THIRD EDITION

Recombinant DNA

GENES AND GENOMES—A SHORT COURSE

THIRD EDITION

Recombinant DNA

GENES AND GENOMES—A SHORT COURSE

James D. Watson
Cold Spring Harbor Laboratory

Richard M. Myers
Stanford University

Amy A. Caudy
Princeton University

Jan A. Witkowski
Cold Spring Harbor Laboratory

W. H. Freeman and Company
New York

Cold Spring Harbor Laboratory Press
Cold Spring Harbor, New York

Recombinant DNA, Third Edition
Genes and Genomes — A Short Course

W. H. Freeman and Company

Publisher	Sara Tenney
Managing Editor	Elaine M. Palucki
Marketing Director	John Britch
Marketing Manager	Debbie Clare
Media and Supplements	Alysia Baker
Manufacturing	RR Donnelley & Sons Company

Cold Spring Harbor Laboratory Press

Publisher	John Inglis
Development Director	Jan Argentine
Project Manager and Developmental Editor	Kaaren Janssen
Art Development Editor	Judy Cuddihy
Production Manager	Denise Weiss
Project Coordinator	Inez Sialiano
Production Editor	Kathleen Bubbeo
Desktop Editor	Susan Schaefer
Art Studio	Dragonfly Media Group
Cover Designer	Ed Atkeson

Front cover illustration: Design concept by James D. Watson and Jan A. Witkowski; rendition by Michael Demaray of Dragonfly Media Group.

Library of Congress Control Number: 2006933192

ISBN: 0-7167-2866-4 (EAN: 9780716728665)

Printed in the United States of America

First printing
W. H. Freeman and Company
41 Madison Avenue
New York, NY 10010
Houndmills, Basingstoke RG21 6XS, England
www.whfreeman.com

Address editorial correspondence to W. H. Freeman and Company, 41 Madison Avenue, New York, NY 10010. Address orders to VHPS/W. H. Freeman and Company, 16365 James Madison Highway, U.S. Route 15, Gordonsville, VA 22942.

For all who contributed to the Human Genome Project

Contents

Detailed Contents

SECTION 1

FOUNDATIONS OF DNA

SECTION 4

HUMAN GENOMICS

Preface

The world of biology has been transformed since the first recombinant molecules were made in the early 1970s. The genetics textbooks of the 1960s described the double helix, the fine genetic mapping of bacteriophage, the operon model of gene control, and the genetic code, but the gene was still an abstract entity, and none had been isolated and characterized. Discussion of the genetics of more complex organisms was limited to the classic breeding studies of *Drosophila* and mice. There was very little of human genetics, and sickle-cell anemia provided the only example of human molecular genetics.

By the time of the first edition of *Recombinant DNA* (1983), genes had been cloned and sequenced, and, by 1992 when the second edition was published, recombinant DNA techniques were providing new insights into the functioning of organisms. We were beginning to understand the molecular control of development based on work in *Drosophila*, we knew how the vertebrate immune system can produce an infinite diversity of antibodies, and inroads were being made into the molecular genetics of cancer. A new field, the molecular genetics of human beings, had been created, leading to the cloning of genes and characterization of mutations responsible for chronic granulomatous disease, Duchenne muscular dystrophy, and cystic fibrosis. Most importantly, DNA-sequencing machines using fluorescent labels were available and the first tentative steps were being taken to do large-scale sequencing. Sequencing of *E. coli* was already under way and a physical map of the *C. elegans* genome was being made in preparation for sequencing. Most impressively, the first eukaryotic chromosome, chromosome III of yeast, had been sequenced in its entirety. These provided tantalizing glimpses of what was to come but nothing could have prepared us for the radical transformation brought about by genome-based studies.

The third edition of *Recombinant DNA* has also undergone a radical transformation, reflected in the book's subtitle, *Genes and Genomes.* New, large-scale techniques, bioinformatic analyses of complete genome sequences, and the insights coming from experiments based on genome data have required substantial revisions of some chapters and the addition of many new ones. Nevertheless, this third edition continues the style of the previous editions.

First, we have not attempted to be encyclopedic in our coverage; there are many excellent textbooks that provide information on (almost) every aspect of cell biology and genetics. Instead, we have selected topics that we think are important—ones that we believe students should know about. These seemingly eclectic topics are linked because they deal with fundamental aspects of biology and because we think that they are exciting. We believe students will, too. Second, as far as possible, we have used real experiments, drawn from papers published in scientific journals, to illustrate topics. We hope that even this glimpse of how experiments are designed and performed will capture the reader's interest. Those who are intrigued can follow up using the citations listed at the end of each chapter. Third, we have written in what we hope is an engaging style and have not refrained from being enthusiastic about discoveries when such enthusiasm is warranted. Biology is wonderful and we believe that textbooks should try to convey that. Fourth, we have continued to make extensive use of figures to illustrate both how experiments were carried out and their results; we hope that these will enhance understanding of the text.

This volume presents what we think are the fundamental concepts of genetics and genomics that form the core of current molecular approaches to studying biological processes. These concepts and topics are what make up the "Short Course" of this book's subtitle. They should be familiar to everyone who, in some way, needs to understand the background to what they may learn in a class on molecular genetics, consider when moving from another science to biology, or read a newspaper story covering some new discovery in biomedical research. The first 13 chapters form a cohesive group and are followed by three chapters that highlight topics of general interest so that the reader can appreciate how what has been covered in the first 13 chapters can make a difference in our lives. We hope that this division will enable a wider audience to become as enthralled by genes and genomes as we are.

There are four sections to the book. The first, "Foundations of DNA," provides a grounding in genes and genetics. It begins with Mendelian genetics before moving to experiments that led to the determination that DNA is the chemical basis of heredity. This is followed by a review of the pathway *DNA to RNA to Protein* and continues with the control of gene expression. These chapters are followed by six chapters that deal with the basic tools of recombinant DNA and what has been learned from their applications to interesting biological problems. There are two completely new chapters. Epigenetics has assumed a much greater importance since 1992 and deserves a chapter to itself. The phenomenon now known as RNA interference (RNAi) did appear in the second edition where it was described as "...not yet understood." Now we understand much of the molecular mechanism of RNAi and a whole new area of biology has been revealed. RNAi is also becoming a technique as important as the polymerase chain reaction.

It is no exaggeration to say that the topics of the second section, "Foundations of Genomics," mark a turning point in the history of biology and provide a pivot in our book as we turn from genes to genomes. "Fundamentals of Whole-Genome Sequencing" introduces many of the basic concepts and techniques for large-scale DNA sequencing, and "How the Human Genome Was Sequenced" tells the story of the science and the politics of the largest project yet undertaken in biological research.

DNA sequence by itself is a meaningless string of As, Ts, Gs, and Cs, and so in "Analyzing Genomes," the third section, we review bioinformatic and experimental approaches to making sense of sequence data. Technical developments have provided the means for performing experiments on a hitherto unimaginable scale and have changed the questions that can be asked about life and the ways in which researchers design their experiments.

Finally, no matter what genome-based biology may tell us about the mysteries of life, we are most intensely interested in ourselves. We have hardly begun to exploit our knowledge of the complete human genome sequence, so "Human Genomics" can be no more than a foretaste of what is to come over the next decades. The potential benefits for biomedical research are evident already in human clinical genetics and cancer research where we now have unprecedented access to those genetic errors that cause such misery. The section, and the book, closes with a chapter on the fascinating science and applications of DNA fingerprinting. What other biological science has so quickly become part of popular culture?

We hope that the topics selected and our treatment of them will make this book suitable for a wide range of readers: undergraduates and graduate students in topics directly related to those in this book—molecular, cell, and developmental biology; biochemistry; genetics; and biotechnology. We also expect that others who wish to learn about the basics of molecular genetics and genomics will find the book useful, for example, medical students and physicians, forensic scientists, patent attorneys, and science journalists.

We have occasionally grumbled, when writing this book, that there was too much to select from—too many topics, too many experiments. But who would wish it otherwise? These are exciting times in biology, and we can be certain that there will be no lack of new discoveries as we explore the new worlds revealed by genomics.

James D. Watson
Amy A. Caudy
Richard M. Myers
Jan A. Witkowski
October 2006

Acknowledgments

A book of this kind is impossible to write without the encouragement, support, and forbearance of our families and colleagues.

A book of this kind is impossible to compose without the help and goodwill of a great many people. We owe special thanks to Neil Lamb who, with contributions from Kate Garber, wrote the first draft of Chapter 14 and to Greg Cooper who provided the first draft for Chapter 12. Their contributions were invaluable. The following found time in their overfilled schedules to review chapters for us and did much to improve our style and correct our errors: Shelley Force Aldred, David Baulcombe, Bruce Budowle, John Butler, Greg Cooper, Abram Gabriel, Alex Gann, Richard Gibbs, Mark Guyer, Gyorgy Hutvagner, Haig Kazazian, Rene Ketting, Jeannie Lee, Scott Lowe, Ron Plasterk, Gavin Sherlock, and Huda Zoghbi.

We are very grateful to the following who read sections, answered queries, made suggestions, and swiftly sent PDFs that helped us on our way: William Anderson, Cecilia Arighi, Michael Ashburner, Art Beaudet, Imre Berger, Fred Bieber, Harald Biessmann, Jon Binkley, David Botstein, Anne Bowcock, Ken Burtis, Angel Carracedo, Tom Caskey, Aravinda Chakravarti, Jeff Chamberlain, Jean-Michel Claverie, Tim Clayton, Michael Coble, Francis Collins, Miguel Constancia, Kay Davies, Kara Dolinski, Lawrence Drayton, Sean Eddy, Argiris Efstratiadis, Steve Elledge, Alain Fischer, David Foran, Tony Frudakis, Fred Gage, Lori Gaglione, Norman Gahn, Mark Gerstein, Jane Gitschier, Aaron Gladman, Ken Goddard, Joel Hagen, Greg Hannon, Stuart Kim, Mary-Claire King,

Bruce Korf, Roger Kornberg, Donna Koslowsky, Adrian Krainer, Fred Kramer, Ilia Leitch, Jim Lupski, Geoff Machin, Stan Maloy, Bruce McCord, Peter Moore, Peter Neufeld, Harry Noller, Robert Nussbaum, Patrick Paddison, Nipam Patel, Norbert Perrimon, Gordon Peters, Lennart Philipson, Craig Pikaard, Joshua Plotkin, Bruce Ponder, Chris Ponting, Daphne Preuss, Mecki Prinz, Jonathan Rees, Christopher Ross, Juan Sanchez, Marco Scarpetta, Bob Shaler, Lincoln Stein, Joan Steitz, Paul Sternberg, Kylie Strong, Bryan Sykes, Ron Taylor, Bill Thompson, Bob Waterston, Victor Weedn, David Werrett, Jon Wetton, Robin Williams, and Michael Yarus.

We wish to thank also the following professors and course instructors who reviewed preliminary drafts in light of their own teaching experiences and expectations: Steven Ackerman, Karen Beemon, Dennis Boygo, Joan Burnside, Kimberly A. Carlson, Christopher Chase, Yvette P. Conley, Shoumita Dasgupta, Mark A. Erhart, Wayne C. Forrester, Shari Freyermuth, Philippe T. Georgel, Cheryl Ingram-Smith, Stephen J. Keller, Stephen T. Kilpatrick, Cindy Klevickis, Carol S. Lin, Pamela A. Marshall, Jacqueline Mooney Andrews, Jessica Moore, Todd P. Primm, Christine Rushlow, Laurie K. Russell, Michael G. Schmidt, Donald Seto, Rebecca Sparks-Thissen, Soichi Tanda, Frans Tax, and David R. Wessner.

Finally, a book of this kind is impossible to produce without the work of a dedicated group of publishing professionals. This book is a joint production of Cold Spring Harbor Laboratory Press and W. H. Freeman and Company Publishers. At Cold Spring Harbor, we owe a special debt of thanks to our Development crew. Kaaren Janssen (Developmental Editor) was very much on the front line with us, spending more time helping us, and polishing and correcting our prose, than we had any right to expect. Judy Cuddihy (Developmental Art Editor) converted our crude figures into forms that made sense and were aesthetically pleasing. Jan Argentine (Development Director) encouraged, exhorted, and cajoled us to keep our noses to the grindstone when the going got tough. We put a great deal of pressure on Inez Sialiano (Project Coordinator), Kathleen Bubbeo (Production Editor), Susan Schaefer (Desktop Editor), and Carol Brown (Permissions Coordinator), who had to deal with innumerable edits in the face of looming deadlines. They turned our manuscript into pages that looked like a proper book. Denise Weiss (Production Manager) guided the look of the pages and her advice and skills are manifest on every page. John Inglis (Executive Director of CSHLP) took on this project and provided us with every means to succeed.

At W. H. Freeman, Sara Tenney (Publisher, Life Sciences) became associated with the project in one of the earlier incarnations of a third edition and has stuck with us through thick and thin. Elaine Palucki (Managing Editor) brought a boundless enthusiasm that buoyed our spirits. They provided valuable guidance on what teachers need and how they teach.

Previous editions of *Recombinant DNA* were noted for their figures and the present edition follows in the tradition. Members of the Dragonfly Media Group—Michael Demaray, Helen Wortham, Rob Fedirko, and Craig Durant—took on the unenviable task of turning our sketches into figures that could be published. We are particularly grateful to Michael Demaray, who produced the beautiful and striking cover illustration from our rough designs.

SECTION 1

FOUNDATIONS OF DNA

This section provides an overview of genetics and of the tools used in recombinant DNA and genomic analyses. We begin with Mendel, whose studies of plant hybridization laid the foundations for one of the most important areas of scientific investigation in the 20th century and, arguably, the most important at the start of the 21st century. Chromosomes were quickly determined to be the physical basis of the Mendelian gene, but our understanding of the chemical nature of the gene did not come until 1953. The discovery of the DNA double helix marked the beginning of the golden age of molecular genetics; it was during this period that the mechanics of translating the information in DNA into proteins and the regulation of gene expression were worked out. By the late 1960s, it seemed that molecular genetics had become mundane and that few new discoveries were yet to be made. This view, however, proved overly pessimistic. In the early 1970s, a set of techniques collectively known as "recombinant DNA" was developed, which opened up new fields of investigation. These techniques revealed surprising and unexpected details of the molecular structure and arrangements of genes and of previously studied phenomena such as movable genes. To this day, extraordinary new discoveries continue to be made in molecular genetics.

CHAPTER 1

DNA Is the Primary Genetic Material

There is no substance so important as DNA. Because it carries within its own structure the hereditary information that determines the structures of proteins, it is the prime molecule of life. The instructions that direct cells to grow and divide are encoded by it; so are the messages that bring about the differentiation of fertilized eggs into the multitude of specialized cells of higher plants and animals. The molecular form of DNA allows a virtually infinite number of structural variations reflected in the variations in hereditary information it transmits. It is the basis for the evolutionary process that has generated the many millions of different life-forms that have occupied the Earth since the first living organisms came into existence 3–4 billion years ago.

The extraordinary capacity of altered DNA molecules to give rise to new life-forms that are better adapted for survival than were their immediate progenitors has made possible the emergence of our own species with our ability to perceive the nature of our environment and to utilize this information to build human civilizations. As a result of our ability for rapid conceptual thought, we have for several centuries been asking ever deeper questions about the nature of inanimate objects like water, rocks, and air, as well as about the stars in surrounding space. And biology, the science of living objects, which only 50 years ago was generally perceived to be a much inferior science, has swiftly come of age. By now there exists a total consensus of informed minds that the essence of life can be explained by the same laws of physics and chemistry that have helped us understand, for example, why apples fall to the ground and why the moon does not, or why water is transformed into gaseous vapor when its boiling point is exceeded.

From DNA issue the commands that regulate the nature and number of virtually all cellular molecules. Through working out the exact structure of a multitude of genetic scripts encoded within DNA molecules, we are now taking giant steps toward eventually understanding the many complex sets of interconnected chemical reactions that cause fertilized eggs to develop into highly complex multicellular organisms. Only in the mid-1980s did it seem possible that we could determine the total genetic information of multicellular organisms, that is, determine the DNA sequences of their genomes. Initially there was much opposition to the huge scale of the Human Genome Project and skepticism as to whether it was technically feasible. Subsequent progress, however, has been extraordinarily rapid. By now the complete genetic scripts of several hundred species, including our own, are known and we will see throughout this book how this new knowledge is transforming our view of the living world and our place in it.

DNA's paramount role in genetics was established only in the middle of the 20th century, but our fascination with heredity stretches back for centuries. In this chapter, we shall look at how patterns of inheritance were determined long before their physical basis was known and then go on to review DNA as the molecule of life.

Mendel's Breeding Experiments with the Pea Plant First Revealed the Patterns of Inheritance

The knowledge that many human traits, such as the color of our eyes and the shapes of our faces, are passed on to us from our parents must go back to the days of early human beings. And from the earliest times that human beings domesticated animals or planted crops, they sought to improve the desirable features of their dogs and goats and grain through selective breeding. However, despite the great successes of breeders over the centuries, theirs was an empirical approach because the patterns of inheritance could not be predicted with certainty. It took the labors of a scientist working alone in Brno (now in the Czech Republic) to build the foundation for understanding how organisms inherit their traits.

That scientist was a monk, Gregor Mendel, who, as early as 1856, had begun breeding experiments by mating parents with specific traits to work out how the characteristics of parents were transmitted to their offspring. Although others had taken a similar approach, Mendel's success was due to his choice of organism (a recurring theme throughout the history of genetics), his experimental design, and his careful tallying and numerical analysis of the results.

Mendel used the pea plant, *Pisum sativum*, which is easy to grow and has distinct traits, or characters (Mendel's term), that are easy to observe. These characters, present in contrasting pairs, include seeds that are yellow or green, and smooth or wrinkled; plants that are short or tall; and flower petals that are purple or white. Mendel bred plants with particular characters for many generations until they showed no variation in that character—these *pure-breeding* plants and their descendants were, for example, either always tall or always short. He then mated plants with contrasting characters and examined all their offspring (the *F_1 generation*), tallying the numbers that showed one or the other character. In the case of plant height, Mendel found that all the offspring of breeding a tall plant with a short plant were tall; none were of medium height. Tallness was considered *dominant* to shortness, which he called *recessive.*

Mendel then bred these tall F_1 offspring with each other to produce an *F_2 generation*. Remarkably, unlike the pure-breeding tall plants, the F_2 progeny of the F_1 tall plants included both tall and short plants: On average, there were three tall plants for every short plant in the F_2 generation. This ratio of 3:1 in the F_2 generation held for other pairs of dominant and recessive characters.

Finally, he analyzed the consequences of cross-breeding plants that had *two* contrasting traits; for example, tall plants with yellow seed were crossed with short plants with green seed. All the F_1 generation offspring were tall plants with yellow seed (tall is dominant to short, and yellow is dominant to green). In the second generation, all four possible combinations—tall plants, yellow seed; tall plants, green seed; short plants, yellow seed; short plants, green seed—appeared. It seemed that different characters were transmitted independently of each other. Furthermore, these four sets of characters always appeared, on average, in the ratio of 9:3:3:1.

Discrete Factors of Inheritance Underlie Mendelian Laws

From these simple experiments, Mendel deduced that there are discrete factors (now called genes) for characters such as seed color or plant height. These factors do not merge, or blend, with each other because crossbreeding produced seeds that were either green or yellow and plants that were either short or tall. The ratios Mendel had found could be accounted for by assuming that each pea plant possesses two factors for each character. During gamete (sex-cell) formation, the two factors for a character separate so that each sperm or egg receives only one factor for each character. A gamete has an equal chance of receiving one or the other of the two factors for each particular character. At fertilization, an offspring receives one factor from the male parent and one from the female parent.

The character shown by a plant depends on the pair of factors it receives from its parents. If both factors are the same, that is, two dominant (yellow seed) or two recessive (green seed), then the plant exhibits those traits—yellow seeds for the former and green seeds for the latter. If one factor is dominant and the other recessive, then the dominant character is expressed; in this case, a plant with a factor for yellow seed and one for green seed will show the dominant trait, yellow seed. But the factor for green seed persists, even if the plant is yellow, and can reappear unchanged in later generations. This was a signal advance. In the 19th century it was believed that the traits of parents were blended in their offspring and, once mixed, the contributing traits could not be recovered. At the time, this was a severe problem for Charles Darwin's theory of evolution through natural selection. His critics pointed out that any favorable variation arising in an individual would have no long-term effect, because it would immediately be lost by blending.

Mendel's seminal contributions, published in 1865, were effectively ignored until 1900 when his research was independently confirmed by the European plant breeders Carl Correns, Hugo deVries, and Erich von Tschermak-Seysenegg. In England, William Bateson became an enthusiast for Mendel and in 1906 gave the name *genetics* (from the Latin *genetikos*, meaning "the produced") to this new field of study, and in 1909, a Danish scientist, Wilhelm Johannsen, named Mendelian factors *genes* (from the Latin *genos*, origin). Each gene can exist in different forms or *alleles* (originally allelomorphs, another contribution by Bateson). For example, there are green and yellow alleles for pea seed color, and when both alleles of a gene are the same, the individual is said to be *homozygous* for that gene; when the alleles are different, the individual is *heterozygous*. In 1911, Johannsen made explicit the important difference between phenotype and genotype. The *genotype* is the complete genetic composition of an organism, whereas its *phenotype* is the physical expression of that genotype.

Chromosomes Are the Cellular Bearers of Heredity

Great advances made in microscopy in the latter part of the 19th century enabled the structure of cells to be seen in detail. Plant and animal cells were shown to have a central body, the nucleus, surrounded by a relatively amorphous cytoplasm. It was assumed that both the nucleus and the whole cell were enveloped in membranes, and in plant cells a thick cell wall was visible. The cytoplasm contained many forms of inclusions—like vacuoles and lipid droplets—as well as rod-shaped granules that are today called mitochondria and chloroplasts. But it was the nucleus that attracted the most attention because of its disappearance at the start of cell division and the concomitant appearance of elongated bodies, called *chromosomes* (they could be stained with special dyes; *chromo* is Greek for "color").

Because embryos arise from the fusion of eggs with sperm, it was evident that heredity is transmitted through these specialized cells. And because a sperm contains very little cytoplasmic material in comparison to the amount of material in its nucleus, the obvious conjecture was that the function of the nucleus was to carry the hereditary determinants of a cell.

By 1902, striking parallels were seen between the behavior of chromosomes and that expected of Mendel's factors. In that year, the American Walter Sutton described the structure and behavior of the chromosomes in the grasshopper *Brachystola magna* and wrote that chromosomes "may constitute the physical basis of Mendelian law of heredity." Within a given cell type, a variety of chromosomes of different sizes were observed, with each distinct chromosome usually being present in two copies (a pair of *homologous chromosomes*). The number of chromosomes was seen to exactly double during cell division (*mitosis*), producing two daughter cells, each of

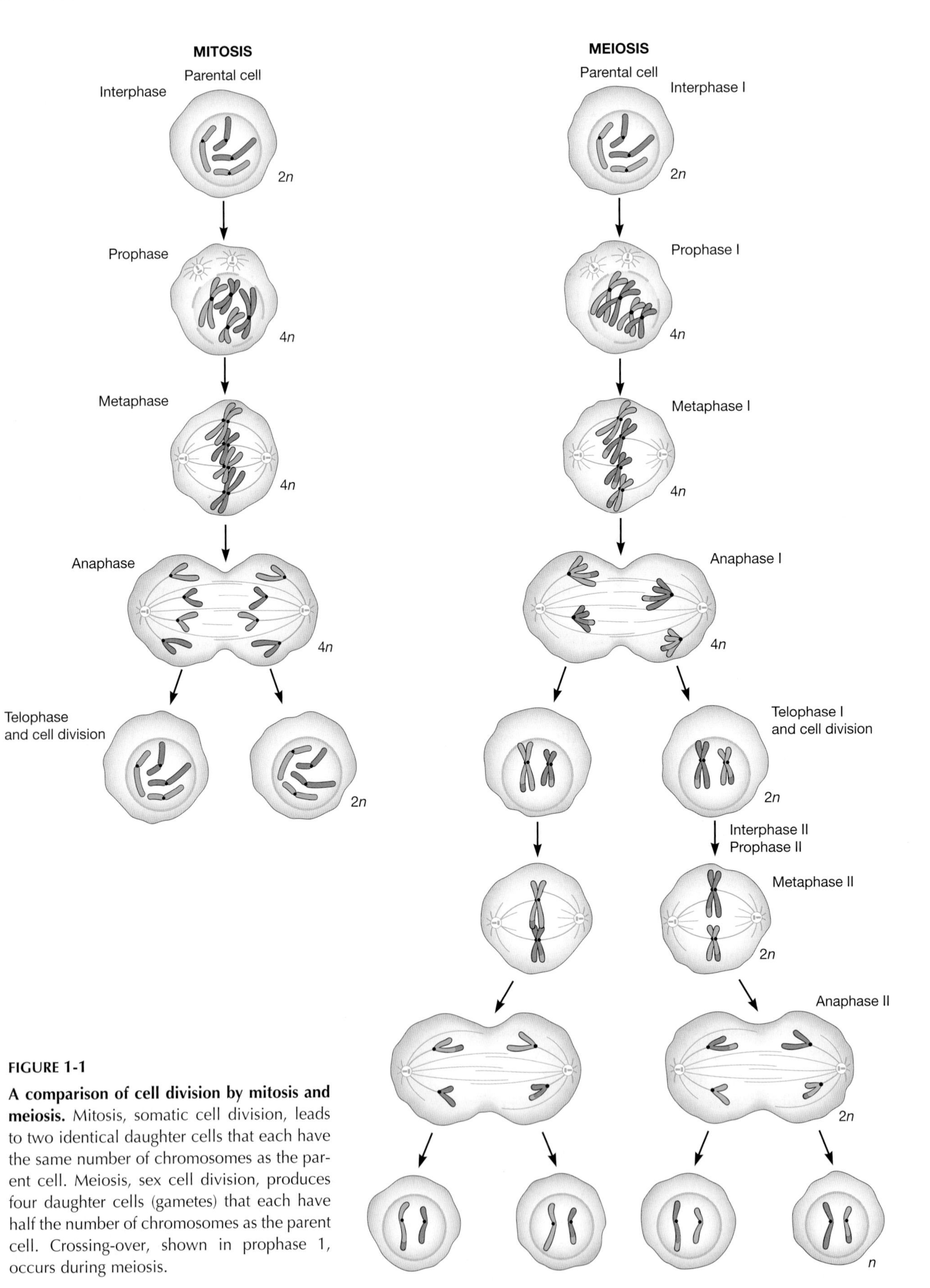

FIGURE 1-1

A comparison of cell division by mitosis and meiosis. Mitosis, somatic cell division, leads to two identical daughter cells that each have the same number of chromosomes as the parent cell. Meiosis, sex cell division, produces four daughter cells (gametes) that each have half the number of chromosomes as the parent cell. Crossing-over, shown in prophase 1, occurs during meiosis.

which received one complete set of chromosomes from the parent cell. The number of chromosomes in the sex cells of sperm and eggs was shown to be exactly half the number found in somatic cells. During the formation of germ cells (*meiosis*), the chromosome count is reduced from the *diploid* number, $2n$, to the *haploid* number, n (Fig. 1-1). The fertilization process between sperm and eggs thus restores the $2n$ chromosome number characteristic of somatic cells, with one chromosome in each pair coming from the male parent and the other from the female parent.

Genes Are Mapped on Chromosomes Using Linkage

If chromosomes are the bearers of genes, then it should be possible to associate the inheritance of specific phenotypes with chromosomes. The first trait to be assigned to a chromosome was sex. The work was done in 1905 at Columbia University in New York City where Nettie Stevens and Edmund Wilson discovered the existence of the so-called *sex chromosomes.* One chromosome, the X, is present in two copies (XX) in all female somatic cells, but there is only one copy in male somatic cells, which also carry a morphologically distinct Y chromosome (XY) (Fig. 1-2). During the halving of chromosome numbers in meiosis, all eggs necessarily receive a single X chromosome, whereas sperm receive either an X or a Y chromosome. Fertilization by a sperm containing an X chromosome generates female (XX) progeny, whereas fertilization by a sperm containing a Y chromosome yields male (XY) progeny. The proposal that sex traits were located on a single pair of chromosomes neatly explained why male and female offspring are produced in a 1:1 ratio; half the fetuses receive an X chromosome from each parent—making them female—and half receive an X chromosome from their mother and a Y chromosome from their father—making them male.

A second example of associating a trait with chromosomes came from Thomas Hunt Morgan's laboratory, which was also at Columbia University. In 1910, while carrying out experimental studies of evolution using the red-eyed fruit fly *Drosophila*, he found a single male with white eyes. This variant appeared to be a typical Mendelian trait when the white-eyed male was bred with red-eyed females—all the progeny had red eyes so that white was recessive to red—but Morgan noticed that all the white-eyed flies in the F_2 generation were male (Fig. 1-3). White eye color is not a trait limited to males, because Morgan could produce white-eyed females by cross-

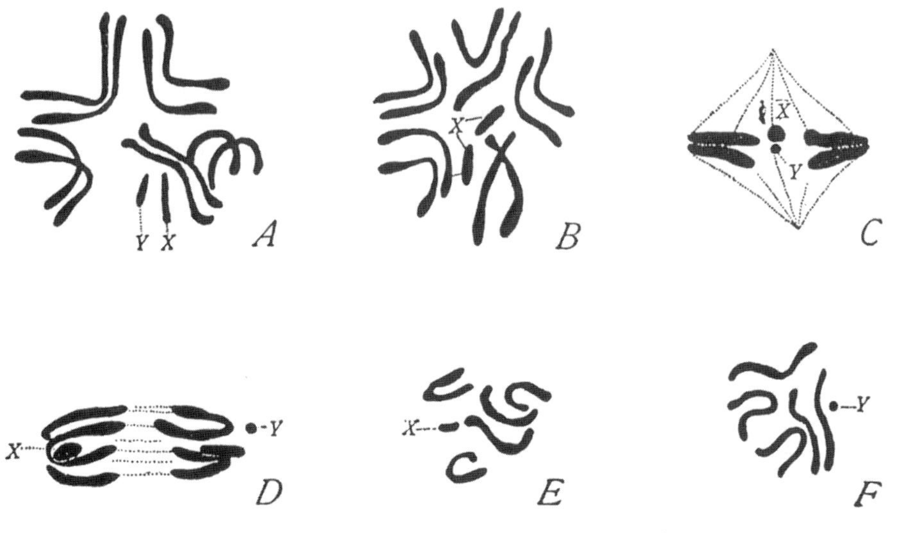

FIGURE 1-2

Facsimile of a figure from Nettie Stevens' 1905 paper on sex chromosomes in the blowfly, *Calliphora vomitoria.* *A* and *B* show the chromosomes in somatic cells of males and females, respectively. There are five pairs of autosomes and one pair of sex chromosomes. The male has one long X chromosome and one shorter Y chromosome, and the female has two X chromosomes. (*C*) During meiosis the X and Y chromosomes pair showing that they are homologs and (*D*) separate so that the spermatocytes contain the haploid number of autosomes and either an X chromosome (*E*) or a Y chromosome (*F*).

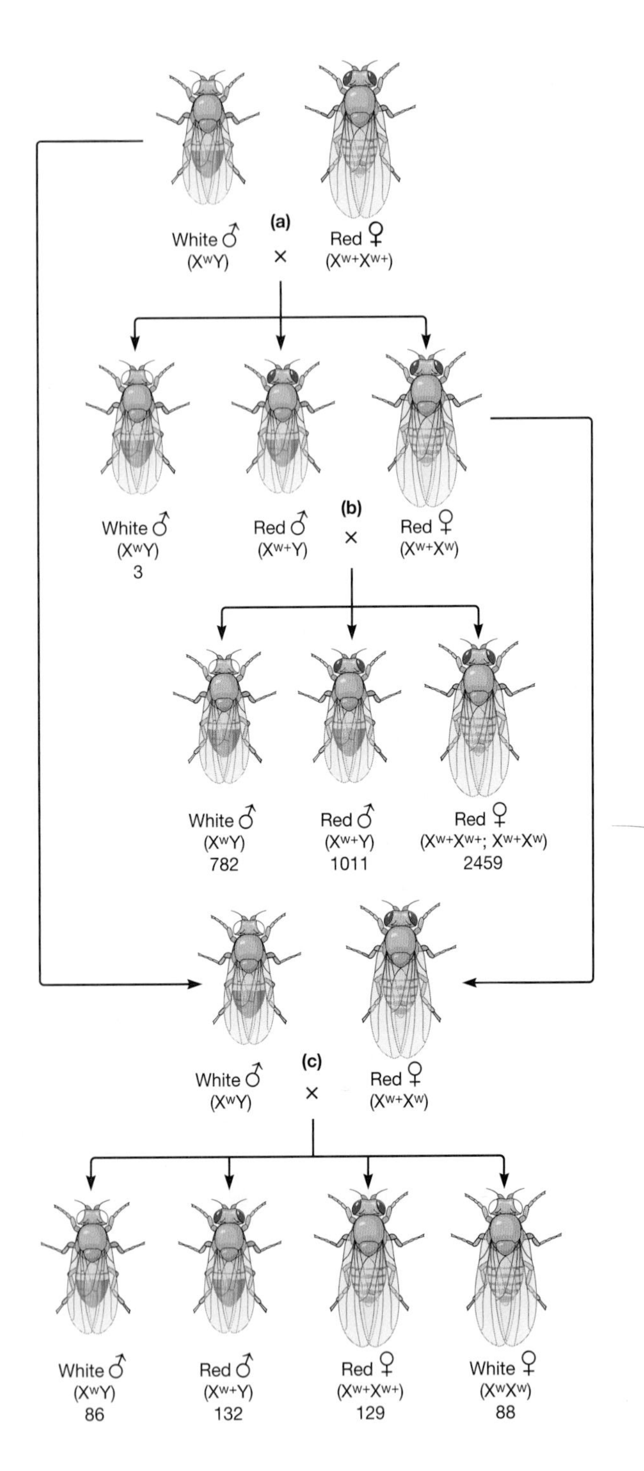

FIGURE 1-3

Morgan's experiments with the white-eyed male *Drosophila*. The male *Drosophila* is smaller than the female, and the wild-type eye color for both sexes is red. The symbolism is modern and not that used by Morgan. X and Y represent X and Y chromosomes, respectively, and the superscripts w+ and w are red eye and white eye, respectively. The numbers of each type of fly are those in Morgan's 1910 *Science* paper. (*a*) In the initial cross, Morgan found that all but three of the offspring of the white-eyed male with red-eyed females had red eyes, a total of 1237 males and females. This indicated that white was recessive to red. There were three white-eyed males that he assumed to be mutants like the original white-eyed male. (*b*) The red-eyed male and female offspring of this cross were then mated to each other. White-eyed flies reappeared in their progeny, but they were always male. This suggested that eye color was somehow associated with maleness. (*c*) However, in a third cross using the original white-eyed male and its red-eyed daughters, white-eyed females were produced, showing that the association was not limited to maleness. All these results and those of other crosses can be explained by assuming that the gene for eye color is carried on the X chromosome and that females are homozygous (XX) whereas males are heterozygous for sex (XY). With only one X chromosome, males show a mutant phenotype even for a recessive trait like white eye.

ing white-eyed males with F_2 red-eyed females. The simplest explanation for these data was that the gene for eye color, like that for sex, must be on the X chromosome. Over the next several years, mutations (or sports as they were then called) in many other *Drosophila* genes were found. When analyzed, many examples were found of two traits tending to be inherited together, or *linked*. After a large number of variants had been studied, it was clear that they formed four *linkage groups*, and that these corresponded to each of *Drosophila's* four chromosomes; that is, linked genes were on the same chromosome. This was a remarkable insight and a powerful argument for the chromosome theory of heredity.

However, linkage between two traits was sometimes broken as a consequence of a physical exchange of chromosome parts (called *crossing-over*) that occurs when homologous chromosomes come together in pairs during meiosis (Fig. 1-4). In 1913, Morgan suggested that the further apart two genes are on the same chromosome, the greater the chance that a crossover will occur and thereby recombine

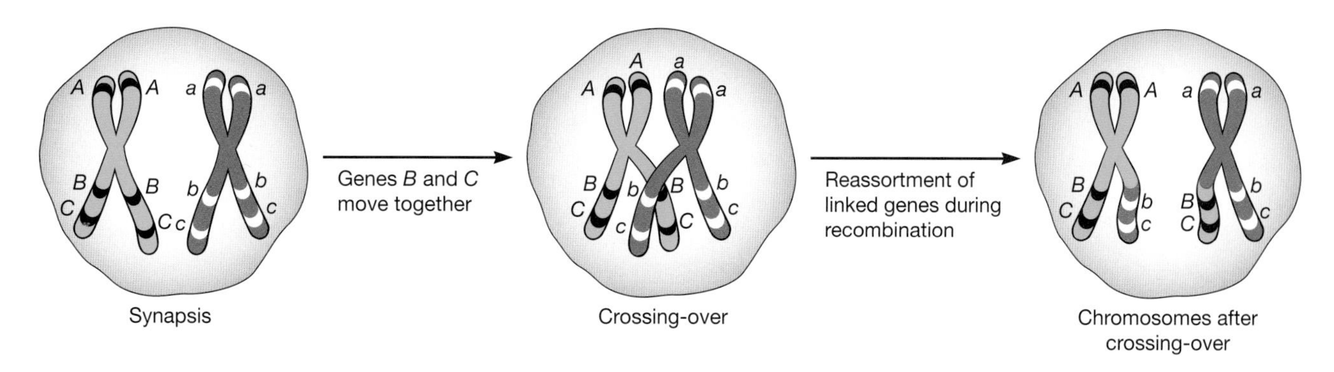

FIGURE 1-4

Reassortment of genes by crossing-over. *B* and *C* (and *b* and *c*) are linked because they move together during recombination. Gene *A* at the other end of the chromosome is seen to be only loosely linked to *B* because a crossover has occurred somewhere between *A* and *B*. Two new combinations have been produced as a consequence of this recombination: *Abc* and *aBC*. In general, the further apart two genes are, the more loosely linked they are and the greater the chance of crossing-over.

genes. Alfred Sturtevant, then an undergraduate student in Morgan's laboratory, realized he could make a quantitative estimate of the distance between a pair of genes by calculating the frequency of crossovers between a pair of genes. Sturtevant went home that evening and returned the next morning with the first genetic map—the positions of five genes on the *Drosophila* X chromosome (Fig. 1-5).

A simple example of mapping calls for determining the distance between the positions (*loci*) of two genes, *pr*, for purple eyes, and *vg*, for vestigial wing (Fig. 1-6). Out of 2839 progeny analyzed, a total of 305, or 10.7%, showed recombination. Sturtevant defined a map unit as the distance between a pair of genes over which crossing-over occurs once in 100 meioses (this map unit has since been named a centimorgan—cM—in honor of Morgan). In this case, the *pr* and *vg* loci are 10.7 cM apart. Analyses become more complicated when more loci are considered, but the principle is the same—determine the number of crossovers between the loci and calculate the frequency of recombinants. Linkage analysis has remained a fundamental strategy for mapping genes in all sexually reproducing organisms, from fruit flies to human beings. However, it is important to remember that recombination frequencies measure genetic distances and that these may differ from the physical distances between genes. This is particularly true for genes separated by large physical distances where multiple crossovers occur. In these cases, even numbers of crossovers will not be detected because these maintain the linkage relationship between the genes and lead to underestimation of recombination frequencies.

Not All Genes Are Found on Chromosomes in Nuclei

Morgan, Sturtevant, and other geneticists in the first part of the 20th century concentrated on the analysis of traits carried on chromosomes in the nucleus, in part because these traits could be studied in terms of Mendelian inheritance and because changes in nuclear genes produced significant changes in phenotypes. However, at the same time some geneticists were studying the role of cytoplasmic inheritance, that is, patterns of inheritance that were not consistent with Mendelian, nuclear inheritance. For many years, the importance of cytoplasmic inheritance was a matter of controversial debate.

One early example, already described by 1909, is *maternal inheritance*, shown by variegated plants such as the four-o'clock plant, *Mirabilis jalapa*. The different parts of a single plant are green, white, or mixed white and green patches, depending on whether the chloroplasts make (green) or do not make (white) chlorophyll. Flowers can be found in all three parts and crossed by transferring pollen. Analysis showed that the source of the pollen had no effect on the color of the progeny, which depended only on the flower that provided the egg. To explain this non-Mendelian inheritance, chloroplasts were early on postulated to have their own genetic material and breed true—green chloroplasts containing chlorophyll make more green chloroplasts, whereas mutant, chlorophyll-free chloroplasts make more colorless chloroplasts. The fertilized zygote receives virtually all of its chloroplasts from the egg cyto-

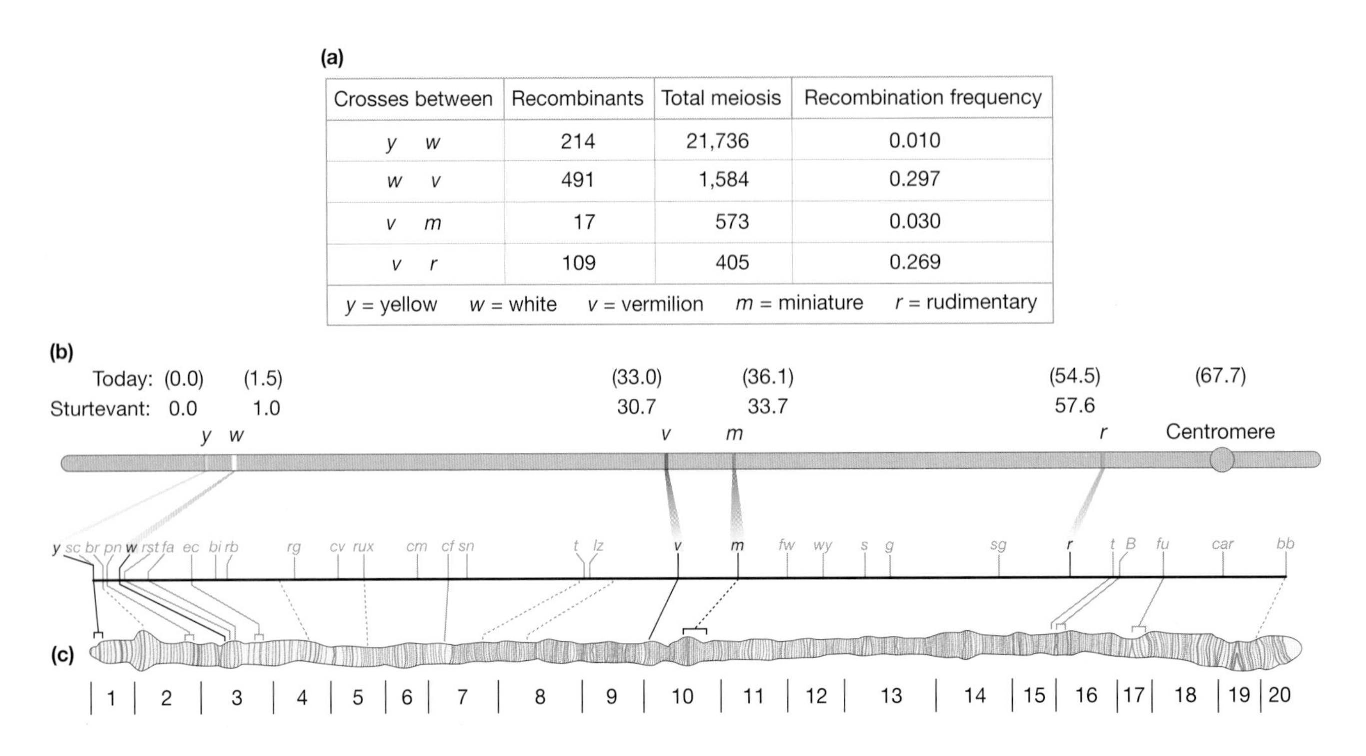

(a)

Crosses between	Recombinants	Total meiosis	Recombination frequency
y *w*	214	21,736	0.010
w *v*	491	1,584	0.297
v *m*	17	573	0.030
v *r*	109	405	0.269
y = yellow *w* = white *v* = vermilion *m* = miniature *r* = rudimentary			

FIGURE 1-5

The first genetic map produced by Sturtevant in 1913. (*a*) A selection of Sturtevant's results of crossing flies and determining the recombination frequencies between pairs of genes. (*b*) The relative positions of Sturtevant's mapped genes are shown on a bar representing *Drosophila* chromosome 1. Sturtevant's values are very close to modern values shown in parentheses above the bar. (*c*) The positions of the same genes are indicated on a drawing of chromosome 1 from the salivary gland. These chromosomes are very large and show remarkable details of chromosomal organization. The map was made in 1935 by Calvin Bridges, another protégé of Morgan and a colleague of Sturtevant. (Bridges' drawings are so accurate that they remain the definitive representation of the polytene chromosomes of *Drosophila melanogaster*.)

plasm; the sperm contributes practically none because it has so little cytoplasm. It is these maternally derived chloroplasts that reproduce and populate the growing plant (Fig. 1-7).

Similarly, mitochondria also have their own genes, and their patterns of cytoplasmic inheritance have been analyzed extensively in fungi, such as *Neurospora*, and the yeast *Saccharomyces cerevisiae*. Like chloroplasts, mitochondria are predominantly inherited from cytoplasm carried with the egg. Mitochondria and chloroplasts have genes because they are the remnants of bacteria-like cells that were assimilated by eukaryotic cells.

Mutations Are Alterations in Genes

Early breeding experiments utilized variants that, like Morgan's white-eyed male, appeared spontaneously at a low frequency. We now know that these spontaneous *mutants* result from changes in single genes and are genetically stable, just like the normal, or *wild-type*, version of the gene. Mutations are essential tools for genetic analysis—the finding of a white-eyed *Drosophila* reveals that there must be a gene that controls eye color, and thus mutant phenotypes act as flags for genes.

However, it is difficult to study spontaneous mutations because they occur at very low frequencies—tens of thousands of individuals may have to be examined to find one. It was with relief and some disbelief that geneticists greeted Herman Muller's claim in 1926 that X rays induced mutations at a very high rate in *Drosophila;* Muller found 100 X ray–induced *Drosophila* mutants in two months, compared to the 200 naturally occurring fruit fly mutants found over the previous 16 years. Muller's findings had profound implications for the health of human beings and he quickly warned against the dangers of unshielded ionizing radiation sources.

Subsequently, it was found that mutations can be

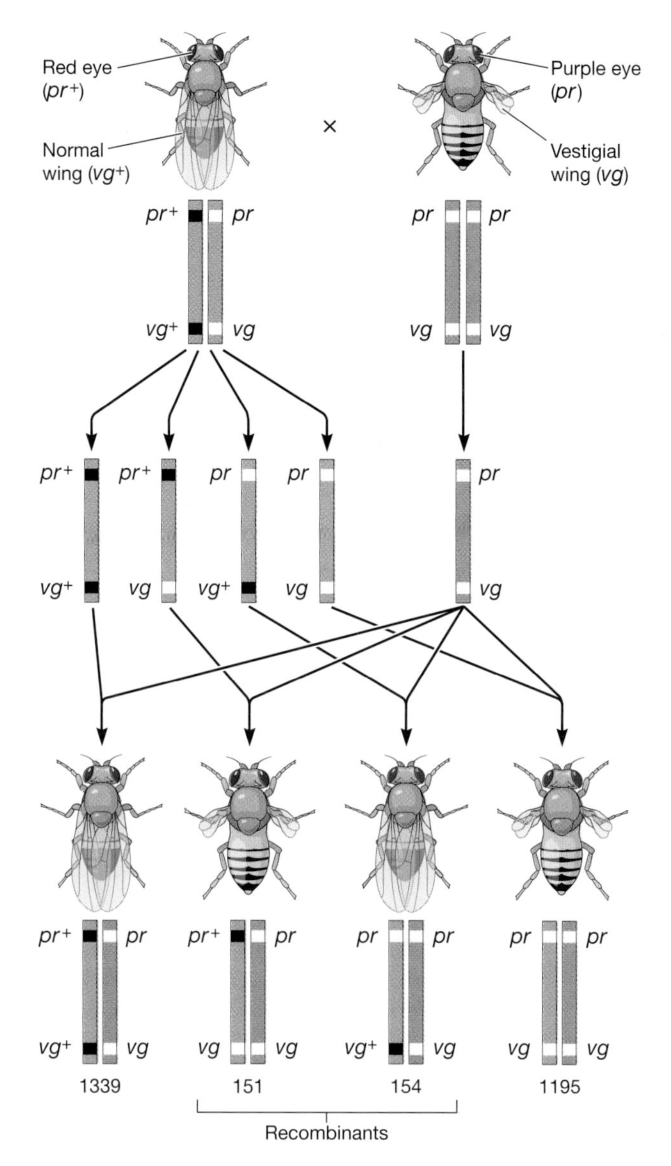

FIGURE 1-6

Determining genetic distances by measuring recombination. The *pr* locus controls eye color: pr^+ is wild-type red eye color, whereas *pr* is recessive purple eye color. The *vg* locus affects wing development: vg^+ produces normal wings, whereas *vg* is for vestigial wings. Because pr^+ and vg^+ are dominant, flies with purple eyes and vestigial wings are *prvg/prvg*. Sturtevant crossed these flies with flies that had red eyes and normal wings but that were heterozygous for these traits, i.e., $pr^+vg^+/prvg$. Four gametes are possible from the heterozygous flies: pr^+vg^+ and *prvg*, and if there has been crossing-over during meiosis, pr^+vg and $prvg^+$. Only one gamete, *prvg*, is possible from the doubly recessive flies. Sturtevant counted the numbers of flies and found that of 2839 flies, there were 305 recombinants: 151 flies that had red eyes but vestigial wings (pr^+vg) and 154 flies with purple eyes and normal wings ($prvg^+$). The recombination frequency between *pr* and *vg* is 305/2839 = 0.107. A recombination frequency of 0.01 is 1 map unit, and so *pr* and *vg* are 10.7 map units apart.

produced by a wide variety of agents—in fact, any agent that in some way disrupts the proper copying of a gene. Chemicals such as mustard gas and ethyl methanesulfonate (EMS) are potent mutagens, and the mutagenic potential of chemicals in the environment continues to be a topic of considerable controversy. Remarkably, it was found that genetic material itself could induce mutations. In the 1940s, at the Cold Spring Harbor Laboratory on Long Island, Barbara McClintock found that the genomes of maize (corn) contain small genetic elements that can move from one chromosomal position to another by inserting themselves at different sites in the genome. Now we know that most organisms have one or more classes of movable (transposable) elements that, when inserted within a gene, can disrupt its functioning (insertional mutagenesis). Retroviruses—viruses that propagate themselves by inserting their genetic material into the chromosomes of their host cells—can similarly cause mutations. Retroviruses have been used in gene therapy as vectors to carry genes into patients, and the risk of insertional mutagensis was thought to be very low based on animal studies. Unfortunately, several children receiving retroviral gene therapy for an immunodeficiency that would have killed them developed leukemia (Chapter 14). (Fortunately, many of these childhood leukemias can be treated successfully.)

Mutagens continue to be used extensively for detecting genes through *mutation* or *genetic screens*; organisms are exposed to a mutagen and their progeny examined for phenotypic abnormalities. In the case of *Drosophila*, mutations have been induced both by chemicals and by a *Drosophila* transposable element, the P element. When the zebrafish was adopted as a model organism, genetic screens were performed using the chemical mutagen ethyl nitrosourea (ENS). Males were exposed to ENS and then crossed to untreated females. The breeding of the second generation required considerable resources—many hundreds of fish tanks—but resulted in the identification of more than 1000 mutations.

Mutations are almost invariably deleterious. The metabolic and developmental pathways of organisms are finely tuned, and throwing a mutational wrench into the cellular works rarely leads to improved performance. However, mutations in a diploid organism are usually recessive because only one of the two copies of a gene is mutated. Yet mutations are essential for evolution. Without the variability that mutations produce, there would be no new forms on which natural selection could act.

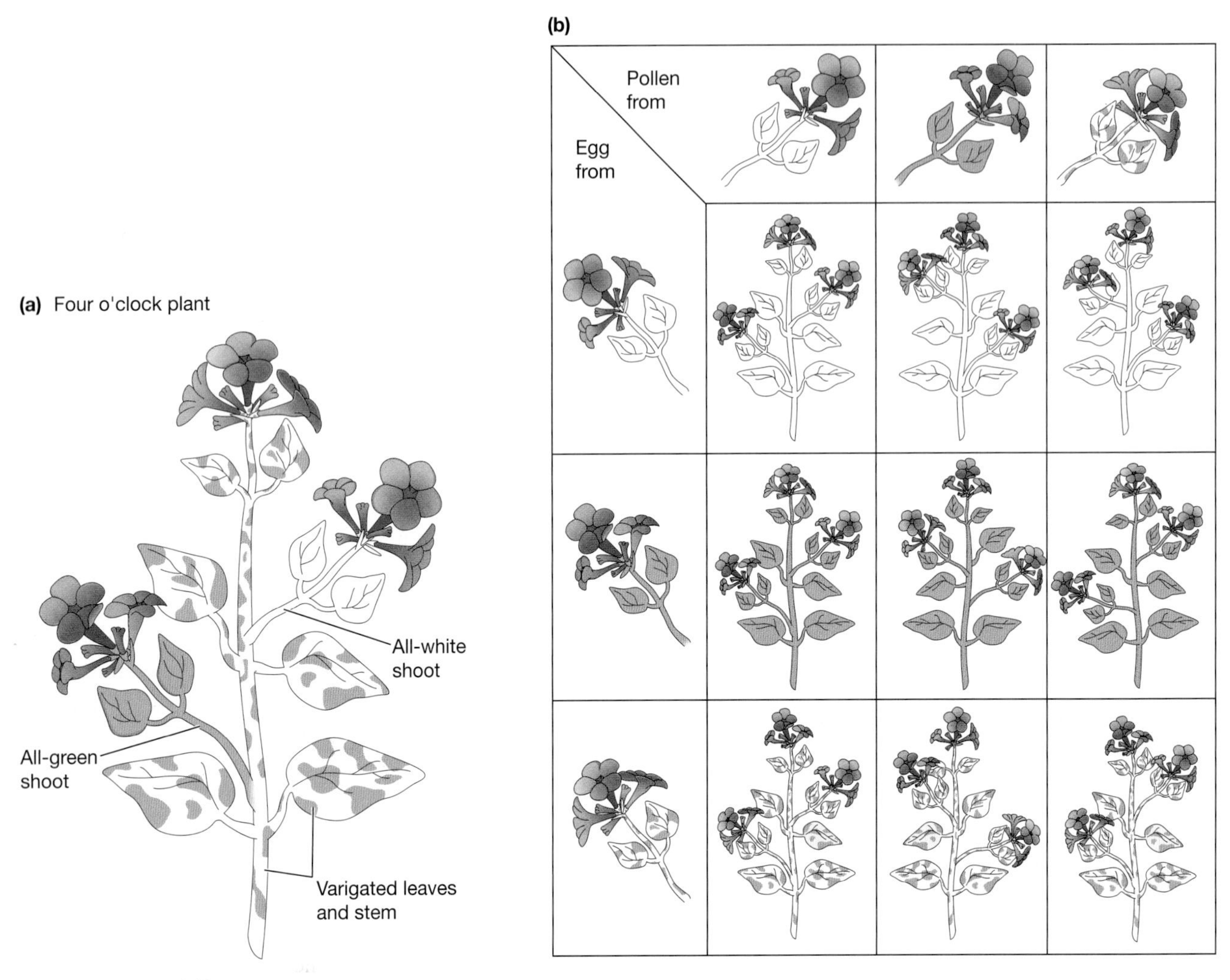

FIGURE 1-7

Extranuclear genes. (*a*) The four o'clock plant is variegated with areas that are green and others that are white, and branch stems can be completely green, completely white, or variegated. (*b*) Pollen taken from flowers on such stems can be used to pollinate flowers on similarly or differently pigmented stems. In all cases, the pigmentation of the F_1 plants is that of the stem bearing the flower that was pollinated; the color of the stem that provided the pollen appears irrelevant. Thus plants derived from eggs borne on a green stem are always green no matter the source of the pollen. This is in marked contrast to Mendelian inheritance where the transmission of a trait is not influenced by the parental origin of a gamete (except in a phenomenon called imprinting; see Chapter 8).

Chromosomes Contain Both Nucleic Acid and Protein

Although it was clear that chromosomes carried genes, early geneticists did not focus on what genes were made of; they did not need to know the physical nature of the gene for their analyses. In fact, analyses of the chemical nature of chromosomes long predated any knowledge of their role in heredity. In 1869, Friedrich Miescher, working in Tübingen, Germany, isolated a phosphorus-containing material he called nuclein from the nuclei of degenerate white blood cells in pus (plentiful in the days before antibiotics). By 1874 he had found that salmon sperm are a much more amenable starting material than pus for isolating nuclein. Miescher showed that nuclein was a mixture of positively charged proteins and an acid material that in 1889 Richard Altman named nucleic acid.

Through analyzing degradation products of nucleic acids from several cellular sources, the German chemist Albrecht Kossel discovered the nitrogen-containing, flat "bases" guanine (1882), adenine (1886), thymine (1893), and cytosine (1894). A fifth base, uracil, was isolated in 1900 (Fig. 1-8). Erroneously it

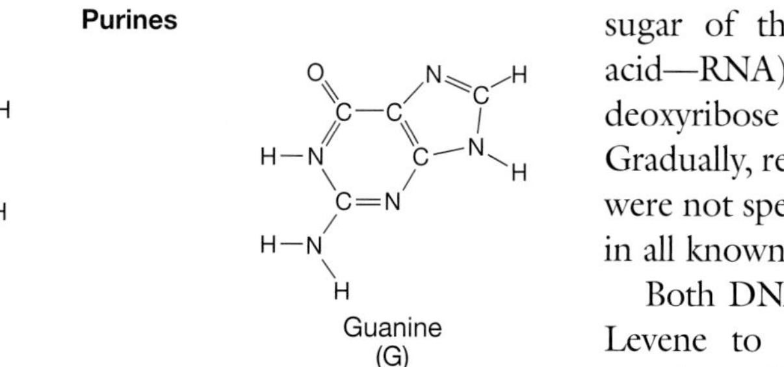

FIGURE 1-8

Bases of nucleic acids. Adenine, guanine, cytosine, and thymine are found in DNA. Uracil is a constituent of RNA where it takes the place of thymine.

was believed that some cells (e.g., those of the calf thymus) had only thymine-containing nucleic acid molecules, whereas other cellular sources like yeast had only uracil-containing nucleic acid molecules. The sugar components of yeast and calf thymus nucleic acids were likewise different, with Phoebus Levene of the Rockefeller Institute finding that the sugar of the former is ribose (hence ribonucleic acid—RNA), whereas the sugar of the latter is deoxyribose (hence deoxyribonucleic acid—DNA). Gradually, researchers realized that these nucleic acids were not special to yeast or animal cells, but occurred in all known cells in differing amounts.

Both DNA and RNA had by then been shown by Levene to be polynucleotides—covalently bonded together collections of nucleotide building blocks. Each nucleotide in turn contains a phosphate group, a sugar moiety (ribose or deoxyribose), and either a purine or pyrimidine base (flat, ring-shaped molecules containing carbon and nitrogen). Figure 1-9 shows the three constituents of a nucleotide. Two purines and two pyrimidines are found in both DNA and RNA. The two purines, adenine and guanine, and the pyrimidine cytosine are used in both DNA and RNA. The pyrimidine thymine is found in only DNA, and the chemically similar pyrimidine uracil appears in only RNA.

In this early period, chemists took no particular pains to use gentle isolation procedures to preserve the macromolecular structures of nucleic acids; they were interested only in determining the constituents of nucleic acids from their degradation products. Levene erroneously believed that a nucleic acid molecule was only four nucleotides long and contained just one of each of the four nucleotides. Later the tetranucleotide hypothesis became increasingly untenable as longer and longer DNA molecules containing hundreds of nucleotides were identified. The possibility, however, could not be logically dismissed that DNA and RNA

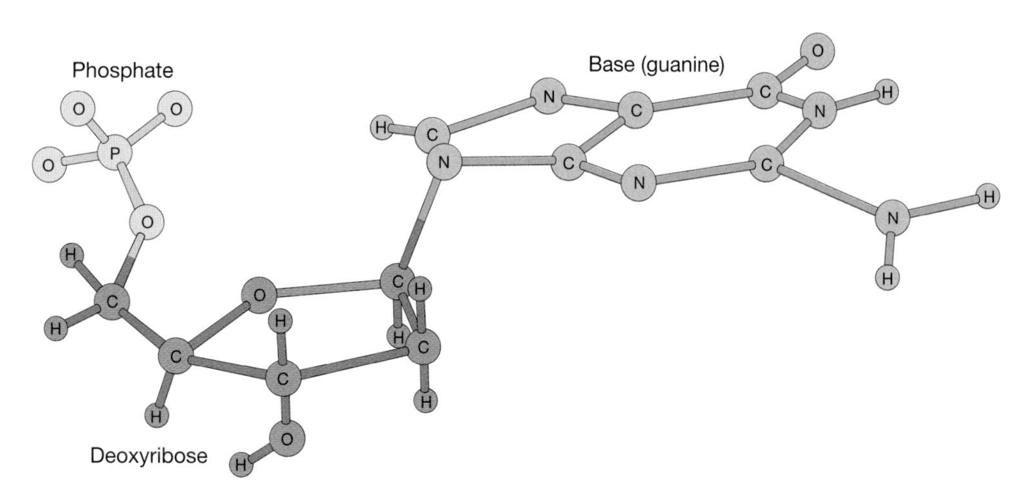

FIGURE 1-9

A DNA nucleotide. The base (*blue*) is attached to a deoxyribose ring (*orange*) that is in turn bonded to a phosphate group (*yellow*). The base shown here, guanine, can be replaced by any one of the other three DNA bases—adenine, cytosine, thymine. In DNA molecules, nucleotides are linked together to form long chains by bonds running from the phosphate group of one nucleotide to the deoxyribose group of the adjacent nucleotide.

were uniform polymers in which each base was repeated every four nucleotides along the entire polynucleotide chain. If so, nucleic acids were effectively nonspecific, unlike proteins of which an effectively unlimited number could in theory be constructed by chaining together the 20 amino acids in different orders.

Quantitating DNA Bases in Different Organisms Hints at Unlimited Variety of DNA Molecules

It was not until the late 1940s that precise quantitation of nucleotides became possible, using the chromatographic techniques developed by A.J.P. Martin and R.L.M. Synge in England for determining the amounts of various amino acids in proteins. The bases of DNA were first quantitatively analyzed by Erwin Chargaff in his laboratory at the College of Physicians and Surgeons of Columbia University. By 1951 it was clear that the four bases were not present in equal amounts (Table 1-1) and so DNA could not be made of the identical tetranucleotide units hypothesized by Levene. It also meant that there were unlikely to be structural constraints on the order of nucleotides in a DNA molecule, and if the sequence of the four bases were indeed irregular, the potential number of different DNA sequences was the astronomically large 4^n, where n is the number of nucleotides in a DNA molecule. Chargaff also noted that the amounts of the four bases did not vary independently (Table 1-1). For all the species he looked at, the amounts of the purine adenosine (A) were very close to the amounts of the pyrimidine thymine (T). Similarly, the amounts of the second purine, guanine (G), were always very similar to those of the second pyrimidine, cytosine (C). However, there were anomalies that may have masked the significance of these ratios. For example, the T2 bacteriophage, a virus that infects bacteria, has no cytosine. The fact remains that Chargaff failed to realize that the A/T and G/C ratios he had discovered held clues to the structure of DNA.

Quantitative determinations of DNA amounts in cells supported the view that DNA could be the genetic material. The same amount of DNA was found in all diploid cells of a given species with half this amount found in the respective sperm and eggs. Equally important, the ratio of A plus T and G plus C was not similar for all organisms. Some have a predominance of A and T, whereas others have a predominance of G and C. Both results fitted with the notion that genes could be made of DNA.

TABLE 1-1. Chargaff's base analyses

	Source of DNA					
Base	Human sperm		Ox thymus		Yeast	
A	0.29		0.28		0.30	
T	0.31	A/T 0.94	0.24	A/T 1.16	0.29	A/T 1.03
G	0.18		0.24		0.18	
C	0.18	G/C 1.00	0.18	G/C 1.33	0.15	G/C 1.20

These are three of Chargaff's most successful analyses where the recovery of material was more than 90% and the base ratios varied between 0.94 and 1.33.

The Very Long Linear Nature and Diameter of DNA Are Determined Using Electron Microscopy

Although in the late 1930s Swedish physical chemists had obtained evidence that DNA molecules were highly asymmetrical from their behavior in solution, direct measurements of their size became possible only when the electron microscope came into general use in the years following the end of World War II. All carefully prepared samples showed extremely elongated molecules many thousands of angstroms (Å = 10^{-10} m) in length and approximately 20 Å thick. All the molecules were unbranched, which confirmed the linear structure proposed by organic chemists. From the lengths of the molecules and the fact that each nucleotide base is just more than 3 Å long, it was clear that most DNA molecules were composed of many thousands of nucleotides and were very possibly much larger than any other natural polymeric molecules.

The Nucleotides of DNA and RNA Are Linked by 5′-3′ Phosphodiester Bonds

But how are the many thousands of nucleotides, the building blocks of DNA, linked together to make these huge molecules? Several years after World War II ended, Alexander Todd, already one of England's most effective chemists, decided to tackle this question. By 1952, his large research group at the Chemical Laboratories of Cambridge University had established the precise phosphate-ester linkages that bound the nucleotides together. Their results were appealingly simple. These linkages were always the same, with the phosphate group connecting the 5′ carbon atom of one deoxyribose residue to the 3′ car-

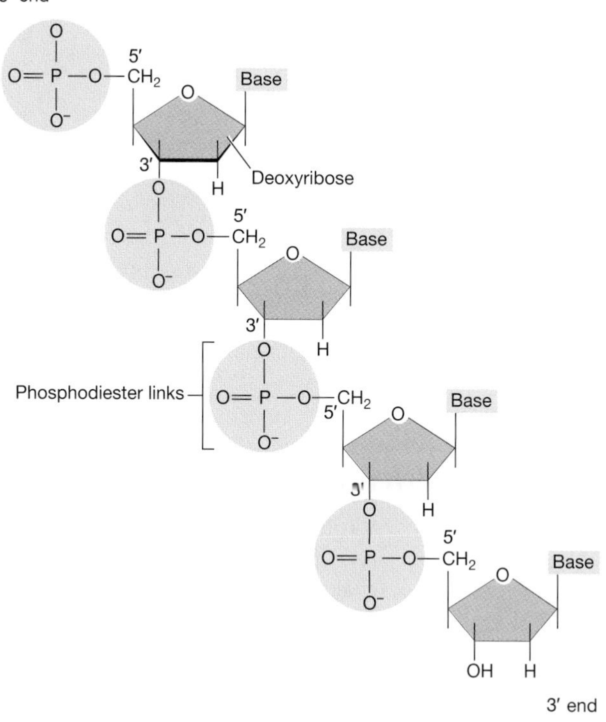

FIGURE 1-10
Individual nucleotides of the DNA molecule are linked together by phosphodiester bonds between the deoxyribose of one nucleotide and the phosphate group of the succeeding nucleotide. This nucleotide-phosphate chain forms the "backbone" of the DNA molecule.

bon atom of the deoxyribose in the adjacent nucleotide (Fig. 1-10). This means that DNA strands end with a 5′-phosphate group at one end and a 3′-hydroxyl group at the other. No traces of any unusual bonds were found, and Todd's group concluded that the polynucleotide chains of DNA, like the polypeptide chains of proteins, are strictly linear molecules. It took longer to settle the question of the nature of the linkages in RNA; but three years later, in 1955, RNA was found to have a highly regular backbone that also employs only 5′-3′ phosphodiester links to hold together its component nucleotides.

A Biological Assay Implicates DNA as the Primary Genetic Molecule

As far back as 1928, the English microbiologist Frederick Griffith, when investigating the pathogenicity of the pneumonia-causing bacterium *Diplococcus pneumoniae*, made a most unexpected observation. Heat-killed pathogenic cells, when mixed with living nonpathogenic cells and injected into mice, were able to transform a small percentage of the nonpathogenic cells into pathogenic cells. In becoming pathogenic, the nonvirulent cells acquired the thick, outer, polysaccharide-rich cell wall (the capsule) that confers pathogenicity to cells that possess it. Griffith thus demonstrated that some substance in the pathogenic cells, not destroyed by heat, could later move into nonpathogenic cells and direct them to make capsules. Furthermore, these newly pathogenic bacteria retained their pathogenicity when isolated and reinjected into other mice. The change appeared heritable.

Griffith himself did not try to identify this substance. This task was taken up by Oswald Avery, whose scientific career at the Rockefeller Institute in New York had been principally spent working out the chemistry of the polysaccharide-containing outer capsules of bacteria. When he began working on the "transforming factor," Avery thought it likely that the active substance would be a complex polysaccharide that in some way primed the synthesis of more polysaccharides of the same kind. Instead, a decade of intensive studies finally ended with Avery and his younger colleagues, Maclyn McCarty and Colin MacLeod, concluding in 1943 that the transforming factor was DNA (Fig. 1-11). Not only was DNA the predominant molecule in their most purified preparation of the transforming factor, but the transforming activity was destroyed by a highly purified preparation of DNase, a then just-discovered enzyme that specifically breaks down DNA. In contrast, the transforming activity was unaffected by exposure to enzymes that degrade proteins or RNA.

Avery's experiments, publicly announced in 1944, had been so carefully done that many scientists accepted his conclusion that DNA was the transforming factor. However, skeptics, including some of Avery's colleagues at the Rockefeller Institute, preferred to believe that Avery had somehow missed the "genetic protein" and that DNA was required for activity in the transformation assay only because it functioned as an unspecific scaffold to which the real gene material—protein—was fixed. There was also the question of the generality of Avery's observation. Perhaps his conclusions drawn from studies of pathogenic bacteria were not true of all genes and might not be applicable to higher organisms. Eight years were to pass before an experiment using bacterial viruses (bacteriophages—phage is from the Greek *phagos* meaning "eating") further pinpointed DNA as the likely genetic molecule.

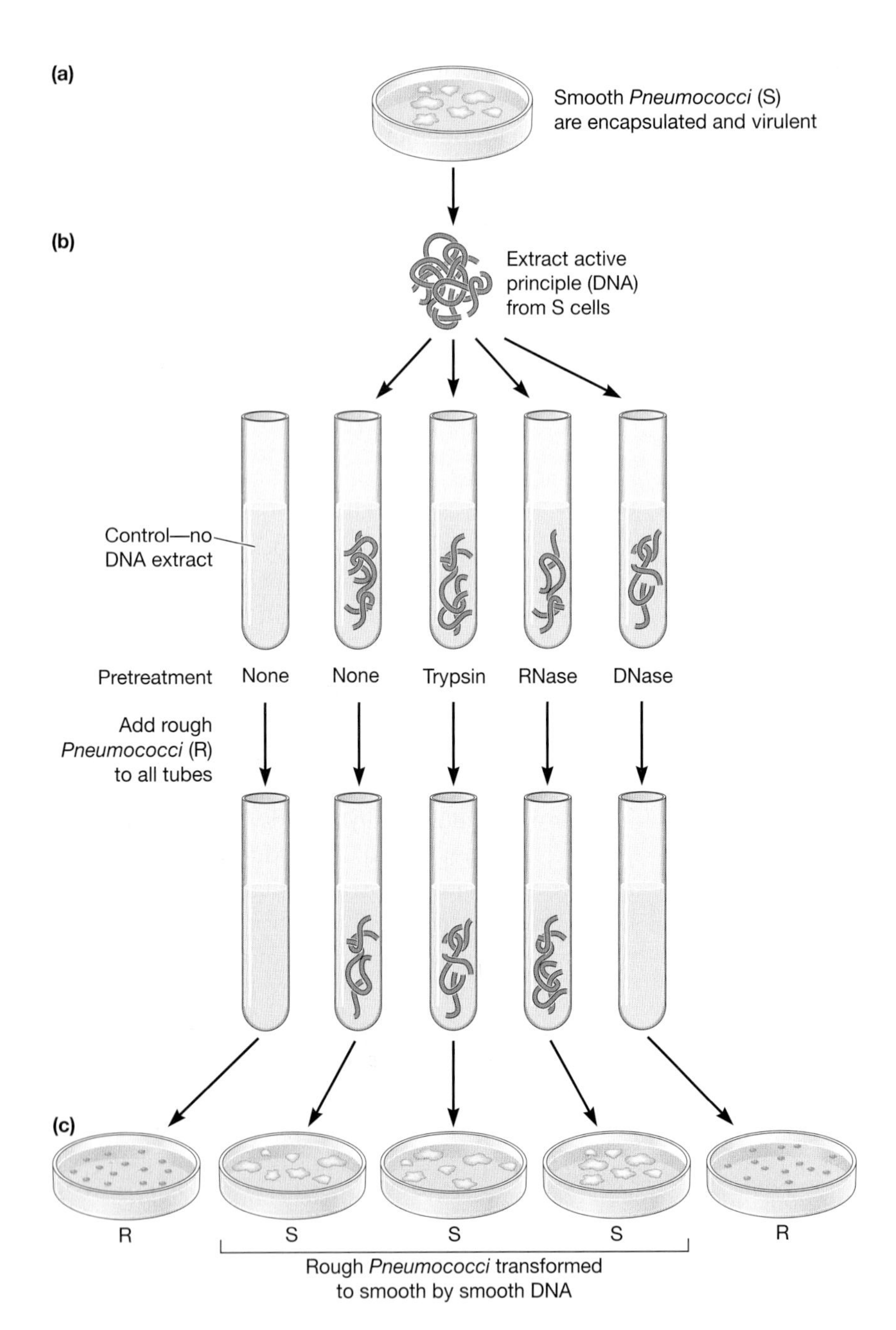

FIGURE 1-11

Avery's transformation experiment. (*a*) Virulent forms of the bacterium *Pneumococcus* have a polysaccharide coat and form large, smooth-surfaced colonies ("S") in culture. The "rough" variants lack the polysaccharide coat and form characteristic rough colonies in culture ("R"). (*b*) Avery prepared DNA from Type III S bacteria. In some cases, he pretreated the DNA with enzymes to determine the nature of the transforming factor. Then Type II R bacteria were added. (*c*) With no extract of S bacteria, the R bacteria continued to produce rough colonies. With S DNA, the R bacteria were transformed and produced a polysaccharide coat, consequently growing as smooth colonies. Pretreatment with proteolytic enzymes (trypsin and chymotrypsin) or ribonuclease had no effect, whereas transforming activity was completely abolished by pretreatment with an extract of dog intestinal mucosa (known to destroy DNA).

Viruses Are Packaged Collections of Genes That Transmit Their Instructions from Cell to Cell

Interest in DNA had also risen as a result of its discovery in the 1930s in several highly purified viruses including bacteriophages. The nature of these tiny disease-causing particles, which multiply only in living cells, was long disputed. Some scientists considered them a sort of naked gene(s); others preferred to think of them as the smallest form of life. Only when it became possible to purify viruses and look at them in the electron microscope did their nature begin to be revealed. They were clearly not minute cells, and as DNA could be readily released from bacteriophage by osmotically shocking them, they appeared to be nothing more than protein packages containing nucleic acid. The best guess, therefore, was that they were obligatory parasites at the genetic level, incapable of multiplying except within living cells. Conceivably by studying such simple viruses to find out how they multiplied, we would understand how genes worked and replicated.

In the 1940s, the physicist Max Delbrück, the physician Salvador Luria, and the physical chemist Alfred Hershey intensively studied a group of phages named T1, T2, T3, T4, etc., which multiply within the common intestinal bacterium *Escherichia coli.* These were eminently suitable for genetic analysis because a single infecting particle needs only some 20 minutes to produce within its host bacterial cell several hundred progeny particles. By 1944, Luria was isolating the first phage mutant with only two more years passing before Hershey demonstrated that phage chromosomes also participate in crossing-over-like events that rearrange parental genotypes. This genetic approach culminated in the experiments done at Purdue University by Seymour Benzer in the mid-1950s. His results took the understanding of the nature of the gene a step farther. He isolated many hundreds of mutations sited within the *rII* gene of the phage T4 and showed that genetic recombination occurs within genes as well as between genes.

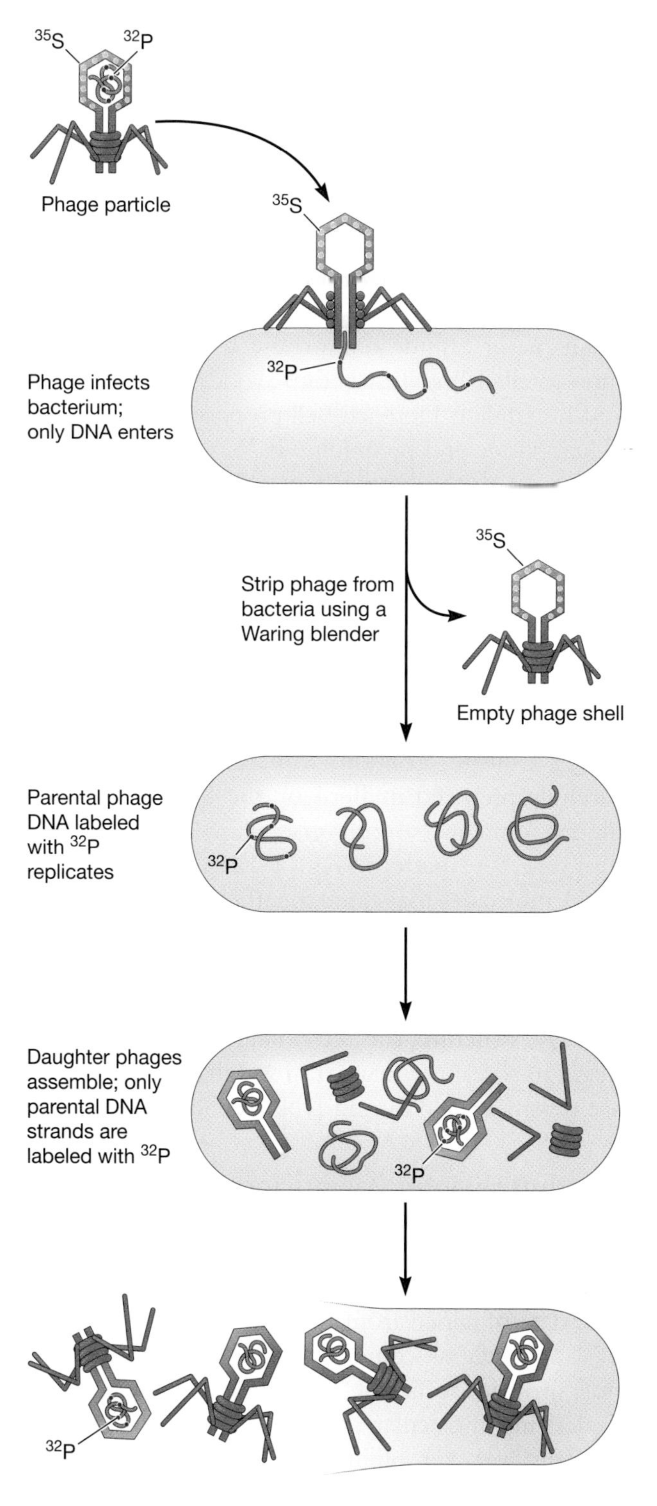

Phage DNA Alone Carries Genetic Information

Purely genetic experiments could not determine whether DNA or one of the protein components of the phage particle carried its genetic specificity, but by 1951, electron microscope pictures of bacteriophage infecting bacteria showed that the phage coats remain attached to the outside of the bacteria. This suggested that phage DNA alone was necessary for the production of new phage. Conceivably the phage protein was simply a protective coat and the means by which the DNA was introduced into the bacterial cell. To test this hypothesis, in 1952, Alfred Hershey and Martha Chase at Cold Spring Harbor, New York, carried out one of the classic experiments of molecular genetics (Fig. 1-12). Using ^{35}S to label phage protein and ^{32}P to label phage DNA, they found that most of the ^{35}S remained outside the infected bacteria, whereas most of the ^{32}P entered the bacteria. Furthermore, as much as 30% of this ^{32}P but less than 1% of the ^{35}S reappeared in the phage progeny, indicating the continuity of the DNA from infecting phage to the progeny. Very likely DNA alone was essential for productive infection.

FIGURE 1-12

The Hershey–Chase experiment. T2 bacteriophages were grown in medium containing ^{32}P to label their DNA and ^{35}S to label the protein coat. Phages were added to a culture of *Escherichia coli* and, at various times after infection, the phages were detached from the bacteria by vigorous shearing in a Waring blender. Analysis showed that most of the ^{35}S activity was removed from the *E. coli,* whereas most of the ^{32}P activity remained with the *E. coli.* Furthermore, if the infection was not interrupted and the progeny phage analyzed, the latter contained ^{32}P, thus indicating that there was continuity of DNA from the infecting phage to their progeny. These results were to be expected if DNA was the hereditary material of the phage.

However, the same objection that was raised against Avery's experiments and his conclusion that DNA was the hereditary material was equally applicable to Hershey and Chase's research. They could not exclude the possibility that a small amount of "genetic protein" was included in the ^{35}S-labeled phage protein that entered the bacterial cells. Nonetheless, the Hershey and Chase experiment, because of its ingenuity and elegance, increased the expectation that DNA was the molecule to study if you wanted to learn the chemical essence of the gene.

DNA Has a Regular, Repeating Shape

To gain a deep understanding of what a biological molecule does, and how it does it, requires knowing its structure. Ultimately, all biology comes down to molecules interacting with each other and these interactions occur at the atomic level. How a hemoglobin molecule takes up oxygen molecules, for example, was only understood once Max Perutz and his colleagues at Cambridge University knew the atomic structure of the hemoglobin molecule. The same is true of DNA with the first attempts to determine the molecular shape of DNA beginning in the 1930s.

In one sense DNA chains are very regular; they contain repeating sugar deoxyribose-phosphate residues that are always linked together by exactly the same chemical bonds. These identical repeating groups form the "backbone" of DNA. However, if the four different bases in DNA are attached in any order along the backbone, this variability would give DNA molecules a high degree of individuality. So, depending on which part of a DNA molecule we focus on, we may view it as regular or as possibly highly irregular. More important, however, is the question of the overall configuration of a DNA molecule: Does the DNA chain fold up into a regular three-dimensional (3D) configuration dominated by its regular backbone? If so, the configuration would most likely be a helical one in which all the sugar-phosphate groups would have identical chemical environments. On the other hand, if the chemistry of the bases dominates the DNA structure, we might fear that no two chains would have identical 3D configurations, because the sequences of the bases are irregular. Were this the case, the task of figuring out how each of these differently shaped molecules could serve as a template for the formation of another DNA chain would have been beyond our reach.

The only direct way to examine the 3D structure of DNA was to see how it diffracts (bends) X rays. Dry DNA has the appearance of irregular white fluffs of cotton, but it becomes highly tacky when it takes on water, and it can then be drawn out into thin fibers. In structural analysis, such fibers are placed in the path of an X-ray beam, and the pattern of the diffracting rays is recorded on photographic film. DNA was first examined this way in 1938 by William Astbury in England, using material prepared in Sweden by Einar Hammarsten, who had used conditions of mild pH and low temperature to isolate DNA of high molecular weight from thymus glands. This DNA did indeed yield a distinctive diffraction pattern, and so many individual DNA molecules must share some preferred orientations. However, the individual diffraction spots were not sharp like those produced by truly crystalline substances, and the possibility remained that DNA chains never assume one precise configuration common to all chains. Just before World War II, Astbury and his student Florence Bell proposed that the individual purine and pyrimidine bases were stacked perpendicular to the long axis of a DNA molecule as if they were a pile of pennies. Better diffraction patterns taken after the war, however, were still difficult to interpret; we now know that Astbury's preparations were mixtures of what came to be called the A and B forms of DNA.

Then in 1950 the physicist Maurice Wilkins and his research student Raymond Gosling, working at King's College, London, with DNA that had been carefully prepared in Bern by the Swiss chemist R. Singer, obtained a truly crystalline diffraction pattern, the "A" pattern (Fig. 1-13a). The individual DNA molecules that came together to form the crystalline fibers must have been very similar in form, or they would not have been able to pack together so regularly. It thus became certain that DNA does have a precise structure, the determination of which might begin to reveal the manner in which DNA molecules are exactly copied during DNA replication.

Two Intertwined Polynucleotide Chains Are the Fundamental Unit of DNA (the Double Helix)

The data obtained from the fiber diffraction pattern of a molecule as complicated as DNA usually does not provide sufficient information to reveal the underlying molecular structure. Inspection of such patterns, however, often provides key parameters that strongly demarcate the outlines of the molecule under investigation. This proved to be the case with DNA. The key X-ray patterns turned out to be not

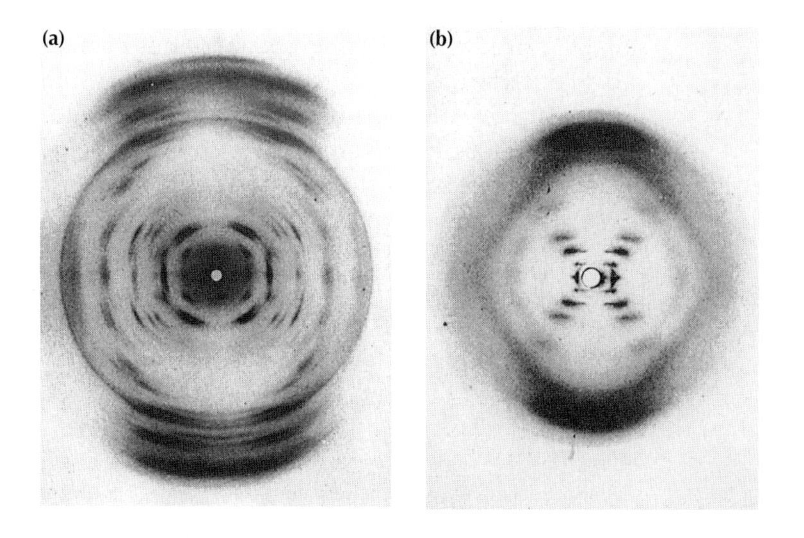

FIGURE 1-13

The X-ray diffraction patterns of DNA. (*a*) The crystalline A form; (*b*) the B form produced at high humidity. The X pattern of diffraction spots is characteristic of a helical structure.

those obtained from the crystalline DNA fibers, but rather those obtained from the less-ordered aggregates that form when DNA fibers are exposed to a higher relative humidity where they can take up more water. These paracrystalline patterns, the "B" form (Fig. 1-13b), were first seen in the summer of 1951 by the English physical chemist, Rosalind Franklin, a colleague of Wilkins at King's College. Her pictures revealed a dominant cross-like pattern, the telltale mark of a helix. Thus, despite the presence of an irregular sequence of bases, the sugar-phosphate backbone of DNA somehow assumed a helical configuration. A separate nucleotide was found every 3.4 Å along its fiber axis, with 10 nucleotides, or 34 Å, being required for every turn of the helix.

A very important inference came from X-ray measurement of the diameter of the helix. Its 20-Å breadth was far too large for a DNA molecule containing only one chain of nucleotides, but how many chains were there? Difficulties in measuring the exact water content of the DNA fiber led to initial confusion about whether two or three chains intertwined. Fortunately X-ray diffraction symmetry arguments favored molecules with two chains that ran in opposite directions. Prior to the use of X-ray diffraction methods, no chemist had ever suspected that DNA was a multichained molecule.

The Two Polynucleotide Chains Are Held Together by Hydrogen Bonds between Complementary Base Pairs

How the two chains are held together in the DNA molecule could not be ascertained from the X-ray data alone. To find out, the biologist James Watson and physicist Francis Crick, working at the Cavendish Laboratory of Cambridge University in February of 1953, built 3D models of DNA to look for the energetically most favorable configurations compatible with the helical parameters provided by the X-ray data at King's College. This approach led them to a molecular model in which the sugar-phosphate backbones are on the outside of the DNA molecule and the purine and pyrimidine bases are on the inside, oriented so that the bases on one strand can form hydrogen bonds to those bases on the opposing chains. Whereas hydrogen bonds are weak noncovalent interactions, there are so many of these bonds between the two chains that the paired chains are held together firmly.

The exact hydrogen-bonding pattern used in DNA emerged with the realization from model building that if a purine on one chain is always hydrogen-bonded to a pyrimidine on the other chain, the dimensions of the paired bases would be the same throughout the length of the DNA molecule. Equally important, the two purines adenine and guanine bond selectively to the two pyrimidines thymine and cytosine, so that adenine (A) can pair only with thymine (T) and guanine (G) can bond only with cytosine (C), providing a structural chemical explanation for Chargaff's rules: A = T and G = C (Fig. 1-14). The earlier anomaly that the T2 bacteriophage did not have cytosine now provided support for the new DNA structure. In 1952, a modified cytosine—5-hydroxymethyl cytosine—had been discovered in T2 DNA. Just like cytosine in other DNAs, the amount of methyl cytosine in T2 was equal to that of T2's guanine, and it formed base pairs just like cytosine.

Each of these base pairs possesses a symmetry that permits it to be inserted into the double helix in two orientations (A:T and T:A; G:C and C:G); thus, along any given DNA chain, all four bases can exist

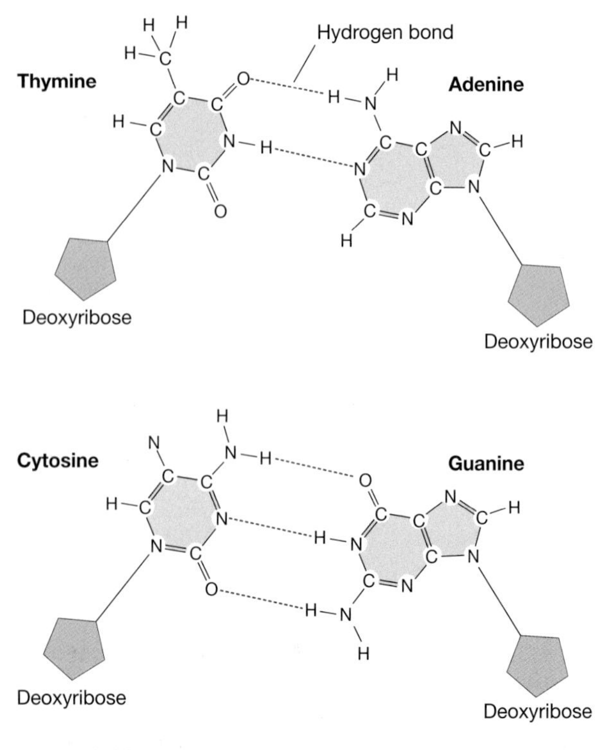

FIGURE 1-14

Hydrogen bonding between the adenine–thymine and guanine–cytosine base pairs. There are two hydrogen bonds holding together the A:T pair and three between G and T.

in all possible permutations of sequence. Because of the specific base pairing, if the sequence of one chain (e.g., TCGCAT) is known, then that of its partner (AGCGTA) is also known. The opposing sequences are referred to as *complementary* and the corresponding polynucleotide partners as *complementary chains.* Only with complementary base pairing could all the backbone sugar-phosphate groups have identical orientations and permit DNA to have the same structure with any sequence of bases (Fig. 1-15). It was also evident from the symmetry of the molecule that the two strands of the double helix ran in opposite directions; the 5′-phosphate at the end of one strand being opposite the 3′-hydroxyl of the other strand. (By convention, single strands of DNA are usually drawn with the 5′ end on the left.)

The Complementary DNA Chains Act as Templates for the Synthesis of New Chains during Replication

Before we knew what genes looked like, it was almost impossible to speculate wisely about how they could be exactly duplicated prior to cell division. Any proposal had to be general. Linus Pauling and Max Delbrück had suggested in 1940 that the surface of the gene somehow acts as a positive mold, or template, for the formation of a molecule of complementary (negative) shape, in just the same way that material can be molded around a piece of sculpture for the purpose of making a cast. The complementary shaped molecule could then serve as the template for the formation of its own complement, thereby producing an identical copy of the original

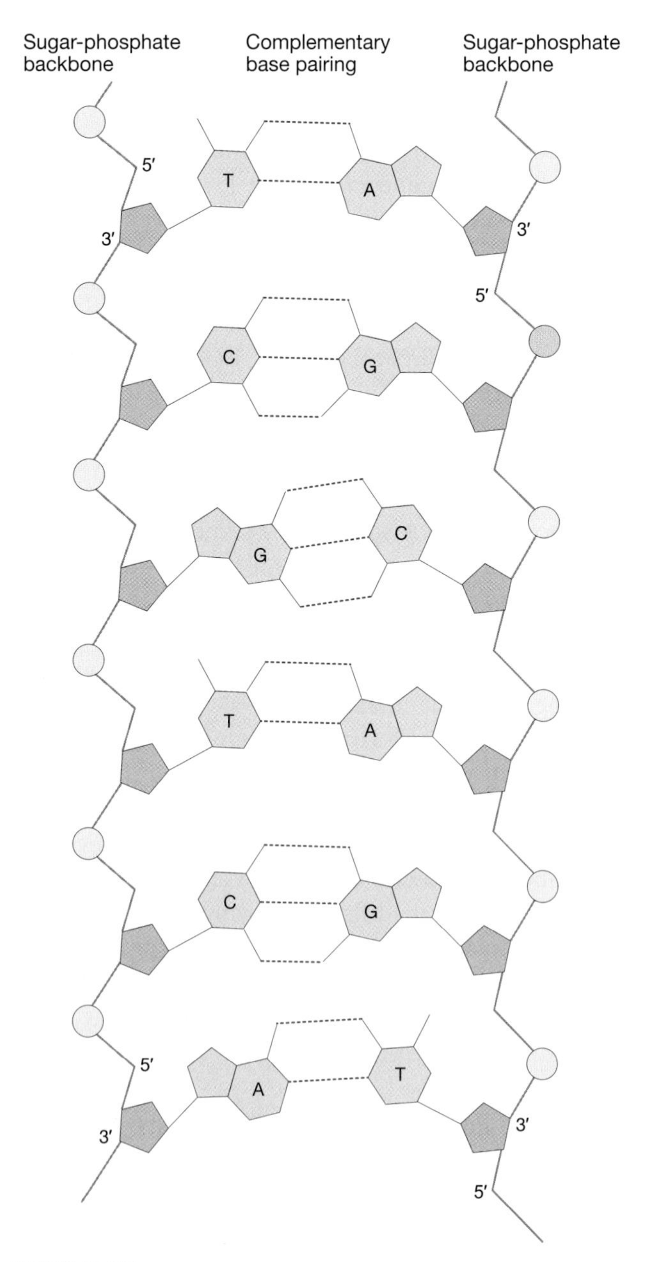

FIGURE 1-15

The DNA molecule has two sugar-phosphate chains held together by hydrogen bonds between complementary base pairs, A:T and G:C.

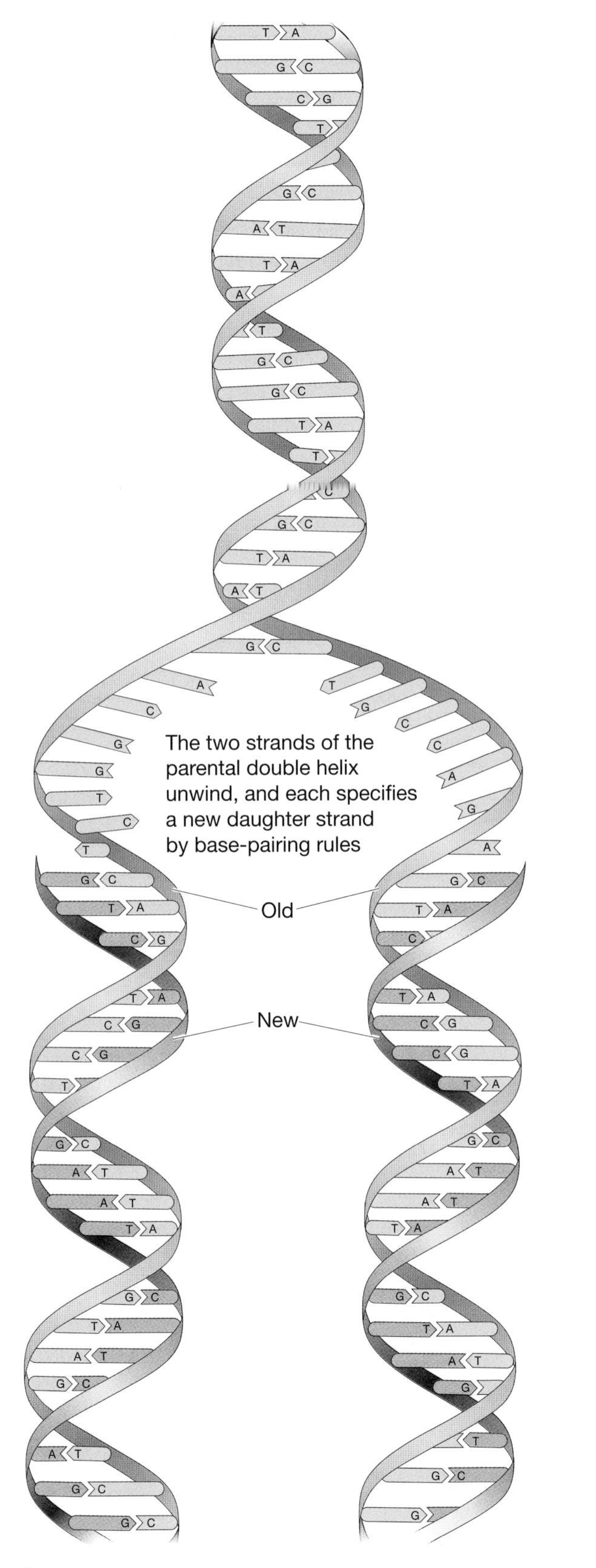

FIGURE 1-16

Identical daughter double helices are generated through semiconservative replication of DNA. The strands of the original helix separate and act as templates for the synthesis of complementary daughter strands.

mold. Thus the realization that the two chains of DNA had complementary shapes caused great excitement, and it was promptly proposed that the two strands of the double helix should be regarded as a pair of positive and negative templates, each specifying its complement and thereby capable of generating two daughter DNA molecules with sequences identical to those of the parental double helix (Fig. 1-16). What was unknown was whether this process was *conservative*, in which case the original double strands stay together and the new complementary strands form a new double helix, or *semiconservative*, where the two original strands separate completely during synthesis of new complementary strands, so that each new double helix contains one parental and one new strand.

DNA Replication Is Semiconservative, Producing One Newly Made Strand Paired with Each Older Strand

It was some five years before there was firm evidence of strand separation. Proof came from the experiments of Matthew Meselson and Franklin Stahl at the California Institute of Technology, in 1958. They had the clever idea of using density differences to separate parental DNA molecules from daughter molecules (Fig. 1-17). They first grew cultures of the bacterium *E. coli* in a medium highly enriched in the heavy isotope ^{15}N. By virtue of its isotopic content, the DNA in these bacteria was much heavier than the DNA produced when cells were grown in the presence of the far more abundant lighter natural isotope ^{14}N. Because of its greater density, the heavier DNA could be clearly separated from the light DNA by high-speed centrifugation in a density gradient of cesium chloride. When heavy DNA–containing cells were transferred to a normal "light" medium and allowed to multiply for one generation, all the heavy DNA was replaced by DNA of a density halfway between heavy and light. The disappearance of the heavy DNA showed that products of replication did not include the original double helix, and thus that the process was semiconservative, with each of the daughter molecules having one heavy (parent) strand and one light (newly synthesized) strand.

Whether the complementary strands completely separate before replication starts was not immediately known. Later, abundant electron microscopic evidence of Y-shaped replication forks indicated that strand separation and replication go hand in hand. As soon as a section of double helix begins to sepa-

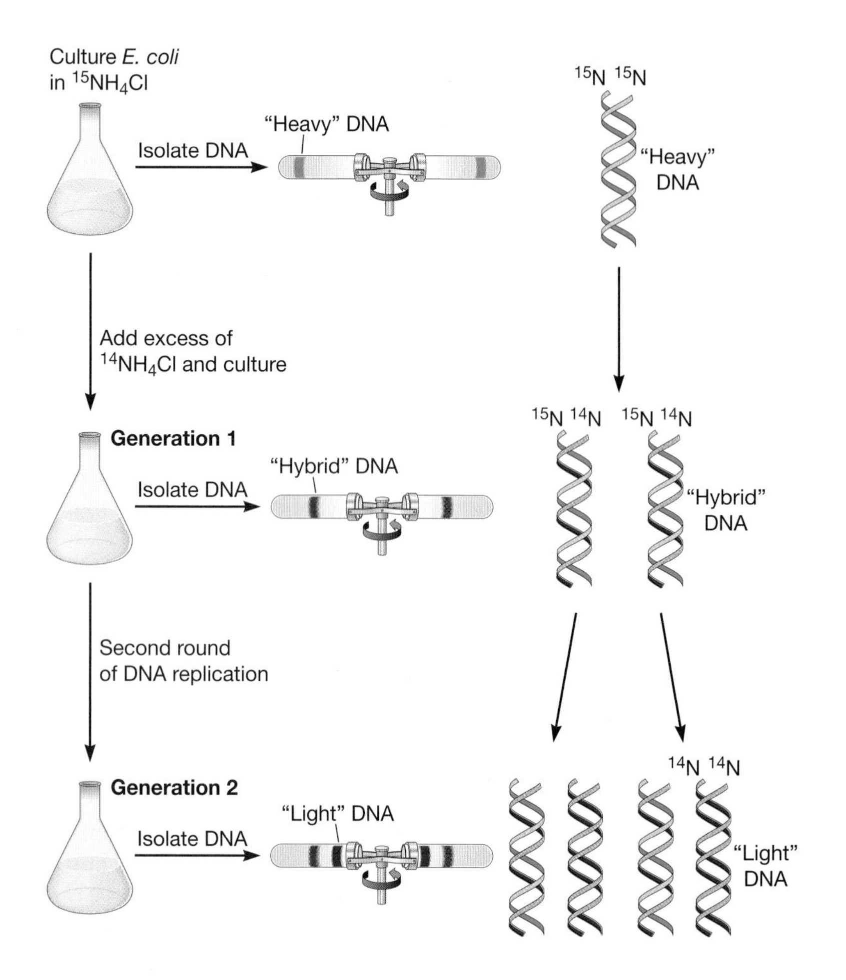

FIGURE 1-17

The Meselson–Stahl experiment demonstrating semiconservative replication of DNA. *Escherichia coli* were grown in medium containing $^{15}NH_4Cl$ to produce "heavy" DNA molecules. Samples were taken from this culture and then at various times after the addition of an excess of $^{14}NH_4Cl$. The ^{14}N was incorporated into replicating DNA, thus producing DNA molecules less dense than those containing ^{15}N. Molecules of differing density can be separated by ultracentrifugation in a gradient of cesium chloride—molecules of higher density forming a band lower in the gradient than molecules of lesser density. After one generation growing in $^{14}NH_4Cl$, a single band of DNA was found, of lower density than the first sample taken prior to addition of the $^{14}NH_4Cl$. The disappearance of the band corresponding to DNA containing only ^{15}N demonstrated that there were no DNA molecules containing two strands labeled with ^{15}N—the single band must contain hybrid DNA molecules with one ^{15}N-labeled and one ^{14}N-labeled strand. After two generations, an even lighter band was found in addition to the band of hybrid DNA. This corresponded to molecules in which both strands are labeled with ^{14}N.

rate for replication, the resulting single-stranded regions are quickly used as templates and become new double-helical regions.

Test Tube Replication of DNA Requires Preexisting DNA Templates

Independent proof that DNA chains serve as direct templates for the formation of progeny DNA chains came from experiments in Arthur Kornberg's lab at Washington University in St. Louis between 1956 and 1958. Working with cell-free extracts of *E. coli*, Kornberg first observed incorporation of radiolabeled deoxyribonucleotides into molecules characteristic of DNA; that is, they were acid insoluble, destroyed by DNase, and resistant to RNase. Through extensive fractionation of these cell-free extracts, the deoxyribonucleotide polymerizing enzyme, DNA polymerase, was discovered, as well as the need for already existing DNA chains to serve as templates for the formation of chains of complementary sequence. Most importantly, the base composition of the test tube–made DNA was identical to the base composition of the template DNA, showing that the latter determined the composition of the former (Table 1-2). This was remarkable to the biochemists because here an enzyme was taking *instructions* from a template, rather than carrying out a straightforward catalytic reaction.

TABLE 1-2. Nucleotide composition of enzymatically synthesized DNA compared with that of the template DNA

DNA	A	T	G	C
Bacteriophage T2				
Template	1.31	1.32	0.67	0.70
Product	1.32	1.29	0.69	0.70
Mycobacterium phlei				
Template	0.65	0.66	1.35	1.34
Product	0.66	0.65	1.34	1.37
Calf thymus				
Template	1.14	1.05	0.90	0.85
Product	1.12	1.08	0.85	0.85
A:T copolymer				
Product	1.99	1.93	<0.05	<0.05

The base composition of the DNA product closely resembles that of the template DNA, whereas the chemically synthesized DNA composed of only A and T directs the synthesis of new DNA molecules with only Ts and As.

Curiously, the Kornberg DNA polymerase is not the polymerase that *E. coli* uses to synthesize DNA in vivo. John Cairns and Paula De Lucia working at Cold Spring Harbor mutagenized *E. coli* and assayed the mutant cells for DNA polymerase activity. After testing more than 3000 mutants, they found one that had no measurable activity yet had normal growth. The strain was more suceptible to DNA damage by UV light, suggesting that perhaps the Kornberg polymerase, now called DNA polymerase I, had a role in DNA repair. Subsequently, two other *E. coli* DNA polymerases were discovered, remarkably by Tom Kornberg, one of Arthur Kornberg's sons! It is DNA polymerase III that is the principal enzyme used to replicate DNA.

Before initiating this work, Kornberg, like most of his fellow enzymatically focused biochemists, regarded the double helix as a hypothesis not yet established. Doubts as to the double helix's reality became intellectually untenable once the Kornberg and Meselson–Stahl experiments became known.

DNA Replication Is Carried Out by a Highly Complex Assembly of Proteins

DNA polymerase III is a very large, molecular machine made up of nine subunits, as befits the number of activities it has to carry out during replication. These include clamping onto the DNA strand, moving along the strand, synthesizing the new strand, checking that the synthesis is correct, and interacting with other proteins essential for replication. These are helicases that unwind the helix, topoisomerases that relieve the stresses created by unwinding, and DNA-binding proteins that stabilize the unwound DNA. The point at which the strands are separated is called the replication fork, which progresses along the DNA as it is unwound and new strands are synthesized.

However, DNA replication is complicated. All DNA polymerases add the new nucleotide to the 3′-OH group of the last nucleotide in the growing chain. This is synthesized from 5′ to 3′, so that the polymerase moves along the template DNA strands from 3′ to 5′. For one of the two DNA strands (*leading*), this works well, but the other strand (*lagging*) runs in the reverse direction. Reiji Okazaki showed that new DNA is synthesized on the lagging strand in short sections, called *Okazaki fragments.* Synthesis on both strands is performed by the same polymerase complex, the DNA of the lagging strand being looped through the complex. The gaps between the Okazaki fragments are filled by DNA polymerase I and the fragments ligated together by DNA ligase. Given the complex choreography of so many enzymes at the replication fork, synthesis of DNA is fairly rapid, about 1000 nucleotides per second.

DNA Molecules Can Be Denatured and Renatured

Each DNA molecule contains so many base pairs that the complementary chains of a DNA helix never spontaneously separate under physiological conditions. If, however, double helices are exposed to near-boiling temperatures or to extremes of pH, they fall apart (i.e., they are denatured or "melt") into their single strands. The temperature at which half of the DNA molecules in a sample have denatured is called the *melting temperature* (T_m); double helices with an excess of G:C base pairs are more stable than helices in which A:T base pairs predominate and have a higher T_m. This is due to the fact that Gs and Cs in DNA have increased *stacking* of bases between adjacent base pairs, which stablizes the double helix, as well as the presence of three hydrogen bonds between G:C base pairs compared to two for A:T pairs. In fact, the proportion of A:T to G:C base pairs in a DNA sample can be directly ascertained by measuring the T_m of the sample.

At first, denaturation was regarded as essentially irreversible, but in 1960 Julius Marmur, Paul Doty, and their coworkers at Harvard showed that complementary single strands recombine to form native double helices when they are kept for several hours at subdenaturing conditions, approximately 65°C (Fig. 1-18). That renaturation of such long molecules occurs at all is remarkable. What is amazing is that renaturation is very specific and will produce perfect double helices when the sequences of the bases of the two combining strands are exactly complementary. These reannealed molecules are biologically functional when transferred into cells.

Annealing is now called *hybridization* and it is an essential and powerful tool in the molecular geneticist's toolbox. For example, given one piece of DNA, that DNA can be used to find its complement, even in a whole genome's worth of DNA. By labeling a DNA fragment with a fluorescent marker, the location of its complement can be found in a section of tissue; and by observing the extent of annealing between DNA strands in solution, researchers can determine the degree of similarity between DNA molecules from different species,

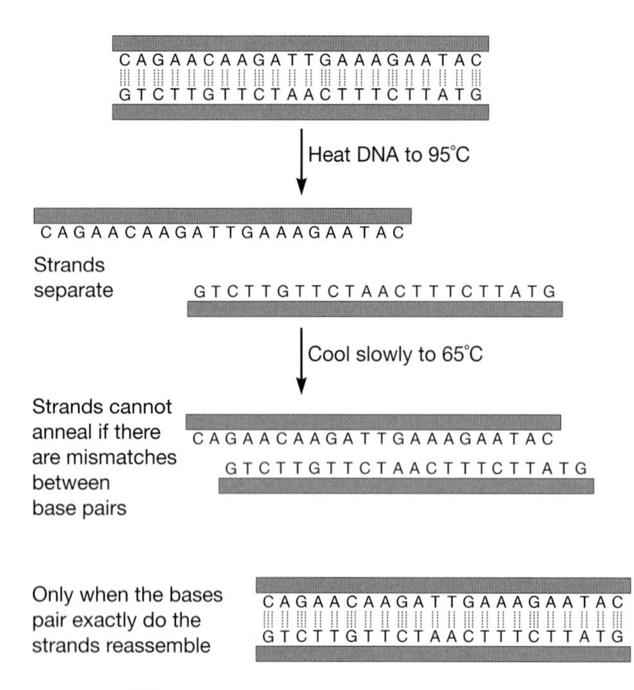

FIGURE 1-18

The base pairing in a small section of a DNA molecule. The three hydrogen bonds between the guanine and cytosine bases and the two hydrogen bonds between adenine and thymine are represented by three and two lines, respectively. When heated to about 95°C, the strands separate. On slow cooling to about 65°C, the strands come together again, but reannealing happens only between perfectly or near-perfectly complementary sequences.

and thus how closely the species are related genetically.

Chromosomes Are Very Long DNA Molecules

When the structure of the double helix was determined, it was not known whether chromosomes of the size of *E. coli*'s contained one or more DNA molecules. Likewise, did a single DNA molecule exist within each of the very much larger chromosomes of higher plants and animals? The latter contain on average some 20 times more DNA than occurs in the *E. coli* chromosome. Initially it was believed that many separate DNA molecules, joined by protein links, were used to construct all but the smallest chromosomes. This picture changed with the realization that long DNA molecules are inherently fragile and easily broken into much smaller fragments. When such shearing was minimized, intact phage DNA chromosomes containing some 200,000 base pairs of phage T2 were seen.

Now that the genomes of many organisms have been sequenced in their entirety, we know that the chromosome of *E. coli* is a single DNA molecule, 4,639,221 bp long, whereas the DNA molecule in human chromosome 6 is 166,880,988 bp long, plus an estimated additional 480,000 bp that could not be sequenced (Table 1-3). A chromosome is thus properly defined as a single, genetically specific DNA molecule to which are attached large numbers of proteins that are involved in maintaining chromosome structure and regulating gene expression.

Some Viruses Have an RNA Genome

Not all organisms use DNA for their genetic material. Indeed, the first virus to be isolated and crystallized—tobacco mosaic virus—was found to contain RNA rather than DNA. Other viruses with RNA genomes include bacteriophage R17, influenza virus, and poliovirus, and among these, there are viruses with both single- and double-stranded RNA. These viruses replicate their RNA genomes in the same way that other organisms replicate their DNA

TABLE 1-3. DNA molecules in chromosomes

Organism and chromosome	Type	Size (bases)	Length (mm)[a]
Bacteriophage λ	Virus	48,502[b]	0.016
Mycoplasma genitalium	Mycoplasma	580,074[b]	0.197
Escherichia coli K12-MG1655	Bacterium	4,639,221[b]	1.577
Saccharomyces cerevisiae chromosome IV[c]	Yeast	1,531,974[d]	0.521
Caenorhabditis elegans chromosome V[c]	Nematode	20,922,238[d]	7.11
Arabidopsis thaliana chromosome 1[c]	Plant	30,494,425[d]	9.90
Drosophila melanogaster chromosome III[c]	Insect	51,243,003[d]	17.42
Homo sapiens chromosome 1[c]	Mammal	246,047,941[d]	83.65

[a]The lengths of the chromosomes were calculated using a value of 0.34 nm per nucleotide. (There are 10 nucleotides per turn of the double helix, and each turn takes 3.4 nm.)

[b]These chromosomes have been sequenced in their entirety.

[c]These are the longest chromosomes of these organisms.

[d]These chromosomes have been sequenced but not completely; some regions that are very difficult to sequence (e.g., centromeres) have not yet been sequenced.

genomes—single RNA chains serve as templates to make chains with complementary sequences.

The retroviruses—for example, human immunodeficiency virus (HIV)—have RNA genomes but replicate very differently from other RNA viruses. The retroviruses were first known for causing cancer in animals; on infecting cells, retroviruses *transform* them into tumorigenic cells. Just as the transformation of pneumococci from one type into another was accompanied by changes in the characteristics of the bacteria, so mammalian cells transformed by RNA tumor viruses take on new characteristics—rapid growth, decreased need for nutrients, and the ability to create tumors when injected into animals. These changes are heritable, and scientists observed that the transformed cells continued to express new proteins, presumably of viral origin. This observation led to the proposal that the RNA tumor viruses were somehow integrated into the DNA chromosomes of the transformed cells. How this was possible—how RNA could be converted into DNA prior to integration—was explained in 1970, when, through the independent research of Howard Temin and Satoshi Mizutani, and of David Baltimore, an enzyme was discovered that carried out the *reverse transcription* of the RNA chromosomes of the infecting viruses into complementary DNA strands.

The enzyme involved, *reverse transcriptase*, is coded for by the viral genomes, and it is present in the viral particles so that it is able to act the moment the infecting RNA chromosome enters an appropriate host cell. The first DNA chain made by reverse transcriptase—complementary to the viral RNA—acts as a template to make a second strand of DNA, thus producing double-helical DNA, which becomes integrated as a *provirus* into host chromosomal DNA. The proviral DNA is replicated along with the host cell DNA. The viral life cycle is completed by transcription of the proviral DNA into an RNA strand identical with that found in the infectious virus and by synthesis of the virus coat proteins. Retroviruses can become integrated into germ cells and passed from generation to generation. These frequently undergo mutations and can no longer replicate, remaining as retroviral "fossils" in the host genome. There are many hundreds of thousands of copies of retroviral-related sequences scattered throughout the chromosomes of most organisms.

Retroviruses have proved to be useful tools for molecular geneticists. The behavior of retroviruses—infecting cells and integrating their DNA into chromosomes—is exploited in gene therapy, where a retroviral vector carries a corrective gene into cells and the gene becomes a permanent addition to the genome (Chapter 6). Reverse transcriptase allows researchers to easily make a DNA copy of any given RNA, and, as we shall see in Chapter 4, this was a key element in developing the techniques for isolating and cloning genes.

By the late 1960s, considerable progress had been made using viruses and bacteria as model systems for studying gene regulation and cancer, but it seemed impossible that we would ever attain a really precise understanding at the molecular level of the crucial workings of the cells of higher organisms. Even though we knew that genes were made of DNA, there seemed to be no way to exploit that knowledge—no way to determine the structure, the sequence of bases, of a particular gene. Furthermore, the idea that it would be possible to isolate and manipulate specific segments of cellular DNA was unthinkable, and understanding how genes are controlled, embryos develop, or cancers arise seemed necessarily to be objectives for the far distant future. All this was to change with the development of recombinant DNA techniques (Chapter 4). As many chapters in this book illustrate, these techniques ushered in a revolution in our understanding of life—a flood of knowledge that continues today and shows no sign of abating.

But in 1953 the first order of business was to work out how genetic information was stored in DNA, and how that information was used by the cell to make proteins. To everyone's surprise, what had been expected by some to be a lifetime's work was achieved within a decade.

Reading List

General

The following three textbooks provide encyclopedic coverage of modern genetics and molecular biology.

Alberts B., Johnson A., Lewis J., Raff M., Roberts K., and Walter P. 2002. *Molecular Biology of the Cell.* Garland, New York.

Lodish H., Berk A., Zipursky S.L., Matsudaira P., Baltimore D., and Darnell J. 2003. *Molecular Cell Biology.* W.H. Freeman, New York.

Watson J.D., Baker T.A., Bell S.P., Gann A., Levine M., and Losick R. 2004. *Molecular Biology of the Gene,* 5th ed. Benjamin Cummings, San Francisco, in conjunction with Cold Spring Harbor Laboratory Press, Cold Spring Harbor, New York.

History

The origins and development of genetics make a fascinating story.

Dunn L.C., ed. 1951. *Genetics in the 20th Century: Essays on the Progress of Genetics during Its First 50 Years.* Macmillan, New York. (A collection of essays written by some of the founders of genetics.)

Sturtevant A.H. 1965. *A History of Genetics.* Cold Spring Harbor Laboratory, Cold Spring Harbor, New York. (Reprinted 2001.)

Judson H.F. 1980. *The Eighth Day of Creation: Makers of the Revolution in Biology.* Simon and Schuster, New York. (Expanded edition, reprinted with a new postscript, Cold Spring Harbor Laboratory Press, 1996; an account of the discovery of the double helix and the development of molecular genetics.)

Carlson E. 2004. *Mendel's Legacy: The Origins of Classical Genetics.* Cold Spring Harbor Laboratory Press, Cold Spring Harbor, New York.

Mendelian Inheritance

Mendel G. 1866. Versuche uber Pflanzen-Hybriden. *Verh. Naturforsch. Ver. Brunn* **4:** 3–47. (This is the citation for Mendel's original paper. German and English versions, annotated with hypertext links, will be found at MendelWeb: http://www.mendelweb.org/)

Olby R.C. 1966. *Origins of Mendelism.* Constable, London. (A discussion of hybridization studies before Mendel.)

Stern C. and Sherwood E.R., eds. 1966. *The Origin of Genetics.* W.H. Freeman, San Francisco. (An English translation of Mendel's paper, together with other interesting documents.)

The Chromosomal Theory of Heredity

Sutton W.S. 1902. On the morphology of the chromosome group in *Brachystola magna. Biol. Bull.* **4:** 124–139.

Stevens N.M. 1905. Studies in spermatogenesis with especial reference to the "accessory" (sex) chromosome. *Carnegie Inst. Wash. Publ.* **36:** 1–33.

Wilson E.B. 1928. *The Cell in Development and Heredity.* Macmillan, New York. (Reprinted, Garland Publishing, New York, 1987; the classic review of cytogenetics.)

Voeller B.R., ed. 1968. *The Chromosome Theory of Inheritance: Classic Papers in Development and Heredity.* Appleton-Century-Crofts, New York. (Extracts from classic papers.)

Linkage

Morgan T.H. 1910. Sex-limited inheritance in *Drosophila. Science* **32:** 120–122.

Sturtevant A.H. 1913. The linear arrangement of six sex-linked factors in *Drosophila,* as shown by their mode of association. *J. Exp. Zool.* **14:** 43–59.

Crow J.F. 1988. A diamond anniversary: The first chromosome map. *Genetics* **118:** 1–3.

Allen G.E. 1978. *Thomas Hunt Morgan: The Man and His Science.* Princeton University Press, Princeton, New Jersey. (A detailed account of Morgan and his research.)

Kohler R.E. 1994. *Lords of the Fly:* Drosophila *Genetics and the Experimental Life.* University of Chicago Press, Chicago. (An interesting account of how *Drosophila* became the organism of choice for genetic analysis.)

Mutations and Genetic Screens

Muller H.J. 1927. Artificial transmutation of the gene. *Science* **46:** 84–87.

Stadler L.J. 1928. Mutations in barley induced by X-rays and radium. *Science* **68:** 186–187.

Nüsslein-Volhard C. and Wieschaus E. 1980. Mutations affecting segment number and polarity in *Drosophila. Nature* **287:** 795–801.

Driever W., Stemple D., Schier A., and Solnica-Krezel L. 1994. Zebra fish: Genetic tools for studying vertebrate development (review). *Trends Genet.* **10:** 152–158.

Carlson E.A. 1981. *Genes, Radiation, and Society: The Life and Work of H.J. Muller.* Cornell University Press, Ithaca, New York. (A biography of a complex scientist.)

Genes outside the Nucleus

Correns C. 1909. Vererbungsversuche mit blaß(gelb) grünen und buntblättrigen sippen bei *Mirabilis jalapa, Urtica pilulifera* und *Lunaria annua. Z. Indukt. Abstammungs-Vererbungsl.* **1:** 291–329.

Gray M.W., Burger G., and Lang B.F. 1999. Mitochondrial evolution. *Science* **283:** 1476–1481.

McFadden G.I. 2001. Chloroplast origin and integration. *Plant Physiol.* **125:** 50–53.

Sapp J. 1987. *Beyond the Gene: Cytoplasmic Inheritance and the Struggle for Authority in Genetics.* Oxford University Press, New York. (An account of research on cytoplasmic inheritance with an interesting sociological analysis.)

DNA as the Primary Genetic Material

Griffith F. 1928. The significance of *Pneumococcal* types. *J. Hyg.* **27:** 13–59.

Avery O.T., MacLeod C.M., and MacCarty M. 1944. Studies on the chemical nature of the substance inducing transformation of *Pneumococcal* types. *J. Exp. Med.* **79:** 137–158. (DNA as the hereditary material.)

Hershey A.D. and Chase M. 1952. Independent functions of viral protein and nucleic acid in growth of bacteriophage. *J. Gen. Physiol.* **36:** 39–56. (A classic experiment.)

Cairns J., Stent G.S., and Watson J.D., eds. 1966. *Phage and the Origins of Molecular Biology.* Cold Spring Harbor Laboratory, Cold Spring Harbor, New York. (A collection of essays celebrating Max Delbrück; reprinted in 1992 with a new introduction and additional materials.)

McCarty M. 1985. *The Transforming Principle: Discovering That Genes Are Made of DNA.* Norton, New York. (An account of the discovery by one of the participants.)

DNA Chemistry and Size

Astbury W.T. and Bell F.O. 1938. X-ray study of thymonucleic acid. *Nature* **141:** 747–748. (The first model for DNA structure.)

Chargaff E. 1950. Chemical specificity of nucleic acids and the mechanism of their enzymatic degradation. *Experientia* **6:** 201–209. (Chargaff draws attention to the base ratios.)

Brown D.M. and Todd A.R. 1952. Nucleotides. Part X. Some observations on structure and chemical behaviour of the nucleic acids. *J. Chem. Soc.* 52–58.

Williams R.C. 1952. Electron microscopy of sodium desoxyribonucleate by use of a new freeze-drying method. *Biochim. Biophys. Acta* **9:** 237–239.

Dekker C.A., Michaelson A.M., and Todd A.R. 1953. Nucleotides. Part XIX. Pyrimidine deoxyribonucleoside diphosphates. *J. Chem. Soc.* 947–951.

Wyatt G.R. and Cohen S.S. 1953. The bases of the nucleic acids of some bacterial and animal viruses: The occurrence of 5-hydroxymethylcytosine. *Biochem. J.* **55:** 774–782.

Portugal F.H. and Cohen J.S. 1977. *A Century of DNA: A History of the Discovery of the Structure and Function of the Genetic Substance.* MIT Press, Cambridge, Massachusetts. (A history beginning with the earliest biochemical studies by Miescher, Kossel, and Levene.)

The Double Helix

Franklin R.E. and Gosling R.G. 1953. Molecular configuration in sodium thymonucleate. *Nature* **171:** 740–741.

Watson J.D. and Crick F.H.C. 1953. Molecular structure of nucleic acids: A structure for deoxyribose nucleic acid. *Nature* **171:** 737–738.

———. 1953. Genetical implications of the structure of deoxyribonucleic acid. *Nature* **171:** 964–967.

Wilkins M.H.F., Stokes A.R., and Wilson H.R. 1953. Molecular structure of deoxypentose nucleic acids. *Nature* **171:** 738–740.

Chambers D.A., ed. 1995. DNA: The double helix—Perspective and prospective at forty years. *Ann. N.Y. Acad. Sci.* **758:** 1–472. (Accounts by key figures in the development of molecular biology, including reprints of the original three *Nature* papers.)

Watson J.D. 1968. *The Double Helix: A Personal Account of the Discovery of the Structure of DNA.* Atheneum, New York. (Reprinted as *The Double Helix* [Norton critical edition, ed. G.S. Stent, Norton, New York, 1980], with several reviews that provide differing perspectives.)

Crick F.C. 1988. *What Mad Pursuit: A Personal View of Scientific Discovery.* Basic Books, New York. (An autobiography covering the whole of Crick's life to the present.)

Maddox B. 2002. *Rosalind Franklin: The Dark Lady of DNA.* HarperCollins, New York.

Wilkins M. 2003. *The Third Man of the Double Helix: The Autobiography of Maurice Wilkins.* Oxford University Press, Oxford.

Ridley M. 2006. *Francis Crick: Discover of the Genetic Code.* HarperCollins, London.

Olby R. 1994. *The Path to Double Helix: The Discovery of DNA.* Dover Publications, New York. (A detailed account of the discovery.)

Replication of DNA

Pauling L. and Delbrück M. 1940. The nature of the intermolecular forces operative in biological processes. *Science* **92:** 77–79.

Meselson M. and Stahl F.W. 1958. The replication of DNA in *Escherichia coli. Proc. Natl. Acad. Sci.* **44:** 671–682.

Kornberg A., Lehman I.R., Bessman M.J., and Simms E.S. 1956. Enzymatic synthesis of deoxyribonucleic acid. *Biochim. Biophys. Acta* **21:** 197–198.

Lehman I.R., Zimmerman S.B., Adler J., Bessman M.J., Simms E.S., and Kornberg A. 1958. Enzymatic synthesis of deoxyribonucleic acid. V. Chemical omposition of enzymatically synthesized deoxyribonucleic acid. *Proc. Natl. Acad. Sci.* **44:** 1191–1196.

De Lucia P. and Cairns J. 1969. Isolation of an *E. coli* strain with a mutation affecting DNA polymerase. *Nature* **224:** 1164–1166.

Kornberg T. and Gefter M.L. 1970. DNA synthesis in cell-free extracts of a DNA polymerase-defective mutant. *Biochem. Biophys. Res. Commun.* **40:** 1348–1355. (This is DNA polymerase II, another repair DNA polymerase.)

———. 1971. Purification and DNA synthesis in cell-free extracts: Properties of DNA polymerase II. *Proc. Natl. Acad. Sci.* 68: 761–764. (This was called *E. coli* DNA polymerase III, which replicates DNA.)

Friedberg E.L. 2006. The eureka enzyme: The discovery of DNA polymerase. *Nat. Rev. Mol. Cell Biol.* **7:** 143–147.

Okazaki R., Okazaki T., Sakabe K., Sugimoto K., and Sugino A. 1968. Mechanism of DNA chain growth. I. Possible discontinuity and unusual secondary structure of newly synthesized chains. *Proc. Natl. Acad. Sci.* **59:** 598–605.

Kornberg A. and Baker T.A. 1992. *DNA Replication.* W.H. Freeman, San Francisco.

Brush G.S. and T.J. Kelly. 1996. Mechanisms for replicating DNA. In *DNA Replication in Eukaryotic Cells* (ed. M.L. DePamphilis), pp. 1–43. Cold Spring Harbor Laboratory Press, Cold Spring Harbor, New York.

DNA Denaturation and Renaturation

Doty P., Marmur J., Eigner J., and Schildkraut C. 1960. Strand separation and specific recombination in deoxyribonucleic acids: Physical chemical studies. *Proc. Natl. Acad. Sci.* **46:** 461–476.

Marmur J. and Lane L. 1960. Strand separation and specific recombination in deoxyribonucleic acids: Biological studies. *Proc. Natl. Acad. Sci.* **46:** 453–461.

Hall B.D. and Spiegelman S. 1961. Sequence complementarity of T2-DNA and T2-specific RNA. *Proc. Natl. Acad. Sci.* **47:** 137–146.

Marmur J. 1994. DNA strand separation, renaturation and hybridization. *Trends Biochem. Sci.* **19:** 343–346. (A historical account.)

Chromosome Sizes

Eukaryote genomes: The European Bioinformatics Institute, Project Ensembl: http://www.ensembl.org/

Prokaryote and viral genomes: The Institute for Genome Research: http://www.tigr.org/tigr-scripts/CMR2/CMRGenomes.spl

Viral genomes: The National Center for Biotechnology Information: http://www.ncbi.nlm.nih.gov/genomes/VIRUSES/viruses.html

Viral Genomes

Baltimore D. 1970. Viral RNA-dependent DNA polymerase. *Nature* **226:** 1209–1211.

Temin H.M. and Mizutam S. 1970. Viral RNA-dependent DNA polymerase. *Nature* **226:** 1211–1213.

Levine A.J. 1992. *Viruses.* Scientific American Library, New York.

CHAPTER 2

Information Flow from DNA to Protein

Once the double helix was found, common sense suggested that the genetic information of genes is embedded within the linear sequences of the four bases A, G, T, and C along DNA molecules. DNA base sequence order is somehow read and acted on by the appropriate cellular machinery to initiate the molecular events that link together in linear fashion the 20 different amino acids of proteins. Now it was possible to examine the geneticist's belief in a one gene–one polypeptide chain connection in molecular terms, restated as the question: How is the four-letter linear language of DNA translated into the 20-letter linear language of polypeptide chains?

Fortuitously, Frederick Sanger and his colleagues were beginning to determine the amino acid sequences of proteins such as insulin at this time, and this raised hopes that the genetic code—the relationship between nucleotides and amino acids—could be solved by treating it as a problem of cryptography (code breaking). Could the knowledge of what amino acids lie next to each other provide clues as to the structure of the DNA language? This approach, however intriguing, quickly went nowhere. Instead, the way genes are first read and then translated came not from theoretical analysis but by doing genetic and biochemical experiments. Gradually these two complementary experimental approaches merged and by 1966 the genetic code was known. But many details remained to be discovered. How were the genes, encoded in the DNA, expressed as protein? How was the genetic code read from DNA? What was the cellular machinery that synthesized proteins? How did cells maintain such a high level of accuracy when making proteins? After 50 years of research, we have a clearer understanding of these processes, and new tools continue to reveal unexpected features of the pathway from DNA to protein.

Proteins Are Key Components of Cells

The crucial role of proteins in the life of the cell was realized early. They play, for example, many structural roles; proteins such as actin and tubulin make up the cytoskeleton of the cell, whereas other proteins such as fibrous collagen aid in building the connective tissue between cells. Structural proteins also have dynamic roles; for example, actin and myosin are components of the molecular machines that make muscles contract. Many other proteins play key roles in health; for example, peptide hormones, such as insulin, regulate metabolism and antibodies are the front line of defense of the body's immune response. And many proteins are enzymes, molecules that act as catalysts of the cell's chemical reactions.

The probability of most chemical reactions occurring unaided at the temperatures at which cells live is very low. Enzymes are the cellular catalysts that bring about chemical reactions that would otherwise be thermodynamically unlikely. Like all other catalysts, enzymes are not used up in the course of chemical reactions, and so a given enzyme molecule may function many thousands of times in a single second. Most enzymes are highly specific and catalyze only a single type of chemical reaction; conversely, each chemical reaction in a cell is catalyzed by only one specific enzyme.

Enzymes act by binding and positioning the potential partners of a chemical reaction (called *substrates*), thus greatly speeding up the rates at which these reactants can interact. In enzyme-catalyzed reactions, certain key atoms of the enzyme often directly participate by temporarily forming chemical bonds with the substrates, thereby forming metastable chemical intermediates that have a higher potential for chemical reactivity. Any change to these *active site* residues can have serious consequences for the activity of the enzyme. The chemical identity of enzymes was initially obscure, with many scientists believing that enzymes might represent a still undiscovered class of biological molecules. By 1935, however, it was increasingly clear that most, if not all, highly purified enzymes were proteins.

A Given Protein Possesses a Unique Sequence of Amino Acids along Its Polypeptide Chain

The realization that enzymes are proteins greatly heightened the interest of chemists in establishing the precise details of protein structure. By 1905, through the work of Emil Fischer in Germany, proteins were already known to be polymeric molecules built up from amino acids linked to each other by peptide bonds to form linear polypeptide chains (Fig. 2-1). The exact number of different amino acids remained

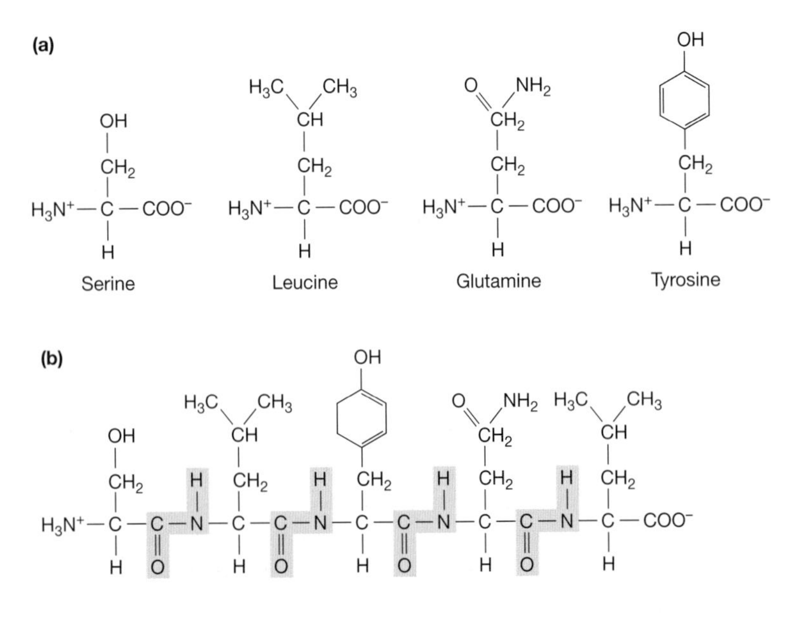

FIGURE 2-1

Amino acids. (*a*) Structural formulas of four amino acids: serine, leucine, tyrosine, and glutamine. These have a common structure of a carboxyl group (COO^-), an amino group (NH_3^+), and a hydrogen atom attached to the same carbon atom. This carbon atom also carries a side chain unique to each amino acid. (*b*) Proteins are polymers of amino acids linked by peptide bonds (*blue shading*) but a short chain of amino acids as shown is called a peptide. This is a pentapeptide with five amino acids: serine, leucine, tyrosine, glutamine, and leucine.

TABLE 2-1. The 20 amino acids in proteins

Amino acid	Three-letter abbreviation	Single-letter code
Glycine	Gly	G
Alanine	Ala	A
Valine	Val	V
Isoleucine	Ile	I
Leucine	Leu	L
Serine	Ser	S
Threonine	Thr	T
Proline	Pro	P
Aspartic acid	Asp	D
Glutamic acid	Glu	E
Lysine	Lys	K
Arginine	Arg	R
Aparagine	Asn	N
Glutamine	Gln	Q
Cysteine	Cys	C
Methionine	Met	M
Tryptophan	Trp	W
Phenylalanine	Phe	F
Tyrosine	Tyr	Y
Histidine	His	H

in question until about 1940, when it was established that the vast majority of proteins were built up using the same set of 20 amino acids (Table 2-1), with the percentage of a given amino acid varying from one protein to another. These findings made it likely that each polypeptide chain was characterized by the unique sequence of its amino acids, a conjecture that was first shown to be correct in 1951, when Sanger in Cambridge, England, reported the order of the amino acids along one of the two polypeptide chains that constitute the hormone insulin.

Most proteins contain only one polypeptide chain. However, there are many other proteins that are formed through the aggregation of polypeptide chains that have different sequences. The oxygen-carrying protein hemoglobin, for example, is formed by the aggregation of four polypeptide chains, two with a specific α sequence and two containing the β sequence. The sizes of different proteins vary greatly. The hemoglobin α chain has 146 amino acids, whereas the giant muscle protein titin has some 27,000 amino acids.

The Functioning of an Enzyme Demands a Precise Folding of Its Polypeptide Chain

Once a polypeptide chain is put together, weak chemical interactions (ionic and hydrogen bonds) between the specific side groups of its amino acids cause it to fold into a unique three-dimensional (3D) form whose exact shape is a function of its amino acid sequence. Most enzymes are globular, with cavities into which their substrates can fit in the way that keys fit into locks. Such cavities, bounded by the appropriate amino acid side groups, enable enzymes to bring their substrates into the close proximity that will permit them to react chemically with one another. The specific amino acid sequence of a given enzyme is thus very important. If inappropriate amino acids are present, then the polypeptide chain cannot fold to form the properly shaped catalytic cavity. By having available 20 of these amino acid building blocks, each with a unique shape and with its own chemical properties, and by being able to form polypeptide chains of variable lengths, a cell has the potential to evolve new proteins possessing catalytic cavities with shapes complementary to virtually any potential substrate.

Proteins, then, through their myriad catalytic functions—both synthetic and degradative—are the molecules that drive all the machinery and activities of the cell. It was clear that there must be some relationship between genes (whatever they were physically) and proteins, and the first observation of such a connection was made very soon after the rediscovery of Mendel's work.

The One Gene–One Protein Hypothesis Is Developed

The first evidence that gene mutations alter proteins—specifically enzymes—came in 1909, when the English physician Archibald Garrod noted that several human hereditary traits were metabolic diseases characterized by the failure of known chemical reactions to take place. Today one of the best understood of such genetic diseases is *phenylketonuria*, in which the amino acid phenylalanine cannot be converted to the related amino acid tyrosine. This "error" leads to a buildup in the blood of the intermediary metabolite phenylpyruvate, which, in still undetermined ways, causes brain damage. Garrod hypothesized that such metabolic diseases, which he called "inborn errors of metabolism," were due to the absence of specific enzymes that were synthesized under the direction of the wild-type genes. If Garrod's hunch was correct, then for every enzyme, and perhaps for each protein that a cell possesses, there must exist a corresponding gene.

Because so few details of cellular metabolism were then understood, some 30 years passed before definitive experiments could be done to prove the one

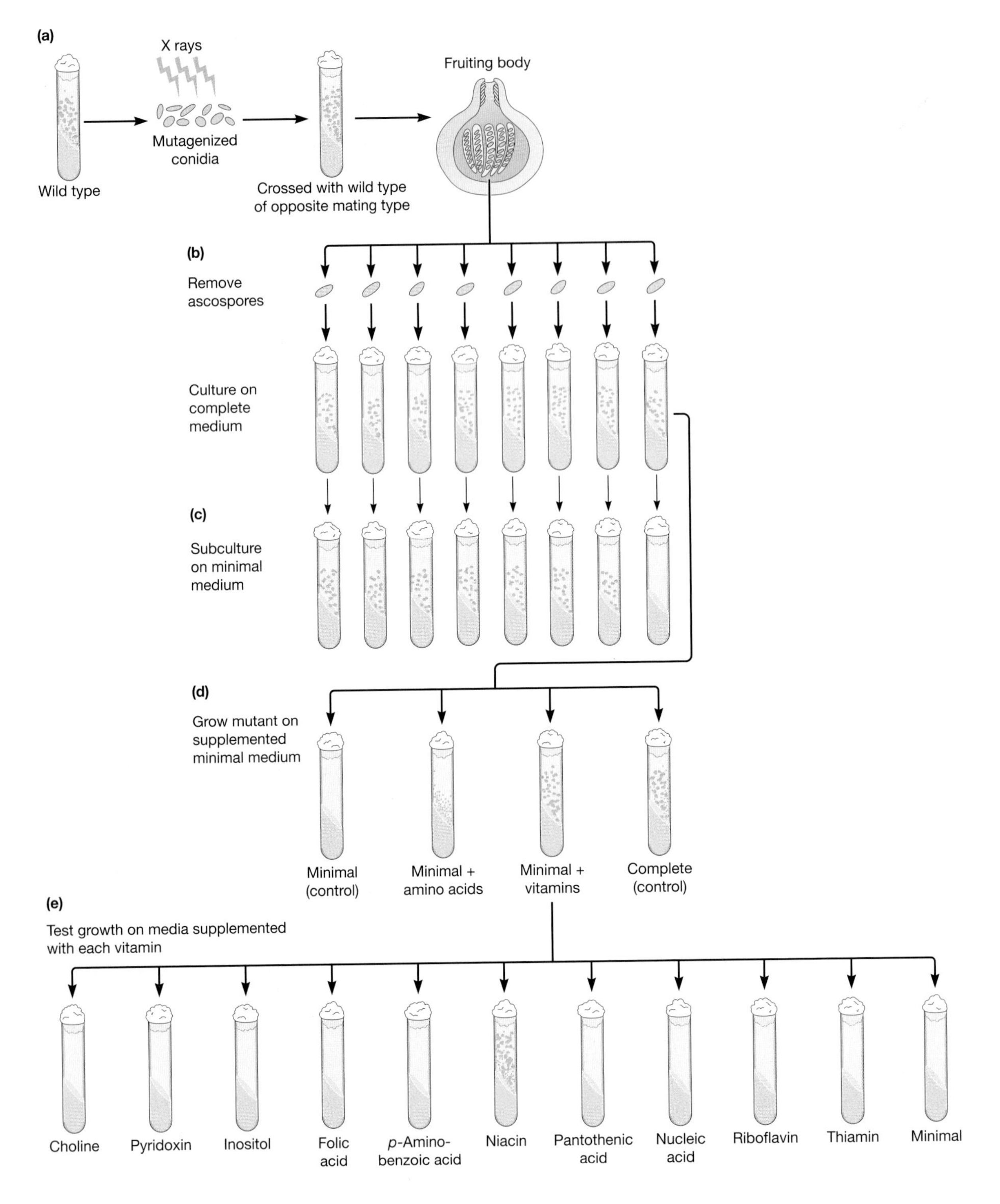

FIGURE 2-2

Determination of biochemical pathways using *Neurospora*. (*a*) Haploid, asexual wild-type conidia (spores) are mutagenized using X rays or UV light and allowed to grow, and the resulting fungus is mated with wild-type fungus. Each diploid nucleus undergoes meiosis followed by mitosis to produce eight ascospores contained within asci. (*b*) Ascospores are isolated and transferred individually to tubes containing a rich culture medium that permits the growth of mutants. (*c*) When these spores have germinated, asexual conidia from each culture are transferred to minimal medium. A few of these conidia are unable to grow in the minimal medium because of mutations. (*d*) Samples are taken from the corresponding culture growing on complete medium and grown on minimal medium supplemented with amino acids or vitamins. In this example, supplying vitamins permits growth, so the defect must be in a pathway involved in the synthesis of a vitamin. (*e*) The precise vitamin involved is determined by using cultures of minimal medium supplemented by a single vitamin. Here, the mutant requires niacin, thus indicating that the mutation is in niacin synthesis. In fact, Beadle and Tatum analyzed many thousands of ascospores; their first mutant was the 250th and their second the 1085th ascospore to be tested.

gene–one enzyme relationship. These studies were done in the early 1940s at Stanford University by the geneticist George Beadle and the biochemist Edward Tatum on the mold *Neurospora.* This microorganism normally grows on a simple diet of glucose and inorganic ions. However, exposure to X rays and UV light, agents that were by then known to increase vastly the rate at which mutant genes arise, led to production of mutant *Neurospora* strains that multiplied only when their normal diets were supplemented with additional nutrients that were specific to each mutant cell (Fig. 2-2). Some of these mutant *Neurospora* cells required specific amino acids like arginine or cysteine, whereas others required a particular vitamin or one of the purine or pyrimidine building blocks of the nucleic acids. In each case, the specific metabolic requirements were afterward shown to be due to the absence of one of the enzymes involved in the specific metabolic pathway that led to the synthesis of the growth factors (Fig. 2-3). A few years later, the one gene–one enzyme concept was generalized to the one gene–one polypeptide chain concept, when Linus Pauling and Harvey Itano at the California Institute of Technology showed that hemoglobin molecules from normal individuals and patients with sickle-cell anemia differed in their electrical charge.

Beadle and Tatum established the relationship between genes and enzymes and, through Sanger's work, the right-hand side of the expression "one gene–one protein" was understood biochemically—a protein was a polymer made up of amino acids linked together in an order specific for each protein. The discovery that DNA is the primary genetic material now provided the biochemical basis for the left-hand side of the equation; in some way, the order of bases along a DNA molecule specified the order of amino acids along a protein molecule. But how?

(a)

Mutant strain number	Minimal medium and ornithine	Minimal medium and citrulline	Minimal medium and arginine	Minimal medium
21502	+	+	+	–
27947	+	+	+	–
34105	+	+	+	–
30300	–	+	+	–
33442	–	+	+	–
36703	–	–	+	–

(b)

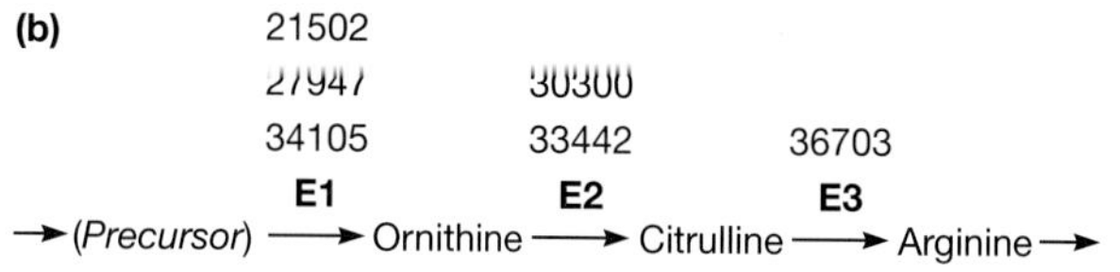

21502
27947
34105
E1
30300
33442
E2
36703
E3

→ (Precursor) → Ornithine → Citrulline → Arginine →

FIGURE 2-3

Biosynthesis of arginine in *Neurospora.* (*a*) None of the six mutants can grow in minimal medium and thus all require supplements. +, Growth; –, no growth. Mutants 21502, 27947, and 34105 require ornithine, citrulline, or arginine; mutants 30300 and 33442 will grow only if provided with citrulline or arginine; and 36703 has an absolute requirement for arginine. (*b*) These data can be accounted for by assuming that ornithine and citrulline are on the biosynthetic pathway to arginine and that each mutant strain has a mutation in an enzyme (enzymes 1–3) required for this pathway. Thus, strain 36703 cannot synthesize arginine from citrulline, so supplying either ornithine or citrulline is ineffective; mutants 30300 and 33442 have mutations that prevent them from making citrulline, but they will grow if supplied with either citrulline or arginine.

Mutant Proteins Have Amino Acid Sequences That Differ from Wild-type Proteins

The first experiments showing mutant proteins with single amino acid replacements came from studies of the abnormal hemoglobin molecules in people suffering from the genetic disease sickle-cell anemia. Vernon Ingram, working in Cambridge, England, in 1957, used Sanger's fingerprinting and peptide sequencing techniques to analyze both the α and β chains that make up the adult form of the hemoglobin molecule. No changes were found in the sickle α chains, but each sickle β chain differed from the normal wild-type hemoglobin β chain through a specific amino acid substitution (glutamic acid to valine) that had occurred at a unique site, position 6, in the β chain. At that time, this was a remarkable finding. It had not been expected that a single amino acid change in such a large and complex molecule as hemoglobin could produce such profound changes in activity. And if change of a single amino acid was sufficient, this could be caused by a change in a single base pair. Probing deeper into the gene–protein relationship in sickle-cell anemia was not possible at that time. There was no conceivable way to isolate the DNA coding for each hemoglobin chain nor any way to determine the nucleotide sequence of the

DNA. It was not until 1976 that the glutamic acid to valine change was shown to be due to a change in a single base, from A to T.

Genetic Mapping Shows Genes and Their Polypeptide Products Are Colinear

The gene could now be precisely defined as the collection of adjacent nucleotides that specify the amino acid sequence of a cellular polypeptide chain. This premise led to the testable prediction that the corresponding nucleotide and amino acid sequences would be colinear, and this hypothesis was soon confirmed by correlating the relative locations of mutations in a gene with the locations of changes in its polypeptide products.

In the mid-1960s, Charles Yanofsky and his collaborators at Stanford University studied mutations in the *Escherichia coli* gene coding for tryptophan synthetase, an enzyme needed to make the amino acid tryptophan. He showed very convincingly that the relative position of each amino acid replacement matched the relative position of its respective mutation along the genetic map (Fig. 2-4). The molecular processes underlying colinearity, however, were not at all obvious, because the 20 different amino acids far exceeded the number of different nucleotides in DNA. Clearly, a one-to-one correspondence between nucleotides and amino acids could not exist. Instead, groups of nucleotides must somehow specify (code for) each amino acid.

No Restrictions Exist as to Possible Amino Acid Sequences

The groups of nucleotides that code for an amino acid came to be called *codons*. From the beginning of codon research it seemed likely that most, if not all, codons were composed of three adjacent nucleotides. Groups of two nucleotides can be arranged in only 16 different permutations ($4 \times 4 = 16$), four too few to code for the 20 different amino acids. Groups of three (AAA, AAC, AAU, etc.), however, would result in 64 independent permutations ($4 \times 4 \times 4 = 64$), many more than are logically needed to specify all the amino acids. So, there was speculation about whether only a subset of these codons specified amino acids or whether the code was redundant in the sense that an amino acid might be specified by more than one codon.

Soon after the DNA helix was found, the Russian-born physicist George Gamow proposed that the codons overlap in such a way that single bases would help specify more than one amino acid. If that were true, then there would be restrictions governing which amino acids could be linked. The first known amino acid sequences were eagerly scanned during 1954 and 1955 by Gamow to see whether some amino acids never occurred next to each other. Initially, too few amino acid sequences were known to be able to determine whether an overlapping code was possible. However, by 1957 Sydney Brenner showed conclusively that no such restrictions of amino acid sequence existed and therefore

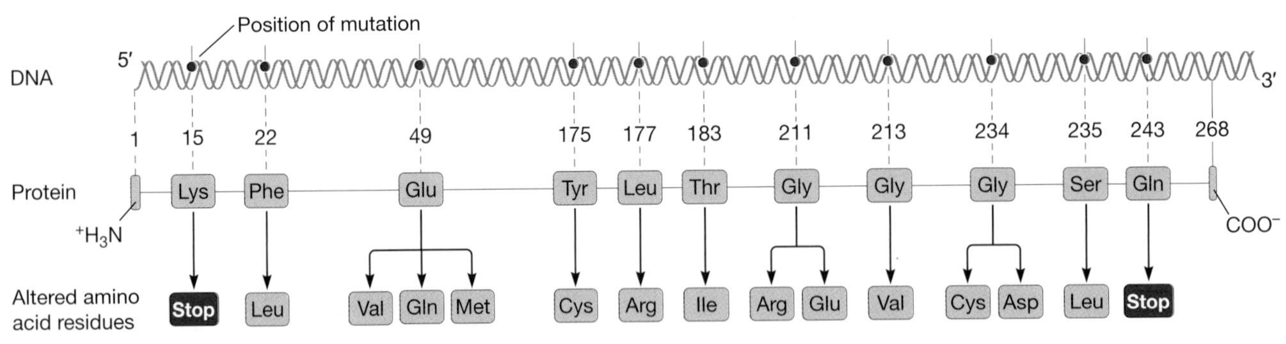

FIGURE 2-4

The order of mutations in a gene coding for tryptophan synthetase is the same as the order of amino acid changes in the gene's polypeptide product. The *red dots* on the DNA strand indicate the positions of the mutations; the numbers below indicate the positions of the changes in the amino acids. The amino acids that appear in these positions in normal chains are in the *lavender boxes* below the numbers, and the amino acids resulting from the mutations appear in the bottom (*peach*) row of boxes. Some mutations cause the synthesis of the protein chain to be terminated (Stop). (Mutations occurring in the same position on the DNA strand in different *E. coli* cells can produce different amino acid changes, as happened here for the 49th, 211th, and 234th amino acids.) Evidence of this kind showed that a gene and its polypeptide product are colinear.

that the successive codons along an RNA chain did not overlap.

Analysis of Triple Mutants in *E. coli* Indicates That There Are Three Nucleotides in a Codon

Confirmation that the genetic code was indeed a triplet code came, in 1961, from a very elegant study performed by Francis Crick, Sydney Brenner, and their colleagues in Cambridge, England. They began with a mutant in the *rII* locus of phage T4, produced using acridines—chemicals that lead to the addition or deletion of single base pairs (Fig. 2-5). Such *rII* mutant phages did not grow on *E. coli* strain K. They then found other mutations that suppressed the *rII* mutation—that is, phage carrying both mutations could grow on *E. coli* strain K. Crick and Brenner put together combinations of these mutations and obtained results that could be best explained by assuming that the acridine-induced mutations involved addition and deletion of bases, and that the genetic code is read in stepwise groups of three base pairs. If one or two bases are added or deleted, the *reading frame* is upset, leading to the use of a completely new collection of codons that invariably code for amino acid sequences that make no functional sense. In contrast, when groups of three base pairs are inserted or deleted, the resulting protein, now containing one more or one less amino acid, remains otherwise unchanged and often retains full biological activity.

The Central Dogma: DNA Makes RNA Makes Protein

A direct template role for DNA in the ordering of amino acids in proteins by then was known to be

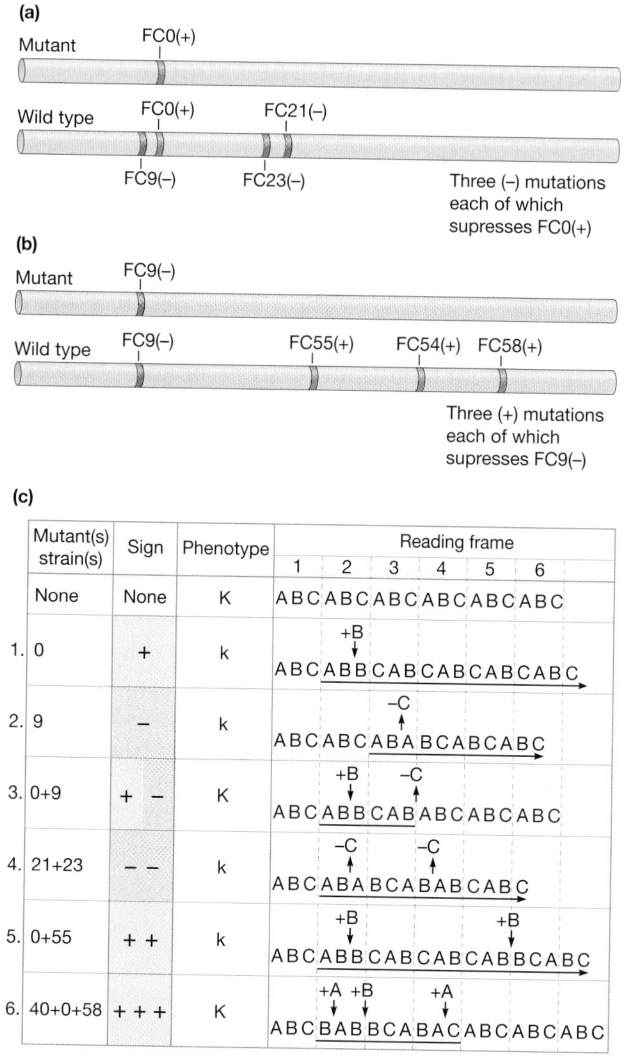

(a)
Mutant FC0(+)
Wild type FC0(+) FC21(–) FC9(–) FC23(–)
Three (–) mutations each of which supresses FC0(+)

(b)
Mutant FC9(–)
Wild type FC9(–) FC55(+) FC54(+) FC58(+)
Three (+) mutations each of which supresses FC9(–)

(c)

	Mutant(s) strain(s)	Sign	Phenotype	Reading frame 1 2 3 4 5 6
	None	None	K	ABCABCABCABCABCABC
1.	0	+	k	+B ABCABBCABCABCABCABC
2.	9	–	k	–C ABCABCABABCABCABC
3.	0+9	+ –	K	+B –C ABCABBCABABCABCABC
4.	21+23	– –	k	–C –C ABCABABCABABCABC
5.	0+55	+ +	k	+B +B ABCABBCABCABCABBCABC
6.	40+0+58	+ + +	K	+A +B +A ABCBABBCABACABCABCABC

FIGURE 2-5

Demonstrating that the genetic code is a triplet code. (*a*) FC0 is a proflavine-induced mutation in the *rII* region of bacteriophage T4. Phage with mutations in this region are unable to grow on *E. coli* strain K12 (λ); the phenotype of the wild-type phage is designated K and that of the mutants as k. Other evidence suggested that proflavine is a mutagen because it adds or deletes a single base—for the purpose of this analysis, it was assumed that FC0 was the consequence of a base insertion and it was designated (+). Several other spontaneously occurring mutations were found that reverted FC0 to wild type—these were designated (–) mutations (FC9, FC21, and FC23). (+ are colored *green* and – colored *red*.) (*b*) These (–) mutants are used in their turn to find a further set of (+) mutants. Here FC9(–) is reverted to wild type by FC55, FC54, and FC58, all designated (+). (*c*) It was now possible to put together (+) and (–) mutations in various combinations and determine which combinations are wild type or mutant. Addition or deletion of single bases (*1* and *2*) or pairs of bases of the same sign (*4* and *5*) disrupt the reading frame (*underline*) so that an abnormal protein results. (The reading frames are indicated by the *dotted lines*. We have used a simple repeating triplet, ABCABC, so that the changes in reading frame can be seen more easily.) Combination of one (+) and one (–) are wild type (*3*), as are three mutants of the same sign (*6*). These combinations produced short disruptions of the reading frame that may have little or no effect on the synthesis or activity of the resulting polypeptide. These results are most easily explained by assuming a triplet code.

impossible. Almost all DNA is located on chromosomes in the nucleus, whereas cell fractionation studies showed that most, if not all, cell protein synthesis occurs in the cytoplasm. The genetic information of DNA (encoded in its nucleotide sequence) thus had to be transferred to an intermediate molecule, which would then move into the cytoplasm where it would order the amino acids of the corresponding protein.

Speculation that this intermediate molecule was RNA became serious as soon as the structure of the double helix was discovered. For one thing, cytochemical studies showed that the cytoplasm of cells making large numbers of proteins always contained large amounts of RNA. Even more importantly, the sugar-phosphate backbones of DNA and RNA were known to be quite similar, and it was easy to imagine the synthesis of single RNA chains upon single-stranded DNA templates. This process of transcription would be analogous to the process of DNA replication (Fig. 2-6). The complementary base pairing would generate hybrid molecules in which one strand is DNA and the other strand is RNA. Cytosine would still pair with guanine, but adenine would base-pair with the unique base of RNA, uracil (U), that is chemically very similar to thymidine (T). The RNA molecules would serve as templates for ordering the amino acids in the polypeptide chains of proteins during the process of translation, so named because the nucleotide language of nucleic acids is translated into the amino acid language of

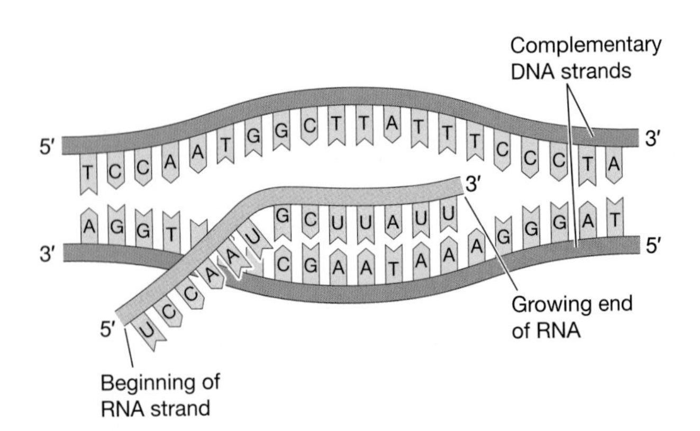

FIGURE 2-6

An RNA strand is synthesized (transcribed) upon a locally single-stranded region of DNA. The sequence of the RNA strand is complementary to that of the DNA strand, G pairing with C and U pairing with A. Ribonucleotides are added at the 3′ end of the mRNA as the RNA polymerase moves 3′ to 5′ along the DNA chain.

proteins. The relationship between DNA, RNA, and protein, as conceived in 1953, was thus

$$\text{DNA} \xrightarrow[\textit{transcription}]{} \text{RNA} \xrightarrow[\textit{translation}]{} \text{Protein}$$

where → indicates the flow of information.

Because the genetic information in DNA had to be conveyed solely by the linear sequences of the four letters (the bases A, T, G, and C) in the DNA alphabet, there had to be a genetic code that when read would tell the protein-synthesizing machinery of the cell how to order the amino acids in a protein. Mutant proteins in turn clearly represented changes in the amino acid sequence, the simplest mutants being proteins in which one amino acid was replaced by another. That information flows unidirectionally from DNA to RNA to protein became known as the Central Dogma of molecular biology, and research over the following decades has established the essential correctness of the concept and established its molecular basis.

RNA Synthesis in Cell-free Extracts

The initial reaction of geneticists to the structure of the double helix was pure delight at seeing how it could function as a template. In a real sense their 50-year quest was over. In contrast, the biochemists, who were then working out how enzymes participated in the synthesis of the nucleotides and amino acids, saw their role as just beginning. They realized that making phosphodiester bonds between nucleotides and peptide bonds between amino acids would also require specific enzymes. The discovery of such enzymes would demand finding conditions in which, say, DNA, RNA, or protein is made in extracts of disrupted cells. These simplified in vitro systems provided the means to assay and isolate the enzymes. These experiments were difficult, however, because cells frequently also possess active enzymes that can break down nucleic acids (*nucleases*) and proteins (*proteases*). These enzymes can destroy newly synthesized molecules, thus making it difficult to determine whether synthesis has, in fact, occurred.

After Arthur Kornberg's mid-1950s success in making DNA using cell-free extracts of *E. coli*, a similar approach was used in 1960 to make RNA. As with DNA synthesis, that of RNA also requires the presence of DNA template and likewise has as its precursors nucleotide triphosphates—nucleotides containing three adjacent phosphate groups. Two of the phosphate groups are split off when nucleotides are linked together, and the energy derived from the broken bonds is used to make the phosphodiester links of the sugar-phosphate backbone. The enzymes that directly make the phosphodiester bonds are called *polymerases* (because of the polymeric nature of the nucleic acids) and named either *DNA polymerases* or *RNA polymerases* depending on which nucleic acid is being synthesized.

DNA and RNA are asymmetrical molecules—their ends are chemically different. Every nucleic acid chain has one end terminating with the 5′ carbon atom (the *5′ end*) and the other end terminating with the 3′ carbon atom (the *3′ end*). By convention, DNA and RNA are drawn with the 5′ end at the left. All nucleic acid chains are synthesized by their polymerases from 5′ to 3′, so that the polymerases move from 3′ to 5′ along the chain that is acting as the template.

The Adaptor Hypothesis of Crick

Following the discovery of the double helix, the ordering of amino acids along polypeptide chains was initially thought to involve the binding of specific amino acids into specific cavities linearly sited along the surfaces of template RNA molecules. No obvious chemical complementarity, however, exists between the specific portions (the side groups) of many amino acids and the purine and pyrimidine bases of RNA. The side groups of the amino acids leucine and valine, for example, cannot form any hydrogen bonds, and, unless somehow modified, they would not be strongly attracted to any RNA template. This chemical insight led Crick, in early 1955, to propose that many, if not all, amino acids had to be attached to some form of adaptor molecules before they could specifically bind to an RNA template. Testing the adaptor hypothesis, however, was impossible until techniques were developed for dissecting biochemically the process by which amino acids become incorporated into growing polypeptides.

In Vitro Systems for Protein Synthesis Reveal Adaptor Molecules (Transfer RNAs)

The first steps toward unraveling the complexities of protein synthesis occurred in the mid-1950s in Boston at the Massachusetts General Hospital. There Paul Zamecnik and Mahlon Hoagland pioneered the use of cell-free extracts to study the incorporation of radioactively labeled amino acids into polypeptide chains. To their surprise they found that amino acids prior to being linked together by peptide bonds became attached to a previously unknown class of small RNA molecules that came to be called transfer RNA (tRNA) (Fig. 2-7). For each of the 20 different amino acids a specific enzyme (an aminoacyl tRNA synthetase) catalyzes its linkage to the 3′ end of its specific tRNA molecule. Soon these tRNA molecules were shown to be the "adaptor molecules" whose existence was predicted the year before by Crick. Remarkably, 50 years after Crick predicted and Zamecnik and Hoagland found transfer RNAs, a new aminoacyl tRNA synthetase has been discovered in an archaebacterium. Examination of the genome sequence of some archaebacteria revealed a previously unknown tRNA synthetase that charges a tRNA (CUA) with pyrrolysine. This is

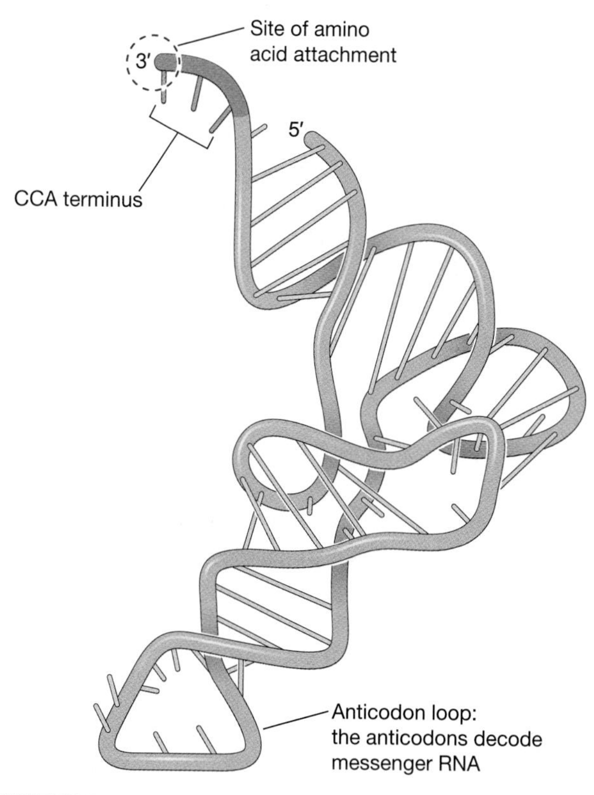

FIGURE 2-7

A transfer RNA molecule. The diagram shows how base pairing within the single-stranded molecule gives it its distinctive shape. The anticodon loop is the portion that decodes messenger RNA. An amino acid attaches to the CCA bases at the 3′ end of the chain.

then incorporated at certain rare UAG codons that are usually stop codons. (Pyrrolysine is the 22nd amino acid. Selenocysteine is the 21st, but it is not encoded in the genome; selenocysteine is synthesized from serine after the latter has been attached to a specialized tRNA.)

Messenger RNA Is the Information Carrier of the Central Dogma

Fractionation of protein-making cell extracts in the Zamecnik lab revealed the attachment of newly made polypeptide chains to cytoplasmically located 200-Å-diameter RNA-containing particles that are today called ribosomes. They are formed by the binding together of two subunits of unequal size with the smaller subunit approximately half the molecular weight of the larger one. In turn, these subunits contain single RNA chains complexed to large sets of ribosomal-specific proteins. When first characterized in 1958, ribosomal RNA (rRNA) molecules were thought to be the RNA templates that ordered amino acids along polypeptide chains. This hypothesis, however, began to fall out of favor when all ribosomes, no matter from which organism, were found to have very similar base compositions, independent of the relative amounts of A:T and G:C in their respective DNA genomes. Equally puzzling was why rRNA chains had invariable base numbers (1500 and 3000) in contrast to the high variable lengths of their putative polypeptide products (250–2000 amino acids). Moreover, experiments at the Institut Pasteur in Paris suggested that bacterial RNA templates for protein synthesis must be very short-lived in contrast to rRNA molecules, which, once made, showed no tendency to break down.

The search for template RNA then became directed to the unstable RNA molecules discovered in 1953 by Alfred Hershey in *E. coli* cells infected by phage T2. In 1956, Elliot Volkin and Lazarus Astrachan showed that this T2 RNA had a base composition very similar to that of T2 DNA. Its role in protein synthesis became clear only in 1960 when independent experiments at Harvard and at CalTech showed that T2 RNA was physically distinct from tRNA or rRNA and represented the long-sought-for class of template RNA. Because they carried the specificity of these genes to the cytoplasm, these RNA molecules soon became called *messenger RNA* (mRNA). Unlike tRNA or rRNA, mRNA is highly variable in length, which is the main reason why its existence took longer to demonstrate. Proof that T2 template RNA molecules were indeed complementary to their DNA templates came from a classic experiment at the University of Illinois by Ben Hall and Sol Spiegelman. *E. coli* cells were infected with T2 phage; T2 DNA was labeled with ^{3}H-thymidine and the newly synthesized RNA was labeled with ^{32}P. Denatured DNA was mixed with RNA and the samples were reannealed. When the samples were analyzed, hybrid DNA–RNA molecules were found (Fig. 2-8).

mRNA is synthesized using only one DNA strand (the sense or template strand), although which of the two DNA chains along a chromosome is used varies from gene to gene. Like DNA polymerase, RNA polymerase synthesizes an RNA molecule from 5′ to 3′, moving along the DNA sense (template) strand from 3′ to 5′. However, by convention, the end of a gene where transcription begins is called the 5′ end of the gene (to correspond to the 5′ end of the mRNA), although chemically the synthesis of the RNA is initiated from the 3′ end of the template DNA strand.

With their putative template roles removed, the role of ribosomes was seen to be that of a highly complex molecular factory whose molecular surfaces brought tRNA and mRNA together in orientations suitable for their roles in protein synthesis. The enzymatic surfaces that catalyzed the formation of peptides had to be located somewhere within ribosomes. Demonstrating the existence of this catalytic site, as we detail below, proved far from straightforward.

In Vitro Translation of Synthetic RNA Reveals Much of the Genetic Code

The realization that ribosomes become programmed to make specific proteins only by binding mRNA molecules led Marshall Nirenberg and Heinrich Matthaei in 1961 to do a historic experiment. The enzyme polynucleotide phosphorylase makes chains of nucleotides joined by phosphodiester bonds without the need for a DNA or RNA template. If the reaction mixture contains uracil diphosphate, the product is polyuridylic acid or poly(U) for short: UUUUUU... . When Nirenberg and Matthaei added poly(U) to cell-free extracts from *E. coli*, they observed that only phenylalanine was synthesized and thus concluded that UUU coded for the amino acid phenylalanine (Table 2-2). poly(A) was found to code for strings of lysine residues, and poly(C) yielded polypeptides containing only proline.

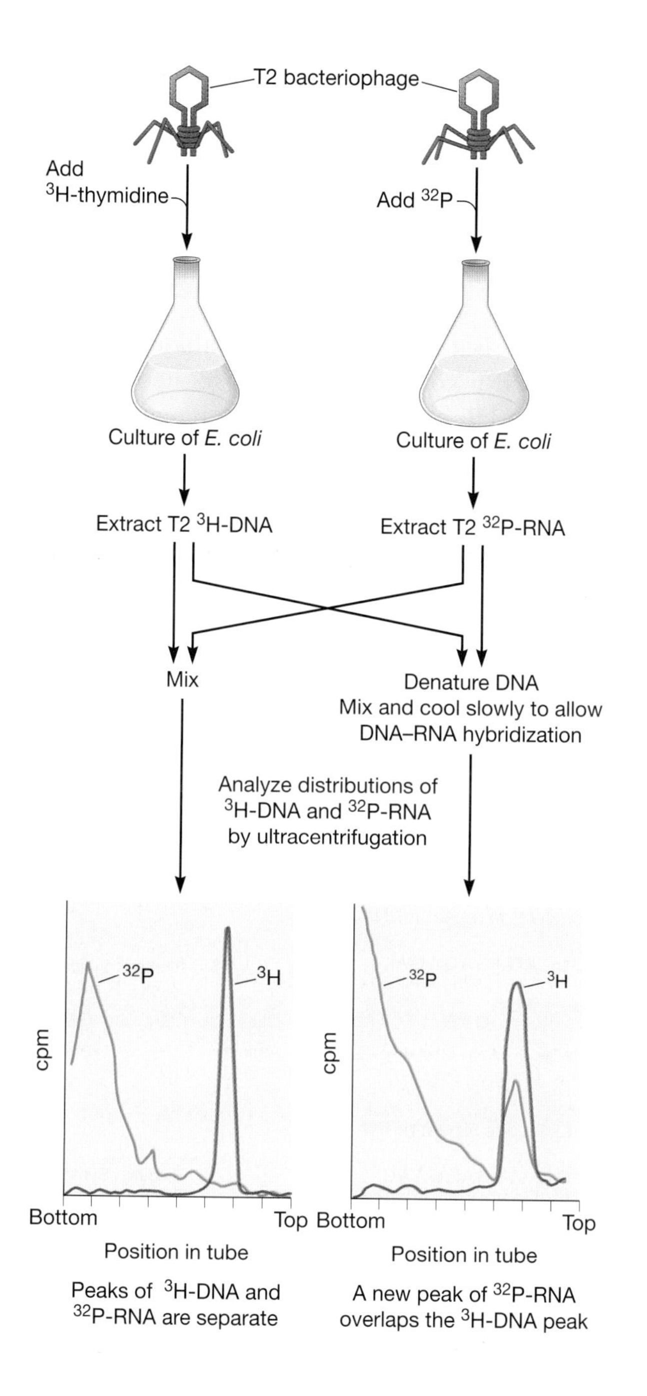

FIGURE 2-8

RNA synthesized in *Escherichia coli* cells after infection by bacteriophage T2 is complementary to phage DNA. Separate cultures of *E. coli* were infected with bacteriophage T2 and with ^{3}H-thymidine or ^{32}P. The former was used for the preparation of ^{3}H-labeled T2 DNA and the latter for the isolation of ^{32}P-labeled T2 RNA. For the control experiment, samples of T2 DNA and T2 RNA were mixed and the distributions of the radioactivity after centrifugation at 33,000 rpm for 5 days in a density gradient were determined. ^{32}P corresponding to T2 RNA formed a broad peak toward the bottom of the gradient, whereas T2 DNA gave a sharp peak toward the top of the gradient. A very different distribution was found when T2 DNA was denatured and mixed with T2 RNA and the mixture was cooled slowly. Now a significant fraction of ^{32}P-RNA is found associated with the ^{3}H-DNA peak, indicating that the RNA is complementary to the DNA. (The peaks of ^{32}P and ^{3}H are coincident because the amount of RNA hybridizing to the DNA is too small to make the RNA–DNA hybrids differ significantly in density from the T2 DNA.)

Over the next several years, synthetic polynucleotides containing random mixtures of two or more nucleotides were used to decode many other codons. Assignations were tentative because coding was indirect, correlating incorporation of amino acids with the calculated statistical distribution of possible codons in these random polynucleotides. Later a more direct method—the filter-binding assay—was devised that unambiguously assigned many more codons to amino acids. The assay involved incubating ribosomes and radiolabeled aminoacyl tRNAs with a polynucleotide. The reaction mixture was filtered through a cellulose acetate filter that retained the ribosomes and any bound aminoacyl tRNAs. For example, ^{14}C-phenylalanine-tRNA was retained 60-fold over controls when poly(U) was included in the reaction mixture. Finally, the brilliant chemical skills of H. Gobind Khorana and his colleagues at the University of British Columbia in Vancouver led to the synthesis of RNAs of known repeating sequences; for example, GUGUGUGU..., AAGAAG..., and TTACTTAC... . These were used not in tRNA filter-binding assays, but as mRNAs for in vitro protein synthesis.

TABLE 2-2. Synthetic polynucleotides are used to determine the genetic code

(a) poly(U) stimulates synthesis of polyphenylalanine	counts/min/mg protein
In vitro system alone	44
+ 10 μg poly(U)	39,800
+ 10 μg poly(A)	50
+ 10 μg poly(C)	38
+ 10 μg poly(I)	57

The in vitro system used in this experiment was made from extracts of *Escherichia coli*. ^{14}C phenylalanine was added, together with a synthetic polynucleotide. After 20 min, the protein was precipitated and the radioactivity in the precipitate was counted. In this example, poly(A), poly(C), and poly(I) did not stimulate incorporation of ^{14}C phenylalanine into polyphenylalanine. However, poly(U) produced a 1000-fold stimulation of incorporation.

(b) Only phenylalanine is incorporated in response to poly(U)	counts/min/mg protein
Experiment 1	
In vitro system alone	68
Phenylalanine + poly(U)	38,300
Experiment 2	
In vitro system alone	20
Mixture (glycine + alanine + serine + glutamine + aspartic acid)	33
Experiment 3	
In vitro system alone	276
Mixture (leucine + isoleucine + threonine + methionine + arginine + histidine + proline + valine + tryptophan + tyrosine + lysine)	899

These experiments show that poly(U) does not lead to the incorporation of amino acids other than phenylalanine. Experiment 1 is a positive control, showing the incorporation of phenylalanine in the presence of poly(U). Experiments 2 and 3 test all other amino acids with poly(U); there was no significant incorporation of any amino acid in response to poly(U).

Comparison of the mRNA codons with the amino acid sequences of the corresponding polypeptides created new assignments and confirmed others (Fig. 2-9).

Not All Base Sequences Code for Amino Acids

By 1966, the search for the genetic code was over and had yielded the following: (1) All codons contain three successive nucleotides, (2) many amino acids are specified by more than one codon (the so-called *degeneracy* of the code), and (3) 61 of the 64 possible combinations of the three bases are used to code for specific amino acids.

The three combinations of bases in mRNA that do not specify any amino acid (UAA, UAG, UGA) were all found to code for stop signals at the ends of mRNA molecules. These sequences bind proteins called *release factors* that make the ribosome fall off the mRNA and dissociate into its two subunits. The finding of stop codons at first created the expectation that specific start codons might also exist, especially because it was becoming more and more certain that all proteins begin with the amino acid methionine. But there is only one methionine codon (AUG), and it codes for internally located methionines as well as for the initiator methionine. However, AUGs that are used to start polypeptides are always present in a specific context. In prokaryotes, the initiating AUG is closely preceded by a purine-rich sequence (e.g., AGGA) that may help to position the ribosome at the start AUG of the mRNA. In eukaryotes, an AUG present at an initiator site has a purine three bases on the 5′ side and is immediately followed by guanine, a sequence named for its discoverer, Marilyn Kozak.

Both DNA and RNA contain many other "codes," although these do not code for amino acids

(a) Dinucleotide repeat mRNA

Poly(AG)$_n$: AGAGAGAGA

Codons: AGA, GAG

Polypeptide: -Arg-Glu-Arg-Glu-

	Binding assay	Synthetic mRNA
Glutamine	GAG	GAG
Arginine	—	AGA

(b) Trinucleotide repeat mRNA

Poly(AAG)$_n$: AAGAAGAAG

Codons: AAG, AGA, GAA

Polypeptides: 1. -Lys-Lys-Lys-Lys-
2. -Arg-Arg-Arg-Arg-
3. -Glu-Glu-Glu-Glu-

	Binding assay	Synthetic mRNA
Lysine	GAG	GAG
Glutamine	GAA	GAA
Arginine	—	AGA

FIGURE 2-9

Synthetic mRNAs with repeated sequences confirmed codon assignments. (*a*) Synthetic mRNAs made up of repeating dinucleotides have two triplet codons that alternate (AGA, GAG). Polypeptides with two alternating amino acids (arginine, glutamine) are made when these RNAs are used as messengers for in vitro protein synthesis. If one of these codons has been assigned to an amino acid (GAG to glutamine), the other codon can be assigned to the second amino acid (AGA to arginine). (*b*) Synthetic mRNAs made with repeating trinucleotide units produce three different polypeptides, each composed of one amino acid. Again, some prior information is needed to assign codons. These experiments were able to assign some codons—for example, AGA to arginine—that did not work in the filter-binding assay.

and are not part of the genetic code. For example, specific nucleotide segments in DNA (called promoters) are recognized by RNA polymerase molecules that start RNA synthesis. In many prokaryotic genes, these sequences—TATAAT and TTGACA—occur 10 and 35 nucleotides before the point where transcription of a gene begins and, in combination with an initiation factor called σ, position the RNA polymerase so that it can begin transcription. We will see repeatedly in the next chapter that DNA-binding proteins play an essential role in regulating when and where genes are expressed.

This simple but intellectually satisfying picture of the relationship between the nucleotide sequence in a gene, its mRNA, and the amino acid sequence in the protein has since undergone profound revisions. There are two processes—RNA splicing and RNA editing (both discussed in Chapter 5)—that lead to modifications in an mRNA sequence so that it is no longer a direct copy of the gene's DNA sequence. Nevertheless, the codon assignments shown in Table 2-3 have withstood the test of time and are used every day in every molecular biology laboratory throughout the world.

TABLE 2-3. The genetic code

First position (5′ end)	Second position U	C	A	G	Third position (3′ end)
U	Phe	Ser	Tyr	Cys	U
	Phe	Ser	Tyr	Cys	C
	Leu	Ser	Stop	Stop	A
	Leu	Ser	Stop	Trp	G
C	Leu	Pro	His	Arg	U
	Leu	Pro	His	Arg	C
	Leu	Pro	Gln	Arg	A
	Leu	Pro	Gln	Arg	G
A	Ile	Thr	Asn	Ser	U
	Ile	Thr	Asn	Ser	C
	Ile	Thr	Lys	Arg	A
	Met	Thr	Lys	Arg	G
G	Val	Ala	Asp	Gly	U
	Val	Ala	Asp	Gly	C
	Val	Ala	Glu	Gly	A
	Val	Ala	Glu	Gly	G

Given the position of the bases in a codon, it is possible to find the corresponding amino acid. For example, the codon 5′-AUG-3′ on mRNA specifies methionine, whereas CAU specifies histidine. UAA, UAG, and UGA are termination signals. AUG is part of the initiation signal, and it codes for internal methionines as well.

Wobble Frequently Allows Single tRNAs to Recognize Multiple Similar Codons

Initially it seemed highly probable that anticodons on tRNAs bind codons by means of three A:U and/or G:C hydrogen bonds that are identical to the A:T and G:C bonds that bind the two strands of the double helix together. But soon experimental results began to show that the anticodon of a single tRNA species, such as the GAA of phenylalanine tRNA, can bind to more than one codon—in this case to both UUU and CUU. This observation suggested to Crick that, although the first two bases in a codon always pair in a DNA-like fashion, the pairing in the third position is less restrictive, so that nonstandard base pairing ("wobble") is allowed for the third base. This idea led to the realization that there need not be a distinct tRNA species for each of the 61 codons corresponding to amino acids. Now that large amounts of DNA sequence can be searched for tRNA genes, we know that this conjecture is true. Yeast, for example, has only 41 tRNAs, and thus many yeast tRNAs recognize more than one codon.

Great variation exists in the relative amounts of particular tRNA species that are present in a given cell. In part, this variation reflects differences in the abundance of the amino acids the tRNAs specify. For example, the amino acids methionine and tryptophan occur relatively rarely in most proteins, and comparatively small amounts of their respective tRNAs are present. Moreover, when more than one tRNA form exists for a given amino acid, the more numerous tRNA form recognizes the more commonly used codons for that amino acid. The rate at which an mRNA message is translated into its corresponding polypeptide chains thus may be controlled in part by whether it contains codons that are recognized by the more abundant tRNA forms for its amino acids.

The Genetic Code Is Nearly Universal

Virtually all the experiments used to decipher the genetic code employed ribosomes and tRNA molecules from *E. coli*. It could thus be asked whether mRNA molecules are always translated into the same amino acid sequences independently of the source of the translation machinery. At the start, the answer was thought to be yes, for it was hard to imagine how the code could change during the course of evolution. By now the initial expectation that the genetic code for chromosomal DNA would prove to be universal has been rigorously confirmed in a large variety of organisms, ranging from the simplest prokaryotes to the most complex eukaryotes. This is demonstrated most dramatically when human proteins are made by bacteria translating human genes. (This observation shows also that the cell machinery for interpreting the genetic code is universal.)

There are exceptions, mainly in assigning one of the stop codons to an amino acid. For example, for the most part the genetic code used by mitochondria is identical with that used by nuclear DNA. However, UGA, a stop codon in nuclear DNA, is read as tryptophan in mitochondria, whereas AGA, coding for arginine in nuclear DNA, is a stop codon in mitochondria. Mitochondria can have a different genetic code because they are closed systems, isolated from the protein-synthesizing machinery of the rest of the cell. These distinctions provide further evidence that mitochondria and chloroplasts are the remnants of primitive bacterial cells that became symbiotically engulfed by the ancestors of present eukaryotic organisms.

Mutations Change the Base Sequence of DNA

Mutations arise as a change in the coding sequence of a gene. *Substitutions* occur when one base is replaced by another. *Insertions* and *deletions* are the addition and removal of one or more bases, respectively (Fig. 2-10). Different mutations have different consequences for the function of the protein. A *nonsense mutation* is a point mutation that converts a codon to a stop codon, producing premature termination of the polypeptide chain and usually a nonfunctional protein. Because of the redundancy of the genetic code, substitutions may not lead to the incorporation of an incorrect amino acid in the protein (*sense mutations*), and even when an incorrect amino acid is used (termed a *missense mutation*), it may have little effect on the function of the protein unless it is a critical portion of the protein. *Deletions* and *insertions* usually have drastic effects on proteins because they alter the reading frame in which the bases are read. Mutations also occur in DNA sequences that do not code for protein. The consequences of these mutations range from none to severe. A mutation in a DNA sequence that affects the regulation of a gene—when it is transcribed and in what cells—may have devastating consequences.

Mutations are important for two reasons. They are responsible for inherited disorders and other diseases such as cancer that involve alterations in genes. At the same time, mutations are the source of phenotypic variation on which natural selection acts. The process by which mutation produces variability and natural selection then favors any resulting advantageous variants is the driving force of evolution.

The Basic Steps in Protein Synthesis Are Determined

By the early 1960s, the process of protein synthesis was well worked out. The mRNA molecules for a par-

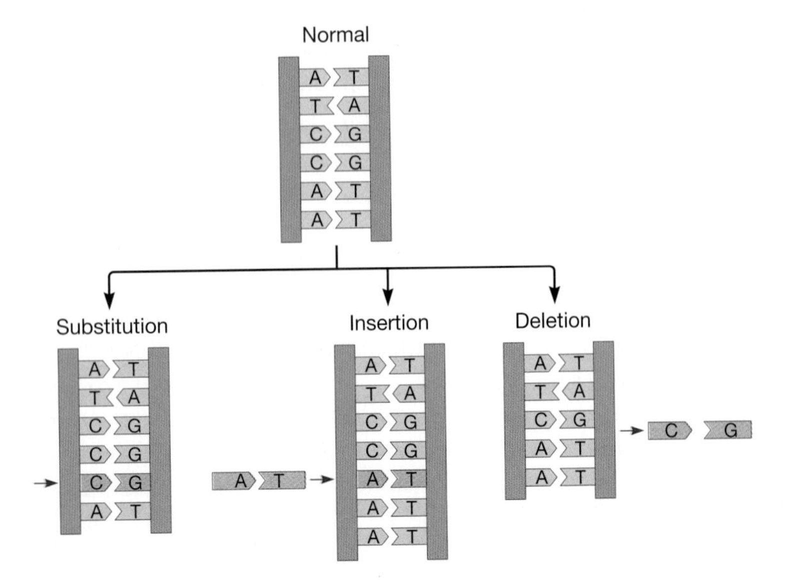

FIGURE 2-10

The three types of mutation. They are substitution, addition, and deletion of a base pair in a DNA strand.

ticular gene are synthesized using DNA of the gene as a template, and those messenger RNAs leave the nucleus for the cytoplasm. There, ribosomes attach to the mRNA where tRNAs carrying their appropriate amino acids line up opposite the codons of mRNA. The binding of tRNA to mRNA is mediated by a set of nucleotides on each tRNA that have sequences complementary to their respective codons on the mRNA; these tRNA sequences are called *anticodons* (Fig. 2-11). Thus, the tRNA molecules mediate the translation of codon sequence into amino acid sequence and, just as Crick had predicted, no contacts exist between the individual amino acids and the mRNA. Ribosomes are thus factories that by themselves have no specificity and therefore can attach to any mRNA molecule. The ribosomes move along the mRNA, so that successive codons are brought into position for ordering the sequence of their respective amino acids. As successive tRNAs enter the ribosome–mRNA complex, the amino acids that they carry are attached by a peptide bond to the growing polypeptide chain. The mRNA molecule is translated from 5′ to 3′, and the polypeptide chain is assembled from the amino to the carboxyl terminus.

Using a wide range of biochemical and genetic techniques, scientists have learned (and continue to learn) many details about the mechanisms of both transcription and translation. Indeed, our understanding of how the molecular machines—RNA polymerase and the ribosome—carry out these processes has undergone a revolution. As happens so often in science, this new knowledge has come through advances in technique, in this case through the increased power of X-ray crystallography. New methods for preparing crystals of molecular complexes, the use of extremely energetic X rays from synchrotron light sources, new algorithms for analyzing the data, and the availability of ever-faster computers have all led to an outpouring of the 3D atomic structures of biological molecules. Seeing these molecules at atomic resolution has led to many new insights into how they function.

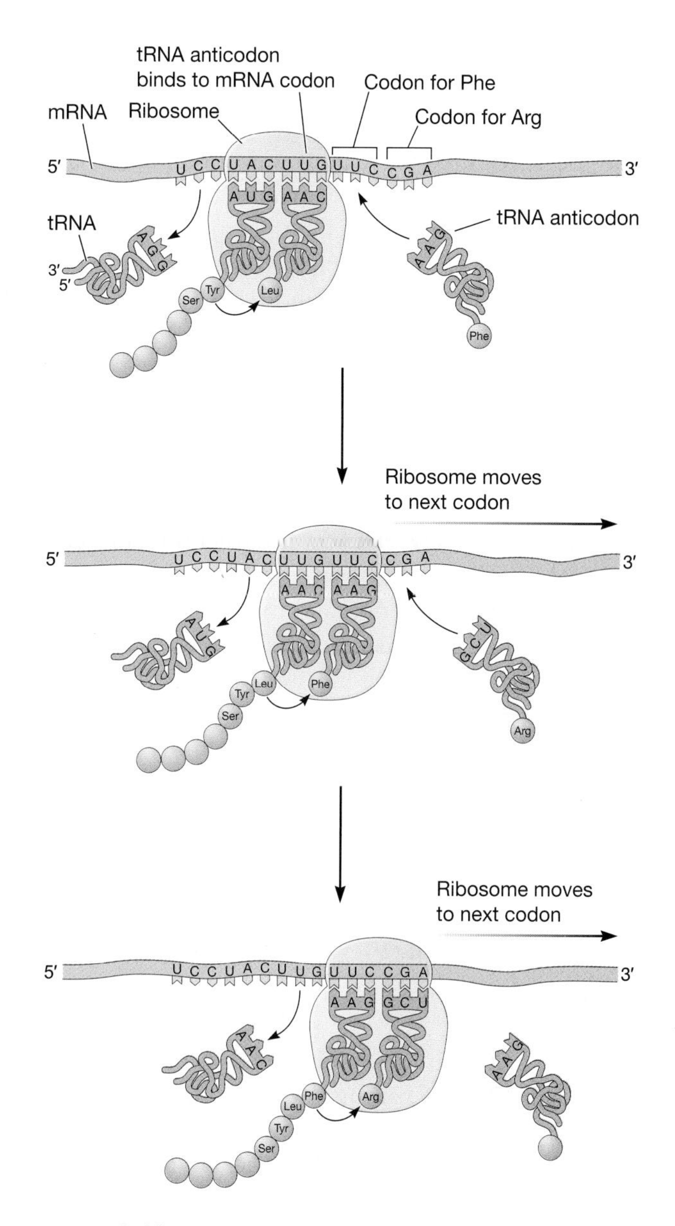

FIGURE 2-11

At the ribosome, the codons of messenger RNA molecules base-pair with the anticodons of transfer RNAs, which are charged with amino acids. The ribosome complex covers two codon triplets at a time (the ribosome has been drawn very small relative to the tRNAs). The 5′ codon (UAC) is paired with the anticodon (3′-AUG-5′) of a tyrosine tRNA that carries the growing polypeptide chain. The 3′ codon is paired with a tRNA carrying the next amino acid to be added—in this case leucine with the anticodon 3′-AAC-5′ paired with the codon UUG. A peptide bond is formed between the amino group of the leucine on the incoming aminoacyl-tRNA and the carboxyl group of the tyrosine, the last amino acid added to the polypeptide chain. Following transfer of the polypeptide chain to the leucine, the tyrosine tRNA is released and the ribosome complex translocates to the next codon, UUC, now occupied by the incoming phenylalanine tRNA with the anticodon 3′-AAG-5′. This cycle is repeated until the ribosome reaches a stop codon.

RNA Polymerase Grasps DNA with a Molecular "Clamp"

The high-resolution crystal structures of bacterial and yeast RNA polymerases were first obtained at Stanford University by Roger Kornberg and his colleagues. Remarkably, there were surprising resemblances between their 3D structures, despite minimal sequence similarity between the individual protein subunits that make up the complexes. These RNA polymerases resemble a crab claw with two pincers, and the active site, where polymerization occurs, is at the base of these pincers. Just how these work was determined by comparing the structure of an empty yeast RNA polymerase complex to the structure of a polymerase actively transcribing RNA. It was found that the claw closes tightly around the DNA, thereby forming a molecular "clamp." The inside of the clamp is lined with positively charged amino acids, thus creating a channel through which the negatively charged DNA and RNA can move smoothly. The clamp is a key feature of the RNA polymerase structure, contacting the DNA as it enters the complex, the DNA–RNA hybrid where nucleotides are added to the growing transcript, and the exiting RNA.

The start site of eukaryotic transcription is established by the binding of initiator proteins to specific DNA sequences at the start of genes. As well as signaling the start of transcription, the initiator factors unwind the DNA and provide a binding site for RNA polymerase. After binding, the polymerase clamp closes around the separated DNA strands and, as it does so, the conformation of the RNA polymerase changes so that several "switch" regions of the molecule can bind DNA. The binding between the switch regions and the DNA stabilizes the closed, active polymerase, and the switch regions bend and twist the DNA so as to point a DNA base into the active site for base pairing with an entering ribonucleotide. This bend permits the careful reading of each DNA base. Once a ribonucleotide has been added to the growing RNA strand, the RNA–DNA hybrid moves so that the next DNA base is moved into position for reading. The just-formed RNA–DNA hybrid is contacted by residues from the RNA polymerase that determine if the correct ribonucleotide has been incorporated. If it has not, the polymerase backtracks on the RNA–DNA hybrid so that the incorrect ribonucleotide can be removed and the correct base added. As transcription proceeds, the RNA is peeled away from the DNA template by a wedge-like portion of the RNA polymerase and guided away from the DNA. When transcription terminates, opening of the clamp changes the polymerase conformation so that the switch regions can no longer bind DNA, and the polymerase readily releases the DNA template. This is a rapid and efficient process—the *E. coli* RNA polymerase adds 10 ribonucleotides per second to an mRNA chain, with an error rate as low as 1 in 100,000 bases. (This is, however, very much slower and less accurate than the *E. coli* DNA polymerase III, which extends a new DNA strand by about 1000 deoxynucleotides per second and with an overall accuracy of 1 in 10^9.)

Making Proteins Is a Highly Accurate Process

The production of a functional protein from an mRNA transcript requires precise translation of the mRNA codons into the corresponding amino acids of the protein (Table 2-4). There are two steps in this process that must be carried out with a high

TABLE 2-4. The accuracy of translation is established through multiple error-checking steps

Stage of protein synthesis	Error-checking step	Mechanism
tRNA charging	Initial binding to tRNA synthetase	tRNA and amino acid bind to acyltransferase pocket.
	Error checking by tRNA synthetase	Aminoacylated tRNA is checked in editing pocket.
tRNA selection at the ribosome	Codon recognition by the ribosome	tRNA anticodon pairs with the codon, but this correct pairing is also sensed by the ribosome.
	GTPase activation and GTP hydrolysis by the translation factor EF-Tu	Ribosomal contacts sensing correct codon–anticodon pairing promote shift from "open" to "closed" conformation that promotes EF-Tu hydrolysis of GTP.
	Accommodation by ribosome allows formation of the peptide bond.	Ribosome senses whether codon–anticodon interactions are correct and shifts to allow tRNA fully into peptide pocket.
	Codon recognition by the ribosome	tRNA anticodon pairs with the codon, but this correct pairing is also sensed by the ribosome.

degree of accuracy. First, the tRNA adaptor molecules must be correctly linked to their corresponding amino acids and, second, the anticodons of each tRNA must bind to the correct codon of the mRNA. There are mechanisms to ensure that both steps are carried out with a minimum of error.

tRNA synthetases, the enzymes that attach amino acids to their correct tRNAs, must do so very accurately. In most organisms, there are 20 tRNA synthetases (some have a few less, and we saw above that archaebacteria have 21), and each must correctly identify a single amino acid from the 20 structurally similar amino acids and also select the corresponding tRNA from the approximately 50 tRNAs. The consequences of not doing so were shown in a classic experiment that demonstrated that the incorporation of amino acids into a polypeptide chain depends entirely on the tRNA anticodon and not the amino acid carried by the aminoacyl-tRNA. Cleverly, the cysteine carried by a $tRNA^{Cys}$ was chemically changed to alanine, whereas the tRNA retained its cysteine anticodon (Fig. 2-12). The alanine was incorporated into protein in the position that should have been cysteine. If then tRNA synthetases make similar mistakes, an incorrect amino acid will be added to the polypeptide chain, which will likely have a disastrous effect on the working of the protein.

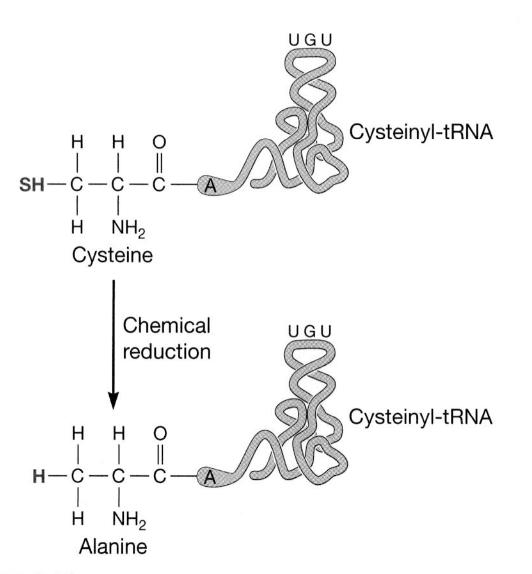

FIGURE 2-12

The incorporation of amino acids into a growing peptide is directed by the codon–anticodon pairing, not by the amino acid carried on the aminoacyl tRNA. A $tRNA^{Cys}$ carrying a cysteine was treated with Raney nickel, a catalyst, so that the cysteine was reduced to alanine, thereby producing a tRNA with the cysteine anticodon but now carrying alanine. The $tRNA^{Cys}$–alanine was added to an in vitro protein synthesis reaction containing random poly(UG) as the synthetic mRNA. Poly(UG) contains many cysteine codons but no alanine codons. Nevertheless, alanine was incorporated in this reaction, thus showing that the amino acid being carried by an aminoacyl-tRNA does not influence the codon–anticodon interaction.

tRNA synthetases face a major challenge in discriminating between structurally similar amino acids. For example, isoleucine and valine differ only in that the latter has an extra methyl group, whereas tyrosine and phenylalanine are distinguished by the former having an additional hydroxyl group. Amino acids that are too large for a particular tRNA are excluded from the acylation site itself, whereas amino acids that are too small are checked for accuracy in an "editing pocket" found in those synthetases where confusion between two related amino acids is possible. The editing pocket contains residues that permit incorrect aminoacyl-tRNAs to fit into its active site where they are hydrolyzed and the amino acid cleaved from the tRNA. On the other hand, aminoacyl-tRNAs with the appropriate amino acid cannot fit into the active site and cannot be cleaved.

Once a tRNA is correctly attached to its corresponding amino acid, the tRNA-amino acid must enter the ribosome and base-pair with the correct mRNA codon. The correct identification of codons is chemically challenging, as there is only a small energy difference between the binding of a correct tRNA, paired to the codons with all three bases, and that of an incorrect tRNA with mismatches at one codon. Calculations based on this energy difference predict an error rate of one in every ten times a tRNA binds, but the actual error rates are two orders of magnitude smaller, ranging from 1 in 2000 to 1 in 20,000 amino acids. It is the ribosome that reduces the chance of codon–anticodon mismatching and the misincorporation of amino acids.

The ribosome accomplishes such amazing accuracy by double-checking the codon–anticodon pairing. tRNA selection is separated into two steps by the irrevocable event of GTP hydrolysis. An aminoacylated tRNA enters the ribosome while bound to the elongation factor EF-Tu. This protein blocks the amino acid from being connected to the growing peptide chain. EF-Tu can leave its aminoacylated tRNA only after hydrolyzing the GTP bound to it. During the initial selection of a tRNA, GTP hydrolysis occurs at a much higher rate for correctly paired tRNAs. Once EF-Tu has hydrolyzed its bound GTP, it can disassociate from the ribosome, which leaves the aminoacylated tRNA available for peptide bond formation. In the second proofreading step, the rate of peptide bond formation is greatly accelerated when the tRNA is correctly paired, but is much slower if pairing is incorrect. This slower rate of peptide bond formation leaves open the opportunity for the incorrect tRNA to leave the ribosome before it is incorporated into the growing chain. This explanation of protein synthesis was made by carefully measuring the rates and binding affinities at each step of translation. Without knowing the precise structure of the ribosome, it was difficult to predict how codon–anticodon pairing could be sensed and how this information was transmitted to affect the rates of EF-Tu GTPase and peptidyltransferase. This required determining the structure of the ribosome at the atomic level.

X-ray Crystallography at Last Reveals the Atomic Structures of Ribosomes

Acquiring the X-ray crystal structure of the ten-subunit RNA polymerase II was a tour de force, but determining the structure of the prokaryotic ribosome was a holy grail for X-ray crystallographers. With more than 50 proteins and more than 3500 nucleotides of RNA, the ribosome was four times larger than any other previously determined structure. For two decades, crystallographers labored to produce increasingly detailed structures of ribosomes and isolated ribosomal components. These low-resolution visions led to useful information on the relative positions of proteins and RNAs in the ribosome, but were inadequate to visualize single atoms. At last, in 2000, the first pictures at atomic resolution were obtained of the 50*S* large ribosomal subunit, both on its own and in complexes with substrates, providing a glimpse of the ribosome in the act of making proteins. One year later, scientists determined the structure of the entire 70*S* ribosome, comprising both the small and large subunits, associated with tRNAs and an mRNA (Fig. 2-13).

These structures gave many insights into the mechanism of translation and the role of the RNA molecules and proteins that make up the ribosome. (Previous attempts to individually crystallize many ribosomal proteins had failed, probably because the 3D structure of the proteins depends on extended interactions with ribosomal RNAs.) The ribosomal proteins are largely present on the surface, although long polypeptides reach in toward the center of each subunit. The small 30*S* subunit is made of three domains that appear mobile and probably move during translation, whereas the 50*S* subunit is an extremely compact structure comprised of a single domain.

The 70*S* structures reveal the interactions between tRNAs, the mRNA, and the two ribosomal subunits. The mRNA is kinked between codons, a conformational change probably key to maintaining a proper reading frame, because the kink separates the codons into distinct segments of the mRNA. The measurements of three tRNAs in the crystal structure showed the large distances (as much as 28 Å) and wide angles of rotation (as much as 46°) through which the tRNAs spiral as they move from one position to the next during translocation. The crystal structure shows clearly the many protein–RNA, RNA–RNA, and protein–protein interactions—intersubunit "bridges"—between the two ribosomal subunits. One such bridge contacts two tRNAs simultaneously, which suggests that these bridges help to coordinate tRNA movement.

The ribosome structure also reveals how translational accuracy is achieved. The shape of the codon–anticodon base pairs at the first two positions is sensed by three nucleotides of the 16*S* ribosomal RNA in the 30*S* ribosomal subunit. Comparison of the structure of an isolated 30*S* ribosomal subunit with the entire 70*S* structure shows that the RNA

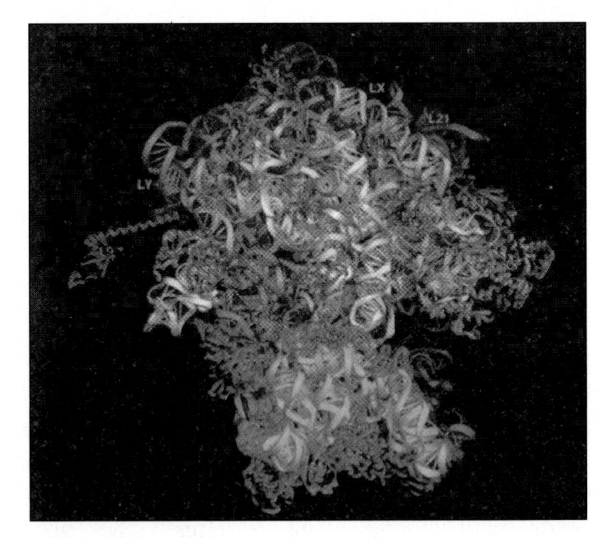

FIGURE 2-13

A three-dimensional representation of a prokaryotic ribosome. This is a ribbon representation of the atomic structure of the ribosome of *Thermus thermophilus*. The 50*S* subunit is above with its 23*S* RNA and 5*S* RNA colored *gray* and *light blue*, respectively. The 50*S* proteins are shown in *magenta*. The 30*S* subunit is below with its 16*S* (*cyan*) and proteins (*dark blue*). Three tRNAs (*gold*, *orange*, and *red*) are shown bound in the interface cavity.

molecules transmit information from the codon–anticodon binding site in the 30*S* subunit to the peptidyltransferase site in the 50*S* subunit and vice versa. Accurate codon–anticodon pairing causes movements of the ribosomal RNAs. These movements are transmitted through the bridge structures and through the tRNAs themselves to the 50*S* subunit. When the pairing is correct, the 50*S* subunit is more likely to assume positions that accelerate the process of initial selection and proofreading. When an incorrect tRNA has bound, the 50*S* subunit remains in positions that slow initial selection and proofreading, so that incorrectly paired tRNA is likely to disassociate.

These high-resolution structures have also settled an important question that had been the subject of debate for decades. Is protein or RNA responsible for catalyzing peptide bond formation? The answer is surprising.

Ribosomal RNA and Not Protein Makes Peptide Bonds

The key reaction catalyzed by ribosomes is the formation of a peptide bond between two amino acids, the *peptidyltransferase reaction*. Experiments had pointed to the possibility that proteins might not be necessary for this reaction, but it was difficult to remove protein molecules entirely from ribosomal RNAs while preserving the 3D structure of the RNAs. The high-resolution structures have settled the issue by providing a detailed view of the active site of the 50*S* subunit where the peptide bond is formed. Remarkably, no protein molecule is within 18 Å of the active site, far too distant for any protein to play a role in catalysis. Instead, it is the 23*S* RNA molecule in the ribosome that is responsible for peptide bond formation.

Enzymes act by positioning the substrates of the reaction so that the reaction proceeds at a much greater rate than would otherwise be the case. The ribosome does this, too, and accelerates the rate of the peptidyltransferase reaction by many orders of magnitude. However, many enzymes also enhance reaction rates by affecting the structure of the substrates and reaction intermediates. They do this either by reacting directly with substrates during the overall reaction, by altering their electronic environments in ways that enhance their reactivity, by stabilizing intermediate states, or by some combination of these. It appears that the ribosome does none of these, although continuing analysis of the intimate workings of the ribosome may reveal unexpected features of the peptidyltransferase reaction.

It would be naive to assume from this that the ribosome is a rather primitive enzyme. After all, the ribosome performs efficiently and is not the rate-limiting step in protein synthesis; the rate-limiting steps are tRNA selection and translocation. The former is limited by diffusion and by the fact that only about 1 tRNA in 50 is acceptable for the next codon. The latter requires major conformational changes in macromolecules and these are relatively slow. As the modern ribosome can achieve rates of peptidyltransferase activity tenfold faster than the rate-limiting steps of tRNA selection and translocation by substrate alignment alone, there may be little evolutionary pressure to improve the performance of the ribosome.

Life May Have Originated in an RNA World

This discovery—that the peptidyltransferase activity of the ribosome is associated with the RNA rather than the protein of the ribosome—provides further support for the hypothesis that RNA rather than protein was the first biologically important molecule. Proteins had been thought to be the first molecules of life since a remarkable discovery made by Stanley Miller in 1953 at the University of Chicago. He found that amino acids are made when electrical discharges occur in an atmosphere of ammonia and hydrogen in the presence of water. These experiments were thought to reproduce the conditions present on the primeval Earth, and it seemed reasonable to assume that ever more complex biological molecules would be formed in the resulting primordial soup. However, proteins cannot reproduce or evolve, and so this vision of life beginning with simple proteins has been supplanted by the speculation that nucleic acids are life's more likely starting materials.

This idea received a great boost when the first *ribozymes*—RNA molecules that can function as enzymes—were discovered in 1982 and 1983. *RNase P* is an enzyme that processes the large precursors of tRNAs. The enzyme has both a protein and an RNA component, but the RNA component alone is able to cut the precursor for the *E. coli* tyrosine tRNA. The second example of a catalytic function for RNA came from studies of the ribosomal RNAs of the ciliate *Tetrahymena*. The precursors of these RNAs contain a segment called an *intron* that is absent from the mature ribosomal RNA. The intron is cut out of the precursor RNA by the action of the molecule itself and, once excised, can catalyze the formation of polynucleotides, act as a site-specific nuclease, and ligate short oligonucleotides! This observation was so

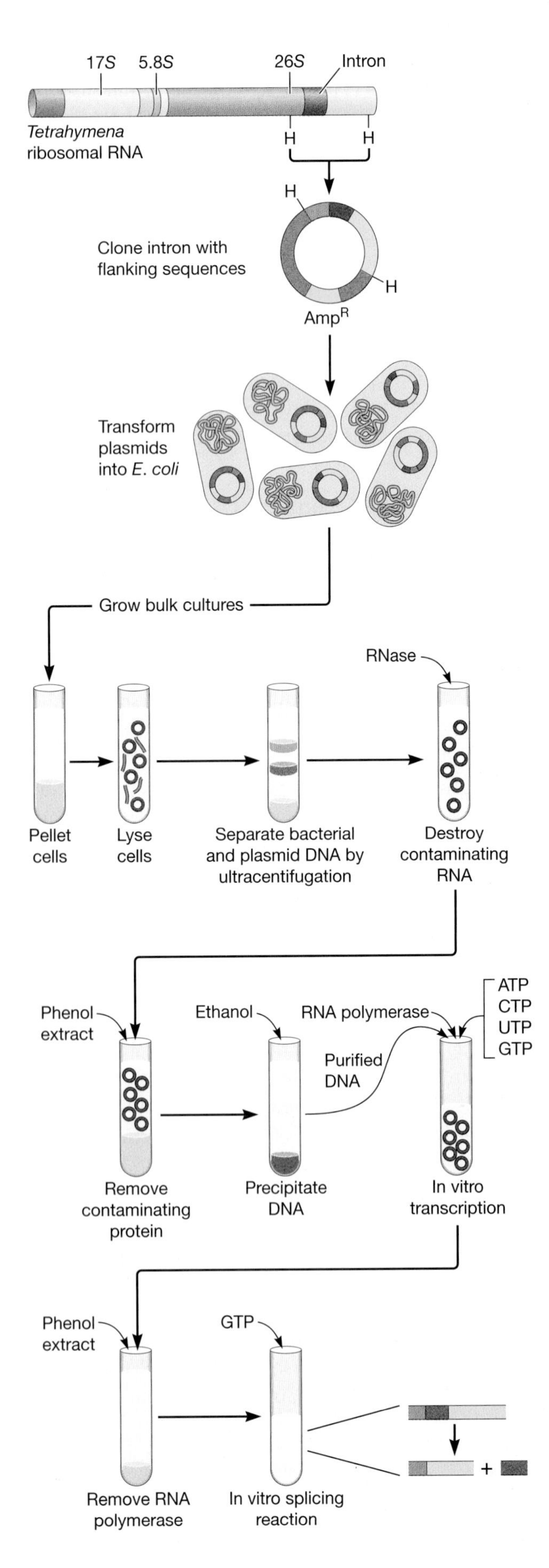

unexpected that especially rigorous proof was required to demonstrate that the RNA was not contaminated by a protein (Fig. 2-14). We will learn much more about introns when we turn to the structure of eukaryotic genes in Chapter 5.

A new class of RNAs called *riboswitches* has recently been identified. These are RNA molecules that sense metabolite concentrations, for example, coenzymes, amino acids, and nucleotides, and alter gene expression to maintain the metabolites at the correct level. Riboswitches usually are found within mRNAs that code for enzymes that produce or transport the molecule that binds to the riboswitch. When riboswitches are bound by their ligand, the shape of the riboswitch changes so that the mRNA forms different base-paired secondary structures that regulate the transcription or translation of the mRNA. Xanthine phosphoribosyltransferase and xanthine permease are key enzymes in the synthesis of purines and in their recycling from nucleic acids; the permease takes up xanthine and the phosphoribosyltransferase converts it into nucleotide precursors. The *xpt-pbuX* operon in *Bacillus subtilis* codes for these enzymes, and a repressor protein had been discovered that inhibited transcription in response to high levels of adenine. However, a corresponding protein had not been found for guanine, leading to the suspicion that an RNA element might be involved in regulating gene expression in response to guanine levels. It was argued that such a riboswitch would likely have a highly conserved sequence because it recognizes a small molecule that has remained unchanged for eons. A bioinformatics comparison of the 5′-untrans-

FIGURE 2-14

Protein-free *Tetrahymena* RNA can self-splice. A portion of the ribosomal RNA genes, including the intron, was cloned into a plasmid. Bacteria containing the plasmid were grown in bulk culture, and plasmid DNA was separated from bacterial DNA by ultracentrifugation of the DNA through a cesium chloride gradient. The plasmid DNA was further purified. RNA was destroyed by treatment with RNase; phenol extractions were used to remove proteins; and pure DNA was precipitated by using ethanol. This DNA was used in an in vitro transcription system to produce RNA. The transcription reaction was treated with phenol to remove the *E. coli* RNA polymerase, and the purified RNA was then used in an in vitro splicing reaction. When the products of this reaction were analyzed by electrophoresis, a band was found that corresponded to the excised intron. Thus, self-splicing occurs in a system scrupulously clean of proteins.

lated region of the *xpt-pbuX* operon with other *xpt* genes in other bacteria identified a conserved, putative guanine-binding sequence, which was called the "G box." The G box emerged as a likely candidate because it created a larger-than-usual space between genes in the bacterial genome. Also, the G boxes from multiple bacteria were predicted to form a similar secondary structure. A 93-nucleotide RNA containing the G box (93 *xpt*) was made by in vitro transcription and labeled with ^{32}P. Experiments showed that guanine induced conformational changes in 93 *xpt*. These conformational changes trigger base-pairing interactions in the RNA so that *xpt* is efficiently transcribed when guanine levels are low (Fig. 2-15).

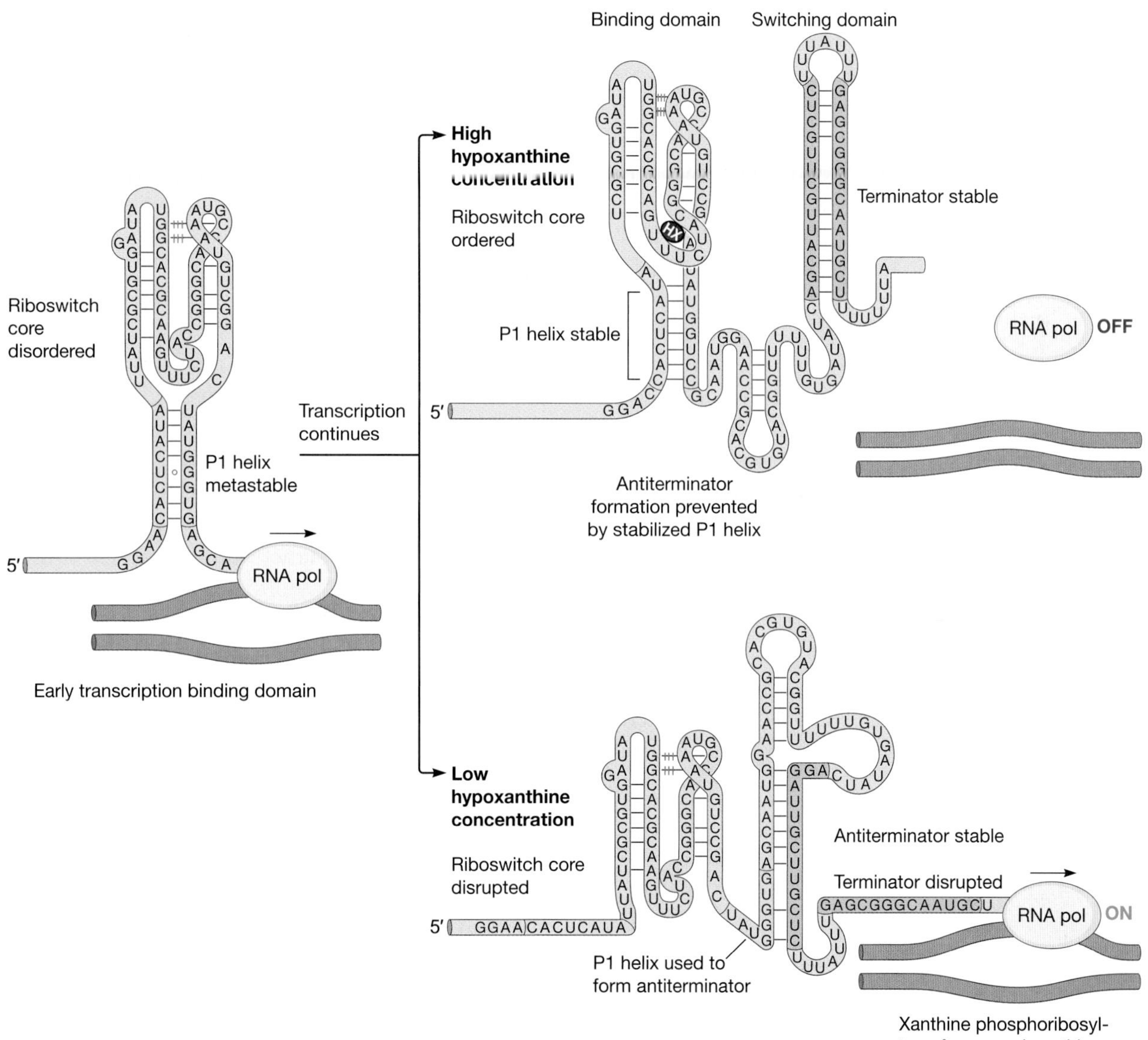

FIGURE 2-15

A riboswitch that binds hypoxanthine and guanine is encoded upstream of the *xpt-pbuX* operon. When hypoxanthine or guanine levels are high, they bind to the riboswitch, thus stabilizing the base pairing in the P1 helix. This conformation allows the mRNA to form a transcription termination loop and transcription stops. When hypoxanthine or guanine are scarce, the P1 helix is not held together so that the 3′ sequences of the P1 helix riboswitch can interact with the following mRNA sequence. The RNA folds to form an antiterminator loop, which disrupts the terminator structure. The RNA polymerase continues synthesizing the mRNA and the operon is transcribed, thereby producing the enzymes necessary for the cell to take up xanthine in preparation for its conversion to guanine.

Riboswitches demonstrate that RNA can specifically recognize many small molecules and in vitro experiments have exploited this by selecting for RNA sequences binding specific molecules or catalyzing unusual chemical reactions.

If RNA was the first molecule and has catalytic functions, then we must assume that it played a central role in the development of the biochemical processes essential to life. Traces of this central role might still be visible in modern cells, and this seems to be the case. Some of the modern functions of RNA are listed in Table 2-5. Furthermore, RNA, ribonucleotides, and ribose are involved in many steps in intermediary metabolism, acting, for example, as cofactors. We will see in Chapter 5 that some RNAs undergo "editing," which may be a relic of an error-correcting process from the time when RNA was the primary genetic material. Although data like these suggest that life may have begun in an RNA world, there are difficulties with this view. The first complex molecules would have emerged in a world that was very inhospitable, particularly for a molecule like RNA, which is very susceptible to hydrolysis, especially at the high temperatures and high concentrations of cations assumed for the prehistoric environment. It has been suggested instead that organic molecules did not arise in a primordial soup, but rather in special and relatively protected environments such as on the surface of clays or pyrites.

Knowing how DNA is copied into RNA to make proteins is not the whole story, any more than knowing how a car assembly line produces cars is the complete story of the functioning of an automobile plant, for there are many factors that determine how many cars are produced each week. There may be demand for a particular model and perhaps a surplus of others; there may be a shortage of parts; or there may be a decline in the economy. The production of cars is *regulated*, and the same is true for proteins, because although many proteins are made in all cells, there are many proteins specific to particular cell types. The production of proteins is often linked to the metabolic activity of the cell, increasing and decreasing as demand requires. In the next chapter we turn to the control of gene expression.

TABLE 2-5. RNA has many functions in the cell

Type of RNA[a]	Function
mRNA	Transfers information from genes to protein-synthesizing machinery
tRNA	Carries activated amino acids for protein synthesis
rRNA	Establishes structure of ribosome to catalyze protein synthesis
U1, U2, U4, U6, U5 snRNAs	mRNA splicing
M1 RNA	Catalytic subunit of RNase P
Telomerase RNA	Template for telomere synthesis
Primer RNA	Initiation of DNA replication
7S RNA	Part of protein secretory complex
ATP	Carrier of energy-rich bonds
Coenzyme A	A key molecule in intermediate metabolism
7SK RNA	Inhibits the transcriptional activator pTEFb
tmRNA	Directs addition of tag to peptides on stalled ribosomes

[a]mRNA, messenger RNA; rRNA, ribosomal RNA; snRNA, small nuclear RNA; tRNA, transfer RNA; tmRNA, transfer messenger RNA.

Reading List

History

Rheinberger H.J. 1997. *Toward a History of Epistemic Things: Synthesizing Proteins in the Test Tube.* Stanford University Press, Stanford, California.

Witkowski J.A., ed. 2005. *The Inside Story: DNA to RNA to Protein.* Cold Spring Harbor Laboratory Press, Cold Spring Harbor, New York. (Contains historical articles on topics covered in this chapter, including many written by the protagonists.)

Protein and Enzymes

Sanger F. and Tuppy H. 1951. The amino acid sequence in the phenylalanyl chain of insulin. *Biochem. J.* **49:** 463–490.

Sanger F. and Thompson E.O.P. 1953. The amino acid sequence in the glycyl chain of insulin. *Biochem. J.* **53:** 353–374.

Sanger F. and Dowding M. 1996. *Selected Papers of Frederick Sanger.* World Scientific, Singapore. (Facsimile printings of 48 of Sanger's most important papers, with fascinating commentaries by Sanger.)

Branden C. and Tooze J. 1999. *Introduction to Protein Structure.* Garland Publishing, New York.

Berg J.M., Stryer L., and Tymoczko J.L. 2002. *Biochemistry.* W.H. Freeman, New York.

The One Gene–One Protein Hypothesis

Garrod A.E. 1908. Inborn errors of metabolism. *Lancet* **2:** 1–7, 73–79, 142–148, 214–220.

Beadle G.W. and Tatum E.I. 1941. Genetic control of biochemical reactions in *Neurospora*. *Proc. Natl. Acad. Sci.* **27:** 499–506.

Pauling L., Itano H.A., Singer S.J., and Wells I.C. 1949. Sickle cell anemia: A molecular disease. *Science* **110:** 543–548.

Bearn A.G. 1993. *Archibald Garrod and the Individuality of Man.* Clarendon Press, Oxford.

Berg P. and Singer M. 2003. *George Beadle: An Uncommon Farmer.* Cold Spring Harbor Laboratory Press, Cold Spring Harbor, New York.

The Central Dogma

Crick F.H.C. 1958. On protein synthesis. Biological replication of macromolecules. *Symp. Soc. Exp. Biol.* **12:** 138–163.

———. 1970. The central dogma of molecular biology. *Nature* **227:** 561–563.

Proteins and Genes

Ingram V.M. 1957. Gene mutations in human hemoglobin: The chemical difference between normal and sickle cell hemoglobin. *Nature* **180:** 326–328.

Yanofsky C., Carlton B.C., Guest J.R., Helinski D.R., and Henning U. 1964. On the colinearity of gene structure and protein structure. *Proc. Natl. Acad. Sci.* **51:** 266–272.

The Coding Problem

Gamow G. 1954. Possible relation between deoxyribonucleic acid and protein structures. *Nature* **173:** 318.

Brenner S. 1957. On the impossibility of all overlapping triplet codes in information transfer from nucleic acid to proteins. *Proc. Natl. Acad. Sci.* **43:** 687–694.

Crick F.H.C., Griffith J.S., and Orgel L.E. 1957. Codes without commas. *Proc. Natl. Acad. Sci.* **43:** 416–421.

Crick F.H.C., Barnett L., Brenner S., and Watts-Tobias R.J. 1961. General nature of the genetic code for proteins. *Nature* **192:** 1227–1232.

Crick F.H.C. 1967. The genetic code—Yesterday, today, and tomorrow. *Cold Spring Harbor Symp. Quant. Biol.* **31:** 3–9. (The Symposium, held in June 1966, marked the end of an era.)

Kay L.E. 2000. *Who Wrote the Book of Life: A History of the Genetic Code.* Stanford University Press, Stanford, California.

Brenner S. 2001. *My Life in Science.* BioMed Central, London.

Knight R.D., Freeland S.J., and Landweber L.F. 2001. Rewiring the keyboard: Evolvability of the genetic code. *Nat. Rev. Genet.* **2:** 49–58.

Synthesizing Nucleic Acids In Vitro

Lehman I.R., Besman M.J., Simms E.S., and Kornberg A. 1958. Enzymatic synthesis of deoxyribonucleic acid. I. Preparation of substrates and partial purification of an enzyme from *Escherichia coli. J. Biol. Chem.* **233:** 163–170.

Hurwitz J., Bresler A., and Diringer R. 1960. The enzymatic incorporation of ribonucleotides into polyribonucleotides and the effect of DNA. *Biochem. Biophys. Res. Commun.* **3:** 15–19.

Stevens A. 1960. Incorporation of the adenine ribonucleotide into RNA by cell fractions from *E. coli* B. *Biochem. Biophys. Res. Commun.* **3:** 92–96.

Weiss S.B. 1960. Enzymatic incorporation of ribonucleotide triphosphates into the interpolynucleotide linkages of ribonucleic acid. *Proc. Natl. Acad. Sci.* **46:** 1020–1030.

Adaptor Molecules and Transfer RNAs

Zamecnik P.C. and Keller E.B. 1954. Relationship between phosphate energy donors and incorporation of labeled amino acids into proteins. *J. Biol. Chem.* **209:** 337–354.

Hoagland M.B., Keller E.B., and Zamecnik P.C. 1956. Enzymatic carboxyl activation of amino acids. *J. Biol. Chem.* **218:** 345–358.

Crick F.H.C. 1957. Discussion. *Biochem. Soc. Symp.* **14:** 25–26.

Hoagland M.B., Stephenson M.L., Scott J.F., Hecht L.I., and Zamecnik P.C. 1958. A soluble ribonucleic acid intermediate in protein synthesis. *J. Biol. Chem.* **231:** 241–257.

Hoagland M. 1996. Biochemistry or molecular biology? The discovery of "soluble RNA." *Trends Biochem. Sci.* **21:** 77–80.

Clark F.C. 2001. The crystallization and structural determination of tRNA. *Trends Biochem. Sci.* **26:** 511–514.

Messenger RNA

Hershey A.D. 1953. Nucleic acid economy in bacteria infected with bacteriophage T2. II. Phage precursor nucleic acid. *J. Gen. Physiol.* **37:** 1–23.

Volkin E. and Astrachan L. 1956. Phosphorus incorporation in *Escherichia coli* ribonucleic acid after infection with bacteriophage T2. *Virology* **2:** 149–161.

Brenner S., Jacob F., and Meselson M. 1961. An unstable intermediate carrying information from genes to ribosomes for protein synthesis. *Nature* **190:** 576–581.

Gros F., Hiatt H., Gilbert W., Kurland C.G., Risebrough R.W., and Watson J.D. 1961. Unstable ribonucleic acid revealed by pulse labelling of *Escherichia coli. Nature* **190:** 581–585.

Jacob F. and Monod J. 1961. Genetic regulatory mechanisms in the synthesis of proteins. *J. Mol. Biol.* **3:** 318–356. (A classic paper of modern biology.)

Volkin E. 1995. What was the message? *Trends Biochem. Sci.* **20:** 206–209.

Solving the Genetic Code

Nirenberg M.W. and Matthaei J.H. 1961. The dependence of cell-free protein synthesis in *E. coli* upon naturally occurring or synthetic polyribonucleotides. *Proc. Natl. Acad. Sci.* **47:** 1588–1602.

Speyer J.F., Lengyel P., Basilio C., and Ochoa S. 1962. Synthetic polynucleotides and the amino acid code, II. *Proc. Natl. Acad. Sci.* **48:** 63–68.

Leder P. and Nirenberg M.W. 1964. RNA code words and protein synthesis II: Nucleotide sequence of a valine RNA code word. *Proc. Natl. Acad. Sci.* **52:** 420–427.

Nishimura S., Jones D.S., and Khorana H.G. 1965. The in vitro synthesis of a co-polypeptide containing two amino acids in alternating sequence dependent upon a DNA like polymer containing two nucleotides in alternating sequence. *J. Mol. Biol.* **13:** 302–324.

Nirenberg M. 2004. Historical review: Deciphering the genetic code—A personal account. *Trends Biochem. Sci.* **29:** 46–54.

Wobble

Crick F.H.C. 1966. Codon–anticodon pairing: The wobble hypothesis. *J. Mol. Biol.* **19:** 548–555.

Ikemura T. 1981. Correlation between the abundance of *Escherichia coli* transfer RNAs and the occurrence of the respective codons in its protein genes. *J. Mol. Biol.* **146:** 1–21.

Sequences That Do Not Code for Amino Acids

Pribnow D. 1975. Bacteriophage T7 early promoters: Nucleotide sequences of two RNA polymerase binding sites. *J. Mol. Biol.* **99:** 419–443.

McKnight S.L. and Yamamoto K.R., eds. 1992. *Transcriptional Regulation.* Cold Spring Harbor Laboratory Press, Cold Spring Harbor, New York.

Ptashne M. and Gann A. 2003. *Genes & Signals.* Cold Spring Harbor Laboratory Press, Cold Spring Harbor, New York.

How Universal Is the Genetic Code?

Bartell B.G., Anderson S., Bankier A.T., deBruijn M.H.L., Chen E., et al. 1980. Different pattern of codon recognition by mammalian mitochondrial tRNAs. *Proc. Natl. Acad. Sci.* **77:** 3164–3166.

Bonitz S.G., Berlani R., Coruzzi G., Li M., Macino G., et al. 1980. Codon recognition rules in yeast mitochondria. *Proc. Natl. Acad. Sci.* **77:** 3167–3170.

Anderson S., Bankier A.T., Bartell B.G., deBruijn M.H.L., Coulson A.R., et al. 1981. Sequence and organization of the human mitochondrial genome. *Nature* **290:** 457–465.

Making Proteins Is a Highly Accurate Process

Chapeville F., Lipmann F., von Ehrenstein G., Weisblum B., Ray W.J., and Benzer S. 1962. On the role of soluble ribonucleic acid in coding for amino acids. *Proc. Natl. Acad. Sci.* **48:** 1086–1092.

Hou Y.M. and Schimmel P. 1988. A simple structural feature is a major determinant of the identity of a transfer RNA. *Nature* **333:** 140–145.

Weisblum B. 1999. Back to Camelot: Defining the specific role of tRNA in protein synthesis. *Trends Biochem. Sci.* **24:** 247–250.

Rodnina M.V. and Wintermeyer W. 2001. Ribosome fidelity: tRNA discrimination, proofreading and induced fit. *Trends Biochem. Sci.* **26:** 124–130.

The Structure of RNA Polymerase and the Ribosome

Ban N., Nissen P., Hansen J., Moore P.B., and Steitz T.A. 2000. The complete atomic structure of the large ribosomal subunit at 2.4 Å resolution. *Science* **289:** 905–920.

Nissen P., Hansen J., Ban N., Moore P.B., and Steitz T.A. 2000. The structural basis of ribosome activity in peptide bond synthesis. *Science* **289:** 920–930.

Wimberly B.T., Brodersen D.E., Clemons W.M., Jr., Morgan-Warren R.J., Carter A.P., et al. 2000. Structure of the 30*S* ribosomal subunit. *Nature* **407:** 327–339.

Cramer P., Bushnell D.A., and Kornberg R.D. 2001. Structural basis of transcription: RNA polymerase II at 2.8 Ångstrom resolution. *Science* **292:** 1863–1876.

Gnatt A.L., Cramer P., Fu J., Bushnell D.A., Kornberg R.D. 2001. Structural basis of transcription: An RNA polymerase II elongation complex at 3.3 Å resolution. *Science* **292:** 1876–1882.

Yusupov M.M., Yusupova G.Z., Baucom A., Lieberman K., Earnest T.N., et al. 2001. Crystal structure of the ribosome at 5.5 Å resolution. *Science* **292:** 883–896.

Nomura M. 1997. Reflections on the days of ribosome reconstitution research. *Trends Biochem. Sci.* **22:** 275–279.

The RNA World

Gesteland R.F., Cech T.R., and Atkins J.F., eds. 2006. *The RNA World.* Cold Spring Harbor Laboratory Press, Cold Spring Harbor, New York.

Kruger K., Grabowshi P.J., Zaug A., Sands A.J., Gottschling D.E., and Cech T.R. 1982. Self-splicing RNA: Autoexcision and autocyclization of the ribosomal RNA intervening sequence of *Tetrahymena*. *Cell* **31:** 147–157.

Guerrier-Takada C., Gardiner K., Marsh T., Pace N., and Altman S. 1983. The RNA moiety of ribonuclease P is the catalytic subunit of the enzyme. *Cell* **35:** 849–857.

Zaug A.J., Been M.D., and Cech T.R. 1986. The *Tetrahymena* ribozyme acts like an RNA restriction endonuclease. *Nature* **324:** 429–433.

Eddy S.R. 2001. Non-coding RNA genes and the modern RNA world (review). *Nat. Rev. Genet.* **2:** 919–929.

Mandal M., Boesc B., Barrick J.E., Winker W.C., and Breaker R.R. 2003. Riboswitches control fundamental biochemical pathways in *Bacillus subtilis* and other bacteria. *Cell* **113:** 577–586.

Orgel L.E. 2004. Prebiotic chemistry and the origin of the RNA world. *Crit. Rev. Biochem. Mol. Biol.* **39:** 99–123.

Tucker B.J. and Breaker R.R. 2005. Riboswitches as versatile gene control elements (review). *Curr. Opin. Struct. Biol.* **15:** 342–348.

Vitreschak A.G., Rodionov D.A., Mironov A.A., and Gelfand M.S. 2004. Riboswitches: The oldest mechanism for the regulation of gene expression? *Trends Genet.* **20:** 44–50.

CHAPTER 3

Control of Gene Expression

Even before the basic outline of the genetic code became established, it was obvious that intricate molecular mechanisms must exist in cells to control the numbers of their many proteins. Within *Escherichia coli*, for example, the relative amounts of the different proteins vary enormously (from <0.01% to ~2% of the total), even though each protein product is coded by a single gene along the *E. coli* chromosome. A priori we can imagine two ways that the cell might achieve this differential synthesis. The first would be by the evolution of molecular signals that control the rates at which specific messenger RNA (mRNA) molecules are transcribed from their DNA templates (*transcriptional control*). The second way would involve molecular devices for controlling the rate at which mRNA molecules, once synthesized, are translated into their polypeptide products (*translational control*). It makes sense for a cell to make only the mRNA molecules that it needs; therefore, molecular biologists initially focused their studies on whether transcriptional control existed. Here, intensive genetic analysis of the so-called induced enzymes of bacteria presented the first key insights into the control of gene expression. However, there are circumstances in which it is necessary to regulate protein synthesis by, for example, regulating the rate of protein synthesis or destroying already synthesized mRNAs—*posttranscriptional control.*

Later, scientists began to study gene expression in eukaryotes, where gene regulation is generally more complex. On average, eukaryotes have larger genes, and these genes are typically controlled by a greater number of regulatory signals than are typical bacterial genes. Also, in eukaryotes, the accessibility of genes to transcription is strongly influenced by the packaging of DNA into chromosomes—around protein complexes called *nucleosomes*—providing yet another level of gene expression not found in prokaryotes. We will begin with examples of gene expression as initially learned from bacteria and later explore how these basic regulatory mechanisms are combined and elaborated in eukaryotes to result in complex patterns of gene expression.

Regulation of Transcription Is the Primary Mechanism Used to Control Protein Levels

Bacteria generally exist in environments that change rapidly; thus, some of the enzymes that they may need at one moment may be useless or even counterproductive at another. Similarly, these cells may suddenly require an enzyme whose presence was previously unnecessary. One way for bacteria (or any other cells) to meet these challenges would be for them to carry on the synthesis of all enzymes simultaneously, whether or not the substrates for these enzymes are present. This approach would be wasteful, however, and, in fact, bacteria do not operate in this way. Instead, most genes function at highly variable rates, so that they make mRNA molecules at appreciable levels only when their genes receive specific signals from outside the cell to go into action.

Cells can, in principle, regulate the levels of a protein at any of four points: by controlling the rate of mRNA synthesis and the stability of mRNAs, and by changing the rate of protein synthesis and the stability of proteins. There are examples of gene regulation occurring at all of these steps, but typically gene regulation occurs most frequently at the beginning of the process, the transcription of mRNAs. This approach is energy efficient, because subsequent regulation steps waste the energy expended in mRNA synthesis. As a result, analyzing the quantity of mRNA present in a cell usually gives a good approximation of the abundance of the corresponding protein. This correlation is the reason that microarray analysis of mRNA levels has been useful in comparing differences in gene expression between different cells. As we shall see in Chapter 13, microarrays are used to analyze gene expression profiles across a genome. Each microarray consists of thousands of DNA sequences immobilized on a solid support and allows the researcher to measure the abundance of every mRNA present in a cell.

Jacob and Monod Proposed a Revolutionary Model Explaining Gene Regulation

When *E. coli* bacteria are exposed to a mixture of nutrients, they first metabolize the simplest sugars, such as glucose. However, when simple sugars are unavailable, the bacteria will produce the enzymes necessary to utilize whatever sugars are available to them. Proteins in the cell sense which sugars are present and activate the appropriate genes. For example, *E. coli* cells normally make the enzyme β-galactosidase (β-gal) at high rates only when lactose is present. (For lactose to be useful as a food source, it must be cleaved into the simpler sugars glucose and galactose; β-galactosidase is the enzyme that catalyzes this splitting.)

Genetic analysis by Jacques Monod and François Jacob of the Institut Pasteur in Paris proved crucial to working out the molecular details of this adaptive phenomenon. Most early work centered on analyzing the β-galactosidase system in *E. coli*; the essential clues emerged from the study of mutants unable to vary the amount of β-galactosidase produced. The key findings were that some mutant bacteria made maximum amounts of enzymes whether or not lactose was present, whereas other mutants produced only traces of enzyme. The use of genetic crosses, similar to those used in the elucidation of the genetic code (Chapter 2), showed that most mutants had a normal gene for galactosidase, indicating that other genetic elements had to be affecting expression levels of the enzyme. Some mutants could be repaired only by a recombination event between the mutant chromosome and the wild-type chromosome, thus demonstrating that the mutation was linked to the β-galactosidase gene on the same DNA molecule. Surprisingly, other mutants could be complemented simply by the presence of a wild-type chromosome in the same cell, which suggests the presence of a diffusible factor that regulated gene expression.

Jacob and Monod explained these observations by proposing that the β-galactosidase gene is regulated by a repressor molecule, the product of another gene, and they described a revolutionary, general model for the control of gene expression. A specific *repressor* molecule exists that binds near the begin-

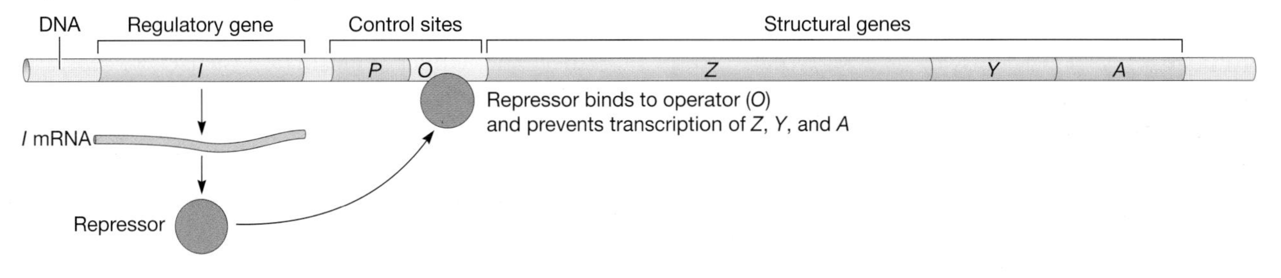

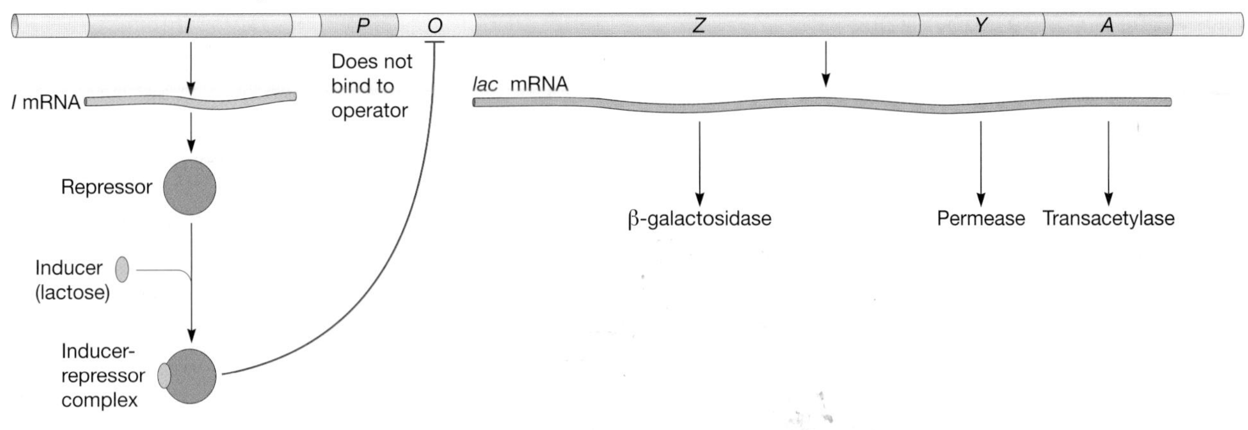

FIGURE 3-1

The regulation of the *lac* operon. The *E. coli* gene for β-galactosidase (*Z*) is located adjacent to two additional genes involved in lactose metabolism. One gene (*Y*) codes for lactose permease, a protein that facilitates the specific entry of lactose into bacteria, whereas the second (*A*) codes for thiogalactosidase transacetylase, an enzyme that may help to remove lactose-like compounds that β-galactosidase cannot split into useful metabolites. The regulatory gene (*I*) codes for the lactose repressor that, in the absence of lactose, binds to the operator sequence (*O*), thus suppressing transcription. When lactose is added to *E. coli* cells, it binds the repressor protein, RNA polymerase binds to the promoter (*P*), and an mRNA transcript encoding all three enzymes—β-galactosidase, permease, and acetylase—is synthesized. Thus, on induction, the amounts of all three proteins rise coordinately. The set of adjacent genes that is transcribed as a single mRNA molecule, together with the control regions, is called an operon. Some operons are quite large—for example, the 11 genes coding for the enzymes in the pathway for the synthesis of the amino acid histidine constitute one operon. The enzymes are translated from one extremely large mRNA molecule containing more than 10,000 nucleotides.

ning of the β-galactosidase gene at a site called the *operator*. By binding to the operator site on the DNA, the repressor prevents RNA polymerase from beginning synthesis of β-galactosidase mRNA. Lactose acts as an *inducer* that, by binding to the repressor, prevents the repressor from interacting with the operator. Thus, in the presence of lactose, the repressor is inactivated and the mRNA is made. Upon removal of lactose, the repressor regains its ability to bind to the *lac* operator DNA and to switch off expression of the β-galactosidase gene (Fig. 3-1). Their model was later extended when it became clear that genes are not only negatively regulated by repressor molecules, but are also positively regulated by activator molecules.

Jacob and Monod's genetic experiments soon showed that the β-galactosidase gene is adjacent to two other genes involved in lactose metabolism—*lacY*, which codes for a lactose permease that permits the uptake of lactose, and *lacA*, which codes for a galactoside transacetylase for which, remarkably, neither a function nor a substrate have yet been identified. These three genes are transcribed as a single mRNA molecule, so that on induction by lactose, the levels of all three proteins rise coordinately. A collection of adjacent genes, together with their control region, that are transcribed into single mRNA molecules is called an *operon*. The organization of genes into operons is an efficient way to ensure that all the enzymes necessary for a particular biochemical process are synthesized when needed and in appropriate amounts. Operons are less commonly found in eukaroytes, although many viruses have adopted the approach, probably to compress the maximum number of genes into a small viral genome.

Jacob and Monod's studies on the β-galactosidase system established the fundamental principle that expression of a gene can be controlled by other genes acting on nearby DNA sequences.

Repressors Bind DNA to Prevent Transcription

Repressors are proteins that interfere with the transcription of their target genes. Genes encoding repressors can be found close to and at a distance from the operons on which they act. The gene for the lactose repressor, for example, lies immediately in front (upstream) of the operon, whereas that for the *trp* operon, which encodes proteins involved in the biosynthesis of tryptophan, is located far away from the operon. Repressors are usually made at a constant rate. Such invariant synthesis is known as *constitutive synthesis*, and the rate of synthesis is a function of the structure of the promoter of the repressor gene. Normally the promoters of repressor genes function at very low rates, leading to the presence of only a few repressor mRNA molecules in the average cell. However, there exist promoter mutants that cause much higher rates of repressor mRNA synthesis and, correspondingly, much higher numbers of repressor molecules per cell. Therefore, even in the presence of high levels of inducer, such mutant cells make smaller-than-usual amounts of the induced proteins.

Because the genetic studies leading to the postulation of repressors were so complete, the role of repressors as key bacterial control elements seemed almost inescapable. Final proof, however, had to await the development of biochemical procedures by which individual repressors could be isolated, chemically identified, and shown to bind specifically to their respective operators. These steps depended on the development of genetic techniques for increasing the few copies of repressor molecules normally present. In 1966, at Harvard University, Walter Gilbert and Benno Müller-Hill created cells with increased numbers of copies of the Lac repressor gene that made large amounts of repressor protein. Gilbert and Müller-Hill isolated the lactose repressor from these cells and demonstrated that it is a protein of molecular weight (MW) 38,000 and that it has two specific binding sites—one for lactose-like compounds and the other for DNA containing the lactose operator sequence (Fig. 3-2). At the same time, Mark Ptashne, also at Harvard, isolated the phage λ repressor that controls the rates at which several classes of λ-specific mRNAs are made. The λ repressor is a 26,000-MW polypeptide chain; it likewise binds to only its specific operator sequence.

In bacteria, repressors interfere directly with RNA polymerase binding to a gene. In eukaryotes, there

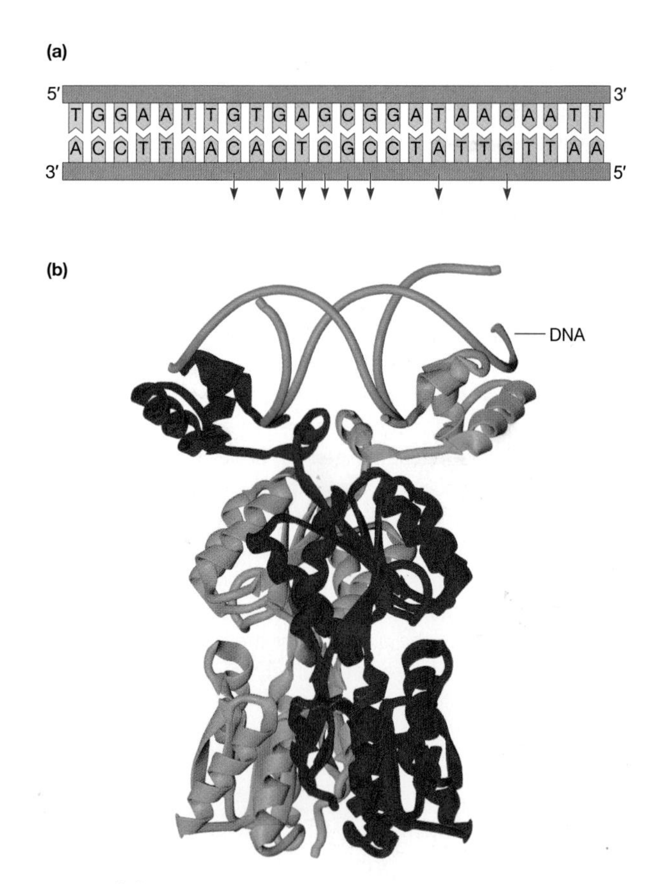

FIGURE 3-2

Lac repressor binds to the operator DNA sequence. (*a*) The sequence of the *lac* operator. Note the twofold symmetry across much of the sequence (palindrome), as highlighted in *pink*. Mutations of the nucleotides indicated by *arrows* lead to constitutive expression (expression in the absence of inducer) because the repressor protein can no longer bind to the operator sequence. (*b*) The structure of a dimer of Lac repressor bound to DNA. The repressor molecules (*orange* and *red*) face one another so that each binds one-half of the near-palindromic operator sequence.

are fewer examples of repressors that act in this way. In addition, various proteins regulate the degree of packaging of the DNA in eukaryotic cells and hence accessibility of the transcription machinery to genes. The degree of compaction has a significant effect on the expression of genes—the more closed the chromatin structure, the less frequently genes tend to be expressed. Chromatin states may be heritable and often affect many genes simultaneously. In Chapter 8 on epigenetics, we will see many examples of the role of chromatin structure in the control of gene expression.

Promoters Are DNA Sequences That Are the Start Signals for RNA Synthesis

Some of the β-galactosidase mutants studied by Jacob and Monod were unable to make β-galactosidase unless they were rescued by recombination with a wild-type *lac* operon. Thus, these mutations somehow affected the synthesis of the mRNA for β-galactosidase. However, many of these mutants had a normal coding sequence for β-galactosidase, suggesting that RNA polymerase was not transcribing the gene. These mutations were mapped to locations upstream of the transcription start site for β-galactosidase—locations where RNA polymerase binds to DNA before initiating transcription.

These sites are now called *promoters.* Because promoter mutations affect only the synthesis of the mRNA molecule immediately downstream, promoters are examples of *cis-acting* control elements (*cis* is a Latin word meaning "on the side of"). There are also *trans-acting* control sequences that can function when physically disconnected from the genes they control—for example, when located on a different chromosome or on an extrachromosomal plasmid. Generally, *trans*-acting genes encode a diffusible molecule that binds *cis*-acting elements and so regulates gene expression—the Lac repressor behaves as a *trans* factor (Fig. 3-3).

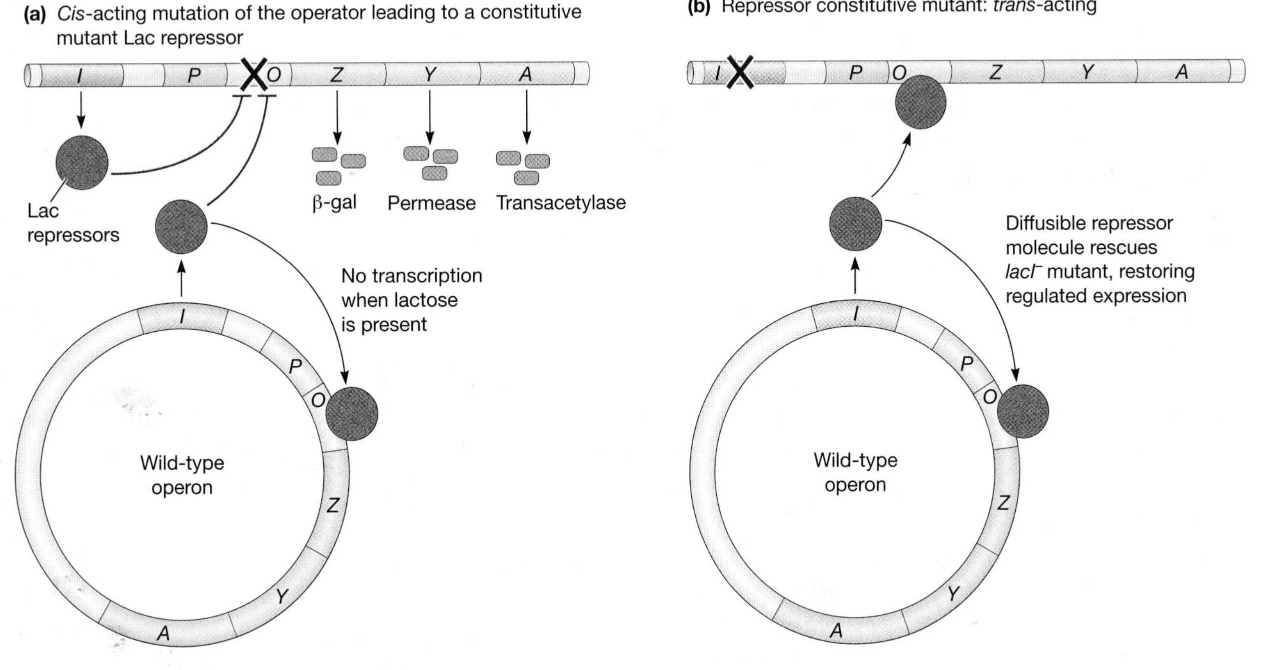

FIGURE 3-3

Mutations can be classified as *cis*- or *trans*-acting. *cis* mutations affect only the DNA molecule to which they are linked, whereas *trans* mutations exert their effects through the action of a factor that can diffuse to other sites. Whether a mutation is *cis* or *trans* can be determined by supplying an extra copy of the wild-type operon, usually cloned into a plasmid. This approach was used to test different constitutively expressed *lac* operon mutants, which expressed the genes of the *lac* operon whether or not lactose was present. (*a*) One mutant was not rescued by the addition of a *lac* operon because of a mutation in the operator sequence that prevented repressor molecules binding to it. (*b*) A second constitutive mutation was caused by a mutation in *I*, the repressor gene. This is a *trans* mutation because it can be rescued by a wild-type copy of the *lac* operon that supplies a functional repressor molecule.

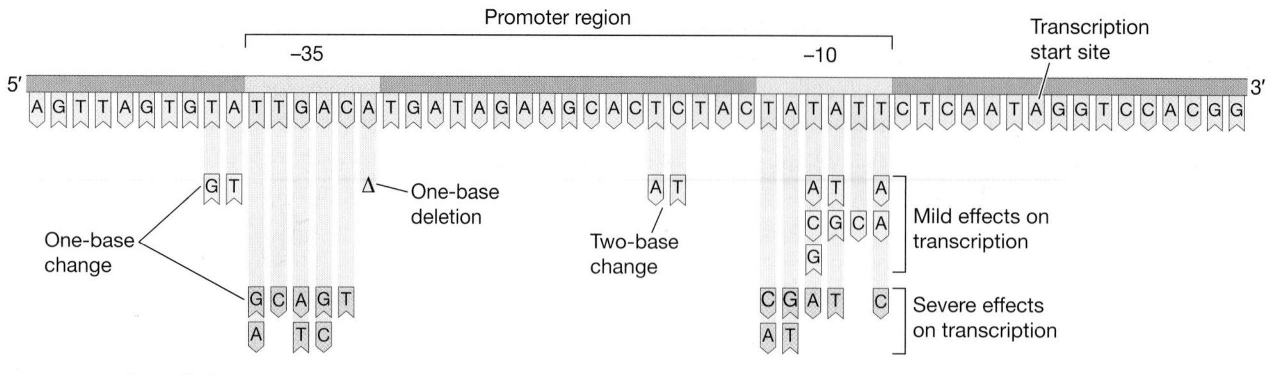

FIGURE 3-4

Specific DNA sequences are important for efficient transcription of *Escherichia coli* genes by RNA polymerase. The boxed sequences approximately 35 and 10 nucleotides upstream of the transcription start site are highly conserved in all *E. coli* promoters. Mutations in these regions have mild (*green*) and severe (*pink*) effects on transcription. The mutations may be changes of single nucleotides or pairs of nucleotides, or a deletion may occur.

Jacob and Monod found other *cis*-acting mutants that produced β-galactosidase continuously even when lactose was not present. When these mutants were mapped and cloned, it was discovered that there were changes in the operator that prevented the binding of the repressor molecule. When the operator sequence is mutated, the Lac repressor can no longer bind and the operon is transcribed continuously.

The frequency of the initiation of transcription at a given promoter can vary by a factor of at least 1000, the maximum rate of transcription from a particular segment of DNA depending on the specific sequence of bases in its promoter. Two highly conserved, separate nucleotide blocks make up the promoter of the *E. coli lac* operon and are characteristic of many other *E. coli* promoters (Fig. 3-4). Both blocks were identified by the existence of *cis*-acting point mutations that blocked RNA synthesis. One block is located about 10 nucleotides upstream of the mRNA start site; the other is located about 25 nucleotides further upstream (the "stream" flows in the direction of transcription). These blocks are designated −10 and −35 relative to the mRNA start site at +1. The initial step in *E. coli* transcription is believed to be the recognition and binding of an RNA polymerase molecule to the −35 region. Subsequently, upon binding of polymerase, the −10 region *melts* (opens up) into its component single strands, thereby allowing transcription to begin at the +1 position. Thus, the sequences required for RNA polymerase binding and initiation are contained within less than 40 nucleotides.

Bacterial RNA polymerase core complexes are bound by a single initiation factor σ (sigma), which directs the polymerase to the promoter sequences. The σ subunits are named according to their molecular weight. It is the σ^{70} subunit that binds to the −35 and −10 sequences in the promoter of the *lac* operon. This is the most common initiation factor in the *E. coli* genome, and many other promoters have similar sequences for recognition by σ^{70}. The activation of σ factors is one way in which *E. coli* makes major changes in its program of gene expression in response to stimuli. For example, when cells are subjected to environmental stress such as a sudden rise in temperature ("heat shock"), the σ^{32} subunit is activated and directs polymerase to transcribe specific promoters to respond to this stress. When cells are starved for nitrogen, σ^{54} is activated, turning on genes needed for the uptake of nitrogen and biosynthesis of glutamine. Some bacteriophages even produce their own σ factors with which they can steal the transcription machinery of the host to transcribe their own genomes.

Genes Are Activated When Transcription Factors Bind Specific DNA Sequences near Promoters

Jacob and Monod's model correctly predicted the mechanisms that prevent β-galactosidase from being expressed in the absence of lactose, but did not take into account the effect of factors that positively affect expression. Promoters can also be under the control of positively acting proteins (activators) that increase the rates at which mRNA chains are made. If both glucose and lactose are present, *E. coli* will first use glucose, a simpler sugar. When this energy source is used up, starvation signals accumulate in the cell, and trigger an activation of the *lac* operon, so that 40-fold more β-galactosidase is produced.

Several proteins that positively regulate RNA polymerase binding, and hence gene expression, have been isolated from *E. coli*. A well-understood positive regulator signals to the appropriate genes that glucose is not available as a food source. When the cell has exhausted its energy supplies, the amount of the intracellular regulator cyclic AMP (cAMP) increases. This cAMP then binds to the DNA-binding protein known as the catabolite gene activator protein (CAP). The resulting cAMP–CAP complexes, by binding to the appropriate promoters, including that of the *lac* operon, help to activate genes whose enzymes can break down alternative sugars such as lactose and galactose.

In bacteria, the regulation of gene transcription is thus controlled in part by the binding of specific regulatory proteins to control sequences situated at the beginnings of their various genes (or operons). These regulatory proteins are called *transcription factors*. We have already seen how repressors can negatively control transcription by blocking the access of RNA polymerase to the promoter. Many other transcription factors such as CAP act positively; they work to recruit RNA polymerase to the promoter of the gene. Still other regulatory proteins promote the initiation of RNA polymerase or help the RNA polymerase separate the DNA strands in the promoter. Because transcription factors are often required for RNA polymerase to transcribe a gene, regulation of the binding of transcription factors to DNA results in the control of gene activity.

Proteins That Regulate Transcription Often Bind DNA Cooperatively

Many transcription factors—repressor and activators—bind DNA as dimers, with one molecule recognizing each half of its operator sequence. Sometimes two dimers will bind to separate sites and at the same time interact with each other. This process is called cooperative binding because the binding of one protein dimer helps the binding of the other to the second site. There are several consequences of cooperative binding, including an increase in affinity for the binding sites, producing stronger repression, and also an increase in specificity. λ repressor binds cooperatively to multiple operator sequences, and that cooperativity is vital for the proper action of the protein, as we shall see.

In another form of cooperative binding, Lac repressor exists as a stable tetramer in the cell. This complex can bind to two operators simultaneously. In both the Lac and λ cases, sometimes DNA has to loop out to accommodate these interactions (Fig. 3-5).

Cooperativity has been found in many other genetic systems, and other genes can carry a number of binding sites for different regulatory proteins. Frequently, the regulatory proteins that bind to DNA sites upstream of a gene interact with one another as well as with the DNA, resulting in cooperativity among the various regulators. As we saw with the *lac* operon, association between regulatory proteins bound to DNA often involves DNA looping. The basic principles established by study of the *lac* operon in *E. coli* have held true for gene regulation in all organisms. Later in this chapter we shall see how the cooperative action of many transcription factors results in complex patterns of gene regulation during the development of organisms.

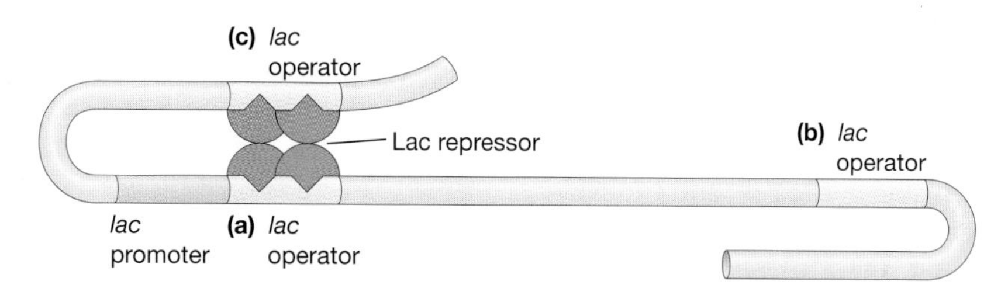

FIGURE 3-5

Lac repressor binds as a tetramer to two operators, looping the DNA. The primary operator (*a*) is located adjacent to the *lac* operon, where bound repressor can block RNA polymerase binding. A second operator (*b*) is located 400 bp downstream, and a third (*c*) lies 90 bp upstream. The combined affinities of Lac repressor for the primary operator and the other nearby operator sequence increase the likelihood that the primary operator is bound by Lac repressor.

Attenuation Is a Method of Regulating Transcription by Preventing Complete mRNA Synthesis

As we have seen, transcriptional activation is the most common mechanism by which gene expression is controlled. However, the expression of many bacterial operons, particularly those involved in amino acid biosynthesis, is also influenced by a process called *attenuation*. This phenomenon was discovered through the elegant experiments by Charles Yanofsky and his coworkers at Stanford University on the tryptophan (*trp*) operon of *E. coli*. This by now exhaustively studied operon consists of a transcriptional control region and five structural genes that encode the enzymes involved in the last steps of tryptophan biosynthesis. The rate at which transcription of *trp* mRNA begins is controlled by a tryptophan-activated repressor molecule that can block the access of RNA polymerase to the *trp* promoter. However, once the repressor leaves the promoter, transcription does not necessarily continue to the end of the operon to produce the very long full-length *trp* mRNA. Instead, incomplete transcription (attenuation) frequently occurs to produce a relatively short 162-base mRNA molecule that codes for a correspondingly small Trp L (leader) protein. Although this tiny protein has no apparent function elsewhere in the cell, it has the very interesting property of being rich in the amino acid tryptophan, whose sole codon is UGG. This allows Trp L to act as a sensor to signal tryptophan levels within the cell.

Whether attenuation takes place depends on the exact folding pattern of the nascent mRNA between the transcribing RNA polymerase and the ribosome translating the *trpL* region. The *trpL* mRNA contains stretches of sequence that can base-pair with each other, forming one of two possible hairpin-like structures (Fig. 3-6). When tryptophan is plentiful, the ribosome translates freely through the leader, and the nascent mRNA folds into a structure that signals RNA polymerase to terminate transcription. When tryptophan is scarce, few activated *trp* transfer RNAs (tRNAs) are available in the cell, and the ribosome therefore stalls at the UGG codons in the leader sequence. Under these circumstances, the mRNA now folds into an alternative structure and RNA polymerase proceeds through it to the end of the operon. Thus, under normal nutritional conditions, most *trp* operon transcripts are "attenuated" or terminated early. It is only when the tryptophan level drops that full-length transcripts of the operon are made efficiently. Tryptophan starvation therefore increases expression of the *trp* operon in two ways: by removing the *trp*-activated repressor, thus increasing transcription by a factor of 70, and by relieving attenuation, increasing transcription a further eight- to tenfold. Overall, then, the expression of the operon can be increased approximately 600-fold. The use of both repression and attenuation allows fine-tuning of the level of intracellular tryptophan. Other operons, all of which control amino acid biosynthesis, have been discovered that rely entirely on attenuation for control. Just as with the tryptophan operon, the leader peptides are rich in the amino acid produced by that operon.

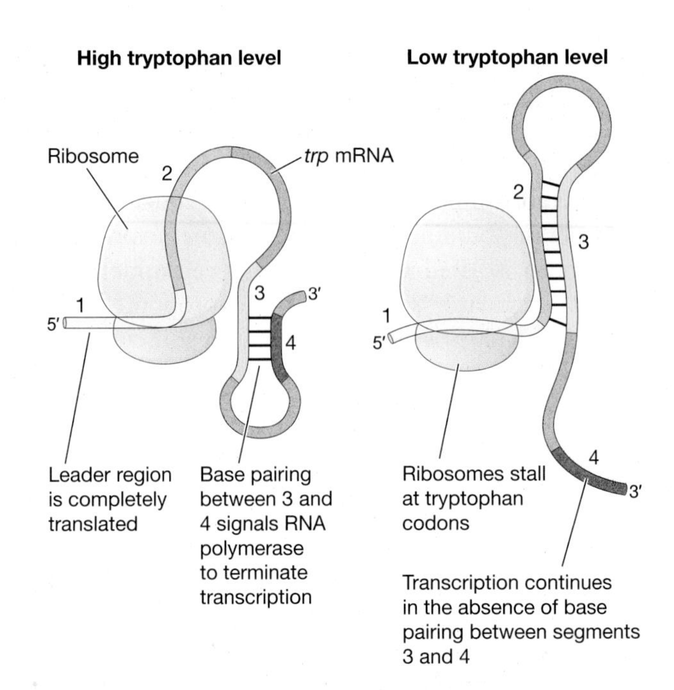

FIGURE 3-6

A model for attenuation at the *trp* operon. The 5′ end of the *trp* operon mRNA, the leader region (region 1), is rich in tryptophan codons. When tryptophan is available, normal translation of this leader sequence occurs. As this happens, the *trp* mRNA forms a stem–loop structure (regions 3 and 4) that apparently does not allow RNA polymerase to continue the transcription of the remainder of the *trp* operon. Note that this "attenuated" state is the *normal* situation; it is only when the tryptophan level drops that the attenuation is relieved. This is believed to occur when a ribosome is stalled trying to translate the leader sequence; the resulting formation of a different stem–loop structure in the next segment of *trp* mRNA (regions 2 and 3) allows the RNA polymerase to continue with the transcription of the remainder of the *trp* operon.

Attenuation is possible because transcription and translation can be coupled in bacteria, which do not have nuclear membranes separating the sites of mRNA and protein synthesis. The mechanism is not possible in eukaryotes, because the nuclear membrane divides transcription in the nucleus from translation in the cytoplasm. However, there are other mechanisms by which the efficiency of transcription elongation and the production of full-length transcripts can be regulated both in eukaryotes and prokaryotes.

Regulation of Translation Is Another Mechanism Used to Control Protein Levels

The expression of proteins is also controlled at the level of translation. Efficient initiation of translation in *E. coli* depends on the presence of a purine-rich sequence of 6–8 nucleotides that lies just upstream from the AUG initiation codon. The existence of this *ribosome-binding site* was first identified in 1974 by John Shine and Lynn Dalgarno in Canberra, Australia, who noticed that the sequence was complementary to the 3′ end of the 16*S* ribosomal RNA (rRNA) molecules found in bacterial ribosomes. This suggested that the initial positioning of mRNA on the smaller ribosomal subunit requires base pairing between the 16*S* rRNA chain and the ribosome-binding (*Shine–Dalgarno*) sequences in the mRNA. In general, in the most efficiently translated mRNAs, the ribosome-binding sequences are centered eight nucleotides upstream from the initiation codon. Mutations in this region, including mutations that place this sequence closer to or farther away from the AUG start codon, can greatly lower the translational efficiency of their respective mRNAs. It is important to note that possession of an appropriately sited Shine–Dalgarno sequence does not guarantee initiation of protein synthesis. Many such sequences are effectively buried in the loop structures and therefore have no way of interacting with 16*S* rRNA.

The efficiency with which a given Shine–Dalgarno sequence works can be modulated by proteins that bind to it and block its availability to the 16*S* ribosomal RNA. The best-understood example involves the ribosomal proteins (*r-proteins*) of *E. coli.* When the rate of r-protein synthesis exceeds the rate at which ribosomal RNA is made, free r-proteins accumulate and certain "key" ones bind to the Shine–Dalgarno sequences on the r-protein mRNA molecules. In this way ribosomal proteins are not synthesized faster than they can be used in making ribosomes, and translational control is achieved by competition between rRNA and r-protein mRNA for the binding of these key proteins.

In eukaryotes, the recognition of the translation start site differs slightly from that in prokaryotes. Instead of a Shine–Dalgarno ribosome-binding sequence, eukaryotic ribosomes bind the 5′ end of an mRNA and scan along the RNA until they find a sequence, first recognized by Marilyn Kozak, which contains the AUG start site in the context of the consensus sequence GCC(G/A)CCAUGG. If the sequence around the initiation codon does not match this consensus Kozak sequence, translation initiation occurs much less frequently. The surrounding structure of the RNA also influences the efficiency of recognition of initiation sites. In the vast majority of eukaryotic genes, a single peptide is translated from each eukaryotic mRNA. In rare cases, a short initial peptide is translated, followed by a longer polypeptide. Also, some RNA viruses and possibly some cellular genes have sequences that act as internal ribosomal entry sites (IRESs), which are RNA motifs that allow internal ribosome binding and translation initiation.

Once translation begins, its rate is determined by the availability of the various tRNAs corresponding to the specific codons encoded in the mRNA molecules. As discussed in Chapter 2, different tRNA molecules are present in quite different amounts in a given cell. Those tRNAs present in larger quantities generally correspond to more commonly used codons. Messages having a high proportion of codons recognized by rare tRNA molecules are thus translated more slowly than those containing codons recognized by abundant tRNAs. Codon usage can affect the expression of cloned genes in other organisms, and so scientists often will adjust the codon usage in a gene that they are attempting to express at high levels.

The Duration of Gene Activity Is Controlled by Protein Modification and Degradation

As long as a protein is present, it will continue to carry out its function. This continued activity is acceptable for many proteins, but what if the state of the cell has changed and continued activity is unnecessary or even counterproductive? We have already seen how the DNA-binding activity of the Lac repressor is regulated by the presence of lactose, but relatively few proteins are regulated by small molecules in this way. Instead, many proteins are controlled by one of several other types of modification.

An irreversible solution to protein regulation is

proteolysis. Normally, proteins turn over at a fairly uniform rate, but pathways exist by which specific proteins can be targeted for destruction. An example in which protein destruction plays a key role in gene regulation is found in the bacteriophage λ. Upon infection of the bacterial host cell, the phage can integrate its DNA into the bacterial genome, thereby forming a *lysogen* that does not reproduce to form more phage. During the lysogenic phase, the λ repressor binds its operator DNA, repressing transcription until the bacterial DNA is damaged (e.g., by irradiation with UV light). At this point, RecA, an *E. coli* protein activated by bacterial DNA damage, clips the repressor between the DNA-binding domains and the dimerization and cooperativity domains. Without the effects of dimerization and cooperativity to stabilize them, the DNA-binding domains are released from the DNA. Transcription of the phage genes begins, resulting in the multiplication of the phage and lysis of the cell (Fig. 3-7). Thus, the inactivation of the λ repressor acts as a switch that allows the phage to multiply inside the cell.

Proteolysis is an irreversible way of regulating the activity of a transcription factor as compared to its being displaced from a complex. Once destroyed, the transcription factor gene must be transcribed and its mRNA translated before it again becomes available to regulate its target. Proteolysis of a transcription factor is like a binary switch, taking a process from one state to another, and cells employ proteolysis of transcription factors in just this way. For example, the transcription factor CtrA in the bacterium *Caulobacter crescentus* blocks transcription of the genes necessary for DNA replication. Only when CtrA is specifically proteolyzed can the cell enter a replication state and synthesize DNA. Immediately after DNA replication is initiated, CtrA is once again synthesized and again blocks DNA replication until the completion of the next cell division cycle. Thus, the level of CtrA is maintained at a high level to prevent repeated rounds of DNA replication between cell divisions; it then drops to zero when replication is needed, before it rises to again block DNA synthesis.

At other times, regulation of transcription should be easily reversible. The activity of transcription factors is frequently regulated through temporary changes, such as the addition of a phosphate group to the transcription factor or the binding of a second protein. When involved in gene regulation, these modifications in some way block either the DNA-binding function or activating interactions of the protein. We shall see a number of examples of flexible types of regulation in the coming chapters.

Eukaryotic Transcription and Gene Regulation Are More Complex Than in Prokaryotes

All bacterial genes are transcribed by the same core polymerase molecule. In contrast, eukaryotic cells have three different RNA polymerase complexes, each with a distinct functional role. rRNA is transcribed off ribosomal DNA genes by the enzyme *RNA polymerase I.* The synthesis of essentially all protein coding mRNAs is catalyzed by *RNA polymerase II.* Transfer RNA and a variety of smaller nuclear and cytoplasmic RNAs are made by *RNA polymerase III.* This multiplicity of RNA polymerases makes it possible for there to be three independently regulated families of promoters.

Not only do eukaryotes have more RNA polymerases, but the polymerases require more factors to initiate RNA synthesis. *E. coli* has about 200 transcription factors and a handful of σ factors required for transcription initiation. On the other hand, although the simple eukaryote, yeast, has a similar number of transcription factors (~160), more than 30 other proteins are needed for transcription initiation. Complex multicellular organisms have many more transcription factors—for example, there are approximately 1500 genes for transcription factors in the human genome. These transcription factors play many different roles, including recognizing other transcription factors, removing chromatin proteins that block polymerase access, and unwinding the DNA at the promoter. Eukaryotic genes typically have more regulatory sites, probably because most eukaryotes require different gene expression patterns during different stages of development and in different types of tissue.

Transcriptional Activator Proteins Have Separable DNA-binding and Activating Functions

Transcription factors must both bind DNA and interact with their targets in the transcriptional machinery—RNA polymerase itself in the case of bacteria, and polymerase or other parts of the transcriptional complex in eukaryotes. These two functions—DNA

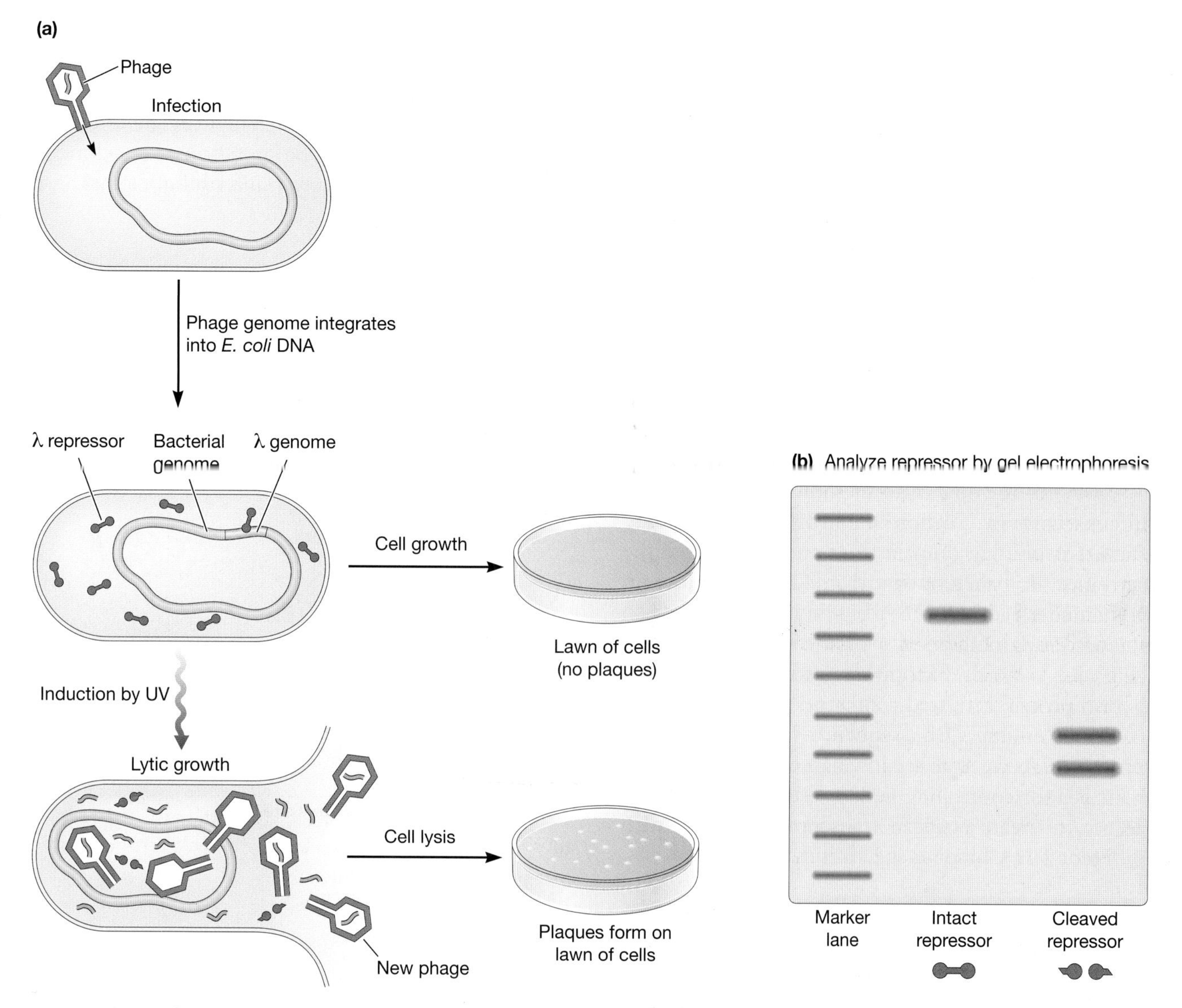

FIGURE 3-7

Cleavage of λ repressor by RecA allows activation of transcription of λ DNA. (*a*) When phage λ infects *Escherichia coli*, the phage generally integrates into the bacterial genome and becomes silent—a state called lysogeny. This silence is maintained by the action of the λ repressor protein, which binds as a dimer to the λ operator sequence and blocks transcription of the λ genome. When the DNA of *E. coli* is damaged by UV light, the RecA DNA damage signaling protease is activated. This protease cuts the λ repressor between the DNA-binding and dimerization/cooperativity domains, thus greatly lowering the affinity of the DNA-binding domains for the λ operator sequence. As a consequence, the repressor DNA-binding domains detach, and transcription of the genes required for λ replication begins. This replication is evident as clear areas (plaques) where the *E. coli* cells have lysed. (*b*) The state of the repressor before and after UV irradiation of *E. coli* cells with lysogenic λ can be monitored on a protein gel, which separates proteins by size. After electrophoresis and transfer of the proteins to a membrane, λ repressor protein is detected using radiolabeled antibodies. Before irradiation, the λ repressor protein is present as a large dimer that moves slowly through the gel; after irradiation and cleavage by RecA, two spots are seen because the subunits differ slightly in size.

binding and activation—are carried on different surfaces of the protein, often even in separate domains. The first demonstration of the separable nature of the DNA binding and activation function came from studies of λ repressor. Despite its name, λ repressor not only acts as a repressor, but for certain genes is a transcriptional activator. Mutant derivatives of λ repressor were isolated that could still bind DNA (and work as a repressor) but were unable to activate transcription. These mutations changed amino acid residues in the surface of the protein that interacted with RNA polymerase—that is, its activating region—but did not affect the region that bound DNA.

This result strongly suggested that the DNA-binding region might be separate from the transcriptional activation region. This was shown to be true also in eukaryotes by creation of a hybrid gene encoding the DNA-binding domain from one protein and the activation domain from another. The hybrid protein contained a bacterial DNA-binding domain fused to the activating portion of a yeast transcription factor (Fig. 3-8). When introduced into yeast cells, this hybrid protein activated transcription only when the bacterial binding DNA sequence was adjacent to a promoter. As we shall see in Chapter 6, experimental methods for controlling gene expression rely on the ability to create artificial transcription factors with different DNA-binding specificities.

Multiple Transcription Factors Work Synergistically to Control Genes

Expression of the bacterial β-galactosidase gene is controlled by a single repressor and a single activator, but most eukaryotic genes are controlled by the combined action of multiple regulators. In this way, one regulator can activate different genes depending on which other regulators are present. A good example of this is found in the process of flower development in the plant *Arabidopsis.* Flowers consist of four concentric whorls of organs: sepals, petals, stamens (the male part), and carpels (the female part). When a plant reaches the point in development at which it should form a flower, the appropriate genes have to be activated to produce this specialized structure. Early on, it became apparent that there were three primary classes of mutations affecting floral development. Each class of mutations affected the development of two adjacent whorls of flower organs. These mutations, termed *homeotic*, lead to transformations of one organ into another (Fig. 3-9).

The ABC model was developed to explain these three classes of mutations. The A activity specifies sepals, A + B specifies petals, B + C specifies stamens, and C alone specifies carpels. When the genes corresponding to these homeotic mutations were cloned, it was found that they were transcription factors and

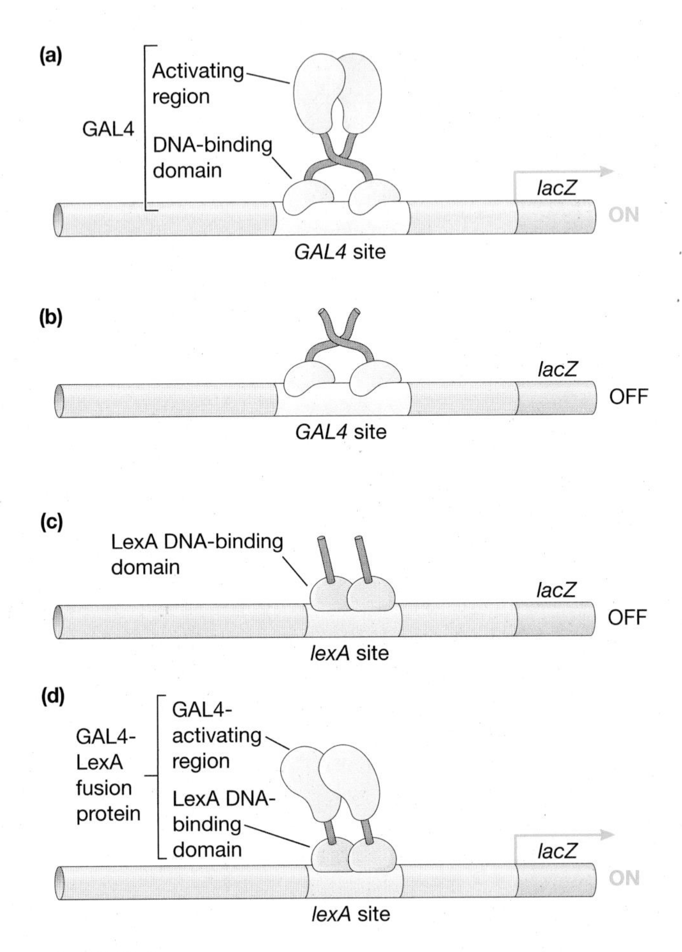

FIGURE 3-8

A domain swap experiment shows that activators have separate activating and DNA-binding domains. (*a*) A β-galactosidase reporter gene was used to test the function of the yeast transcriptional activator, GAL4. GAL4 turned on the reporter gene when a *GAL4* recognition sequence was present upstream of the reporter. (*b*) However, when a truncated GAL4 protein containing only the DNA-binding domain was tested, no activation of the reporter occurred. (*c*) Similarly, the DNA-binding domain of the bacterial protein LexA was tested, using a reporter in which the *GAL4* recognition sequence had been replaced with the DNA sequence bound by LexA. The LexA DNA-binding domain did not activate transcription. (*d*) Then, a "domain swap" experiment was done. A hybrid protein was created by attaching the activation domain from GAL4 to the DNA-binding domain of LexA. In this case, transcriptional activation occurred—the LexA DNA-binding domain of the hybrid molecule binding to the *lexA* sequence, and the GAL4 activation domain initiating transcription.

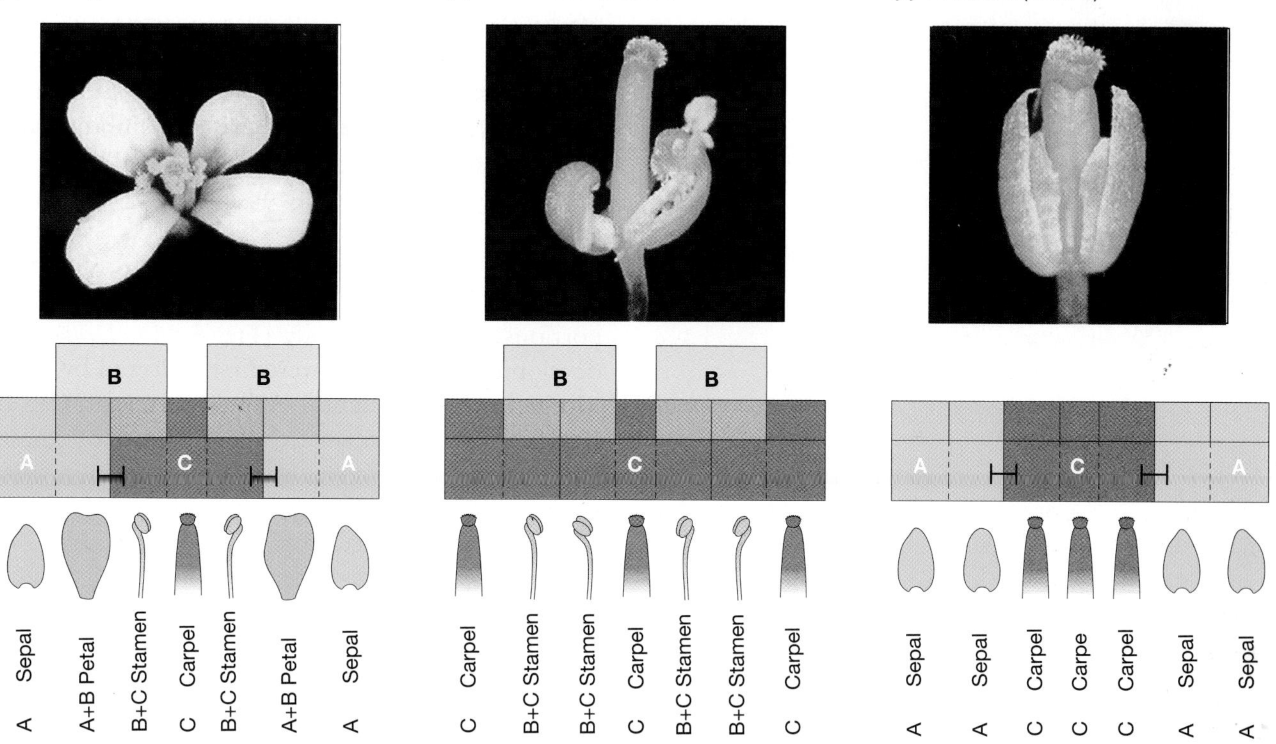

FIGURE 3-9

ABC homeotic mutations change the identity of the tissues in a flower. (*a*) Photograph of a wild-type *Arabidopsis* flower, with diagram *below* indicating the regions of gene activity. The outermost whorl of sepals (hidden by the petals in the photograph) is specified only by the A class (*green*) genes, whereas the combination of A and B (*yellow*) creates petals. The activities of both B and C (*red*) genes create stamens, the site of pollen production. In the center of the flower, the C genes cause the formation of the carpels, which contain the ovules. (*b*) *AP2* (*APETALA-2*), a class A mutation, causes the transformation of petals into stamens and sepals into carpels. The expression of C genes spreads throughout the flower because A genes such as *AP2*, which normally restrict the zone of C expression, are no longer present. (*c*) In *PI* (*PISTILLATA*), a class B mutation, stamens are transformed into carpels and petals are transformed into sepals, the default states of A-only and C-only gene expression in the wild-type flower.

that each was expressed in two sets of organs. That is, each homeotic gene promotes the development of two completely different organs; the combined actions of the pairs of transcription factors result in different developmental outcomes. This was verified by studying the interaction between the various flower-patterning homeotic genes. The B group genes *APETALA-3* (*AP3*) and *PISTILLATA* (*PI*), in combination with the general flower specification gene *SEP3*, can interact with either the Class A gene *APETALA-1* (*AP1*) or the Class C gene *AGAMOUS* (*AG*) to activate transcription (Fig. 3-10). The targets of these genes that lead to organ specification are still under study. This is one of many examples in which a transcription factor can activate different targets depending on what other transcription factors are present.

Development Is Controlled by Cascades of Transcription Factors

Few phenomena are more fascinating than embryonic development—the intricate and precise program by which a large and complex animal arises from a single fertilized egg cell. Studying development in mammals is difficult, because the process occurs tucked out of sight in the uterus and spans weeks or months, and because the genetic analysis is slow and complicated. Instead, many developmental biologists have focused their efforts on organisms like fish, flies, frogs, and worms, in which development is easier to study. The molecular basis of development has emerged from studies of these organisms, and the general principles involved have helped us understand mammalian development.

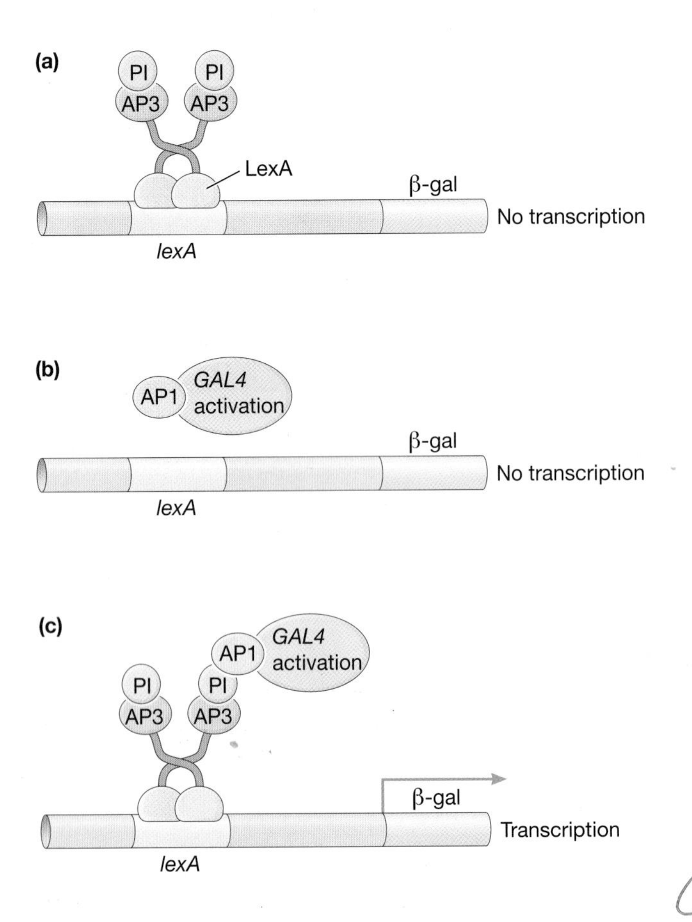

(d)

AP3-LexA	PI	AP1	SEP3	AG	Transcription
+	+	−	−	−	−
+	−	+	−	−	−
−	−	+	−	−	−
+	+	+	−	−	+
−	−	−	−	+	−
+	+	−	+	−	−
+	+	−	+	+	−

FIGURE 3-10

Tetrameric complexes determine floral organ identity. (*a*) APETALA-3 (AP3) was fused to the *lexA* DNA-binding domain. When this was tested in yeast cells carrying a β-galactosidase gene with an upstream *lexA* binding site, the AP3-LexA fusion did not activate transcription. (*b*) An APETALA-1 (AP1)-GAL4 activating domain fusion could not activate the reporter. (*c*) Only when AP3-LexA, AP1-GAL4, and PISTILLATA (PI) were coexpressed was the reporter activated. (*d*) This table shows the various combinations of transcription factors needed for transcription. For activation by the class C gene *AGAMOUS* (*AG*), the floral specification gene *SEP3* was also required. Class B *AP2* and *PI* floral identity genes can form complexes with both class A and class C genes.

Development in the fruit fly, *Drosophila*, is particularly amenable to molecular analysis, in part because of the many developmental mutants that are known and because of the relative ease of examining the fly larva. The life cycle consists of embryogenesis, three larval stages, a pupal stage, and the adult stage; the total development period from egg to adult lasts about 10 days. The embryo is divided through a process of regulated gene expression into 15 unique segments, each destined to form specific portions of the adult body (Fig. 3-11). *Drosophila* developmental mutants were first collected by T.H. Morgan and his students—Alfred Sturtevant, Calvin Bridges, and Herman Muller. Some of these mutants are remarkable; *bicoid* produces two tails, *antennapedia* turns a leg into an antenna, and *Ultrabithorax* produces two pairs of wings.

A major advance occurred in 1985, when Christiane Nüsslein-Volhard and Eric Wieschaus, working in Heidelberg, Germany at the European Molecular Biology Laboratory, embarked on a major genetic screen for mutants that affected *Drosophila* development; they eventually identified 139 mutations. The genes fell into several general classes. The *gap* genes, for example, were recognized because mutations resulted in the loss of several segments of the embryo, thus forming a large gap in the normal segmentation pattern. Three of the principal gap genes, *hunchback*, *Krüppel*, and *giant*, have been extensively studied. The genes were cloned and shown to encode transcription factors; this suggests that they transmit signaling information to the segmentation genes by directly modulating their transcription. In situ hybridization of the gap gene clones to embryos revealed that they were expressed in broad bands across the embryos, which was the first sign of segmentation in the embryo. After the gap genes establish stripes of gene expression across the embryo, pair-rule genes such as *even-skipped*, *fushi tarazu*, and *odd-paired* further divide these regions into individual segments. Mutations in the pair-rule genes led to loss of segments in an alternating pattern, so that mutant embryos end up with half the number of segments. Other unusual mutant alleles were found that lost only one or a few segments. Analyses of these mutations found deletions in putative regulatory sequences upstream of the genes, which suggests that these elements direct expression to individual stripes. Are these regulatory sequences where the transcription factors produced by the gap genes act?

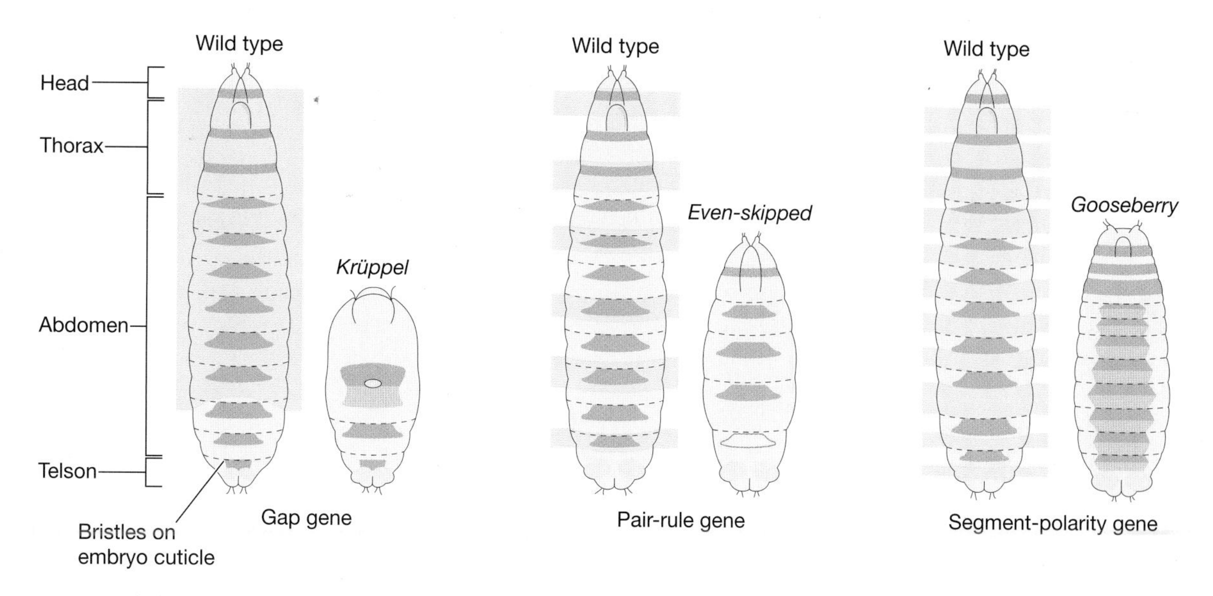

FIGURE 3-11

Patterning of the *Drosophila* embryo and phenotypes of developmental mutants. Shown at the left of each pair is a normal *Drosophila* larva, with the wild-type segmentation pattern of head, thorax, abdomen, and telson. Shown at the right of each pair are the phenotypes of selected gap (*left pair*), pair-rule (*middle pair*), and segment polarity (*right pair*) mutants. In the gap mutants, several segments are lost (*pale lavender boxes*). In pair-rule mutants, such as *even-skipped*, every other segment is missing. In segment polarity mutants, a portion of each segment is deleted and is replaced by a mirror image of the portion of the segment that remains.

Drosophila geneticists linked these elements with a reporter gene to determine where they are active in the embryo. The *E. coli* β-galactosidase gene from the *lac* operon is a commonly used reporter gene. The expression of the enzyme is detected by adding a synthetic substrate such as X-gal that produces an intense blue precipitate when hydrolyzed by β-galactosidase. The precipitate cannot diffuse and individual stained cells can be seen easily. Different sections of the region upstream of the pair-rule gene *even-skipped* were linked to the β-galactosidase gene and introduced into developing embryos. The different sequences activated β-galactosidase expression and directed it to different stripes along the body of the embryo, so that one sequence led to blue cells in only stripe 1, whereas another directed expression in only stripe 5, yet another in only stripe 2, and a fourth in both stripes 4 and 6 (Fig. 3-12).

To determine how the gap genes influence expression of *even-skipped* in stripe 2, the reporter gene construct containing the fragment that produced only stripe 2 was introduced into embryos having mutations in each of the gap genes *hunchback*, *Krüppel*, and *giant*. In the *hunchback* mutant, stripe 2 disappeared, suggesting that *hunchback* is normally an activator of *even-skipped* in stripe 2. On the other hand, in both *Krüppel* and *giant* mutants, stripe 2 expanded, which suggests that the products of these gap genes normally repress expression of *even-skipped*. Furthermore, *Krüppel* and *giant* regulate *even-skipped* differentially within stripe 2; the *Krüppel* mutation expanded stripe 2 expression in the posterior edge and *giant* mutation expanded stripe 2 expression in the anterior edge (Fig. 3-13). Thus the combination of activating and repressing regulatory sites establishes the tight boundaries of *even-skipped* expression. Correct expression patterns of *even-skipped* are very important for the further development of the embryo, because *even-skipped* is itself a transcriptional regulator of several other patterning genes.

Enhancers Contain Recognition Sites for Transcription Factors

The DNA elements upstream of *even-skipped* are termed *enhancers*, because they can enhance transcription of a promoter. Enhancers contain DNA sequences that are recognized and bound by tran-

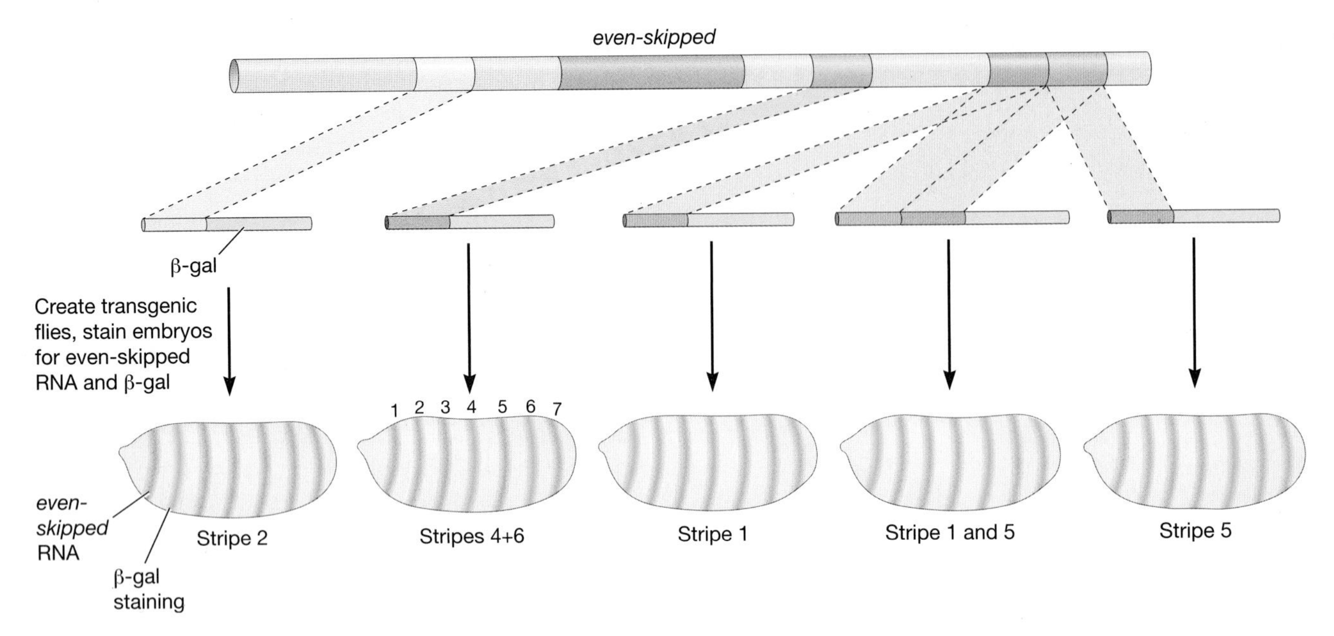

FIGURE 3-12

DNA sequences near the *even-skipped* gene direct its expression in distinct stripes in the *Drosophila* embryo. DNA sequences surrounding the *even-skipped* gene were fused to a β-galactosidase reporter gene. The embryos were simultaneously stained with a probe complementary to β-galactosidase mRNA (shown in *blue*) and another probe complementary to the endogenous *even-skipped* mRNA (shown in *orange*). The former showed where the sequence was directing expression of β-galactosidase, and the latter showed the normal pattern of stripes. Several sequences were discovered that directed the expression of the reporter gene in specific stripes corresponding to the natural pattern of *even-skipped* expression. Note that these sequences are characteristic of enhancers in that they function whether they are located upstream or downstream of *even-skipped*. Remarkably, when two enhancers were tested together, the expression pattern reflected the combination of those two enhancers.

scription factors. Then, the transcription factors promote transcription from the nearby promoter, usually by recruiting RNA polymerase and the associated factors necessary for transcription.

Once the *hunchback*, *Krüppel*, and *giant* genes had been cloned, the DNA-binding specificities of the transcription factors they encode was determined by testing the ability of the proteins to bind different DNA sequences. The *even-skipped* stripe 2 enhancer contains multiple binding sites for the repressor proteins *giant* and *Krüppel*, and for the activators *hunchback* and *bicoid*, the former competing with the latter for binding to the enhancer (Fig. 3-14). All these sites contribute to regulating *even-skipped* expression; if even one binding site is mutated, it dramatically decreases *even-skipped* transcription. This demonstrates that the cooperative activity of multiple activators binding at multiple sites is required for gene activation, and how the boundaries of gene expression can be established through competition for repressors.

Eukaryotic Enhancers Act over Long Distances

By injecting genomic DNA containing the *even-skipped* gene into fly embryos, scientists found that sequences both upstream and downstream of the coding sequence are required for correct expression of all of the stripes. Some stripe-specific enhancers are upstream of the gene, including the stripe 2 and stripe 3 and 7 enhancers described above, but several other stripe-specific enhancers are located downstream. Furthermore, these enhancers function properly whether placed upstream or downstream of a reporter gene and often at significant distances from the transcription start site. Indeed, enhancers can be very far from the genes they affect. For example, the enhancer of the *Sonic hedgehog* (*Shh*) gene, which controls pattern formation in the embryo, is regulated by an enhancer 1,000,000 bases away from the *Shh* promoter! This ability to activate transcription at a distance, and when placed on either side of a gene,

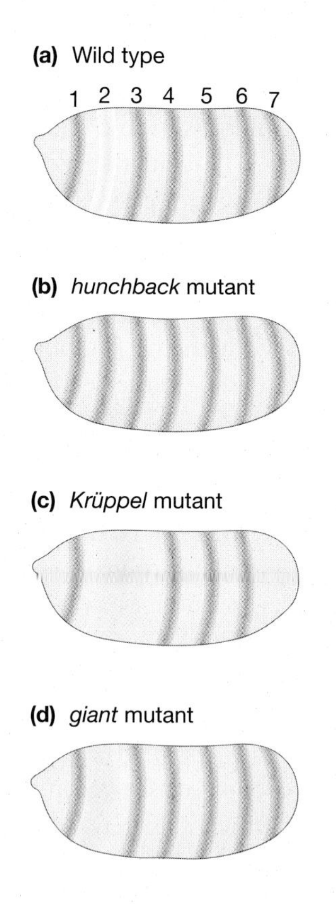

FIGURE 3-13

Mutations of gap genes affect the expression of pair-rule genes. The sequence upstream of *even-skipped* that directs expression of the β-galactosidase reporter gene to stripe 2 (*a*; see also Fig. 3-12) was introduced into flies mutant for the gap genes *hunchback* (*b*), *Krüppel* (*c*), and *giant* (*d*). The boundaries of reporter gene expression were altered by these mutations, which leads to the prediction of a model for the effect of the gap genes on controlling the expression pattern in these stripes. Note that the pattern of endogenous *even-skipped* expression is also altered in these mutants. The use of the reporter constructs helps to clarify the effect on stripe 2 in isolation.

is one of the defining properties of an enhancer element (Fig. 3-15a). It is thought that enhancers exert their effects at such great distances by being brought in close proximity to a promoter by looping of the DNA strand.

The effect of enhancers is constrained by *insulator* sequences, DNA elements that set physical boundaries on enhancer activity. Insertion of an insulator sequence from another location in *Drosophila* between the enhancer and promoter of *even-skipped* abolishes activation (Fig. 3-15b). Many insulators also block the effect of suppressive signals. Geneticists have made use of insulator sequences when trying to ensure expression of a transgene. Flanking a transgene with insulator sequences frequently leads to higher levels of transgene transcription than if it is introduced into the genome without insulator sequences. The mechanism by which insulators block enhancers is still mysterious, although the presence of an insulator often alters chromatin conformation.

Locus Control Regions Regulate Some Groups of Genes

The α and β subunits of the key oxygen transport protein hemoglobin are encoded by two gene families whose members are closely related structurally and whose expressions are regulated during development (Table 3-1). The human α-globin gene cluster is located on chromosome 16 and the β-globin gene cluster is on chromosome 11. Each gene in these

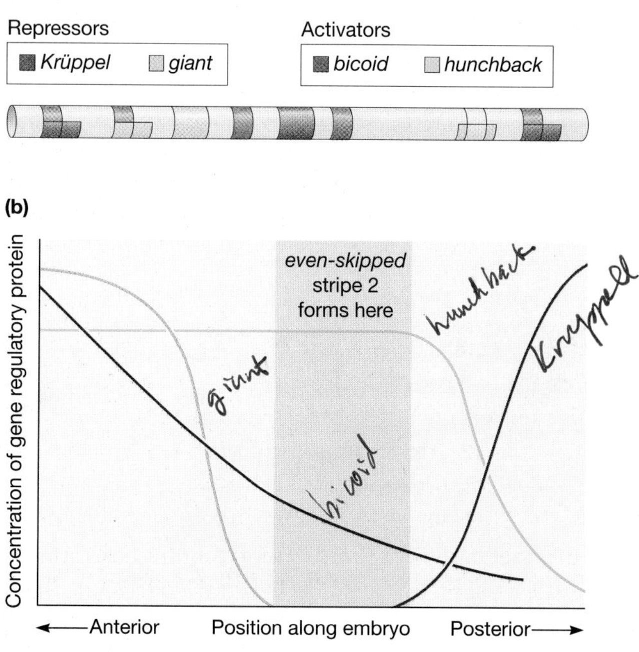

FIGURE 3-14

The *even-skipped* stripe 2 enhancer contains binding sites for the gap genes *Krüppel, giant,* and *hunchback* as well as the maternal effect gene *bicoid.* (*a*) The binding sites that have been mapped within the *even-skipped* stripe 2 enhancer. *Krüppel* and *giant* act as repressors, whereas *bicoid* and *hunchback* are activators. (*b*) Stripe 2 is characterized by a high level of *hunchback* and a decreased but significant expression of *bicoid* in the cells that express *even-skipped* stripe 2. The boundaries of the stripe are set by the presence of the repressors *Krüppel* and *giant*, which interfere with *bicoid* and *hunchback* binding sites in cells anterior and posterior to stripe 2.

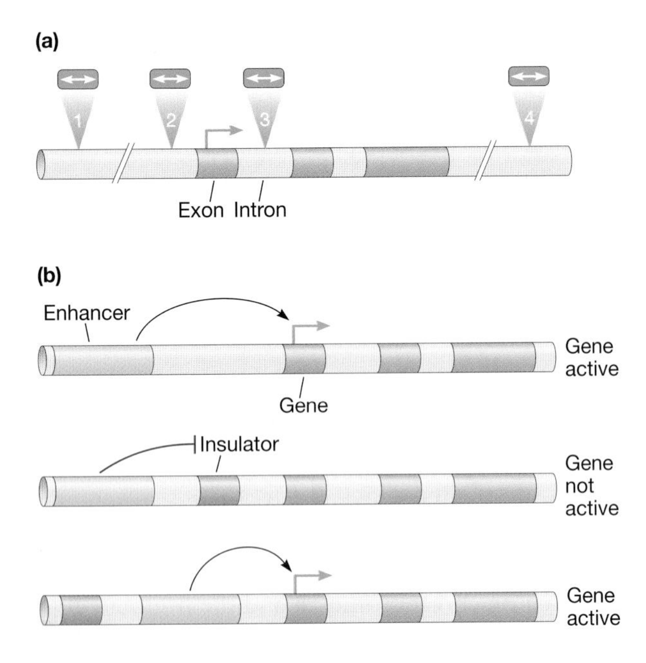

FIGURE 3-15

Enhancers increase gene expression independently of their position relative to the gene, but their action can be constrained by insulator sequences. (*a*) The *purple boxes* indicate the coding exons of a gene. The *light green boxes* above the gene indicate positions where enhancers function. In addition to functioning close to the promoter (2), enhancers can work from thousands of base pairs upstream (1) or downstream (4) of a gene. In some cases (e.g., the immunoglobulins), the enhancer is located within an intron (3). In most cases, enhancer sequences function in either orientation relative to the gene (shown by *arrows* in both directions). (*b*) When inserted between an enhancer and a gene, an insulator blocks the activating activity of the enhancer. However, if the insulator element lies outside a gene-enhancer pair, the activation of the gene by the enhancer is not affected.

clusters is a separate transcriptional unit consisting of upstream enhancers and the coding sequence, followed by transcriptional termination signals.

Anemia results from a deficiency in the oxygen-carrying ability of the blood and one form, thalassemia, is caused by a lack of hemoglobin production. More than 150 mutations are known that can cause thalassemia, many of which affect the promoters, enhancers, or RNA processing of the gene. However, a group of mutations that affect a region upstream of the entire β-globin locus was discovered in a Hispanic population. Although these genes could be transcribed normally in vitro, the patients had no expression from any genes within the β-globin locus. The function of this region was examined experimentally by transferring the human β-globin gene locus into mice. It was found that this *locus control region* (LCR) was required for full activation of the individual genes in the locus (Fig. 3-16). Furthermore, the LCR was unusual in that transgenes carrying it were expressed at the same level regardless of their site of integration in the mouse genome. The importance of LCRs for regulating gene expression was demonstrated by making transgenic mice with the β-globin locus and two copies of the LCRs. These expressed twice the amount of β-globin.

When isolated, some parts of the LCR have enhancer-like function, others have insulator-like function, and still others have the properties of promoters. LCRs control entire groups of nearby genes, whereas enhancer sequences typically control just one gene. The ability of LCRs to regulate large regions of the genome means that the order and position of eukaryotic genes is important for their regulation.

Careful microscopic observations of the β-globin locus suggest that the LCR may help to organize the region into loops of DNA. In embryonic red blood cell precursors, the active embryonic γ-globin genes were closely associated with the LCR. Later in development, the now-active adult β-globin genes were located near the LCR, whereas the silent embryonic genes were physically distant. This has led to a model in which the LCR brings the active genes close to it, but it loops out genes that are inactive. It is still a mystery how the physical organization of large segments of DNA alters gene expression.

Gene Expression in Eukaryotes Is Also Controlled by DNA Packaging

Another distinctive feature of eukaryotic cells is the way DNA molecules are tightly wrapped around *histone* proteins to form structures called *nucleosomes.* The strings of nucleosomes are then further looped and wrapped into a compact structure called *chro-*

TABLE 3-1. Expression of globin genes during human development

	α-globin cluster		β-globin cluster			
	α	ζ	ε	γ	δ	β
Embryo	✓	✓	✓	✓		
Fetus	✓			✓		✓[a]
Adult	✓				✓	✓

[a]Less than 10%.

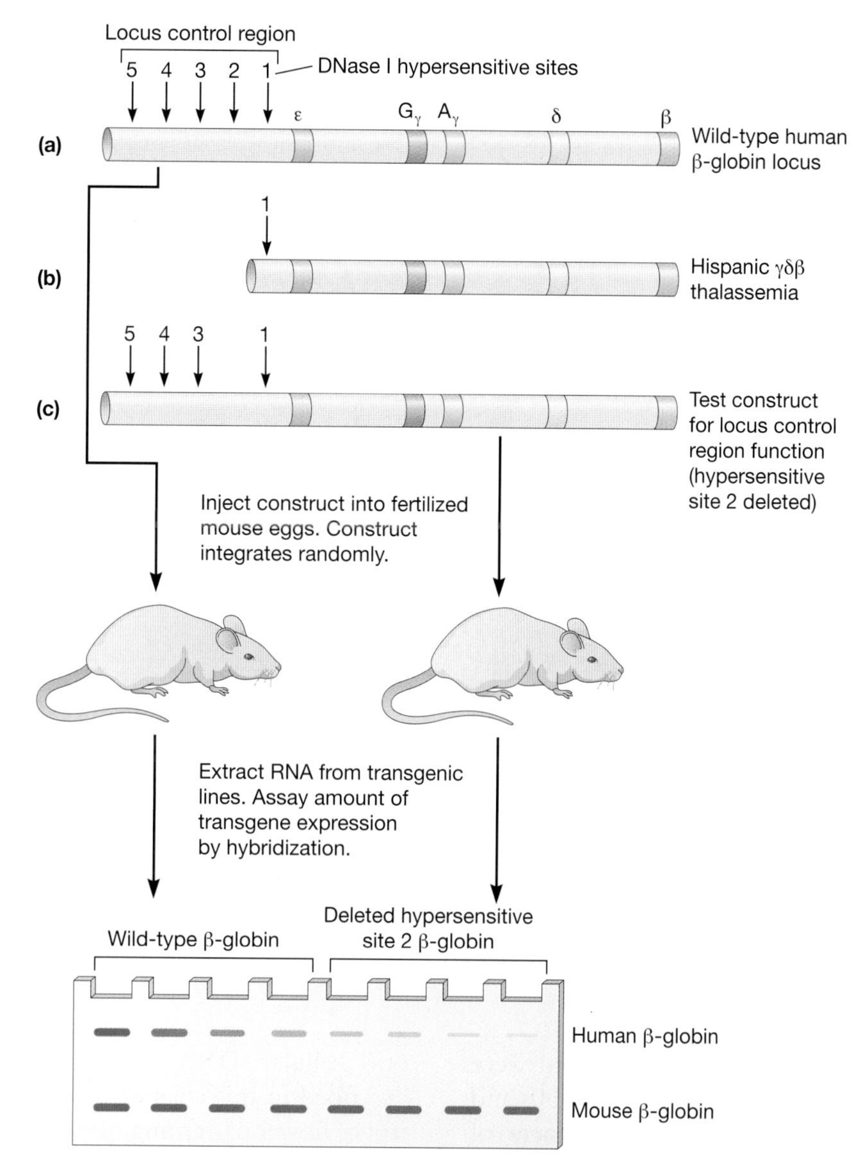

FIGURE 3-16

The human β-globin gene cluster is regulated by a locus control region. (*a*) The human β-globin locus showing the five genes in the region and the upstream locus control region. The five DNase-hypersensitive sites are indicated by *arrows*. (*b*) One form of thalassemia removes a large part of the locus control region greatly diminishing expression of β-globin, leaving just the DNase-hypersensitive site closest to the gene cluster. (*c*) Transgenic mouse lines were made with the wild-type human β-globin locus and the LCR-deleted locus. RNA was extracted from these lines, and the abundance of the human β-globin transgenes and the mouse β-globin gene was measured by a method called an *S1* nuclease protection assay. The transgenes with the entire locus control region were faithfully expressed, whereas the constructs with a deleted LCR were expressed in only a fraction of the lines.

matin (Fig. 3-17). Clearly, the basic nucleosome structure must be modified as RNA polymerase passes through individual nucleosomes as it transcribes mRNA. As soon as chromatin began to be examined with the electron microscope, observers claimed that especially "active" chromatin (e.g., that making rRNA) lacked the normal appearance of chromatin. In contrast, inactive chromatin in some eukaryotic cells is visible as tightly condensed DNA structures. At the molecular level, molecular biologists commonly find that actively transcribed DNA is more accessible to nucleases such as restriction enzymes than is inactive chromatin. In Chapter 8 we will learn more about how chromatin structure is controlled.

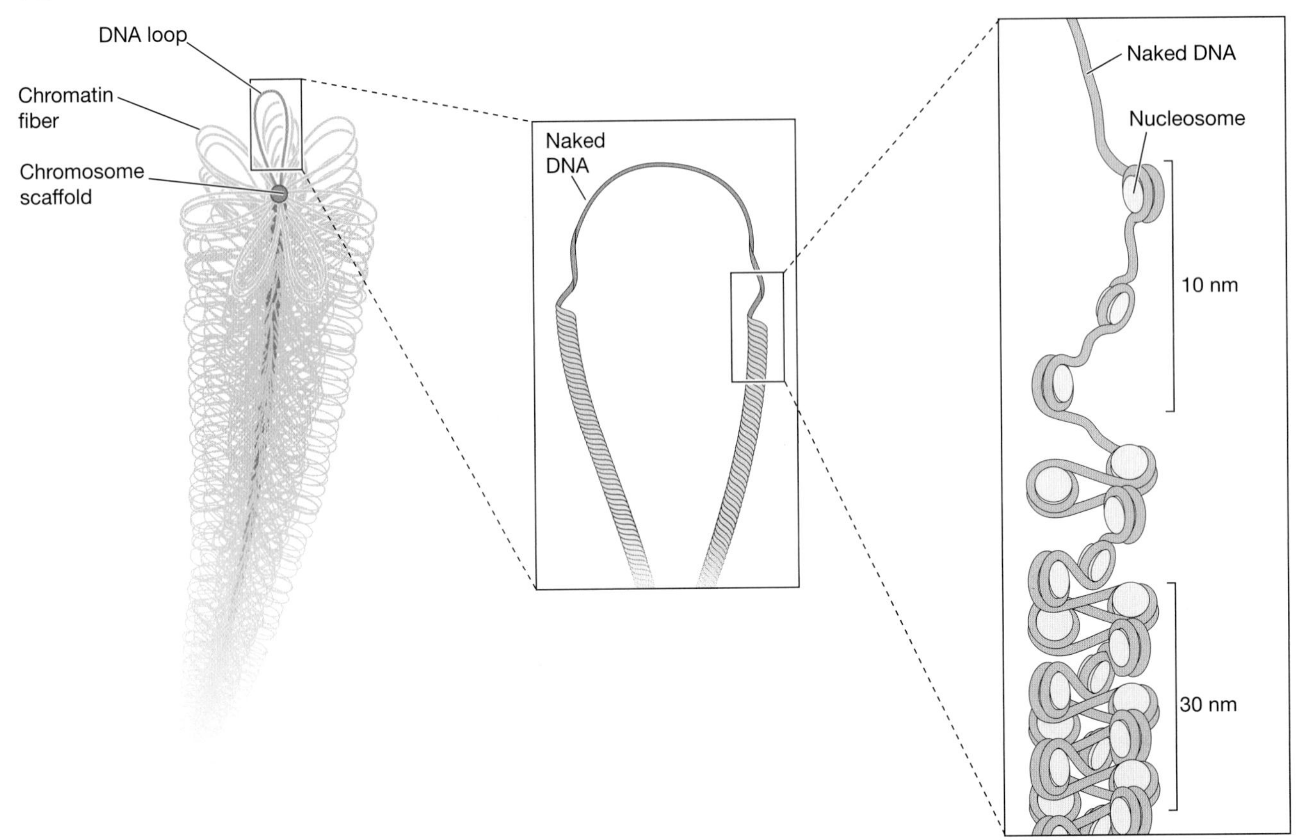

FIGURE 3-17

DNA is highly organized in chromosomes. DNA is wrapped around a complex of histone proteins to form structures called nucleosomes. These nucleosomes are clustered together into 30-nm-wide fibers, which are further packed into other dense structures. Typically, sites of active transcription loop out from the chromosome and contain either naked DNA or single nucleosomes.

The fundamentals of gene expression were worked out in bacteria and viruses in the 1950s and 1960s. At that time, prior to the development of recombinant DNA techniques, learning how the genes of animal and plant cells are controlled seemed next to impossible. These cells would never be as simple to work with as bacteria or their phages, and the technical means to analyze the molecular genetics of eukaryotic cells were not available. Furthermore, the idea that it would be possible to isolate and manipulate specific segments of eukaryotic cellular DNA was unthinkable. The work described in this chapter on the combinatorial control of flower patterning, the molecular regulation of gene expression during *Drosophila* development, and the control of hemoglobin expression had to await the development of recombinant DNA techniques. In the next chapter, we turn to a review of the basic tools of recombinant DNA and in subsequent chapters we will cover further complexities of eukaryotic genes—complexities that could not have been learned from bacteria or viruses.

Reading List

General

Beckwith J., Davies J., and Gallant J.A. 1983. *Gene Function in Prokaryotes.* Cold Spring Harbor Laboratory, Cold Spring Harbor, New York.

Ptashne M. 2004. *A Genetic Switch: Phage Lambda Revisited*, 3rd ed. Cold Spring Harbor Laboratory Press, Cold Spring Harbor, New York.

Muller-Hill B. 1996. *The* lac *Operon: A Short History of a Genetic Paradigm.* Walter de Gruyter, Berlin.

Ptashne M. and Gann A. 2002. *Genes and Signals.* Cold Spring Harbor Laboratory Press, Cold Spring Harbor, New York.

Repressors and Operators

Jacob F. and Monod J. 1961. Genetic regulatory mechanisms in the synthesis of proteins. *J. Mol. Biol.* **3:** 318–356.

Beckwith J.R. and Signer E.R. 1966. Transposition of the lac region of *Escherichia coli.* I. Inversion of the lac operon and transduction of lac by ϕ80. *J. Mol. Biol.* **19:** 254–265.

Gilbert W. and Müller-Hill B. 1966. Isolation of the lac repressor. *Proc. Natl. Acad. Sci.* **56:** 1891–1898.

Ptashne M. 1967. Isolation of the lambda phage repressor. *Proc. Natl. Acad. Sci.* **57:** 306–313.

Ptashne M. and Hopkins N. 1968. The operators controlled by the lambda phage repressor. *Proc. Natl. Acad. Sci.* **60:** 1282–1287.

Shapiro J., Machattie L., Eron L., Ihler G., Ippen K., and Beckwith J. 1969. Isolation of pure lac operon DNA. *Nature* **224:** 768–774.

Wang J.C., Barkley M.D., and Bourgeois S. 1974. Measurements of unwinding of lac operator by repressor. *Nature* **251:** 247–249.

Anderson W.F., Ohlendorf D.H., Takeda Y., and Matthews B.W. 1981. Structure of the cro repressor from bacteriophage lambda and its interaction with DNA. *Nature* **290:** 754–758.

Pabo C.O. and Lewis M. 1982. The operator-binding domain of lambda repressor: Structure and DNA recognition. *Nature* **298:** 443–447.

Steitz T.A., Ohlendorf D.H., McKay D.B., Anderson W.F., and Matthews B.W. 1982. Structural similarity in the DNA-binding domains of catabolite gene activator and cro repressor proteins. *Proc. Natl. Acad. Sci.* **79:** 3097–3100.

Kim R. and Kim S.H. 1983. Direct measurement of DNA unwinding angle in specific interaction between *lac* operator and repressor. *Cold Spring Harbor Symp. Quant. Biol.* **47:** 451–454.

Anderson W.F., Ohlendorf D.H., Takeda Y., and Matthews B.W. 1981. Structure of the cro repressor from bacteriophage lambda and its interaction with DNA. *Nature* **290:** 754–758.

Moreau P.L., Pelico J.V., and Devoret R. 1982. Cleavage of lambda repressor and synthesis of RecA protein induced by transferred UV-damaged F sex factor. *Mol. Gen. Genet.* **186:** 170–179.

Promoters

Pribnow D. 1975. Bacteriophage T7 early promoters: Nucleotide sequences of two RNA polymerase binding sites. *J. Mol. Biol.* **99:** 419–443.

Wang J.C., Jacobsen J.H., and Saucier J.M. 1977. Physiochemical studies on interactions between DNA and RNA polymerase. Unwinding of the DNA helix by *Escherichia coli* RNA polymerase. *Nucleic Acids Res.* **4:** 1225–1241.

Chamberlin M.J., Nierman W.C., Wiggs J., and Neff N. 1979. A quantitative assay for bacterial RNA polymerases. *J. Biol. Chem.* **254:** 10061–10069.

McClure W.R. 1980. Rate-limiting steps in RNA chain initiation. *Proc. Natl. Acad. Sci.* **77:** 5634–5638.

Youderian P., Bouvier S., and Susskind M.M. 1982. Sequence determinants of promoter activity. *Cell* **30:** 843–853.

Ackerson J.W. and Gralla J.D. 1983. In vivo expression of *lac* promoter variants with altered –10, –35, and spacer sequences. *Cold Spring Harbor Symp. Quant. Biol.* **47:** 473–476.

Burgess R.R., Travers A.A., Dunn J.J., and Bautz E.K. 1969. Factor stimulating transcription by RNA polymerase. *Nature* **221:** 43–46.

Ishihama A. 2000. Functional modulation of *Escherichia coli* RNA polymerase (review). *Annu. Rev. Microbiol.* **54:** 499–518.

Attenuators

Yanofsky C. 1981. Attenuation in the control of expression of bacterial operons (review). *Nature* **289:** 751–758.

Bertrand K., Korn L., Lee F., Platt T., Squires C.L., et al. 1975. New features of the regulation of the tryptophan operon. *Science* **189:** 22–26.

Oxender D.L., Zurawski G., and Yanofsky C. 1979. Attenuation in the *Escherichia coli* tryptophan operon: Role of RNA secondary structure involving the tryptophan codon region. *Proc. Natl. Acad. Sci.* **76:** 5524–5528.

Positive Control

Zubay G., Schwartz D., and Beckwith J. 1970. Mechanism of activation of catabolite-sensitive genes: A positive control system. *Proc. Natl. Acad. Sci.* **66:** 104–110.

Epstein W., Rothman-Denes L.B., and Hesse J. 1975. Adenosine 3′:5′-cyclic monophosphate as mediator of catabolite repression in *Escherichia coli. Proc. Natl. Acad. Sci.* **72:** 2300–2304.

Simpson R.B. 1980. Interaction of the cAMP receptor protein with the *lac* promoter. *Nucleic Acids Res.* **8:** 759–766.

McKay D.B., Weber I.T., and Steitz T.A. 1982. Structure of catabolite gene activator protein at 2.9-Å resolution. Incorporation of amino acid sequence and interactions with cyclic AMP. *J. Biol. Chem.* **257:** 9518–9524.

Translational Control in Bacteria

Shine J. and Dalgarno L. 1974. The 3′-terminal sequence of *Escherichia coli* 16*S* ribosomal RNA: Complementarity to nonsense triplets and ribosome binding sites. *Proc. Natl. Acad. Sci.* **71:** 1342–1346.

Nomura M., Yates J.L., Dean D., and Post L.E. 1980. Feedback regulation of ribosomal protein gene expression in *Escherichia coli:* Structural homology of ribosomal RNA and ribosomal protein mRNA. *Proc. Natl. Acad. Sci.* **12:** 7084–7088.

Modular Activators

Ogata R.T. and Gilbert W. 1978. An amino-terminal fragment of lac repressor binds specifically to lac operator. *Proc. Natl. Acad. Sci.* **75:** 5851–5854.

Pabo C.O., Sauer R.T., Sturtevant J.M., and Ptashne M. 1979. The lambda repressor contains two domains. *Proc. Natl. Acad. Sci.* **76:** 1608–1612.

Brent R. and Ptashne M. 1985. A eukaryotic transcriptional activator bearing the DNA specificity of a prokaryotic repressor. *Cell* **43:** 729–736.

Murre C., McCaw P.S., Vaessin H., Caudy M., Jan L.Y., et al. 1989. Interactions between heterologous helix-loop-helix proteins generate complexes that bind specifically to a common DNA sequence. *Cell* **58:** 537–544.

ABC Model of Floral Development

Krizek B.A. and Fletcher J.C. 2005. Molecular mechanisms of flower development: An armchair guide (review). *Nat. Rev. Genet.* **6:** 688–698.

Honma T. and Goto K. 2001. Complexes of MADS-box proteins are sufficient to convert leaves into floral organs. *Nature* **409:** 525–529.

Enhancers in Drosophila *Development*

Stanojevic D., Small S., and Levine M. 1991. Regulation of a segmentation stripe by overlapping activators and repressors in the *Drosophila* embryo. *Science* **254:** 1385–1387.

Fujioka M., Emi-Sarker Y., Yusibova G.L., Goto T., and Jaynes J.B. 1999. Analysis of an *even-skipped* rescue transgene reveals both composite and discrete neuronal and early blastoderm enhancers, and multi-stripe positioning by gap gene repressor gradients. *Development* **126:** 2527–2538.

Berman B.P., Nibu Y., Pfeiffer B.D., Tomancak P., Celniker S.E., et al. 2002. Exploiting transcription factor binding site clustering to identify *cis*-regulatory modules involved in pattern formation in the *Drosophila* genome. *Proc. Natl. Acad. Sci.* **99:** 757–762.

Eukaryotic Enhancers

Church G.M., Ephrussi A., Gilbert W., and Tonegawa S. 1985. Cell-type-specific contacts to immunoglobulin enhancers in nuclei. *Nature* **313:** 798–801.

Marriott S.J. and Brady J.N. 1989. Enhancer function in viral and cellular gene regulation (review). *Biochim. Biophys. Acta* **989:** 97–110.

Lettice L.A., Heaney S.J.H., Purdie L.A., Li L., de Beer P., et al. 2003. A long-range *Shh* enhancer regulates expression in the developing limb and fin and is associated with preaxial polydactyly. *Hum. Mol. Genet.* **12:** 1725–1735.

Locus Control Regions

Kioussis D., Vanin E., deLange T., Flavell R.A., and Grosveld F.G. 1983. Beta-globin gene inactivation by DNA translocation in gamma beta-thalassaemia. *Nature* **306:** 662–666.

Milot E., Strouboulis J., Trimborn T., Wijgerde M., de Boer E., et al. 1996. Heterochromatin effects on the frequency and duration of LCR-mediated gene transcription. *Cell* **87:** 105–114.

Tolhuis B., Palstra R.J., Splinter E., Grosveld F., and de Laat W. 2002. Looping and interaction between hypersensitive sites in the active β-globin locus. *Mol. Cell.* **10:** 1453–1465.

CHAPTER 4

Basic Tools of Recombinant DNA

The deciphering of the genetic code was the crowning achievement of the decade of research following the discovery of the DNA double helix. That research program had been set out clearly in the Central Dogma, and by the mid-1960s, it had been confirmed experimentally. Semiconservative DNA replication as suggested by the double helix had been demonstrated and an enzyme carrying out DNA synthesis had been found. The role of an RNA intermediary—messenger RNA (mRNA)—had been confirmed and its existence demonstrated. Protein synthesis had been studied intensively by biochemists and many of the components involved—mRNA, activated amino acids, transfer RNA (tRNA), and ribosomes—had been isolated and their functions elucidated. With the deciphering of the genetic code, it was possible now in theory to understand the information contained in any DNA or RNA molecule. A new discipline, molecular genetics, had been born.

To many of the biologists who had played a part in that birth, it seemed that molecular genetics was entering a period of consolidation. They felt there would be further growth, but that this would be incremental; new conceptual advances seemed unlikely as the fine details of the old problems were worked out and biochemists dissected in ever more minute detail the pathways involved. At that time, in the 1960s, molecular genetics was almost entirely based on the study of bacteria and phages. New advances would come in working out the functioning of genes in higher organisms, but the experimental tools were lacking and it seemed as far in the future as putting a man on the moon. So to avoid possibly marking time, several distinguished contributors to our primary knowledge on the storage of genetic information in DNA left molecular genetics to start up new careers in neurobiology. What they could not have foreseen was the very rapid development over the next decade of the enzymological and chemical techniques that gave rise to recombinant DNA, which by now has come to include any technique for manipulating DNA or RNA.

Recombinant DNA techniques are so powerful because they provide the tools to study the genetics of any organism by isolating the DNA of virtually any gene. A particular gene can be isolated and produced in large quantities through *cloning* and its genetic information can be read by *sequencing*. The functions of that gene can then be analyzed by using *in vitro mutagenesis* to make a specific alteration in that information before reintroducing the mutated DNA into the organism to determine the effects of the mutation. By the late 1970s, as it became clear that these tools offered the fastest and surest route to understanding the molecular mechanisms of formerly intractable processes such as development and cell division, they were seized eagerly by biologists in almost every field. The subsequent period of scientific excitement and achievement has seen few, if any, parallels in the history of biological research.

In this and the next two chapters, we examine some of the tools on the molecular biologist's workbench. In this chapter, we look at the basic techniques of recombinant DNA: enzymes that cut, join, and synthesize DNA; electrophoresis for separating DNA molecules; DNA sequencing and chemical synthesis; the polymerase chain reaction; and the use of hybridization to select specific DNA sequences. In Chapter 5, we will turn to what techniques used for isolating genes have taught us about the structures of genomes, and in Chapter 6 we review the current technologies for studying the function of DNA sequences within cells.

Restriction Enzymes Cut DNA at Specific Sequences

Protein chemists had been able to put *proteases*, in particular, those enzymes that cut polypeptide chains at specific sites, to good use in determining the structure of proteins. By using different proteases, cutting at different specific sites, different smaller fragments (peptides) can be obtained and their amino acid sequences determined. The original order of the peptides in the protein can then be established by looking for overlaps in amino acid sequence between the peptides.

Unfortunately for early nucleic acid biochemists, all the known *nucleases*, the enzymes that were first found to break the phosphodiester bonds of nucleic acids, showed very little sequence dependency. The prevailing opinion was that, in contrast to proteases, highly specific nucleases would never be found and therefore the isolation of discrete DNA fragments, even from viral DNA, would not be possible. Doubt was cast on this view with the observations, beginning as early as 1953, that when DNA molecules from one strain of *Escherichia coli* were introduced into a different *E. coli* strain (e.g., *E. coli* strain B and *E. coli* strain C), the DNA was nearly always quickly fragmented into smaller pieces. Occasionally, however, the infecting DNA molecule somehow became modified and remained intact, so that it could now replicate in the new bacterial strain. This phenomenon was recognized as being similar to host restriction when bacteriophages infect bacteria. Most of the bacteriophage DNA is destroyed by bacterial enzymes, but a small amount escapes destruction and phages containing this DNA can efficiently infect that same bacterial strain. It was clear that the bacterial DNA must in some way be modified to escape destruction by its own enzymes, and that occasionally the phage DNA was modified in the same way.

In 1966 chemical analysis of such modified DNA molecules revealed the presence of methylated bases not present in the unmodified DNA. Methylated bases are not inserted as such into growing DNA chains; rather, they arise through the enzymatically catalyzed addition of methyl groups to newly synthesized DNA chains. The stage was thus set. In the 1960s, Stuart Linn and Werner Arber, working in Geneva, found in extracts of cells of *E. coli* strain B both a specific *modification enzyme* that methylated unmethylated DNA and a *restriction nuclease* that broke down unmethylated DNA. The term "restriction" was applied to the nuclease because this enzyme recognized specific unmethylated sequences on the DNA molecule. Over the next several years, other restriction nucleases and their companion modification methylases were identified, suggesting that many site-specific nucleases might exist. None of these early *E. coli* restriction enzymes lived up to these hopes, however; although the enzymes recognized specific unmethylated sequences, they all cleaved the DNA at random locations far removed from these sites.

The first restriction nuclease that did cleave at a specific site in DNA was discovered in 1970 by Hamilton Smith of Johns Hopkins University, who followed up his accidental finding that the bacterium *Haemophilus influenzae* rapidly broke down foreign phage DNA. This degradative activity was subsequently observed in cell-free extracts and shown to be due to a true restriction nuclease. Thomas Kelly and Smith determined that HindII, as this enzyme is called, bound to the following sequence, in which the arrows indicate the exact cleavage sites on each DNA strand, and "Py" and "Pu" represent any pyrimidine or purine residue:

5′-G T Py ↓ Pu A C-3′
3′-C A Pu ↑ Py T G-5′

Since then, restriction enzymes that cut specific sequences have been isolated from several hundred bacterial strains, and more than 150 different specific cleavage sites have been found (Table 4-1). Nearly all restriction enzymes recognize *palindromic sites*—that is, the sequences from 5′ to 3′ on each strand are the same. We know now that most restriction enzymes are dimers of two identical subunits that fit on the DNA so that their twofold axis of symmetry is the same as that of the sequence. The enzyme moves along the DNA strand, and when it reaches the target sequence, the DNA is distorted and the strands are cut.

Restriction enzymes recognize specific sequences of four to eight base pairs. A given 4-bp site occurs on average every 256 bp, so an enzyme cutting a four-base sequence will produce many more, much smaller DNA fragments than an enzyme cutting an 8-bp sequence, which will occur on average every 65,536 bp. However, predicting the size of DNA fragments produced by a restriction enzyme is complicated by many factors. One factor that affects the frequency with which DNA is cut is its base composition. Take, for example, the enzyme NotI, which has an eight-base recognition sequence that includes two cytidine–guanine dinucleotides (usually abbreviated as CpG). This dinucleotide sequence occurs much less frequently in vertebrate genomes than would be expected statistically—at about 20% of the expected frequency—and thus NotI cuts less frequently than expected. Furthermore, cytidine is

TABLE 4-1. Some restriction enzymes and their cleavage sequences

<table>
<tr><th>Microorganism</th><th>Enzyme abbreviation</th><th>Sequence</th><th>Notes</th></tr>
<tr><td>Haemophilus aegyptius</td><td>HaeIII</td><td>5′ . . . G G|C C. . . 3′
3′ . . . C C|G G. . . 5′</td><td>1</td></tr>
<tr><td>Thermus aquaticus</td><td>TaqI</td><td>5′ . . . T|C G A . . . 3′
3′ . . . A G C|T . . . 5′</td><td>2</td></tr>
<tr><td>Haemophilus haemolyticus</td><td>HhaI</td><td>5′ . . . G C G|C . . . 3′
3′ . . . C|G C G . . . 5′</td><td>3</td></tr>
<tr><td>Desulfovibrio desulfuricans</td><td>DdeI</td><td>5′ . . . C|T N A G . . . 3′
3′ . . . G A N T|C . . . 5′</td><td>4</td></tr>
<tr><td>Moraxella bovis</td><td>MboII</td><td>5′ . . . G A A G A $(N)_8$| . . . 3′
3′ . . . C T T C T $(N)_7$| . . . 5′</td><td>5</td></tr>
<tr><td>Escherichia coli</td><td>EcoRV</td><td>5′ . . . G A T|A T C . . . 3′
3′ . . . C T A|T A G . . . 5′</td><td>1</td></tr>
<tr><td></td><td>EcoRI</td><td>5′ . . . G|A A T T C . . . 3′
3′ . . . C T T A A|G . . . 5′</td><td>2</td></tr>
<tr><td>Providencia stuarti</td><td>PstI</td><td>5′ . . . C T G C A|G . . . 3′
3′ . . . G|A C G T C . . . 5′</td><td>3</td></tr>
<tr><td>Microcoleus</td><td>MstII</td><td>5′ . . . C C|T N A G G . . . 3′
3′ . . . G G A N T|C C . . . 5′</td><td>4</td></tr>
<tr><td>Nocardia otitidiscaviarum</td><td>NotI</td><td>5′ . . . G C|G G C C G C . . . 3′
3′ . . . C G C C G G|C G . . . 5′</td><td>6</td></tr>
</table>

Notes
1. Enzyme produces blunt ends.
2. The single strand is the 5′ strand.
3. The single strand is the 3′ strand.
4. The base pair *N* can be any purine or pyrimidine pair.
5. The enzyme does not cut within the recognition sequence, but at whatever sequence lies eight nucleotides 3′ to the recognition site.
6. NotI has an eight-base recognition sequence and cuts mammalian DNA very infrequently.

often chemically modified by methylation and, as NotI cannot cut methylated CpG sites, it will produce even larger fragments. Moreover, the CpG sites are not distributed randomly throughout the genome—they tend to be clustered in the regions 5′ to genes. NotI produces fragments between 1 million and 1.5 million base pairs in size instead of the 65,536 base pairs expected of an "eight-base cutter." Enzymes like NotI that produce very large DNA fragments have proved invaluable for long-range physical mapping of mammalian DNA.

Electrophoresis Is Used to Separate Mixtures of Nucleic Acid Fragments

Proteins and the nucleic acid fragments produced by restriction enzymes can be analyzed by *electrophoresis.* Such molecules, differing in electrical charge, size, and shape, move at different rates in an electrical field and so can be separated under conditions in which they retain their biological activity. It was Arne Tiselius and Theodor Svedberg who, in 1926, developed moving boundary electrophoresis. Subsequently robust solid supports were used, most commonly filter paper and gels made of polymers such as agarose or polyacrylamide.

The rate at which the fragments migrate through the gel is a function of their lengths, charge, and shape, with small fragments generally moving much faster than large fragments (Fig. 4-1). Different supports are used depending on the absolute sizes of the fragments to be separated as well as on the relative differences in the sizes of these fragments. DNA fragments ranging in size from 100 base pairs to several thousand base pairs can be efficiently separated in a gel made of 1% agarose, provided these fragments differ in size from each other by a few hundred base pairs. When it is necessary to resolve fragments of DNA differing from one another by only a single nucleotide (e.g., in DNA sequencing), polyacrylamide gels are needed.

At first, when electrophoresis was used to analyze viral DNA fragments produced by restriction enzymes, the viral DNA was first radioactively labeled and the fragments were detected by autoradiography. A most useful advance came with the realization that DNA-binding dyes such as ethidium bromide could be used to stain DNA fragments directly in the gel. The fragments appear as a series of bands when a stained gel is illuminated with UV light, each band corresponding to a restriction frag-

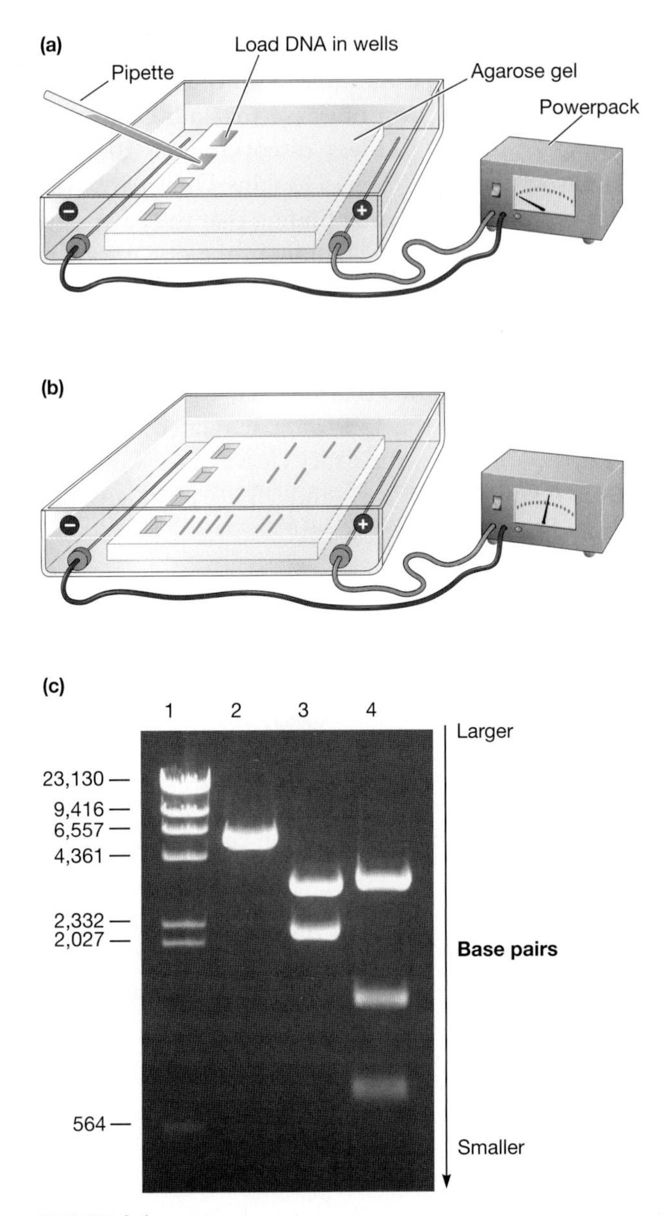

FIGURE 4-1

DNA molecules of different sizes can be separated by electrophoresis. Electrophoresis is a process in which an electrical field is used to move the negatively charged DNA molecules through porous agarose gels. (*a*) A slab of agarose is cast with slots to take the samples. These usually include a blue dye to make it easier to pipette the samples into the slots. (*b*) The current is turned on and the DNA molecules (negatively charged) move toward the positive electrode. Small DNA molecules move faster than larger molecules, and molecules of the same size move at the same speed so that they become separated into bands. (Other factors, including charge and shape, also influence the movement of molecules through the gel.) (*c*) These become visible when the gel is removed from the electrophoresis tank and stained with ethidium bromide. This intercalates between the nucleotides and fluoresces when illuminated with UV light. In the gel shown, the plasmid pCDNA3.1 has been digested with EcoRI (lane *2*), SalI (lane *3*), and NcoI (lane *4*). Lane *1* contains HindIII-digested λ DNA as size markers.

ment of specific size. The sizes can be established by calibration with DNA molecules of known sizes. Bands containing as little as 0.005 μg of DNA can be visualized. Most importantly, DNA molecules move unharmed through these gels and can be recovered (eluted) from the gel as biologically intact double helices for use in experiments.

Restriction Enzyme Sites Are Used to Make Maps of DNA

Different restriction enzymes, having different target sequences, necessarily give different restriction fragments for the same DNA molecule. If restriction enzymes are used singly and in combination, and the lengths of the fragments are carefully determined by electrophoresis, then the fragments can be reassembled to make a *restriction map* of the molecule showing the positions of sites cut by the enzymes. The first such map was obtained in 1971 by Daniel Nathans, a colleague of Hamilton Smith at Johns Hopkins. Nathans used the HindII enzyme to cut ("digest," in the jargon of the field) the circular DNA of the SV40 virus into 11 specific fragments. By measuring the sizes of fragments produced when the DNA was completely cut, and also the overlapping intermediate-sized fragments produced by brief treatments with HindII (resulting therefore in only partial digestion of the DNA), Nathans located the sites on the circular viral DNA that are attacked by the restriction enzyme. With this information, it became possible to determine the regions of biological importance on the circular viral DNA. For example, brief radioactive labeling of replicating viral DNA followed by digestion with HindII allowed Nathans to prove that the replication of SV40 DNA always begins in one specific HindII fragment and proceeds bidirectionally around the circular DNA molecule. Restriction enzymes thus provide a critical tool for molecular biologists to predictably fragment a given DNA molecule and to analyze its structure and features in fine detail. This becomes especially powerful when combined with techniques that detect DNA fragments with specific sequences ("Southern blotting," see Fig. 4-11).

DNA Ligase Joins the Ends of DNA Fragments Produced by Restriction Enzymes

Restriction enzymes together with gel electrophoresis provide a convenient means to purify specific DNA fragments and also, because of the way they cut double-stranded DNA, a way to recombine fragments. Some restriction enzymes, such as HindII, cut DNA at the center of their recognition site to produce blunt-ended fragments; these fragments are fully base-paired out to their ends and have no tendency to stick together (Fig. 4-2). In contrast, many restriction enzymes make staggered or asymmetric cuts on the two strands of the DNA to create short complementary single-stranded tails on the ends of each fragment. The EcoRI recognition sequence is GAATTC and it is cut to generate a four-base-long tail, AATT (Fig. 4-2b). The single-stranded tail sequences left by different restriction enzymes can be similar even though the recognition sites are different. MfeI recognizes the sequence CAATTG and cuts this to produce the same AATT tail as does EcoRI (Fig. 4-2c).

Complementary single-stranded tails tend to associate by base pairing and thus are often called *cohesive*, or *sticky, ends.* For example, the linear molecules that EcoRI generates by cutting circular SV40 DNA often temporarily recyclize because of base pairing between their tails. Base pairing occurs only between complementary base sequences, so the cohesive AATT ends produced by EcoRI will not, for example, pair with the AGCT ends produced by HindIII. However, any two DNA fragments with complementary ends (regardless of the organism from which they originate) can associate by base pairing and temporarily anneal. What was needed was a way to permanently join the associated ends.

In fact, such enzymes had already been discovered in 1967. The *DNA ligases* play important roles in DNA repair and DNA replication, where a ligase links the DNA Okazaki fragments generated on the lagging strand. By combining restriction enzymes and ligases, scientists had the tools to cut DNA and to paste the fragments together. Such experiments were first done at Stanford University in 1972 by Janet Mertz and Ronald Davis, who realized that EcoRI in conjunction with DNA ligase provided a general way to achieve in vitro, site-specific genetic recombination.

Thus, by the early 1970s, techniques for manipulating and isolating genes were available and plans were under way to clone genes and produce large quantities of the cloned genes in bacteria. But even as the first plans were being made to carry out gene cloning experiments, concerns began to be raised about the safety of such experiments.

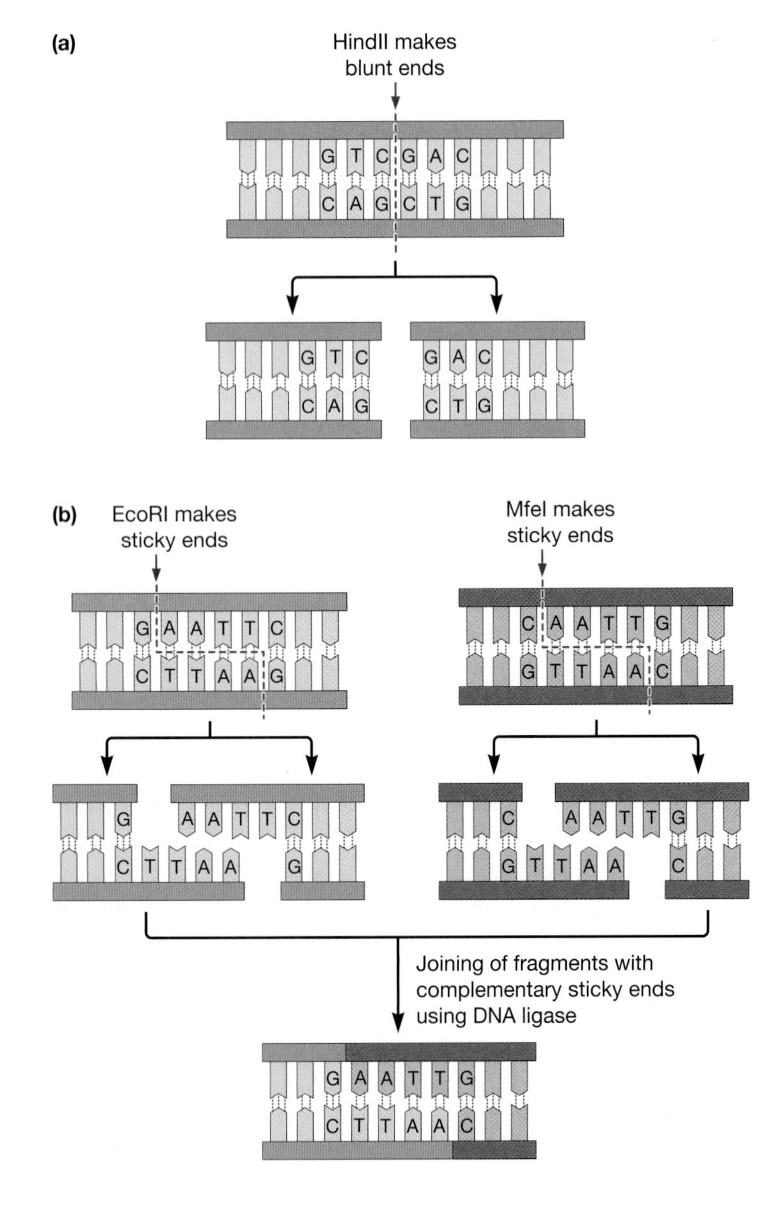

FIGURE 4-2

Restriction enzymes can produce blunt ends or "sticky" ends. (*a*) The HindII restriction enzyme cuts DNA at the center of its recognition site, leaving blunt ends. These blunt ends can be ligated to blunt ends produced by any other process. (*b*) The EcoRI and MfeI restriction enzymes make staggered, symmetrical cuts in DNA away from the center of the recognition site, leaving cohesive, or sticky, ends. A sticky end produced by EcoRI digestion can anneal to any other sticky end produced by EcoRI cleavage or any other sticky end with complementary sequence produced through other methods. For example, the restriction enzyme MfeI produces identical overhangs to EcoRI, although the flanking sequences are different. MfeI sticky ends can be ligated to those from EcoRI, although the ligated product cannot be recut with either enzyme because the resulting sequence is recognized by neither enzyme.

Scientists Voice Concerns about the Dangers of Gene Cloning at the Asilomar Conference

These first concerns—prior to the cloning of any gene—centered on proposals to clone the genomes of DNA tumor viruses and replicate the recombinant molecules in bacterial cells like *E. coli*. The question was: Could *E. coli*, present in the intestinal tract of all human beings, act as a vector to transmit the cancer-causing genes of tumor viruses to human beings? In 1972, it was pointed out that the DNA of mice, and possibly that of human beings, harbored the DNA genomes (proviruses) of latent RNA tumor viruses. Many of the recombinant plasmids that scientists would isolate in their search for specific genes might contain as well these DNA proviruses that might be able to cause human cancer. If so, should all recombinant DNA experiments using vertebrate DNA be regarded as possibly dangerous? The answers were mixed. This was hardly surprising given that the necessary data were not yet available to allow rational assessment of the risks. There was virtually unanimous agreement, however, that researchers should not have unrestricted freedom to do experiments that might have military consequences.

Thus the first discussions of the conjectured hazards of recombinant DNA techniques took place before any experiments had been carried out. However, the announcement of the first successful cloning experiments brought matters to a head as it became clear that the future of studying the molecular basis of genetics lay in using recombinant DNA techniques. The question then was whether to move ahead as fast as possible or to try to devise methods that would allay worries about the possible risk of these experiments. In July of 1974, a letter appeared in *Science* urging that scientists considering recombinant DNA experiments should pause until there had been further evaluation of the risks involved. This evaluation took place in February of 1975, when a group of more than 100 internationally respected molecular biologists gathered at the Asilomar Conference Center located near Monterey, California. In the absence of clear knowledge whether any danger might exist, a nearly unanimous consensus emerged that some restrictions on DNA cloning were appropriate.

Afterward, the "Asilomar recommendations" were considered by a special committee appointed by the National Institutes of Health (NIH). In its deliberations, the committee recommended guidelines that effectively precluded the use of recombinant DNA techniques for studying the genes of cancer viruses and required the use of genetically disabled bacteria that would not grow well outside the laboratory. These recommendations became codified in official federal regulatory guidelines that took effect in July 1976. These regulations in the United States were paralleled by the establishment of guidelines developed by similar bodies in Europe.

A remarkable feature of the discussions about the safety of recombinant DNA was the unprecedented participation of nonscientists in communities like Cambridge, Massachusetts, in the debates. This led to the appointment of lay members of the Recombinant DNA Advisory Committee and ensured an increasing involvement of nonscientists in discussions about the societal impact of genetics. The most notable example is the Ethical, Legal, and Social Issues (ELSI) Research Program of the Human Genome Project (Chapter 11).

Many scientists thought that the NIH regulations were too restrictive and in some cases scientifically unsound. As more experiments were performed, and more data accumulated, it became increasingly apparent that the actual hazards of recombinant DNA experiments were extremely small. Discussions throughout 1978 led to new, less restrictive NIH regulations that took effect in January 1979 and permitted cloning of viral cancer genes.

Plasmids and Viruses Are Used as Vectors to Carry DNA Sequences

The goal of cloning is not only to isolate a specific gene, but also to produce large quantities of the gene for analysis and further manipulations. The obvious way to do this was to use bacteria to replicate the gene, even if the gene was not bacterial in origin. This requires linking the cloned gene to DNA sequences that have the appropriate signals for directing replication of the hybrid molecule in bacteria. Fortunately, bacteria possess *plasmids*, which have just the right properties to act as *vectors* for carrying foreign DNA fragments into bacteria. Plasmids are tiny, circular DNA molecules only a few thousand base pairs long—found in bacteria. There may be as few as two or as many as several hundred in each bacterial cell (referred to as *copy number*) and they replicate independently of the host bacterial chromosome. Plasmids were first noticed as genetic elements that were not linked to the main chromosome and that carried genes that made the bacteria resistant to antibiotics such as tetracycline or kanamycin. That these genes were found on plasmids as opposed to main-chromosomal DNA was not a matter of chance. Antibiotic resistance requires relatively large amounts of the enzymes that chemically destroy the antibiotics, and because the genes encoding these enzymes are carried on plasmids, they are present in much higher numbers than would be the case if they were located on the single bacterial chromosome. The small size of plasmids (which makes it easier to manipulate them) and the presence of antibiotic resistance genes (which provides a means to select for bacteria containing the plasmids) made them ideal vectors for cloning.

In the first cloning experiments, Stanley Cohen and Herbert Boyer and their collaborators at Stanford and the University of California, San Francisco, used two plasmids—pSC101 and pSC102—found in *E. coli* (Fig. 4-3). pSC101 was chosen as the vector because it contained only a single EcoRI recognition site that would accept DNA fragments cut with EcoRI, and it could be selected for by using the antibiotic tetracycline because it carried a gene for tetracycline resistance. pSC102 contained a gene for resistance to the drug kanamycin, and this was the gene to be cloned. The two plasmids were mixed and converted into linear molecules with EcoRI sticky ends by treatment with EcoRI. After allowing the DNA fragments to link and using DNA

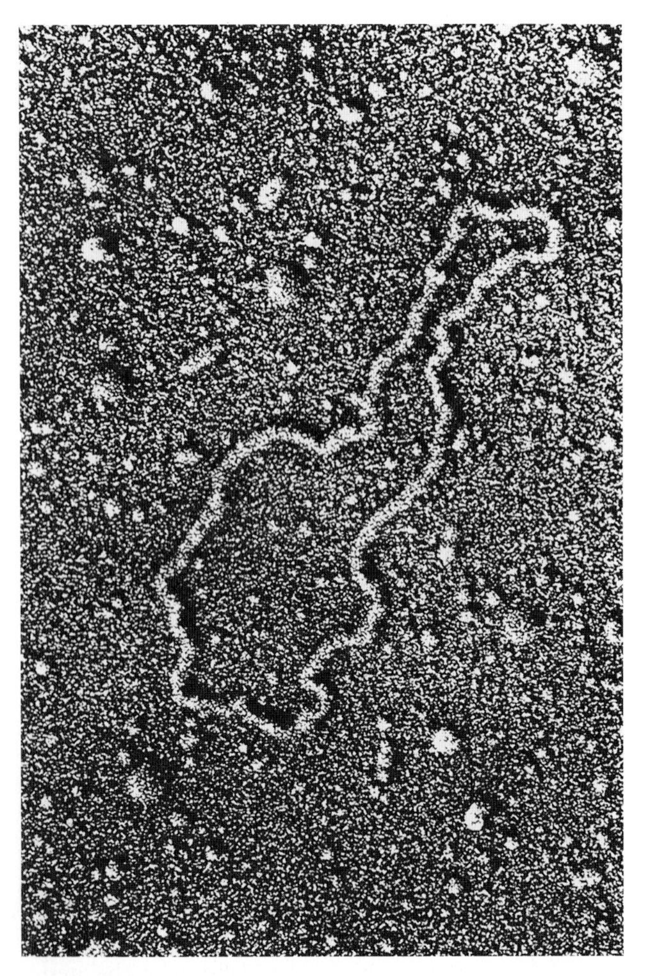

FIGURE 4-3

Electron micrograph of plasmid pSC101. The plasmid is a circle of double-stranded DNA, of 9263 base pairs approximately 1.3 μm long.

ligase to make the links permanent, the recombined DNA molecules were introduced into the bacteria. These were plated on media containing both tetracycline and kanamycin. Those bacteria that survived and grew were resistant to both antibiotics because they contained recombinant plasmids carrying tetracycline resistance from the pSC101 plasmid and kanamycin resistance from pSC102.

Although in the original Cohen–Boyer experiment the two DNAs were both from plasmids, it was clear that all sorts of foreign DNA, from microbes and, most importantly, from higher plants and animals, could be cloned into these plasmids. For example, the *E. coli* chromosome with its 4 million base pairs contains about 500 different recognition sites for EcoRI. By random insertion of individual EcoRI fragments into pSC101, it would be possible to clone all the *E. coli* genes. Plasmids were engineered to be ever more useful and versatile; an early example was pBR322 and elements of this were used to create others, including the popular pUC18 and pUC19. These have additional enzyme cleavage sites for cloning, have different drug-resistance genes, and are present at much higher copy number than other plasmids. All plasmids have one great advantage—they are easily purified from *E. coli* (Fig. 4-4).

Encouraged by their success using plasmids, scientists looked to nature to find other vectors suitable for cloning in bacteria, and bacteriophages were an obvious choice. *Transducing bacteriophages* are naturally occurring vectors that carry parts of bacterial chromosomes from one bacterial cell to another. (Transducing phages integrate into the host bacterial genome to wait for the proper conditions to excise themselves. If this excision process is not precise, the phage may carry a segment of the bacterial genome with it. Such transduction events had long been used for fine-scale mapping of bacterial genes.) Bacteriophage λ was rapidly adapted as a general cloning vector and engineered, by in vitro recombination methods, to carry virtually any fragment of foreign DNA. Much larger fragments of DNA can be cloned in bacteriophages than in plasmids, although there is an upper limit set by the need to package the DNA in the phage protein coat; a maximum of 25 kb of DNA can be cloned in bacteriophage λ. As for plasmids, genetic engineers created more sophisticated phage vectors, including, for example, λgt and λZAP. Subsequently, other vectors with much larger capacity have been developed; bacterial artificial chromosomes (BACs) and yeast artificial chromosomes (YACs) can hold inserts as long as 100–500 kb and 250–1000 kb, respectively (Table 4-2). (Large-scale vectors are discussed in Chapter 11.)

There Are Five Basic Steps in Cloning

There are five steps in the basic process of cloning DNA, described here for cloning in plasmid vectors (Fig. 4-5). The first is to choose the appropriate DNA to be cloned. Sources of DNA for cloning are genomic DNA or a so-called complementary DNA (cDNA), a DNA copy of an mRNA made using the enzyme reverse transcriptase.

The second step is to produce a collection of DNA fragments of a size suitable for inserting into appropriate vectors carrying one or more antibiotic resistance genes. Although these fragments are pre-

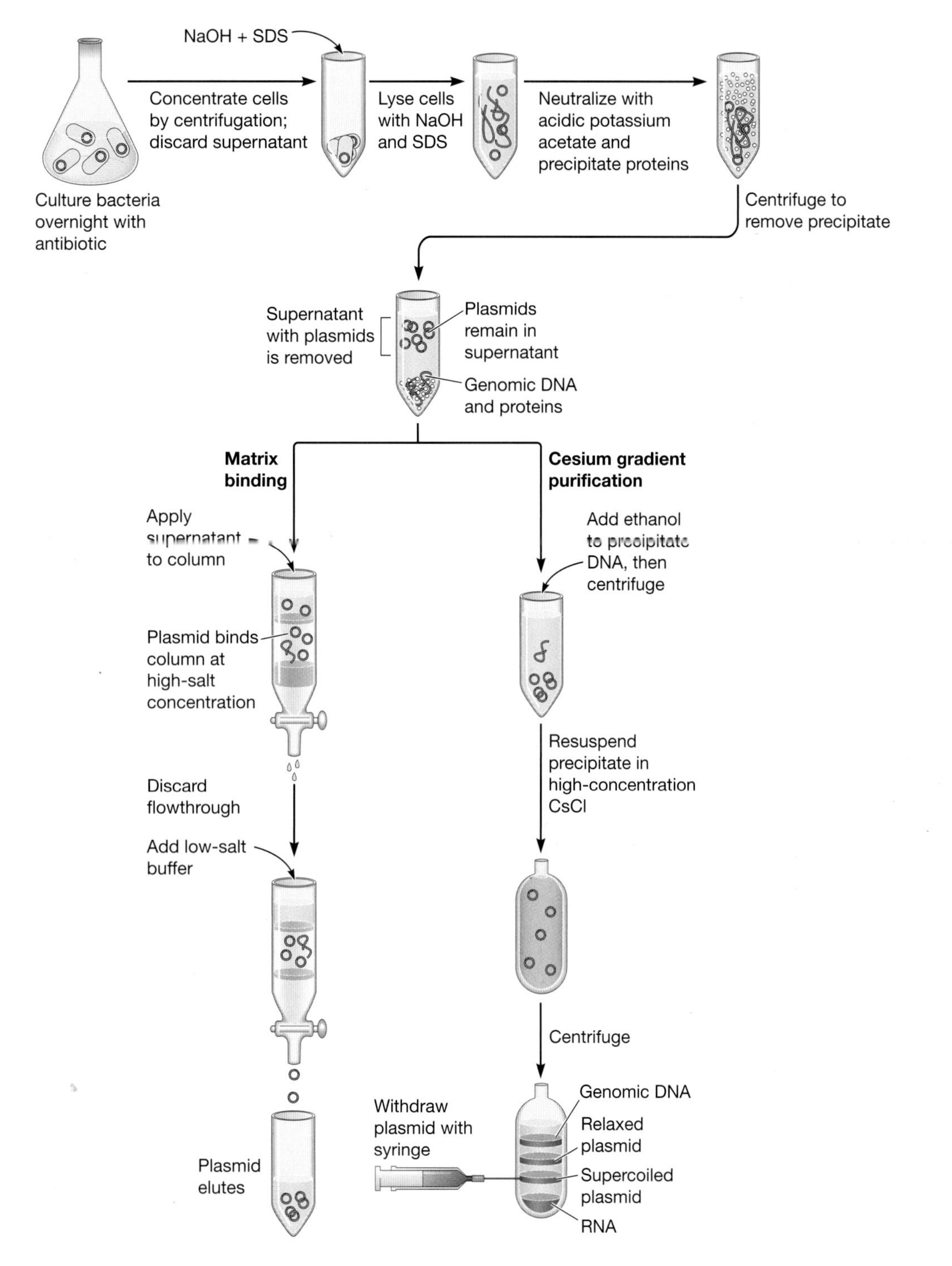

FIGURE 4-4

Purification of plasmid DNA from *Escherichia coli*. Bacteria containing a plasmid with a cloned DNA insert and an antibiotic resistance gene are grown in medium containing the antibiotic to select for bacteria carrying the plasmid. Most plasmids used for molecular biology are *multicopy* (i.e., present in many copies per cell). The medium is centrifuged to pellet the *E. coli* cells and the supernatant is discarded. Sodium hydroxide and the detergent sodium dodecyl sulfate are added to lyse the bacteria, thus freeing the plasmids. Potassium acetate is added, which forms an insoluble precipitate of potassium dodecyl sulfate that contains the *E. coli* genomic DNA and many proteins. This precipitate is removed by centrifugation; the small plasmid DNA remains in the supernatant, which also contains *E. coli* RNA. Two methods are shown for further purification of the plasmids. One method uses density centrifugation in a cell to separate out the plasmid DNA from other nucleic acids. (Plasmid DNA can be very tightly coiled ["supercoiled"], or one strand can be nicked, thus allowing it to adopt a "relaxed shape.") Another approach is to use a resin that specifically binds DNA in buffers of high salt concentration. The column with the bound plasmid DNA is washed several times before eluting the plasmid DNA with a low salt buffer.

TABLE 4-2. Cloning capacity of commonly used vectors

Vector	Insert size range (kb)
Plasmid	<10
Phage	<23
Cosmid	30–46
P1 artificial chromosome (PAC)	130–150
Bacterial artificial chromosome (BAC)	<300
Yeast artificial chromosome (YAC)	200–2000

pared from chromosomal DNA by using a restriction endonuclease, most cDNAs are of a suitable size for cloning without any further manipulation.

The third step is to insert the DNA into the vector and use DNA ligase to covalently link the DNA fragment to the vector DNA. In the fourth step, the vectors with inserted DNA fragments are introduced into a population of bacteria; this step is called *transformation.* The transformed bacteria are plated on agar containing an antibiotic. Only those bacteria containing plasmids will grow because the plasmids confer antibiotic resistance on the bacteria. The bacteria are plated at low density so that as each resistant bacterial cell divides, it gives rise to a colony of bacteria, all derived from a single cell containing a single molecule of recombinant DNA. This is the cloning step proper. Such a collection of cloned DNA fragments propagated in bacteria is called a *library.*

This library is a library without a catalog to tell us which clone contains a particular sequence, so the fifth step is to determine which colonies contain the desired sequence. In the next sections we will review each of these steps in more detail.

Choosing the Right Starting Material Is Essential in Cloning

The first decision to be made is whether to clone a gene starting with chromosomal DNA or a cDNA copy of an mRNA. The choice depends on the particular problem to be tackled. If one is interested in the amino acid sequence of a protein, this information can be obtained most readily from the nucleotide sequence of a cloned cDNA. On the other hand, if one is interested in the regions of a gene that regulate its expression, or in gene sequences not contained within the mRNA, then this information can be obtained only from genes cloned from chromosomal DNA.

The purity of the starting DNA sample is important because if DNA from other organisms is present it will be represented in the final library. This type of

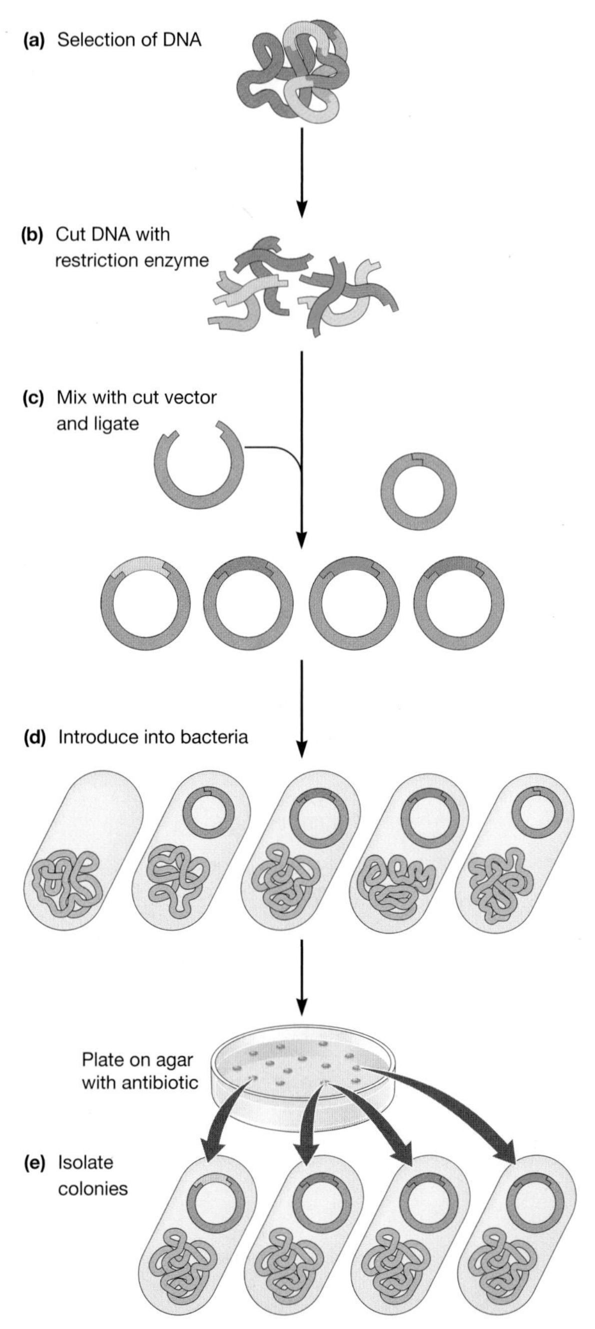

FIGURE 4-5

The five basic steps of cloning DNA in a plasmid. (*a*) DNA containing the gene of interest must be prepared. (*b*) The DNA is treated with a restriction enzyme to cut the DNA into fragments of a size suitable for inserting into a vector that has been cut with the same restriction enzyme. (*c*) Third, the DNA fragments are mixed with vector molecules and DNA ligase is used to covalently link them. (*d*) In the fourth step, the recombinant molecules are introduced into bacteria that are plated on agar containing an antibiotic. This kills any bacteria that have not taken up a plasmid (which carries an antibiotic resistance gene). The bacteria are plated at low density so that each colony arises from a single bacterial cell carrying a plasmid. (*e*) Finally, those very few clones containing the gene of interest are selected from the library.

DNA mixing could occur if, for example, plants harvested for DNA extraction had bacteria and fungi growing on the surface of the leaves and roots. It is also important to ensure that the DNA sample contains the gene of interest. It would not be a good idea to try to clone human Y-chromosome genes beginning with a sample of DNA from a woman.

The mRNA that will serve as the template for the synthesis of a cDNA must be prepared from cells that express the gene of interest. As any given cell type expresses only a subset of its chromosomal genes, proper selection of the starting tissues or cells is especially important. This may not be easy, because many genes are expressed in only a limited number of cell types or under certain growth conditions or at particular stages in development. The first eukaryotic gene to be cloned was the rabbit β-globin gene because there was a readily available and plentiful source of β-globin mRNA—reticulocytes in the blood; 50–90% of the total mRNA in a reticulocyte is globin mRNA. Similarly, pancreatic cells were used as the source of mRNA for cDNA cloning of the insulin gene.

Techniques have improved so that we are now able to clone cDNAs from rare mRNAs, but the relative abundance of mRNAs from different genes affects the relative number of copies of each cDNA present in the library. There will be many more cDNAs derived from an abundant mRNA than from a rare mRNA, so that there will be many more clones of the former than the latter in a library, making it difficult to find clones of a rare transcript. Several techniques have been developed for *normalization* of cDNA libraries, which reduce the proportion of clones corresponding to abundant mRNAs.

mRNA Is Converted to cDNA by Enzymatic Reactions

The first step in constructing a cDNA library involves isolating total cellular RNA. A fraction that contains mostly mRNA is then isolated from the total cellular RNA. As we will describe in detail in Chapter 5, the majority of eukaryotic mRNA molecules have a run of adenine nucleotide residues called a *poly(A) tail* at their 3′ ends. This tail is important for the function of the mRNA, but also the poly(A) provides a very convenient way to isolate mRNAs from total cellular RNA, the bulk of which is ribosomal RNA and tRNA, which do not have poly(A) tails (Fig. 4-6). (The fraction of poly(A)-containing mRNA is usually only ~1–2% of total cellular RNA.) The selection relies on the use of oligonucleotides composed only of deoxythymidine (oligo(dT)), which can be linked to a solid support such as beads of cellulose. These beads are packed into small columns. When a preparation of total cellular RNA is passed through such a column, the poly(A) tails of the mRNA molecules bind to the oligo(dT), thereby trapping the mRNAs to the support while the rest of the RNA flows through the column. The bound mRNAs are then eluted from the column and collected.

The poly(A) tails of the mRNA molecules are also used for the next step of cDNA cloning, the preparation of a DNA copy of the RNA (Fig. 4-7). Short oligonucleotides containing 12–20 deoxythymidines (oligo(dT)) are mixed with the purified mRNA and hybridized to the poly(A) tails, where they act as primers for reverse transcriptase. This enzyme, which is isolated from certain RNA tumor viruses (see Chapter 5), can use RNA as a template to synthesize a DNA strand. (Its name comes from its ability to reverse the normal first step of gene expression.) The product of the reaction is an RNA–DNA hybrid. Using oligo(dT) as the primer has the disadvantage that, because the reverse transcriptase must begin at the 3′ end of its mRNA, it may not reach the 5′ end of the molecule. This is a particular problem for very long mRNAs. To circumvent this difficulty, a second method, *randomly primed* cDNA synthesis, is used. Oligonucleotide fragments, 6–10 nucleotides long and made up of many possible sequences, are used as primers for the cDNA synthesis. In this way, priming of the mRNA occurs from many positions, not only from the 3′ end. Sequences close to the 5′ end of long mRNAs are more readily cloned using this method. Again the product of reverse transcription is an RNA–DNA hybrid. At this point, the RNA–DNA hybrid molecules must be converted into double-stranded DNA molecules that can be cloned into appropriate vectors.

The most common way to synthesize a double-stranded cDNA from an mRNA–cDNA hybrid makes use of an enzyme from *E. coli*, RNase H, that recognizes RNA–DNA hybrid molecules and digests the RNA strand into many short pieces. These RNA pieces remain hybridized to the first cDNA strand and serve as primers for *E. coli* DNA polymerase, which uses the original cDNA as a template to synthesize the complementary strand of DNA. Eventually, this process completely replaces the original RNA with DNA, except for a small piece of RNA at the extreme 5′ end. The new DNA strand is not entirely contiguous but rather contains breaks ("nicks"). These breaks

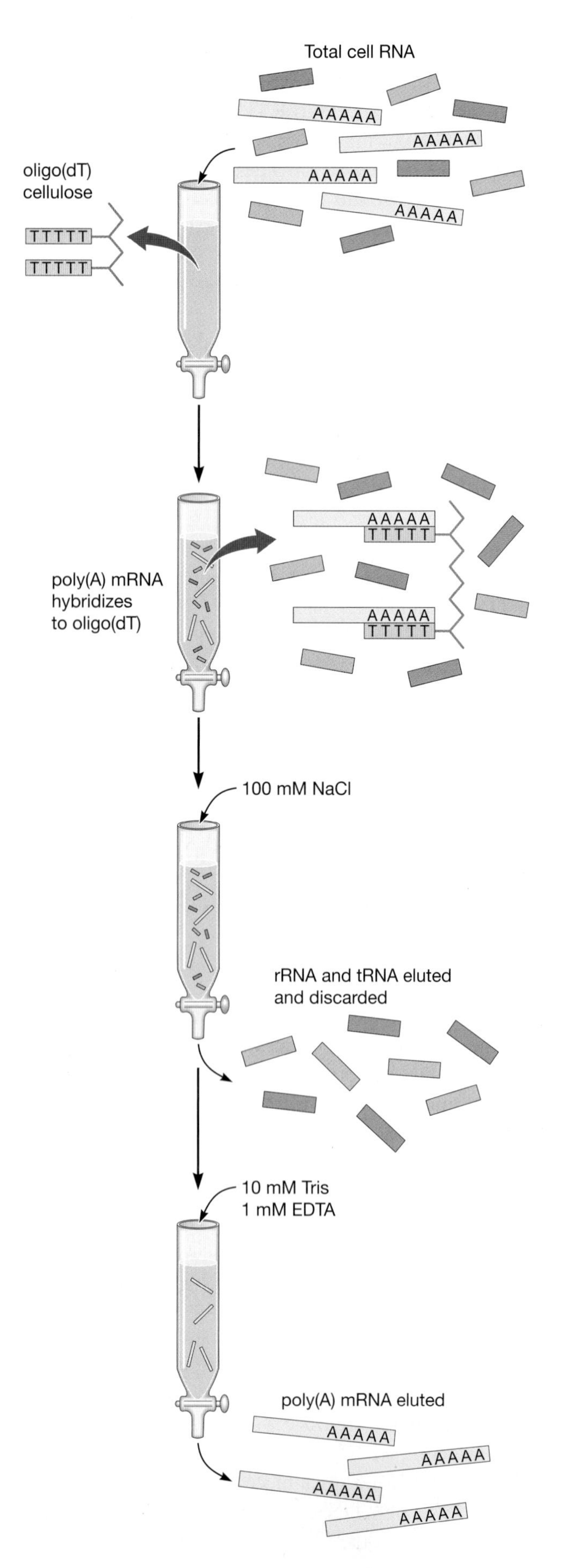

are joined (ligated) by the action of DNA ligase, thus forming a double-stranded DNA molecule.

cDNA Molecules Are Joined to Vector DNA to Create a Clone Library

The double-stranded cDNA molecules obtained by these procedures are then inserted into a plasmid or a phage vector. Modern plasmid vectors have been extensively engineered to contain a variety of features to aid the cloning process, including, for example, *polylinker regions* containing many restriction enzyme sites. For library construction, artificial restriction enzyme sites are added to the ends of the cDNAs (Fig. 4-8). The restriction enzyme sites are contained in 8- to 12-bp oligonucleotides, synthesized chemically, called *linkers* or *adaptors*. These adaptors are added to the double-stranded cDNAs using DNA ligase and then cut with the appropriate restriction enzyme to produce single-stranded sticky ends.

The cDNA, now carrying sticky ends generated by the restriction enzyme, is combined with a vector that has been cleaved with the same enzyme, and the ends are sealed together using ligase. The recombinant molecules are now ready to be introduced into bacterial cells using a method appropriate for the vector type. Plasmids are introduced by a variety of transformation procedures (detailed in Chapter 6), whereas λ phage vectors are first packaged in vitro to form infectious phage particles that can replicate in bacteria. Although plasmid vectors offer the advantage of ease of manipulation of the inserted cDNA, phage libraries contain a greater number of clones and can be screened in much larger numbers.

FIGURE 4-6

Isolation of poly(A) mRNA. Most eukaryotic mRNAs carry a poly(A) tail, which can be used to purify the mRNA fraction from the bulk of cellular RNA. Cellular RNA is passed over a column consisting of an inert material, often cellulose or agarose, to which oligonucleotides consisting entirely of deoxythymidine (dT) residues have been attached. The poly(A) tails hybridize to this oligo(dT), causing the mRNA to stick to the column, whereas the rest of the RNA runs through. After extensive washing of the column to remove the last traces of contaminating material, the column is washed with a buffer of low ionic strength. Under these conditions the poly(A)–oligo(dT) hybrids dissociate, and the purified mRNA washes off the column.

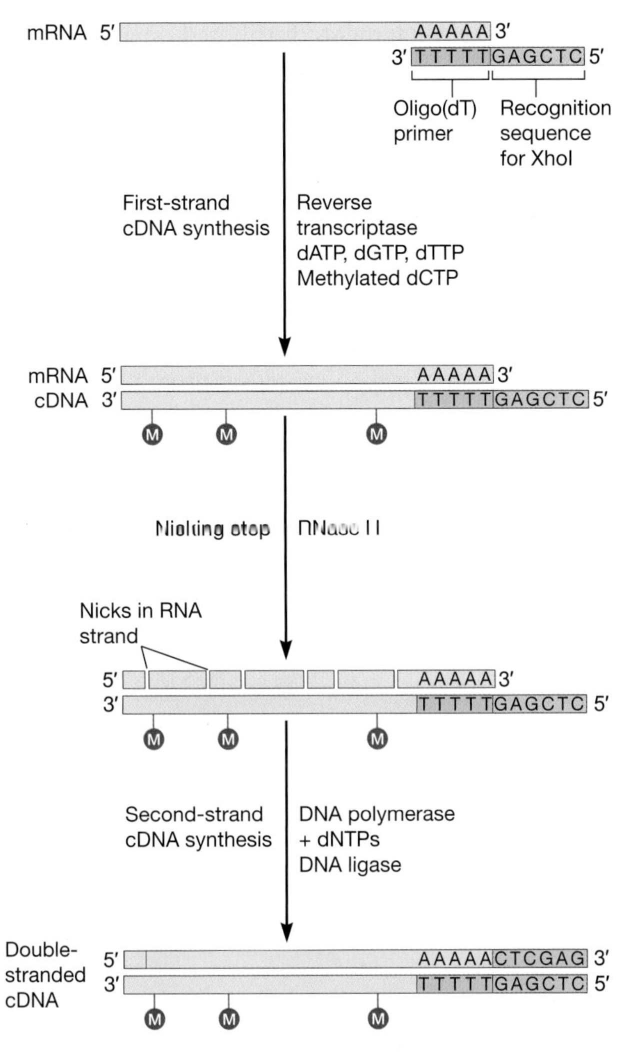

FIGURE 4-7

Synthesis of cDNA. poly(A) mRNA is incubated with deoxythymidine-containing oligonucleotides, which hybridize to the poly(A) tails, forming primed templates for the enzyme reverse transcriptase. When the cDNA will be used to construct a library, the oligonucleotide primers are designed so that they contain a restriction site, such as XhoI, which will be used to create sticky ends for ligation to the vector. Deoxynucleotide triphosphates are added to the mRNA but with methylated deoxycytosine instead of the usual unmethylated deoxycytosine. On addition of reverse transcriptase, cDNA molecules are synthesized on the mRNA templates; this methylated cDNA cannot be cut at internal XhoI sites. The result of the reverse transcriptase reaction is a collection of RNA–DNA hybrids and the RNA strands must be destroyed and replaced with DNA. RNase H nicks the RNA strands and these serve as initiation sites for DNA synthesis by *E. coli* DNA polymerase. This DNA synthesis step is done using unmethylated deoxycytosine to ensure that the XhoI site in the oligonucleotide at the ends of the cDNAs can be digested by XhoI. Eventually, most of the RNA fragments are replaced by DNA. The double-stranded DNA molecules are finally treated with DNA ligase, which seals up any remaining nicks in the new DNA strand.

How can the experimenter be certain that the process of cloning to this point has resulted in bacteria containing cloned DNA? One simple approach, called *blue/white screening,* relies on the β-galactosidase gene from the *lac* operon (Chapter 3). The enzyme β-galactosidase hydrolyzes the chemical X-gal to produce an insoluble blue dye. The cloning site of many vectors is within a copy of the β-galactosidase gene, so that the insertion of DNA disrupts the gene and bacterial colonies or phage plaques remain colorless ("white") if X-gal is included in the culture medium (Fig. 4-9). In contrast, colonies or plaques containing vectors without DNA inserts turn blue. Blue/white screening can be used to determine whether recombinant vectors are present at high frequency in the library and picking white colonies or plaques excludes those containing empty vectors.

Libraries of Genomic DNA Represent the Complete Sequence of Organisms

cDNA libraries are useful because they represent the proteins expressed in a given tissue and can be directly used for the expression and study of proteins. However, mRNAs only represent a snapshot of the complete genome of an organism. To study the entire genome (much of which is not expressed as mRNA), it is necessary to create a library containing all segments of the genome of an organism. This is possible by similar techniques to those described above for making cDNA libraries (in fact, it is simpler in that one begins with double-stranded DNA rather than having to make cDNA). One limitation of the plasmid and λ vectors is that they can hold only small DNA fragments. This means not only that many clones are needed to ensure complete coverage of genomic DNA, but also that many large genes cannot be contained on a single clone. Other vectors with much larger capacity, such as BACs and YACs, are indispensable for genome-scale cloning. Genomic libraries have long been used to study specific parts of the genome, and creation of high-quality libraries is essential for genome sequencing, as we shall see in Chapter 11.

Nucleic Acid Probes Are Used to Locate Clones Carrying a Desired DNA Sequence

The net result of plating out a clone library is hundreds of thousands to a million phage plaques or bacterial colonies (for plasmid vectors), each containing

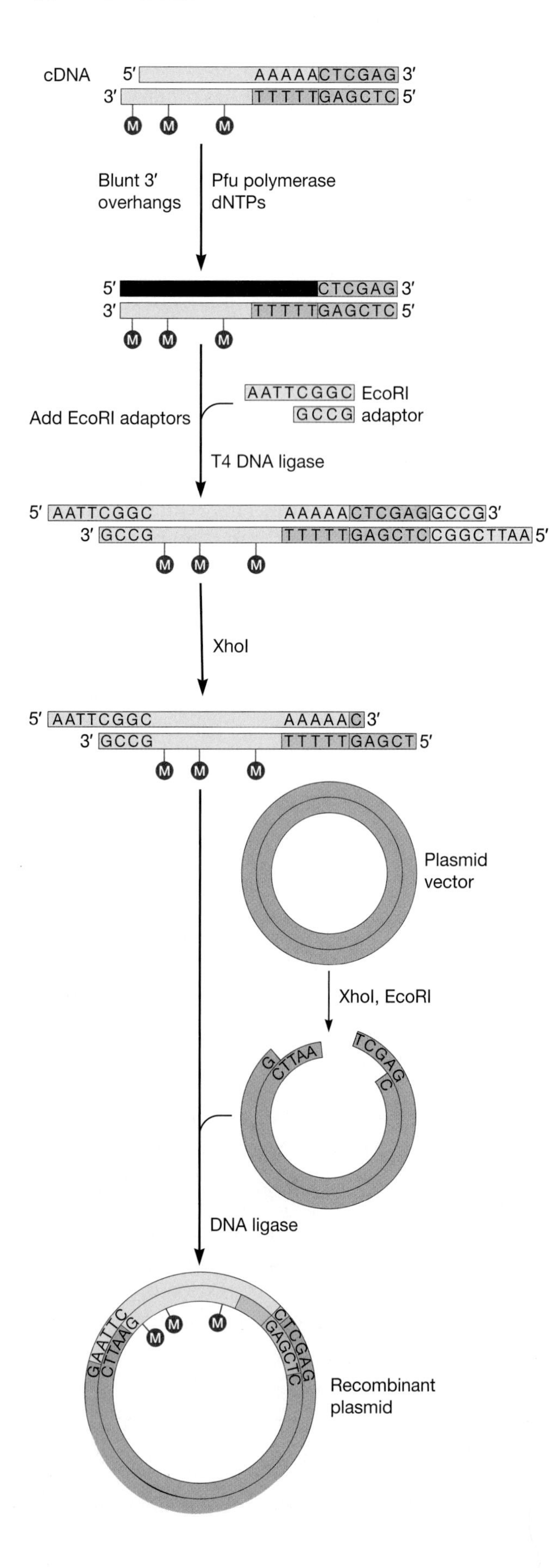

a cloned DNA fragment, distributed on a set of agar plates. Once the library has been plated out, a copy, or *replica*, is prepared on nitrocellulose filters or nylon membranes (Fig. 4-10). This process transfers a portion of each plaque or colony to the nitrocellulose and is done in such a way that the pattern of plaques or colonies on the original plates is maintained on the filters. Screening is carried out by incubating these nitrocellulose replicas with a nucleic acid probe to detect sequences of interest or with an antibody that can detect an expressed protein of interest.

The most direct method of screening is to use nucleic acid hybridization, a very sensitive means for detecting DNA sequences with small pieces of DNA (probes) that are complementary to those sequences. This requires knowledge of the sequences being sought. In some cases, part of the gene may already have been cloned, and this can be used to search for clones that contain additional sequences flanking the starting clone. In other cases, the gene of interest may be selected based on genome sequence information from that organism or from another organism. If the sequence of interest is not precisely known, it is still possible to use hybridization to find partially

FIGURE 4-8

Directional cloning of double-stranded cDNA in a plasmid vector. The cDNA is first manipulated to give it different single-stranded tails, complementary to restriction sites on the vector DNA. The cDNA is treated with *Pfu* DNA polymerase, which has 3′ to 5′ exonuclease activity and 5′ to 3′ polymerase activity. These combined activities fill in 5′ overhangs and remove 3′ overhangs, producing blunt ends on the cDNA product. Next, the EcoRI adaptors are attached to the ends of the cDNA using DNA ligase. Ligation requires a 5′-phosphate and, because the linkers are not phosphorylated, only one linker can be attached at each end of the cDNA, to the free 5′-phosphates. This produces a sticky end compatible with EcoRI at one end of the cDNA. The cDNA is then digested with XhoI. Because methylated cytosines were used for the cDNA synthesis, the only XhoI site that can be cut is that in the reverse transcriptase primer at the 3′ end of the original mRNA. The end product is a cDNA molecule with a sticky EcoRI at one end and a XhoI at the other. Next, the vector DNA is treated with XhoI and EcoRI, and the cut vector is purified. The vector and cDNAs now have compatible ends and they are covalently joined with DNA ligase. The result is a population of circular molecules, with the cDNA directionally cloned with the original 5′ end at the EcoRI site and the 3′ end at the XhoI site.

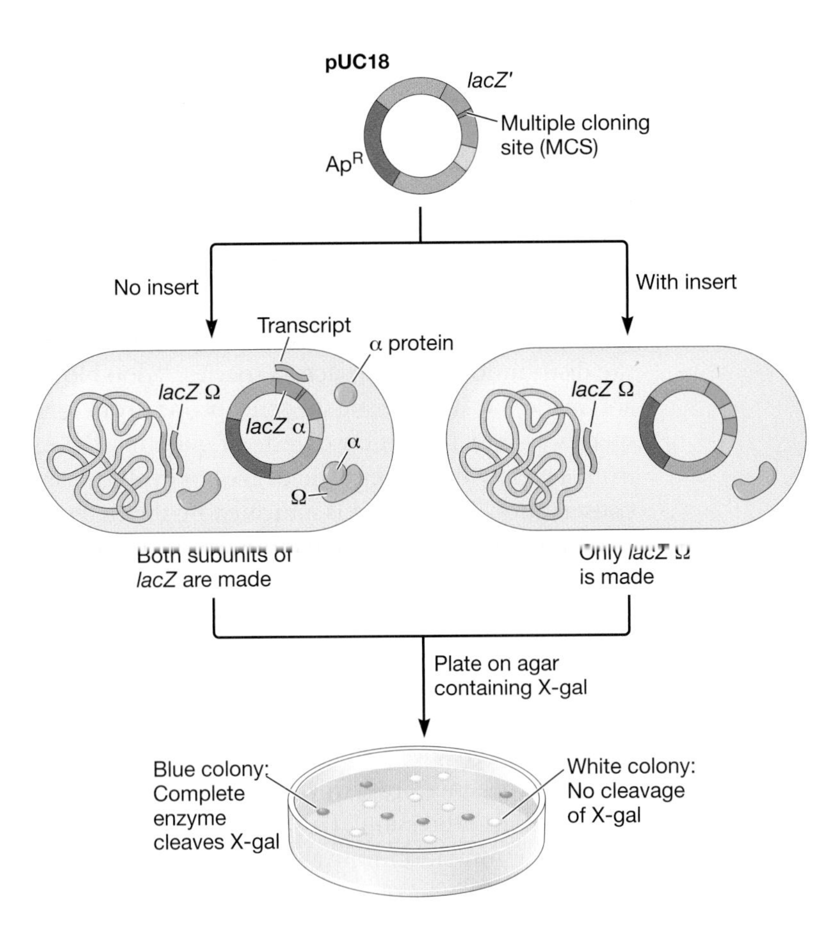

FIGURE 4-9

Blue/white screening to recognize vectors containing inserts. The *Escherichia coli lacZ* gene encodes the enzyme β-galactosidase. When bacteria making β-galactosidase are grown on agar containing the colorless chemical X-gal, cleavage of the X-gal by the enzyme produces an insoluble blue product. In many modern vectors, the multiple cloning site is situated within a small 150-nucleotide fragment of the *lacZ* gene (*lacZ*′), which produces a 50-amino-acid-long α fragment. The multiple cloning site does not disrupt the *lacZ*′ reading frame. When the α fragment is combined with the *lacZ* Ω fragment, they assemble to form an active enzyme. So, expression of the α fragment from a plasmid in an *E. coli* host producing the Ω fragment leads to the formation of active β-galactosidase, and colonies grown on X-gal containing agar are blue. In contrast, if there is a cDNA insert in the multiple cloning site, the α fragment cannot be synthesized, functional β-galactosidase is not made, and the colonies are colorless ("white"). The blue and white colonies (or plaques if a phage vector has been used) are easily distinguishable, providing a ready means for identifying clones with inserts.

matching sequences. Hybridization between probe and cloned gene sequences containing mismatched bases will occur if screening is performed at lower temperatures (e.g., 42°C instead of 65°C) and higher salt concentrations. The filter with spots showing where the probe hybridized to clones is then compared to the original plate to select clones containing the sequence of interest. These clones are removed from the plates and the phages or plasmids are grown in bulk to produce large quantities of the cloned DNA for further study.

Southern and Northern Blotting Procedures Analyze DNA and RNA by Hybridization

Once a DNA segment of interest has been cloned, it is useful to study its position in the genome and expression in the organism. In 1975 Ed Southern combined gel electrophoresis with hybridization to create an extremely powerful tool for detecting sequences, the so-called *Southern blotting* procedure (Fig. 4-11). Genomic DNA is cut with one or several restriction enzymes. The resultant fragments are denatured to separate the DNA strands and separated by electrophoresis on an agarose gel. The gel is then overlaid with a sheet of nitrocellulose or nylon membrane, and a flow of buffer is set up through the gel, carrying the DNA fragments onto the filter, where they bind. Thus, a replica of the distribution of DNA in the gel is created on the filter. A radioactively labeled nucleic acid probe, complementary to the sequences being sought, is then hybridized to the DNA molecules bound on the filter. Unbound probe is washed off, and the dried filter is exposed to X-ray film. The end product is an *autoradiograph* with a pattern of bands indicating the positions of DNA fragments complementary to the probe. The sizes of the fragments can be estimated by comparisons with size standards run on the same gel.

Southern blotting made possible the production of detailed restriction maps of complex genomes, in

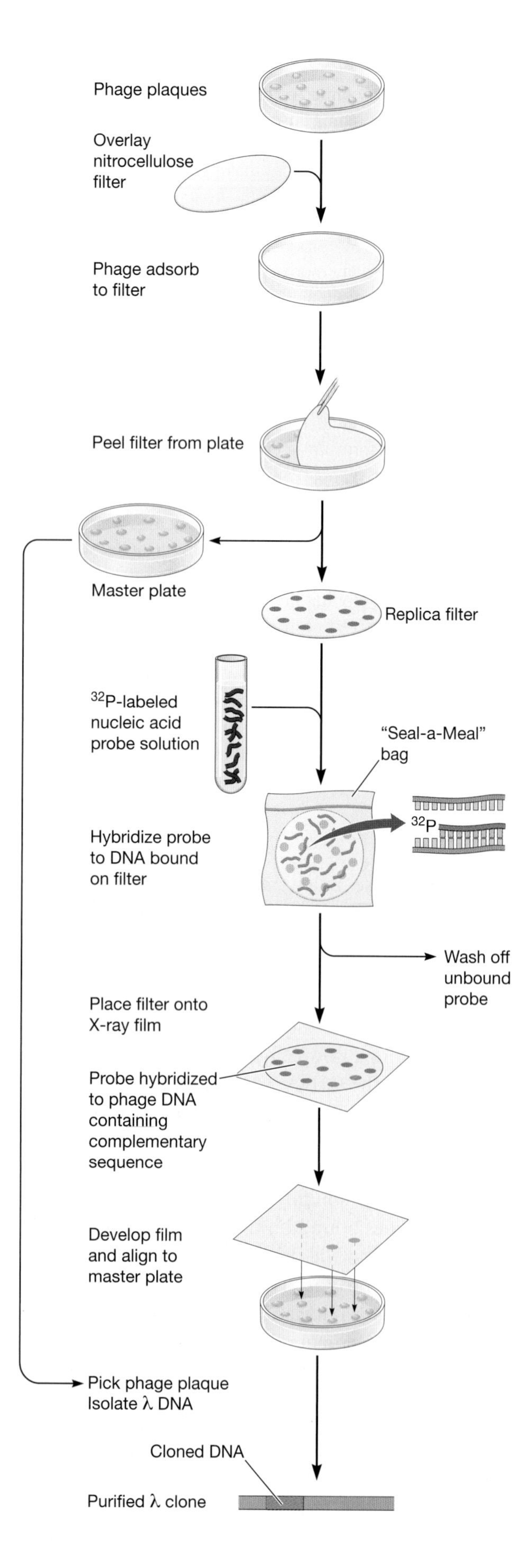

which specific fragments can be detected among the millions of other fragments produced by restriction enzyme digests. It has been used in conjunction with electrophoretic separation of very large DNA molecules to prepare restriction maps covering distances of hundreds of kilobases. In this way genes can be ordered along chromosomes, revealing the clustering of genes into functionally related groups. Southern blotting has also been useful for looking at the distribution of a gene across species. "Zoo blots" are made by hybridizing a gene probe from one species against restriction digests of genomic DNAs from a variety of species. The degree of evolutionary conservation of the gene is indicated by the range of DNAs to which the probe hybridizes.

The analogous technique for analyzing RNA is called *Northern blotting.* Instead of separating DNA on a gel, an RNA sample is electrophoresed, blotted, and hybridized with a labeled probe of interest. Northern blotting is particularly valuable for determining which tissues or cell types express a particular gene, when genes are turned on, and what factors regulate their expression. The availability of extensive sequence information now allows the reverse experiment, called a *microarray.* To construct a microarray, several hundred to hundreds of thousands of known DNA sequences (equivalent to the probes used in

FIGURE 4-10

Screening a library with a nucleic acid probe to find a clone. Libraries are typically screened by spreading several hundred thousand phages on 10–20 large agar plates covered with the host bacterial culture (lawn). After the phage plaques have grown to visible size, nitrocellulose filters are carefully laid onto the surface of the plates. Phage particles from the plate adhere to the filter, creating a precise replica on the filter of the pattern of plaques on the plate. The filters are treated to strip off phage proteins and bind the phage DNA to the filter surface. The filter is incubated in a solution containing a radioactively labeled DNA or RNA probe complementary in sequence to a portion of the gene being sought. This hybridization reaction is often carried out in sealed plastic bags. The filters are carefully washed to remove unbound probe, which leaves behind only the probe molecules tightly bound to complementary sequences within phage DNA. The location of the bound probe is determined by exposing the filters to X-ray film (autoradiography). The position is represented by a spot of exposure on the film. By orienting the film with the original agar plate, the phage plaque carrying the complementary sequence can be identified and the desired clone can be isolated.

Southern blotting) are immobilized in tiny grids and hybridized with labeled DNA or RNA. We shall describe this extremely powerful, recently developed method in more detail in Chapter 13.

Although the basics of Southern and Northern blotting have remained unchanged, the method for detecting DNA and RNA fragments in the gels has changed dramatically. Instead of using photographic film to record the positions of radioactively labeled probes, researchers use phosphorimagers, which have screens containing crystals that respond to radioactive emissions. When these are illuminated by a laser, the crystals emit light, which is recorded by a photomultiplier. This is much more rapid than using X-ray film—exposures are measured in hours rather than days—and much more sensitive. Furthermore, quantitation is more accurate because the response of the imager is linear over a wide range of radioactivity. Alternatively, probes can be labeled using non-radioactive labels. Indirect methods incorporate nucleotides carrying a small molecule such as biotin into a probe and detect this biotin using an antibody carrying alkaline phosphatase. This in turn acts on a

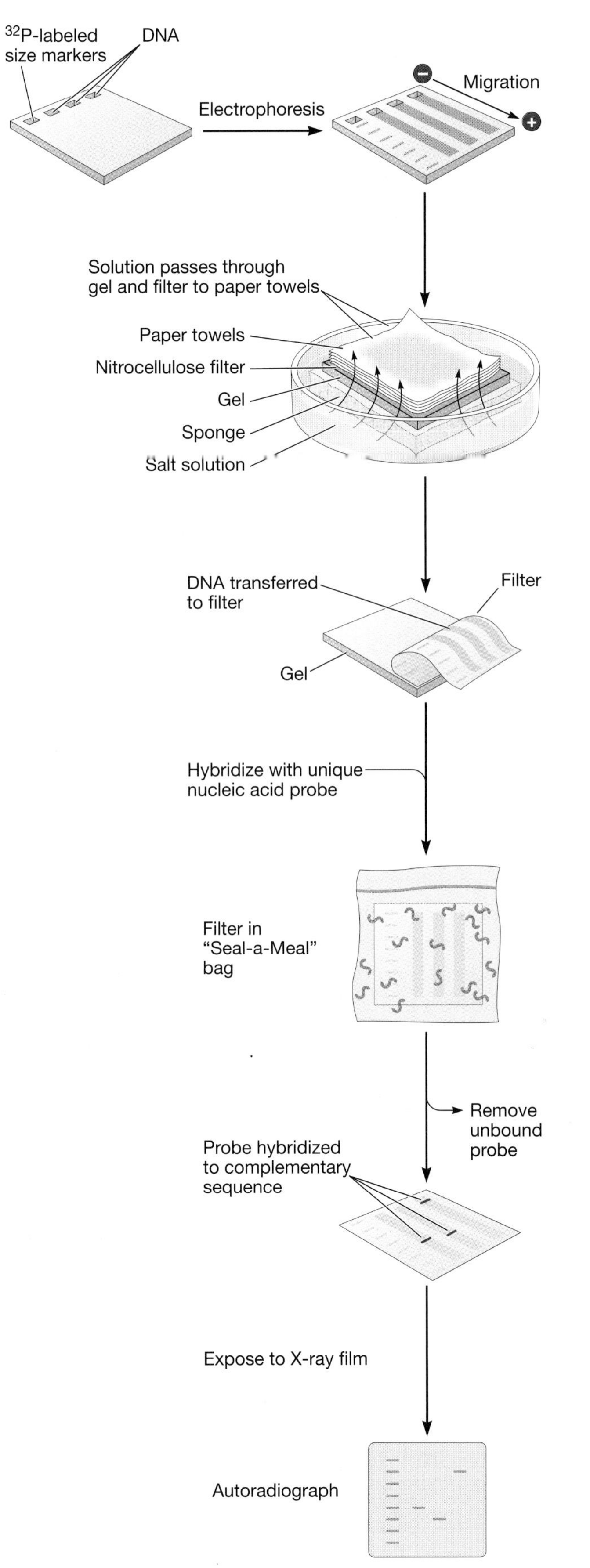

FIGURE 4-11

Southern blotting: analyzing DNA by gel electrophoresis, blotting, and hybridization. DNA cleaved with restriction enzymes is applied to an agarose gel and electrophoretically separated by size. The DNA in the gel is transferred to a nitrocellulose filter to make a precise replica of distribution of DNA as it was in the gel. This is usually done by placing the gel atop a sponge sitting in a tray of buffer. The filter is laid over the gel and covered with a stack of paper towels that acts as a wick, pulling buffer up through the sponge, gel, and filter. DNA fragments from the gel are carried up onto the filter, where they stick tightly. The filter is removed and hybridized with a radioactively labeled probe. Hybridization specifically tags the DNA fragment of interest, even though it may constitute only a minute fraction of the nucleic acids on the filter—this is the basis of the exquisite selectivity of the method. Unbound probe is washed off, and the filter is exposed to X-ray film. The position of a DNA fragment complementary to the probe appears as a band on the film. This procedure is termed Southern blotting when DNA is transferred to nitrocellulose. A very similar procedure called Northern blotting can be performed using RNA. The process is called Western blotting when protein is transferred, usually from an SDS-polyacrylamide gel rather than an agarose gel. In Western blotting, the protein of interest is visualized using an antibody that specifically recognizes it within the background of other cellular proteins.

substrate that emits light when hydrolyzed by the enzyme. Direct methods label a probe with dyes that fluoresce when stimulated by a laser.

Hybridization analogous to Southern and Northern blotting can also be performed directly on cells, a procedure called fluorescence in situ hybridization (FISH; Chapter 14). In FISH, fluorescently labeled probes are hybridized to "fixed" (chemically preserved) cells to determine the localization of a DNA or RNA sequence within the cell.

Only one other point of the compass has come into common usage, although not dealing with nucleic acids. In *Western blotting*, proteins in complex mixtures are separated by electrophoresis, usually on polyacrylamide gels, and then transferred to nitrocellulose or polyvinylidene fluoride (PVDF) filters. Specific proteins are detected by using antibodies.

Powerful Methods Are Used to Sequence DNA

Although by the mid-1970s, genes could be isolated and produced in large quantities for analysis, there was no way to read the nucleotide sequence of cloned DNA. In fact, the first nucleic acid sequences had been determined in the 1960s, but these were of the relatively small tRNA molecules, only 75–80 nucleotides in length. In the late 1960s, Frederick Sanger turned his attention from sequencing proteins to developing fast, simple procedures for sequencing larger RNA molecules. These provided a roundabout way to determine the sequence of DNA by first using RNA polymerase to synthesize a complementary RNA chain, and then sequencing this RNA. A breakthrough came a few years later with the advent of methods that allowed direct sequencing of DNA fragments between 100 and 500 nucleotides in length. The first of these, the *plus–minus* method, was developed by Sanger in 1975, and, in conjunction with using polyacrylamide gel electrophoresis (another Sanger innovation), a large part of the 5386-bp sequence of the small DNA phage ϕX174 was quickly determined (Chapter 10). Two years later, Allan Maxam and Walter Gilbert at Harvard, and Sanger and his colleagues at Cambridge, England, developed new methods that were significant advances on the plus–minus method.

Maxam and Gilbert's method was based on the chemical degradation of DNA chains. In this method, DNA fragments are labeled at one end and then divided into four separate samples. Each sample undergoes a different treatment, producing DNA fragments ending in G, C, a combination of A and G, and a combination of C and T. The DNA fragments in the completed reactions are separated by electrophoresis in polyacrylamide gels and detected by autoradiography. Merging the data from all the samples gives the complete sequence. The sequence of all the 5243 base pairs of SV40 DNA was completed quickly using this method, and that of the small recombinant plasmid pBR322 (4362 bp) was determined in less than a year by Greg Sutcliffe in Gilbert's laboratory.

It is Sanger's second method for sequencing DNA that has become the standard technique (Fig. 4-12). Specific terminators of DNA chain elongation—2′,3′-dideoxynucleoside triphosphates (ddNTPs)—were synthesized. These ddNTPs are incorporated normally into a growing DNA chain through their 5′-triphosphate groups, but they cannot join with the next incoming deoxynucleotide triphosphate (dNTP) because they lack the 3′-OH group needed to make the phosphodiester bond. When a small amount of a specific ddNTP (say, ddATP) is included in the reaction mixture along with the other three dNTPs required for DNA synthesis by DNA polymerase, the products are a series of chains that are specifically terminated wherever a ddATP is incorporated. Four reactions are set up, each with a different ddNTP; the DNA chains of each reaction are separated by electrophoresis; and the sequence is read off.

The next major advance in DNA sequencing came when Michael Hunkapiller and Leroy Hood developed a method for labeling each ddNTP with a differently colored dye, so that all four reactions could be run in a single lane on a gel. Later, semiautomated sequencing machines became available and determining the sequence of a segment of DNA became a standard procedure in every molecular biology laboratory. Continuing developments, including robots for preparing reaction samples and fully automatic capillary-based sequencing machines, have led to sequencing the entire genomes of higher organisms. At the date of this printing, the genomes of more than 1000 viruses and 300 organisms, including human, mouse, and rat, have been determined. Chapters 10, 11, and 12 provide many more details of the strategies and technologies that have enabled us to achieve feats regarded as impossible only a few years ago.

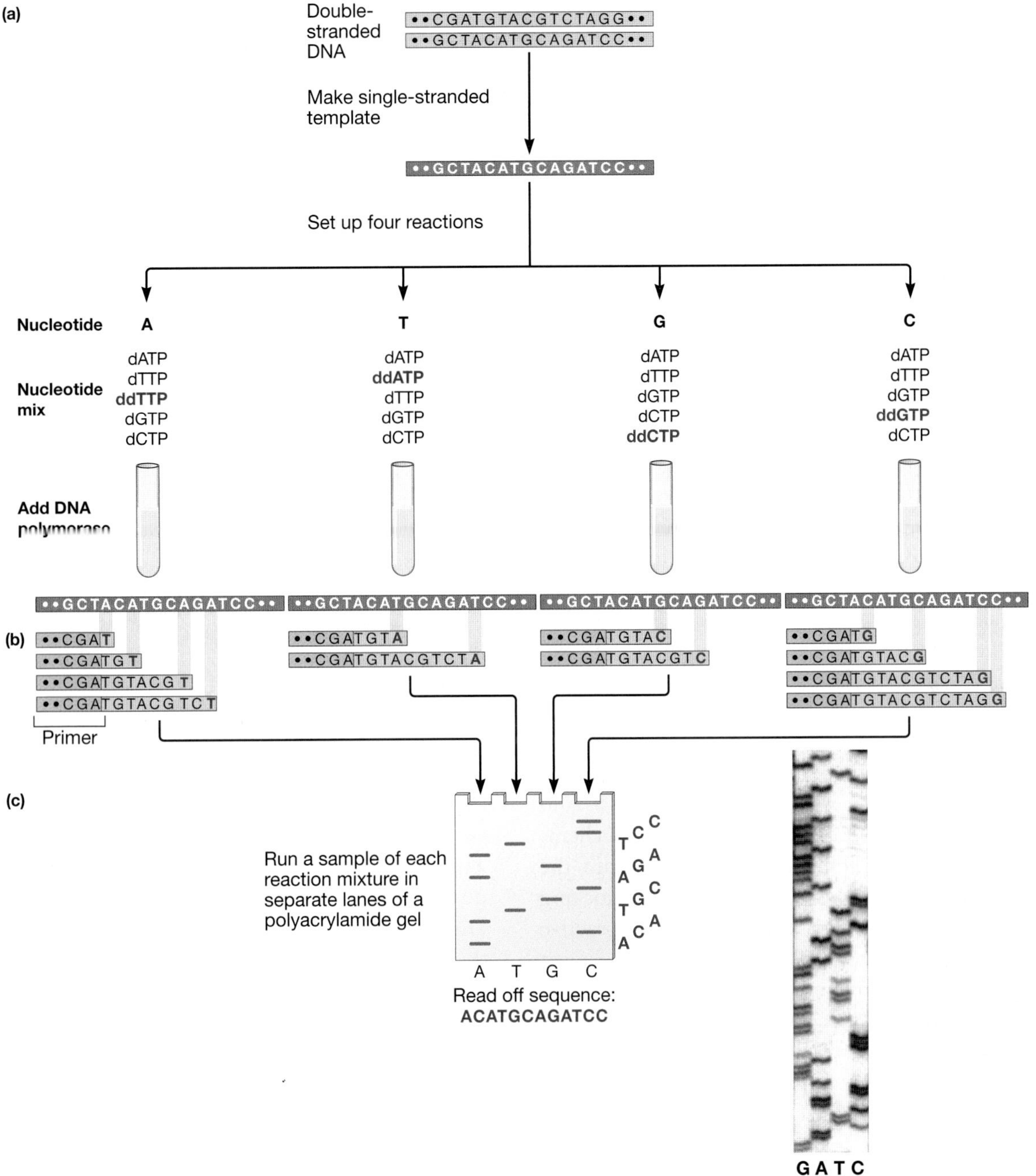

FIGURE 4-12

The Sanger dideoxy DNA sequencing procedure. 2´,3´-dideoxynucleotides of each of the four bases are prepared. These molecules can be incorporated into a growing DNA strand by DNA polymerase because they have a normal 5´-triphosphate; however, once incorporated into a growing DNA strand, the dideoxynucleotide (ddNTP) cannot form a phosphodiester bond with the next incoming dNTP. Growth of that particular DNA chain stops. (*a*) A Sanger sequencing reaction consists of a DNA strand to be sequenced, a short piece of DNA (the primer) that is complementary to the end of that strand, a carefully controlled ratio of one particular ddNTP with its normal dNTP, and the other three dNTPs. A small amount of one or more radioactive dNTPs is also included so that DNA molecules can be visualized later by autoradiography. (*b*) When DNA polymerase is added, normal polymerization will begin from the primer; when a ddNTP is incorporated by chance, the growth of that chain will stop. If the correct ratio of ddNTP:dNTP is chosen, a series of labeled strands will result, the lengths of which are dependent on the location of a particular base relative to the end of the DNA. (*c*) The resulting labeled fragments are separated by size on an acrylamide gel, and autoradiography is performed; the pattern of the fragments gives the DNA sequence. Generally, in modern sequencing methods the products are visualized by detecting fluorescence from dyes conjugated to the chain terminating ddNTPs. By using a different dye for each ddNTP, a single reaction can be done and analyzed on a single lane of a gel.

Oligonucleotides Are Synthesized Chemically

The ready availability of oligonucleotides of defined sequence is critical for many recombinant DNA techniques. Methods for linking nucleotides were worked out during the 1970s and this early phase culminated with the synthesis of a complete alanine tRNA by H. Gobind Khorana's laboratory in 1972.

The emergence of quick, convenient methods for the synthesis of moderately long oligonucleotides with defined sequences depended on three technical advances. The first followed the example of peptide synthesis and attached the growing nucleotide chain to a solid support. The polymerases in cells synthesize DNA and RNA in a 5′ to 3′ direction, whereas it is most frequently used for in vitro oligonucleotide synthesis done in reverse, from 3′ to 5′. Thus, the 3′-hydroxyl of the first base is tethered to the solid support. This leaves just one end of the molecule available for reactions, and adding and washing off reagents is much easier with a tethered molecule.

The second advance was the development of improved chemistries for protecting both the 5′ and the 3′ end of a mono- or oligonucleotide from chemical reactions by attaching *blocking groups* onto the nucleotides used in DNA synthesis. These blocking groups are additional atoms linked to the reactive hydroxyl and amine groups on the nucleotides, which prevent unwanted chemical reactions between the next added base and the growing oligonucleotide chain. After synthesis is complete, the blocking groups are removed, leaving a DNA molecule chemically identical to those synthesized in cells.

The third advance came with the development of programmable machines that synthesize oligonucleotides using phosphoramidite chemistry and solid supports. Until then, synthesis of oligonucleotides was a time-consuming process that was limited to linking together fewer than 20 nucleotides. Now, completely automated machines synthesize oligonucleotides as long as 100 bases in a few hours. The limiting factor is the progressively lower yield of full-length oligonucleotides with increasing length.

Genes Can Be Synthesized from Oligonucleotides

Oligonucleotides are essential for many techniques and they have also been used to synthesize genes. Because the yield of each reaction step of oligonucleotide synthesis is less than 100%, there are limits on the maximum length of synthetic oligonucleotides. Complete genes have to be made by stitching together multiple oligonucleotides and the first gene thus synthesized, in 1970, was for a tRNA. Synthesis of genes encoding small peptide hormones such as the 42-bp human hormone somatostatin and the two chains of human insulin, 63 and 90 bp, soon followed. At that time, chemical synthesis provided a way for researchers to bypass the prohibitions against the insertion of human DNA into a bacterium. Although those prohibitions were soon lifted, chemical synthesis continued to be a practical, although somewhat expensive, way to create genes and, by the 1980s, genes such as tissue plasminogen activator, 1610 bp in length, were being made. Further developments led to the synthesis of the complete genomes of poliovirus (7440 bp) and of the bacteriophage ϕX174 (5386 bp).

The Polymerase Chain Reaction Amplifies Specific Regions of a DNA Target

The *polymerase chain reaction* technique (PCR) was devised by Kary Mullis in the mid-1980s and, like DNA sequencing and synthesis, has revolutionized the practice of molecular genetics; those who learn the craft today find it inconceivable that any research was done prior to 1988 when PCR became a reliable and convenient tool. There are so many variations of the PCR technique, and so many applications, that we can provide only a sampling of them in this chapter. Here, we will go over the basic features of PCR, whereas many of the experiments described in later chapters will provide examples of the applications of PCR.

A major problem in analyzing genes is that they are rare targets in a complex genome—the human genome, for example, has about 23,000 genes. Various procedures were devised to isolate and detect these rare targets, but such techniques were generally complex and time consuming. PCR changed this by enabling us to produce enormous numbers of a specified DNA sequence without resorting to cloning. PCR uses DNA polymerase to synthesize multiple copies of a DNA sequence ("amplify") determined by the primers needed to initiate DNA synthesis.

The starting material for a PCR is a DNA sample containing the sequence to be amplified. DNA preparation for a PCR is relatively easy except for the most demanding of samples. The target DNA sequences do not have to be isolated from other DNA because they are defined by the oligonucleotide primers used in the reaction. In fact, DNA released by disrupting cells by boiling or by detergents can be used directly without any purification. Because PCR amplifies DNA, the amount of DNA needed to initiate a PCR is very small—even a single DNA molecule will suffice. Remarkable sources of DNA for PCR include biopsies embedded in paraffin for more than 40 years; blood samples, taken by heel prick of newborns, for the neonatal detection of phenylketonuria and stored as dried spots on cards; and even the tooth of a 10- to 11-year-old Neanderthal child dating from 100,000 years ago.

DNA polymerase uses single-stranded DNA as a template for the synthesis of a complementary new strand. These single-stranded templates are produced simply by heating double-stranded DNA to near boiling. DNA polymerase requires a small section of double-stranded DNA to initiate ("prime") synthesis. Therefore, the starting point for DNA synthesis in vitro can be specified by supplying an oligonucleotide primer that anneals to the template at that point (Fig. 4-13).

Both DNA strands of a double helix can serve as templates for synthesis, provided an oligonucleotide primer is supplied for each strand. The primers are chosen to flank the region of DNA that is to be amplified, so that the newly synthesized strands of DNA, starting at each primer, extend beyond the position of the primer on the opposite strand. Therefore new primer binding sites are generated on each newly synthesized DNA strand. The reaction mixture is heated to separate the strands, the primers anneal to the newly available binding sites, and new chains are synthesized. The cycle of heating, primer binding, and extension is repeated (Figs. 4-14 and 4-15), so that at the end of n cycles, the reaction mixture contains a theoretical maximum of 2^n double-stranded DNA molecules that are copies of the sequence between the primers (Table 4-3). In practice, the typical yield is 10–30% of the theoretical maximum. Important factors include the length and

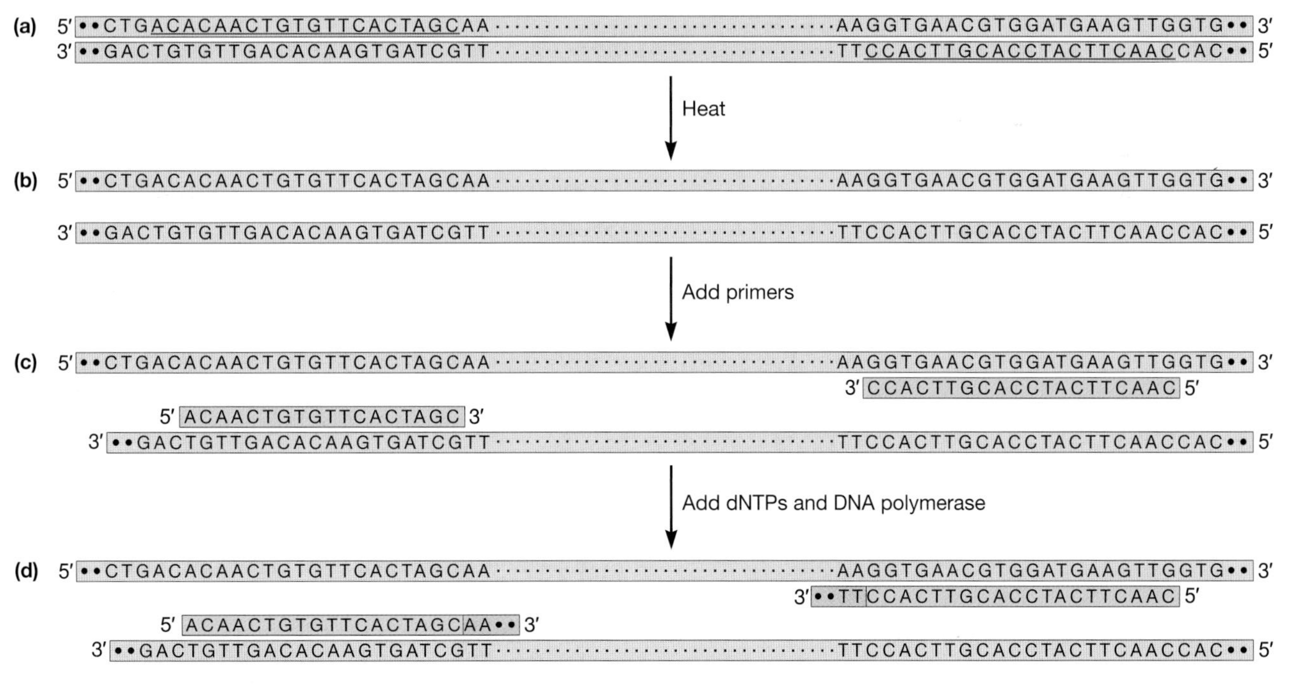

FIGURE 4-13

Primers for DNA polymerase. (*a*) The target for amplification, a small section covering 110 bp of the β-globin gene, is shown. Two sequences separated by 60 nucleotides are detailed. Oligonucleotide primers for the PCR are made to hybridize to the two sets of 20 nucleotides (*underlined*). (*b*) When the DNA is heated, the strands separate. (*c*) The oligonucleotide primers (shown in *green*) hybridize specifically to their complementary sequences at the 3′ ends of each strand of the target sequence. (*d*) DNA polymerase uses these primers to begin synthesis of new strands (shown in *pink*) complementary to the target DNA sequences in the 5′ to 3′ directions.

composition of the sequence to be amplified, because DNA polymerase can falter on long sequences or in regions of high GC content. Nevertheless, PCR can produce so many copies that the DNA can be easily visualized on an agarose gel stained with ethidium bromide rather than having to resort to Southern blotting.

Optimizing a PCR requires the adjustment of many parameters, including temperature, salt concentrations, and cycle durations, but the primers are the key factor—it is they that direct where synthesis of new DNA strands will begin. The specificity of primer binding depends on the length of the primers and their nucleotide composition. Oligonucleotides between 18 and 24 nucleotides long tend to be highly specific under standard conditions where the temperature of the primer-binding step (called *annealing*) is close to the temperature at which the primers dissociate from the template DNA (the *melting temperature*, T_m). It is also important to choose primer pairs with similar T_ms, so that both anneal at the chosen temperature. Increasing primer length increases specificity, but at the price of increasing the T_m. If the T_m becomes too high, the required annealing temperature will exceed the temperature for polymerase extension and may risk inaccurate priming.

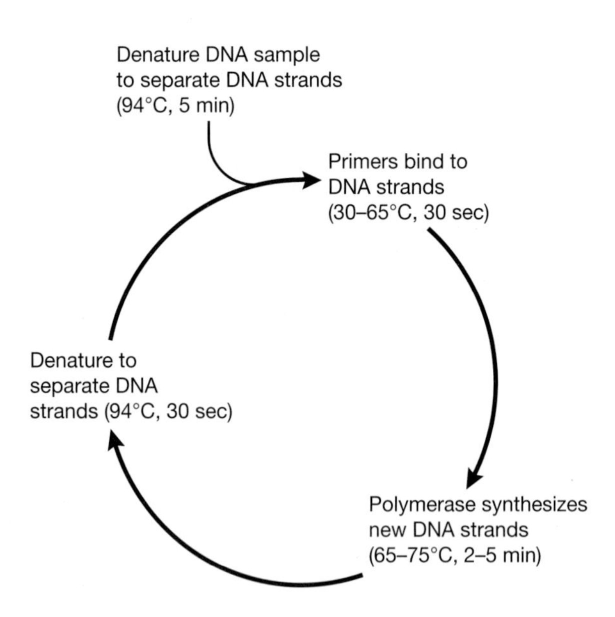

FIGURE 4-14

The PCR cycle. The DNA sample is heated to separate the DNA strands (initial denaturation), and then the reaction mixture goes through repeated cycles of primer annealing, DNA synthesis, and denaturation. The target sequence doubles in concentration for each cycle.

Thermostable Polymerases Simplify and Improve PCR

Once the extraordinary power of PCR had been recognized, many improvements were made in the basic technique. Undoubtedly, the most significant was the use of DNA polymerases from thermophilic bacteria that live at very high temperatures. Because *E. coli* DNA polymerase is destroyed at the temperatures needed to separate double-stranded DNA, fresh enzyme had to be added for each cycle of the reaction. However, the DNA polymerase from the bacterium *Thermus aquaticus*, which lives in water at a temperature of 75°C, has a temperature optimum of 72°C and is reasonably stable even at 94°C. Called Taq polymerase, this enzyme remains active through a complete set of amplification cycles and has enabled automation of PCR using *thermocyclers*, machines designed to carry out the time and temperature cycles for a PCR. For processing very large numbers of samples, the PCRs are set up using robots. This is essential for genome projects where millions of PCRs are carried out without any manual intervention.

Contamination Can Be a Problem in PCR Studies

An unwelcome corollary of the amplification power of the PCR is that minor contamination of the starting material can have serious consequences. This is true especially for critical applications, such as amplifying DNA isolated in forensic cases where the samples taken at a crime scene are rarely in the pristine state of samples used to prepare DNA in the laboratory, as we will see in Chapter 16.

An essential control is to carry out a PCR amplification without template DNA to test for contaminating DNA in the reaction reagents. One source of contamination is the products of previous amplification reactions. A completed PCR mixture may contain as many as 10^{13} amplified fragments, so that even minute volumes, such as droplets in an aerosol from a pipette tip, contain very large numbers of amplifiable molecules. A simple and very effective

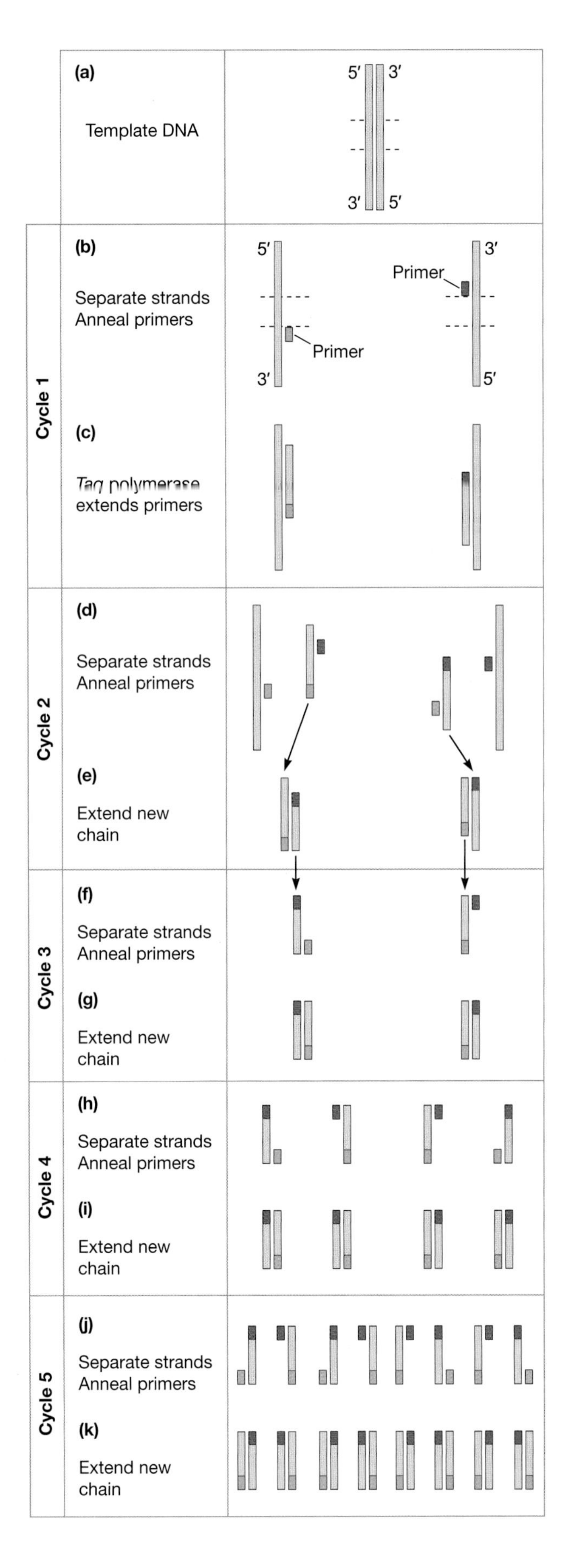

way to reduce contamination is to physically separate pre- and postamplification steps so that amplified reactions are never in the same area that is used for setting up the reactions. DNA can also be made inactive as a template for PCR. Often, areas in the lab set aside for PCR are treated with UV light. This cross-links DNA, thus creating intrastrand bridges that stop the polymerase as well as breaking DNA strands.

More sophisticated methods can be used for especially critical studies or to deal with persistent contamination. A clever method is based on the incorporation of deoxyuracil instead of thymidine into PCR products. (Remarkably, this has little effect on the properties of the amplified DNA except that many restriction enzyme sites are lost.) When a new reaction is set up, the enzyme uracil-*N*-glycosylase is included and this degrades any previously amplified DNA contaminating the new reaction by destroying the uracils. dUTP can be used in the new reaction because uracil-*N*-glycosylase enzyme is inactivated by the high temperature in the new PCR.

FIGURE 4-15

The exponential increase of DNA copies during PCR. (*a*) The starting material is a double-stranded DNA molecule. The region to be amplified is marked by *dashed lines.* (*b*) The strands are separated by heating the reaction mixture and then cooling so that the primers anneal to the two primer-binding sites flanking the target region, one on each strand. (*c*) *Taq* polymerase synthesizes new strands of DNA, complementary to the template, that extend a variable distance beyond the position of the primer-binding site on the other template. (*d*) The reaction mixture is heated again; the original and newly synthesized DNA strands separate. Four binding sites are now available to the primers, one on each of the two original strands and the two new DNA strands. (To simplify the diagram, subsequent events involving the original strands are omitted.) (*e*) *Taq* polymerase synthesizes new complementary strands, but the extension of these chains is limited precisely to the target sequence. The two newly synthesized chains thus span exactly the region specified by the primers. (*f*) The process is repeated, and primers anneal to the newly synthesized strands (and also to the variable-length strands, but these are omitted from the figure). (*g*) *Taq* polymerase synthesizes complementary strands, producing two double-stranded DNA fragments that are identical to the target sequence. (*h–k*) The process is repeated and the number of target fragments doubles for each subsequent cycle of the reaction.

TABLE 4-3. Amplification of molecules during polymerase chain reaction

Cycle number	Number of double-stranded target molecules
1	0
2	0
3	2
4	4
5	8
6	16
7	32
8	64
9	128
10	256
11	512
12	1,024
13	2,048
14	4,096
15	8,192
16	16,384
17	32,768
18	65,536
19	131,072
20	262,144
21	524,288
22	1,048,576
23	2,097,152
24	4,194,304
25	8,388,608
26	16,777,216
27	33,554,432
28	67,108,864
29	134,217,728
30	268,435,456
31	536,870,912
32	1,073,741,824

Fidelity of DNA Synthesis Determines the Accuracy of PCR Amplification

Contamination is not the only possible source of error in PCR. Like all other biochemical processes, DNA replication is not a perfect process, and occasionally DNA polymerase will add an incorrect nucleotide to the growing DNA chain. The rate of misincorporation measured in a naturally replicated DNA molecule is approximately 1 in 10^9 nucleotides. Cells achieve such extraordinary accuracy because the DNA replication machinery removes mismatched nucleotides added to the DNA chain. In vitro, *Taq* polymerase does not have this "proofreading" capability, and using the temperatures and salt concentrations typical in a PCR, the enzyme incorporates one incorrect nucleotide for about every 2×10^4 nucleotides incorporated. This is not a serious matter for bulk analysis of PCR products because molecules with the same misincorporated nucleotide will form a minute proportion of the total number of molecules synthesized. But misincorporation is important if PCR fragments are to be used for cloning, where each clone is derived from a single amplified molecule. If this molecule contains one or more misincorporated nucleotides, then all DNA isolated from that clone will carry the identical mutation. This problem can be reduced by beginning the PCR with a large, rather than a small, number of template molecules—fewer cycles of amplification are needed and less total DNA synthesis takes place. Another way to reduce the error rate is to use different DNA polymerases. Polymerases from many thermophilic bacterial species have been isolated for use in PCR, and some of these, such as *Pfu* DNA polymerase, do have proofreading 3′ to 5′ exonuclease activity. The exonuclease activity removes the mismatched nucleotides, which allows the polymerase another opportunity to synthesize the correct sequence.

PCR Will Amplify Sequences from a Single DNA Molecule

One of the most remarkable applications of PCR is in performing linkage analysis using sperm. Human genetic linkage analysis is difficult because it has to be carried out using families where the number of available offspring is often much lower than the geneticist would like for statistical analysis. However, examining 1000 sperm is like studying a family with 1000 children, because each sperm is the result of a meiotic division. Linkage analysis can be done by choosing a male heterozygous for the loci to be tested and determining what proportion of his sperm cells show recombination between the loci. But a sperm cell is haploid, containing only one copy of each chromosome, and each chromosome is a single molecule of DNA. Linkage analysis using sperm requires PCR to work effectively and reliably in a reaction mixture containing a single target DNA molecule!

In the example given here, alleles at three loci were analyzed—two on chromosome 19 (DS19S49 and *APOC2*) and one on chromosome 9 (D9S52) (Fig. 4-16). These loci contain microsatellite repeats, stretches of DNA where a sequence is repeated over and over again. D19S49 has repeats

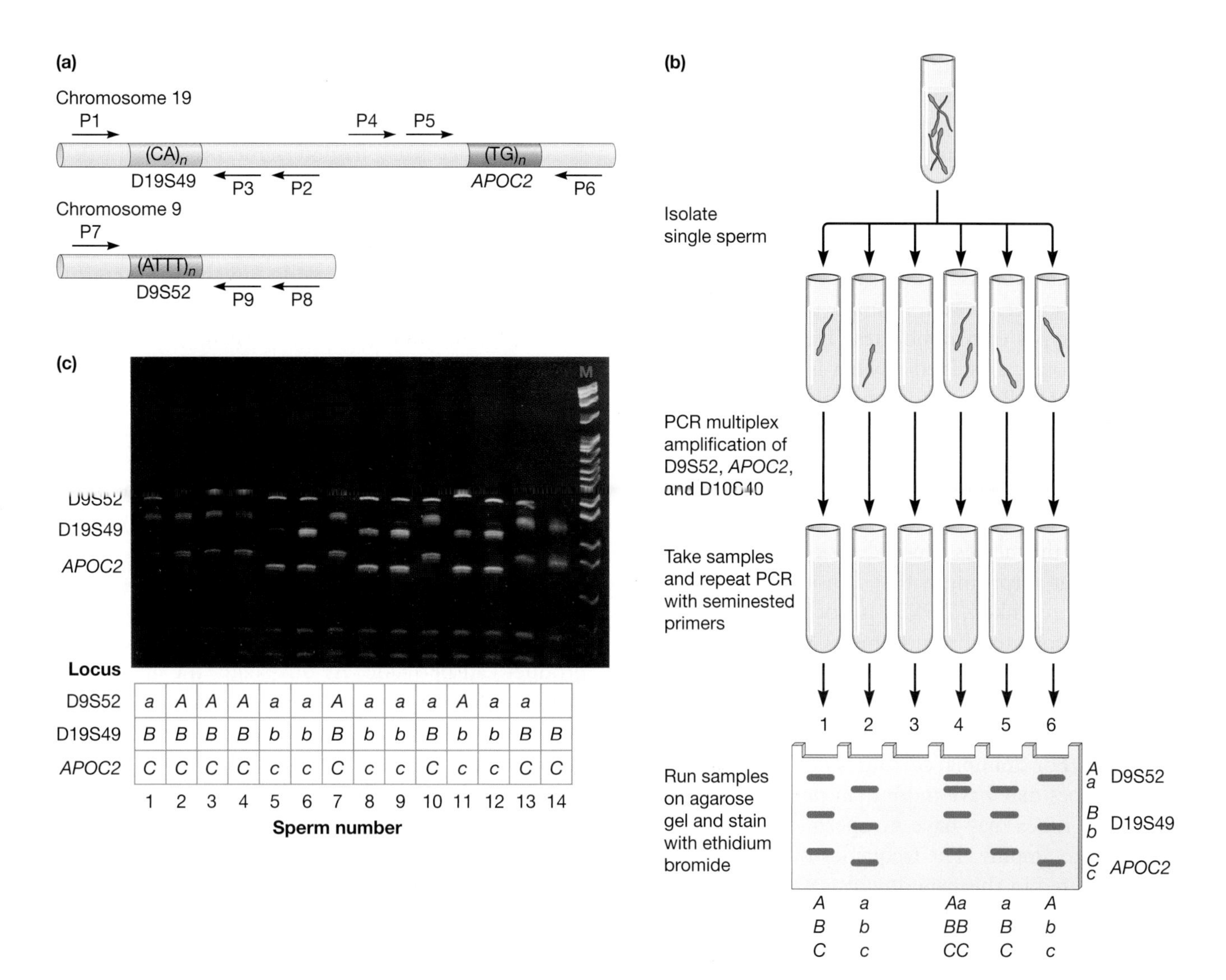

Locus	1	2	3	4	5	6	7	8	9	10	11	12	13	14
D9S52	*a*	*A*	*A*	*A*	*a*	*a*	*A*	*a*	*a*	*a*	*A*	*a*	*a*	
D19S49	*B*	*B*	*B*	*B*	*b*	*b*	*B*	*b*	*b*	*B*	*b*	*b*	*B*	*B*
APOC2	*C*	*C*	*C*	*C*	*c*	*c*	*C*	*c*	*c*	*C*	*c*	*c*	*C*	*C*

FIGURE 4-16

Amplifying single DNA molecules in linkage analysis using single sperm. (*a*) Three polymorphic regions were used: D19S49 and *APOC2* on chromosome 19 and D9S52 on chromosome 9. The polymorphisms arise from differing numbers of repeat units at each locus: (CA) at D19S49; (TG) at *APOC2*; and (ATTT) at D9S52. Sets of three primers specific for each locus were synthesized. (*b*) A sperm sample was obtained from a man who was heterozygous for each locus—$(CA)_{15}$ and $(CA)_{21}$; $(TG)_{18}$ and $(TG)_{22}$; and $(ATTT)_{12}$ and $(ATTT)_{13}$. Single sperm were transferred to tubes, and amplification of the three loci was carried out simultaneously using the outer primers for each locus (P1 and P2; P4 and P6; P7 and P8). Samples were taken and the amplification repeated using one outer and one inner primer (P1 and P3; P5 and P6; P7 and P9). Samples were run on an acrylamide gel and stained with ethidium bromide. There should be three bands in each lane, corresponding to DNA amplified at each locus, and the haplotypes of each sperm for these loci can be read off from the gel. Thus, the sperm in lane 1 is *ABC*, whereas that in lane 2 is *abc*. However, lane 3 has no bands, presumably because the tube contained no sperm, whereas there are two bands corresponding to both alleles of the D9S52 locus in lane 4. This sample presumably had two sperm in it—*ABC* and *aBC*. (*c*) Data from an experiment. The gel and its analysis are shown. One amplification product—for D9S52—is missing in lane 14. It is not unusual for such "dropout" to occur in multiplex reactions. The data show linkage between D19S49 and *APOC2*. For this man, chromosome 19 has allele *B* of D19S49 associated with allele *C* of *APOC2*, and allele *b* with allele *c*. There is no linkage between alleles of D9S52 and D19S49-*APOC2* because they are on different chromosomes.

of CAs, *APOC2* has repeats of TG, and D9S52 has repeats of ATTT. The alleles at these sites result from differing numbers of repeats, the minimum difference being one repeat, equivalent to two nucleotides at *APOC2* and D19S49 and four nucleotides at D9S52. Sperm was obtained from men who were heterozygous at all three loci, and single sperm were isolated using a fluorescence-activated cell sorter, more commonly used for preparing different types of lymphocytes. The three loci were simultaneously amplified in what is termed a *multiplex reaction*. Primers for multiplex PCR must be carefully designed so that each PCR product has a distinct size and so that the different primer sets anneal to the template and not to one another. Samples were electrophoresed using polyacrylamide gels and the alleles were determined. Inspection shows, not surprisingly, that although *APOC2* and D19S49 are linked, D9S52 is not linked because it is on a separate chromosome. Calculation of the recombination frequency between the two linked loci requires statistical treatment of the data because technical complications have to be taken into account. For example, sorting is not perfect, so that some tubes may have more than one sperm, whereas other tubes may have no sperm, and sometimes loci fail to amplify. The frequency of recombination as calculated in this experiment was 0.083, which is in good agreement with the value determined from similar analysis performed on 40 families. These estimates of genetic distances are for *male* chromosomes, which is an important point, given that recombination frequencies for genes differ for male and female meioses.

Real-Time PCR Is Used for the Rapid and Accurate Quantitation of RNA and DNA

It is often necessary to determine the amount of a particular DNA or RNA molecule in a sample. This can be done by measuring the intensities of bands on Southern or Northern blots, but this is not a very sensitive method; PCR provides much more versatile and accurate methods.

In general, the greater the number of target molecules in the starting sample, the more amplified molecules will be present at each step of the reaction. By measuring the final PCR products on a gel, it is possible to get some idea of how much of the target molecule was present at the beginning of the reaction. However, this is not so easy to do reliably. If the quantities of primers and nucleotides are too low, or there are too many cycles of PCR, the primers and nucleotides in the reaction will be exhausted and the reaction will reach a plateau after which no more product can be made. At this stage, the amount of DNA made no longer reflects how much of the target molecule was present initially. A much better strategy is to use quantitative real-time PCR methods that measure the amplification of the target molecule continuously during a PCR. After an initial phase where amplification is undetectable, the PCR enters an exponential phase in which the product is almost doubling at each step. If more target molecules are present at the start, fewer cycles will be required to reach the exponential phase. Comparison of the number of cycles required to reach this point allows the determination of the initial template concentration of the reaction (Fig. 4-17).

FIGURE 4-17

Real-time PCR is used to measure the quantity of target sequences in a sample. (*a,b*) Plots of the theoretical accumulation of DNA molecules during a PCR. In *a*, the scale is linear, and the curve is close to the baseline of cycle 25 because the number of molecules is low compared to later cycles. In *b*, the accumulation of molecules is plotted on a logarithmic scale, producing a linear plot because of the exponential increase. In both cases the curves flatten by cycle 32. (*c,d*) Plots of experimental data. The PCR products were detected by the dye SYBR Green, a nonspecific dye that fluoresces when intercalated into double-stranded DNA. The amount of PCR product is so low in the early cycles that it cannot be detected above background fluorescence. Note that the best fit to the exponential accumulation shown in the theoretical logarithmic graph (*b*) is during cycles 18–24 and it is in this period, when the reaction is behaving truly exponentially, that comparison of template concentrations should be made. Note that the linear portion of the plot in *c*—cycles 26–30—is *not* the correct time to do quantitative determinations! (*e*) Tenfold dilutions of the starting DNA concentrations have been compared using real-time PCR and plotted on a linear scale. The greater the amount of template, the fewer the number of cycles are needed to first be able to detect the product. Note that the reactions reach different steady-state levels that do not reflect the initial differences in template concentration; the reaction corresponding to the *light blue* curve has 1000-fold more template than the *red* reaction, yet produces less final product. This reflects the effects of small, random variations during the initial stages of the reaction, which become amplified during the later stages, and emphasizes the importance of making comparisons at the correct times.

Several different approaches are available for determining the amount of PCR product present at the end of each cycle, but all rely on measuring the amount of a fluorescent tag that is associated with each newly synthesized molecule. Early tags included ethidium bromide and SYBR Green I, both of which insert in the grooves of double-stranded DNA. A more recent approach uses oligonucleotide probes that hybridize to an internal sequence of the target molecule. One type of probe is called a *molecular beacon*, which is an oligonucleotide carrying a fluorophore at one end and a nonfluorescent quencher at the other. The molecular beacon is designed so that the fluorophore and quencher come together, thus forming a hairpin. When illuminated in this configuration with UV

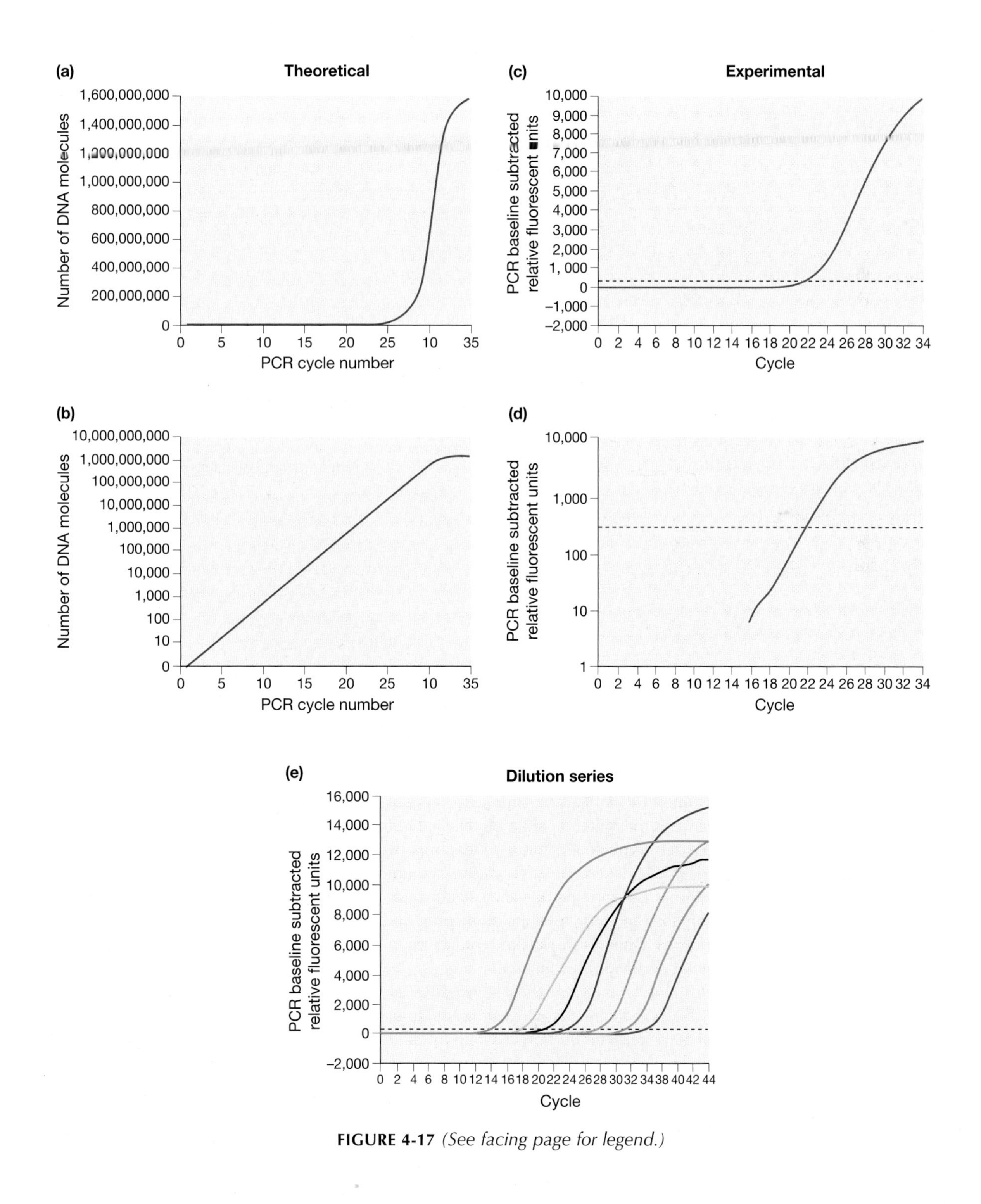

FIGURE 4-17 *(See facing page for legend.)*

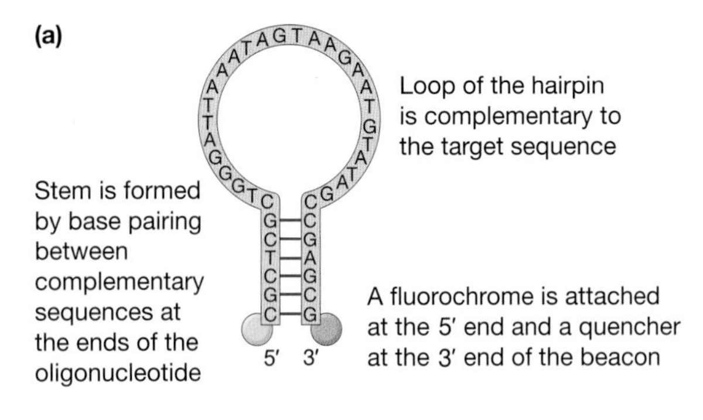

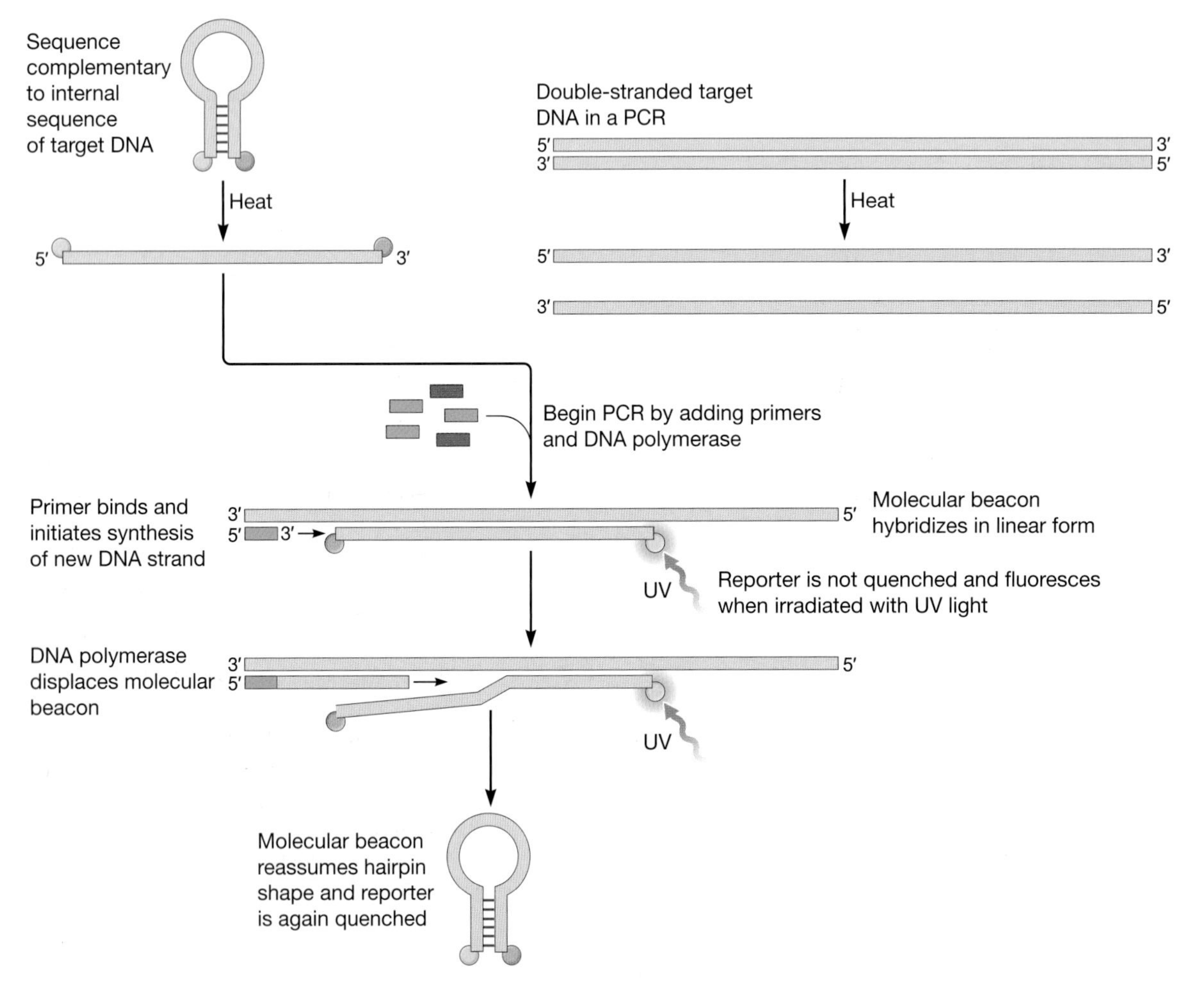

FIGURE 4-18

Molecular beacons for real-time PCR. (*a*) A molecular beacon is a short (30-bp to 40-bp) oligonucleotide synthesized to be complementary to an internal site of the segment to be amplified. The beacon has a fluorescent reporter dye attached to its 5´ nucleotide and a quenching dye attached to its 3´ nucleotide. The first few, usually six, 5´ nucleotides and the last 3´ nucleotides are complementary so that the beacon folds onto itself, forming a hairpin structure that brings the dye and quencher into close proximity. This association permits the transfer of energy from the reporter dye to the quenching dye, so that the quenching dye absorbs the energy emitted by the reporter dye when it is irradiated, preventing any emission of the energy as a fluorescent signal. (*b*) When the PCR is heated to dissociate the target DNA strands, the beacon hairpins also melt, producing a linear molecule. During the annealing step of PCR, when the reaction is cooled to allow hybridization of the primers, the molecular beacon hybridizes to the target strand. The reporter and quencher dyes are now separated, so that the reporter fluoresces when irradiated. Because the reporter fluoresces only when bound to the target sequence, the fluorescence signal is proportional to the quantity of target molecules in the PCR. The beacon does not interfere with the amplification process because it is displaced from the target as the DNA polymerase moves along its template strand. It is then available for the next cycle.

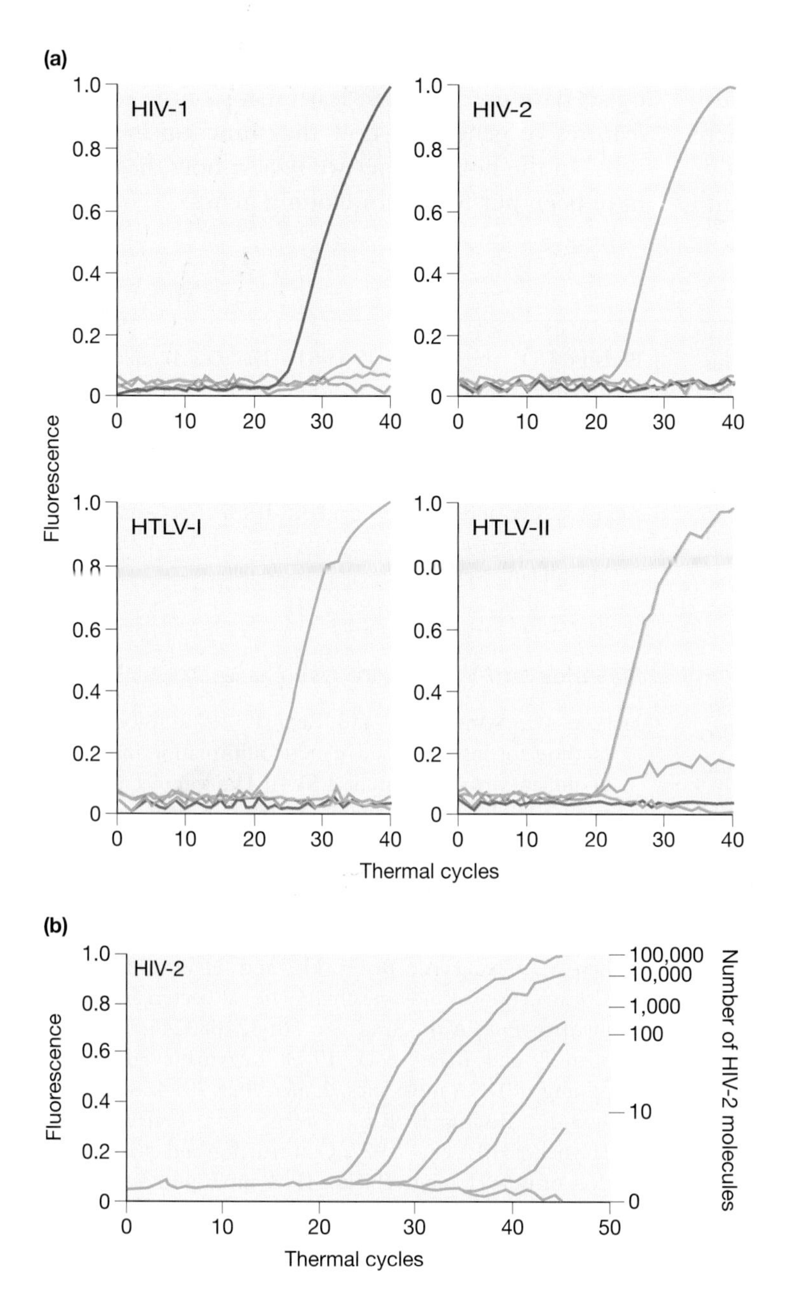

FIGURE 4-19

Detecting different human retroviruses. Primers were designed to amplify the *gag* gene of HIV-1, the *env* gene of HIV-2, the *tax* gene of HTLV-I, and the *pol* gene of HTLV-II. Oligonucleotides specific for conserved internal sequences of each amplified viral sequence were synthesized and each labeled with a different fluorochrome and the same non-fluorescent quencher. (*a*) The four primer–molecular beacon combinations were used in the same reaction with one virus. The specificity of the combinations is clear; the HIV-1 primers (in *purple*) amplify only HIV-1. (*b*) The technique can detect very small numbers of one virus in the presence of large numbers of a second. Here the PCR contained a constant number of HTLV-I molecules (100,000) and tenfold dilutions of HIV-2 molecules, ranging from 100,000 to 10. The number of PCR cycles needed to develop a measurable signal is inversely proportional to the starting number of HIV-2 molecules, and as few as ten HIV-2 molecules can be detected in the presence of 100,000 HTLV-I molecules.

light, energy from the fluorophore is absorbed by the quencher in a process called fluorescence resonance energy transfer (FRET), and no light is emitted. However, when the probe is heated and hybridizes to the amplified DNA sequence, fluorophore and quencher are separated, and now the former emits light when irradiated (Fig. 4-18). The increasing fluorescence measured as the reaction proceeds is a measure of the increase in the number of amplified molecules.

Multiplex PCR has been combined with molecular beacon technology for a very sensitive means of detecting and quantitating human retroviruses—HIV-1, HIV-2, HTLV-I, and HTLV-II. By using different fluorochromes, multiple beacons can be used to report on PCR products amplified from the different viruses in a single reaction. As few as ten molecules of one virus can be detected in the presence of 100,000 molecules of a second virus (Fig. 4-19).

Recombinant DNA Comes of Age

Recombinant DNA technology has fulfilled its promise many times over. What were once esoteric and difficult techniques, limited to a small number of laboratories, now can be learned readily from one of many manuals and there is a service industry to supply the enzymes, oligonucleotides, and other reagents that once had to be prepared painstakingly by individual researchers. The tools of recombinant DNA are used routinely around the world and biologists have used them to learn many of the essential features of genes and how they function in organisms. In the next chapter we review how these tools have been put to use for cloning genes.

Reading List

The best resource for much of the material covered in this chapter is Sambrook and Russell (2001). Although it is a laboratory manual, the introductions to each chapter contain extensive descriptions of the underlying principles of cloning.

Sambrook J. and Russell D.W. 2001. *Molecular Cloning: A Laboratory Manual*, 3rd ed. Cold Spring Harbor Laboratory Press, Cold Spring Harbor, New York.

Restriction Enzymes and Maps

Linn S. and Arber W. 1968. Host specificity of DNA produced by *Escherichia coli*, X. In vitro restriction of phage fd replicative form. *Proc. Natl. Acad. Sci.* **59:** 1300–1306.

Meselson M. and Yuan R. 1968. DNA restriction enzyme from *E. coli*. *Nature* **217:** 1110–1114.

Kelly T.J., Jr., and H.O. Smith. 1970. A restriction enzyme from *Hemophilus influenzae*, II. Base sequence of the recognition site. *J. Mol. Biol.* **51:** 393–409.

Smith H.O. and Wilox K.W. 1970. A restriction enzyme from *Hemophilus influenzae*, I. Purification and general properties. *J. Mol. Biol.* **51:** 379–391.

Danna K. and Nathans D. 1971. Specific cleavage of simian virus 40 DNA by restriction endonuclease of *Hemophilus influenzae*. *Proc. Natl. Acad. Sci.* **68:** 2913–2917.

Sharp P.A., Sugden B., and Sambrook J. 1973. Detection of two restriction endonuclease activities in *Hemophilus parainfluenza* using analytical agarose-ethidium bromide electrophoresis. *Biochemistry* **12:** 3055–3062.

Roberts R.J., Vincze T., Posfai J., and Macelis D. 2003. REBASE: Restriction enzymes and methyltransferases. *Nucleic Acids Res.* **31:** 418–420.

The Recombinant DNA Debate

Singer M. and Soll D. 1973. Guidelines for DNA hybrid molecules. *Science* **181:** 1174.

Berg P., Baltimore D., Boyer H.W., Cohen S.N., Davis R.W., et al. 1974. Potential biohazards of recombinant DNA molecules. *Science* **185:** 303.

Berg P., Baltimore D., Brenner S., Roblin R.O., and Singer M.F. 1975. Asilomar conference on recombinant DNA molecules. *Science* **188:** 991–994.

Norman C. 1976. Genetic manipulation: Guidelines issued. *Nature* **262:** 2–4.

Rogers M. 1976. *Biohazard*. Knopf, New York.

Department of Health, Education, and Welfare. 1980. Guidelines for research involving recombinant DNA molecules. *Federal Register*, Tuesday, January 29.

Watson J.D. and Tooze J. 1981. *The DNA Story*. W.H. Freeman, San Francisco. (A sourcebook for documents on the recombinant DNA debate.)

Krimsky S. 1983. *Genetic Alchemy*. MIT Press, Cambridge, Massachusetts.

Zilinskas R.A. and Zimmerman B.K. 1986. *The Gene-splicing Wars: Reflections on the Recombinant DNA Controversy*. Macmillan, New York.

Plasmids and Viruses Are Used as Vectors

Jackson D., Symons R., and Berg P. 1972. Biochemical method for inserting new genetic information into DNA of simian virus 40: Circular SV40 DNA molecules containing lambda phage genes and the galactose operon of *Escherichia coli*. *Proc. Natl. Acad. Sci.* **69:** 2904–2909.

Mertz J.E. and Davis R.W. 1972. Cleavage of DNA by RI restriction endonuclease generates cohesive ends. *Proc. Natl. Acad. Sci.* **69:** 3370–3374.

Cohen S., Chang A., Boyer H., and Helfing R. 1973. Construction of biologically functional bacterial plasmids in vitro. *Proc. Natl. Acad. Sci.* **70:** 3240–3244.

Lobban P. and Kaiser A.D. 1973. Enzymatic end-to-end joining of DNA molecules. *J. Mol. Biol.* **79:** 453–471.

Bolivar F., Rodrigues R.L., Greene P.J., Betlach M.C., Heyneker H.L., et al. 1977. Construction and characterization of new cloning vehicles. II. A multi-purpose cloning system. *Gene* **2:** 95–113.

Yanisch-Peron C., Vieira J., and Messing J. 1985. Improved M13 phage cloning vectors and host strains: Nucleotide sequences of the M13mpl8 and pUC19 vectors. *Gene* **33:** 103–119.

Hohn B. and Murray K. 1977. Packaging recombinant DNA molecules into bacteriophage particles in vitro. *Proc. Natl. Acad. Sci.* 74: 3259–3263.

Blattner F.R., Williams B.G., Blechl A.E., Denniston-Thompson K., Faber H.E., et al. 1977. Charon phages: Safer derivatives of bacteriophage lambda for DNA cloning. *Science* **196:** 161–169.

Huynh T.V., Young R.A., and Davis R.W. 1985. Constructing and screening cDNA libraries in lgt10 and lgt11. In *DNA Cloning: A Practical Approach* (ed. D.M. Glover), vol. 1, pp. 49–78. IRL Press, Oxford.

Short J.M., Fernandez J.M., Sorge J.A., and Huse W.D. 1988. Lambda ZAP: A bacteriophage lambda expression vector with in vivo excision properties. *Nucleic Acids Res.* **16:** 7583–7600.

Cloning

Aviv H. and Leder P. 1972. Purification of biologically active globin messenger RNA by chromatography on oligothymidylic acid-cellulose. *Proc. Natl. Acad. Sci.* **69:** 1408–1412.

Maniatis T., Kee S.G., Efstratiadis A., and Kafatos F.C. 1976. Amplification and characterization of a β-globin gene synthesized in vitro. *Cell* **8:** 163–182.

Okayama H. and Berg P. 1982. High-efficiency cloning of full length cDNA. *Mol. Cell. Biol.* **2:** 161–170.

Soares M.B., Bonaldo M.D.F., Jelene P., Su L., Lawton L., and Efstratiadis A. 1994. Construction and characterization of a normalized cDNA library. *Proc. Natl. Acad. Sci.* **91:** 9228–9232.

Carninci P., Shibata Y., Hayatsu N., Sugahara Y., Shibata K., et al. 2000. Normalization and subtraction of cap-trapper-selected cDNAs to prepare full-length cDNA libraries for rapid discovery of new genes. *Genome Res.* **10:** 1617–1630.

Grunstein M. and Hogness D.S. 1975. Colony hybridization: A method for the isolation of cloned DNAs that contain a specific gene. *Proc. Natl. Acad. Sci.* **72:** 3961–3965.

Southern and Northern Blotting

Southern E.M. 1975. Detection of specific sequences among DNA fragments separated by gel electrophoresis. *J. Mol. Biol.* **98:** 503–517.

Southern E.M. 2000. Blotting at 25. *Trends Biochem. Sci.* **25:** 585–588. (Southern provides the inside story of how he devised his eponymous blot.)

Alwine J.C., Kemp D.J., and Stark G.R. 1977. Method for detection of specific RNAs in agarose gels by transfer to diazobenzyloxymethyl-paper and hybridization with DNA probes. *Proc. Natl. Acad. Sci.* **74:** 5350–5354.

Sequencing DNA

Sanger F. and Dowding M. 1996. *Selected Papers of Frederick Sanger*. World Scientific, Singapore. (Facsimile printings of 48 of Sanger's most important papers, with fascinating commentaries by Sanger.)

Sanger F., Nicklen S., and Coulson A.R. 1977. DNA sequencing with chain-terminating inhibitors. *Proc. Natl. Acad. Sci.* **74:** 5463–5467.

Sanger F. and A.R. Coulson. 1975. A rapid method for determining sequences in DNA by primed synthesis with DNA polymerase. *J. Mol. Biol.* **94:** 444–448.

Sanger F., Air G.M., Barrel B.G., Brown N.L., Coulson A.R., et al. 1977. Nucleotide sequence of bacteriophage ϕX174. *Nature* **265:** 678–695.

Maxam A.M. and Gilbert W. 1977. A new method of sequencing DNA. *Proc. Natl. Acad. Sci.* **74:** 560–564.

Sutcliffe J.G. 1979. Complete nucleotide sequence of the *Escherichia coli* plasmid pBR322. *Cold Spring Harbor Symp. Quant. Biol.* **43:** 77–90.

Sutcliffe J.G. 1995. pBR322 and the advent of rapid DNA sequencing. *Trends Biochem. Sci.* **20:** 87–90. (A personal history of the early days of sequencing.)

Oligonucleotide Synthesis and Making Genes

Gait M.J. and Sheppard R.C. 1977. Rapid synthesis of oligodeoxy-ribonucleotides: A new solid-phase method. *Nucleic Acids Res.* **4:** 1135–1158.

Beaucage S.L. and Caruthers M.H. 1981. Deoxynucleotide phosphoramidites—A new class of key intermediates for deoxypolynucleotide synthesis. *Tetrahedron Lett.* **22:** 1859–1862.

Heyneker H.L., Shine J., Goodman H.M., Boyer H., Rosenberg J., et al. 1976. Synthetic *lac* operator is functional in vivo. *Nature* **263:** 748–752.

Khorana H.G. 1979. Total synthesis of a gene. *Science* **203:** 614–625.

Cello J., Pau A.V., and Wimmer E. 2002. Chemical synthesis of poliovirus cDNA: Generation of infectious virus in the absence of natural template. *Science* **297:** 1016–1018

Polymerase Chain Reaction

Mullis K.B. 1990. The unusual origin of the polymerase chain reaction. *Sci. Am.* **262:** 56–65.

Saiki R.K., Gelfand D.H., Stoffel S., Scharf S.J., Higuchi R., et al. 1988. Primer-directed enzymatic amplification of DNA with a thermostable DNA polymerase. *Science* **239:** 487–491.

Mullis K. and Faloona F. 1987. Specific synthesis of DNA in vitro via a polymerase catalyzed chain reaction. *Methods Enzymol.* **55:** 335–350.

McCabe E.R.B., Huang S.-Z., Seltzer W.K., and Law M.L. 1987. DNA microextraction from dried blood spots on filter paper blotters: Potential applications to newborn screening. *Hum. Genet.* **75:** 213–216.

Higuchi R., von Beroldingen C.H., Sensabaugh G.F., and Erlich H.A. 1988. DNA typing from single hairs. *Nature* **332:** 543–546.

Krings M., Stone A., Schmitz R.W., Krainitzki H., Stoneking M., and Pääbo S. 1997. Neandertal DNA sequences and the origin of modern humans. *Cell* **90:** 19–30.

Chien A., Edgar D.B., and Trela J.M. 1976. Deoxyribonucleic acid polymerase from the extreme thermophile *Thermus aquaticus*. *J. Bacteriol.* **127:** 1550–1557.

Saiki R.K., Scharf S.J., Faloona F., Mullis K.B., Horn G.T., Erlich H.A., and Arnheim N. 1985. Enzymatic amplification of beta-globin sequences and restriction site analysis for diagnosis of sickle cell anemia. *Science* **230:** 1350–1354.

Kwok S. and Higuchi R. 1989. Avoiding false positives with PCR (review). *Nature* **339:** 237–238.

Longo M.C., Berringer M.S., and Hartley J.L. 1990. Use of uracil DNA glycosylase to control carry-over contamination in polymerase chain reactions. *Gene* **93:** 125–128.

Keohavong P. and Thilly W.G. 1989. Fidelity of DNA polymerases in DNA amplification. *Proc. Natl. Acad. Sci.* **86:** 9253–9257.

Eckert K.A. and Kunkel T.A. 1990. High fidelity DNA synthesis by the *Thermus aquaticus* DNA polymerase. *Nucleic Acids Res.* **18:** 3739–3744.

Hubert R., Weber J.L., Schmitt K., Zhang L., and Arnheim N. 1992. A new source of polymorphic DNA markers for sperm typing: Analysis of microsatellite repeats in single cells. *Am. J. Hum. Genet.* **51:** 985–991.

Vet J.A. and Marras S.A. 2005. Design and optimization of molecular beacon real-time polymerase chain reaction assays. *Methods Mol. Biol.* **288:** 273–290.

Kramer R. and Tyagi S. 1996. Molecular beacons: Probes that fluoresce upon hybridization. *Nat. Biotechnol.* **14:** 303–308.

Vet J.A.M., Majitha A.R., Marras S.A., Tyagi S., Dube S., Poiesz B.J., and Kramer F.R. 1999. Multiplex detection of four pathogenic retroviruses using molecular beacons. *Proc. Natl. Acad. Sci.* **96:** 6394–6399.

CHAPTER 5

Fundamental Features of Eukaryotic Genes

The picture of the gene largely remained unchanged during the 1960s. It was based essentially on the model proposed by François Jacob and Jacques Monod for prokaryotes; there was a sequence of DNA that carried the genetic code for a protein, and there were associated sequences that did not code for a protein but that bound regulatory molecules and modified transcription of the gene. This model was derived from analyses of bacterial gene expression, but Jacob and Monod believed that it was likely to apply to the cells of higher organisms. The scheme of repressors, operators, and structural genes could account for the regulation and specificity of gene expression that are key features of eukaryotic cells. As Jacob remarked, "What is true of *E. coli* is true of elephants."

However, even before recombinant DNA techniques were used to dissect genes, evidence was accumulating that prokaryotic and eukaryotic genes were not the same. Density gradient ultracentrifugation of eukaryotic DNA revealed that in addition to the main band of genomic DNA, there were thin bands of DNA that were named "satellite" DNA. Because this DNA reannealed very rapidly, it was assumed to be made up of very large numbers of repeated sequences. Prokaryotic and eukaryotic cells also differed in the properties of their RNAs. It was found that RNAs in the nucleus—heterogeneous nuclear RNAs—were much larger than mRNAs found in the cytoplasm, which suggested that the primary RNA transcript of a eukaryotic gene underwent some form of processing before being translated.

These were inferences from indirect analysis of DNA and RNA, but recombinant DNA and other techniques that were developed in the early 1970s provided the means for analyzing the molecular structures of eukaryotic genes and genomes. It is not surprising that these revealed many unexpected and important features of gene structure and regulation of gene expression. In this chapter, we bring together a variety of studies that illustrate how the concept of the gene has changed from the clear definition of Jacob and Monod to a much more complex and subtle understanding.

Split Genes Are Discovered

The first indication that all genes did not conform to the Jacob and Monod model based on prokaryotic genetics came from studies of gene expression and gene mapping in adenovirus. During adenovirus replication, precursors of viral RNA transcripts (pre-mRNAs) within the nucleus of an infected cell were found to be shortened by removal of one or more internal sections to produce smaller mRNA molecules. These processed or "spliced" mRNAs moved to the cytoplasm, where they served as templates for viral protein synthesis. The gene segments missing from cytoplasmic mRNAs were identified in electron micrographs of RNA–DNA heteroduplexes between adenovirus DNA and isolated RNA transcripts. The surprising and exciting discovery of mRNA splicing was first announced at the 1977 Cold Spring Harbor Laboratory Symposium on Chromatin. Quickly, splicing was shown to occur in another virus, SV40, and the question immediately arose whether splicing might also be involved in the processing of cellular RNA (Fig. 5-1). For several years it had been known that many eukaryotic mRNAs are first synthesized as

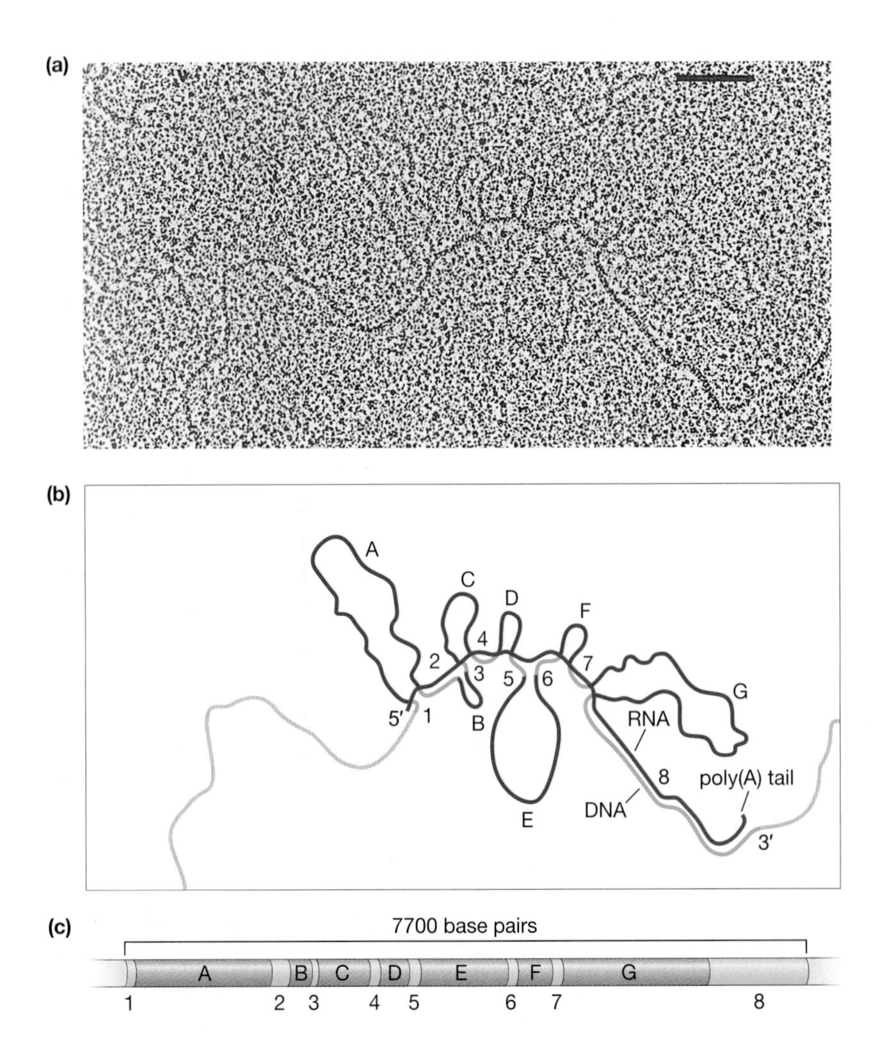

FIGURE 5-1

Examination of gene structure by electron microscopy of DNA–mRNA hybrids. (*a*) DNA containing the gene for ovalbumin mRNA and prepared for electron microscopy with uranyl acetate. Regions where genomic DNA hybridized to mRNA form a thicker line than do the single-stranded genomic DNA loops. The scale bar represents 0.086 μm. (*b*) Interpretation of the electron micrograph based on subsequent sequencing of the exon–intron boundaries in genomic DNA and comparison with cDNA sequence. The 5′ and 3′ ends of the mRNA are labeled. The ovalbumin mRNA (*purple*) hybridizes to the eight exons of the ovalbumin gene DNA (1–8, *blue*). The DNA loops out where there are introns in the gene (A–G, *red*) because there are no complementary sequences in the mRNA. The poly(A) tail does not hybridize to the DNA. (*c*) Arrangement of exons and introns in the ovalbumin gene.

large pre-mRNAs that are later processed in the nucleus to much smaller products. However, until the announcement of adenovirus splicing, it had always been assumed that this processing necessarily and exclusively involved removal of long sections at the 5′ and 3′ ends of pre-mRNA. Excited by the results on viral RNAs, investigators working on the structure of eukaryotic genes searched for evidence of splicing of cellular RNAs.

Eukaryotic Cells Splice Pre-mRNAs to Remove Introns and Produce Mature mRNAs

At the time that the electron microscopy experiments on adenovirus were done, few cellular genes had been cloned, so proof that chromosomal genes were spliced also came initially from these studies. For example, studies of heteroduplexes of *Drosophila* ribosomal genes and mRNAs showed that these coding sequences were also interrupted by noncoding DNA. The regions of the chromosomal DNA not present in the mature mRNA were given the name *introns*. The coding sequences were called *exons*, because the processed mRNAs, without the introns, "exit" from the nucleus to the cytoplasm. Once the first genes were cloned, introns could be identified by comparing the cloned genomic DNA with the corresponding cloned complementary DNA (cDNA). Early on, these comparisons had to be made by Southern blot hybridization of probes to restriction fragments of cDNA and chromosomal DNA. For example, HaeIII restriction enzyme digestion of rabbit β-globin cDNA produced one restriction fragment, 333 bp long, but a similar digest of chromosomal DNA produced a fragment about 800 bp long. The "missing" base pairs come from an intron between the second and third exons. The precise size of this rabbit β-globin intron was known because the rabbit β-globin gene was one of the first to be sequenced. As DNA sequencing became more widespread, the sizes of the exons and introns in many other eukaryotic genes and the locations of the intron–exon boundaries were determined precisely by sequencing cloned genomic DNA and comparing the sequence with the amino acid sequence of the corresponding protein.

Introns exist in genes from all eukaryotic animals, in plant genes, and, surprisingly, in some genes of the *Escherichia coli* bacteriophage T4. However, some genes, such as the genes coding for the α and β forms of interferon and most of the genes from the yeast *Saccharomyces cerevisiae*, do not contain introns. Often the introns of a gene contain many more nucleotides than coding exons, thus accounting for the previously unexplained large sizes of many primary mRNA transcripts. The lengths of introns can vary widely; the smallest human introns known are less than 100 nucleotides, whereas the human dystrophin gene has an intron more than 210,000 nucleotides long. Although the numbers and sizes of introns vary widely from one gene to another, the arrangement of introns within a gene is often similar, or *conserved*, between closely related organisms. For instance, two introns are present in all genes of the human β-globin family, which encode the adult form of the β chain of hemoglobin. The sizes of introns in the β-globin gene from different species differ slightly, but their positions are always the same relative to the coding sequence (Fig. 5-2).

Intronless mammalian genes can also be generated using recombinant DNA methods and tested to see how they function in vivo. For example, the introns of pre-mRNAs for adenovirus major late gene, *fushi tarazu* (a *Drosophila* development gene), and other pre-mRNAs were deleted precisely, and both the pre-mRNAs and their deleted counterparts were injected into *Xenopus* oocytes. After incubation the nuclei were dissected from the oocytes, and the distribution of the RNAs between nucleus and cytoplasm determined. (The very large size of *Xenopus* oocytes makes this possible.) Remarkably, although the mRNAs spliced in vivo from the pre-mRNA and deleted pre-mRNA are identical in sequence, the former were much more efficiently transported to the cytoplasm. Further analysis showed that after incubation in in vitro splicing reactions, pre-mRNA and deleted mRNA formed very different complexes, suggesting that proteins involved in export were loaded onto the pre-mRNA during splicing. It is now clear that splicing is an essential step in preparing many mRNAs for export. During splicing, a number of proteins such as YRA1 in yeast and Aly in higher eukaryotes become associated with the mRNA. These proteins interact with the transport mechanism that moves processed mRNAs through the nuclear pore and into the cytoplasm.

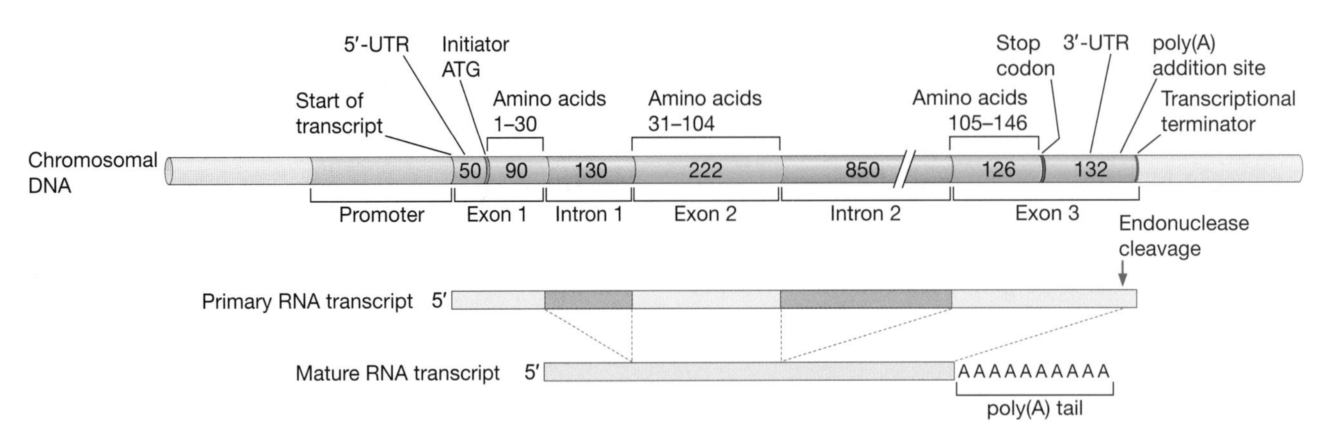

FIGURE 5-2

Organization of the human β-globin gene. The human β-globin gene encodes the adult form of the β chain of hemoglobin. The β-globin gene from all mammals consists of three exons interrupted by two introns. The numbers within the *colored* segments indicate the number of nucleotides in each exon or intron. The primary transcript (pre-mRNA) contains the exon and intron sequences. The introns are removed by splicing enzymes to form the mature mRNA. The 5′ and 3′ exons each contain an untranslated region (UTR). The 5′- and 3′-UTRs contain sequences important for specifying RNA processing and translation events. The sequence AAUAAA, near the 3′ end of the primary transcript, directs an endonuclease to cleave the RNA 15–30 nucleotides farther along the molecule. The end generated by this cleavage is the site for the addition of a string of As, the "poly(A) tail."

Specific Sequences and Protein Factors Ensure Very Accurate Splicing

By the summer of 1978, just a year after the first split genes were discovered, the sequences of many exon–intron boundaries from cellular genes had been determined. Researchers had expected to find that the sequence at the upstream (5′) and downstream (3′) ends of an intron would be complementary. These sequences would therefore be expected to hybridize and form a stretch of double-stranded RNA, which would be recognized and precisely excised by specific splicing enzymes. However, this idea was soon discounted when it was shown that the sequences at the upstream and downstream splice sites were not complementary. Yet the base sequences at the boundaries between exons and introns were not random and, after many boundaries had been sequenced, a pattern emerged. The sequences at the 5′ ends of introns were similar to each other, as were the sequences at the 3′ ends of introns (Fig. 5-3). There is also a third, less well-defined, sequence essential for splicing, found within the intron. This is the branch point site found 20–50 nucleotides 5′ to the 3′ splice site, within the intron. The branch point sequence in yeast is UACUAAC, but it is very variable between species, and the consensus sequence is YNYURAC, where Y is a C or T; R is a G or A; and N can be any base. However, the A is invariant because this A takes part in the splicing reaction that removes introns and joins exons; mutations of this branch site A prevent splicing.

The number of sequences involved in splicing regulation continues to grow. The most recent additions are splicing enhancers and silencers. The former, exonic splicing enhancers (ESEs), are found in almost every exon and may be especially important in regulating alternative splicing—that is, determining which exons are selected for splicing among the several that may be present in a gene (see the following section). Putative ESEs were identified by comparing exon and nonexon sequences. Site-directed mutagenesis was used to make single-base-pair changes in 22 of these sequences and in 24 control sequences. The mutations in the control sequences had no effect on splicing, but splicing was significantly reduced in 18 out of the ESE sequences. These bind a class of splicing factors called SR proteins. (These proteins are characterized by having serine-arginine dipeptide repeats; the single-letter codes for serine and arginine are S and R.) The SR proteins have an RNA-binding domain that binds the mRNA, while the SR domain in turn binds components of the splicing machinery and determines which exons are included or excluded by preventing exon skipping. In some way, the complex of SR and other proteins at an ESE inhibits the splicing of the upstream and downstream flanking exons. Conversely, there are exonic splicing

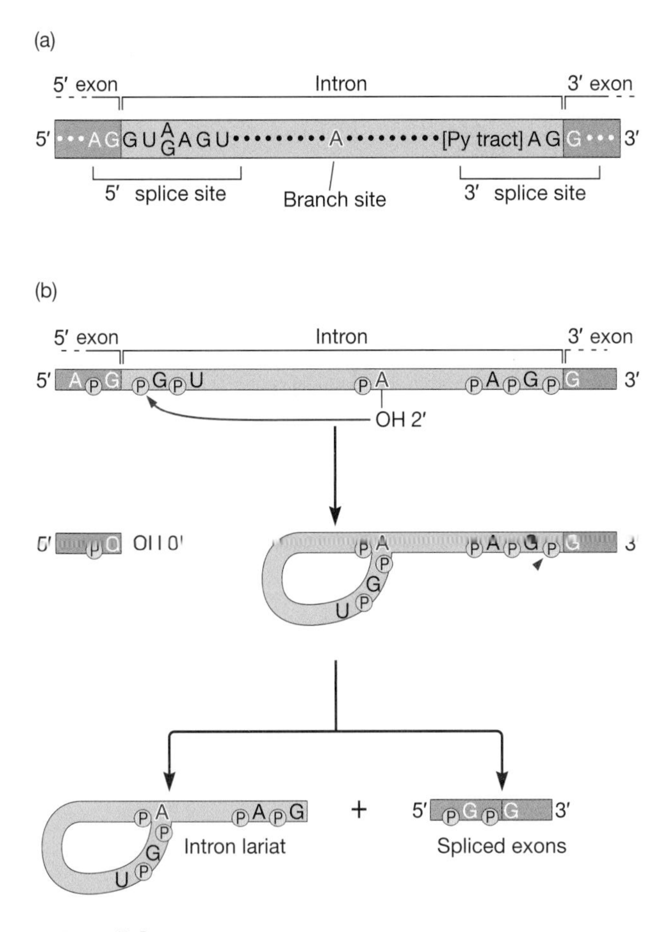

FIGURE 5-3

Sequences at exon–intron boundaries, and within the intron, determine the sites of splicing. (*a*) Consensus sequences of 5′ and 3′ splice junctions in eukaryotic mRNAs. Almost all introns begin with G:U and end with A:G. From the analysis of many exon–intron boundaries, extended consensus sequences of preferred nucleotides at the 5′ and 3′ ends have been established. In addition to exon–intron boundary sequences, a less well-defined sequence—the branch site—just upstream of the 3′ splice junction is important for splicing. The key nucleotide in the branch site sequence is an A. (*b*) This A is important because it takes part in the splicing reaction. The hydroxyl group from the ribose of the branch site A attacks the phosphoryl group of the G, the last nucleotide of the exon, leading to cleavage of the sugar-phosphate backbone of the DNA. Next, the newly exposed hydroxyl group of the exonic G reacts with the phosphoryl group of the G that is the first nucleotide of the next exon. This joins the two exons and releases the intron in a lariat-like shape.

silencers that in some, as yet not well-understood, fashion suppress splicing. It is possible that the default state of some exons is to be silenced (i.e., excluded from the mRNA) unless the silencing is relieved by the binding of splicing factors to splicing enhancers.

Splicing is carried out by a complex of proteins and RNAs that make up a molecular machine called the spliceosome. The involvement of RNA in eukaryotic splicing had been suspected from studies of transfer RNA (tRNA) processing in *E. coli*, which was shown to be accomplished by a complex of RNA and protein. The nuclei of eukaryotic cells contain very large numbers of small nuclear RNAs (snRNAs), five of which (U1, U2, U4, U5, and U6) associate with proteins to form small nucleoprotein particles (snRNPs), each named for the snRNA it contains. It is these snRNPs, together with other proteins, that constitute the spliceosome. The U1 snRNP is the first component to bind to the splice site. The nucleotide sequence of the 5′ end of the U1 RNA is complementary to the 5′ consensus sequence of a splice site and it binds there, to be followed by the U2 snRNP binding to the branch site. The other three snRNPs assemble, and the complex brings together the 3′ and 5′ ends of the exons to be spliced. It is the U6 snRNP that removes the intron and joins the exons (Fig. 5-3).

Alternative Splicing Creates Different, but Related, mRNAs from a Single Gene

Many primary gene transcripts can be spliced in different ways to produce distinct RNA molecules that each encode a different protein. Differential splicing was first demonstrated in mRNAs from adenovirus, SV40, and polyoma virus and subsequently in mRNAs from many cellular genes. Alternative splicing often produces two forms of the same protein that are necessary at different stages of development or in different cell types. For example, immunoglobulins of the IgM class exist as either a membrane-bound protein displayed on the cell surface or as a soluble protein secreted into the blood. The membrane-bound form is expressed first during B-cell development and then, as the B cell differentiates into a plasma cell, expression of the membrane-bound form ceases and the secreted form is produced. Immunoglobulins are a complex of four protein molecules, two "heavy" and two "light." By direct analysis of the secreted and membrane-bound antibody proteins, it was demonstrated that the two forms of antibodies contain different heavy chains. Analysis of cloned antibody cDNAs showed that this difference is due to alternative splicing of the heavy-chain gene transcript. The two heavy-chain mRNAs differ only at their 3′ ends (Fig. 5-4). The B cell–specific mRNA contains two exons that encode

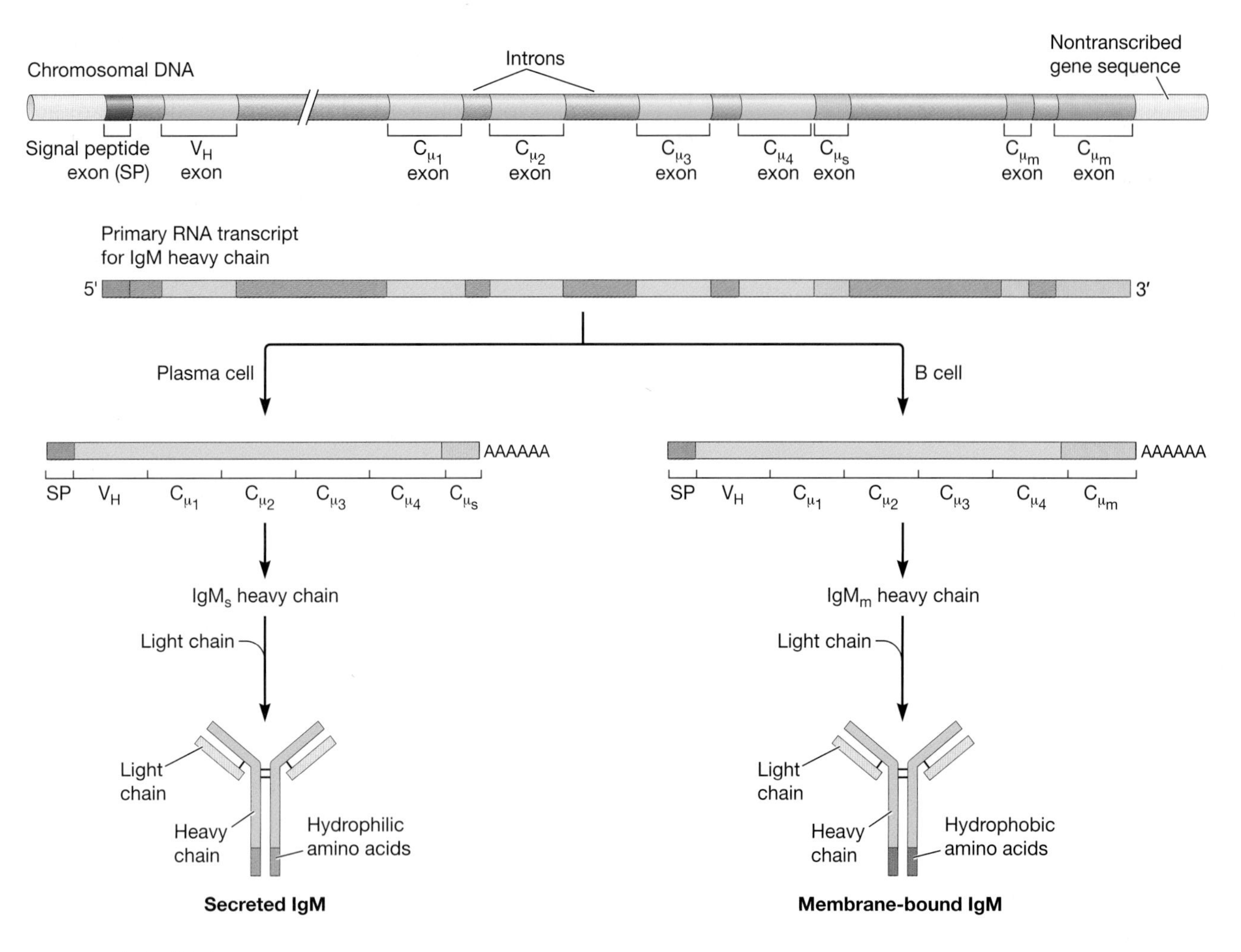

FIGURE 5-4

Alternative splicing produces secreted and membrane-associated forms of IgM from a single gene. The μ gene encoding the heavy chain for an IgM molecule is shown. The heavy and light chains of an antibody are composed of a series of structural *domains*. The organization of an immunoglobulin gene parallels this domain structure of the protein. For example, in the heavy-chain gene shown here, the coding sequences for the signal peptide (SP)—amino acids at the amino terminus that target the antibody for secretion—are contained within the first exon. Similarly, the sequences encoding the variable (V_H) and constant (C_μ) domains reside within individual exons. The same pre-mRNA is produced in both B cells and plasma cells, but each cell type processes the primary transcript in a different way. In a plasma cell (which secretes immunoglobulin molecules into the blood), the mature mRNA is spliced so that it includes the C_μ exon, encoding hydrophilic amino acids. In a B cell (which displays immunoglobulin molecules on its surface), the pre-mRNA is spliced so that it includes the two C_{μ_m} exons, encoding hydrophobic amino acids, thus allowing the immunoglobulin to be anchored in the plasma membrane.

very hydrophobic amino acids that anchor the protein in the membrane. These exons are missing from the plasma cell IgM heavy-chain mRNA. Instead, this mRNA contains a different exon that is shorter and encodes less hydrophobic amino acids appropriate for secretion of the protein.

Alternative splicing can be extremely complicated. For example, the gene encoding the protein α-tropomyosin contains 14 exons. Different combinations of exons are used to form mature tropomyosin mRNAs in skeletal muscle, smooth muscle, and non-muscle cells (Fig. 5-5). This complex process most likely evolved to produce a tropomyosin protein with a particular structure that is necessary for each cell type. Although the overall structure of each tropomyosin protein is similar, the cell type–specific amino acids may function as binding sites for other proteins.

The mechanism by which cell type–specific splicing occurs is beginning to be explained for some genes.

Pre-mRNAs can be bound by enhancers and repressors of splicing. The enhancer proteins have domains that can recruit the splicing machinery, whereas splicing repressors bind RNA to block the splicing machinery near specific splice junctions. This has been best demonstrated in the process of sex determination in *Drosophila*, where a splicing event directs the developing embryo to become either male or female, because a protein, Transformer, produced only in females, directs alternative splicing of the *doublesex* gene. In males, *doublesex* is spliced into a form that encodes a repressor of female genes, permitting male development. In females, the alternatively spliced form of *doublesex* encodes a protein of entirely different function—a repressor of male genes and an activator of female genes.

The importance of alternative splicing for generating protein diversity has been highlighted by data from the Human Genome Project. For many years it was thought that the complexity of an organism would be reflected in the number of its genes. A "back of the envelope" calculation had estimated that there might be as many as 100,000 human genes, a figure that suited our view of ourselves. However, there was considerable surprise when, as the sequence data came in from the Human Genome Project, the estimate fell first to 35,000 and then to 20,000–25,000 genes when 99% of the euchromatic portion of the genome had been sequenced. This is about the same number of genes as in the nematode *Caenorhabditis elegans* (19,500) and less than the number in the plant *Arabidopsis thaliana* (27,000). The current (August 2006) number of genes so far identified in the Ensembl Project analysis of the human genome sequence is 22,000, whereas the number of gene transcripts is 35,845. The difference between these two numbers is accounted for by alternative splicing. Furthermore, our ideas about transcription may have to be revised. Determinations of the 5′ and 3′ ends of full-length mouse cDNAs, combined with other data, have revealed more than 180,000 transcripts, yet the mouse genome has only about 20,000 protein-coding genes. The assumption has been that transcription is concerned with the production of coding mRNAs, but this study revealed that about one-half of the transcripts are noncoding RNAs.

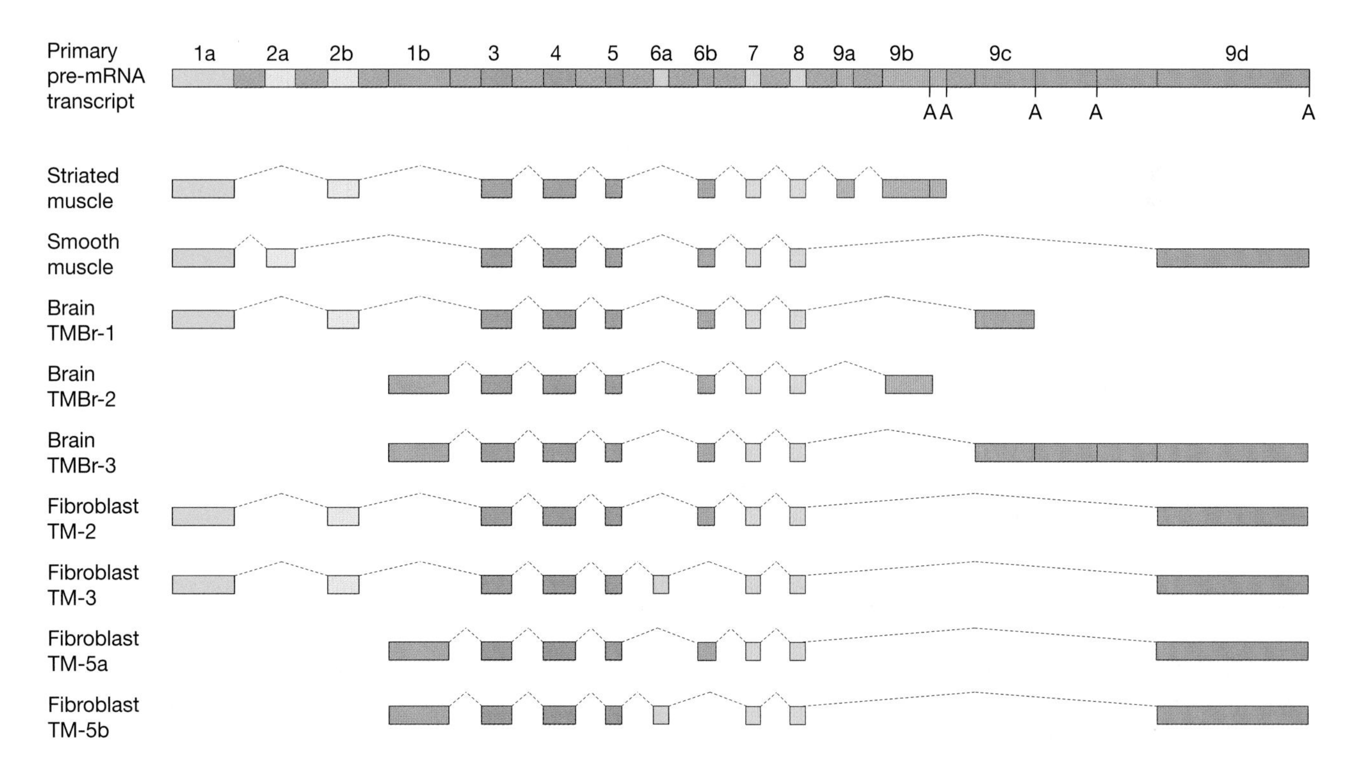

FIGURE 5-5

Complex patterns of eukaryotic mRNA splicing. The pre-mRNA transcript of the α-tropomyosin gene is alternatively spliced in different cell types. The *red-orange boxes* represent introns; the *other colors* represent exons. Polyadenylation signals are indicated by an A. *Dashed lines* in the mature mRNAs indicate regions that have been removed by splicing. TM, tropomyosin.

Errors in Splicing Cause Disease

Given the ubiquity and importance of splicing, it is not surprising that mutations affecting splicing can have a devastating effect. As more and more is learned about the DNA sequences determining splicing patterns and the myriad of protein factors regulating splicing, increasing numbers of human genetic diseases are found to result from splicing defects. It is estimated that at least 15% of human genetic disorders are of this nature. Mutations in consensus splicing sequences may lead to exon skipping, the omission of the following exon. Alternatively, a mutation may create a new splice site and lead to the inclusion of sequences that should not be part of the mRNA. Mutations in the β-globin gene that disrupt correct splicing of mRNA transcripts cause many cases of β-thalassemia, a disease in humans that affects the oxygen-carrying capability of hemoglobin. Such mutations in a splice site can lead to multiple mutant mRNAs as cryptic splice sites (sequences that can act as splice sites but are usually not used) are now used (Fig. 5-6). The mutations need not affect the consensus splicing sequences; it is now recognized that mutations in ESEs can have profound effects. One such mutation—a G to T transversion—occurs in exon 18 of the breast cancer susceptibility gene *BRCA1* and causes exon skipping. Experimental studies with an in vitro splicing system and an engineered *BRCA1* minigene showed that exon skipping correlated with disruption of ESEs, no matter the nature of the mutation. Even a silent mutation—one not producing an amino acid change—can disrupt an ESE and cause splicing problems.

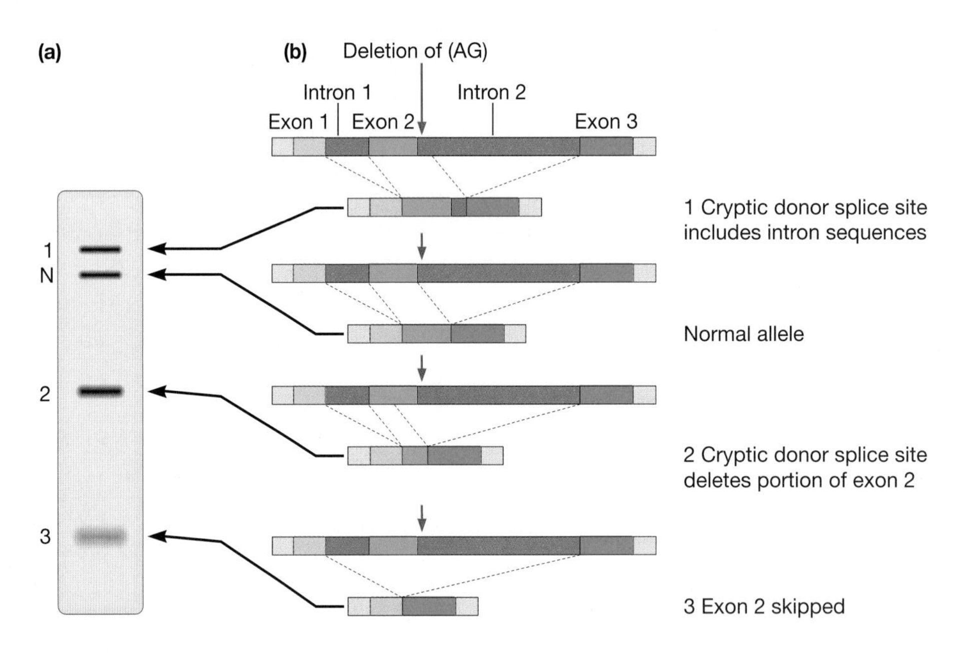

FIGURE 5-6

Mutations in the splicing sites of the β-globin gene disrupt splicing and cause β-thalassemia. (*a*) Electrophoresis of reverse transcriptase polymerase chain reaction (PCR) products of mRNAs from a carrier of thalassemia. The patient has one normal allele of the β-globin gene and one with a deletion of two nucleotides (AG) in the 5′ splice site of the second intron. There is a band corresponding to the normal allele (N) in the carrier, and three products derived by variant splicing of the mRNA from the mutant allele. (*b*) Splicing patterns that give rise to the mRNAs. The site of the deletion is marked by the *arrows*. mRNA produced by the normal allele (N) is spliced correctly—the two introns (*reds*) are removed and the three exons (*blues*) joined together. (1) Loss of the splice site leads to activation of a cryptic donor splice site 3′ to the lost site, so that a portion of intron 2 is included in this mRNA. (2) Activation of another cryptic donor splice site in the second exon produces a shorter mRNA lacking the 3′ end of exon 2. (3) Finally, by mechanisms not well understood, a mutation at one splice site can affect splicing at nearby splice site. In this case, the 5′ acceptor site of exon 2 is not used, so that exon 2 is skipped.

Introns Have Been Acquired and Lost during Evolution

Fascinating questions arose immediately following the discovery that genes contain noncoding regions (introns): Where did they come from (Are introns ancient partners of genes or newly acquired interlopers?) and why are they there (Does this arrangement of genes into introns and exons have a functional significance?). These have not been easy questions to answer because genomes change and it is hard to recognize the traces of molecular events from millions of years ago. Indeed, the picture is complicated and the simplistic "introns old" or "introns new" argument has been replaced by a more subtle approach, which recognizes that some introns are ancient and others are acquired.

One approach is to survey the distributions of introns in the organisms. The living world is divided into three great domains: archaea, bacteria, and eukarya. Archaea have introns in some genes, but because these are self-splicing and different from eukaryotic introns, they do not shed much light on the origins of the latter. Bacteria were thought to have no introns at all, because, as the "introns are ancient" hypothesis assumed, bacteria lost their introns while developing the "streamlined" genomes characteristic of modern bacteria. On the other hand, the "introns are new" hypothesis proposes that modern bacteria lack introns because they never had them; introns entered the world after the separation of the eubacterial and eukaryotic domains. An intron has been discovered in the leucine tRNA of eight species of cyanobacteria, which suggests that introns were present in the genomes of organisms predating the development of eukarya. The fact that the same intron is found in eight different species suggests that it could be very ancient; indeed, it must have been present early in the evolution of the cyanobacteria some 2.5 billion years ago.

However, data accumulated suggesting that introns are both lost and gained. For example, there is a single insulin gene in the hagfish, chicken, and human, and this gene has two introns. The rat has two insulin genes, I and II; gene II has two introns, but gene I has only one. This suggests that the single, original rat insulin gene duplicated to produce genes I and II, and that the former has lost one of its introns. Evidence for the insertion of introns comes from the α- and β-tubulin genes. These are very large families of genes present in all eukaryotes, and the two families arose by duplication of an ancestral gene. Comparisons of the distributions of introns in the tubulin gene of a variety of organisms show that they are found at 35 different places in the α and β families, with only one intron in common. This suggests that the introns were gained by the genes after the families separated. Furthermore, the distribution of introns within a family is best explained by the acquisition of a particular intron rather than by the multiple losses of introns (Fig. 5-7).

If introns can be gained, where do they come from? One analysis suggests that some may come from other introns. The nematode worms *C. elegans* and *Caenorhabditis briggsae* each have more than 100,000 introns in their genes. Introns unique to *C. elegans* or *C. briggsae* were found by comparing them with introns in the orthologous genes in other species of nematodes, and in *Drosophila*, mosquito, mouse, and human. This search revealed 81 introns found only in *C. elegans* and 41 introns found only in *C. briggsae*. These were assumed to be new introns inserted into the two genomes after divergence of the two species. The sequences of these introns were compared with those of all other introns in each species. Twenty-eight introns had significant DNA sequence similarity to other introns. Furthermore, there were three genes in which a new intron was similar to another intron in that same gene. This finding, and the fact that genes with new introns tended to be expressed in the germ line, suggested that the new introns arose through reverse splicing of the original intron.

Exons Often Define Protein Functional Domains and Exon Shuffling May Contribute to Protein Evolution

Introns do not encode proteins, yet they have been maintained in eukaryotic genomes for many millions of years. Why? What is the selective advantage of a genome that contains introns and has to replicate them along with coding sequences?

It has been suggested that introns might facilitate genetic variation (the substrate for natural selection) if exons encode functional domains of proteins, and introns promote recombination of exons. This process, called *exon shuffling*, contrasts with gene duplication, a process in which an entire gene becomes available for evolutionary changes. The concept of exon shuffling was inspired by studying the sequences of proteins such as the low-density-

(a) Distribution of introns at various codons in members of the α- and β-tubulin families

α-tubulins	2	4	5	9	13	16	17	19	20	21	33	35	41	56	58	59	62	76	*90	*90	95	126	134	177	208	211	257	319	327	351	353	407	412	437	448
Homo sapiens	+	–	–	–	–	–	–	–	–	–	–	–	–	–	–	–	–	+	–	–	–	+	–	–	–	–	–	–	–	–	–	–	–	–	–
Rattus norvegicus	+	–	–	–	–	–	–	–	–	–	–	–	–	–	–	–	–	+	–	–	–	+	–	–	–	–	–	–	–	–	–	–	–	–	–
Drosophila melanogaster	+	–	–	–	–	–	–	–	–	–	–	–	–	–	–	–	–	–	–	–	–	–	–	–	–	–	–	–	–	–	–	–	–	–	–
Physarum polycephalum	–	+	–	–	–	–	–	–	–	–	–	–	+	–	–	–	–	–	+	–	–	–	–	+	–	–	+	–	+	–	–	–	+	–	–
Chlamydomonas reinhardtii	–	–	–	–	–	+	–	–	–	–	–	–	–	–	–	–	–	–	–	+	–	–	–	–	–	–	–	–	–	–	–	–	–	–	–
β-tubulins																																			
Homo sapiens	–	–	–	–	–	–	–	–	+	–	–	–	–	–	+	–	–	–	–	–	+	–	–	–	–	–	–	–	–	–	–	–	–	–	–
Neurospora crassa	–	–	+	–	+	–	–	–	–	+	–	+	–	+	–	–	–	–	–	–	–	–	–	–	–	–	–	+	–	–	–	–	–	–	–
Schizosaccharomyces pombe	–	–	+	–	–	–	–	–	–	+	–	+	–	+	–	–	–	–	–	–	–	–	–	–	–	–	–	–	–	+	–	–	–	–	–
Chlamydomonas reinhardtii	–	–	–	+	–	–	–	–	–	–	–	–	–	–	–	+	–	–	–	–	–	–	+	–	–	–	–	–	–	–	–	–	–	–	–
Toxoplasma gondii	–	–	–	–	–	–	–	–	–	–	+	–	–	–	–	–	–	–	–	–	–	–	–	–	–	–	–	–	–	–	–	+	–	+	–

*An intron occurs at two different positions within codon 90

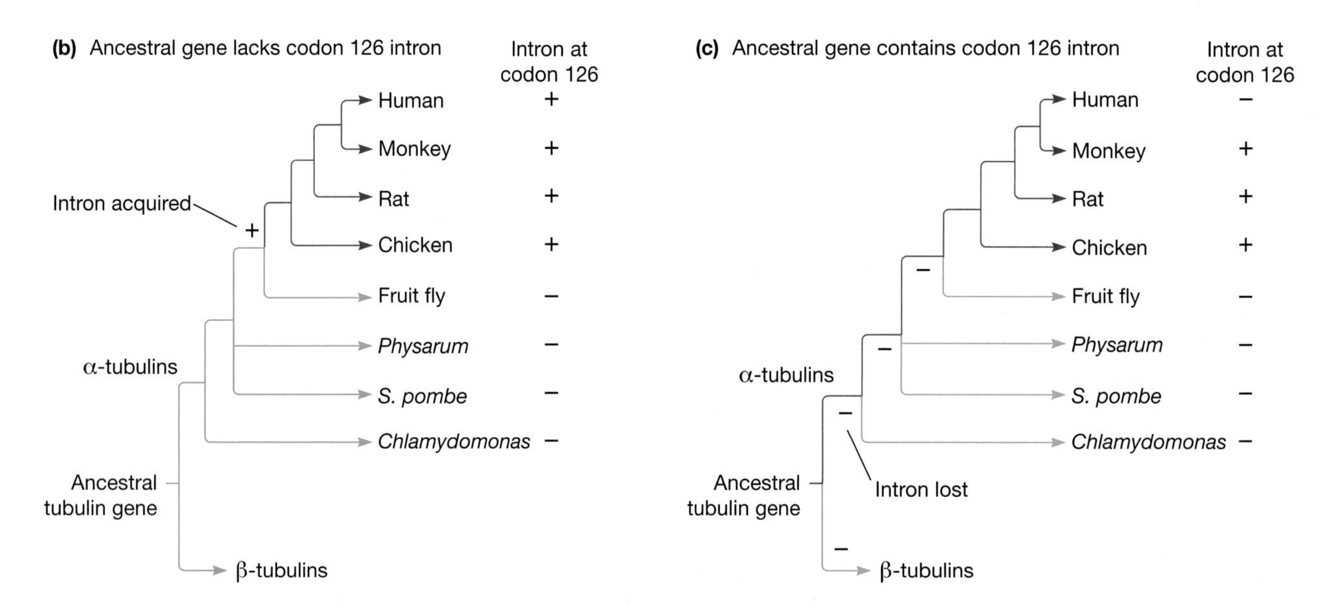

FIGURE 5-7

The tubulin family of genes provides evidence for insertion of introns. (*a*) Analysis of a large number of tubulin genes from a wide variety of organisms shows that introns are present in at least 35 different locations in these genes. Most tubulins have three introns, some have none, and a tubulin gene from *Aspergillus nidulans* has eight (not shown). The present-day distribution of these introns can be best explained by the hypothesis that the tubulins have *acquired* introns during evolution. (*b*) The distribution of the intron found at codon 126 in α-tubulin can be explained if the gene acquired the intron after the divergence of the vertebrates and invertebrates, as shown. (*c*) If it is assumed that this intron was present in the ancestral gene tubulin, then repeated losses (–) of the intron would have had to occur to generate the present distribution of this intron. Repeat losses are much less likely than a single insertion.

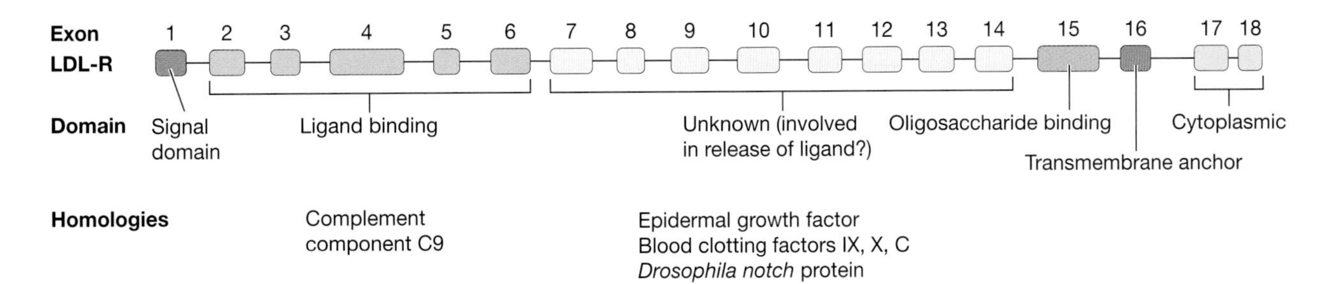

FIGURE 5-8

Evidence of exon shuffling in the low-density lipoprotein receptor (LDL-R) gene. The structure of the LDL-R exemplifies the modular construction of proteins, a feature that may have played an important role in evolution. The receptor contains six functionally distinct domains. Four of these are motifs common to many membrane proteins. There is an amino-terminal signal sequence that is needed for the protein to move to the cell surface; exon 16 codes for the hydrophobic transmembrane domain; exon 15 codes for a region that binds oligosaccharides and may keep the receptor away from the cell membrane; and exons 17 and 18 code for a cytoplasmic domain. LDL binds to the ligand-binding domain (exons 2–6). This region has a striking similarity to the complement component factor C9. Exons 7–14 code for a region of unknown function that may be involved in mediating release of LDL once the LDL-LDL-R complex has been internalized by the cell. This domain is homologous to domains in a large number of other proteins, including epidermal growth factor.

lipoprotein receptor (LDL-R), which appears to be made up of bits of other proteins stitched together to make a new protein. The LDL-R gene has 18 exons that can be grouped into six blocks that encode protein domains with different functions (Fig. 5-8).

There is some evidence that introns do divide proteins into important domains, which would be expected if exons tend to encode self-contained protein domains. For example, the different triosephosphate isomerases of chick and maize have six introns in common, and the introns occur at the boundaries of α helices and β-pleated sheets in the protein. Unfortunately, there are exceptions to the idea that exons always contain complete protein domains. As more triosephosphate isomerase genes from other species have been cloned and sequenced, more introns have been discovered. Some of these break protein domains and the 21 different intron sites (across seven species) appear to be distributed randomly.

Nevertheless, exon shuffling remains an attractive idea. The advantage of constructing proteins in this modular fashion is that only a limited number of units are needed to generate the enormous diversity of proteins found in cells. "Useful" units are evolutionarily successful and will be kept and reused. The number of units has been estimated by first identifying exons and then determining which have been reused by comparing their sequences. The result of this calculation is that all the proteins so far sequenced are made up of only some 1000–7000 different exons! This estimate is highly controversial, but it indicates that the reassortment of exons to produce new proteins may be an important factor in protein evolution.

Introns Are Found to Have a Function

There is one fascinating example of introns performing a function, where they rather than exons are the functional product of a gene. Almost all of the genes for small, stable nucleolar RNA genes (snoRNAs) that are involved in ribosomal RNA maturation in vertebrates are found within the introns of genes that code for proteins. After splicing and excision of the introns, nucleotides are removed from the introns' 5′ and 3′ ends to produce functional snoRNAs. The human U22 snoRNA is in an intron of the U22 host gene (UHG) that does not appear to code for a protein. The spliced UHG mRNA has numerous stop codons in all three reading frames, and, while it is expressed, polyadenylated, and exported to the cytoplasm, it is rapidly degraded. Comparisons of the human and mouse UHGs showed that regions within eight of the nine introns were highly conserved (including U22)—about 90% similarity between the two species. In contrast, the exons were poorly conserved. Northern blotting

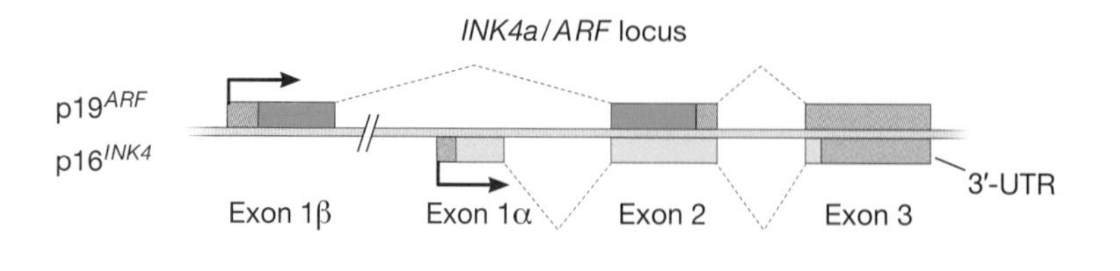

FIGURE 5-9

The tumor suppressor genes p16^{INK4a} and p19^{ARF} are encoded by a single locus in the human genome. These genes are transcribed from different promoters and have different first exons (E1α and E1β), but share two exons (E2 and E3) that are read in different frames. The resulting mRNAs encode completely unrelated proteins.

showed that the newly discovered regions in the seven introns were expressed as small RNAs. These were associated with fibrillarin, a nucleolar ribonuclear protein involved in ribosomal RNA (rRNA) processing, and sequence analysis showed that they could base-pair with 18*S* and 28*S* rRNAs, all characteristics of snoRNAs. Thus, in a curious reversal of the conventional pattern, it is the intronic sequences of UHG that are "useful."

Genes Are Sometimes Encoded within Other Genes

Overlapping coding sequences, or *nested genes*, were initially known to exist in bacteria, bacteriophages, and viruses. It was argued that these overlapping sequences were a peculiarity of the very small genomes of these organisms, thus making maximum use of their genetic material. With the discovery of introns and the realization that these can be very large, it was not surprising to find that eukaryotes have not wasted this DNA and that genes can be found within the introns of other genes. The first example of this was identified in a gene from *Drosophila*. During the analysis of the GART (glycinamide ribonucleotide formyltransferase) gene, which encodes an enzyme important for the biosynthesis of purines, an unrelated gene for a pupal cutile protein was found embedded within one of the introns. Surprisingly, the direction of transcription of the cuticle gene was shown to be opposite that of the GART gene transcript. Many other examples of genes embedded within introns of other genes are now known. For instance, intron 22 of the human factor VIII gene is very large (32 kb) and harbors two genes, F8A and F8B. Both are transcribed, the former in the opposite direction to the factor VIII gene. (F8A is responsible for ~45% of hemophilia A. There are two other copies of F8A 400 kb away from the factor VIII gene; homologous recombination between the intronic F8A and one of the other copies causes a large inversion and translocation of exons 1–22 away from the remaining exons, which produces severe hemophilia.) The *INK4a* locus provides a remarkable example of two genes, p16^{INK4a} and p19^{ARF}, encoded by the same segment of DNA. The two transcripts begin with different exons that are spliced to a common second exon; however, the reading frames are different so that two different proteins are made from the same DNA (Fig. 5-9). This locus is found to be mutated in many different kinds of cancer as both p16^{INK4a} and p19^{ARF} are involved in cell cycle control.

Many DNA Sequences Are Tandemly Repeated in the Genome

Although most genes are present only once per haploid genome, the genes for histones, tRNAs, and rRNAs are present many times within the genome and are often clustered together. *C. elegans*, for example, has 659 tRNA and 275 rRNA genes. These genes are all very abundantly expressed, and organisms probably need multiple copies of the genes so as to transcribe sufficient mRNA molecules. For example, histones—proteins that maintain and regulate chromosome structure—are required in great abundance during periods of rapid DNA replication in embryonic development of lower eukaryotes. The histone gene family consists of five major genes in most eukaryotes (H1, H2A, H2B, H3, and H4). In sea urchin and *Drosophila*, these five genes occur in a cluster of about 5000–6000 bp, and each cluster is tandemly repeated between 100 and 1000 times. A cluster of histone genes exists in higher eukaryotes but at only 10–40 copies per genome.

However, most of the repetitive DNA in eukaryotes does not correspond to genes; that is, it is not coding DNA (Chapter 11). *Microsatellite DNAs* range in size from two to six nucleotides, and are present in tandem arrays of five to about 30 copies. *Minisatellite repeats* are 15- to 100-bp long and repeated 20–50 times. Analysis of a particular microsatellite repeat from a number of people showed that the number of repeats and sometimes the specific sequence differed for each person. These differences form the basis for DNA fingerprinting now used in forensics. The genetic characterization of individuals by this and other methods will be discussed further in Chapter 16.

The largest proportion of repetitive DNA in most organisms consists of *transposons*, segments of DNA that can move (transpose) from one position in the genome to another. In contrast to the simple sequence repeats, transposons are generally not tandemly repeated but, rather, exist as isolated elements that may be present in many thousands of copies per genome. *Short interspersed nucleotide elements* (SINEs) range in length from 130 to 300 bp, and the most abundant family of SINEs in mammals, the *Alu* repeats, are present at about a million copies per genome, thus constituting some 10% of the mass of the human genome. (These are called the "Alu family" because they usually contain the recognition sequence AGCT for the restriction enzyme AluI.) We will discuss the mechanisms of transposition, and the role of transposons in genome evolution, in Chapter 7.

Protein Families Arise from Gene Duplication and Subsequent Sequence Divergence

The most important way in which eukaryotic genomes increase in size and complexity is through gene duplication and subsequent sequence divergence. Duplication of a gene allows one copy of the duplicated gene to undergo mutation in the absence of selection, because the other copy supplies the protein needed to sustain cell function. This gradual accumulation of mutations in the absence of selective pressure is called *genetic drift*. The evolutionary significance of genetic drift is that the mutations may lead to a protein acquiring new functions—an enzyme acting on a different substrate, for example. This is particularly clear in the case of the β-globin cluster, where five β-globins have evolved from an ancestral β-globin gene (Fig. 5-10). These different β-globin subunits have different physiological properties and are expressed at different times during development. For example, fetal hemoglobin F is composed of two α chains and two γ chains. It has a higher affinity for oxygen than does hemoglobin A, the adult form, which has two α chains and two β chains. Thus, transport of oxygen across the placen-

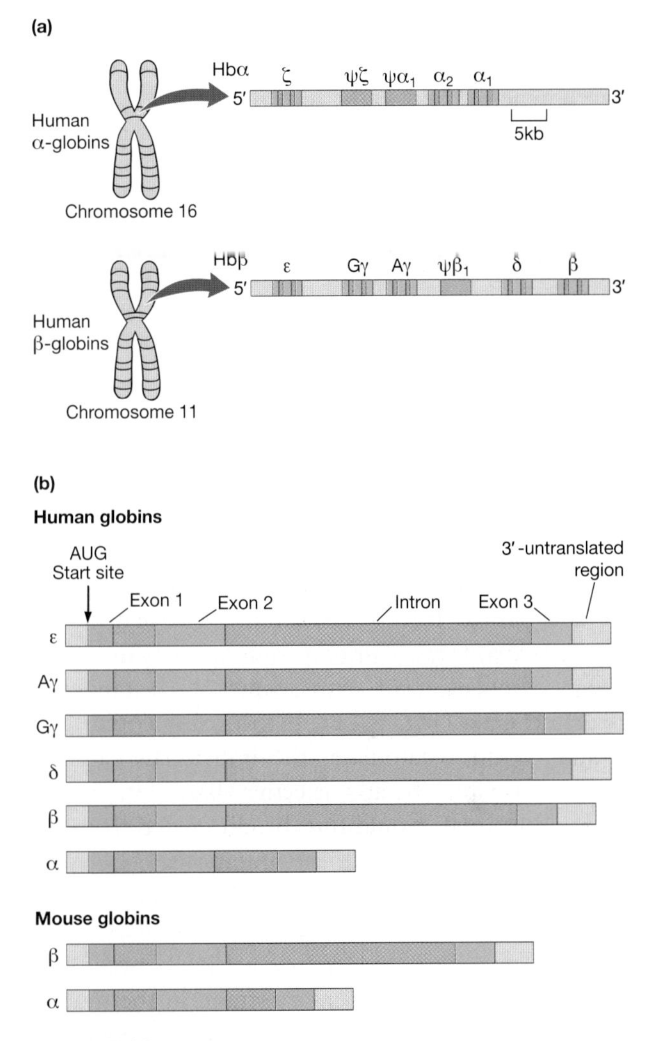

FIGURE 5-10

Gene duplication leads to the development of gene variants with new properties. The globin gene family has provided a wealth of information about gene evolution. (*a*) The present-day human globin genes are in two clusters. Following duplication of the ancestral globin gene and the translocation of one copy, the α-globin cluster covers about 30 kb on chromosome 16, and the β-globin genes are spread over ~50 kb on chromosome 11. The order of the genes along the chromosomes (from *left* to *right*) is the order in which they are expressed during development. Pseudogenes (ψζ, $\psi\alpha_1$, $\psi\beta_1$) are nonfunctional. (*b*) The overall morphology of the genes is remarkably constant. Comparisons of the exon–intron boundaries of the five β-globin genes, and of their sequences, show clearly that they are related. Exons, *blue*; introns, *red*.

ta is from the low-affinity hemoglobin A in the maternal circulation to the higher-affinity hemoglobin F of the fetus.

A similar divergence in the functions of duplicated genes is evident in plant genes encoding enzymes for synthesizing terpenes. Plants make terpenes for signaling to other plants and to repel pathogens and animals. Terpenes are made by the enzyme terpene synthase in the upper parts of the plant. Bioinformatic analysis of the *Arabidopsis* genome revealed seven genes related to the terpene synthases, three of which were closely related. Indeed, two of these—*25820* and *25830*—were identical to one another and the third—*25810*—is 80% identical to these two. Curiously, the latter gene was expressed exclusively in roots and does not synthesize one of the terpenes made by *25820* and *25830*. It is clear that an ancestral gene underwent a duplication; one gene diverged in expression and function to give rise to *25810*, whereas the other underwent a second duplication to produce *25820* and *25830*.

Genetic drift in duplicated genes may be desirable for generating genetic diversity, but in other situations it needs to be counteracted. For example, tandemly repeated genes like those for histones must be maintained free of mutations. This may come about in two ways. *Unequal crossing-over* can lead to the accumulation of extra copies of a tandemly repeated gene (Fig. 5-11). If too many copies of a mutant form accumulate, the individuals carrying these copies may be at a selective disadvantage, and those individuals would thus be eliminated from the population together with the mutations.

Mutations may be corrected rather than eliminated by *gene conversion*. Here a break in a DNA strand of the normal allele leads to base pairing between that strand and the complementary strand in the mutant allele. Following repair and replication, the mutant sequence is replaced by the normal sequence. This outcome is more likely for duplicated genes than the converse, in which the normal sequence is replaced by the mutant sequence, because there are more wild-type alleles than mutant alleles. Therefore, the break that initiates the process is more likely to occur in the wild-type allele. Unequal crossing-over and gene conversion work against the process of gene duplication and divergence through genetic drift. For a drifted mutation to become common (*fixed*) in a population, there must be a selective advantage associated with the variant gene; although, if the population is small enough, mutations can be fixed by chance.

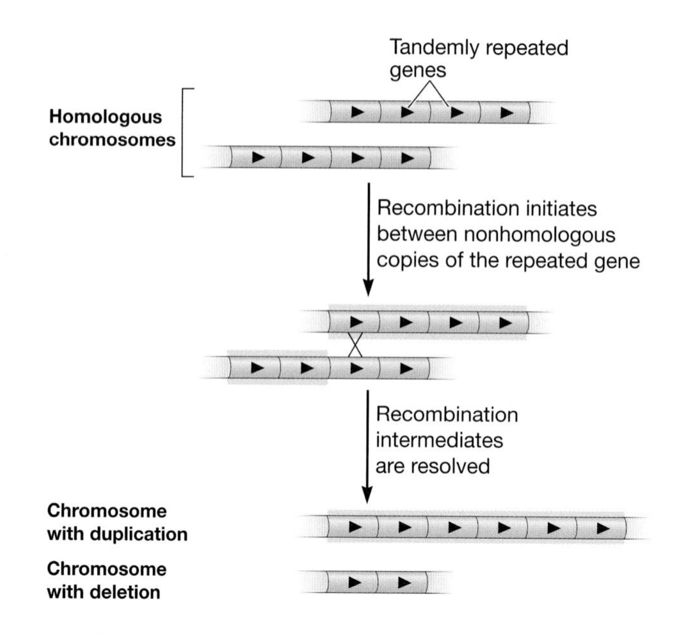

FIGURE 5-11

Unequal crossing-over between similar sequences located at nonhomologous sites on a chromosome generates one chromosome with a duplication. The other chromosome carries the corresponding deletion. Through this recombination-based mechanism, the copy number of repeated genes can increase or decrease.

It is not only single genes that can be duplicated—the duplication of entire genomes is common in plants, leading to *polyploidy*, the presence of more than one genome in a cell. At some stage in their evolutionary history, an ancestor in each of the lineages that gave rise to grasses, maize, tomato, potato, soybean, tobacco, and cotton underwent a massive natural form of genetic engineering. Cotton and tobacco are tetraploids, with 52 and 48 chromosomes, respectively, derived from ancestral plants with haploid numbers of 13 and 12. Many of the duplicated genes are lost; polyploidy in maize arose some 11 million years ago and about half of the duplicated genes have been lost over that period. Others diverge in function or patterns of gene expression. For example, 62% of recently duplicated genes in *Arabidopsis* have changed function. Polyploidy is much rarer in animals but there is some evidence for whole-genome duplication from comparative analyses of the pufferfish and the human genome sequences. There was not only synteny between regions of the two genomes, but there were two such regions in the pufferfish genome for each syntenic region in the human sequence. This is an ancient duplication in the ancestor of all bony fish and is estimated to have occurred 230 million years ago.

Pseudogenes Arise by Duplication of Functional Genes and Accumulate Mutations during Evolution

Although one of a pair of duplicated genes may diverge to acquire new functions, it may instead acquire deleterious mutations. As researchers continued to clone and sequence genes, they found DNA sequences that were highly related to known genes, but that included inserted or deleted sequences or other mutations that prevented the production of a functional protein. These nonfunctional sequences, closely related to functional genes, were termed *pseudogenes.* The first identified pseudogene was in the 5*S* rRNA gene cluster of the frog *Xenopus laevis.* Soon after this finding, pseudogenes were identified in the globin gene loci of various organisms, including in the human α-globin locus. The $\psi\alpha_1$ pseudogene is closely related to the three functional α-globin genes, but contains many mutations and no longer encodes a functional globin protein (Fig. 5-12). For example, the initiator methionine codon ATG of the functional α-globin gene has been mutated to GTG, which codes for valine. The consensus sequences at the 5′ ends of both introns also contain mutations expected to disrupt splicing. Furthermore, there are numerous base changes and several deletions within the coding region. The sequence of the $\psi\alpha_1$ pseudogene suggests that this gene arose by duplication of an α_2 globin gene. It is thought that, initially, this duplicated gene was functional, but at some time during evolution acquired an inactivating mutation. Because the gene is duplicated, this mutation in one of the copies did not affect the organism's survival. Subsequently, additional mutations accumulated in this gene, which yielded the sequence of the modern pseudogene. Pseudogenes have been found in many other gene families, such as β-globin and the genes encoding the cytoskeletal proteins actin and tubulin.

Another type of nonfunctional gene called a *processed pseudogene* is present in eukaryotic chromosomes. These processed pseudogenes resemble mRNA molecules in that they lack promoter and intron sequences and contain a stretch of adenines at the 3′ end of the gene, similar to those found on the end of an mRNA ("poly(A) tails"—see the following section). These pseudogenes appear to have arisen by reverse transcription of an mRNA molecule into DNA, followed by its insertion into the genome. In Chapter 7, we will discuss the ways in which genomes are much more dynamic than was once thought.

How many pseudogenes are there? Genome-scale sequencing and bioinformatic analysis have shown that pseudogenes are very common in vertebrate genomes, although the estimates vary rather widely depending on the criteria used to identify a pseudogene. There are thought to be more than 19,000 pseudogenes in the human genome, of which approximately 8000 are processed pseudogenes. The mouse has about the same number, whereas *C. elegans* has 2168.

3′ poly(A) Tails and 5′ Methylguanosine Caps Are Added to the Ends of Eukaryotic mRNAs

So far, we have been discussing complexities of the genome as reflected at the level of genes and their sequences. However, there are added complexities once the mRNA transcripts have been synthesized on their DNA templates. (RNA splicing occurs after transcription; we discussed this earlier in this chapter because it reflects the structure of the gene.) The first modifications of mRNAs were detected during stud-

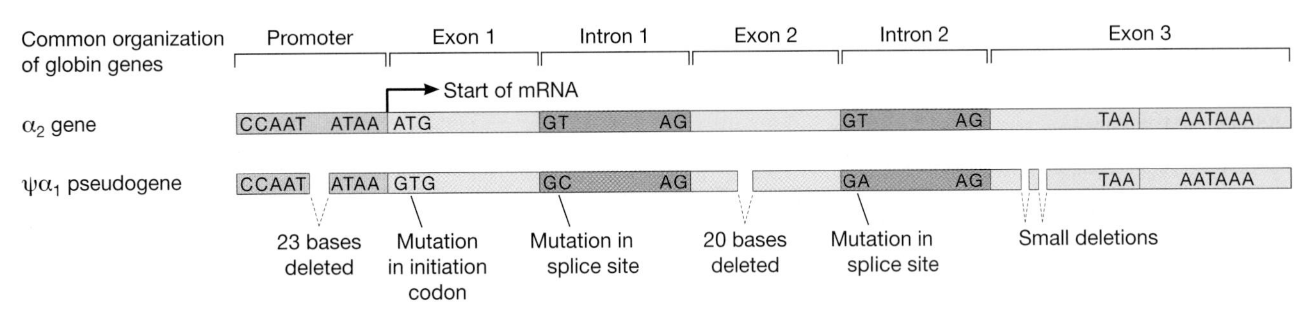

FIGURE 5-12

Structure of a human α-globin pseudogene. The organization of the human $\psi\alpha_1$ pseudogene is similar to that of functional globin genes, and its DNA sequence is ~70% identical to the α_2-globin gene; however, mutations have accumulated throughout the gene so that it does not encode a functional protein.

ies of the degradation of RNAs by ribonucleic acid nucleases (RNases). It was found that the mRNAs transcribed by RNA polymerase II had sequences that were resistant to RNases that digest RNA only at G and C nucleotides; analysis of the nucleotide content of the resistant portion revealed that it was composed of only A. These *poly(A) tails* at the 3′ ends of mRNAs do not come from sequences encoded in the DNA but are added enzymatically to the ends of transcribed RNAs. The presence of the poly(A) sequence increases translation and stabilizes mRNAs, and also appears to promote export of some mRNAs from the nucleus. Only those mRNAs transcribed by RNA polymerase II, one of three eukaroytic RNA polymerases, are polyadenylated. Poly(A) tails have also been observed on mRNAs from bacteria and archaea, but they are not required for translation and, in fact, can destabilize mRNAs. Whatever its function, the poly(A) tail provides a very convenient way to isolate mRNAs from total cellular RNA (as we discussed in Chapter 4).

The types of nucleotide analysis that led to the discovery of the poly(A) portion of RNA similarly led to the discovery of another unique component of eukaroytic mRNAs, the 5′ 7-methylguanosine cap. This cap was first discovered in the course of analyzing viral RNAs from reovirus-infected cells. The RNA molecules produced by the virus were found to contain an unusual nucleotide that was subsequently identified as a methylated guanosine residue linked to the mRNA through an unusual chemical bond. (Most other RNAs, such as tRNAs and rRNAs, are not modified by capping.) The capping of mRNAs was soon shown not to be unique to viruses, but to be a common feature of most eukaroytic mRNAs, from yeast to humans. The cap structure plays an important role in initiating translation. The initiation factor eIF4E binds to the cap, and in so doing leads to the assembly of a complex of other initiation factors, as well as an RNA helicase. The complex binds to the 40*S* ribosome subunit and begins unwinding the secondary structure of the 5′-untranslated region of the mRNA, thus allowing the codon–anticodon interaction at the initiation codon. The presence of the cap stabilizes mRNA and seems to act as a signal for cells that an RNA should be translated into protein; uncapped mRNAs are not efficiently translated.

RNA Editing Modifies Sequence Information at the mRNA Level

Alternative splicing produces several variations of a protein from a single gene by selecting exons, but the information, as reflected in nucleotide sequence, flows unchanged from DNA through RNA to protein. Remarkably, *posttranscriptional* RNA editing, in which the sequence of an RNA transcript is altered, has been discovered in a variety of genes from several organisms. Comparisons of the genomic and mRNA sequences for the cytochrome oxidase gene in two trypanosomes showed that the mRNA molecules had nucleotides added or subtracted to produce the correct reading frame for translation (Fig. 5-13). This editing requires small RNAs called guide RNAs

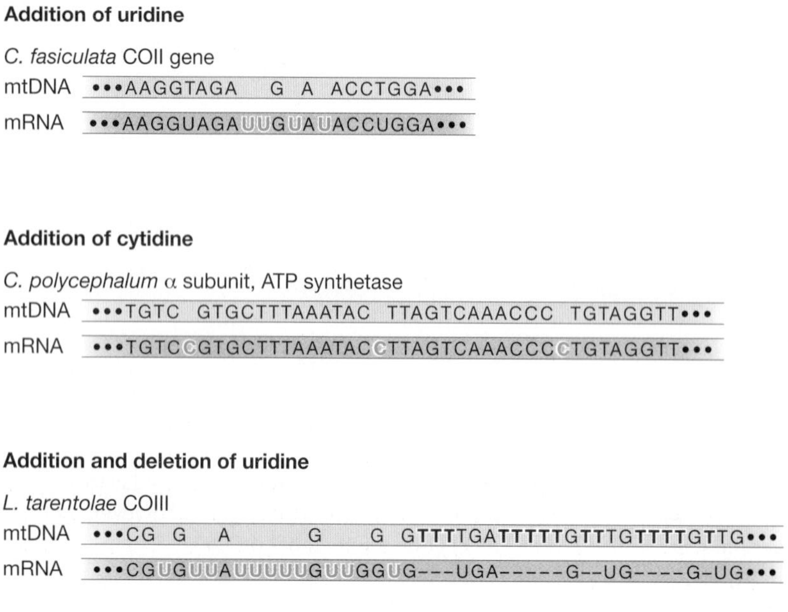

Addition of uridine

C. fasiculata COII gene
mtDNA •••AAGGTAGA   G A ACCTGGA•••
mRNA •••AAGGUAGAUUGUAUACCUGGA•••

Addition of cytidine

C. polycephalum α subunit, ATP synthetase
mtDNA •••TGTC GTGCTTTAAATAC TTAGTCAAACCC TGTAGGTT•••
mRNA •••TGTCCGTGCTTTAAATACCTTAGTCAAACCCCTGTAGGTT•••

Addition and deletion of uridine

L. tarentolae COIII
mtDNA •••CG G A G G GTTTTGATTTTTGTTTGTTTTGTTG•••
mRNA •••CGUGUUAUUUUUGUUGGUG---UGA-----G--UG----G-UG•••

FIGURE 5-13

RNA editing adds (or removes) uridines and cytidines from transcribed RNA molecules. The examples show posttranscriptional modifications of RNA transcribed from the mitochondrial genes (mtDNA) of two organisms. The phenomenon was described first in the cytochrome oxidase subunit II gene (COII) of the protozoan *Crithidia fasiculata*. Comparisons of the RNA sequence with the genomic DNA sequence showed that several uridines, for which there were no corresponding thymidines in the DNA sequence, had been inserted in the mRNA. In some cases, both addition (*green*) and removal (*red*) of uridines occur in the same mRNA, as shown by the sequence from the COIII gene of the protozoan *Leishmania tarentolae*.

(gRNAs), which hybridize to the sequence to be edited and lead to nuclease cleavage of the target sequence (Fig 5-14). The gRNA holds together the 5′ and 3′ parts of the cleaved RNA sequence while Us are removed by an exonuclease or are added by terminal uridylyl transferase. However, trypanosomes are strange organisms with many special features to enable them to survive in their host cells and it was conceivable that RNA editing was unique to them. Thus it was the subsequent discoveries of RNA editing in plants, insects, mollusks, and mammals that demonstrated that this is a fundamental process in living organisms.

The second example of RNA editing to be discovered was that of the mammalian apolipoprotein B gene transcript. Apolipoprotein B, an essential component of the various forms of plasma lipoproteins, is found in two forms: apo-B100 is synthesized by the liver, and apo-B48 is synthesized in the intestine. apo-B48 is about one-half the size of apo-B100. When cDNA clones were compared, a stop codon (TAA) was found in the apo-B48 cDNA sequence where there was a glutamine codon (CAA) in apo-B100. However, when the apo-B48 gene was sequenced, it was found that the genomic DNA contained a glutamine codon and not a stop codon (Fig. 5-15). The conclusion is that in the intestine, the CAA codon in the apo-B48 mRNA is edited to UAA. In this case, the editing is carried out without cutting the phosphodiester backbone of the mRNA; the Apobec-1 cytidine deaminase enzyme, in a complex with other proteins, simply removes an amine group from the cytidine, converting it to uridine. (There are several members of the APOBEC protein family, some of which play a role in fighting retroviral infections by deaminating cytidines in newly replicated viral RNA.)

There are also adenosine deaminases that convert adenosine to inosine, which base-pairs with cytidine. A notable example is the editing of the 2c subtype of mammalian serotonin 5-HT receptor mRNA. These are G protein–coupled receptors that, when they bind serotonin, activate phospholipase C. This in turn activates a cascade of reactions leading to the formation of second messenger molecules that alter neuronal signaling. The former mobilizes calcium release from intracellular stores, whereas the latter activates protein kinase C. Comparison of the nucleotide sequences of 2c subtype mRNAs with genomic DNA found that 11 different RNAs are generated by editing of four nucleotides, producing three amino acid changes (two of the nucleotide changes are within the same

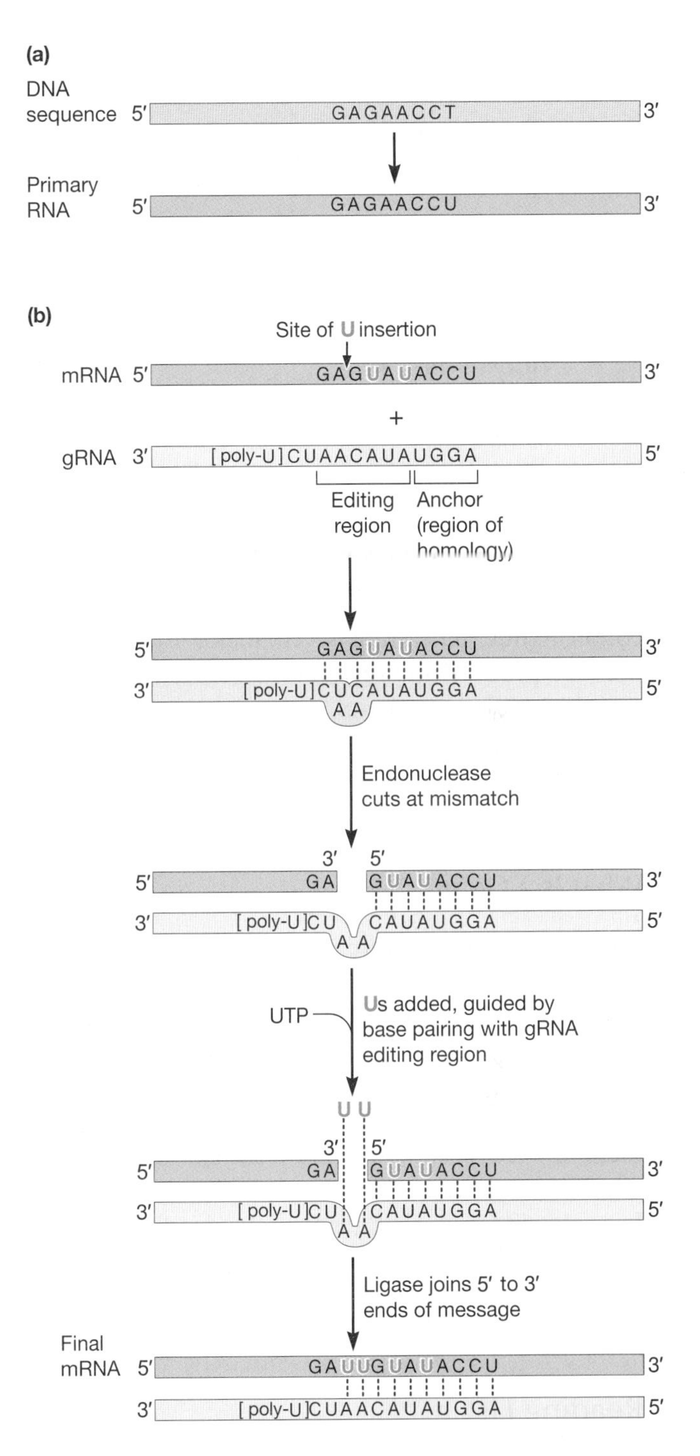

FIGURE 5-14

RNA editing of trypanosome *coxII* genes requires a *guide RNA* that hybridizes imperfectly to the target sequence. (*a*) The *coxII* gene is transcribed. (*b*) The primary transcript is modified by insertion of four uridine residues at three separate sites. After the first two uridine residues are inserted, a guide RNA (gRNA) directs insertion of the third and fourth uridine residues. A gRNA-directed endonuclease cuts the mRNA at the mismatch. The unpaired nucleotides in the guide RNA then act as the template for the addition of Us to the mRNA. Finally, an RNA ligase makes the chain complete.

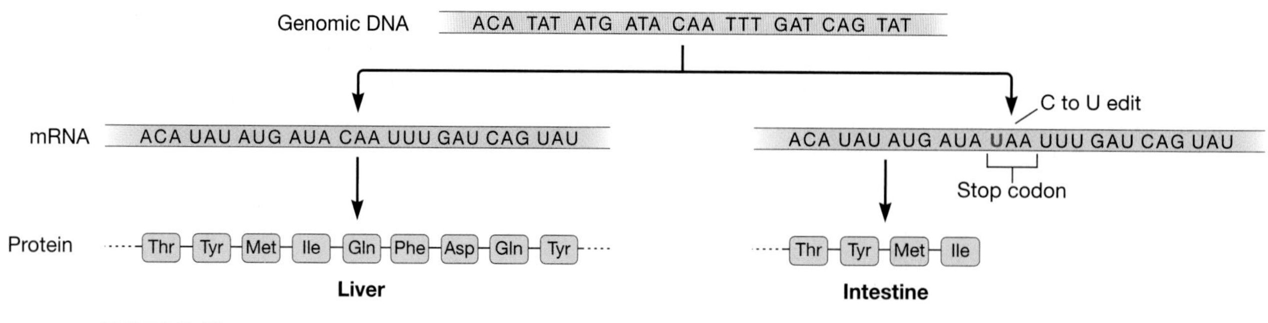

FIGURE 5-15

RNA editing produces tissue-specific forms of the human apolipoprotein B mRNA. A portion of the human apolipoprotein B mRNA is shown. In the intestine, a cytidine at position 6457 is deaminated to become a U. This changes a CAA codon coding for glutamine to a UAA termination codon. The resulting protein is much shorter than the protein encoded by mRNA produced in other tissues, where editing does not occur.

codon), thus generating seven major isoforms of the receptor. Moreover, these isoforms showed region-specific patterns of expression in the brain, which suggests that the RNA editing provides an unexpected regulatory control.

What Is a Gene?

Mendel's "factors" were abstract entities, recognized by the contrasting phenotypes they produced—tall or short, green or yellow, smooth or wrinkled—and by the patterns of their transmission from one generation to the next. The discovery that genes are carried on chromosomes did not make them any less abstract—genes were still defined by genetic analysis rather than by their chemical nature. Even the discovery that genes are made of DNA did not answer the question: What is a gene? A few years later, Jacob and Monod's description of structural genes and operators—still derived from genetic analysis—showed that a gene was going to be more complex than had been expected. As we have seen in this chapter, recombinant DNA techniques and genomic analysis have contributed to the realization that the gene is, in fact, much more complex than we could have imagined. Taken together with modifications of mRNA and protein by posttranscriptional and posttranslational processes, the relationship between gene and protein is much more indirect than was envisaged 40 years ago. This is hardly surprising and we can be sure that continuing analyses of genomes will reveal ever more interesting complications. We may never be able to provide a succinct definition of a gene that satisfies all circumstances. Although this state of affairs may alarm philosophers of science, geneticists have done very well with a flexible, operational concept of the gene, which has accommodated each successive discovery made over the past century.

Reading List

Split Genes Are Discovered

Cold Spring Harbor Symp. Quant. Biol., 1978, vol. 42. Chromatin. Cold Spring Harbor Laboratory, Cold Spring Harbor, New York.

Witkowski J.A. 1988. The discovery of "split" genes: A scientific revolution (review). *Trends Biochem. Sci.* **13:** 110–113.

Berget S.M., Moore C., and Sharp P. 1977. Spliced segments at the 5′ termini of adenovirus-2 late mRNA. *Proc. Natl. Acad. Sci.* **74:** 3171–3175.

Chow L.T., Gelinas R.E., Broker T.R., and Roberts R.J. 1977. An amazing sequence arrangement at the 5′ ends of adenovirus 2 messenger RNA. *Cell* **12:** 1–8.

Eukaryotic Cells Splice Pre-mRNAs to Remove Introns and Produce Mature mRNAs

Jeffreys A.J. and R.A. Flavell. 1977. The rabbit β-globin gene contains a large insert in the coding sequence. *Cell* **12:** 1097–1108.

Luo M.-J. and Reed R. 1999. Splicing is required for rapid and efficient mRNA export in metazoans. *Proc. Natl. Acad. Sci.* **96:** 14937–14942.

Specific Sequences and Protein Factors Ensure Very Accurate Splicing

Zhang X.H., Kangsamaksin T., Chao M.S., Banerjee J.K., and Chasin L.A. 2005. Exon inclusion is dependent on predictable exonic splicing enhancers. *Mol. Cell. Biol.* **25:** 7323–7332.

Ibrahim E.C., Schaal T.D., Hertel K.J., Reed R., and Maniatis T. 2005. Serine/arginine-rich protein-dependent suppression of exon skipping by exonic splicing enhancers. *Proc. Natl. Acad. Sci.* **102:** 5002–5007.

Nilsen T.W. 2003. The spliceosome: The most complex macromolecular machine in the cell? *BioEssays* **25:** 1147–1149.

Alternate Splicing Creates Different, but Related, mRNAs from a Single Gene

Early P., Rogers J., Davis M., Calame K., Bond M., Wall R., and Hood L. 1980. Two mRNAs can be produced from a single immunoglobulin μ gene by alternative RNA processing pathways. *Cell* **20:** 313–319.

Lees-Miller J.P., Goodwin L.O., and Helfman D.M. 1990. Three novel brain tropomyosin isoforms are expressed from the rat α-tropomyosin gene through the use of alternative promoters and alternative RNA processing. *Mol. Cell. Biol.* **10:** 1729–1742.

Sosnowski B.A., Belote J.M., and Mckeown M. 1989. Sex-specific alternative splicing of RNA from the *transformer* gene results from sequence-dependent splice site blockage. *Cell* **58:** 449–459.

Inoue K., Hoshijima K., Sakamoto H., and Shimura Y. 1990. Binding of the *Drosophila Sex-lethal* gene product to the alternative splice site of transformer primary transcript. *Nature* **344:** 461–463.

Bell L.R., Horabin J.L., Schedl P., and Cline T.W. 1991. Positive autoregulation of *Sex-lethal* by alternative splicing maintains the female determined state in *Drosophila*. *Cell* **65:** 229–239.

International Human Genome Sequencing Consortium. 2001. Initial sequencing and analysis of the human genome. *Nature* **409:** 860–921.

Venter J.C., Adams M.D., Myers E.W., Li P.W., Mural R.J., et al. 2001. The sequence of the human genome. *Science* **291:** 1304–1351.

International Human Genome Sequencing Consortium. 2004. Finishing the euchromatic sequence of the human genome. *Nature* **431:** 931–945.

Hubbard T., Andrews D., Caccamo M., Cameron G., Chen Y., et al. 2005. Ensembl 2005. *Nucleic Acids Res.* **33:** D447–D453.

Ensembl Web site: http://www.ensembl.org/index.html

Claverie J.M. 2005. Fewer genes, more noncoding RNA. *Science* **309:** 1529–1530.

The FANTOM Consortium and Riken Genome Exploration Research Group & Genome Science Group. 2005. The transcriptional landscape of the mammalian genome. *Science* **309:** 1559–1563.

Errors in Splicing Cause Disease

Cartegni L., Chew S.L., and Krainer A.R. 2002. Listening to silence and understanding nonsense: Exonic mutations that affect splicing. *Nat. Rev. Genet.* **3:** 285–298.

Faustino P., Osorio-Almeida L., Romao L., Barbot J., Fernandes B., Justica B., and Lavinha J. 1998. Dominantly transmitted β-thalassemia arising from the production of several aberrant mRNA species and one abnormal peptide. *Blood* **91:** 685–690.

Liu H.-X., Cartegni L., Zhang M.Q., and Krainer A.R. 2001. A mechanism for exon skipping caused by nonsense or missense mutations in *BRCA1* and other genes. *Nat. Genet.* **27:** 55–58.

Robberson B.L., Cote G.J., and Berget S.M. 1990. Exon definition may facilitate splice site selection in RNAs with multiple exons. *Mol. Cell. Biol.* **10:** 84–94.

Introns Have Been Acquired and Lost during Evolution

Roy S.W. 2003. Recent evidence for the exon theory of genes. *Genetica* **118:** 251–266.

Woese C.R., Kandler O., and Wheelis M.L. 1990. Towards a natural system of organisms: Proposal for the domains Archaea, Bacteria, and Eucarya. *Proc. Natl. Acad. Sci.* **87:** 4576–4579.

Kuhsel M.G., Strickland R., and Palmer J.D. 1990. An ancient group I intron shared by eubacteria and chloroplasts. *Science* **250:** 1570–1573.

Bell G.I., Pictet R.L., Rutter W.J., Cordell B., Tischer E., and Goodman H.M. 1980. Sequence of the human insulin gene. *Nature* **284:** 26–32.

Perler F., Efstratiadis A., Lomedico P., Gilbert W., Kolodner R., and Dodgson J. 1980. The evolution of genes: The chicken proinsulin gene. *Cell* **20:** 555–556.

Dibb J. and Newman A.J. 1989. Evidence that introns arose at proto-splice sites. *EMBO J.* **8:** 2015–2021. (Tubulin gene family.)

Coghlan A. and Wolfe K.H. 2004. Origins of recently gained introns in *Caenorhabditis*. *Proc. Natl. Acad. Sci.* **101:** 11362–11367.

Exons Often Define Protein Functional Domains and Exon Shuffling May Contribute to Protein Evolution

Gilbert W. 1978. Why genes in pieces? *Nature* **271:** 501.

Sudhof T.C., Russell D.W., Goldstein J.L., Brown M.S., Sanchez-Pescador R., and Bell G.I. 1985. Cassette of eight exons shared by genes for LDL receptor and EGF precursor. *Science* **228:** 893–895.

Dorit R.L., Schoenbach L., and Gilbert W. 1990. How big is the universe of exons? *Science* **250:** 1377–1382.

Marchionni M. and Gilbert W. 1986. The triosephosphate isomerase gene from maize: Introns antedate the plant–animal divergence. *Cell* **46:** 133–141.

Logsdon J.M., Jr., Tyshenko M.G., Dixon C., D.-Jafari J., Walker V.K., and Palmer J.D. 1995. Seven newly discovered intron positions in the triose-phosphate isomerase gene: Evidence for the introns-late theory. *Proc. Natl. Acad. Sci.* **92:** 8507–8511.

Liu M. and Grigoriev A. 2004. Protein domains correlate strongly with exons in multiple eukaryotic genomes—Evidence of exon shuffling? *Trends Genet.* **20:** 399–403.

Fedorova L. and Fedorov A. 2003. Introns in gene evolution. *Genetica* **118:** 123–131.

Functions Are Found for Intron Sequences

Tycowski K.T., Shu M.D., and Steitz J.A. 1996. A mammalian gene with introns instead of exons generating stable RNA products. *Nature* **379:** 464–466.

Genes Are Sometimes Encoded within Genes

Henikoff S., Keene M.A., Fechtel K., and Fristron J.W. 1986. Gene within a gene: Nested *Drosophila* genes encode unrelated proteins on opposite DNA strands. *Cell* **44:** 33–42.

Levinson B., Kenwrick S., Lakich D., Hammonds G., Jr., and Gitschier J. 1990. A transcribed gene in an intron of the human factor VIII gene. *Genomics* **7:** 1–11.

Quelle D.E., Zindy F., Ashmun R.A., and Sherr C.J. 1995. Alternative reading frames of the *INK4a* tumor suppressor gene encode two unrelated proteins capable of inducing cell cycle arrest. *Cell* **83:** 993–1000.

Sharpless N.E. and DePinho R.A. 1999. The *INK4A/ARF* locus and its two gene products. *Curr. Opin. Genet. Dev.* **9:** 22–30.

Many DNA Sequences Are Tandemly Repeated in Genomes

Britten R.J. and Kohne D.E. 1968. Repeated sequences in DNA. *Science* **161:** 529–540.

Schmid C.W. and Jelinek W.R. 1982. The Alu family of dispersed repetitive sequences. *Science* **216:** 1065–1070.

Singer M.F. 1982. SINEs and LINEs: Highly repeated short and long interspersed sequences in mammalian genomes. *Cell* **28:** 433–434.

Jeffreys A.J., Wilson V., and Then S.C. 1985. Individual-specific fingerprints of human DNA. *Nature* **316:** 76–78.

Hentschel C.C. and Birnstiel M.L. 1981. The organization and expression of histone gene families. *Cell* **25:** 301–305.

Protein Families Arise from Gene Duplication and Subsequent Sequence Divergence

Ohno S. 1970. *Evolution by Gene Duplication.* Springer-Verlag, New York.

Prince V.E. and Pickett F.B. 2002. Splitting pairs: The diverging fates of duplicated genes. *Nat. Rev. Genet.* **3:** 827–837.

Chen F., Dae-Kyun R., Petri J., Gershenzon J., Bohlmann J., Pichersky E., and Tholl D. 2004. Characterization of a root-specific *Arabidopsis* terpene synthase responsible for the formation of the volatile monoterpene 1,8-cineole. *Plant Physiol.* **135:** 1956–1966.

Mazet F. and Shimeld S.M. 2002. Gene duplication and divergence in the early evolution of vertebrates. *Curr. Opin. Genet. Dev.* **12:** 393–396.

Efstratiadis A., Posakony J.W., Maniatis T., Lawn R.M., O'Connell C., et al. 1980. The structure and evolution of the human β-globin gene family. *Cell* **21:** 653–668.

Hentschel C.C. and Birnstiel M.L. 1981. The organization and expression of histone gene families. *Cell* **25:** 301–305.

Jaillon O., Aury J.-M., Brunet F., Petit J.-L., Strange-Thomann N., et al. 2004. Genome duplication in the teleost fish *Tetraodon nigroviridis* reveals the early vertebrate proto-karyotype. *Nature* **431:** 946–957.

Moore R.C. and Purugganan M.D. 2005. The evolutionary dynamics of plant duplicate genes. *Curr. Opin. Plant Biol.* **8:** 122–128.

Ober D. 2005. Seeing double: Gene duplication and diversification in plant secondary metabolism. *Trends Plant Sci.* **10:** 444–449.

Adams K.L. and Wendel J.F. 2005. Polyploidy and genome evolution in plants. *Curr. Opin. Plant Biol.* **8:** 135–141.

Lai J., Ma J., Swigonova Z., Ramakrishna W., Linton E., et al. 2004. Gene loss and movement in the maize genome. *Genome Res.* **14:** 1924–1931.

Blanc G. and Wolfe K.H. 2004. Widespread paleopolyploidy in model plant species inferred from age distributions of duplicate genes. *Plant Cell* **16:** 1667–1678.

Pseudogenes Arise by Duplication of Functional Genes and Accumulate Mutations during Evolution

Proudfoot N.J. and Maniatis T. 1980. The structure of a human α-globin pseudogene and its relationship to α-globin gene duplication. *Cell* **21:** 537–544.

Jacq C., Miller J.R., and Brownlee G.G. 1977. A pseudogene structure in 5*S* DNA of *Xenopus laevis. Cell* **12:** 109–120.

Zhang Z. and Gerstein M. 2004. Large-scale analysis of pseudogenes in the human genome. *Curr. Opin. Genet. Dev.* **14:** 328–335.

Zhang Z., Harrison P.M., Liu Y., and Gerstein M. 2003. Millions of years of evolution preserved: A comprehensive catalog of the processed pseudogenes in the human genome. *Genome Res.* **13:** 2541–2558.

3′ Tails of Repeated Adenine Bases and 5′ Methylguanosine Caps Are Added to Eukaryotic mRNAs

Shatkin A.J. and Manley J.L. 2000. The ends of the affair: Capping and polyadenylation. *Nat. Struct. Biol.* **8:** 838–842.

Furuichi Y., Morgan M., Muthukrishnan S., and Shatkin A.J. 1975. Reovirus messenger RNA contains a methylated, blocked 5′-terminal structure: m-7G(5′)ppp(5′)G-MpCp-. *Proc. Natl. Acad. Sci.* **72:** 362–366.

Lim L. and Canellakis E.S. 1970. Adenine-rich polymer associated with rabbit reticulocyte messenger RNA. *Nature* **227:** 710–712.

Darnell J.E., Wall R., and Tushinski R.J. 1971. An adenylic acid-rich sequence in messenger RNA of HeLa cells and its possible relationship to reiterated sites in DNA. *Proc. Natl. Acad. Sci.* **68:** 1321–1325.

Edmonds M., Vaughan M.H., Jr., and Nakazato H. 1971. Polyadenylic acid sequences in the heterogeneous nuclear RNA and rapidly-labeled polyribosomal RNA of HeLa cells: Possible evidence for a precursor relationship. *Proc. Natl. Acad. Sci.* **68:** 1336–1340.

Lee S.Y., Mendecki J., and Brawerman G. 1971. A polynucleotide segment rich in adenylic acid in the rapidly labeled polyribosomal RNA component of mouse sarcoma 180 ascites cells. *Proc. Natl. Acad. Sci.* **68:** 1331–1335.

Aviv H. and Leder P. 1972. Purification of biologically active globin messenger RNA by chromatography on oligothymidylic acid-cellulose. *Proc. Natl. Acad. Sci.* **69:** 1408–1412.

RNA Editing Modifies Sequence Information at the mRNA Level

Koslowsky D.J. 2004. A historical perspective on RNA editing: How the peculiar and bizarre became mainstream. *Methods Mol. Biol.* **265:** 161–198.

Benne R., van den Burg J., Brakenhoff J.P., Sloof P., Van Boom J.H., and Tromp M.C. 1986. Major transcript of the frame-shifted *coxII* gene from trypanosome mitochondria contains four nucleotides that are not encoded in the DNA. *Cell* **46:** 819–826.

Shaw J. M., Feagin J.E., Stuart K., and Simpson L. 1988. Editing of kinetoplastid mitochondrial mRNAs by uridine addition and deletion generates conserved amino acid sequences and AUG initiation codons. *Cell* **53:** 401–411.

Mahendran R., Spottswood M.R., and Miller D.L. 1991. RNA editing by cytidine insertion in mitochondria of *Physarum polycephalum. Nature* **349:** 434–438.

Chen S.-H., Habib G., Yang C.-Y., Gu Z.-W., Lee B.R., et al. 1987. Apolipoprotein B-48 is the product of a messenger RNA with an organ-specific in-frame stop codon. *Science* **238:** 363–366.

Powell L.M., Wallis S.C., Pease R.J., Edwards Y.H., Knott T.J., and Scott J. 1987. A novel form of tissue-specific RNA processing produces apolipoprotein B-48 in intestine. *Cell* **50:** 831–840.

Hadjiagapiou C., Giannoni F., Funahashi T., Skarosi S.F., and Davidson N.O. 1994. Molecular cloning of a human small intestinal apolipoprotein B mRNA editing protein. *Nucleic Acids Res.* **25:** 1874–1879.

Harris R.S. and Liddament M.T. 2004. Retroviral restriction by APOBEC proteins. *Nat. Rev. Immunol.* **4:** 868–877.

Burns C.M., Chu H., Rueter S.M., Hutchinson L.K., Canton H., Sanders-Bush E., and Emeson R.B. 1997. Regulation of serotonin-2C receptor G-protein coupling by RNA editing. *Nature* **387:** 303–308.

What Is a Gene?

Stadler L.J. 1954. The gene. *Science* **120:** 811–819. (A discussion of the nature of the gene pre–double helix.)

Carlson E.A. 1966. *The Gene: A Critical History.* Saunders, Philadelphia.

Carlson E.A. 1991. Defining the gene: An evolving concept. *Am. J. Hum. Genet.* **49:** 475–487.

Waters K.C. 1994. Genes made molecular. *Philos. Sci.* **61:** 163–185.

Beurton P., Falk R., and Rheinberger H.-J., eds. 2000. *The Concept of the Gene in Development and Evolution. Historical and Epistemological Perspectives.* Cambridge University Press, Cambridge.

CHAPTER 6

A New Toolbox for Recombinant DNA

Cloning a gene is only the first step in understanding what a gene does. Even sequencing the gene and predicting features of the protein it encodes using bioinformatics provides only limited knowledge of the gene's function. To know what a gene does—where and when—requires carrying out experiments. There are three prerequisites for a detailed analysis of a gene. First, we must isolate the gene by cloning. Second, we must be able to manipulate its sequence so as to explore the functions of the various parts of the protein or RNA it encodes. Third, we must be able to return the altered gene to cells growing in culture or in the body to determine how the gene functions in its proper environment. We have already seen how genes can be isolated by cloning. This was soon followed by the development of methods for manipulating DNA sequences and persuading cells to take up the foreign DNA, despite millions of years of evolutionary barriers to DNA uptake. However, investigators soon discovered that it was not enough to get a gene into cells; the gene had to be retained and function properly. So, a great deal of work and energy has been devoted to developing methods for retaining the correct functioning of transferred genes. Methods have also been developed to inactivate genes already present in cells; "knocking out" a cellular gene provides invaluable information on its function.

With many of these methods now in place, gene transfer has become a routine tool for studying gene structure and function. It is used to identify the regulatory sequences that control gene expression. It provides a means to determine gene function by the transfer of new or altered genes into new cellular environments. And it provides the basis for high-level protein expression, which is used by researchers to produce proteins for analysis and by the biotechnology industry to produce new protein drugs. Sophisticated methods allow investigators to introduce new genes or alter existing ones in intact animals, which makes possible useful animal models of complex human diseases. The ability to isolate genes as molecular clones, the development of tools to modify gene sequences in the test tube, and the power to return altered genes to the organism to test their function have revolutionized the way genetics is done in higher organisms.

Useful Sequences Can Be Appended to a DNA Sequence by Polymerase Chain Reaction

Before the polymerase chain reaction (PCR) was invented, investigators had to be content with whichever restriction sites happened to occur naturally in a sequence. PCR makes it easy to append any sequence to the ends of an amplified product, provided that the 3′ tail of a primer is complementary to the target sequence so that the polymerase can initiate DNA synthesis. Fortunately, 5′ extensions of the primer carrying the sequence to be added do not hinder extension of the new strand (Fig. 6-1). 5′ extensions have many uses—one common practice is to add restriction sites so that the PCR product can be conveniently cloned into a digested vector. (Caution is required in the choice of site, however, because the site must not be contained within the amplified sequence itself, or the insert will be cut not only at the termini but also at the internal site.) Another frequent application of this method is the addition of a DNA sequence for an *epitope tag*, an amino acid sequence that can be recognized by an antibody. When the epitope tag (usually 7–10 amino acids) is added to a protein, an antibody that binds to the epitope tag can be used to detect and isolate the protein.

In Vitro Mutagenesis Is Used to Study Gene Function

There are few achievements in molecular biology quite as remarkable as our ability to make specific and precise modifications in the nucleotide sequence of a gene and in so doing modify the functions of

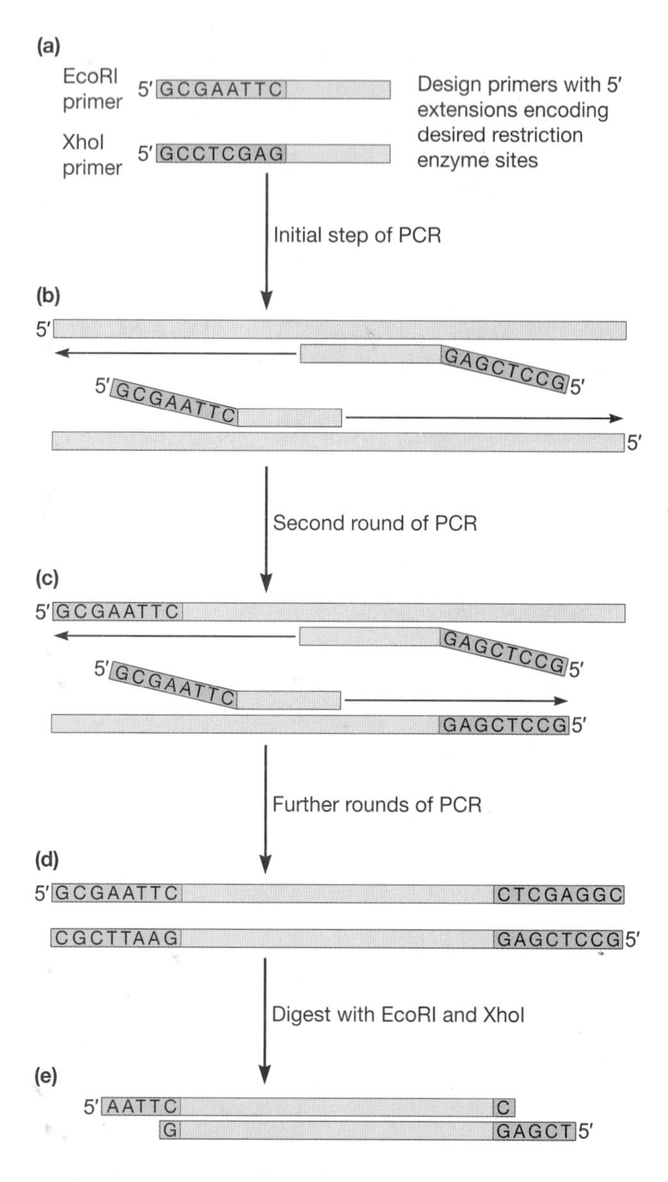

FIGURE 6-1

Addition of novel sequences to the 5′ end of polymerase chain reaction (PCR) products. (*a*) A pair of PCR primers is designed so that their 3′ sequences hybridize to the target sequence, whereas their 5′ ends contain sequences encoding a restriction enzyme site. In this case, one primer adds an EcoRI restriction site (GAATTC) and the other an XhoI site (CTCGAG). Two additional nucleotides are added to each because most restriction enzymes need an extension of a few nucleotides to cut a sequence. (*b*) The target DNA is denatured so that the primers can anneal. In the first cycle of PCR, the 5′ ends of the primers are single stranded and do not hybridize to the target sequence. (*c*) In the next round of PCR—only the newly synthesized strands are shown—the DNA primers anneal again and this time synthesis produces double-stranded DNA molecules, just as in a conventional PCR. However, these molecules now have a restriction site at one end. (*d*) The products of the second round and all the subsequent rounds of PCR have restriction sites at both ends. (*e*) When these are cut with EcoRI and XhoI, sticky ends are produced.

the protein it encodes. In vitro mutagenesis of cloned genes has become a standard tool in the functional analysis of nucleic acids and proteins: DNA containing the gene of interest is treated in vitro by some mutagenesis procedure (or agent) that alters the DNA either chemically or enzymatically.

The various approaches to mutagenesis can be grouped broadly into random and site-directed methods. Random methods introduce mutations (including deletions, insertions, or base changes) anywhere in a segment of DNA. These random approaches are best used to identify the location and boundaries of a particular function within a cloned DNA fragment and are most readily and successfully used for this purpose when a simple genetic screen (or selection) is available. A genetic screen or selection consists of a system to test the function of the DNA in cells without having to isolate and test each plasmid individually (Fig. 6-2). Random mutagenesis is often used as a first step when little is known about the function encoded by a particular DNA fragment. The value of a random strategy is that it quickly narrows the focus of attention from a large gene to a smaller region that can be studied subsequently in greater detail. However, analysis of random mutants generally provides only an approximate identification of the key functional regions of the gene; it does not explain how things work on a molecular level.

Once an important functional region of a gene has been identified by random mutagenesis, site-directed methods—placing or targeting mutations precisely where they are needed—are used to define the role(s) of specific sequences. In addition, directed mutagenesis provides a powerful tool for the analysis of protein function, by allowing researchers to make specific changes in the structure of the protein. A number of strategies have been developed to construct site-directed mutants in vitro. For example, the simplest approaches involve creating directed deletions of a sequence using available restriction sites or by producing specific PCR products. However, to be most powerful, mutagenesis must allow the experimenter to place *any* modification at *any* position desired in cloned DNA. The easiest way to do this is with synthetic DNA oligonucleotides.

Synthetic Oligonucleotides Can Be Used to Create Specific Mutations

Oligonucleotides provide the means to design a particular mutation and then to place it precisely within

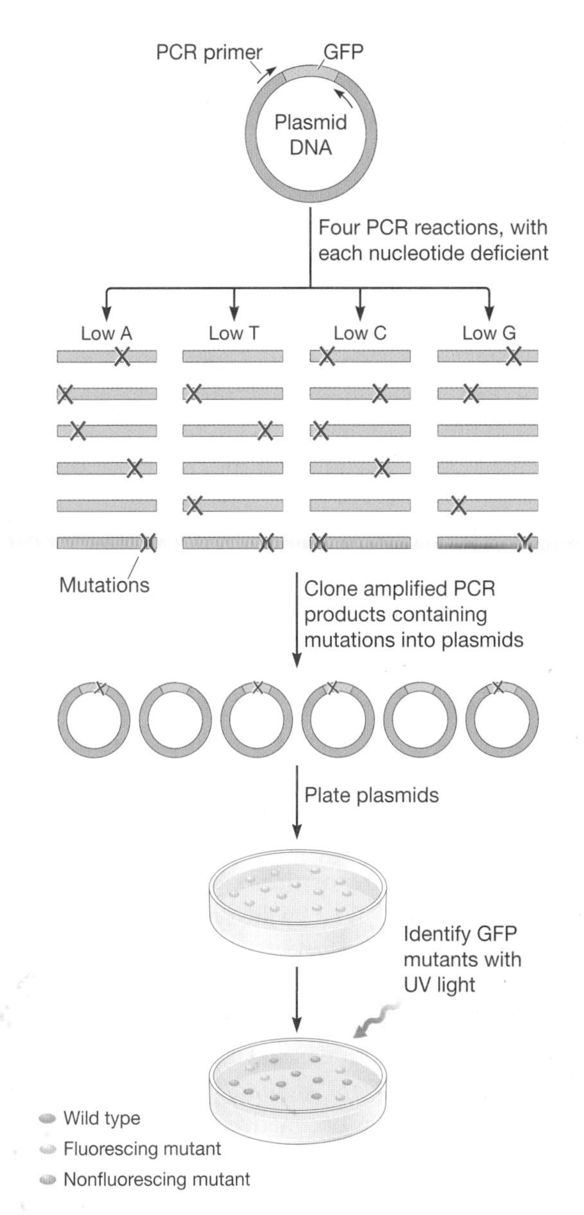

FIGURE 6-2

Random mutagenesis of a gene and screening for mutants. To create new variants of the green fluorescent protein (GFP), the gene was amplified under mutagenic PCR conditions. To increase the frequency of mutations in PCR, a non-proofreading polymerase is used, $MnCl_2$ rather than $MgCl_2$ is included, and nucleotide concentrations are imbalanced so that one nucleotide is at low concentration. Four PCR assays are performed, each deficient in a different nucleotide. The amplified PCR products containing mutations are cloned into plasmids for expression. The plasmids are plated out so that each colony grows from a single cell containing a single mutant plasmid. GFP mutants that were brighter or had different colors could be identified by shining an ultraviolet light on the colonies. Note that the mutagenic PCR destroyed GFP function in many of the colonies. The precise mutations in each colony of interest are identified by sequencing the plasmid. In the actual experiment, only a few colonies of several thousand had the desired properties.

the desired sequence. The simplest method for oligonucleotide-directed mutagenesis relies on enzymatic primer extension on a DNA template (Fig. 6-3). In this method, an oligonucleotide is designed that carries the mutation flanked by 10–15 nucleotides of wild-type sequence. The oligonucleotide and its reverse complement are synthesized, and these "mutagenic" oligonucleotides are hybridized to their complementary sequence in wild-type DNA, thus forming two heteroduplexes with mismatched nucleotides at the site of the mutation. Although the oligonucleotides are not perfectly complementary, they will anneal if the hybridization conditions are not too stringent. The annealed oligonucleotides then serve as primers for in vitro enzymatic DNA synthesis by a DNA polymerase. The polymerase synthesizes DNA all the way around the plasmid, which results in a plasmid incorporating the mutant primers and containing staggered nicks at the primer termini. In this way, all regions of the plasmid except the region containing the mutagenic oligonucleotide will be wild type in sequence. Multiple rounds of synthesis are performed by thermocycling. After amplification, the mutated DNA must be selected from the mixture of the original wild-type and mutant plasmids. The wild-type DNA can be removed by a clever approach that takes advantage of a natural process in the cell. It happens that DNA prepared from most strains of *Escherichia coli* is methylated at guanosine bases in the sequence GATC (a process called *Dam methylation*), but the DNA synthesized in vitro is not methylated. Thus, a restriction enzyme that cuts only methylated DNA selectively degrades the parental methylated plasmids. The remaining plasmids carrying the mutations are then transformed into *E. coli*, where the nicks are sealed by bacterial enzymes. The types of mutations that can be made by this approach (and a number of other similar methods) range from single-nucleotide substitutions to deletions or insertions, which are limited only by the size of the oligonucleotide needed.

These methods for site-directed mutagenesis are so simple that it is possible to explore the functions of a large number of mutant proteins. Detailed studies of the blood clotting protein factor VIII pinpointed a single amino acid important for binding of factor VIII to a calcium ion, which is needed for factor VIII to cleave its target and activate the clotting cascade. By making all 19 possible substitutions at the key calcium-binding amino acid, researchers found one change, from glutamic acid to alanine, that doubled the activity of the protein. Deficiency of factor VIII causes hemophilia A, and the development of a recombinant factor VIII with higher potency could be a more effective treatment for hemophiliacs than the wild-type protein.

Systems Are Developed for Inducible Gene Expression

The *lac* operon, described in detail in Chapter 3, was one of the first systems adopted for inducible gene expression. Expression of the genes in this operon is under control of the Lac repressor and the activator CAP (catabolic gene activator protein). Thus, if the

FIGURE 6-3

Production of oligonucleotide-directed mutants. (*a*) Plasmids containing the cloned DNA to be mutated are produced in a strain of *Escherichia coli* that has *dam* methylase activity so that GATC sites are methylated at G, indicated by the letter Ms on the target DNA. Plasmids are isolated and heated to separate the double-stranded DNA. (*b*) Oligonucleotides complementary to the cloned DNA but containing mutated bases (indicated by the X) are annealed to the cloned DNA. (*c*) Thermostable polymerase is added so that the mutagenic oligonucleotides prime synthesis of DNA strands to extend around the plasmid. The reaction contains four standard dNTPs, which produce a new unmethylated strand. (*d*) Cycles of annealing and extension are repeated, generating a population of mutagenized, unmethylated plasmids in addition to the methylated template strands. (*e*) After DNA synthesis, the mixture is digested with a methylation-specific restriction enzyme such as DpnII, which cleaves only the original, methylated DNA strand. (*f*) The uncut mutated cloned DNA is used to transform *E. coli*. Once inside the bacterium, the nicks remaining from the original nucleotide are repaired. When this procedure is used, ≥80% of colonies contain mutant plasmids. These mutant plasmids can be confirmed by DNA sequencing or, if a convenient site is available, by restriction digestion.

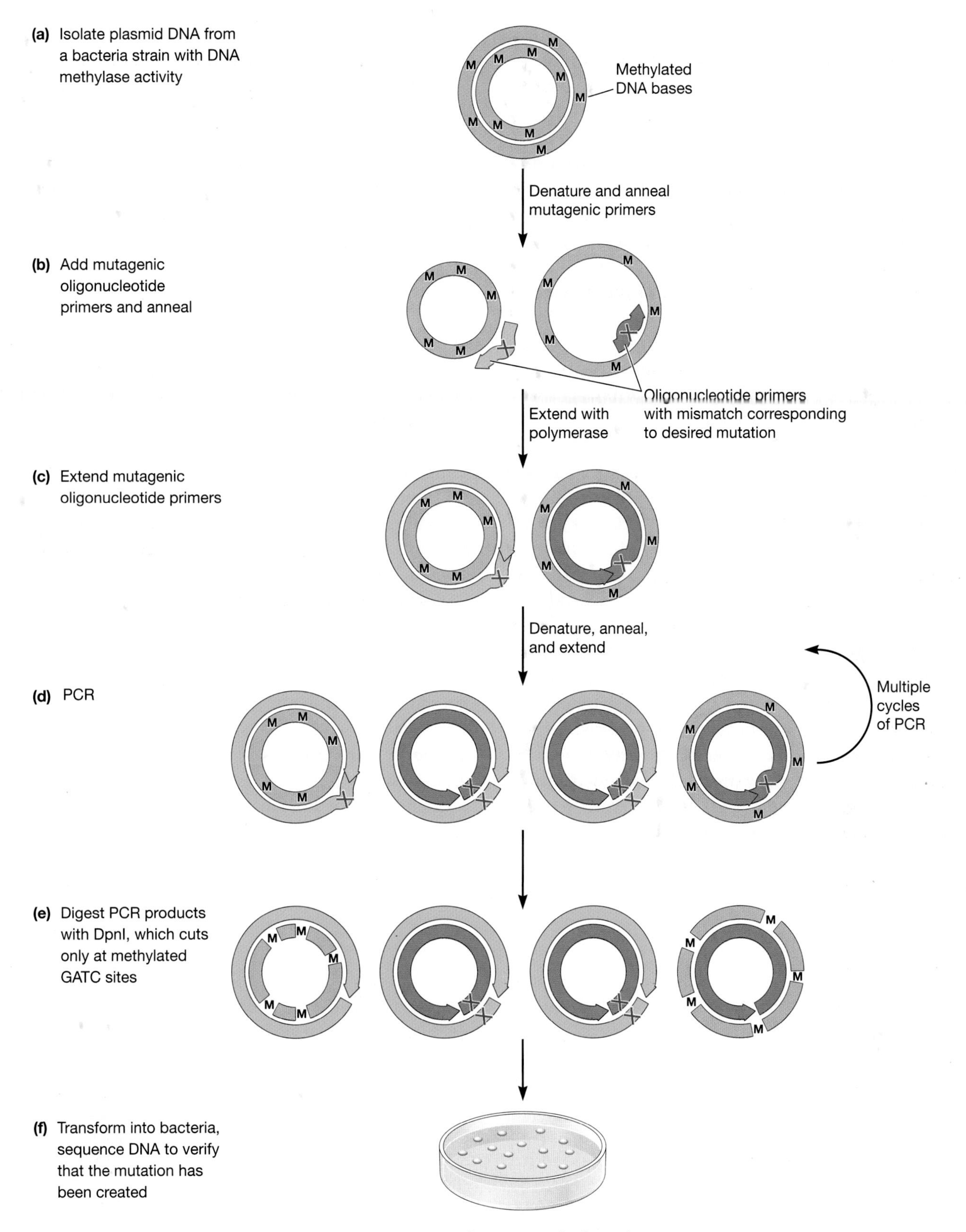

FIGURE 6-3 *(See facing page for legend.)*

β-galactosidase gene, the first gene in the *lac* operon, is replaced by a gene of interest, expression of that gene can be controlled by the addition of lactose (or synthetic inducers) to the medium. The proper regulation of a gene under control of the *lac* operator depends on Lac repressor, CAP, and the interaction of RNA polymerase with the proteins bound at the operator. Although the system is dependent on these multiple bacterial factors, the *lac* operator sequence can be inserted downstream of eukaryotic promoters. When Lac repressor is then expressed, it binds, which prevents transcription of the target gene. To induce gene expression, an inducer is added, which causes Lac repressor to dissociate from the promoter, thereby permitting complete transcription of the gene.

The most common method for inducible gene expression in eukaryotic cells is the more versatile *tet* operon system from *E. coli*. The tetracycline-resistance protein is encoded in the TN10 transposon. In the absence of tetracycline, Tet repressor binds tightly to the operator of the Tet-resistance gene, *tetO*, preventing transcription. When tetracycline is present, the repressor dissociates and transcription begins. The *tet* system was adapted for mammalian cells by creating a novel transcription factor that combines modules from two different transcription factors. The new transcription factor is made up of the DNA-binding domain from the Tet repressor and the activation domain of the herpesvirus protein VP16. The resulting activator protein binds to the *tet* operator only when tetracycline is absent. Therefore, when the *tet* operator is inserted upstream of a promoter, transcription is activated by the binding of the Tet-VP16 hybrid at the operator site. Addition of tetracycline causes the activator to dissociate from the operator sequence(s), so that transcription is no longer activated. This system is called the "Tet-off" system because it allows the experimenter to turn off a gene by adding tetracycline.

Because the binding can be introduced upstream of any promoter by careful placement of the *tet* operator sequences, *tet* systems have been developed for use in yeast, plants, and animals. For example, the Tet-off system was used to investigate the role of a tumor-promoting gene, *myc*, in cancer. Overexpression of Myc protein in white blood cells causes cancer. However, biologists did not know whether the continued expression of *myc* was required for the growth of tumors. Using a Tet-off system, the researchers expressed *myc* in mice. As expected, tumors developed in these mice. Amazingly, when *myc* was turned off by administering tetracycline to the mice, the tumors were blocked and the mice recovered—even when *myc* expression was subsequently restored (Fig. 6-4).

Extensive studies and mutagenesis experiments using the Tet repressor DNA-binding domain yielded a mutant tetracycline repressor protein that does the reverse—it *binds* DNA only when tetracycline is present. This mutant allowed the creation of an oppositely controlled system, the Tet-on system, which is active only when tetracycline is present. Both systems utilize the same *tet*-binding operator sequences, so by introducing either the *tTA* or *rtTA* gene, it is possible to switch from suppression by tetracycline to activation by it.

FIGURE 6-4

The Tet-off system is used to test whether continued Myc expression is necessary to maintain tumor growth. The DNA-binding domain of the bacterial Tet repressor gene (*tetR*) is fused to the herpesvirus transcriptional activation domain VP16. This hybrid activator is produced under the control of the promoter for the heavy chain of antibodies, which is active only in antibody-producing white blood cells (B cells). The cancer-promoting gene *myc* is also introduced by using a promoter that requires an activator protein for expression. There are Tet-responsive sequence elements (TREs) upstream of this promoter that can be bound by the Tet repressor. When tetracycline is absent, the Tet-repressor–activator hybrid binds the TREs and activates transcription of *myc*. When tetracycline is present, the Tet repressor protein cannot bind, and so transcription halts. Transgenic mice were prepared with both genes. These transgenic mice died rapidly from lymphoid tumors. However, if, having developed tumors, they were treated with tetracycline so that the *myc* gene was no longer activated by tTA, the tumors regressed and as many as 80% of the mice survived, which shows clearly that tumors required continued Myc protein for growth. A "Tet-on" system is also available, in which a gene is activated by the addition of tetracycline. In the Tet-on system, the Tet repressor portion of the activator protein has been mutated to create what is called the reverse Tet repressor. In contrast to the normal protein, this mutant protein binds DNA only in the presence of tetracycline, so that the target gene is activated only in the presence of tetracycline.

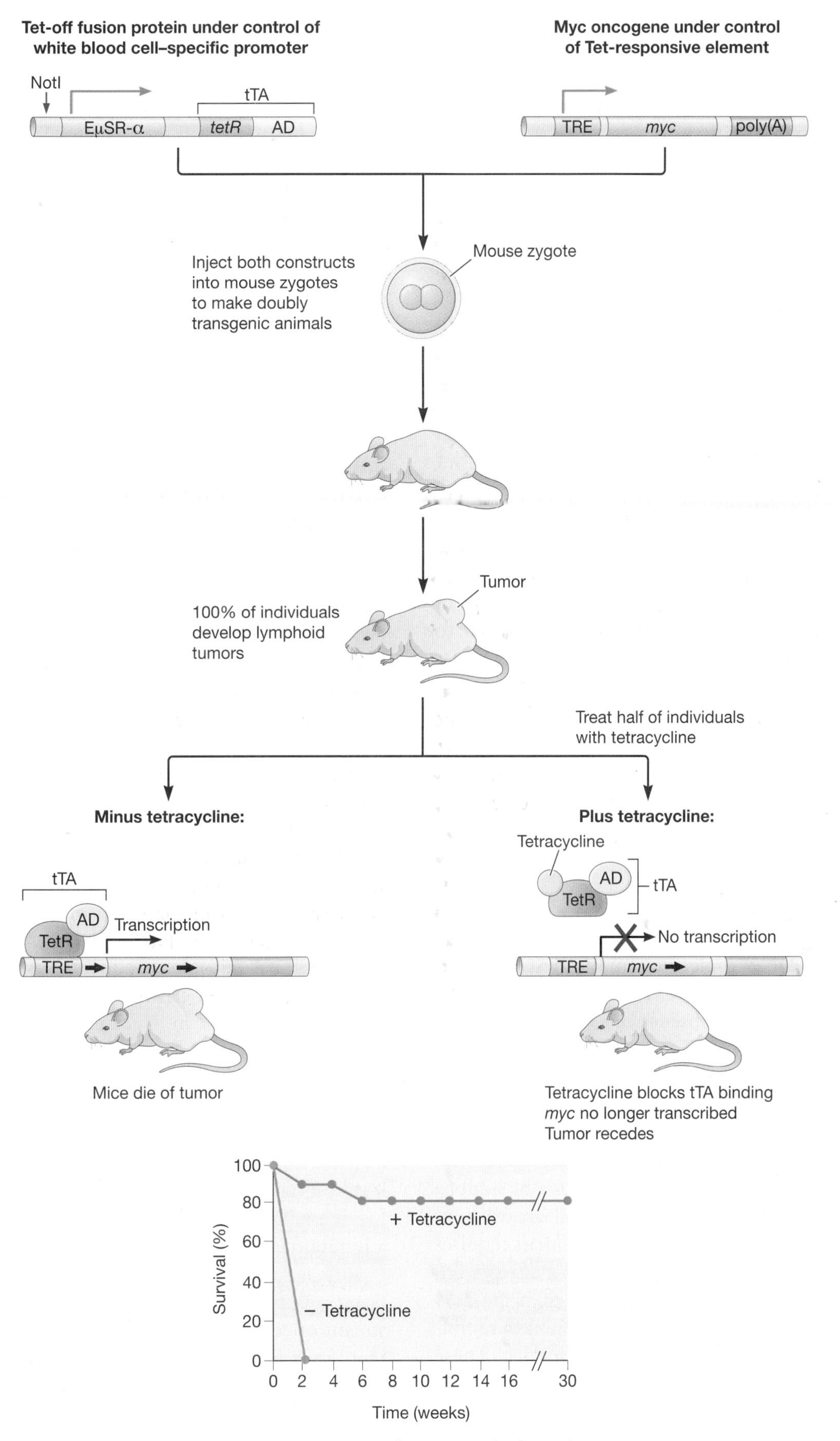

FIGURE 6-4 *(See facing page for legend.)*

Different Promoters Have Different Transcription Levels and Are Active in Different Cell Types

Generally, promoters function properly only in related cell types—bacterial promoters work well in bacteria (often even between very different bacterial species) but work poorly in eukaryotic cells, and the converse is true as well. It is important, then, to choose promoters carefully when constructing expression vectors. For example, "strong" promoters from viruses or promoters for highly expressed cellular genes such as actin will be used for maximum gene expression, whereas in other cases, lower levels of expression may be needed and a "weaker" promoter that is transcribed at a lower rate is appropriate. Many promoters have been discovered that are *tissue specific*, that is, are active only in a particular cell type. When genes are placed under control of such promoters, their expression can be limited to specific tissues within an organism. These promoters are especially useful in the production of transgenic animals, as we will see later in this chapter.

Selectable Markers Are Developed for Eukaryotic Cells to Allow the Creation of Stable Cell Lines

Once a piece of DNA is introduced into a population of cells, one must be able to identify the cells bearing the DNA molecule. In the description of gene cloning in Chapter 4, we saw that genes permitting growth in the absence of a key nutrient or in the presence of an antibiotic can be used for selection of bacteria. Similar approaches are used for yeast, mammalian cells, and plants, although different drugs are used because many antibiotics developed to kill bacteria do not affect eukaryotic cells. For example, antibiotics that kill bacteria by inhibiting synthesis of the bacterial cell wall are without effect on eukaryotic cells. If an experiment requires multiple plasmids, a different selectable marker can be incorporated in each plasmid and then the cells will be selected on medium containing each antibiotic (or lacking each applicable nutrient). These selectable markers permit the growth of only those cells bearing the marker and generally require the extended growth of cells in selective medium to eliminate those lacking the piece of DNA with the marker.

Selectable markers are genes that add a new trait to cells. Those we have just seen confer survival, but it is possible to isolate cells using other traits. In bacteria, the gene for β-galactosidase is often used in conjunction with a special dye that is processed into a blue precipitate in cells expressing the enzyme. Colonies of bacteria containing the vector can be picked on the basis of their color (as we describe in Chapter 4). A similar selection system for plants makes use of another *E. coli* gene, that for the enzyme β-glucuronidase (GUS), which uses glucuronides as substrates. Its advantage as a reporter gene for plants is that plants have virtually undetectable levels of GUS, because the enzyme is restricted to vertebrates and their microorganisms. When plant cells expressing β-glucuronidase are incubated with X-glucuronide (X-gluc, analogous to X-gal), a blue color is produced that can be detected histochemically; with a different substrate, GUS can be measured quantitatively with a fluorimeter.

However, both β-galactosidase and β-glucuronidase have a significant disadvantage as marker systems for cells other than bacteria—the histochemical reaction to produce the blue color kills the cells. Fortunately, researchers soon found even more versatile markers that could be used with living cells. The first visual marker developed for living eukaryotic cells was green fluorescent protein (GFP) found in a jellyfish. The protein absorbs blue light and emits green light. Cells expressing GFP glow green when observed in a microscope using ultraviolet light. GFP has become a frequently used marker for promoter activity in cells. It can even be imaged in whole animals to see in which cells the gene is active and can be fused to other proteins to determine their localization within cells. GFP and other fluorescent proteins are now available in a rainbow of colors—ranging from blue to red. Many biologists use the *luciferase* gene in a similar manner. When luciferin and ATP are added to cells expressing luciferase, light is produced, which is measured with a luminometer.

Bulk isolation of GFP-expressing cells can be done using a fluorescence-activated cell sorter (FACS). A stream of cells passes, one cell at a time, past a laser that emits light of the correct wavelength to excite GFP in cells expressing GFP. These cells are given a small positive electrostatic charge and, as they pass a charged plate, they are deflected and collected. Another approach can be used to select cells expressing a gene of interest from a mixed culture, if an anti-

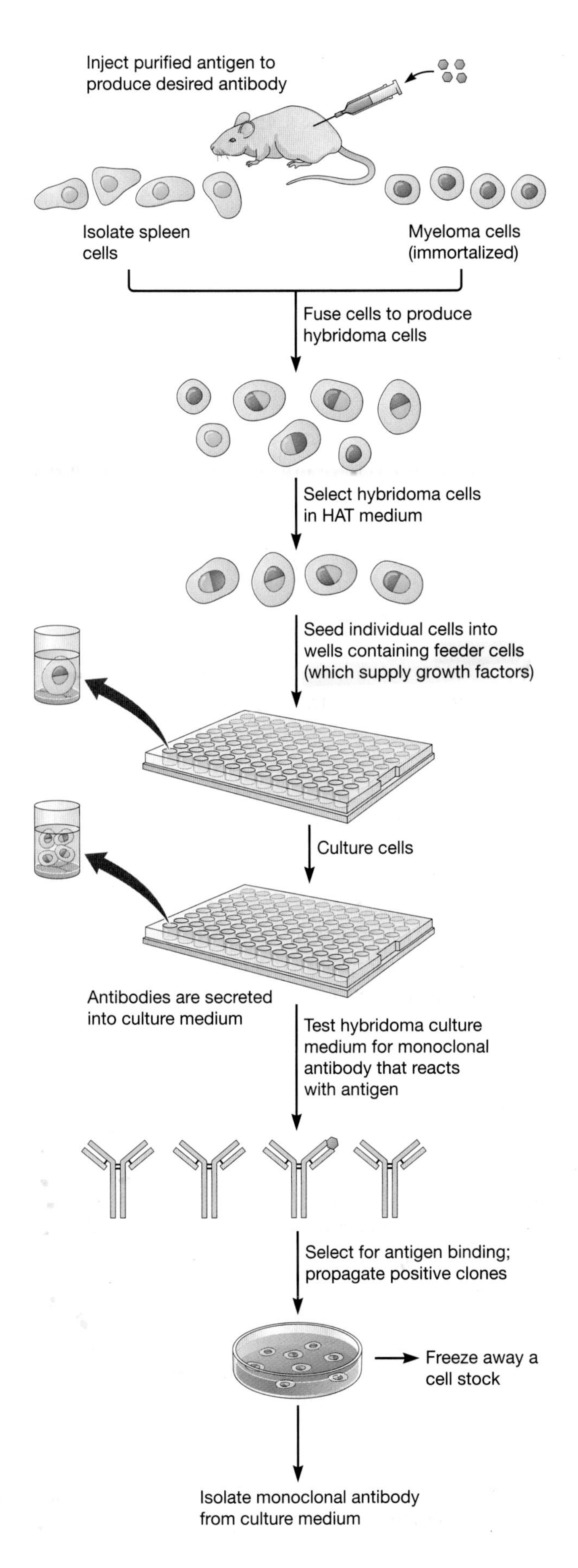

body is available that recognizes the protein being expressed. Once the target gene is cloned, it can serve as a marker by using the antibody to detect the cells expressing that gene.

Monoclonal Antibodies Can Be Produced to Recognize Any Molecule

Antibodies are exquisitely selective proteins produced by the immune system that can bind to a single target among millions of irrelevant sites. Researchers long dreamed of harnessing the specificity of antibodies for a variety of uses that require the targeting of drugs and other treatments to particular sites in the body. It is this use of antibodies as targeting devices that led to the concept of the "magic bullet," a treatment that could effectively seek and destroy tumor cells and infectious agents wherever they resided.

The major limitation in the therapeutic use of antibodies was the difficulty of producing a useful antibody in large quantities. Initially, researchers screened myelomas, which are antibody-secreting tumors, for the production of useful antibodies. But it was not possible to program a myeloma to produce a specific antibody. This situation changed dramatically with the development of monoclonal antibody technology. The procedure for producing monoclonal antibodies, or mAbs, is shown in Figure 6-5. First, a mouse or rat is inoculated with the antigen to

FIGURE 6-5

Production of a monoclonal antibody (mAb). A mouse is inoculated with an antigen. This stimulates the proliferation of lymphocytes expressing antibodies against the antigen. Lymphocytes are taken from the spleen and fused to myeloma cells by treatment with polyethylene glycol. Hybrid cells that express hypoxanthine phosphoribosyltransferase (HPRT) are selected for growth in hypoxanthine (HAT) medium. Myeloma cells lack HPRT and thus die in this medium unless they become fused with a lymphocyte, which expresses the missing enzyme. Unfused lymphocyte cells soon die off as well, because they do not grow for long in culture. Individual hybrid cells are transferred to the wells of a microtiter dish and cultured for several days. Aliquots of the culture fluids are removed and tested for the presence of antibody (Ab) that binds the antigen. Cells that test positive are cultured for monoclonal antibody production. Hybridoma cell lines can be cultured indefinitely. Antibody-producing cell lines are stored frozen in liquid nitrogen.

which an antibody is desired. After the animal mounts an immune response to the antigen, its spleen, which houses antibody-producing cells (B lymphocytes), is removed, and the spleen cells are fused en masse to a specialized myeloma cell line that no longer produces an antibody of its own. The resulting fused cells, or hybridomas, retain properties of both parents. They grow continuously and rapidly in culture like the myeloma cell, yet they produce antibodies specified by the lymphocyte from the immunized animal. Hundreds of hybridomas can be produced from a single fusion experiment, and they are systematically screened to identify those producing large amounts of the desired antibody. Once the appropriate hybridoma line is identified, the antibody becomes available in limitless quantities. Monoclonal antibodies are already widely used for the diagnosis of infections and cancer and for the imaging of tumors for radiotherapy. In Chapter 15, we will see how several mAbs are used to treat cancer.

Recombination Technologies Allow Rapid Exchange of DNA Fragments between Vectors

A gene will have been isolated in a vector specifically designed for cloning cDNAs or large genomic fragments, but this may not be a suitable vector for subsequent studies. For example, the researcher might subsequently want to examine the gene's expression in bacteria or in yeast, analyze its function in animal cells, and simultaneously use insect cells to prepare large quantities of the protein for structural studies. Each of these applications requires the use of different promoters, and most projects require repeatedly transferring the gene into different vectors with different selectable markers ("subcloning"). Although molecular cloning often seems straightforward on paper, in practice subcloning can be rather tricky, and producing multiple versions of a gene often consumes a great deal of time and effort. Subcloning a gene between vectors is especially complicated when different vectors have different cloning restriction sites, and, conversely, it becomes very complicated if several genes need to be cloned into the same vector as each gene will require a unique cloning strategy based on available restriction enzyme sites. Fortunately, several methods have been developed to move sequences from one vector to another without having to use restriction enzymes. In principle, all of these methods are similar to restriction enzyme approaches, because a compatible sequence has to be established in both the vector and the insert, thus allowing them to be joined.

One of the most straightforward approaches to transferring DNA between vectors exploits a peculiar property of *Taq* DNA polymerase. When *Taq* polymerase is used in PCR, the enzyme tends to add an extra adenosine residue at the 3′ terminus of the PCR product. This single-stranded overhang can be used as if it were a sticky end produced by a restriction enzyme and can be ligated into a vector that has been digested so that it contains a 5′ T overhang. This approach is called TA subcloning, and it can be used to clone PCR products without needing to digest them with restriction enzymes. There are even methods for TA subcloning that avoid the use of ligase, which instead use a *recombinase* enzyme that cuts and then reconnects DNA strands. One type of recombinase is called topoisomerase, which untangles DNA strands and relaxes over- or undertwisting of DNA molecules during DNA replication. Topoisomerase can be used in TA cloning by cutting the vector and allowing topoisomerase to covalently attach to the ends. When the PCR product is added, the topoisomerase covalently links the strands to the ends of vector and PCR fragments (Fig. 6-6). Topoisomerase cloning can also be used to recombine types of ends other than TA.

Some recombinases will exchange strands between any two DNA molecules containing a region of identical sequence. This process, called *homologous recombination*, occurs naturally during meiosis, when sister chromatids pair and exchange genetic information. Other than in germ-line cells undergoing meiosis, homologous recombination is a relatively rare event. However, the budding yeast, *Saccharomyces cerevisae*, is particularly adept at homologous recombination and this opened the door to a new range of genetic experimentation in this yeast: Each of the 6000 genes in the yeast genome has been systematically knocked out using homologous recombination.

Homologous recombination techniques have also been developed for manipulating DNA in *E. coli*. Although *E. coli* has its own mechanism for homologous recombination, expressing a group of recombinase genes from phage λ in *E. coli* significantly increases the recombination efficiency of *E. coli*. Such recombinogenic *E. coli* strains can be used to efficiently swap pieces of DNA between different plasmids (Fig. 6-7). Standard restriction enzyme–based methods for cloning require a different strategy for

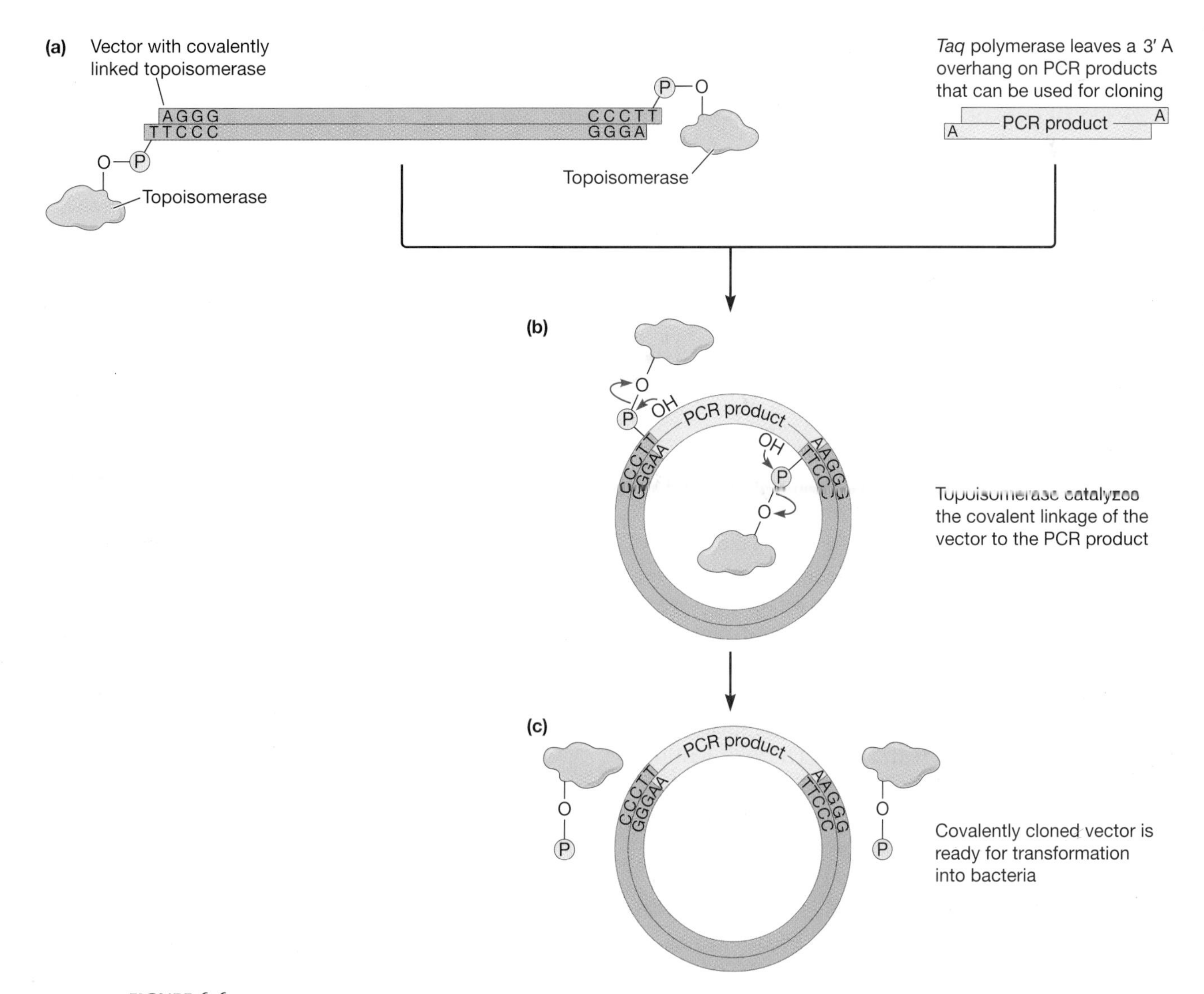

FIGURE 6-6

Topo TA cloning substitutes DNA ligase with topoisomerase. (*a*) A PCR product is prepared using *Taq* polymerase, which leaves 3´ A overhangs at both ends. The vector is treated with topoisomerase I from *Vaccinia* virus, which binds to DNA at specific sites and cleaves the DNA phosphodiester backbone after 5´CCCTT. The topoisomerase remains covalently linked to the DNA strand through a 3´-phosphate residue. The vector, which has overhanging 3´ T residues, is mixed with the PCR product. (*b*) The 5´-OH group of the PCR product attacks the bond between the phosphate group and the topoisomerase. (*c*) After catalyzing the reaction, topoisomerase is no longer bound to the vector, leaving the covalently closed vector and insert. The requirement for a 5´-OH terminus makes the reaction extremely selective for PCR products, because primers are typically synthesized without a 5´-phosphate. Other fragments of DNA in the reaction (for instance, the other vector molecules) cannot be linked by topoisomerase unless they too have a 5´-OH.

each gene, which depends on the sites that are available. Homologous recombination methods are an extremely useful technological advance because they can be adapted to work with any vector. This allows the investigator to use the same tool kit to simultaneously swap many different genes between several vectors of interest. Bacterial strains capable of homologous recombination have also proven particularly useful for the manipulation of large pieces of DNA. Cloning vectors called bacterial artificial chromosomes (BACs) have been developed, which can accommodate hundreds of kilobases of DNA. BACs are ideal for studying large segments of the genome, but they contain inserts so large that there are few if any unique restriction sites. Instead, sequences are replaced or inserted at a specific site using homologous recombination in a manner very similar to that used for constructing knockouts in yeast.

A special class of recombinases is *site specific*, recognizing specific DNA sequences and catalyzing rearrangement between them. At the end of this chapter we will see how the site-specific recombinase Cre is used to generate rearrangements within the mouse genome.

Foreign Genes Can Be Transferred into Cells by Physical Methods

Manipulating the sequence of a gene in a test tube is not sufficient for discovering its function—it remains to study the functioning of the gene in a living cell and this requires persuading a cell to take up the gene. Cells do not take up DNA easily—this could have dire consequences should a cell take up a fragment of a harmful piece of DNA from its environment. So, geneticists have had to devise clever methods to bypass natural barriers to gene transfer.

Fortunately, many bacterial and yeast species can be directly *transformed* with DNA. The first example was the transformation experiments of Oswald Avery that demonstrated that DNA was most likely the genetic material (Fig. 1-11). The efficiency of transformation has been improved by first treating bacteria with salts such as calcium chloride, adding plasmid DNA, and then exposing the cells to a brief heat shock, for example, 42°C for 40 seconds. Another effective approach for gene transfer is *electroporation*, in which cells are placed in a solution containing DNA and subjected to a brief electrical pulse that causes holes to open transiently in their membranes. DNA enters through the holes directly into the cytoplasm and then passes into the nucleus (Fig. 6-8).

Electroporation is also an effective means for introducing DNA into most animal cells, but the earliest gene transfer experiments into animal cells were done with DNA tumor viruses (i.e., viruses whose genes are encoded in DNA). DNA was isolated from purified viruses and introduced into cultures of uninfected cells. These cells eventually produced fully infectious viruses. This DNA-mediated transfer of infectious virus was dubbed *transfection*, to distinguish it from *infection*, the natural route of entry for viruses.

There are a wide variety of methods used to transfer DNA directly into cells. Perhaps the simplest to understand, although quite challenging to do, is microinjection, in which DNA is injected directly into the nucleus of cells through fine glass needles. Microinjection is an efficient process on a per cell basis—that is, a large fraction of the injected cells actually receive the DNA. When first developed, microinjection was a time-consuming process and only a few cells could be injected in a single experiment. Once again, computers have revolutionized a technique; in

FIGURE 6-7

Recombinogenic strains of *Escherichia coli* can be used to swap segments of DNA between plasmids in *E. coli*. Two different strains of bacteria each contain a different plasmid. (*a*) In the donor strain, the *pir*-116 allele of the Π replication protein permits replication of the donor plasmid, which has a RK6γ origin of replication. This donor strain also contains the F plasmid (F for fertility), which contains the transfer operon, Tra. This operon allows the transfer via mating from oriT on the donor plasmid into the mating strain. (*b*) In the recipient strain, λ recombinase is under the control of the inducible *lac* operon. The I-SceI restriction enzyme is under the control of the arabinose (Ara)-inducible promoter. (I-Sce I is a *Saccharomyces cerevisiae* restriction endonuclease. It has an 18-bp recognition site that occurs very rarely in DNA.) The donor and recipient vectors contain regions of homology. When the donor and recipient strains are mixed, conjugation occurs and the donor plasmid is transferred into the recipient strain. Then, arabinose is added to the media, which triggers expression of the I-SceI restriction enzyme. The donor segment of DNA and the target location on the recipient vector are flanked by I-Sce I restriction enzyme sites. After the induction and expression of I-SceI, the restriction enzyme cleaves the plasmid, releasing the target gene for recombination with the recipient vector. Recombination is triggered by addition of the *lac* inducer IPTG, which activates the λRed recombination genes. The recombinase catalyzes strand exchange between the regions of homology at the termini of the donor insert and the recipient plasmid. Recombinants are selected by selecting for ampicillin resistance, Ap, present on the recipient plasmid. The donor plasmid contains only a kanamycin resistance gene and cannot grow on medium containing ampicillin. Nonrecombinants are selected against because the insert on the original plasmid contains a gene called *pheS* that is toxic to *E. coli* when the phenylalanine analog D,L-*p*-Cl-phenylalanine is added to the culture medium. Also, the RK6γ origin of replication on the donor plasmid is only functional in the donor cells that contain the *pir*-116 gene. Thus, the donor plasmid is only capable of replication in the donor strain and not in the non-*pir* recipient strain.

this case, computer-controlled, automated injection means that hundreds of cells can be injected.

The first method used for introducing DNA into animal cells en masse was to incubate the DNA with an inert carbohydrate polymer (dextran) to which a positively charged chemical group (DEAE [diethylaminoethyl]) has been coupled. The DNA sticks to the DEAE–dextran via its negatively charged phosphate groups. These large DNA-containing particles in turn stick to the surfaces of cells, which are thought to take them in by a process known as *endocytosis*, a normal feature of membrane turnover. Some of the DNA evades destruction in the cytoplasm of the cell and escapes intact into the nucleus,

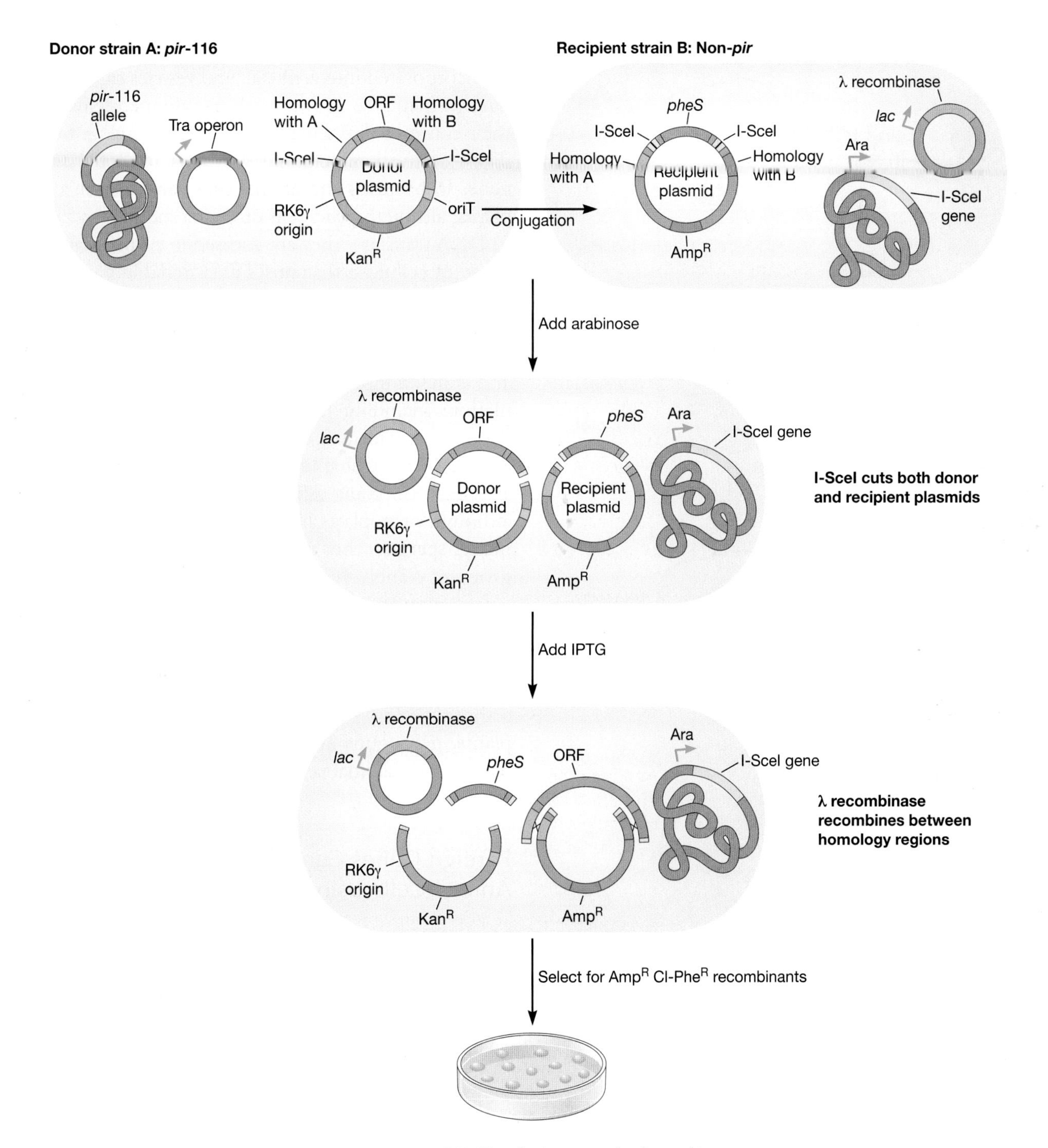

FIGURE 6-7 *(See facing page for legend.)*

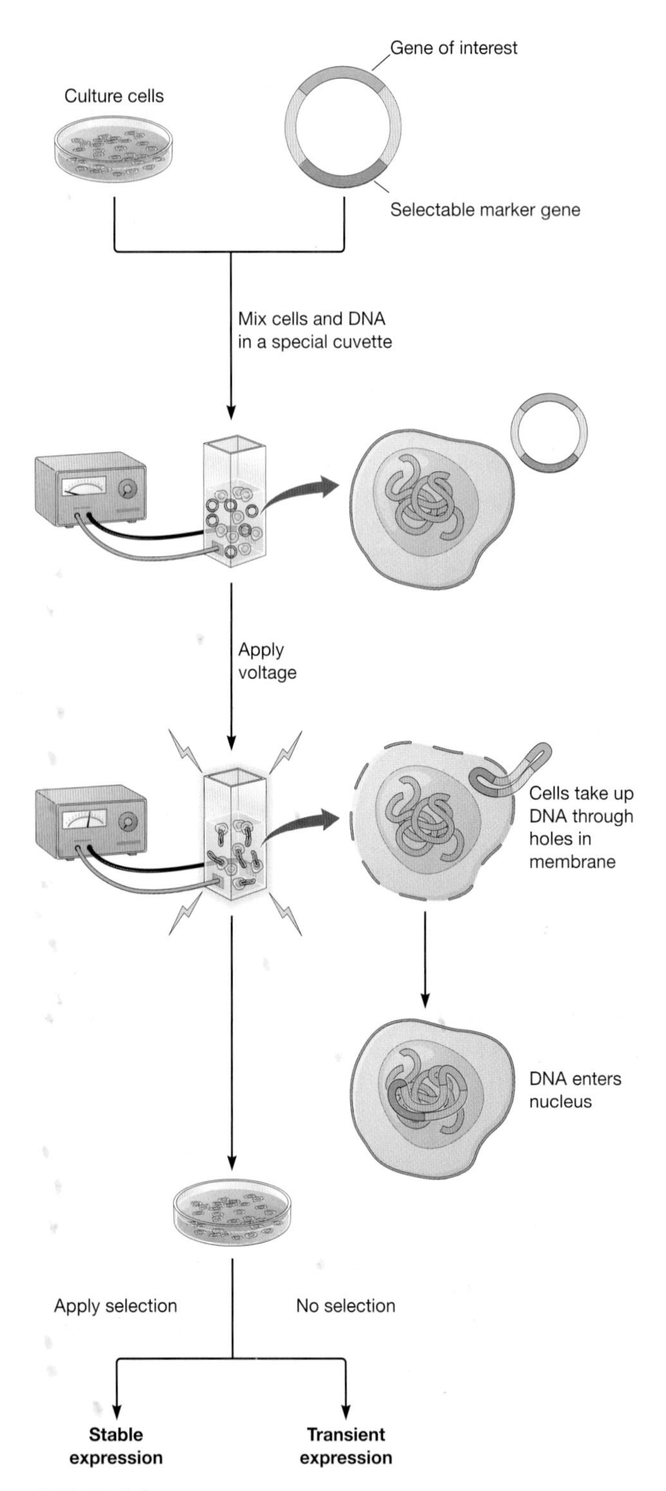

FIGURE 6-8

Gene transfer by electroporation. Cells are mixed with the DNA to be transfected and placed in a small chamber with electrodes connected to a specialized power supply. A brief electric pulse is discharged across the electrodes, which transiently opens holes in cell membranes. DNA enters the cells, which are removed and plated in fresh medium. The cultures can be harvested for experiments during the transient expression phase, or selection can be applied to isolate stably transfected clones.

where it can be transcribed into RNA like any other gene in the cell.

The DEAE–dextran method, although relatively simple, was very inefficient for many types of cells. The breakthrough that eventually made gene transfer a routine tool for workers studying animal cells was the discovery that cells efficiently took in DNA in the form of a precipitate with calcium phosphate. This discovery arose from prior work showing that divalent cations such as calcium and magnesium promoted the uptake of DNA into bacteria. DNA can also be incorporated into artificial lipid vesicles called *liposomes*, which fuse with the cell membrane, delivering their contents directly into the cytoplasm (Fig. 6-9).

The same gene-cloning methods used to study genes in bacterial and animal cells can be applied in plants, although plant cells present a special challenge to DNA transfer—they are encased in a thick cell wall made of cellulose that poses a formidable barrier to efficient gene transfer. One approach is to remove this wall using fungal cellulase enzymes. The resulting ***protoplast*** is enclosed only by a plasma membrane and is much more amenable to experimental manipulations including transfection and electroporation. Protoplasts will take up macromolecules like DNA, and they are capable of regenerating whole plants, as are nearly all plant cells (Fig. 6-10). Unfortunately, although protoplasts have been used successfully for many species, the most agriculturally important group of plants—cereals—are very difficult to regenerate from protoplasts. In a stunningly direct approach, DNA can be absorbed on the surfaces of tungsten or gold microprojectiles, 1 μm in diameter, which are fired into intact cells as though pellets from a shotgun (Fig. 6-11). Although first developed for plants, gene guns are also used to introduce DNA directly into animal cells in tissues.

Foreign Genes Can Be Transferred into Animal Cells Using Viruses

Many DNA transfer methods have low efficiency, and some cell types are completely refractory to DNA transfer using these artificial methods. Instead, researchers have turned to viruses, natural gene delivery systems that work very efficiently. Viruses have to grow in cells and they have to get their genomes into cells. Evolution and natural selection have led viruses to develop clever ways to do so. In fact, the first gene transfer methods for bacteria used

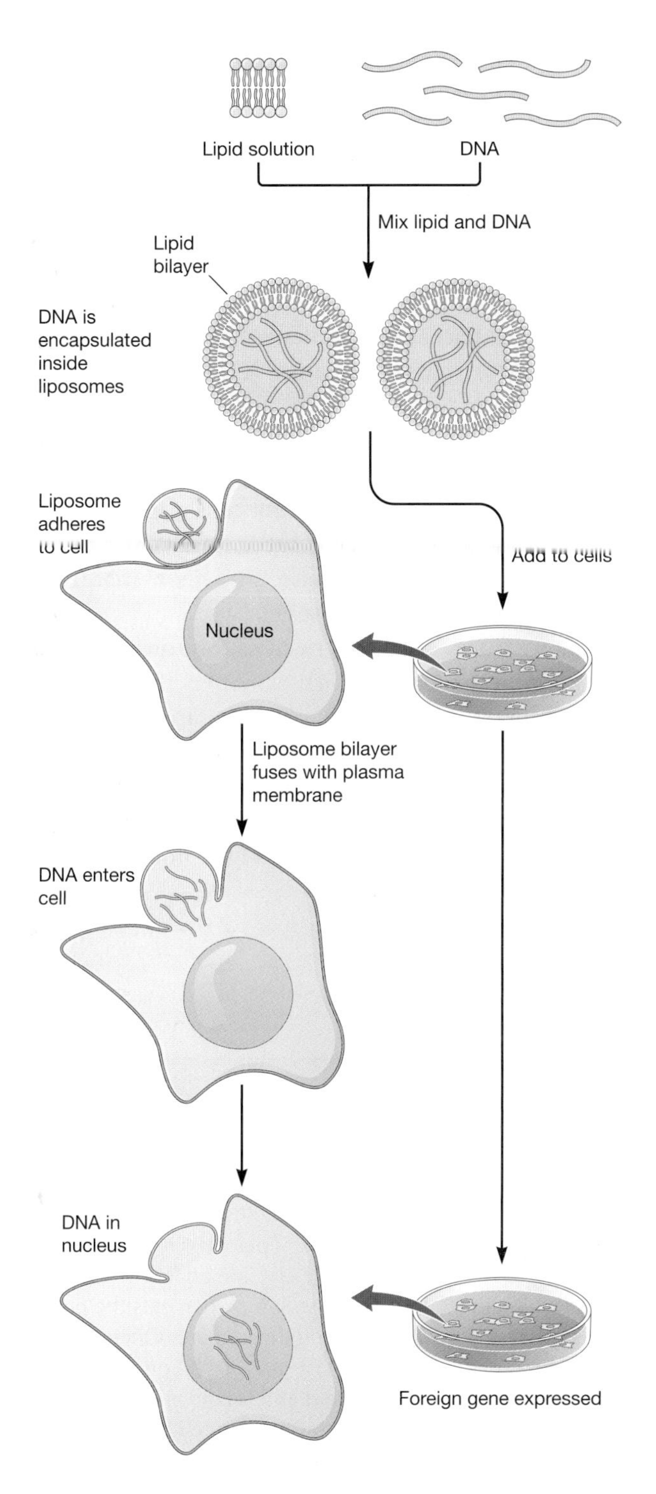

FIGURE 6-9

Transfection using liposomes (*lipofection*). DNA, which is negatively charged at near-neutral pH because of its phosphodiester backbone, is mixed with lipid molecules with positively charged (cationic) head groups. The lipid molecules form a bilayer around the DNA molecules, which creates liposomes that are mixed with cells. Most mammalian cells are negatively charged at their surface, so the positively charged liposomes interact with the cells. Cells take up the lipid–DNA complexes, and some of the transfected DNA enters the nucleus.

bacteriophages, which are still in common use for gene transfer. The strategy is simple: The target gene is incorporated into the phage genome and transferred into the cell upon infection by the phage.

The earliest viral vectors for mammalian cells were based on the monkey tumor virus SV40, where it was possible to substitute some of the viral genes with foreign genes. The use of viruses like SV40, however, was limited because only small genes can be inserted into the SV40 genome before the DNA becomes too large to be packaged in the viral coat proteins. Furthermore, SV40 genomes are commonly rearranged or deleted after infecting cells. Many other types of viruses have now been adapted for use in gene transfer.

Many viruses are *lytic* viruses—after replicating in cells, they have to kill and disrupt the cells to be released and propagate the infection. So, these viruses cannot be used as vectors to stably introduce a gene into cells. However, a number of nonlytic viruses drawn from the *retrovirus, adenovirus,* and *adeno-associated* virus groups have been modified to transfer genes to animal cells. Retroviruses (including lentiviruses like HIV), after infecting cells, use reverse transcriptase to convert their RNA genome to DNA. This viral DNA is efficiently integrated into the host genome as a "provirus," where it permanently resides, replicating along with host DNA at each cell division. Adenoviruses and adeno-associated viruses are DNA viruses. (Adeno-associated viruses are so named not because they are related to adenoviruses but because they need the assistance of an adenovirus for replication.) Adeno-associated viruses, like retroviruses, integrate into the genome, whereas adenoviruses are maintained as extrachromosomal DNA molecules. These viruses make attractive vectors because they can infect nearly any cell, and although most retroviruses can infect only dividing cells, lentiviruses, adenoviruses, and adeno-associated viruses can infect nondividing cells. As many differentiated cells in the brain, muscles, and other tissues no longer divide, the development of the latter viral vectors has greatly expanded the range of experiments that can be carried out with these specialized cell types.

Viral vectors are prepared by removing most of the viral genes, leaving room for insertions of the genes of interest (Fig. 6-12). This means that these viral vectors cannot replicate. The vectors usually contain a selectable marker as well as the foreign gene to be expressed. To prepare virus stocks,

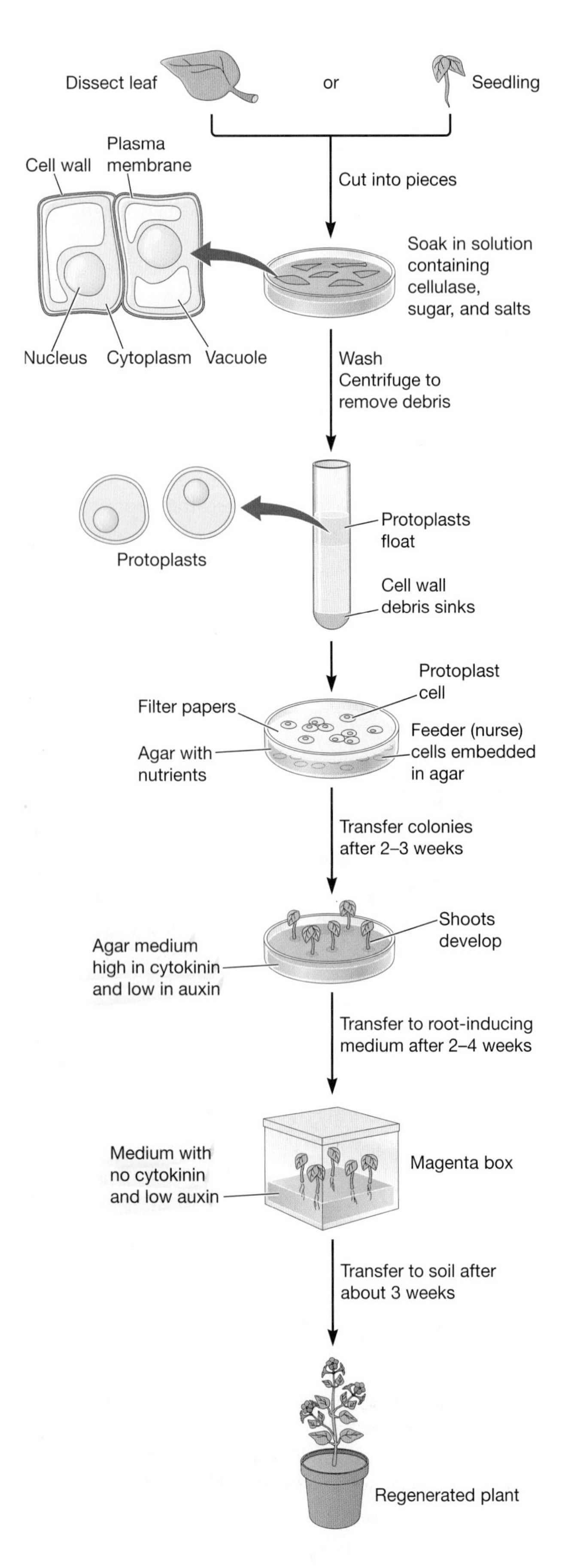

cloned proviral DNA is transfected into a *packaging cell.* These cells contain an integrated provirus with all its genes intact, but lacking the sequence that directs packaging of the viral genome in the viral coat proteins. Thus, the packaging provirus produces all the proteins required for assembling infectious virus particles, but it cannot package itself. Retroviral vectors carry the sequence of interest between the packaging sequences, so that the retroviral DNA is transcribed into RNA and packaged into infectious virus particles for release from the cell. The resulting virus stock, which carries the gene of interest but lacks wild-type replication-competent virus, can be used to infect target cells. On infection, the recombinant genome is reverse transcribed into DNA by reverse transcriptase, and then integrates into the genome of the host cell.

Adenoviruses enter the nucleus, but do not replicate. Adeno-associated viruses enter the nucleus, their single-stranded DNA genomes are made double stranded, and they integrate into the host cell's DNA. In both cases, the cells express the new virally introduced gene, but never produce any virus because the recombinant virus genome lacks the necessary viral genes.

FIGURE 6-10

Regeneration of plants from protoplasts. Leaf cells have a cytoplasmic compartment containing numerous chloroplasts, a large vacuole, and a nucleus. The plasma membrane is surrounded by a tough cellulose cell wall that can be removed by incubating pieces of plant tissue in a solution containing cellulase. Sugars and salts are added to the solution to maintain osmotic balance, which prevents the protoplast from lysing because of osmotic pressure. Once the cell debris is removed, the protoplasts are placed on solid nutrient-rich media high in cytokinin and low in auxin. The protoplasts rapidly regenerate a protective cell wall, and the nutrient-rich medium encourages them to grow. Shoots appear in a number of weeks. Then, the cultured cells are transferred to a container called a Magenta box, which contains root-inducing medium lacking cytokinin and low in auxin. Once roots appear, the plantlets can be placed in soil, where they develop into regenerated plants. Regeneration from a handful of cells can also be done from fragments of plants without going through the protoplasting process—in that case, cells transformed by other means (such as biolistics; see Fig. 6-11) are placed on growth medium and regenerated in the same way.

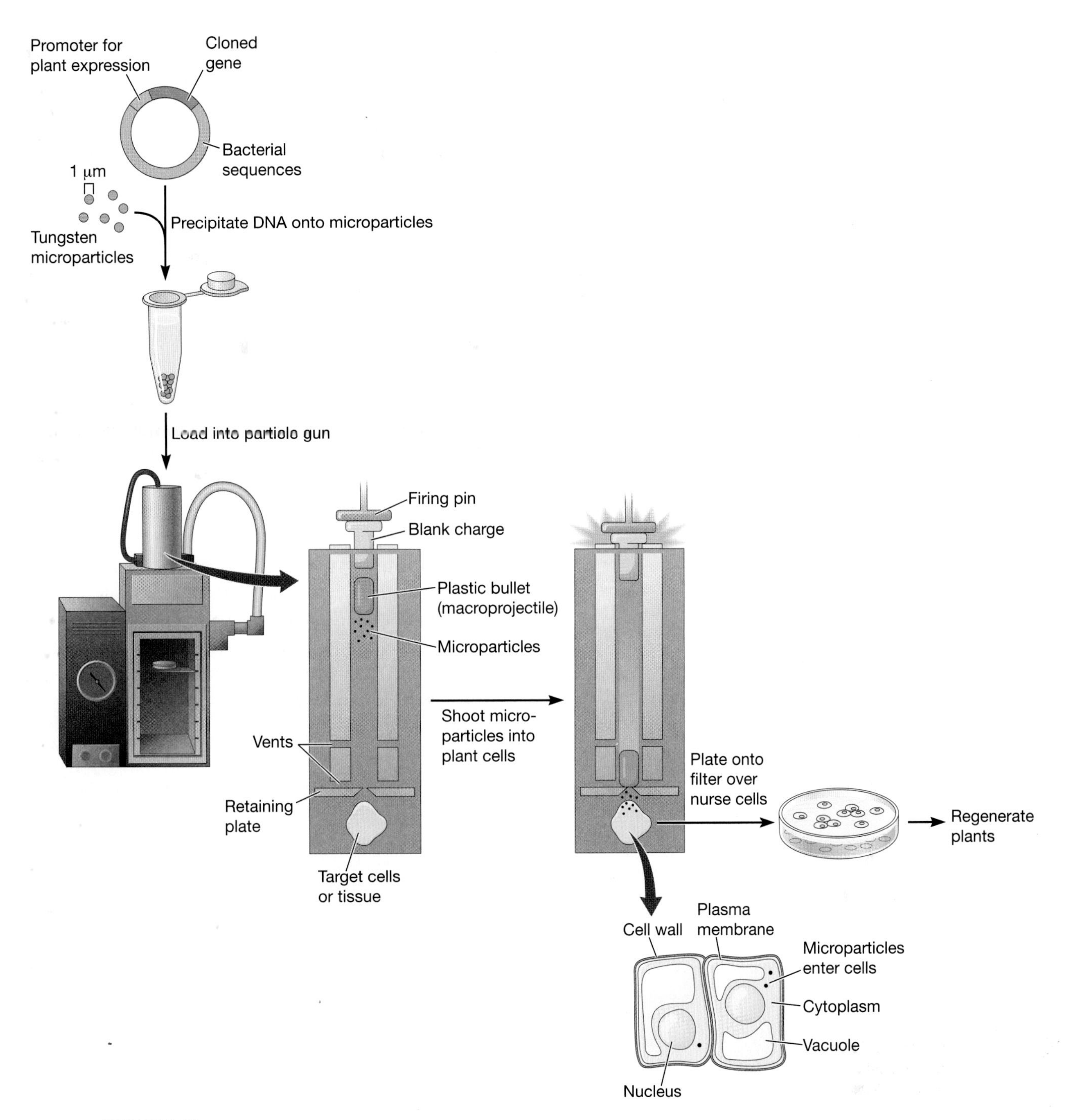

FIGURE 6-11

Direct transfer of DNA into plant cells by microprojectile bombardment ("shotgun"). A thick coat of DNA is deposited onto the surface of a 1-μm-diameter tungsten or gold particle by precipitation with calcium chloride. The beads are placed on the end of a plastic bullet (the "macroprojectile") in the barrel of a particle gun designed especially for this purpose. The target plant tissue or suspension cells are placed next to a small opening at the end of the barrel. The macroprojectile is propelled toward the cells by an explosive charge and, as it slams into the retaining plate, the particles it carries pass through the aperture and hit the cells. The barrel of the gun and the specimen chamber are usually evacuated, otherwise the air resistance slows the velocity of the microparticles. Plant cells can withstand a vacuum for as long as 2 min. Following bombardment, the cells are transferred to a cell culture plate and plants are regenerated as described in Fig. 6-10.

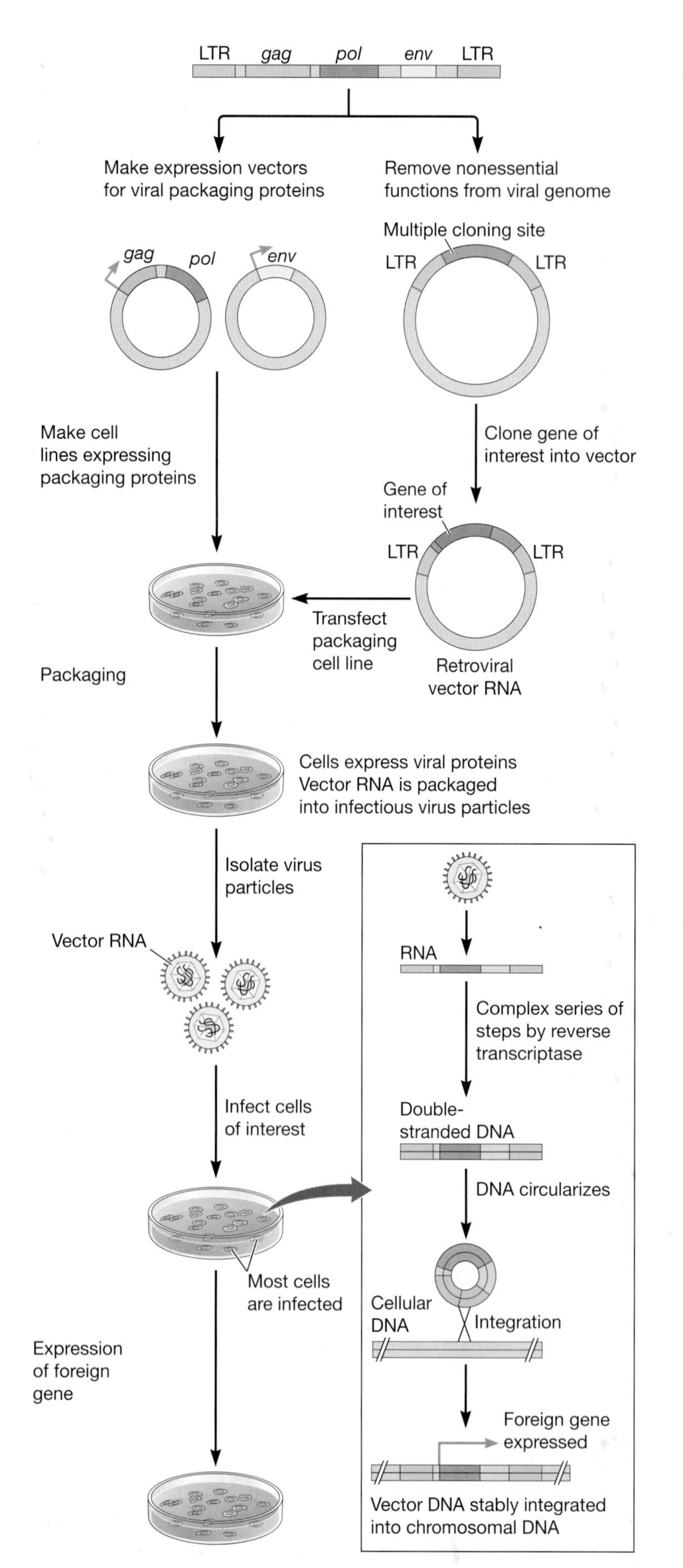

FIGURE 6-12

Use of a retrovirus vector for stable, long-term expression of a foreign gene. All retroviruses contain the three genes *gag, pol,* and *env,* which encode the main structural components of retroviruses—*gag* encodes a large protein that is proteolytically processed into components of the virion core; *pol* encodes, among other functions, the viral reverse transcriptase necessary to convert the RNA genome into DNA after infection; and *env* encodes the proteins that will form the virion envelope. All these are required for packaging their RNA genomes into viral particles. The genome is flanked by long terminal repeats (LTRs), which are needed for the retroviral genome to integrate into the infected cell's genome. The packaging proteins can be supplied in *trans* by expressing the *gag, pol,* and *env* genes from plasmids. These are stably transfected into cells to create a packaging cell line. The gene of interest can be cloned between the LTR sequences and be expressed under the control of the LTR, which has promoter activity. The retroviral vector also contains the Ψ sequence that directs packaging of the RNA. The recombinant viral plasmid is transfected into the packaging cell line and the RNA produced from the recombinant viral plasmid is packaged. Virus released from the packaging cells contains only the gene of interest flanked by LTRs. This virus can be used to infect many other cell types, which results in the integration of the gene of interest and the stable production of the protein encoded by the gene. Because the recombinant viruses do not contain *gag, pol,* or *env,* the infected target cells do not produce virus.

Foreign Genes Are Introduced into Plants Using the Ti Plasmid of *Agrobacterium*

The most popular method for DNA transfer into many types of plants uses *Agrobacterium tumefaciens*, a bacterium that infects plant cells and drives the infected cells to multiply wildly, thus producing the plant tumors known as *crown galls*. The tumor-inducing agent in *Agrobacterium* is a plasmid, called Ti, that integrates some of its DNA into the chromosome of its host plant cells, causing them to divide repeatedly. Ti plasmids are large, circular double-stranded DNA molecules of about 200 kb, and, like other bacterial plasmids, they exist in *Agrobacterium* cells as independently replicating genetic units.

When *Agrobacterium* infects plant cells, part of the plasmid, called T-DNA, is transferred to the plant cell. The T-DNA is flanked by Ti plasmid sequences, each 25 bp long. These flanking sequences are called *borders*. Excision is a two-stage process in which the right-hand border is nicked within the 25-bp repeat. A second nick in the left-hand border releases the T-DNA as a single strand. The process of transfer from the bacterial cell to the plant cell is mechanistically similar to the process of bacterial conjugation; it is as though the *Agrobacterium* is mating with the plant cell! Once inside the plant, the T-DNA enters the nucleus and integrates into the plant cell DNA. The T-DNA carries genes for an auxin (a plant hormone) and it is this that causes the infected cell to divide, as well as to diffuse to adjacent cells and induce their division, thus producing the crown gall.

The standard method for using T-DNA as a vector is the *binary system* (Fig. 6-13). This method was devised when investigators realized that the essential functions for transfer are supplied separately by the T-DNA itself and by the Ti plasmid, and that the components can be carried on separate vectors. The *binary vector* contains the T-DNA sequences that are needed for excision and integration. The hormone genes of the T-DNA are removed to create room for the insertion of foreign DNA, which will be transferred to the plant cell. (Deleting the hormone genes has the added advantage of preventing the uncontrolled growth of the recipient cells.) The other essential genes from the Ti needed for transfer are supplied in *trans* on a *helper* plasmid. A very important factor in the development of T-DNA-based vectors is the availability of selectable markers such as neomycin phosphotransferase II (NPTII) and dihydrofolate reductase. These markers are included within the 25-bp repeats of the binary vector, so they too are transferred into the plant cell and are used to select for cells that have taken up the T-DNA. The binary system was used to try to produce tomato plants that ripened more slowly and so would not bruise during transport to supermarkets. Disappointingly, these fruits were as soft as wild-type tomatoes, presumably because the targeted enzyme is just one of several factors that contribute to fruit softening. Inhibiting synthesis of ethylene, which leads to ripening, has been more successful and growers can ship such tomatoes without bruising them and then use ethylene to ripen them. However, despite this initial setback, *Agrobacterium* transformation of plant cells has led to the development of crops that have the potential to transform the food supply. The most notable of these is corn carrying the Bt toxin, which is produced by the common soil bacterium *Bacillus thuringiensis*. This protein is nontoxic to humans, but kills caterpillars that eat it. For many years, the bacteria were sprayed directly on crops to control caterpillars, but transgenic corn has been developed that expresses the toxin directly within the leaves. Bt-expressing plants are more effectively protected than by previous methods, and the use of Bt-expressing plants greatly reduces the use of other insecticides.

Many Organisms Can Be Used for Making "Foreign" Proteins

The Lac repressor was isolated by overexpressing it (Chapter 3) and this has become standard procedure for producing large quantities of protein. Many proteins can be successfully expressed and purified from bacteria (usually *E. coli*) and a host of different vectors and inducible systems have been developed for use in bacteria. The latter can sometimes have the advantage of growing rapidly in inexpensive media but they are generally best suited to produce small proteins—it can be difficult to express large proteins efficiently. The latter can sometimes be made in yeast cells, which grow nearly as rapidly and economically as bacteria.

One major drawback of protein expression in bacteria and yeast is that these organisms do not modify proteins in precisely the same manner as do animal cells. Proteins from higher organisms can have a

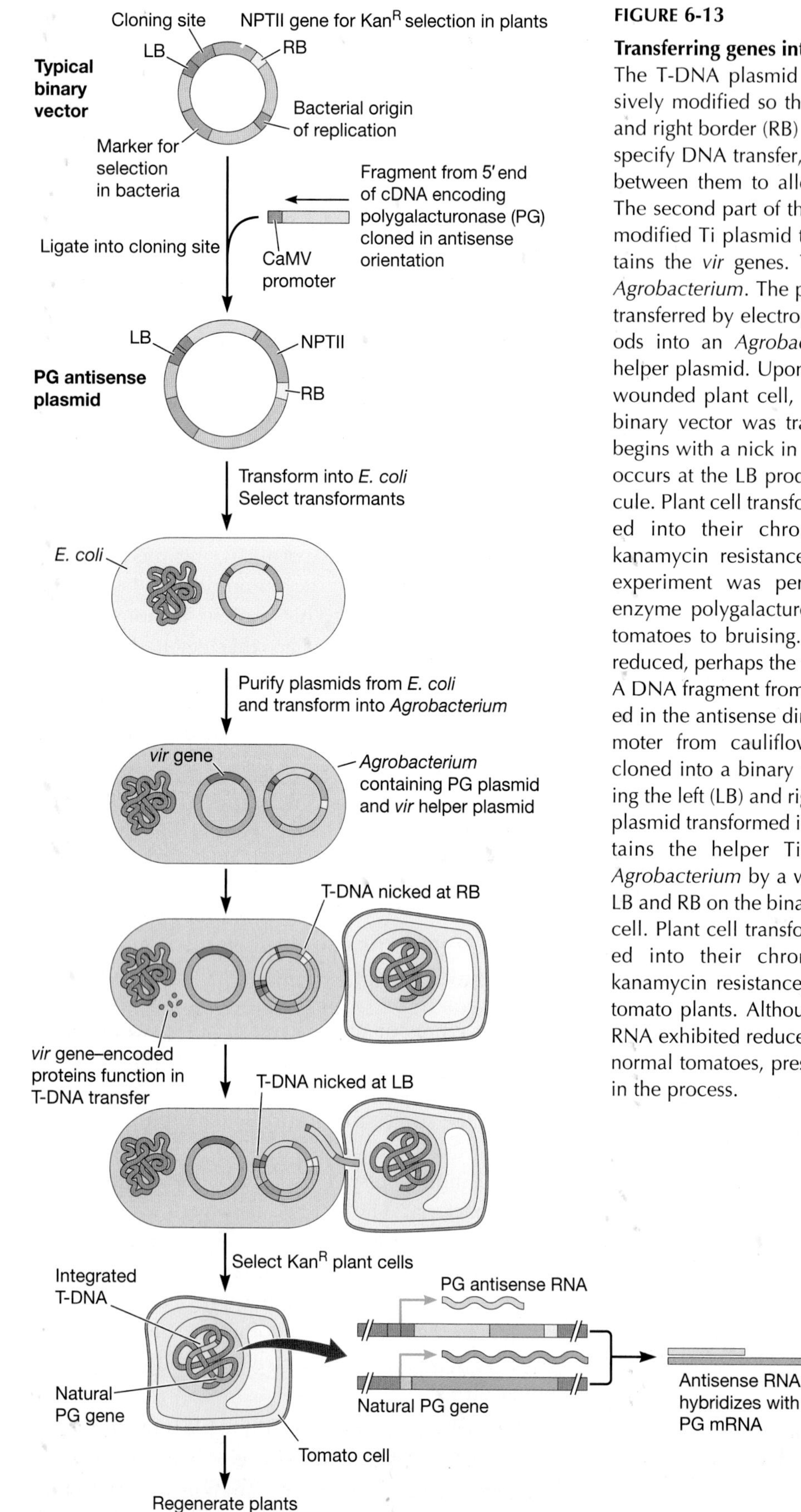

FIGURE 6-13

Transferring genes into plant cells by use of a binary vector. The T-DNA plasmid from *Agrobacterium* has been extensively modified so that it contains only the left border (LB) and right border (RB) 25-bp imperfect repeat sequences that specify DNA transfer, and the NPTII gene has been inserted between them to allow antibiotic selection in plant cells. The second part of the binary system is a helper plasmid, a modified Ti plasmid that is missing its T-DNA but still contains the *vir* genes. This helper plasmid is maintained in *Agrobacterium*. The plasmid constructed in *E. coli* was then transferred by electroporation or other transformation methods into an *Agrobacterium* strain that also contains the helper plasmid. Upon activation of the *Agrobacterium* by a wounded plant cell, the DNA between LB and RB on the binary vector was transferred into the plant cell. Transfer begins with a nick in the DNA strand in the RB; then a nick occurs at the LB producing a single-stranded T-DNA molecule. Plant cell transformants, which have this DNA integrated into their chromosomal DNA, were selected by kanamycin resistance and used to regenerate plants. An experiment was performed to investigate the role the enzyme polygalacturonase (PG) plays in the sensitivity of tomatoes to bruising. If the cellular levels of PG could be reduced, perhaps the fruit would be hardier during shipping. A DNA fragment from the 5′ end of the PG cDNA was ligated in the antisense direction to the constitutively active promoter from cauliflower mosaic virus (CaMV) and then cloned into a binary vector between the sequences encoding the left (LB) and right (RB) T-DNA borders. The antisense plasmid transformed into a strain of *Agrobacterium* that contains the helper Ti plasmid. Upon activation of the *Agrobacterium* by a wounded plant cell, the DNA between LB and RB on the binary vector was transferred into the plant cell. Plant cell transformants, which had this DNA integrated into their chromosomal DNA, were selected by kanamycin resistance and used to regenerate fruit-bearing tomato plants. Although tomatoes expressing the antisense RNA exhibited reduced PG activity, they were just as soft as normal tomatoes, presumably because PG is just one factor in the process.

variety of different sugars and other chemical groups added after translation, and these modifications are frequently essential for proper functioning of the protein. Protein expression in an insect cell or animal cell is often the best solution for the production of functional protein, although these cells are much more expensive to grow than are bacteria or yeast.

An insect virus called baculovirus that infects cells from a moth is widely used for protein production. Baculoviruses attracted attention because they produce proteins at very high levels in the later stages of an infection. These proteins are not needed when the virus is used to infect cells in tissue culture, and the genes can be replaced by foreign genes. These too are produced at a high level—milligrams of nonviral proteins can be produced. In addition, proteins synthesized in baculovirus usually undergo posttranslational modifications such as glycosylation. Very large DNA inserts can be cloned in baculovirus and these are introduced into the viral genome using homologous recombination. The large capacity of baculoviruses has been exploited to synthesize large multiprotein complexes from a single baculovirus. All five proteins of the human TFIID transcription factor were expressed appropriately in the baculovirus system. They were found to form a functional complex of 700 kD. This has required the construction of sophisticated vectors for use in *E. coli* for the assembly of multiple genes at a time. These are inserted into a bacmid, a baculovirus shuttle vector that grows as a large plasmid in *E. coli* and can be transfected into insect cells.

Several plant systems have been developed for protein expression, because growing acres of plants could be an inexpensive solution for producing very large quantities of a protein. A large number of proteins have been made in plants—for example, human serum albumin, α-interferon, α_1-antitrypsin, and collagen. Especially promising are recombinant antibodies and subunit vaccines, and several of these have entered clinical trials. One of the most interesting is a chimeric secretory IgA/IgG antibody to the major adhesion molecule of *Streptococcus mutans*. This is the leading cause of tooth decay, and topical application of an antibody against the adhesion molecule that binds the bacteria to teeth is an effective preventative treatment. Production of the antibody required both recombinant DNA engineering and conventional plant breeding. Four separate lines of transgenic tobacco plants had to be produced, each expressing a different portion of the antibody molecule. These lines were then crossed to produce a plant that synthesized the complete antibody. This antibody was purified from the plants and successfully prevented the colonization of *S. mutans* in the mouth.

Many nuclear transgenes are expressed at low levels in plants. Instead it is possible to use microprojectile bombardment to produce plants that have transgenic chloroplasts. These have the great advantage that they are present in thousands of copies in each cell. This means that very high levels of protein can be achieved; fragment C of tetanus toxin can reach levels as high as 25% of the soluble proteins in transgenic tobacco plants. Also, this approach allays concern about the escape of transgenes into the environment—because chloroplasts are not transmitted in pollen, the transgenes cannot escape on the wind.

Efficient purification of the transgene protein is essential and various tricks have been devised to make this easy (and cheap) to achieve. However, it would be more convenient if pharmaceutical proteins did not have to be purified—instead the plants would be eaten. Lettuce, bananas, and potatoes have all been used in this way. The harmless, laboratory version of *E. coli* appears repeatedly in these pages, but there is also an enterotoxigenic form that makes an enterotoxin that causes severe gastrointestinal illness in human beings and is a leading cause of diarrhea in childen in developing countries. The LTB subunits of the enterotoxin attach the toxin to the gut cells, thus enabling the entry of the enzymatically active LTA subunit, which leads to water loss from cells. The LTB subunit has been expressed in transgenic potatoes and when these are eaten (raw!) they induce localized antibodies against LTB from the gut mucosa as well as systemic antibodies. The economic advantages of being able to deliver vaccines in this way in developing countries are obvious.

Transgenic Animals Can Be Made by Injecting DNA Directly into Embryos

Eukaryotic gene expression can be studied by introducing genes into yeast cells or mammalian cells in tissue culture, but what if we want to study the genetic control of development in higher organisms, where cell interactions play a crucial role? Fortunately, a set of techniques has been developed that can be used to introduce genes into early animal embryos. Genes can be injected directly into the embryos of fruit flies, worms, zebrafish, and mice to create animals carrying integrated genes. For the production of transgenic mice, the most effective technique has proved to be microinjection of the cloned genes into fertilized eggs, which contain two pronuclei, one from the sperm (male) and one from the egg (female), that ultimately form the nucleus of the one-celled embryo. In this approach, a few hundred copies of the foreign DNA in about 2 picoliters of solution are microinjected into one of the two pronuclei. The injected embryos are then transferred to the oviduct of a foster mother and, upon subsequent implantation in the uterus, many develop to term (Fig. 6-14). The percentage of eggs that survive the manipulation and develop to term varies, but usually ranges between 10% and 30%. Of the survivors, between a few percent and 40% have foreign DNA integrated into their chromosomes. As with microinjection, transfection, or infection, the introduced DNA appears to integrate randomly without preference for a particular chromosomal location, usually as a tandem array of many copies at a single locus. Mice that carry the foreign gene are referred to as *transgenic*, and the foreign DNA is termed a *transgene*. Lentivirus infection has been more recently adopted as a method of transgenesis for early embryos.

DNA introduced into one-celled embryos can become stably integrated in both somatic and germ-line cells. Mice derived from embryos injected with cloned human interferon DNA or with rabbit β-globin DNA were shown to transmit these transgenes to their offspring as a Mendelian trait just as with their own genes. All the mice derived from such a single founder mouse form a *line* of mice; every member of a line of mice has the same transgene at the identical position in its genome.

Transgene Expression Can Be Targeted to Specific Tissues

Although a transgene integrates in a chromosomal location different from that of its endogenous counterpart, it is often expressed in a pattern that mimics the expression of the endogenous gene. To determine the pattern of expression, various tissues are analyzed for the presence of RNA or protein products encoded by the transgene. For example, when transgenic mice harboring the human insulin gene were analyzed, human insulin RNA was found in the pancreas but not in other tissues. Transcription of the human insulin transgene had been induced by the same signals that induced the endogenous mouse insulin genes. Thus, not only can a foreign

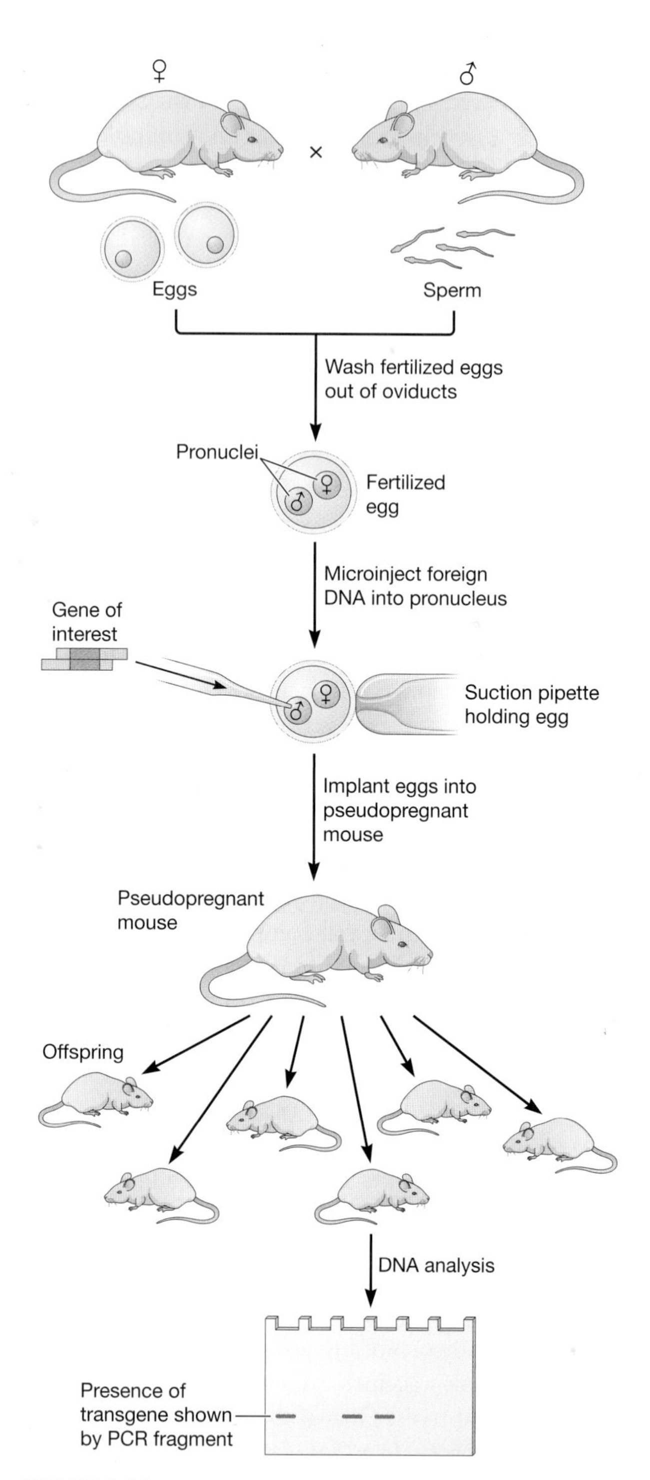

FIGURE 6-14

Producing transgenic mice by microinjection. Fertilized eggs are collected by washing out the oviducts of mated females, and the gene of interest is injected into one of the two pronuclei. The injected eggs are transferred to foster mothers (female mice made pseudopregnant by mating with vasectomized males). Three weeks after the birth, the offspring are checked for the presence of the transgene by Southern blotting or PCR of DNA extracted from a small piece of the tail. In the example shown, three of the offspring carry the transgene.

transgene be expressed in the correct tissue, but it may be subject to the same regulatory signals as the endogenous genes. Similar results have been obtained using many genes introduced into transgenic mice by *embryonic stem* (ES) *cell* transfer.

If the sequences responsible for tissue-specific regulation of a gene are known, they can be used to target expression of a protein to a tissue in which it is not normally expressed. For example, the islets of Langerhans in the pancreas are composed of four cell types—α, β, δ, and PP—characterized by the hormone that they produce—glucagon, insulin, somatostatin, and pancreatic polypeptide, respectively. The promoter-enhancer control region of the rat insulin gene (RIP) was used to target expression of SV40 large T antigen (Tag), a viral oncogene, to the β cells of the islets (Fig. 6-15). Mice carrying this RIP-Tag transgene died at 9–12 weeks of age, and pathological analysis showed hyperplasia (abnormal proliferation of cells) and tumors of the islets of Langerhans. All other tissues of the transgenic mice were normal despite containing the transgene, which showed that tissue-specific expression was occurring in the islets. Immunohistochemical analysis of tissue sections from the transgenic mice showed that the tumor cells expressed T antigen and that they were exclusively β cells. That is, the insulin gene control region had directed expression of the T antigen to precisely the appropriate cell type. These RIP-Tag mice have been used extensively in studies of cancer (Chapter 15).

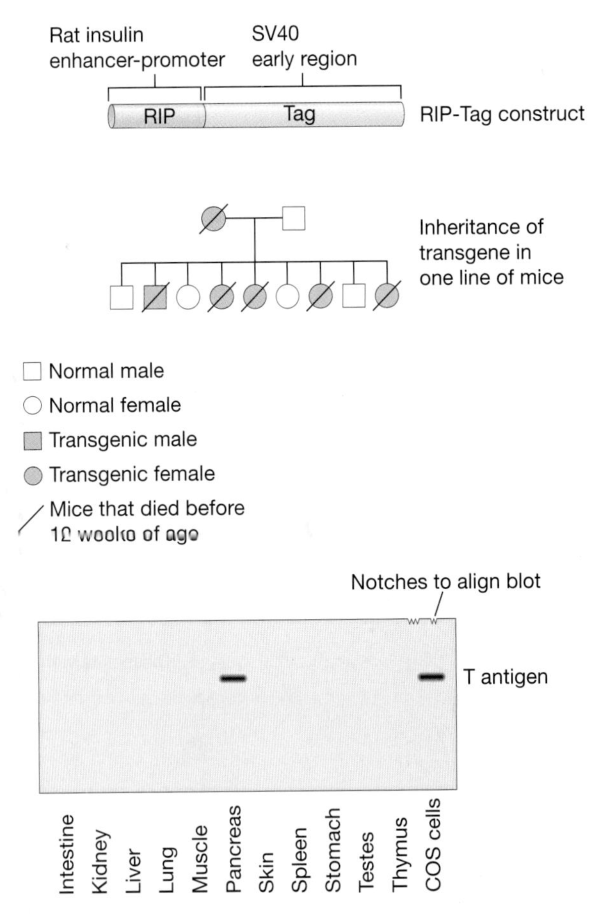

FIGURE 6-15

Gene targeting to specific tissues. The insulin gene is expressed exclusively in the β cells of the pancreas. The enhancer-promoter region of the insulin gene was linked to the gene for the large T antigen (Tag) of the DNA tumor virus SV40. It was expected that the resulting transgene would be expressed only in the pancreas of transgenic animals. The transgene integrated into the germ line, as shown by the inheritance pattern. The most striking phenotypic expression of the transgene was death of all transgenic animals 9–12 weeks after birth with tumors of the islets of Langerhans. The tissue specificity of large T antigen expression was determined by using an antibody to large T antigen to immunoprecipitate tissue homogenates, carrying out electrophoresis, and detecting large T antigen with a second, radiolabeled antibody. Large T antigen was found only in the pancreas and in extracts of COS cells that constitutively express large T antigen.

Embryonic Stem Cell Lines Can Be Injected into Embryos to Create Mice with Altered Genomes

Direct transfer of DNA into the pronuclei of fertilized mouse eggs is an efficient method of producing transgenic mice, but there is no opportunity to manipulate or otherwise control DNA integration. However, this can be done by first introducing the DNA into embryonic stem cells (ES cells) and then injecting the transfected cells into embryos, where they become incorporated into the developing embryo (Fig. 6-16). The ES cells are obtained by culturing the inner cell mass of mouse blastocysts, which gives rise to the embryo. (The rest of the cells of the blastocyst make the extraembryonic membranes that give rise to the placenta.) The ES cells are grown in tissue culture just like other cells, except that they must be prevented from differentiating by growing them on a feeder layer of fibroblasts, or by adding leukemia inhibitory factor (LIF) to the culture medium. Under these conditions, ES cells can be grown for many weeks. These extraordinary ES cells can be regarded as the equivalent of unicellular mice, and when they are injected into mouse blastocysts, they are able to participate in the

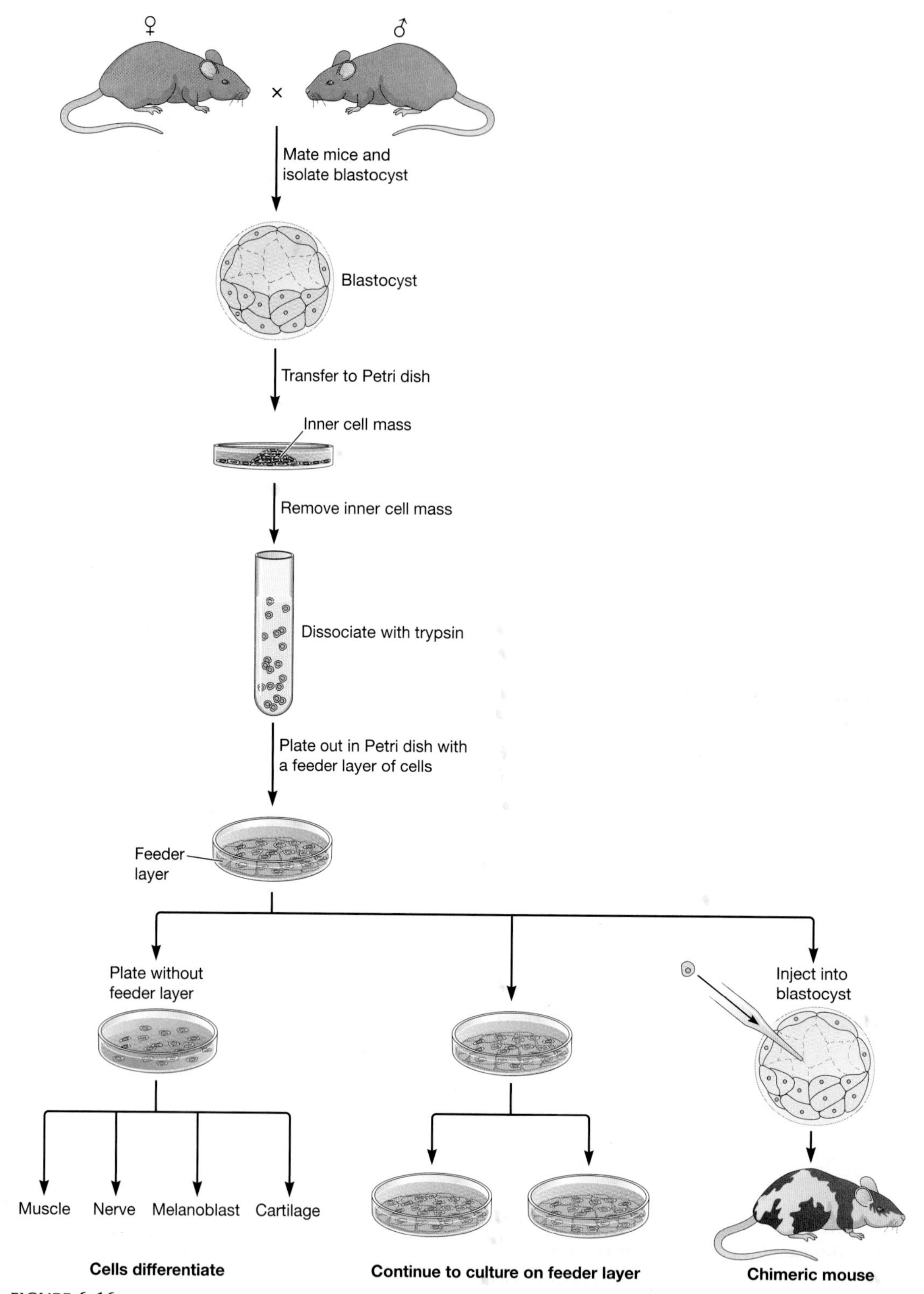

FIGURE 6-16

Embryonic stem (ES) cells produced from mouse blastocysts. Mice are mated, and 3 days later, blastocysts are isolated and cultured in Petri dishes. The cells spread out over the surface of the dish so that the clump of cells forming the inner cells mass, and corresponding to the future embryo, can be removed. The clump of cells is dissociated into single cells using trypsin, a proteolytic enzyme. If the cells are replated out on a plain culture-dish surface, they will differentiate into a variety of tissues, but if they are grown on a feeder layer of fibroblasts, they will continue to proliferate and can be subcultured repeatedly. (A feeder layer is a monolayer of cells that has been treated so the cells can no longer divide. They continue to metabolize, and in so doing "condition" the culture medium so that the cells seeded on top of them survive and grow better). The cells can be microinjected into a blastocyst, where they will become assimilated into the inner cell mass and take part in the formation of many tissues of the chimeric mouse. It is usual to use ES cells and recipient blastocysts derived from mice with different coat colors so that the contribution of the ES cells to the chimeric offspring can be assessed by simply looking at their coat color.

formation of all tissues, including germ cells.

The most important advantage of ES cells for gene transfer into mice is that the transgene can be manipulated in vitro before the cells are injected into the embryo. DNA can be introduced into ES cells by transfection, retroviral infection, or electroporation, and cells carrying properly expressed transgenes can be selected for before injection into a blastocyst. Homologous recombination techniques can be applied in ES cells to produce transgenic mice with mutations in specific genes, to replace a mutant gene with the normal equivalent, or to introduce a gene of interest at a specific location within the genome.

Homologous Recombination Can Be Used to Knock Out Genes in Mice

The most significant limitation of many gene transfer methods for animal and plant cells is that once foreign DNA is taken up by cells, it is lost from 90% or more of the cells that originally take it up. In the rare instances that it becomes stably integrated into the host cell genome, the integration occurs at random. Random integration results in wide fluctuations in the expression levels of the transferred gene and in the pattern of its regulation. Thus, in some cases, the foreign gene integrates in the vicinity of a highly expressed cellular gene and comes under the influence of that gene's regulatory apparatus, leading to high levels of foreign-gene expression. In other cases, foreign genes may insert into quiet areas of the genome where their expression is likely to be suppressed. For the experimenter, this often means tedious and lengthy sifting of dozens of clones of transfected or infected cells to find the cell clone with the desired level of foreign-gene expression.

Investigators have little control over where the gene recombines into the host genome. In the vast majority of cases, this occurs by *heterologous recombination*, when the gene recombines into an unrelated sequence. But in some instances, a gene can recombine precisely into the identical sequence in the genome by *homologous recombination*. Homologous recombination is a powerful tool for studying gene function because it can be used for *gene targeting*—that is, placing foreign DNA at a precise locus in the genome. Gene targeting is now used to inactivate mammalian genes by homologous recombination of a disrupted transgene into the target gene of interest (a process called "gene knockout"), and it can also be used to introduce new sequences into a defined location in the genome.

Homologous recombination is a relatively rare event in mammalian cells. Some targeting vectors recombine homologously with a frequency of only once per thousand heterologous insertions, although most vectors have a much higher frequency of homologous insertion, ranging from 5% to as high as 20%. Nonetheless, effective strategies are needed to enrich the population of transfected cells for those in which homologous recombination has occurred. One way to do this is depicted in Figure 6-17. The gene to be knocked out is interrupted by the incorporation of a region of DNA that carries a selectable marker gene such as *neo*, followed immediately downstream by the thymidine kinase (*tk*) gene, which can be selected

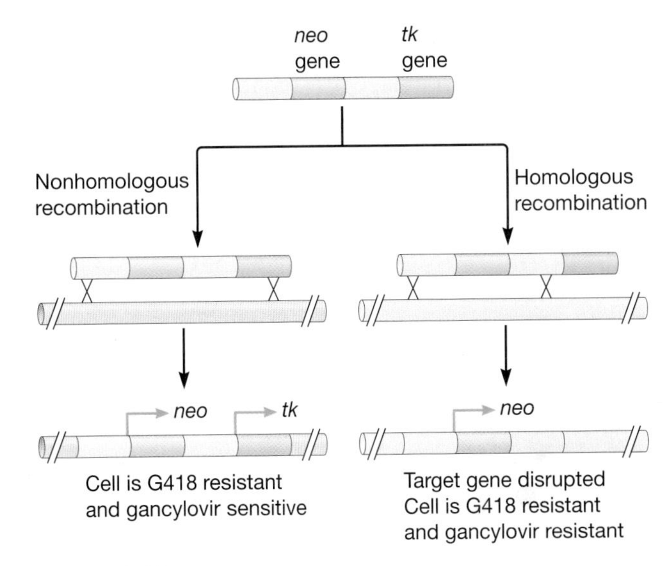

FIGURE 6-17

***Positive–negative selection* is a commonly used strategy for enriching for gene-targeting or homologous recombination events.** In this scheme, an exon from a targeted gene is disrupted with a *neo* gene containing its own promoter (encoding resistance to the antibiotic G418). Downstream from the exon is a *tk* (thymidine kinase) marker. In most random integrations, which occur near the ends of the transfected fragments, both markers become incorporated into the genome. A homologous recombination, however, requires recombination within the homologous cellular DNA fragments. In these homologous recombination events, the *neo* gene, but not the *tk* marker, is retained. Thus, after transfection, a double selection is applied with G418 and gancyclovir. G418 ensures that only cells with an integrated *neo* gene grow up, and gancyclovir kills the cells that also express thymidine kinase. Thus, only neo^+ tk^- cells survive this selection, and these colonies frequently carry the desired targeted mutation. There are other possible ways in which *tk* function can be lost—for example, through small deletions or by silencing from an adjoining region of the genome. Consequently, neo^+ tk^- clones must still be verified for the correct integration event.

against. After transfection, a dual selection is applied: selection for the *neo* gene (resistant to the drug neomycin, called G418) and against expression of *tk*. Heterologous recombination usually results in the insertion of both marker genes, so these cells do not survive the double selection. In a homologous recombination, only the first marker is retained, and the cells survive double selection. Using methods of this type, the enrichment for homologous recombinants can be 100-fold or more.

Precise Rearrangements in the Mouse Genome Can Be Made Using Homologous Recombination

Very effective and efficient techniques for homologous recombination in mouse cells are based on the site-specific recombinase Cre, originally identified in studies of bacteriophage P1 replication. This phage has a circular genome that can become tangled during replication. To separate these entangled copies, the phage expresses Cre that acts at sequences called *lox* sites. Cre acts at pairs of *lox* sites, cutting each *lox* site in half. The DNA between the *lox* sites is degraded and Cre joins the cut ends. Site-specific recombinases enable a geneticist to make alterations at will in the mouse genome: inserting a gene at a specified location, replacing one or several exons, or even rearranging entire chromosome segments.

Chromosome rearrangements are accomplished by introducing *lox* sequences at locations flanking the target DNA. Then, when Cre recombinase is expressed in the cell, the two sites recombine to produce a chromosomal rearrangement. Cells containing the correct rearrangement can be selected, because the two targeting vectors are designed so that recombination between the *lox* sites produces a functional selectable marker (Fig. 6-18). One use of this strategy was to create a mouse model of DiGeorge syndrome, the most common deletion syndrome that occurs in humans, caused by a 3-Mb deletion within chromosome 22. Patients with this syndrome have an array of symptoms including heart defects, defects in the hypoparathyroid gland, an abnormal facial appearance, and learning difficulties. The lack of a natural mouse mutant meant that genetic engineering had to be performed to create such a mutant. Mouse ES cells were prepared in which the region of the mouse genome homologous with human 22q11 was flanked by *lox* sequences, and Cre was used to generate the corresponding deletion. These cell lines were used to create mice. Mice homozygous for the deletion died during gestation, but those heterozygous for the deletion survived. About one-quarter of these heterozygotes had heart defects similar to those observed in humans. Just as in humans with DiGeorge syndrome, *haploinsufficiency* (i.e., having a single copy of the deleted genes) is sufficient to cause the syndrome. Mice were designed that carried a duplication of the genes elsewhere in the genome, and this rescued the defects. Diseases caused by haploinsufficiency are relatively unusual, and the fact that the mouse model recapitulates this feature of DiGeorge syndrome suggests that it will be useful for studying this disorder.

Gene Therapy May Permit the Manipulation of Gene Expression in Humans

It takes many years to develop effective drugs, even when we have detailed knowledge of the mutated gene and the function of its encoded protein. An alternative strategy for treating human genetic disorders is *gene therapy*, the restoration of a functional gene through modification of the mutant gene or by introducing a functional gene. Unfortunately, although gene therapy has been considered the holy grail for correcting genetic disorders, it continues to pose many challenges. Quite apart from the technical challenges of delivering a gene to a specific cell type and expressing it in a physiologically responsive matter, there are serious issues of safety. For example, Jesse Gelsinger, a volunteer in a trial for treating ornithine transcarbamylase deficiency (OTCD), died from an immune response to the adenovirus viral vector being used to deliver the OTC gene.

One long-standing concern of using viral vectors to deliver genes is that they might induce cancers by integrating into the host cell DNA and activating or forcing inappropriate expression of a cellular gene. We will discuss an example of this, the treatment of children suffering from severe combined immunodeficiency (SCID) in Chapter 14.

The suite of recombinant DNA tools—cloning, sequencing, and modifying genes, and introducing modified genes into cells and animals—has transformed biology. Our current knowledge of the wonderfully intricate processes of life was unimaginable to previous generations of biologists. That knowledge continues to increase at a remarkable rate and, as we shall see in the following chapters, these tools continue to reveal unsuspected facets of life.

235 @ 9 am

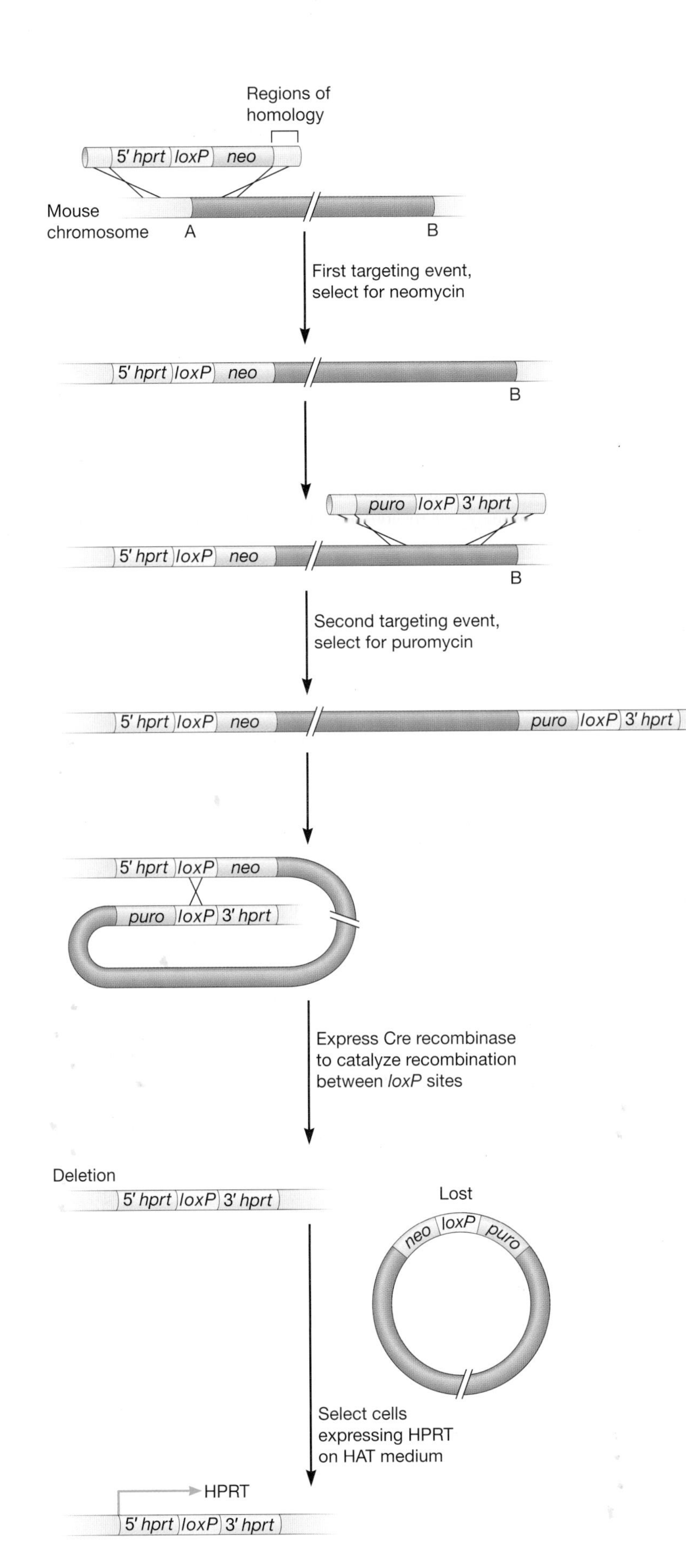

FIGURE 6-18

Chromosome deletions in the mouse using ES cells, Cre-*lox* recombination, and positive selection. Two rounds of targeting in ES cells are used to insert two *loxP* sites at a large distance apart onto the mouse chromosome. A different positive marker is used to insert each *loxP* site: a neomycin-resistance gene in the first targeting and a puromycin-resistance gene in the second round. Furthermore, the 5′ targeting vector contains the 5′ portion of the hypoxanthine phosphoribosyltransferase gene, *hprt*, and the 3′ vector contains the 3′ portion of *hprt*. When Cre is transiently expressed in the ES cells, it catalyzes recombination between two *loxP* sites, excising the intervening sequence as a circle and joining the two ends of the chromosome. The excised circular molecule lacks a centromere and is not stably inherited in subsequent cell divisions. The strength of this technique comes from the power of the selection for those ES clones in which the correct rearrangement has occurred. The separate portions of *hprt* in the targeting vector are not functional. HAT is added to the culture medium of the ES cell clones and the only clones that survive are those in which *loxP* recombination has occurred precisely, thus bringing together the 5′ and 3′ portions of the gene and producing functional *hprt*. (Targeting is done in *hprt*-deficient ES cells.)

Reading List

Targeted Mutagenesis

Kunkel T.A. 1985. Rapid and efficient site-specific mutagenesis without phenotypic selection. *Proc. Natl. Acad. Sci.* **82:** 488–492.

Taylor J.W., Ott J., and Eckstein F. 1985. The rapid generation of oligonucleotide-directed mutations at high frequency using phosphorthioate-modified DNA. *Nucleic Acids Res.* **13:** 8765–8785.

Vandeyar M.A., Weiner M.P., Hutton C.J., and Batt C.A. 1988. A simple and rapid method for the selection of oligodeoxynucleotide-directed mutants. *Gene* **65:** 129–133.

Wakabayashi H., Su Y.-C., Ahmad S.S., Walsh P.N., and Fay P.J. 2005. A Glu113Ala mutation within a factor VIII Ca^{2+}-binding site enhances cofactor interactions in factor Xase. *Biochemistry* **44:** 10298–10304.

Inducible Expression

Gossen M. and Bujard H. 1992. Tight control of gene expression in mammalian cells by tetracycline-responsive promoters. *Proc. Natl. Acad. Sci.* **89:** 5547–5551.

Gossen M., Freundlieb S., Bender G., Muller G., Hillen W., and Bujard H. 1995. Transcriptional activation by tetracyclines in mammalian cells. *Science* **268:** 1766–1769.

Felsher D.W. and Bishop J.M. 1999. Reversible tumorigenesis by MYC in hematopoietic lineages. *Mol. Cell* **4:** 199–207.

Promoters for Gene Expression

Cheng L., Ziegelhoffer P.R., and Yang N.S. 1993. In vivo promoter activity and transgene expression in mammalian somatic tissues evaluated by using particle bombardment. *Proc. Natl. Acad. Sci.* **90:** 4455–4459.

Colored and Fluorescent Proteins

Hull G.A. and Devic M. 1995. The beta-glucuronidase (gus) reporter gene system. Gene fusions; spectrophotometric, fluorometric, and histochemical detection. *Methods Mol. Biol.* **49:** 125–141.

Chalfie M., Tu Y., Euskirchen G., Ward W.W., and Prasher D.C. 1994. Green fluorescent protein as a marker for gene expression. *Science* **263:** 802–805.

Shaner N.C., Campbell R.E., Steinbach P.A., Giepmans B.N.G., Palmer A.E., and Tsien R.Y. 2004. Improved monomeric red, orange and yellow fluorescent proteins derived from *Discosoma* sp. red fluorescent protein. *Nat. Biotechnol.* **22:** 1567–1572.

Cloning without Ligase

Mead D.A., Pey N.K., Herrnstadt C., Marcil R.A., and Smith L.M. 1991. A universal method for the direct cloning of PCR amplified nucleic acid. *Biotechnology* **9:** 657–663.

Shuman S. 1994. Novel approach to molecular cloning and polynucleotide synthesis using vaccinia DNA topoisomerase. *J. Biol. Chem.* **269:** 32678–32684.

Baudin A., Ozier-Kalogeropoulos O., Denouel A., Lacroute F., and Cullin C. 1993. A simple and efficient method for direct gene deletion in *Saccharomyces cerevisiae*. *Nucleic Acids Res.* **21:** 3329–3330.

Li M.Z. and Elledge S.J. 2005. MAGIC, an in vivo genetic method for the rapid construction of recombinant DNA molecules. *Nat. Genet.* **37:** 311–319.

Physical Methods of DNA Transfer

Mercenier A. and Chassy B.M. 1988. Strategies for the development of bacterial transformation systems (review). *Biochimie* **70:** 503–517.

Shigekawa K. and Dower W.J. 1988. Electroporation of eukaryotes and prokaryotes: A general approach to the introduction of macromolecules into cells (review). *Biotechniques* **6:** 742–751.

Capecchi M.R. 1980. High efficiency transformation by direct microinjection of DNA into cultured mammalian cells. *Cell* **22:** 479–488.

Pellicer A., Robins D., Wold B., Sweet R., Jackson J., et al. 1980. Altering genotype and phenotype by DNA-mediated gene transfer (review). *Science* **209:** 1414–1422.

Wigler M., Silverstein S., Lee L.S., Pellicer A., Cheng Y., and Axel R. 1977. Transfer of purified herpes virus thymidine kinase gene to cultured mouse cells. *Cell* **11:** 223–232.

Felgner P.L., Gadek T.R., Holm M., Roman R., Chan H.W., et al. 1987. Lipofection: A highly efficient, lipid-mediated DNA-transfection procedure. *Proc. Natl. Acad. Sci.* **84:** 7413–7417.

Horsch R.B., Fry J.E., Hoffman N.L., Eichholts D., Rogers S.G., and Fraley R.T. 1985. A simple and general method for transferring genes into plants. *Science* **227:** 1229–1231.

Klein T.M., Wolff E.D., Wu R., and Sanford J.C. 1987. High-velocity microprojectiles for delivering nucleic acids into living cells. *Nature* **327:** 70–73.

Viral Vectors

Goff S.P. and Berg P. 1976. Construction of hybrid viruses containing SV40 and lambda phage DNA segments and their propagation in cultured monkey cells. *Cell* **9:** 695–705.

Mann R., Mulligan R.C., and Baltimore D. 1983. Construction of a retrovirus packaging mutant and its use to produce helper-free defective retrovirus. *Cell* **33:** 153–159.

Cepko C.L., Roberts B.E., and Mulligan R.C. 1984. Construction and applications of a highly transmissible murine retrovirus shuttle vector. *Cell* **37:** 1053–1062.

Lai C.M., Lai Y.K., and Rakoczy P.E. 2002. Adenovirus and adeno-associated virus vectors (review). *DNA Cell Biol.* **21:** 895–913.

Transfer of Genes Using Agrobacterium

Zambryski P., Tempe J., and Schell J. 1989. Transfer and function of T-DNA genes from Agrobacterium Ti and Ri plasmids in plants (review). *Cell* **56:** 193–201.

Smaith C.J.S., Watson C.F., Ray J., Bird C.R., Morris P.C., Schuch W., and Grierson D. 1988. Antisense RNA inhibition of polygalacturonase gene expression in transgenic tomatoes. *Nature* **334:** 724–726.

Estruch J.J., Carozzi N.B., Desai N., Duck N.B., Warren G.W., and Koziel M.G. 1997. Transgenic plants: An emerging approach to pest control (review). *Nat. Biotechnol.* **15:** 137–141.

Hanson B., Engler D., Moy Y., Newman B., Ralston E., and Gutterson N. A simple method to enrich an Agrobacterium-transformed population for plants containing only T-DNA sequences 1999. *Plant J.* **19:** 727–734.

Protein Expression Systems

Jones I. and Morikawa Y. 1996. Baculovirus vectors for expression in insect cells (review). *Curr. Opin. Biotechnol.* **7:** 512–516.

Haq T.A., Mason H.S., Clements J.D., and Arntzen C.J. 1995. Oral immunization with a recombinant bacterial antigen produced in transgenic plants. *Science* **268:** 714–716.

Daniell H., Kumar S., and Dufourmantel N. 2005. Breakthrough in chloroplast genetic engineering of agronomically important crops. *Trends Biotechnol.* **23:** 238–245.

Ma J.K.C., Hiatt A., Hein M., Vine N.D., Wang F., et al. 1995. Generation and assembly of secretory antibodies in plants. *Science* **268:** 716–719.

Germ-Line Transmission of Transferred Genes

Jaenisch R. and Mintz B. 1974. Simian virus 40 DNA sequences in DNA of healthy adult mice derived from preimplantation blastocysts injected with viral DNA. *Proc. Natl. Acad. Sci.* **71:** 1250–1254.

Gordon J.W., Scangos G.A., Plotkin D.J., Barbosa J.A., and Ruddle F.H. 1980. Genetic transformation of mouse embryos by microinjection of purified DNA. *Proc. Natl. Acad. Sci.* **77:** 7380–7384.

Costantini F. and Lacy E. 1981. Introduction of a rabbit beta-globin gene into the mouse germ line. *Nature* **294:** 92–94.

Gordon J.W. and Ruddle F.H. 1981. Integration and stable germ line transmission of genes injected into mouse pronuclei. *Science* **214:** 1244–1246.

Embryonic Stem Cells

Evans M.J. and Kaufman M.H. 1981. Establishment in culture of pluripotential cells from mouse embryos. *Nature* **292:** 154–156.

Martin G.R. 1981. Isolation of a pluripotent cell line from early mouse embryos cultured in medium conditioned by teratocarcinoma stem cells. *Proc. Natl. Acad. Sci.* **78:** 7634–7638.

Gossler A., Doetschman T., Korn R., Serfling E., and Kemler R. 1986. Transgenesis by means of blastocyst-derived embryonic stem cell lines. *Proc. Natl. Acad. Sci.* **83:** 9065–9069.

Robertson E., Bradley A., Kuehn M., and Evans M. 1986. Germ-line transmission of genes introduced into cultured pluripotential cells by retroviral vector. *Nature* **323:** 445–448.

Sternberg N. and Hamilton D. 1981. Bacteriophage P1 site-specific recombination. I. Recombination between *loxP* sites. *J. Mol. Biol.* **150:** 467–486.

Swift G.H., Hammer R.E., MacDonald R.J., and Brinster R.L. 1984. Tissue-specific expression of the rat pancreatic elastase I gene in transgenic mice. *Cell* **38:** 639–646.

Gene Targeting by Homologous Recombination

Mansour S.L., Thomas K.R., and Capecchi M.R. 1988. Disruption of the proto-oncogene *int-2* in mouse embryo–derived stem cells: A general strategy for targeting mutations to non-selectable genes. *Nature* **336:** 348–352.

Zijlstra M., Li E., Sajjadi F., Subramani S., and Jaenisch R. 1989. Germ-line transmission of a disrupted β_2 microglobulin gene produced by homologous recombination in embryonic stem cells. *Nature* **342:** 435–438.

Lindsay E.A., Botta A., Jurecic V., Carattini-Rivera S., Cheac Y.C., et al. 1999. Congenital heart disease in mice deficient for the DiGeorge syndrome region. *Nature* **401:** 379–383.

Gene Therapy

Somia N. and Verma I.M. 2000. Gene therapy: Trials and tribulations (review). *Nat. Rev. Genet.* **1:** 91–99.

Hacein-Bey-Abina S., Von Kalle C., Schmidt M., McCormack M.P., Wulffraat N., et al. 2003. *LMO2*-associated clonal T cell proliferation in two patients after gene therapy for SCID-X1. *Science* **302:** 415–419.

CHAPTER 7

Mobile DNA Sequences in the Genome

For many years, DNA was regarded as the safe deposit box of the cell's hereditary information. Changes (mutations) do occur and the cell has devised a variety of mechanisms to repair corrupted DNA or, as a fail-safe measure, to kill cells whose DNA has undergone irredeemable damage. Large-scale rearrangements occur in meiosis when recombination between homologous chromosomes ensures that genes are redistributed, but it was thought that DNA remained largely undisturbed. So, the discovery, by Barbara McClintock in the late 1940s, that there are genetic elements in maize that can direct their own movement throughout the genome was startling. Other examples were found in other species and it gradually came to be realized that, far from being an exception, these movable genetic elements are ubiquitous. Furthermore, they and their inactive derivatives are present in huge numbers; they make up, for example, as much as 50% of the human genome.

Initially, these sequences were believed not to have a function; they merely cluttered a genome and imposed a replicative burden on the cell. They were dismissed as garbage or "junk DNA." However, that they are maintained in genomes despite the burden they put on the cell suggests that they do have a function. In fact, we now know that these sequences have important roles—affecting gene expression, generating mutations, and reorganizing genome structure. In large part because of the activity of *transposons*, as these movable DNA elements came to be called, genomes are not fixed entities. Rather, transposons shape major changes in the genome and play an important role in genome evolution.

Genetic Analysis Reveals the Existence of "Movable Elements" in Maize and Bacteria

McClintock named the strange genetic elements she found in maize "controlling elements." These genetic elements were first noticed because they inhibited the expression of other maize genes with which they came into close contact. For example, she studied the pattern of expression of the genes for the pigment anthocyanin, which gives a deep purple color to maize kernels. There were mutations in these genes that were stable and gave rise to white kernels, whereas other mutations reverted at high frequency, thus producing patches or streaks of color where the mutation had reverted in otherwise white kernels. If plants with unstable mutations were crossed with plants with stable mutations, the latter also began to revert at high frequency. McClintock could account for the results of her genetic experiments—meticulous and exhaustive crosses of thousands of maize plants—by assuming that there were two genetic elements that she called *Activator* (Ac) and *Dissociation* (Ds). It appeared that the Ac and Ds could be inserted into DNA and later excised; after excision, the function of a previously dormant gene often returned. Remarkably, these elements did not have fixed chromosomal locations but seemed to move about the maize genome.

For many years, the corn plant provided the only evidence for movable genetic elements, but similar elements later were proposed to account for several highly mutable genetic loci in *Drosophila*. Nevertheless, most geneticists paid little attention to such oddities until the discovery, in the late 1960s, that some polar mutations in *Escherichia coli* (mutations affecting several genes in an operon) behaved like similar loci in maize and *Drosophila*; that is, they spontaneously reverted at high frequency, even though conventional point mutagens could not revert the mutations. Molecular analysis showed that these resulted from the insertion and excision of large segments of DNA called *insertion sequences* (ISs). By examining heteroduplexes between genes carrying these insertions, it became clear that many presumably independent insertion events involved exactly the same DNA sequences (Fig. 7-1).

With the advent of recombinant DNA techniques a few years later, these insertions could be cloned and studied. These studies elucidated the organization of IS elements and showed how they could "jump" from one place to another, a phenomenon called *transposition*. Moreover, as eukaryotic genomes were explored by cloning and sequencing, it became apparent that virtually all genomes were littered with sequences that appeared to be relics of transposition events long past. Cloning also revealed that many organisms have adopted segments of transposable elements to solve difficult problems in gene regulation and diversity. Other transposable elements have no obvious function. Why are they a part of our DNA? Are they simply molecular parasites or do they serve a useful, as-yet-undiscovered function? Our growing understanding of genome sequences is beginning to provide clues, but many of these questions are still unresolved.

Sequencing Reveals the Organization of Transposable Elements

The *E. coli* insertion sequences were among the first transposons to be sequenced. They range in size from 750 to 2500 bp and are typically (but not always) flanked by inverted repeats—closely related sequences in opposite orientations—from 10 to 40 bp in length. Enclosed by the inverted repeats are one or more open reading frames (ORFs) that encode proteins required for transposition of the element. In the simplest ISs, a single ORF encodes a transposase enzyme that recognizes the inverted repeats and catalyzes excision of the element. Other IS elements are more complex, resulting from the close juxtaposition of two or more transposons. These associated transposons can move as a unit, bringing along the DNA between them. These *complex transposons* were recognized because the mobilized DNA carried genes conferring antibiotic resistance. Complex transposons can easily jump from the bacterial chromosome to phage genomes or conjugative plasmids, and when this happens, the transposons can then spread to other bacteria. Transposons of this sort are a major source of antibiotic resistance in naturally occurring bacteria.

The "controlling elements" of maize, painstakingly studied by McClintock, turned out to be remarkably similar to the transposable elements of bacteria (Fig. 7-2). Cloning and sequencing of these elements also offered a clear explanation for the two

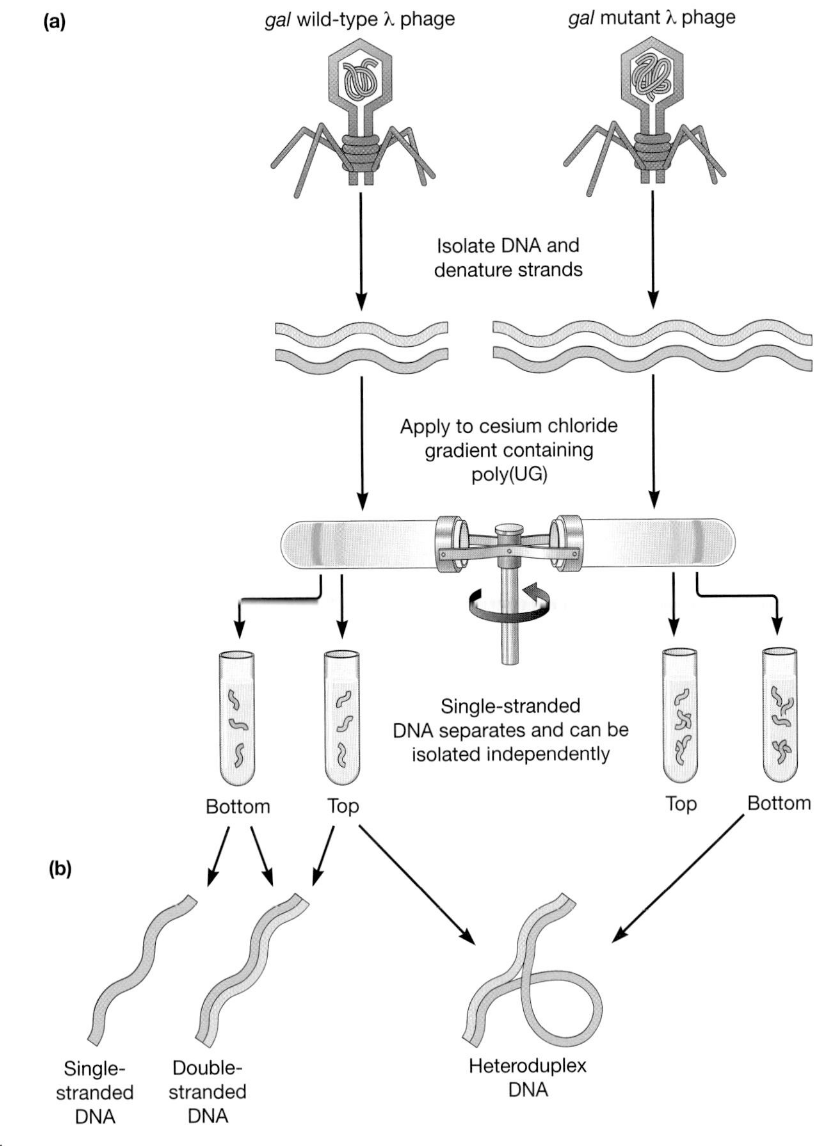

FIGURE 7-1

Polar mutations in different bacterial genes are caused by a small number of insertion sequences. (*a*) Insertion mutations were analyzed by density gradient centrifugation. The mutations of interest at the *gal* operon were moved from the bacterial chromosome into transducing λ phage. The *gal* mutations were compared to other polar mutations in λ phage, and wild-type phages were used as control molecules. The top and bottom strands of phage DNA could be separated in a cesium chloride density gradient. To better separate the strands, poly(U) RNA was added, which tends to preferentially bind one strand more than the other, thus increasing the separation of the two DNA strands in the gradient. Careful comparison of the gradients revealed that phage-bearing insertion sequences contained more DNA than wild-type phage. However, it was not known whether there was any homology between the various insertion mutations. (*b*) Electron microscopy was used to address this question by testing whether the different phage mutants shared regions of homology other than the phage sequences themselves. To do this, the single-stranded DNA isolated from the CsCl gradients was mixed together, allowed to hybridize, and prepared for analysis by electron microcopy. The hybridized molecules were examined. The electron micrographs can be measured to determine whether the DNA is single or double stranded, and the length of the DNA can be estimated based on the micrograph. The top or bottom strand alone failed to hybridize, so that only single-stranded DNA was visible in the micrograph. When the top and bottom strands of wild-type phage were mixed (or, indeed, if the top and bottom strands of any one phage were mixed), the result was a single, completely double-stranded DNA. However, when the top strand of wild-type phage was mixed with the bottom strand of any insertion mutant, a double-stranded DNA molecule containing a single-stranded loop was formed. This loop corresponded to the insertion sequence, and the length of the sequence could be confirmed by measuring the micrograph. Similar experiments on other insertion mutations established that there were several different kinds of insertion sequences, which were not homologous to one another.

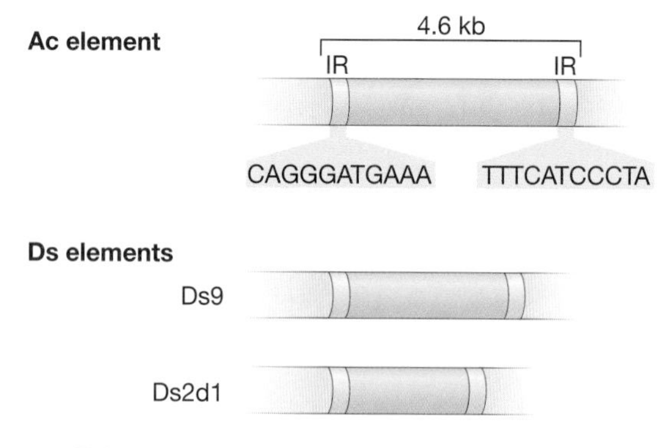

FIGURE 7-2

The Ac/Ds transposable elements of maize. The Ac element is an intact transposon organized much like bacterial transposons, with short inverted repeats (IRs) flanking sequences encoding a transposase enzyme. Because it carries an active transposase gene, Ac moves autonomously. The Ds elements are Ac elements that have deletions in the transposase gene. They cannot move on their own, but when Ac is present, the transposon can bind the IRs of the Ds element and catalyze its movement. In the absence of Ac, Ds insertion mutations are stable.

types of elements identified by McClintock. Ac and Ds are derivatives of the same transposable element. Ac is 4500 bp in length and encodes a transposase that catalyzes its movement. Ds elements are defective Ac elements, usually with deletions so that they no longer express transposase. When Ac is present and providing transposase activity in *trans*, Ds freely hops out of the anthocyanin gene, which allows "recovery" (reversion) of pigment production in those cells of the kernel where successful Ds excisions occur. In the absence of Ac, Ds is immobile, and Ds insertion mutations cannot revert.

There are Three Categories of Mobile Elements

As more and more repetitive elements were cloned and sequenced, it became apparent that there are three distinct families of transposons distinguished by their mechanism of jumping and overall sequence organization. The bacterial ISs and maize Ac and Ds elements are *DNA transposons*, which remain as DNA as they are excised and inserted elsewhere in the genome. There are two groups of *retrotransposons*, so named because their transposition involves an RNA transcript from which DNA is synthesized, thereby reversing the normal flow of genetic information. One group of retrotransposons closely resembles retroviruses, and on this basis are called *viral-like retrotransposons* or *LTR retrotransposons.* (Long terminal repeats [LTRs] are directly repeated sequences at each terminus of these elements, essential for the process of reverse transcription and integration.) Retrotransposons of the second group lack LTRs but carry a 3′ poly(A) sequence, apparently derived from a messenger RNA (mRNA) transcript—these are called *poly(A) retrotransposons* or *non-LTR retrotransposons.*

All of these types of transposons have been cloned and studied, and we are beginning to understand the mechanisms by which they move through the genome. To mobilize, transposons require specific DNA sequences (which may be transcribed into RNA) that are recognized by proteins that catalyze the mobilization of the transposase complex. The core catalytic domains encoded by DNA transposons and the integrases encoded by LTR retrotransposons are extremely similar, and structural and biochemical data reveal that the core catalytic mechanism has been conserved throughout evolution. Transposons that can synthesize transposase and the proteins necessary for mobility are called *autonomous* transposons. Many transposons are *nonautonomous*; that is, the proteins necessary for transposition are supplied from transposons elsewhere in the genome. Also, host proteins are often involved in transposon mobility—for example, retrotransposons are transcribed by host RNA polymerases.

DNA Transposons Move Using Two Distinct Mechanisms

The bacterial transposon Tn*3* does not actually jump at all; rather, it copies itself to a new location and leaves the parental element intact. This was first demonstrated in experiments studying plasmid-to-plasmid transfer, in which the surprising observation was made that the donor plasmid never lost its transposon. Transposition of Tn*3* is a multistep process (Fig. 7-3), which was established by isolating mutant Tn*3* elements that got stuck between transposon replication and excision of the original element. This intermediate structure, termed a *cointegrate*, consists of two full Tn*3* elements tandemly

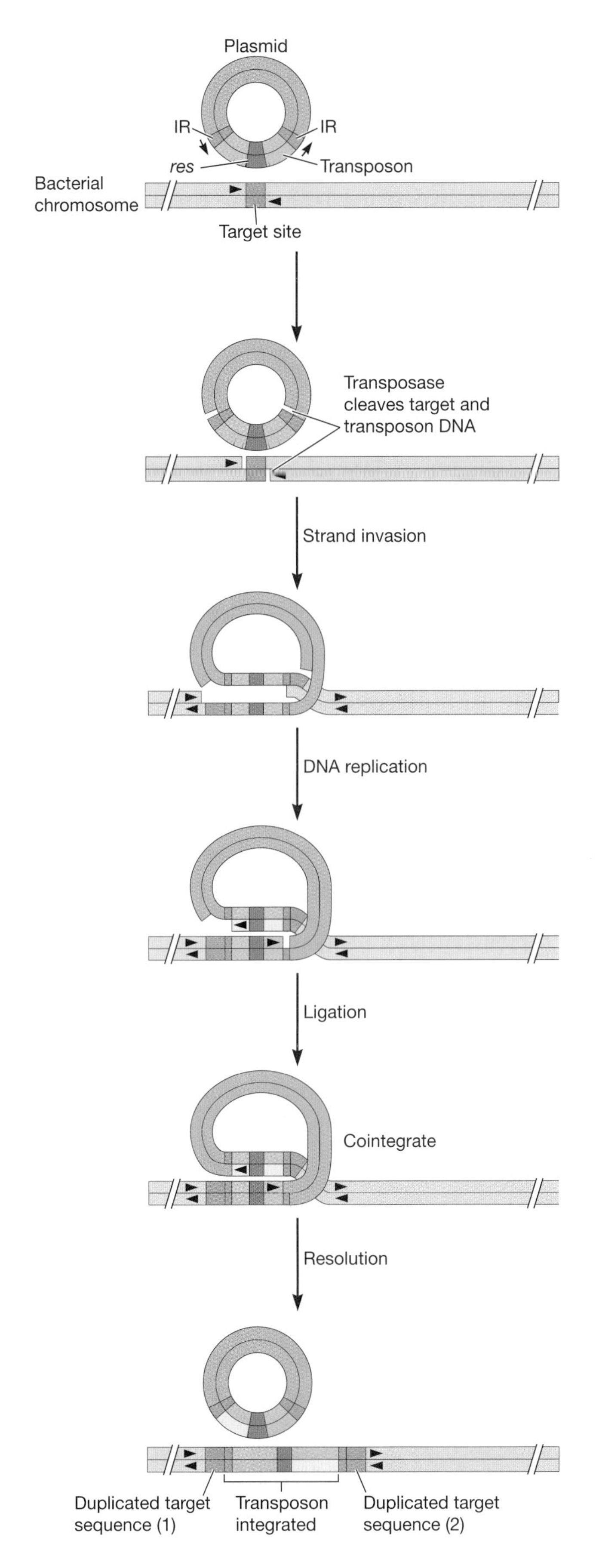

repeated. Formation of the cointegrate requires activity of the Tn*3* transposase as well as replication of the transposon. The cointegrate is resolved into two molecules by the action of the transposon-encoded resolvase protein at specific target sequences, called *res*. That replication of the transposon occurs during cointegrate formation was shown by analysis of cointegrates formed in strains carrying two distinguishable forms of Tn*3*. Cointegrates in these strains always contained two identical elements, which indicates that cointegrates are formed by copying of the parental element rather than by joining of two preexisting ones.

The cointegrate is normally resolved into two separate molecules by recombination between the two elements. Mutants that get stuck in the cointegrate state carry mutations in either the *res* element or the resolvase gene. In the latter case, the mutants could be induced to transpose correctly if resolvase was supplied from another source, whereas mutants lacking the *res* sequence remained as cointegrates. Together with biochemical studies of the resolvase enzyme, these data suggest that resolvase binds to the two *res* sequences in the adjacent transposons

FIGURE 7-3

Tn*3* transposition occurs by a combination of replication and recombination. The diagram shows transposition of Tn*3* from a plasmid to the bacterial chromosome. The transposase enzyme initiates transposition by putting single-stranded nicks in the IRs at each end of the transposon. Staggered nicks are also made in the chromosomal DNA, and the free ends of the transposon DNA strands invade the chromosomal molecule. This unstable structure is held together by the transposase protein. Note that each strand of the original transposon is now a free single strand. The transposase promotes the formation of a replication fork at one of the sites of strand invasion, so that complementary strands are newly synthesized by DNA polymerase. Ligation of the newly synthesized DNA to the parental molecules yields a cointegrate in which the donor and recipient molecules are covalently joined with a transposon at each junction. Each transposon consists of one strand from the original molecule and one newly synthesized strand. The cointegrate is resolved by recombination to regenerate the donor plasmid molecule and the chromosomal DNA molecules now carrying a transposon insertion. Resolution occurs by alignment of a short DNA segment, *res*, in both transposons, and recombination is catalyzed by the resolvase molecule encoded by the transposon. Note that transposition also results in a small duplication of the target sequence into which the transposon is inserted.

and catalyzes recombination between the elements, which leaves behind a new Tn*3* insertion in the recipient DNA and returns the parental element to its original location.

In contrast to Tn*3*, which must replicate when it moves to a new site, Tn*10* performs a more conventional jump without any DNA replication. This was elegantly demonstrated by constructing two λ phages carrying Tn*10* insertions (Fig. 7-4). Each had been engineered to carry a *lacZ* gene encoding β-galactosidase and a gene for tetracycline resistance. The only difference between the two phage genomes was that the *lacZ* gene in one of the phages had been mutated so that the β-galactosidase was inactive. The phage genomes were mixed, denatured, and renatured to form heteroduplex phage molecules consisting of one strand from each parent. The phage molecules carrying heteroduplex transposons were introduced into *E. coli*, and bacterial colonies resistant to tetracycline were selected. Because the phages could neither replicate nor integrate in the bacteria, tetracycline-resistant colonies had to have arisen by transposition of Tn*10* from phage DNA to the bacterial chromosome. If Tn*10* transposition required DNA replication as for Tn*3*, information from only one strand of the heteroduplex transposon would be integrated into the *E. coli* chromosome, with the other strand remaining at the donor site. The tetracycline-resistant colonies would be either *lac*$^+$ or *lac*$^-$, depending on which replicated strand had integrated into the chromosome. If, on the other hand, the transposon moved nonreplicatively—cut from the donor site and pasted into the recipient chromosome—then both strands of the heteroduplex would be transferred, and the resulting colonies would contain a mixture of *lac*$^+$ and *lac*$^-$ cells. In this experiment, mixed colonies were recovered, thus providing strong evidence that transposition of Tn*10* is nonreplicative.

Many DNA transposons coordinate their transposition with DNA replication sequences to become more abundant in the genome. A transposon is replicated so each of the daughter double helices contains a complete transposon (Fig. 7-5). One of the transposons then excises itself and the double-strand break left by the excision of this transposon is repaired by homologous recombination. The transposon integrates into the as-yet-unreplicated DNA, ahead of the replication fork, and is replicated when the replication fork passes through its sequence. The net result is that there are now four transposons where there had been two. Coupling transposition and replication means that this type of transposon is active only in dividing cells.

Retrotransposons Transpose through RNA Intermediates

Yeast *Ty* elements were first identified by the genetic properties of the mutations they cause. In addition to extinguishing gene function by insertion into coding or regulatory sequences, *Ty* insertions have the unusual property of sometimes activating genes that were previously silent. Cloning and sequencing the transposons revealed the reason. *Ty* elements are organized differently from DNA transposons. Instead of having inverted repeats at their ends, the *Ty1* element carries 334-bp direct repeats, termed δ *elements*. These repeats carry strong promoters for RNA polymerase; in fact, *Ty* element transcription accounts for 5–10% of total yeast mRNA. Because these LTRs are direct repeats, each element has one LTR that is oriented so that it promotes transcription of chromosomal DNA located downstream of the insertion. If, then, a *Ty1* element inserts 5′ to a silenced gene, it can activate that gene. LTRs are also found in the *copia* element of *Drosophila*, which has many of the same properties as the *Ty* element. For example, *copia* LTRs drive high levels of transcription (the transposon's name derives from the copious quantity of transcripts made by these elements). Because of this strong promoter activity, retroviruses or retrotransposons can cause cancer in higher animals when they insert near growth-promoting genes and activate them. As we will see in Chapter 15, several cancer-promoting genes have been discovered by mapping the location of an LTR transposon insertion.

Because the sequence organization of retrotransposons like *Ty* and *copia* is so different from that of DNA transposons such as Tn*3* and Tn*10*, it seemed likely that they transpose by a different mechanism. In fact, *Ty* elements transpose through an RNA intermediate. This was established in a series of clever experiments in which *Ty* genomes were engineered by recombinant DNA techniques to test whether the *Ty* RNA or DNA was the entity that transposed. First, a *Ty* element was placed under the control of the inducible Gal promoter. Increasing the rate of transcription of the *Ty* element increased the rate at which it transposed, showing that tran-

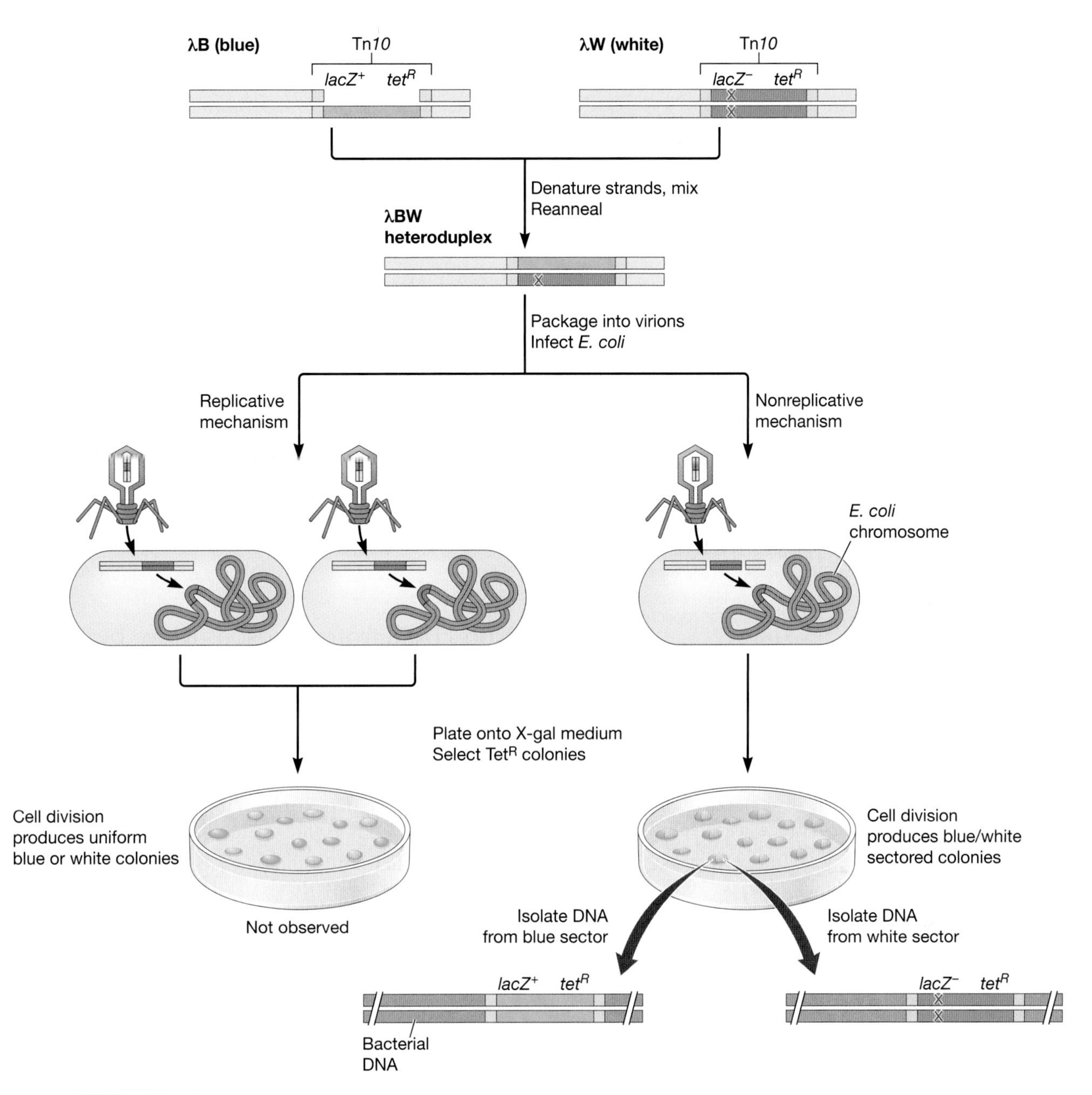

FIGURE 7-4

Demonstration that movement of Tn*10* occurs by a nonreplicative "cut-and-paste" mechanism. Two λ phage derivatives were constructed by insertion of a Tn*10* molecule carrying a *lacZ* gene (which encodes β-galactosidase). One phage, λBlue, carried a wild-type *lacZ* gene; the other, λWhite, carried a mutant gene. The two phage DNA molecules were denatured, mixed, and reannealed, which yielded heteroduplexes carrying one strand from each parent. The phage DNA was packaged into virion particles and used to infect *E. coli*. The transposon hops to the *E. coli* chromosome, yielding tetracycline-resistant (Tet^R) colonies. The *lac* phenotype of the strains was examined using X-gal, which stains lac^+ colonies blue. If transposition occurs by a replicative mechanism as seen in Tn*3* (see Fig. 7-3), only a single strand of the heteroduplex is copied to the chromosome, thus yielding colonies that are either uniformly blue (if they received the λBlue strand) or white (if they received the λWhite strand). If the transposon is cut from the phage and pasted into the chromosome, then both the λBlue and the λWhite strands move to the chromosome. Replication of the chromosome and cell division segregate the two strands, which leads to mixed colonies with blue and white cells. This is what happened in the experiment. Comparison of the transposon in blue and white cells from a single colony showed that, as expected, the lac^+ and lac^- transposons were integrated at the same location.

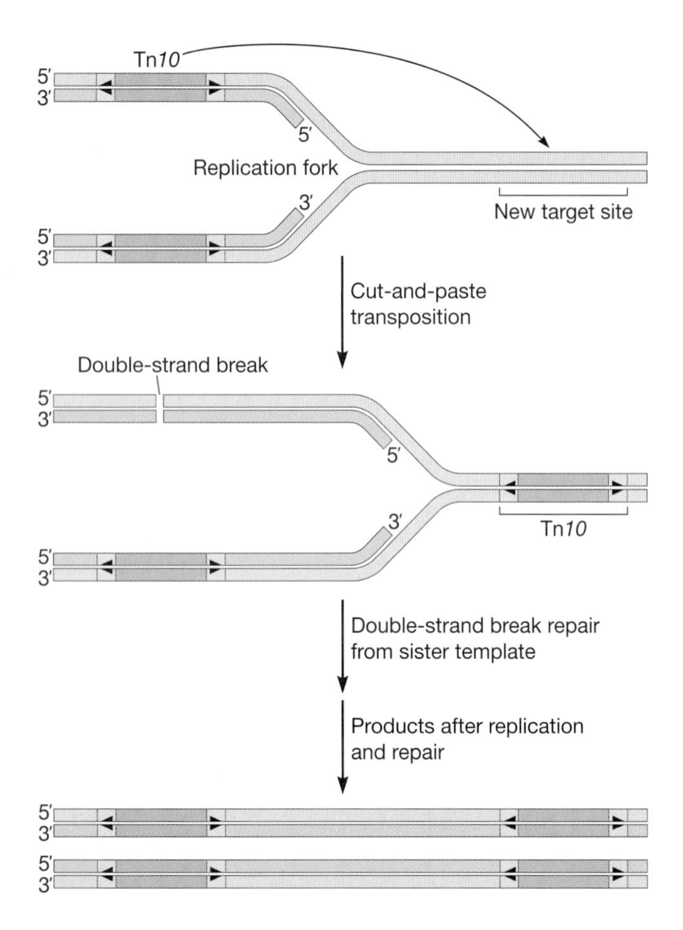

FIGURE 7-5

Transposition of Tn*10* after passage of a replication fork. Transposition is activated by the hemimethylated DNA that exists just after DNA replication, because it is only in this situation that the transposase can efficiently bind. During transposition, a double-stranded break is made in the chromosomal DNA where the element excised. This break can be repaired, but because the repair pathway uses the other strand to direct repair, a copy of Tn*10* is regenerated at the site of excision. By this mechanism, transposition may appear to be "replicative" in nature, although the actual recombination process goes through the cut-and-paste (nonreplicative) pathway.

scription is important for transposition. However, was the act of transcription important, or was it the RNA transcript itself that was transposing? This was tested in a very clever experiment. An intron was placed in the donor *Ty* element. After transposition it was found that the transposed elements lacked this intron. This showed that the newly transposed elements must be DNA copies of spliced RNA molecules transcribed from the donor element. Other LTR retrotransposons such as the *copia* element have a similar mechanism of retrotransposition.

Transcription initiates within the upstream LTR and terminates within the downstream LTR, so that neither end of the transcript contains a complete repeat. Yet, intact LTRs are required for the transposon-encoded integrase to recognize DNA integration. This problem is solved by multiple steps of reverse transcription that piece together the portions of the LTR from each end of the virus. This produces a DNA copy containing complete copies of the LTR at both ends. Retroviruses use an identical mechanism for their reverse transcription and integration into provirus form. The main characteristic differentiating LTR retrotransposons from retroviruses is that retroviruses are packaged and released to infect other cells, whereas LTR retrotransposons usually have lost the genes necessary for packaging into viral particles and instead move within the genome of a single cell.

Non-LTR Retrotransposons Move by Target-primed Reverse Transcription

The third class of transposable element, known as poly(A) retrotransposons or non-LTR retrotransposons, includes the human LINES and SINES (long and short interspersed elements). The LINE-1 (L1) element makes up 21% of the total DNA content of the human genome. The high copy number led to its discovery long before genome sequencing. It appeared as a highly abundant group of 1.2-, 1.5-, 1.8-, and 1.9-kb fragments when genomic DNA was digested with KpnI (the 6-kb L1 element contains several internal KpnI sites). When the L1 element was cloned and sequenced, it was found to contain two ORFs, ORF1 and ORF2. ORF1 encodes a RNA-binding protein. ORF2 is a multifunctional protein that encodes both endonuclease and reverse transcriptase activities. The L1 element and the many elements related to it are integrated into the genome by the process of *target-primed reverse transcription* in which the ORF2 protein makes a single-stranded nick in the DNA at a T-rich site. Then, the 3′-OH of the DNA strand hybridizes to the poly(A) region of the retroelement RNA and acts as a primer for the reverse transcriptase activity of ORF2. Although the details of the process are unclear, after reverse transcription, the 5′ end of the element integrates into the chromosome and the second strand of the retroelement is replicated. Integrated retrotransposons, whether LTR or non-LTR, and DNA transposons typically result in a *target site duplication* in the genome, in which an identical genomic DNA sequence is present at each end of the element. This duplication is caused by integration of the element at a staggered break in the DNA. The duplication results from filling in the gap across the staggered break.

Reverse transcription of non-LTR retrotransposons commonly stops before the element has been fully transcribed. In fact, the vast majority of L1 elements in the human genome are truncated. Without the promoter regions in their 5′ ends, these elements are unable to transpose.

Like the L1 element, another non-LTR retrotransposon, the *Alu* element, was identified even before genome sequencing. They are called "Alu" because they were recognized by virtue of an internal AluI recognition sequence (AGCT), which produces an abundant band in AluI digested human genomic DNA. This family of 300-bp sequences is the most abundant repetitive element in primate genomes; there are some 1.1 million *Alu* elements in the human genome, which make up of 11% of the total genome sequence.

The *Alu* sequence is similar to the 7SL RNA gene, which is transcribed by RNA polymerase III. 7SL RNA is highly abundant in cells as part of the signal-recognition particle (SRP), which aids in transporting proteins to the endoplasmic reticulum for secretion. It is suggested that an ancestral derivative of 7SL RNA suffered a large deletion, duplicated, and then diverged, which gave rise to short segments called "free left" and "free right" *Alu* monomers. These fused and, with the addition of some other sequences, gave rise to the *Alu* sequence. Subsequent divergence from this ancestral sequence has led to the development of an "*Alu* family" with some 16 subfamilies. The *Alu*s are ancient; they are believed to have arisen as long as 60 million years ago. Like the 7SL RNA, most, if not all, *Alu*s are transcribed by RNA polymerase III.

SINES such as *Alu* do not encode their own reverse transcriptase and so are nonautonomous transposons, which require reverse transcriptase supplied from other elements. These are likely to be LINES, as the LINES and SINES have a common 3′ sequence. Transactivation of an *Alu* element by LINE proteins was demonstrated by a clever experiment using an *Alu* with a modified sequence (Fig. 7-6). An *Alu* element was modified to contain a gene-encoding resistance to the antibiotic neomycin. The neomycin-resistance gene was interrupted by the introduction of a self-splicing intron and so could confer resistance only when the RNA was spliced, reverse transcribed, and integrated into the genome. The self-splicing intron was used because it would be processed even if the RNA was not recognized by the normal splicing machinery of the cell. When the *Alu* reporter construct was transfected into cells, an extremely small number of cells became resistant to neomycin by integrating the modified *Alu* element. However, when the construct was co-transfected with a vector expressing the L1 ORF2 protein, which contains reverse transcriptase and endonuclease activities, the *Alu* reporter transposed 1500 times more frequently. A retroviral reverse transcriptase had no effect on *Alu* transposition, thus demonstrating that the L1 ORF2 protein specifically enhanced retrotransposition.

The *Alu* element diverged long ago, but it still retains enough 7SL sequences that it is recognized by several of the proteins that make up the signal recognition particle. By virtue of associating with SRP proteins, *Alu* RNAs are localized to the ribosome. Consequently, they are present at the site where reverse transcriptases are translated and are present for immediate reverse transcription and integration.

DNA copies of other RNAs, such as transfer RNAs (tRNAs), are found scattered throughout the genome. The most spectacular examples of apparent movement of RNA molecules into genomic DNA are the *processed pseudogenes*, which we discussed in Chapter 5. These are copies of complete coding genes that lack introns and frequently end with a poly(A) tail. These pseudogenes clearly represent gene sequences that were transcribed and processed before being reverse transcribed and inserted into genomic DNA. Pseudogenes may result from reverse transcriptase activity supplied by non-LTR retrotransposons or perhaps by LTR retrotransposons or retroviruses.

Transposons Are Abundant in Most Genomes

Different types of transposable elements are abundant in different organisms. We have just seen that copies of the L1 non-LTR retrotransposon make up more of the mass of the human genome than any other sequence. In all, about 50% of the human genome consists of transposable element sequences.

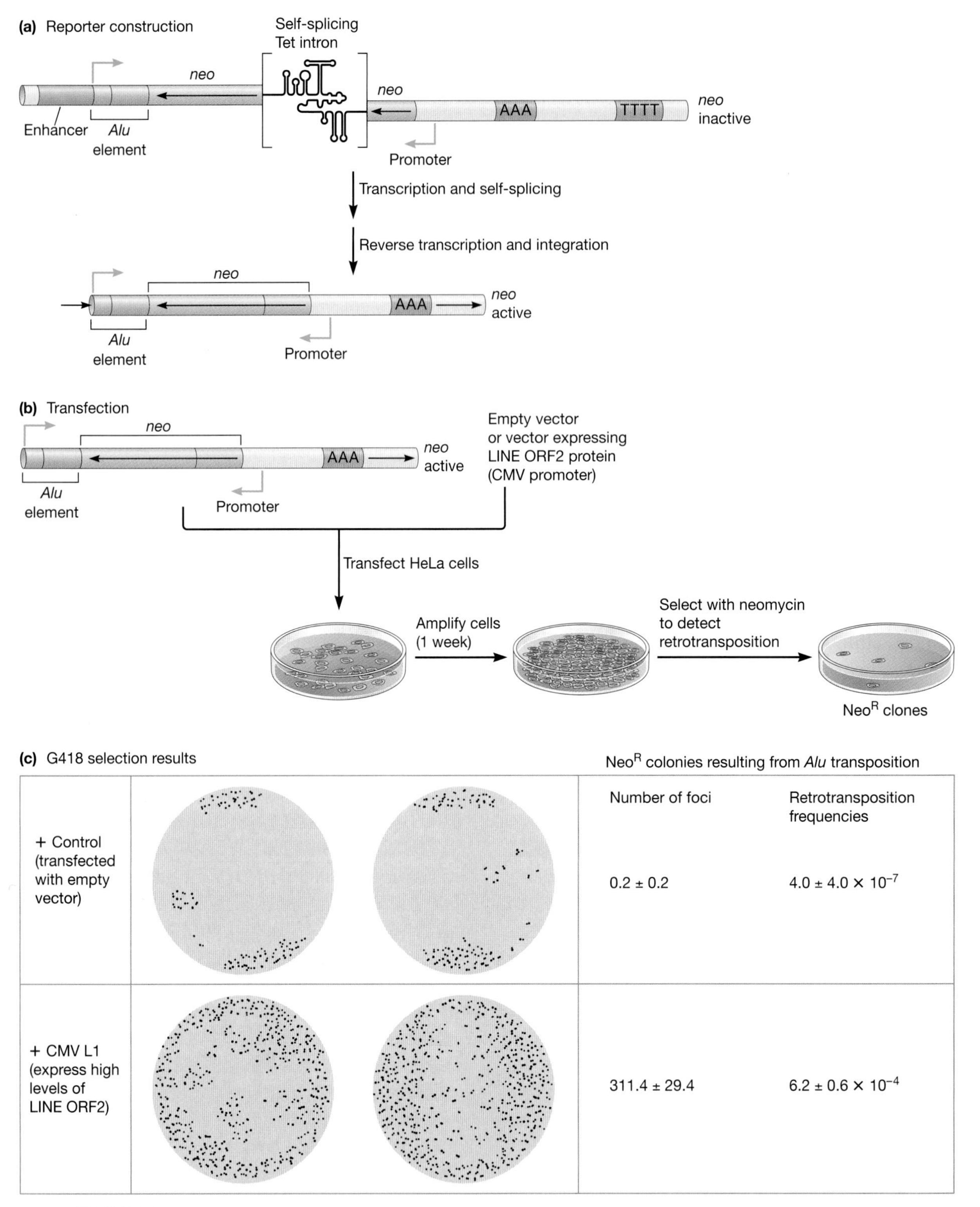

FIGURE 7-6

The transposition of *Alu* elements is promoted by the reverse transcriptase encoded by LINE elements. (*a*) A reporter of *Alu* transposition was constructed by using a gene encoding neomycin resistance interrupted by a self-splicing intron. This type of intron was used because *Alu* elements are transcribed by RNA Pol III, which does not recruit the splicing machinery to its transcripts. Self-splicing introns catalyze their own excision and do not need other splicing factors. Transcription, self-splicing, and reverse transcription will confer neomycin resistance on cells. (*b*) The marked *Alu* element was transfected into tissue culture cells along with an expression vector. The expression vector was either empty or expressed the LINE ORF2 protein, which encodes reverse transcriptase and integrase activities. (*c*) In this experiment, neomycin-resistant colonies resulting from *Alu* transposition were 1500 times more common in cells expressing high levels of LINE ORF2 than in cells transfected with an empty vector.

TABLE 7-1. Proportions of different types of transposable elements in different organisms

Species	Common name	LTR retrotransposons	Non-LTR retrotransposons	DNA transposons
Saccharomyces cerevisiae	Baker's yeast	100.0	0.0	0.0
Dictyostelium discoideum	Slime mold	45.8	38.5	15.6
Caenorhabditis elegans	Nematode worm	1.7	6.9	91.4
Arabidopsis thaliana	Thale cress	46.0	5.0	48.9
Drosophila melanogaster	Fruit fly	69.2	22.7	8.1
Homo sapiens	Humans	18.6	75.2	6.3
Oryza sativa	Rice	56.2	3.7	40.1
Zea mays	Maize (corn)	95.0	1.7	3.3

The contributions of each class of transposon to the total transposons in each organism is given. Note that different classes of transposons are abundant in different genomes. For instance, LTR (long terminal repeat) retrotransposons are rare in *C. elegans* but abundant in *Z. mays*.

The vast majority of these elements are no longer active, either as a consequence of defective integration events or of mutations occurring over time. In many genomes, the contribution from mobile elements is even larger than in the human genome. In maize, more than 60% of the genome consists of transposable elements, although in the maize genome (as in many other plants) the most abundant elements are LTR retrotransposons, rather than the non-LTR retrotransposons that are so abundant in primate genomes. It is not known why the relative proportion of different transposable elements varies between different species, but large differences are indeed present (Table 7-1). In general, RNA transposons are more abundant than DNA transposons, probably because RNA transposons have the potential to produce many RNA copies for retrotransposition, whereas DNA transposons are limited to producing at most a single copy per transposition event.

Transposons explain, in part, the *C-value paradox*, the long-known fact that genome size—estimated by hybridization kinetics or measured as picograms of DNA—does not seem to be correlated with the complexity of organisms or the expected number of genes present in organisms. For example, humans have a similar number of genes to fruit flies, yet our genome is 20 times larger (Table 7-2). Much of the difference in genome size between species results from the number of transposable elements present—in most organisms, genome size increases linearly with the number of transposons present. It is also important to realize that, in many organisms, a

TABLE 7-2. Transposable elements make up a large proportion of genomes

Species	Common name	Genome (Mb)	Number of protein-coding genes	% of genome occupied by transposable elements
Saccharomyces cerevisiae	Baker's yeast	12	5,773	3
Dictyostelium discoideum	Slime mold	34	9,000	10
Caenorhabditis elegans	Nematode worm	100	18,400	6
Arabidopsis thaliana	Thale cress	125	25,498	14
Drosophila melanogaster	Fruit fly	180	13,600	15
Anopheles gambiae	Malaria mosquito	278	13,000	16
Takifugu rubripes	Pufferfish	400	38,000	2
Oryza sativa	Rice	400	37,544	35
Zea mays	Maize (corn)	3,200	50,000	60
Homo sapiens	Humans	3,000	25,000	44
Mus musculus	Mouse	2,500	30,000	40
Hordeum vulgare	Barley	4,800	40,000	70

The number of megabases (Mb) indicated in each genome is representative of the average haploid genome size. Plants, in particular, are notorious for having widely varying genome sizes even within the same species, in large part because plants often become polyploid. The protein-coding gene estimates are contentious. More than a decade after the completion of the yeast genome sequence, it is likely that the gene number of 5773 is correct within 1%. The exact gene numbers present in the larger genomes are estimates. Some of the larger genomes, such as *Z. mays* or *H. vulgare*, have not even been sequenced in their entirety. Even in fully sequenced genomes, the exact number of protein-coding genes is not known accurately, because gene prediction is difficult, as we will see in Chapter 12. The number of transposons in the genome is also uncertain. For instance, various methods had estimated the transposon content of the rice genome at 16%, but after sequencing was complete, it was apparent that 35% or possibly more of the sequence is transposon derived.

large percentage of the genome is occupied by highly repetitive sequences called satellite repeats, which are not mobile elements.

The large numbers of transposons and other similar elements pose difficulties for genome sequencers. Transposon-rich regions are difficult to sequence because the sequences are so repetitive, which makes it difficult to assemble the short sequence reads correctly (as we shall discuss in Chapter 11). However, we now know with high accuracy the numbers and distributions of transposons in the completely sequenced genomes of the rat, mouse, and human.

Transposable Elements Maintain the Ends of *Drosophila* Chromosomes

The ends of chromosomes are special. In 1938, the *Drosophila* geneticist Hermann Muller used X rays to induce chromosome breakage. He found that flies did not survive if there were deletions at the ends of their chromosomes. This led Muller to propose that there were special structures at the ends of chromosomes, which he called *telomeres.* Later, McClintock noted that the telomeric ends of broken chromosomes were "sticky"—once broken, they quickly stuck together or to the ends of other chromosomes. This suggested that telomeres had rather special properties.

Once the detailed mechanism of DNA replication was understood, it was clear that replication of the ends of chromosomes posed a special problem. DNA replication complexes move 3′ to 5′ along the DNA of a chromosome but cannot synthesize the last 50–100 base pairs at the 3′ end, because the replication machinery sitting on the lagging strand template blocks synthesis of the final Okazaki fragments (Chapter 1). If cells did not have a mechanism for maintaining the ends of chromosomes, these would gradually get shorter until genes would be lost. The telomeres solve this problem. In most organisms, telomeres consist of a simple sequence about 6 bp in length repeated hundreds or thousands of times, producing extensions several kilobases long at the ends of chromosomes. These sequences are added by *telomerase*, a specialized enzyme complex containing a short RNA and a specialized reverse transcriptase. The latter uses the short RNA as a template for reverse transcription so that DNA sequences can be added to the single-stranded overhangs left at the telomeres after DNA replication. The protein component of telomerase is itself related to the reverse transcriptases encoded by retrotransposons, although it is a normal cellular gene and not encoded within a mobile element.

Telomerase maintains telomeres in many organisms, but at least two groups of organisms have abandoned telomerase in favor of a different solution. The parasite *Giardia lamblia* and dipteran insects, such as *Drosophila*, have co-opted retrotransposons for telomere maintenance. *Drosophila* lost its telomerase gene at some point during evolution and its telomeres now consist of two non-LTR retrotransposons, called *HeT-A* and *TART.* These are related to other retroelements found elsewhere in the genome, but by some unknown mechanisms, *HeT-A* and *TART* transpose only to the ends of chromosomes and manage to maintain the appropriate numbers of transposons to provide correct telomere length. *HeT-A* and *TART* can even create telomeres on broken DNA ends. This was observed by treating *Drosophila* deficient in double-strand break repair with X rays. The absence of repair mechanisms prevented broken ends from being repaired. The male flies were examined for mutations in the *yellow* gene, which is responsible for the wild-type dark pigment on the fly's outer cuticle and which lies near the end of the X chromosome. (If females had been used, repair might have occurred through homologous recombination with the female's second X chromosome.) The broken X chromosomes in the resulting flies were unstable, so that mutant alleles exhibited progressively more severe phenotypes. Southern blot analysis of the chromosome ends revealed that the ends lost about 75 bp of DNA per generation. Suddenly, some of these chromosome ends gained DNA. When the X chromosomes from these flies were analyzed using probes for telomere transposons, it was apparent that *HeT-A* elements had transposed onto the ends. Over successive generations, the number of *HeT-A* elements increased.

It is still unclear how insects managed to tame two retroelements for the maintenance of telomeres. One clue is the unusual behavior of the proteins encoded by the retroelements. Although many retroelements in *Drosophila* are actively transcribed, the proteins they encode only rarely enter the nucleus. The Gag proteins encoded by *HeT-A* and *TART* were found localized at the telomeres, but only when both were expressed at the same time. Perhaps *HeT-A* and *TART* coevolved to maintain the telomeric sequences.

Retrotransposons Are Important Components of Many Centromeres

Telomeres are not the only part of a chromosome maintained by transposable elements. When mitosis is observed under the microscope (see Fig. 1-1), the condensed chromosomes are paired tightly at a single region called the *centromere*. Specific proteins bind to these specialized regions of DNA and link the chromosome to the mitotic spindle. If a centromere lacks the proper sequences, it cannot bind to the spindle and the chromosomes will not be properly segregated into the two daughter cells.

Proper segregation of chromosomes is of paramount importance, but, surprisingly, the DNA sequences present at centromeres vary greatly between organisms and even among different chromosomes of a single organism. Despite these sequence differences, the same set of proteins binds to the centromeres of most eukaryotes. The CENH3 protein is closely related to histone H3 and replaces histone H3 in the nucleosomes at the centromere. It is found only at centromeres and so is a useful marker for them. To understand better the composition of the maize centromere, an antibody to the CENH3 protein was used to precipitate CENH3 and the DNA bound to it from lysed maize plants (Fig. 7-7). This type of experiment is called a chromatin immunoprecipitation (ChIP) experiment, and we will see further examples of it throughout the book. The isolated DNA was hybridized with probes corresponding to known repeats from the maize genome. The only repeats from the genome found associated with CENH3 were the centromeric satellite repeat and CRM, a retrotransposon found only at the centromere of maize. The same family of retrotransposons was then found to be specifically located within the centromeres of a broad range of other grains, including rice, wheat, barley, rye, and oat. The presence of this element at the centromeres of so many plants suggests that this family of retrotransposons is important for centromere function in some as-yet-unknown way.

Another centromeric protein suggests a further link between centromeres and transposable elements. The CENP-B protein forms an essential part of the kinetochore, the complex that attaches the chromatids of a chromosome to the microtubules of the spindle during cell division. The sequence of the protein is closely related to transposases of the Tc*1/Mariner* family of transposable elements. CENP-B binds DNA and is required for efficient chromatin segregation. It is still not known exactly what functions CENP-B performs, but the presence of a transposase-like protein at the kinetochore is intriguing. Transposons of all kinds are found near and often within the centromeres of many organisms. It is still unclear whether this occurs because transposons tend to transpose there and are not eliminated, or they are there because they are important for centromere function.

Immunoglobulin Genes Are Rearranged to Generate Antibody Diversity

The various forms of transposons are not the only DNA sequences to move, and in contrast to the early views of DNA as a static repository of genetic information, we have come to recognize the dynamic character of genomes. Some cells reorganize their own DNA, most notably when generating antibodies. A long-standing problem in immunology had been to understand how the vertebrate immune system is able to generate molecules to recognize millions of different antigens. Structural analysis of antibody proteins showed that each B lymphocyte produced a unique antibody molecule. But if each possible antibody molecule were individually encoded in the genome, coding for the immune system alone would account for a large share of all the DNA in our cells, and limiting expression to one and only one antibody per B cell would pose a tricky problem.

Antibody genes, because of the interest in this problem and because they are very highly expressed in certain B cells, were among the first genes cloned. Antibody proteins are tetramers that contain two identical heavy chains and two smaller light chains. In 1976, Susumu Tonegawa was able to purify and label the mRNA encoding an antibody light chain from a myeloma cell line. This RNA preparation was radioactively labeled and hybridized to DNA to determine the structure and abundance of the gene from which the mRNA came. He found that in embryonic DNA there were two different restriction fragments that hybridized to the probe, but in the myeloma cell DNA there was only one and its size differed from both embryonic fragments. Thus, the DNA encoding the antibody gene had been rearranged in the myeloma cell.

The nature of these rearrangements was dramatically revealed when the first antibody genes were cloned. Physical characterization and, eventually,

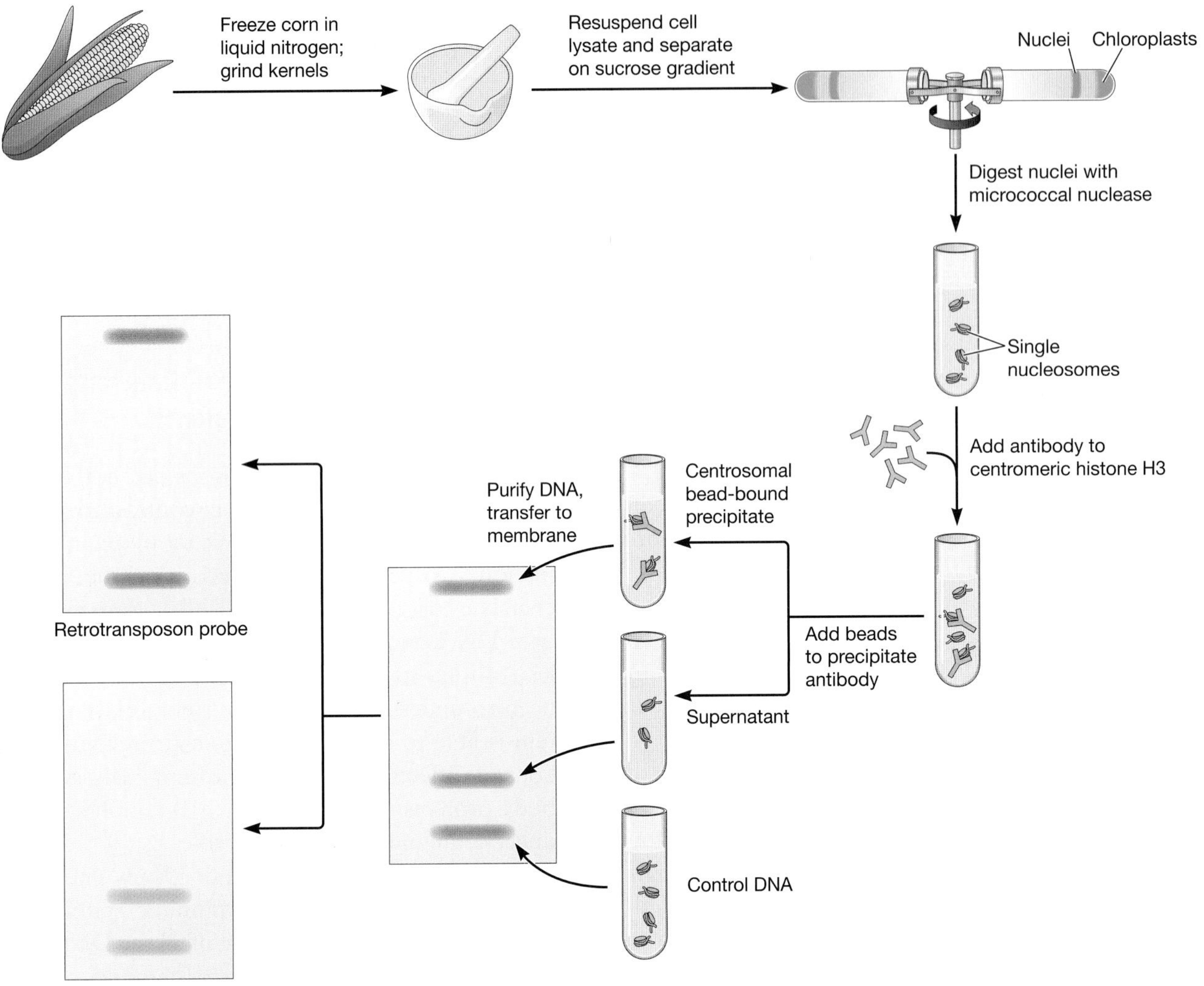

FIGURE 7-7

Identification of a retrotransposon at the maize centromere. Growing ears of maize were collected (many cells in this tissue are actively engaged in mitosis, which makes it a good choice for studies of centromere function) and quick-frozen in liquid nitrogen. The frozen tissue was ground in a mortar and pestle. The grinding action ruptures the tough plant cell walls. The cell lysate was resuspended and fractionated in a sucrose density gradient to separate the nuclei from other cellular components, such as chloroplasts and cytoplasm. The nuclear fraction from the sucrose gradients was then digested with micrococcal nuclease, so that single nucleosomes were present. An antibody to the centromeric histone CENH3 was added to the lysate. Then, agarose beads bearing the antibody-binding *Staphylococcus aureus* protein A were added to the mixture to precipitate the antibody-CENH3 nucleosomal complexes. DNA was separately prepared from the centrosomal bead-bound precipitate and the supernatant and applied to a membrane. The membrane was hybridized with a probe to the centromeric retroelements CentC and CRM and with probes to the Tekay retroelement. The CentC and CRM probes were preferentially found in the centromeric fraction, whereas the Tekay retroelement, although abundant in the genome, was not present on nucleosomes containing the centromeric histone CENH3.

direct DNA sequence analysis of these clones showed that each antibody gene was indeed created from a large family of DNA segments (Fig. 7-8). The light-chain locus, for example, contains a single sequence encoding the constant region (C) shared by all antibodies, and multiple segments encoding different variable regions (V) and joining regions (J), each with distinct sequences. During the development of each B cell, the immunoglobulin locus undergoes a programmed rearrangement that brings one (and only one) of the V genes and one of the J genes to a position immediately upstream of the C gene, which generates a sequence encoding a functional antibody chain. This happens through a cut-and-paste mechanism that deletes intervening DNA and joins the V, J, and C segments in a somewhat imprecise manner, adding to the diversity of potential antibody sequences. A similar process occurs at the heavy-chain locus, except that an additional diversity segment (D) is incorporated between the V and J segments, thus producing a molecule of the form VDJC.

The fascinating process of DNA rearrangement that produces antibody molecules has been the subject of intense study, and the molecular mechanisms of the reaction have been studied by introducing artificial recombination substrate sequences into B-cell lines growing in culture. A DNA molecule was created in which a selectable marker such as a drug-resistance gene was separated from a promoter by a stretch of spacer DNA. Correct rearrangement of the DNA resulted in expression of the drug-resistance gene and the consequent formation of drug-resistant cells. If the appropriate sequences are provided, B-cell lines will readily recombine them, whereas other cell types, such as fibroblasts, will not rearrange the substrates, because fibroblasts do not express the necessary enzymes. This difference was used to hunt for the genes that control the recombination reaction. DNA from an antibody-expressing B-cell line was transfected into a population of fibroblast cells containing an artificial recombination substrate (Fig. 7-9). After several rounds of selection, the segment of human B-cell DNA responsible for this recombination activity was cloned and found to contain two neighboring genes, *RAG1* and *RAG2*, which cooperate to carry out recombination with high efficiency.

What sort of genes are *RAG1* and *RAG2*? They have an unusual structure—no introns, a rarity among mammalian genes. The proteins they produce structurally resemble DNA transposases and catalyze similar DNA nicking events as carried out by transposases encoded by the hAT family of DNA transposons. Recently, biochemical analysis of hAT transposition helped to explain the imperfect palindromic sequences characteristic of the junctions of the rearranged antibody genes. When a hAT transposon excises itself, two covalently closed hairpin ends are left behind in the DNA. This loop must be nicked in order to produce ends that allow the DNA strand to be reassembled. Most commonly, the nick is not precisely in the middle, and the overhanging end is filled in by DNA repair enzymes. This repair process pro-

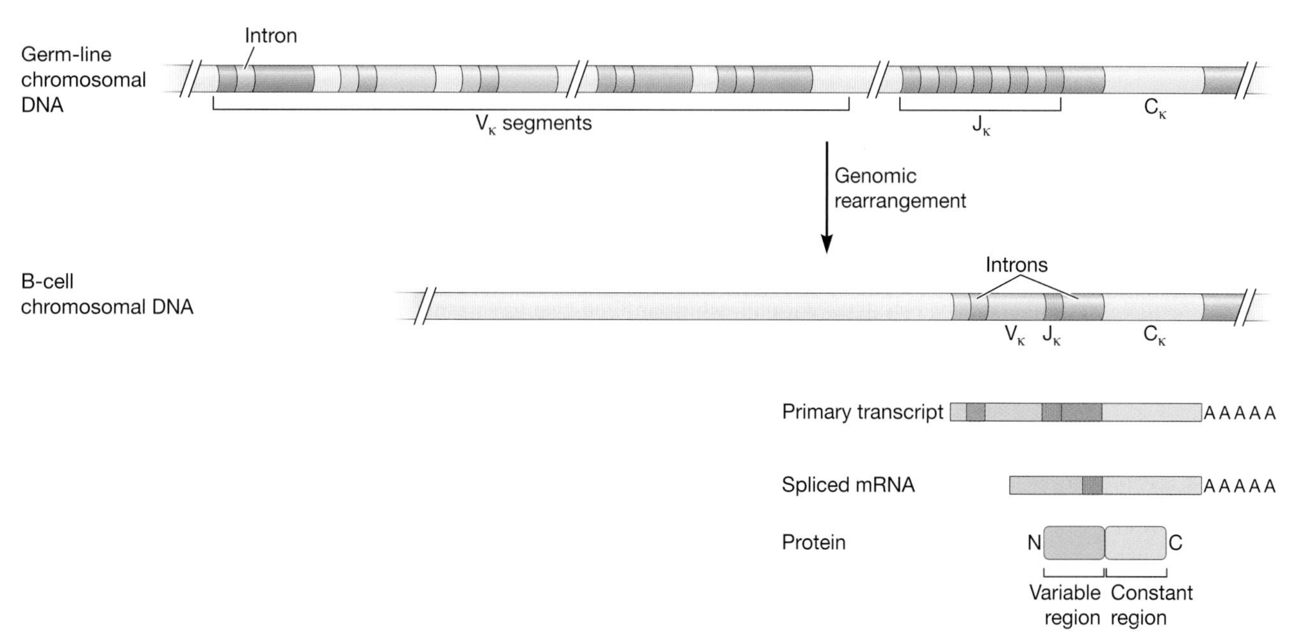

FIGURE 7-8

Rearrangement of the immunoglobulin (Ig) gene locus to form a functional antibody-coding gene. Antibody molecules consist of a constant region (C) that is shared among all antibodies and a variable region (V) that is unique for each antibody and determines its antigen-recognition specificity. In the κ-light-chain gene, about 100 V-region specificities are stored as unexpressed gene segments (V_κ) far upstream of the sequences encoding the constant region (C_κ). In the development of a particular B cell, a selected V_κ is transposed to a "joining" region (J_κ) immediately upstream of C_κ. This occurs by a cut-and-paste mechanism that joins V_κ to J_κ. There are several J_κ segments in the κ locus, thereby increasing the number of possible junction sequences in the resulting antibody gene.

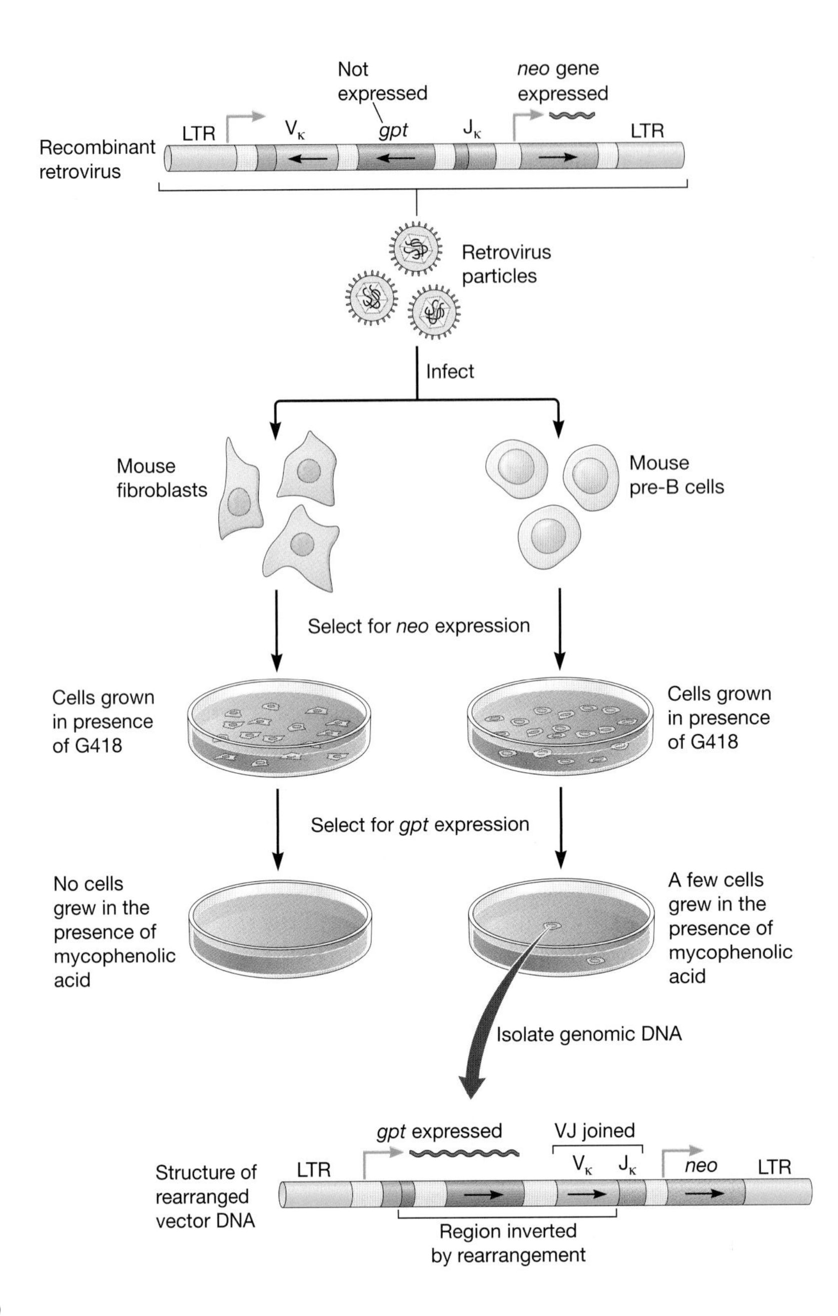

FIGURE 7-9

Artificial recombination substrates are used to study antibody gene rearrangement. A retrovirus vector was designed as a recombination substrate with the structure shown at the top of the figure. It contained V and J gene segments together with their recombination signals (*teal* and *brown boxes*) in opposite orientation to one other. Between the segments was placed a drug-resistance gene (*gpt*) in the reverse orientation relative to the strong viral promoter located in the LTR. Because the V and J are oriented in opposite direction, VJ joining requires inversion of the intervening DNA segment. This inversion also flips the *gpt* gene, placing it in the correct orientation for transcription from the LTR promoter. Expression of the *gpt* gene allows cells to grow in the presence of mycophenolic acid. The experiment was performed by packaging the substrate molecule into retrovirus particles, which were used to infect cultures of recipient cells. Infected cells were selected for resistance to the drug G418, a property conferred by a second gene in the vector (*neo*) that is expressed even in the absence of recombination. Next, the cells were treated with mycophenolic acid. The only cells that grew were those able to flip the *gpt* gene by performing VJ recombination. Resistant cells were obtained only in cell lines of the B lineage known to be undergoing antibody gene rearrangements. Introduction of the construct into fibroblast cells, for example, never yielded mycophenolic acid–resistant cells. In another set of experiments, fibroblast cells bearing a recombination construct like this were used to identify the genes responsible for recombination. As described in the text, the *gpt* recombination substrate was used in combination with transfection of genomic DNA from a B-cell line to identify the genes expressed in B cells. In actuality, any genomic library would have worked for this experiment. Although the *RAG* genes are usually silent in fibroblasts, introducing them at high copy number in a cloned context permitted expression in fibroblasts at levels sufficient to promote V(D)J recombination.

duces a short palindrome in the sequence. The two blunt ends are then fused by other DNA repair enzymes. The sequences between the VJ segments of antibody genes are nicked and repaired in the same way. The similarity between hAT transposition and VJ rearrangement underscores the relationship between the cellular process of antibody rearrangement and the action of invasive transposons.

Transposons Can Cause Mutations

We saw that transposons were discovered as "unstable" mutant alleles in maize and later in bacteria. Transposon-induced mutations are widespread and some of the most striking examples come from plants where variation in pigment synthesis can produce beautiful patterns. For example, the *pallida* gene is required for the synthesis of red pigment in the garden flower *Antirrhinum majus*, commonly known as snapdragon. A mutant allele of *pallida*, called *pallida*^{recurrens}, was discovered that had ivory-colored flowers with red spots and streaks. Further examination of the *pallida* gene revealed a copy of the DNA transposon *Tam3* just upstream of the *pallida* transcription start site in the *recurrens* allele. The pigmented spots on the petals are the result of transposon excision during the process of flower development, similar to the speckled maize kernels observed by McClintock.

Surprisingly, *pallida*^{recurrens} produces five types of offspring. Most commonly, the plant produces offspring with random spots just like itself. However, four other kinds of progeny have been found. The second type of progeny is stably red (full reversion); the third group is a paler red, ranging from nearly ivory to nearly full red (partial reversion); the fourth group produces stably patterned flowers that have both red and ivory regions; and the fifth shows a new pattern of transposon distribution that produces spotted flowers (Fig. 7-10). The *pallida* gene

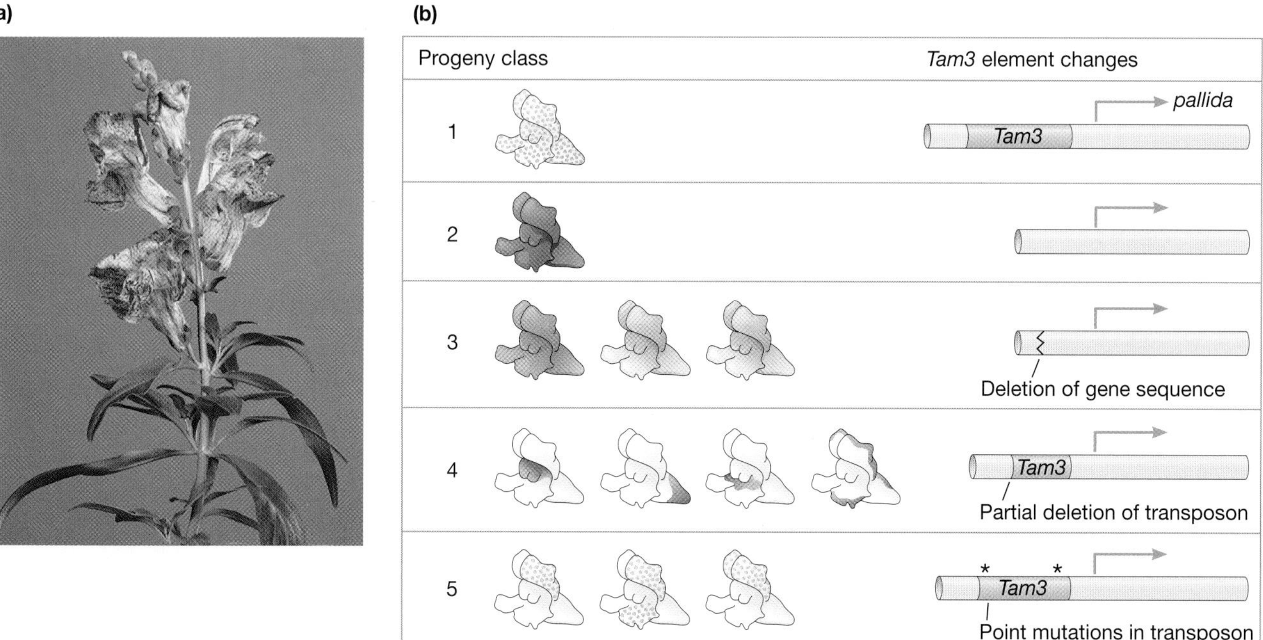

FIGURE 7-10

The *pallida* mutation of snapdragon illustrates the gene expression changes that can be caused by transposon excision. (*a*) The *pallida*^{recurrens} mutation causes spots in the snapdragon flower. (*b*) There are five types of progeny of *pallida* mutants. Most progeny are just like the parent, indicated here as class 1. Class 2 progeny revert to a full red pigment throughout the flower, whereas class 3 progeny are partial revertants that have less than full pigment expression. The fourth class of progeny have restricted patterns of pigment expression, but the pigmented areas are even and show no evidence of spotting as seen in flowers where *Tam3* transposition is active. The fifth group of progeny shows new patterns of transposition. At *right*, diagrams illustrate the changes to the *Tam3* element in the different progeny classes. Class 2 progeny have clean excisions, whereas classes 3 and 4 have partial excisions. In particular, new sequences from sloppy DNA repair are often present in the class 4 progeny. Class 5 progeny have changes within the sequence of the *Tam3* element itself.

was analyzed in these four groups. In all of the bright red, full revertants, *Tam3* had cleanly excised, leaving the gene just 1 bp different from the wild-type sequence. In the pale progeny and those plants showing new, stable patterns of pigment, *Tam3* had excised less cleanly, deleting or leaving behind extra DNA sequences. The remaining sequences apparently affected the enhancers that control gene expression. Several varieties of snapdragon from the fourth group are sold commercially because they have unique and beautiful pigment patterns. In the group of unstable alleles showing novel patterns and frequencies of transposition, the sequence of the *Tam3* element itself had mutated, probably affecting its recognition by the transposase.

Clearly, transposons can alter the genome to produce beautiful phenotypes, but most transpositions are deleterious rather than pretty. New retrotransposon insertions have been discovered in about 48 isolated cases of 34 different human diseases, including hemophilia, muscular dystrophy, β-thalassemia, and severe combined immunodeficiency. For example, two unrelated hemophilia patients were found to have insertions in exon 14 of the hemophilia gene, which presumably disrupted normal functioning of the factor VIII gene (Fig. 7-11). When the inserted DNA was cloned and sequenced, it was found to be derived from L1 elements. In these cases, the parents of these patients did not have the insertion, which implies that the transposable element likely jumped in the germ

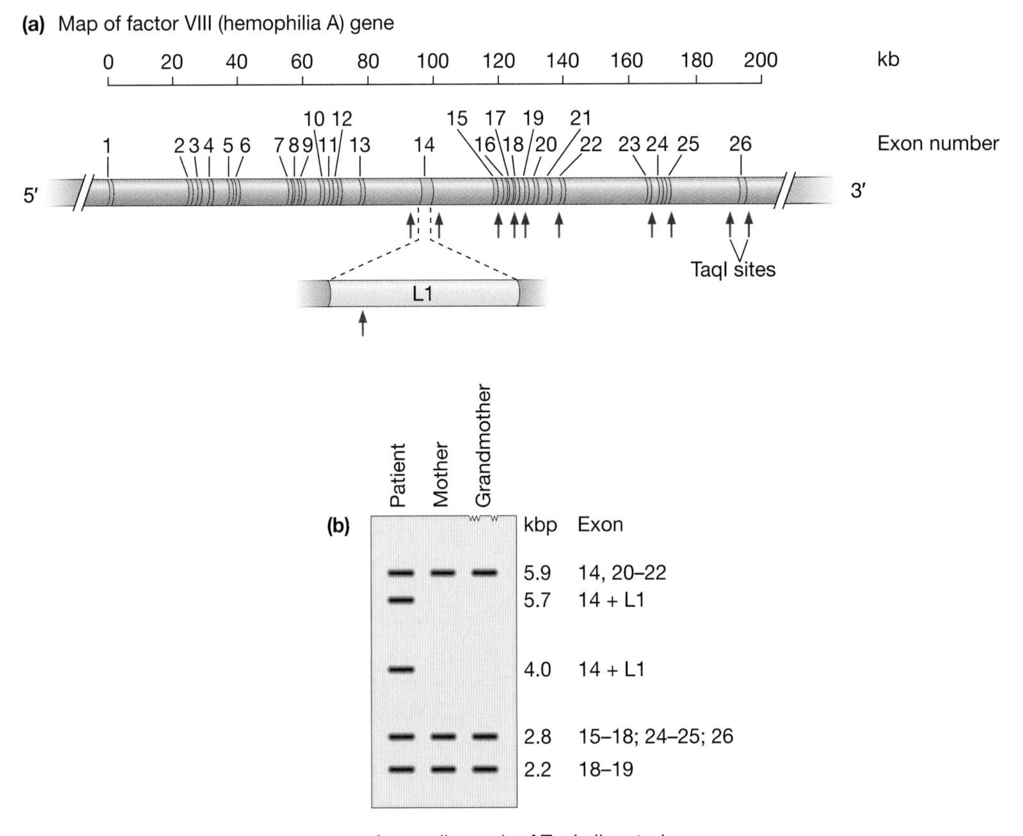

FIGURE 7-11

The mutation in the hemophilia A (factor VIII) gene caused by the insertion of an L1 element. (*a*) A map of the hemophilia gene, showing the locations of its exons (numbered *1–26*) and the TaqI sites (*arrows*). The L1 element lies within exon 14 and contains one TaqI site (*arrow*). (*b*) Digestion of normal DNAs (from mother and grandmother) with TaqI, followed by hybridization with a cDNA probe for exons 14–26, reveals three fragments from the factor VIII gene; their sizes are 5.9, 2.8, and 2.2 kbp. The band at 5.9 kbp actually contains two fragments of similar size, one including exon 14 and one covering exons 20–22. The patient's DNA has the latter fragment, but the fragment containing exon 14 is cut into two fragments because the L1 element inserted in his hemophilia gene contains a TaqI site. The L1 element increases the size of this exon 14 fragment from 5.9 to 9.7 kbp, and its two fragments are 5.7 and 4.0 kbp.

line or in the embryo. Current estimates are that as many as 1 in 600 mutations in humans are caused by retrotransposon insertions. This is certainly an underestimate because of the difficulty of detecting insertions with more subtle effects. This sampling bias is shown by the fact that one-half of the insertions were found in X chromosomes. These act as dominant mutations in males and so are more easily discovered.

In the coming years, better DNA-sequencing technologies will allow us to estimate more accurately the rates of transposon activity in the genome. Whole genome analysis, for example, indicates that there are many more *Alu* families than had been thought and that the evolutionary history of *Alu* is complex. A bioinformatics analysis of the human genome sequence found 480,000 *Alu* sequences, which clustered into 213 subfamilies, compared with early estimates of just 31 subfamilies. The rate of mutagenesis by transposons differs widely among organisms—in *Drosophila,* transposable element jumps cause more than 50% of new mutations, whereas in the mouse about 10% of new mutations are due to transposable element activity. It is not known why some organisms have much more transposable element activity than others. Sequence comparisons indicate that when transposable elements invade a genome, they initially multiply rapidly but then their rate of spread slows.

Mobile Elements Are Used as Tools for Mutagenesis and Transgenesis

Because transposable elements cause mutations, they have been used to carry out genetic screens in many organisms. Transposons are useful mutagens because they can be used to clone the gene that they have disrupted. One common way to do this is by inverse polymerase chain reaction (PCR) (Fig. 7-12). Genomic DNA of the mutant is digested with a restriction enzyme that cuts within the transposon. Among the many fragments from the host genome, there will be fragments containing the ends of the transposon adjacent to host DNA. The digested DNA is ligated under dilute conditions, so that the molecules recircularize instead of ligating to one another. Now, the unknown flanking host DNA can be amplified with primers that point outward from the transposon sequences, and the resulting PCR products can be sequenced. This sequence information can be used to determine the precise location of the insertion and to identify the host gene disrupted by the transposon.

P elements (transposable DNA segments that consist of a transposase gene flanked by inverted

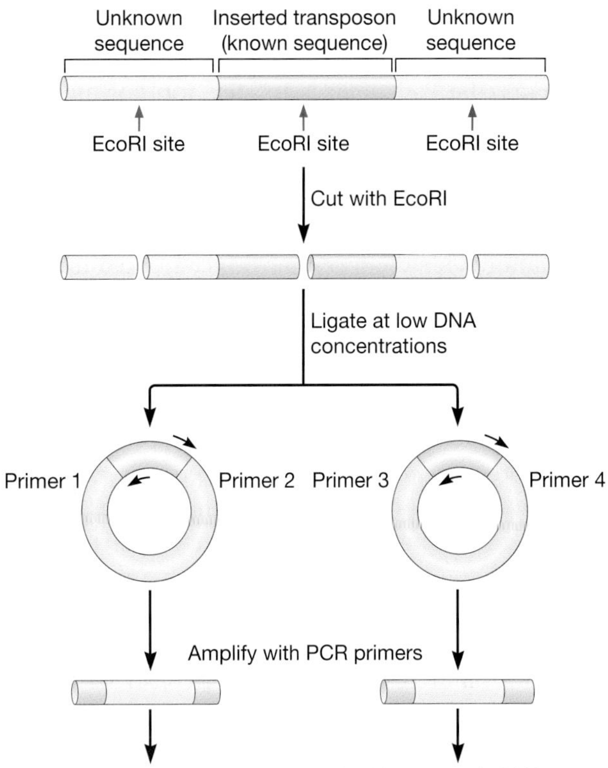

FIGURE 7-12

Inverse PCR allows amplification of unknown sequences flanking a known insertion sequence. Genomic DNA is prepared from an organism with an inserted transposable element of known sequence. The DNA is digested with a restriction enzyme known to cut within the inserted sequence (in this example, EcoRI is used). By chance, the restriction enzyme will also have recognition sites within the unknown, flanking DNA. The digested DNA is diluted to a low concentration and ligated, so that the formation of self-annealed circular fragments is favored (rather than the formation of ligation products between different fragments). The mixture is then amplified by PCR using primers facing outward from the known transposon sequence. Amplified products are sequenced to reveal the genomic DNA flanking the insertion site.

repeats) entered *Drosophila melanogaster* from a related *Drosophila* species sometime within the past 50 years. Biologists suspect that the P element hitched a ride in a tiny mite that fed first on a species carrying the P element and then on a *D. melanogaster*. (P-element DNA has been detected in mites feeding on flies.) The P element expanded rapidly throughout the *D. melanogaster* population and is likely to be present in any fly found in the wild. Fortunately for geneticists, T.H. Morgan and his colleagues in the Fly Room at Columbia established stocks of *D. melanogaster* in the early years of the

20th century, long before the P element entered wild *Drosophila*. Because laboratory strains lack P elements, the P element is an ideal tool for mutagenesis of laboratory *Drosophila*—there can be no confusion with endogenous elements.

As might be guessed by its rapid spread throughout wild *Drosophila*, the P element is very active and can multiply rapidly when introduced into a fly. The P element is an autonomous retrotransposon encoding its own transposase. To restrict its activity, a nonautonomous P element was engineered by deleting its transposase sequence, leaving the terminal inverted repeats that are necessary for transposition. The modified P element must be provided with an exogenous transposase if it is to move. This is supplied in *trans* from an expression vector, thus making P-element transposition dependent on a transposase provided from a separate gene. P elements are commonly used to introduce transgenes into *Drosophila*. The gene of interest is cloned into a construct so that it is flanked by the recognition sequences for the P-element transposase. The gene is then microinjected into *Drosophila* embryos along with a separate expression plasmid that encodes the P-element transposase. The transposase is transiently expressed during early development, which allows transposition of the P-element construct into the genome. As the embryo grows, the nonintegrating transposase plasmid is lost and transposition stops.

An early use of the nonautonomous P element was as an "enhancer trap" to find transcriptional enhancers within the genome. A P element was modified to carry two marker genes. The first, *rosy*, under control of a promoter derived from the *white* gene that drives expression in the eye, confers a red eye color on the offspring of flies with white eyes. The second gene encodes β-galactosidase under the control of a promoter that is very weakly expressed in nearly every cell type. This modified P element was injected into fly embryos. Transgenic individuals were identified by their red eye phenotype, and these were crossed with flies expressing the P-element transposase to mobilize the P element. In some of the progeny of this cross, the P element had moved to new locations and when these flies were stained for β-galactosidase, individuals displayed an amazing variety of expression patterns in different cells and tissues (Fig. 7-13). These varying expres-

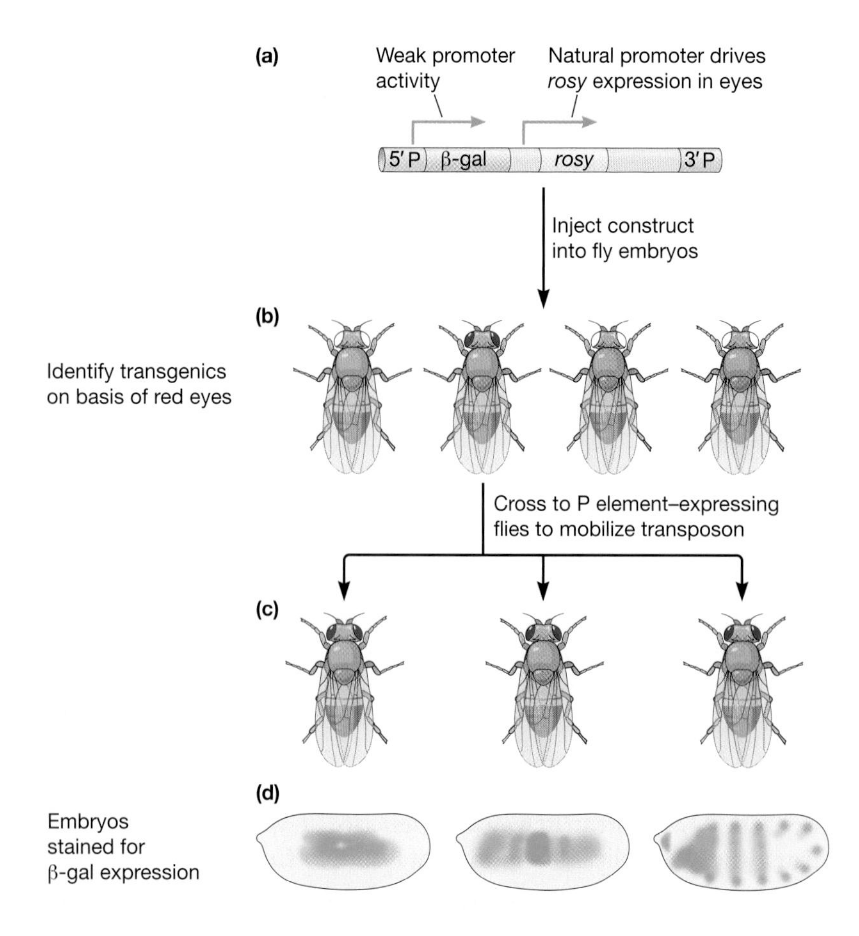

FIGURE 7-13

A modified P element containing marker genes allows identification of enhancer sequences throughout the genome. (*a*) The enhancer trap transposon used as a reporter of enhancer activity. A *rosy* gene is included, which allows identification of transgenic lines by eye color. The *rosy* gene was expressed using an eye-specific promoter. The β-galactosidase gene was expressed under control of the weak P-element promoter. The expression level of the P element on its own is low. Therefore, strong expression of *lacZ* requires that the transgene integrates near to an enhancer sequence in the genome. The vector also contains a gene for ampicillin resistance within the P element. (*b*) Transgenic flies are prepared by injecting the P element construct into embryos from a line of flies with white eyes. Transgenic flies are identified by looking for individuals with red eyes, and individuals carrying the transgene were identified. (*c*) These transgenic flies were then crossed to males carrying a copy of the P-element transposase. (*d*) Embryos from each line are collected and stained to reveal the distribution and expression of the enhancer trap.

sion patterns are the consequence of the different insertion sites of the reporter. High levels of expression indicate that the reporter has integrated close to the enhancer of a gene. Strains in which β-galactosidase was expressed particularly strongly in a specific organ or tissue have been useful for identifying genes that are regulated in a tissue- or stage-specific manner. In addition, the enhancer trap P element acts as an insertional mutagen, so these enhancer trap lines not only mark interesting locations in the genome that control gene expressions but are also a source for new mutants.

A Transposon Is Resurrected for Mutation Experiments in Mammalian Cells

In marked contrast to the highly active P elements of *Drosophila*, all of the DNA transposons in the human genome (and indeed in most vertebrates) have been rendered inactive through mutation. Although about 2% of the human genome consists of DNA transposon sequences, none has been active in the past 50 million years. This is unfortunate from the standpoint of the experimental geneticist wanting to use DNA transposons for experiments in vertebrates. It is also unfortunate that, in contrast to the Ac/Ds and *Spm* transposons that are active in many plant species, animal transposons tend to be species specific. Because no active vertebrate transposons exist, a group of biologists decided to make one by comparing the sequences of a number of inactive transposons to deduce what sequences were needed to reconstruct an active vertebrate transposon.

The Tc*1/mariner* family of DNA transposable elements are found in a wide variety of organisms, including protozoa, fish, the African clawed frog *Xenopus*, and human beings. This broad distribution suggested that it might be useful as a transposon for mutagenesis experiments. The Tc*1/mariner* transposons found in members of the salmon family appear to be the youngest and so are likely to have accrued the fewest disabling mutations. Twelve individual elements were cloned from eight species and their sequences were compared. Site-directed mutagenesis was used to restore ORFs and to recreate what was hoped would be an active transposase (Fig. 7-14). Each modification was assayed for its transposase activity in in vitro assays. The end result of this painstaking work was the awakening of a transposon of a kind that had probably not been active for 10 million years. Appropriately, it was dubbed *Sleeping Beauty*.

The *Sleeping Beauty* transposon system—the transposase gene together with terminal inverted repeats from a nonautonomous salmon element—can be used to generate germ-line insertions similar to those generated by *Drosophila* enhancer traps. Other recent work has taken advantage of the ability of *Sleeping Beauty* to insert into the DNA of somatic cells. Most cancers arise as a consequence of mutations in somatic cells. Therefore, experimentally inducing cancers by making somatic cell mutations using a tagged retrotransposon would be a useful way to determine which genes are required for tumor development. A version of *Sleeping Beauty* was constructed with sequences that could either inactivate or activate a gene, depending on whether the element integrated within a transcribed region or outside it. First, the construct contained splicing signals flanking an exon filled with stop codons in all reading frames, so that it would, if inserted within a gene, create a prematurely stopped protein. This should reveal a gene whose activity is required to inhibit cancer; in its absence cancer develops. Second, *Sleeping Beauty* was engineered to contain a strong enhancer and promoter, so that integration upstream of genes will induce overexpression, thus revealing genes whose overexpression promotes cancer. These constructs were introduced into somatic cells, and, when cancerous tumors subsequently developed in the animals, the tumors were analyzed. The sites of insertion of the constructs were determined to identify which genes or sequences had been disrupted. Several of these were known cancer genes. In addition, insertions into a number of unknown genes were recovered, all candidates for new tumor-suppressor genes. Other insertions produced new protein products from existing genes, transcribed by the promoter present in *Sleeping Beauty* and spliced to exons in the genome. These transcripts produced new protein products, such as a constitutively active kinase missing the regulatory domain that normally limits its phosphorylation activity.

Transposable Elements Influence Gene Expression

Most transposons have promoters that drive their own transcription, and these promoters often remain functional even when internal sequences are deleted. Transcripts from these promoters may extend into neighboring genes, turning them on.

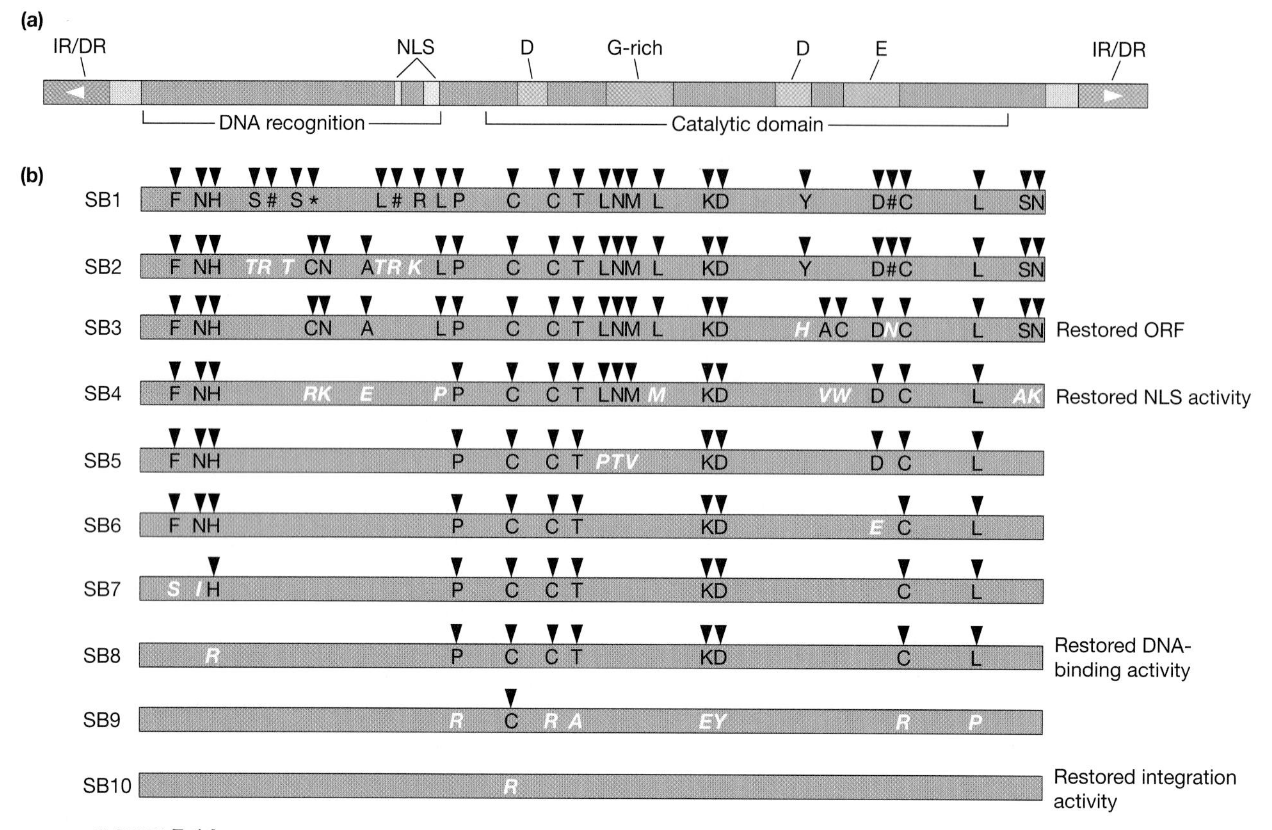

FIGURE 7-14

The active transposon, *Sleeping Beauty*, was created from silent transposable elements. (*a*) Diagram of a salmonid Tc*1*-like transposable element. IR/DR, inverted repeat/direct repeat; NLS, nuclear localization signal; DDE, domain that catalyzes transposition. (*b*) Twelve transposable elements from eight species were compared to generate a consensus sequence for the active ancestor. An element found in salmon closest to this sequence (SB1, Sleeping Beauty 1) was chosen as the starting point for multiple rounds of site-directed mutagenesis to replace amino acids. There were many amino acid differences between the starting and consensus sequence (*black letters* and *arrows*), as well as termination codons (*) and frame shifts (#) (SB1). The open reading frame was restored and consensus amino acids were substituted (*white letters*) (SB3). Then the nuclear localization signal was corrected (SB4) and other amino acids changed. The DNA-binding region was corrected in SB8 and a fully active transposon was achieved in SB10.

Also, as transposons move through the genome, they may carry additional flanking DNA along with them, analogous to transduction by phage (Chapter 4). This is especially true for retrotransposons, because a transcript that reads into sequences 3′ of the retrotransposon can be reverse transcribed and integrated along with the retrotransposon. Thus, transposons can move sequences other than themselves around the genome, thereby possibly moving elements that influence gene expression. There are a growing number of examples in which transposons are known to influence gene expression in ways that alter the phenotype of the organism.

LINE insertions within genes are rarely observed because they are likely to disrupt gene function, which typically makes organisms less well-adapted to survival. However, a survey of insertions of the Doc family of LINE elements in *Drosophila* revealed surprising findings, suggesting that in fact a LINE insertion that disrupts a gene may provide a selective advantage to the organism. In this study, one particular element—*Doc1420*—was found in many natural populations. (Only a few populations lacked the element, and most of these were from Africa, where the *Drosophila* insect family originated.) This Doc element occurred at surprisingly high frequency (~80% in populations tested in eight different countries) and, perhaps more surprisingly, had inserted into and remained within a particular sequence in these populations—a gene the authors called *CHKov1*. Comparative sequence analysis suggested that the original insertion event likely occurred about 90,000 years ago, and the element was now much more common than would be expected by chance. This was evidence that selective pressure had promoted the reproductive success of flies carrying *Doc1420* so that the insertion became widespread. This *selective sweep* occurred within an extraordinarily brief evolutionary time frame—between the last 25 to 240 years (an estimated 500–2400 *Drosophila* generations of 10–20 generations per year).

What could this (relatively recent) selective pressure have been? Organophosphates were introduced as pesticides in the 1940s and are the most widely used pesticides. They paralyze and kill insects by inhibiting acetylcholinesterase, an enzyme that breaks down acetylcholine at nerve synapses. It was suspected that the loss of *CHKov1*, a choline kinase, would affect choline metabolism and perhaps the function of acetylcholinesterase itself, conferring pesticide resistance on those flies with the *Doc1420* insertion. To examine whether this was so, two lines of *Drosophila* were bred, one with and one without the *Doc1420* insertion, and both lines were exposed to organophosphate insecticides. The flies carrying the *Doc1420* insertion were observed to be more resistant to organophosphates (81% survived in the presence of the pesticides compared to 32% of those lines without the insertion). It therefore seems likely that the *Doc1420* insertion into *CHKov1* suddenly spread throughout the world because it helped flies evolve resistance to a new environmental selective pressure—organophosphate pesticides.

Another example of a transposon changing gene expression was found in a study of *Drosophila* resistant to the pesticide DDT. Although *Drosophila* species are not important crop pests, they, like many other insects around the world, have developed resistance to DDT, the first modern pesticide. (*Drosophila* laboratory strains established in the 1930s prior to the introduction of DDT are not resistant.) To better understand the molecular mechanisms of pesticide resistance in insects, DDT-resistant *Drosophila* were collected from around the world and microarrays were used to compare the expression of approximately 90 genes suspected to be involved in pesticide detoxification compared to the susceptible laboratory strains. The resistant strains all showed high levels of expression of a gene called *Cyp6g1*, which is a member of the cytochrome P450 gene family that can detoxify many compounds. DDT-sensitive flies were made that express high levels of *Cyp6g1*, and this was sufficient to make susceptible strains resistant to DDT. When the sequences of *Cyp6g1* were compared among the strains, all of the resistant strains were found to have an LTR retrotransposon called *Accord* inserted just upstream of *Cyp6g1*. The transcription from the LTR increases the expression of *Cyp6g1* so that the flies can more efficiently break down DDT. Although subsequent studies in *Drosophila* and other insects have implicated other genes in addition to *Cyp6g1* as important for DDT resistance, this example highlights the ability of transposable elements to modify gene expression and to affect the evolution of a species.

Another role for transposable elements in regulating gene expression was revealed from studies of the early development of the mouse embryo. Analysis of a cDNA library, prepared from poly(A) RNA from early mouse embryos, showed that 15% of the clones included retroelement sequences, thereby demonstrating that retrotransposons are very transcriptionally active in these early stages. In contrast, fewer than 1% of the clones in a cDNA library prepared from later-stage embryos came from retroelements. When these sequences were analyzed, it was found that the transposable elements were acting as alternative promoters for a number of protein-coding genes. Some of the transposon-directed mRNAs were predicted to encode novel proteins, and one of these proteins could be detected. Thus, transposable elements may have been co-opted for regulating a set of early embryonic genes.

Transposable Elements Help Generate Diversity for Evolution

Mobile elements increase genetic variability by changing gene expression and promoting large-scale rearrangements of DNA. In so doing, they generate new phenotypes on which natural selection can act. Many mobile elements are activated during stressful conditions—times when an organism might wish to increase its repertoire of variations. Broken chromosomes create stress for a cell, and McClintock observed that the presence of a broken chromosome in maize cells activated transposons, not just near the broken DNA, but throughout the genome. In some cases, this transposon activation helps to stitch the genome back together at the DNA double-strand breaks, but McClintock also saw that the global activation of transposons often produced profound DNA rearrangements. Such changes could quickly lead to the formation of reproductively isolated populations of a species, because major rearrangements can interfere with the pairing of homologous chromosomes during meiosis. When chromosomes fail to pair, they do not segregate correctly into gametes, which greatly decreases the likelihood of forming a healthy embryo. In this way, transposon activation could promote the formation of new species, because reproductively isolated groups accumulate other changes that push them further and further apart.

Environmental stresses can alter the movement of transposable elements. For example, the abundance of the BARE-1 retrotransposon was observed to vary in natural populations of barley. Plants growing in stressful, hot, dry environments had three times as many copies of BARE-1 as did plants from cooler, wetter locations. The plants in the stressful environment had more whole retrotransposons than did plants growing in more favorable locations. The latter also had smaller genomes, not only because there were fewer retroelements but also because there were many isolated LTRs, evidence that intervening retrotransposon sequences had been removed by recombination. These and a number of other observations have led to the suggestion that transposition is promoted by environmental stress so that genome variability increases during difficult times. Although many variants are probably less successful in their new environment, by chance a few individuals could be better suited to the new conditions.

We have focused on the act of transposition as an agent of change. However, the enormous number of transposons throughout the genome has another important consequence. The widespread distribution of transposons means that recombination at transposon sequences can bring together distant parts of the genome. Furthermore, because the recombination depends only on the presence of very similar sequences, rearrangements can occur equally well between transposons that have become inactivated by point mutations as between transposons that are active.

Crossing-over between transposable elements located at nonhomologous chromosomal sites can result in the formation of duplications or deletions. Examples of both have been observed at the human low-density-lipoprotein receptor gene (Fig. 7-15). Different rearrangments between *Alu* insertions result in the production of mutant receptors that cannot remove cholesterol from the bloodstream. Individuals carrying these *Alu*-triggered mutations were identified because they had extremely high cholesterol levels.

The human genome has an exceptionally large number of duplications (Chapter 11); about 5% of the human genome consists of segmental duplications, blocks of duplicated sequence 200–400 kb in length. Nearly one-third of these duplications are flanked by *Alu* elements, which suggests that many were generated by *Alu* mispairing and recombination. The fact that so many of these duplications are apparently the result of *Alu* sequences is likely due to the stunning abundance of *Alu* elements in our genome, rather than to some mysterious capacity for directing duplication. Although many duplication and deletion events have deleterious consequences, occasionally such events may be evolutionarily useful. In the human globin locus, there are two copies of the fetally expressed γ-globin gene. This duplication event occurred as a consequence of an unequal crossover between L1 elements flanking the single γ-globin gene in our primate ancestor. The function of the gene evolved from a gene expressed only in the embryo to a fetal gene with coding changes that allow the fetus to efficiently capture oxygen in the placenta from the mother. This duplication event was successful enough to be present in all modern primates.

Transposons Are Usually Silenced

Genomes are replete with mobile elements, which, if left unchecked, could cause havoc, turning genes on and off at random or by promoting dangerous recombination. Organisms with high transposon activity have developed strategies to control and minimize transposition.

In *Drosophila*, certain crosses give a surprising result—the offspring of apparently normal individuals are sterile and bear numerous spontaneous mutations and chromosomal aberrations. This phenomenon, termed *hybrid dysgenesis*, occurs only in one direction—females of one line will show hybrid dysgenesis with males of another line, but if the cross is reversed, using males from the first line and females from the second, the offspring are normal.

The first discovered example of hybrid dysgenesis

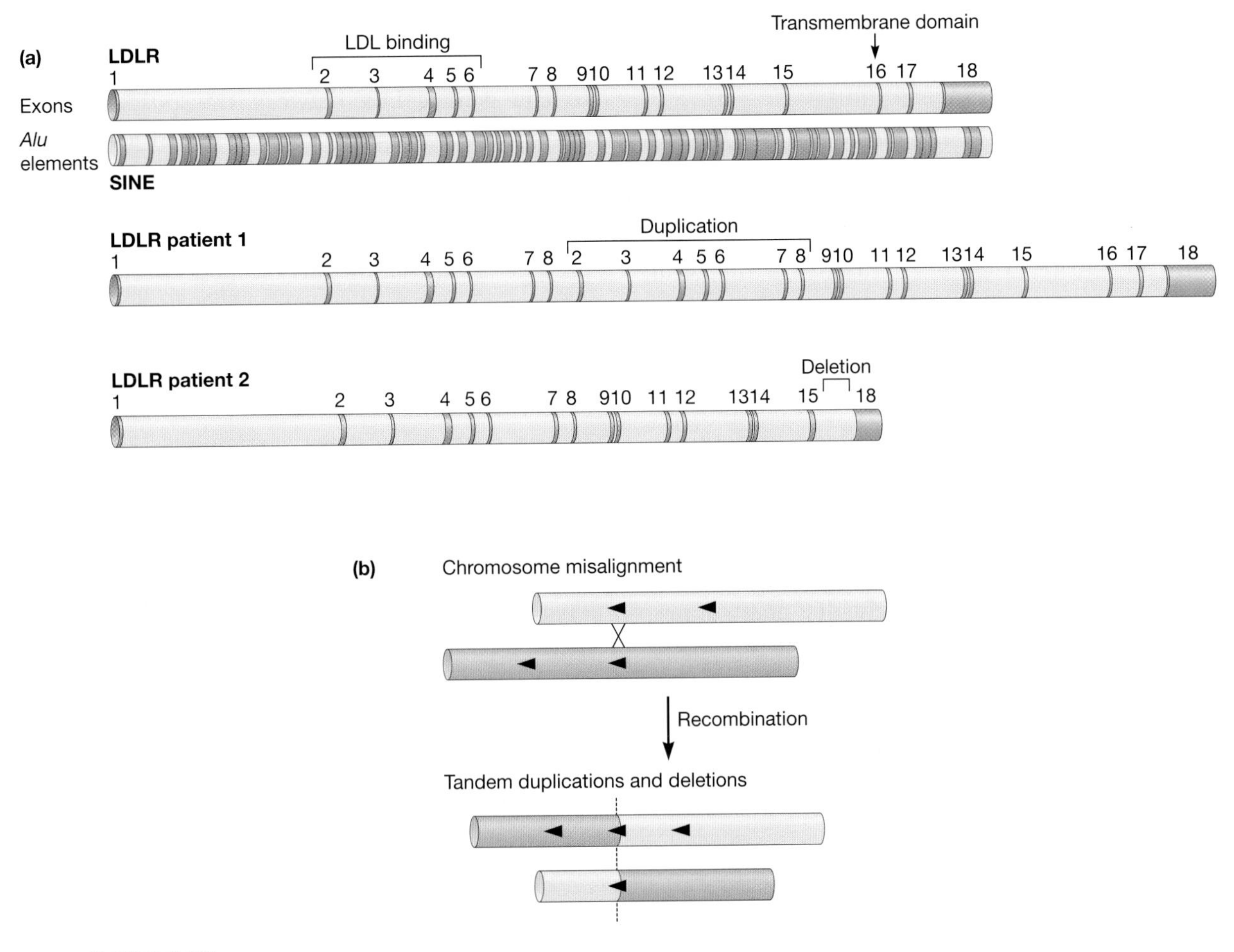

FIGURE 7-15

Recombination between *Alu* elements causes mutations in the low-density lipoprotein receptor (LDLR) gene. (*a*) A map of the intron/exon structure of the gene encoding the low-density lipoprotein receptor. The receptor binds low-density lipoprotein (LDL), which transports cholesterol in the blood. When the receptor binds LDL, a signal is transmitted to the inside of the cell so that the cholesterol can be taken up by cells and used. Below the intron/exon diagram, the position of the many *Alu* elements present in the gene is illustrated. Below, two examples of *Alu*-flanked duplications and deletions found among patients with hypocholesterolemia. In the first example, exons 2–8 are duplicated. Although this mutant molecule binds LDL, it does not internalize it correctly and LDL is not removed from the blood. This causes extremely high cholesterol levels to accumulate in the bloodstream, which increases the risk of heart disease. In the second example, deletion of exons 16, 17, and part of 18 results in the removal of the transmembrane domain. Consequently, the receptor is synthesized and exported from the cell. Without the transmembrane domain, it cannot be recognized by the uptake machinery and again LDL is not removed from the blood. (*b*) A diagram of how misalignment between *Alu* elements in the LDL receptor locus could lead to duplications and deletions during recombination. Another way that deletions may form is by recombination between two *Alu* elements on the same strand of DNA.

in *Drosophila* is caused by the P element. In addition to encoding a transposase, the P element encodes an inhibitor of transposition, an RNA-binding protein that alters the splicing of the transposase. In most cells, expression of the inhibitor protein results in the production of an internally deleted, defective transposase from the P element. When oocytes develop in the body of female *Drosophila* with P elements, the inhibitor accumulates in the oocyte cytoplasm and inhibits transmission of active P-element transposase. If a female is P element–free, she produces oocytes without the inhibitor. Should she mate with a male carrying P elements, there is a burst of transposon activity in the embryo because the inhibitor present in the very small volume of sperm cytoplasm is diluted in the very large volume of oocyte cytoplasm.

Laboratory strains of *Caenorhabditis elegans* provide another example of the regulation of transposon activity. In the worm, "cut-and-paste" DNA transposons from the Tc*1*/*mariner* family hop frequently in somatic cells but are inactive in the germ line. This prevents transposons from creating heritable mutations. However, in natural isolates of *C. elegans*, transposons actively hop in the germ line, and, when lab lines are mutagenized, it is possible to identify mutants with active germ-line transposition. These observations suggest that germ-line transposition is normally inhibited in the commonly used laboratory lines such as Bristol N2. When the mutations that allow germ-line transposition were mapped, it was found, surprisingly, that many mutations also affected two other silencing processes, *cosuppression* and *RNA interference* (RNAi). As we will see in detail in Chapter 9, cosuppression and RNA interference are silencing phenomena in which small RNAs direct the silencing of homologous genes either at the transcriptional level or by degrading an mRNA once transcribed.

The genetic link between transposon silencing, cosuppression, and RNAi led investigators to look for small RNAs homologous to the DNA transposons. Indeed, small RNAs homologous to the TIR (terminal inverted repeat) sequences of the Tc*1* element are present in the germ line of these worms. To test whether Tc*1* homology could trigger silencing, a fusion transcript was prepared in which the TIR from Tc*1* was placed in the 3′ untranslated region of GFP (Fig. 7-16). This construct was silenced, although a similar control construct containing DNA derived from the *unc-22* gene was not. The GFP construct was silenced only when Tc*1* was part of the transcript—when the Tc*1* TIR was placed after a transcription termination signal, no silencing occurred. This result suggests that the silencing signal acts on the RNA transcript of Tc*1* rather than directly on the DNA. Mutator genes like *mut-16* are needed for germ-line silencing of Tc*1*. When the GFP-Tc*1* fusion construct was tested in worms in which the mutator gene *mut-16* was silenced, GFP was

(a)

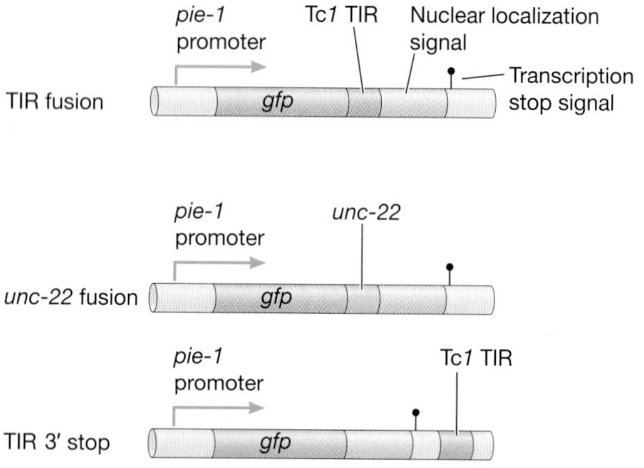

TIR fusion: *pie-1* promoter, *gfp*, Tc*1* TIR, Nuclear localization signal, Transcription stop signal

unc-22 fusion: *pie-1* promoter, *gfp*, *unc-22*

TIR 3′ stop: *pie-1* promoter, *gfp*, Tc*1* TIR

(b)

Transgene	Allele	Transgene copy number	GFP expression	
			Wild-type background	*mut-16* feeding
TIR fusion	0	3	+/−	++
	1	4	−	+/−
	2	2	+	+++
	3	1	−	+/−
	4	1	−	+/−
unc-22 fusion	5	1	++++	++++
	6	4	++	++
	7	1	++++	++++
	8	1	++	++
	9	1	+++	+++
TIR 3′ stop	11	3	+++	+++
	12	2	+++	+++

FIGURE 7-16

Sequences from *Caenorhabditis elegans* Tc*1* DNA transposons can trigger silencing when present in other transcripts. (*a*) Transgenic lines of *C. elegans* were created that contained GFP under the control of a germ line–specific promoter followed by either the 54-bp terminal inverted repeat sequence from the Tc*1* DNA transposon or by a 54-bp fragment randomly selected from the unrelated *unc-22* gene. The Tc*1* sequence was tested in two positions, both within the transcribed portion of the transgene and beyond the transcription stop site. (*b*) Multiple transgenic lines were prepared carrying each construct. GFP expression in the germ line was scored in wild-type lines, where the Tc*1* TIR sequence prevented GFP expression, but only when it was present within the transcribed portion of the transgene. The suppression by the TIR was relieved in worms in which the *mut-16* mutator gene was inactivated by RNA interference.

expressed. That the silencing of GFP-Tc*1* depends, just like germ-line silencing of Tc*1*, on *mut-16* indicates that germ-line silencing is also brought about by an RNA-dependent mechanism that recognizes the Tc*1* element. The link between transposon silencing and RNAi suggests that one function of RNAi is to control the activity of repetitive elements in the genome. As we will see in Chapter 9, a link between RNAi and transposon silencing has now been made in many other organisms.

One of the wonderful things about biology is that so much remains to be discovered about how cells function. For many years, the control of gene expression was believed to occur largely as outlined in Chapter 3, but recently a much wider array of mechanisms controlling gene expression have been discovered. We have just seen how RNA-dependent events control the amount of transposon RNA in *C. elegans*. From this and many other examples, we now know that mutations are not the only means by which a gene can be modified. In fact, there is growing evidence that the structure of the chromatin itself is modified to control gene expression and DNA rearrangments, even as the DNA sequence itself remains unchanged. Even more remarkably, unlike mutations, these chromatin modifications are *reversible*. These *epigenetic* changes of DNA are the subject of the next chapter.

Reading List

General

Curcio M.J. and Derbyshire K.M. 2003. The outs and ins of transposition: From mu to kangaroo (review). *Nat. Rev. Mol. Cell. Biol.* **4:** 865–877.

Discovery of Transposable Elements

McClintock B. 1950. The origin and behavior of mutable loci in maize. *Proc. Natl. Acad. Sci.* **36:** 344–355.

Shapiro J.A. 1969. Mutations caused by the insertion of genetic material into the galactose operon of *Escherichia coli*. *J. Mol. Biol.* **40:** 93–105.

Jordan E., Saedler H., and Starlinger P. 1968. O^0 and strong-polar mutations in the gal operon are insertions. *Mol. Gen. Genet.* **102:** 353–363.

McClintock B. 1987. *The Discovery and Characterization of Transposable Elements: The Collected Papers of Barbara McClintock*. Garland Publishing, New York.

Comfort N.C. 2001. *The Tangled Field: Barbara McClintock's Search for the Patterns of Genetic Control.* Harvard University Press, Cambridge.

DNA Transposons

Nevers P. and Saedler H. 1977. Transposable genetic elements as agents of gene instability and chromosomal rearrangements (review). *Nature* **268:** 109–115.

Fiandt M., Szybalski W., and Malamy M.H. 1972. Polar mutations in lac, gal and phage lambda consist of a few IS-DNA sequences inserted with either orientation. *Mol. Gen. Genet.* **119:** 223–231.

Hirsch H.J., Starlinger P., and Brachet P. 1972. Two kinds of insertions in bacterial genes. *Mol. Gen. Genet.* **119:** 191–206.

Malamy M.H., Fiandt M., and Szybalski W. 1972. Electron microscopy of polar insertions in the lac operon of *Escherichia coli*. *Mol. Gen. Genet.* **119:** 207–222.

Gill R., Heffron F., Dougan G., and Falkow S. 1978. Analysis of sequences transposed by complementation of two classes of transposition-deficient mutants of transposition element Tn*3*. *J. Bacteriol.* **136:** 742–756.

Foster T.J., Davis M.A., Roberts D.E., Takeshita K., and Kleckner N. 1981. Genetic organization of transposon Tn*10*. *Cell* **23:** 201–213.

Fedoroff N., Wessler S., and Shure M. 1983. Isolation of the transposable maize controlling elements Ac and Ds. *Cell* **35:** 235–242.

Bender J. and Kleckner N. 1986. Genetic evidence that Tn*10* transposes by a nonreplicative mechanism. *Cell* **45:** 801–815.

LTR Retrotransposons

Cameron J.R., Loh E.Y., and Davis R.W. 1979. Evidence for transposition of dispersed repetitive DNA families in yeast. *Cell* **16:** 739–751.

Boeke J.D., Garfinkel D.J., Styles C.A., and Fink G.R. 1985. Ty elements transpose through an RNA intermediate. *Cell* **40:** 491–500.

Non-LTR Retrotransposons

Shafit-Zagardo B., Maio J.J., and Brown F.L. 1982. KpnI families of long, interspersed repetitive DNAs in human and other primate genomes. *Nucleic Acids Res.* **10:** 3175–3193.

Boeke J.D. and Stoye M.P. 1997. Retrotransposons, endogenous retroviruses, and the evolution of retroelements. In *Retroviruses* (ed. J.M. Coffin et al.), pp. 343–435. Cold Spring Harbor Laboratory Press, Cold Spring Harbor, New York.

Cost G.J., Feng Q., Jacquier A., and Boeke J.D. 2002. Human L1 element target-primed reverse transcription in vitro. *EMBO J.* **21:** 5899–5910.

Haynes S.R. and Jelinek W.R. 1981. Low molecular weight RNAs transcribed in vitro by RNA polymerase III from Alu-type dispersed repeats in Chinese hamster DNA are also found in vivo. *Proc. Natl. Acad. Sci.* **78:** 6130–6134.

Lee M.G., Lewis S.A., Wilde C.D., and Cowan N.J. 1983.

Evolutionary history of a multigene family: An expressed human beta-tubulin gene and three processed pseudogenes. *Cell* **33:** 477–487.

Ullu E. and Tschudi C. 1984. Alu sequences are processed 7SL RNA genes. *Nature* **312:** 171–172.

Dewannieux M., Esnault C., and Heidmann T. 2003. LINE-mediated retrotransposition of marked Alu sequences. *Nat. Genet.* **35:** 41–48.

Transposons and Genome Size

Kidwell M.G. 2002. Transposable elements and the evolution of genome size in eukaryotes (review). *Genetica* **115:** 49–63.

Examples of Endogenous Recombination

Tonegawa S. 1988. Nobel lecture in physiology or medicine—1987. Somatic generation of immune diversity (review). *In Vitro Cell Dev. Biol.* **24:** 253–265.

Jones J.M. and Gellert M. 2004. The taming of a transposon: V(D)J recombination and the immune system (review). *Immunol. Rev.* **200:** 233–248.

Hozumi N. and Tonegawa S. 1976. Evidence for somatic rearrangement of immunoglobulin genes coding for variable and constant regions. *Proc. Natl. Acad. Sci.* **73:** 3628–3632.

Brack C., Hirama M., Lenhard-Schuller R., and Tonegawa S. 1978. A complete immunoglobulin gene is created by somatic recombination. *Cell* **15:** 1–14.

Early P., Huang H., Davis M., Calame K., and Hood L. 1980. An immunoglobulin heavy chain variable region gene is generated from three segments of DNA: VH, D and JH. *Cell* **19:** 981–992.

Matsunami N., Hamaguchi Y., Yamamoto Y., Kuze K., Kangawa K., et al. 1989. A protein binding to the J_κ recombination sequence of immunoglobulin genes contains a sequence related to the integrase motif. *Nature* **342:** 934–937.

Schatz D.G., Oettinger M.A., and Baltimore D. 1989. The V(D)J recombination activating gene, RAG-1. *Cell* **59:** 1035–1048.

Oettinger M.A., Schatz D.G., Gorka C., and Baltimore D. 1990. RAG-1 and RAG-2, adjacent genes that synergistically activate V(D)J recombination. *Science* **248:** 1517–1523.

Zhou L., Mitra R., Atkinson P.W., Hickman A.B., Dyda F., and Craig N.L. 2004. Transposition of hAT elements links transposable elements and V(D)J recombination. *Nature* **432:** 995–1001.

Transposons at Telomeres

Biessmann H., Mason J.M., Ferry K., d'Hulst M., Valgiersdottir K., Traverse K.L., and Pardue M.L. 1990. Addition of telomerase-associated HeT DNA sequences "heals" broken chromosome ends in *Drosophila*. *Cell* **61:** 663–673.

Transposons at Centromeres

Henikoff S., Ahmad K., and Malik H.S. 2001. The centromere paradox: Stable inheritance with rapidly evolving DNA (review). *Science* **293:** 1098–1102.

Zhong C.X., Marshall J.B., Topp C., Mroczek R., Kato A., et al. 2002. Centromeric retroelements and satellites interact with maize kinetochore protein CENH3. *Plant Cell* **14:** 2825–2836.

Transposons Cause Mutations

Coen E.S., Carpenter R., and Martin C. 1986. Transposable elements generate novel spatial patterns of gene expression in *Antirrhinum majus*. *Cell* **47:** 285–296.

Kazazian H.H., Jr., Wong C., Youssoufian H., Scott A.F., Phillips D.G., and Antonarakis S.E. 1988. Haemophilia A resulting from de novo insertion of L1 sequences represents a novel mechanism for mutation in man. *Nature* **332:** 164–166.

Ostertag E.A. and Kazazian H.H., Jr. 2001. Biology of mammalian L1 retrotransposons (review). *Annu. Rev. Genet.* **35:** 501–538.

Transposons as Genetic Tools

Rubin G.M. and Spradling A.C. 1982. Genetic transformation of *Drosophila* with transposable element vectors. *Science* **218:** 348–353.

O'Kane C.J. and Gehring W.J. 1987. Detection in situ of genomic regulatory elements in *Drosophila*. *Proc. Natl. Acad. Sci.* **84:** 9123–9127.

Bellen H.J., O'Kane C.J., Wilson C., Grossniklaus U., Pearson R.K., and Gehring W.J. 1989. P-element-mediated enhancer detection: A versatile method to study development in *Drosophila*. *Genes Dev.* **3:** 1288–1300.

Bier E., Vaessin H., Shepherd S., Lee K., McCall K., et al. 1989. Searching for pattern and mutation in the *Drosophila* genome with a P-lacZ vector. *Genes Dev.* **3:** 1273–1287.

Wilson C., Pearson R.K., Bellen H.J., O'Kane C.J., Grossniklaus U., and Gehring W.J. 1989. P-element-mediated enhancer detection: An efficient method for isolating and characterizing developmentally regulated genes in *Drosophila*. *Genes Dev.* **3:** 1301–1313.

Houck M.A., Clark J.B., Peterson K.R., and Kidwell M.G. 1991. Possible horizontal transfer of *Drosophila* genes by the mite *Proctolaelaps regalis*. *Science* **253:** 1125–1128.

Ivics Z., Hackett P.B., Plasterk R.H., and Izsvák Z. 1997. Molecular reconstruction of *Sleeping Beauty*, a *Tc1*-like transposon from fish, and its transposition in human cells. *Cell* **91:** 501–510.

Transposons Influence Gene Expression

Peaston A.E., Evisikov A.V., Graber J.H., de Vries W.N., Holbrook A.E., Solter D., and Knowles B.B. 2004. Retrotransposons regulate host genes in mouse oocytes and preimplantation embryos. *Dev. Cell* **7:** 597–606.

Aminetzach Y.T., Macpherson J.M., and Petrov D.A. 2005.

Pesticide resistance via transposition-mediated adaptive gene truncation in *Drosophila*. *Science* **309:** 764–767.

ffrench-Constant R., Daborn P., and Feyereisen R. 2006. Resistance and the jumping gene. *BioEssays* **28:** 6–8.

Lippman Z., Gendrel A.V., Black M., Vaughn M.W., Dedhia N., et al. 2004. Role of transposable elements in heterochromatin and epigenetic control. *Nature* **430:** 471–476.

Daborn P.J., Yen J.L., Bogwitz M.R., Le Goff G., Feil E., et al. 2002. A single P450 allele associated with insecticide resistance in *Drosophila*. *Science* **297:** 2253–2256.

Pedra J.H., McIntyre L.M., Scharf M.E., and Pittendrigh B.R. 2004. Genome-wide transcription profile of field- and laboratory-selected dichlorodiphenyltrichloroethane (DDT)-resistant *Drosophila*. *Proc. Natl. Acad. Sci.* **101:** 7034–7039.

Transposons and the Evolution of Genomes

McClintock B. 1984. The significance of responses of the genome to challenge. *Science* **226:** 792–801.

Kazazian H.H., Jr. 2004. Mobile elements: Drivers of genome evolution (review). *Science* **303:** 1626–1632.

Kalendar R., Tanskanen J., Immonen S., Nevo E., and Schulman A.H. 2000. Genome evolution of wild barley (*Hordeum spontaneum*) by *BARE*-1 retrotransposon dynamics in response to sharp microclimatic divergence. *Proc. Natl. Acad. Sci.* **97:** 6603–6607.

Lehrman M.A., Schneider W.J., Sudhof T.C., Brown M.S., Goldstein J.L., and Russell D.W. 1985. Mutation in LDL receptor: Alu-Alu recombination deletes exons encoding transmembrane and cytoplasmic domains. *Science* **227:** 140–146.

Lehrman M.A., Goldstein J.L., Russell D.W., and Brown M.S. 1987. Duplication of seven exons in LDL receptor gene caused by Alu-Alu recombination in a subject with familial hypercholesterolemia. *Cell* **48:** 827–835.

Bailey J.A., Liu G., and Eichler E.E. 2003. An Alu transposition model for the origin and expansion of human segmental duplications. *Am. J. Hum. Genet.* **73:** 823–834.

Fitch D.H., Bailey W.J., Tagle D.A., Goodman M., Sieu L., and Slightom J.L. 1991. Duplication of the γ-globin gene mediated by L1 long interspersed repetitive elements in an early ancestor of simian primates. *Proc. Natl. Acad. Sci.* **88:** 7396–7400.

Silencing of Transposable Elements

Bregliano J.C., Picard G., Bucheton A., Pelisson A., Lavige J.M., and L'Heritier P. 1980. Hybrid dysgenesis in *Drosophila melanogaster* (review). *Science* **207:** 606–611.

Kidwell M.G., Kidwell J.F., and Sved J.A. 1977. Hybrid dysgenesis in *Drosophila melanogaster*: A syndrome of aberrant traits including mutation, sterility, and male *recombination*. *Genetics* **86:** 813–833.

Bingham P.M., Kidwell M.G., and Rubin G.M. 1982. The molecular basis of P-M hybrid dysgenesis: The role of the P element, a P-strain-specific transposon family. *Cell* **29:** 995–1004.

Robert V.J., Vastenhouw N.L., and Plasterk R.H. 2004. RNA interference, transposon silencing, and cosuppression in the *Caenorhabditis elegans* germ line: Similarities and differences (review). *Cold Spring Harbor Symp. Quant. Biol.* **69:** 397–402.

Sijen T. and Plasterk R.H. 2003. Transposon silencing in the *Caenorhabditis elegans* germ line by natural RNAi. *Nature* **426:** 310–314.

CHAPTER 8

Epigenetic Modifications of the Genome

Mendelian genetics is one of the most successful explanatory frameworks in biology, but an early criticism was that it was concerned with the *transmission* of hereditary information and not with how the information was used in development. Indeed, the developmental biologists thought that geneticists were so caught up with genes and chromosomes that they ignored other mechanisms and processes that were essential for the proper development and functioning of organisms. For example, a key problem for embryology was that differentiation occurred even though all the cells are derived from the fertilized egg and have the same nuclear genome. It seemed self-evident to embryologists that there must be factors in the cytoplasm—maternally inherited—that contributed to differentiation. As well, examples of gene expression and inheritance were found that did not fit the Mendelian pattern—for example, Barbara McClintock's observations of variegation in maize kernels and the inheritance of traits associated with cytoplasmic organelles such as chloroplasts.

The term "epigenetics," coined originally to describe how genetic information was used in development to produce an organism, came to cover a diverse set of phenomena but was regarded as peripheral to mainstream genetics. However, as the tools of recombinant DNA began to be applied, it became increasingly clear that these non-Mendelian patterns of inheritance and control of gene expression play an extremely important role in the life of the cell. Epigenetics is now taken to mean heritable modifications to DNA that alter gene expression but not the genome DNA sequence (mutations). These modifications are typically inherited through mitosis, and often through meiosis, and although they are heritable, they are not permanent and can be reversed. The field of epigenetics now includes some of the most fascinating biological phenomena. In this chapter, we will describe such epigenetic phenomena as X-chromosome inactivation and imprinting. The mechanisms of the control of gene expression we consider here will point the way toward Chapter 9, where we will examine a newly revealed world in which RNA plays wholly unsuspected roles.

The Different Dosage of Sex Chromosomes Demands a Solution to the Problem of Gene Expression

Having the right number of genes is vital to the well-being of many organisms, especially, it seems, of mammals. Human beings are diploid (we have pairs of chromosomes) and chromosomal imbalances have severe consequences. These imbalances arise during meiosis when chromosomes fail to separate ("nondisjunction"), thereby producing gametes missing a chromosome or having an extra copy of a chromosome. In the fertilized egg, these result, respectively, in *monosomy* or *trisomy* of that chromosome. In humans, monosomy of any of the autosomes (the non–sex chromosomes) is lethal before birth, and although children with trisomies may be born, they suffer from severe deformities. Trisomy 21 (Down syndrome) is an exception and it is probably a consequence of the very small number of genes (~330) present on chromosome 21. However, the sex chromosomes are anomalous; human males are monosomic for two chromosomes (the X and Y), whereas human females are diploid for the X chromosome and have no copy of the Y. This situation reflects the evolutionary origin of the Y chromosome and provides a remarkable example of epigenetics.

During male meiosis, even though the X and Y chromosomes are very different in size, they come together and pair, albeit only over a short distance, just like the other chromosomes. Such pairing requires substantial regions of similarity between chromosomes, and this observation suggests that the X and Y chromosomes, despite their differences, are in fact homologs, which gradually began to diverge beginning some 300 million years ago. The two sexes are decribed as the *homogametic sex* and the *heterogametic sex*. The *homogametic* sex possesses two identical chromosomes and so produces identical gametes, all containing the same sex chromosome. The *heterogametic* sex has one chromosome identical to those in the homogametic sex, as well as a second unique chromosome. When these two sex chromosomes are segregated into gametes, the result is two different types of gametes. This system of sex determination is a straightforward mechanism for producing equal numbers of male and female offspring. Mating the heterogametic sex (XY) to the homogametic sex (XX) produces four possible embryos, two each of XX and XY, so that the offspring have an overall 1:1 sex ratio (Table 8-1). Sexual differentiation requires a large number of genes, and localization of these genes on sex chromosomes ensures that they segregate as a group during meiosis. So that these genes remain together, crossing-over that normally occurs during meiosis between chromosome homologs is suppressed for the sex chromosomes in the heterogametic sex. In humans, there is a limited exception to this rule, which occurs in the pseudoautosomal region, where homology is sufficient to allow pairing and crossing-over between the X and the Y. In fact, if a crossover event fails to occur in this region, the X and Y do not segregate properly during meiosis.

Sex determination using XX and XY has two major genetic costs. First, the heterogametic sex is haploid for both sex chromosomes, and so mutant alleles on a male X chromosome cannot be compensated by a normal allele on a second X chromosome. As a consequence there are many X-linked diseases, such as hemophilia and Duchenne muscular dystrophy, that predominantly affect males because they have only a single copy of each gene on the X. Women with a mutation on one X chromosome are termed *carriers* and generally have few or no symptoms. Second, there is a difference in gene dosage between the homogametic and heterogametic sex. How is it that the homogametic sex can have a double dose of each gene carried (in the case of mammals) on the two X chromosomes?

TABLE 8-1. Genotypes of offspring produced in a system of heterogametic and homogametic sex determination

		Heterogametic parent	
		X	Y
Homogametic parent	X	XX	XY
	X	XX	XY

That there is a special mechanism to correct gene dosage is evident from the fact that human beings with unusual numbers of sex chromosomes are viable. Women with Turner syndrome, who have only one X chromosome (typically called XO, where O represents the absence of a chromosome), are relatively normal, although they are sterile and have minor developmental changes. Individuals with an XXY chromosome complement (Klinefelter syndrome) are male and, although sterile, they develop fairly normally. Other X polysomies, including XXX and XXXX (females) and XXXY and XXXXY (males), have been observed. Males with XYY constitutions occur as frequently as 1 in 1000 male births. These men develop normally and are fertile.

Only One X Chromosome Is Active in Female Mammals

Dosage compensation was noted first by Hermann Muller for the *w* locus on the *Drosophila* X chromosome. The wild-type w^+ eye color is red, whereas the mutant allele w^a produces orange eyes. Muller noticed that females homozygous for w^a/w^a had the same intensity of eye color as did males who had a single copy of w^a (and a Y chromosome [Fig. 8-1]). The implication was that there was some form of regulation that made one copy of w^a equivalent to two copies of w^a. The mechanism by which this occurs is significantly different from that in mammals, but the discovery of dosage compensation in *Drosophila* led the way for the explanation of sex chromosome dosage compensation in mammals.

In 1949, Murray Barr and Ewart Bertram found a surprising difference between the nuclei of neurons in male and female cats. In the nuclei of females, but not males, they found a heterochromatic body that stained deeply with DNA-sensitive dyes. This observation of female-specific heterochromatin was corroborated in many other mammals, including humans, and became known as the "Barr body." It was assumed that the Barr body represented an X chromosome, but it was Susumo Ohno's studies of chromosomes in regenerating rat liver cells that demonstrated the identity of the Barr body as an X chromosome. In the prophase nuclei of female animals, a very condensed X chromosome was observed paired to a less-condensed X. Particularly convincing evidence came from studies of nuclei from individuals with differing numbers of X chromosomes—the number of Barr bodies was always one less than the number of X chromosomes (Table 8-2).

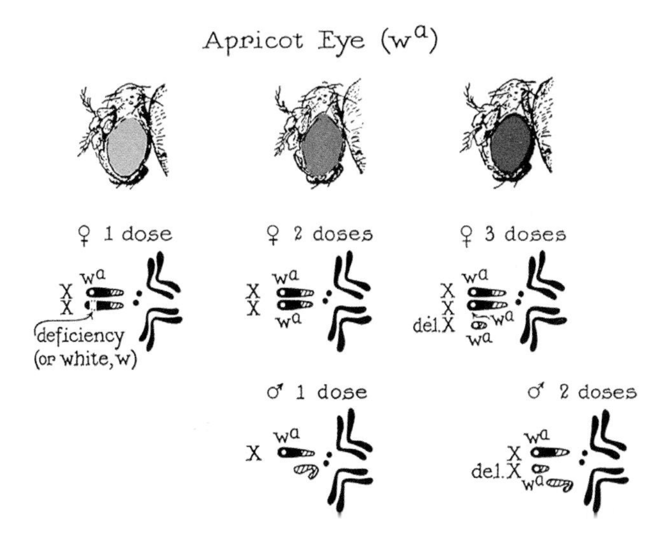

FIGURE 8-1

Dosage compensation of genes on the *Drosophila* X chromosome. (*Left column*) The apricot allele (w^a) of the *white* gene (responsible for pigmentation in the *Drosophila* eye) produces a pale orange color rather than the wild-type red color in females with a single copy of the gene. This gene is located on the *Drosophila* X chromosome, and females bearing a single copy can be prepared by crossing a line with a deficiency in the region of the X carrying *white* with a fly carrying the w^a allele. Surprisingly, the eyes of males with one copy of w^a (on the single X chromosome) are a deeper shade of orange, identical to females homozygous for w^a/w^a. Thus, the activity of the *white* gene is adjusted so that a single copy of w^a in males is equivalent to two copies in females. (*Right column*) This dosage effect on the X chromosome was further extended by studying male and female flies with an extra copy of w^a, carried on a deleted X chromosome (delX). (This fragment of X was used because flies triploid for an entire X are not viable and XXY flies are female.) The eyes of females with three doses of w^a were nearly red, and those of males showed even more pigmentation. Careful measurements of the pigment demonstrated that the ratio of these three-dose females to two-dose males was approximately 1.5 to 2, as expected for a compensation mechanism that equalizes two doses of X-linked genes in females to one dose in males.

The mammalian X chromosome was not the first condensed chromosome observed. As long ago as 1928, the embryologist Emil Heitz developed techniques for cell staining that revealed that portions of chromosomes were highly condensed throughout the cell cycle. He termed this condensed material *heterochromatin*, to distinguish it from the euchromatin that cycled between a condensed form during mitosis and a decondensed state during interphase. Major portions of the X chromosome in *Drosophila* are permanently condensed, and it was shown that the X-chromosome heterochromatin had a silencing

TABLE 8-2. Barr bodies and X chromosomes

Chromosome complement	Number of Barr bodies
XO, XY, XYY	0
XX, XXY, XXYY	1
XXX, XXXY, XXXYY	2
XXXX, XXXXY	3
XXXXX	4

influence on genes translocated nearby (Fig. 8-2). For example, deletions and rearrangements that brought the *white* gene w^+ (making red eyes) close to the heterochromatic region of the X caused silencing, so that w^+ was turned off in some cells. The eyes of these flies were colored in a variegated pattern of red and white. These observations in *Drosophila* correlated heterochromatin with silencing and suggested how dosage compensation in mammals might be brought about. If chromosomes functioned similarly in mammals, the heavily condensed X chromosome that forms the Barr body was also likely to be inactive.

In 1961, Mary Lyon advanced a hypothesis to explain how X-chromosome dosage was controlled in mammals. While studying genes affecting coat color in mice, she observed that mutations in X-linked coat-color genes were neither dominant nor recessive. Instead, females heterozygous for these mutant coat-color genes had mottled or dappled coats (Fig. 8-3). However, males never displayed these patterns, nor did rare XO females. Similar variegation of coat color was also observed in animals with translocations between the X chromosome and autosomes carrying coat-color genes. This indicated that the X was influencing the expression of the translocated genes in a manner similar to that observed in *Drosophila* X chromosome translocations. Lyon proposed that one of the two X chromosomes was randomly inactivated in female animals early during embryogenesis, so that in some cells one X chromosome was active and in some cells the other was active. The outcome of this random inactivation was a random distribution of hair cells expressing different pigments, which produced randomly patterned coat coloration in females. It also made the active X chromosome complements of male and female cells the same—both having a single active X chromosome.

Lyon's X-inactivation hypothesis was soon validated by Ernest Beutler's studies of another X-linked gene, the enzyme glucose-6-phosphate dehydrogenase (G6PD). Deficiency of this enzyme produces a type of anemia, and the activity of the enzyme can be easily assayed. Women heterozygous for a G6PD mutation had two populations of red blood cells, one population expressing G6PD and the other deficient in the enzyme. This demonstrated that the X chromosome with the wild-type G6PD gene was active in the cells expressing G6PD, where-

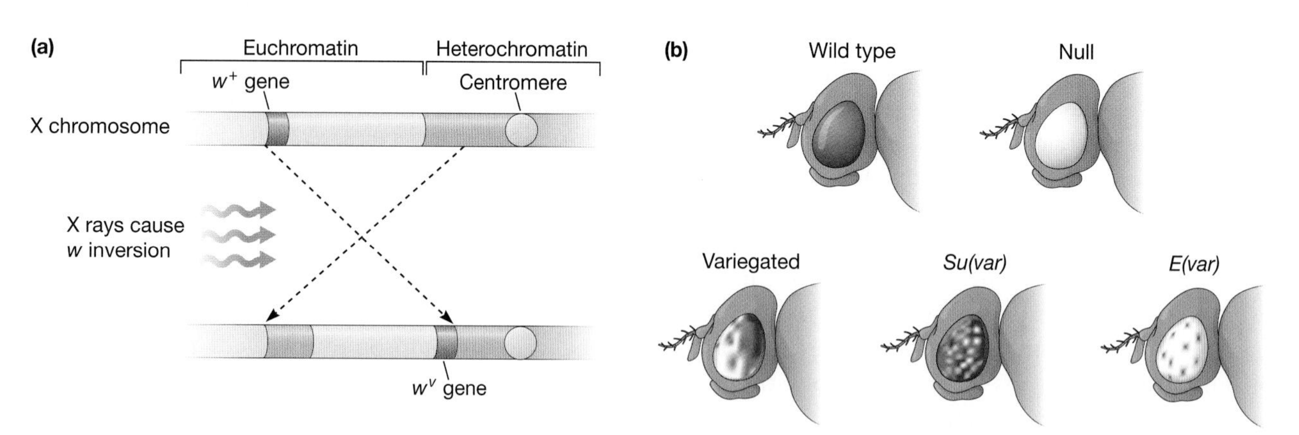

FIGURE 8-2

Heterochromatin on the *Drosophila* X chromosome silences nearby genes. Flies bearing a variegating eye color were discovered. (*a*) When their chromosomes were examined, it was apparent that the X chromosome had been rearranged, so that the *white* gene had been brought closer to the highly condensed (heterochromatic) portion of the X chromosome. (*b*) Similar rearrangements of the *white* gene near to the centromeres or ribosomal DNA locus of other chromosomes showed a similar variegated silencing pattern. This phenomenon has been termed *position effect variegation* (PEV), because the marker gene (in this case, *white*) shows varying levels of gene expression when it is placed near silent regions of the genome.

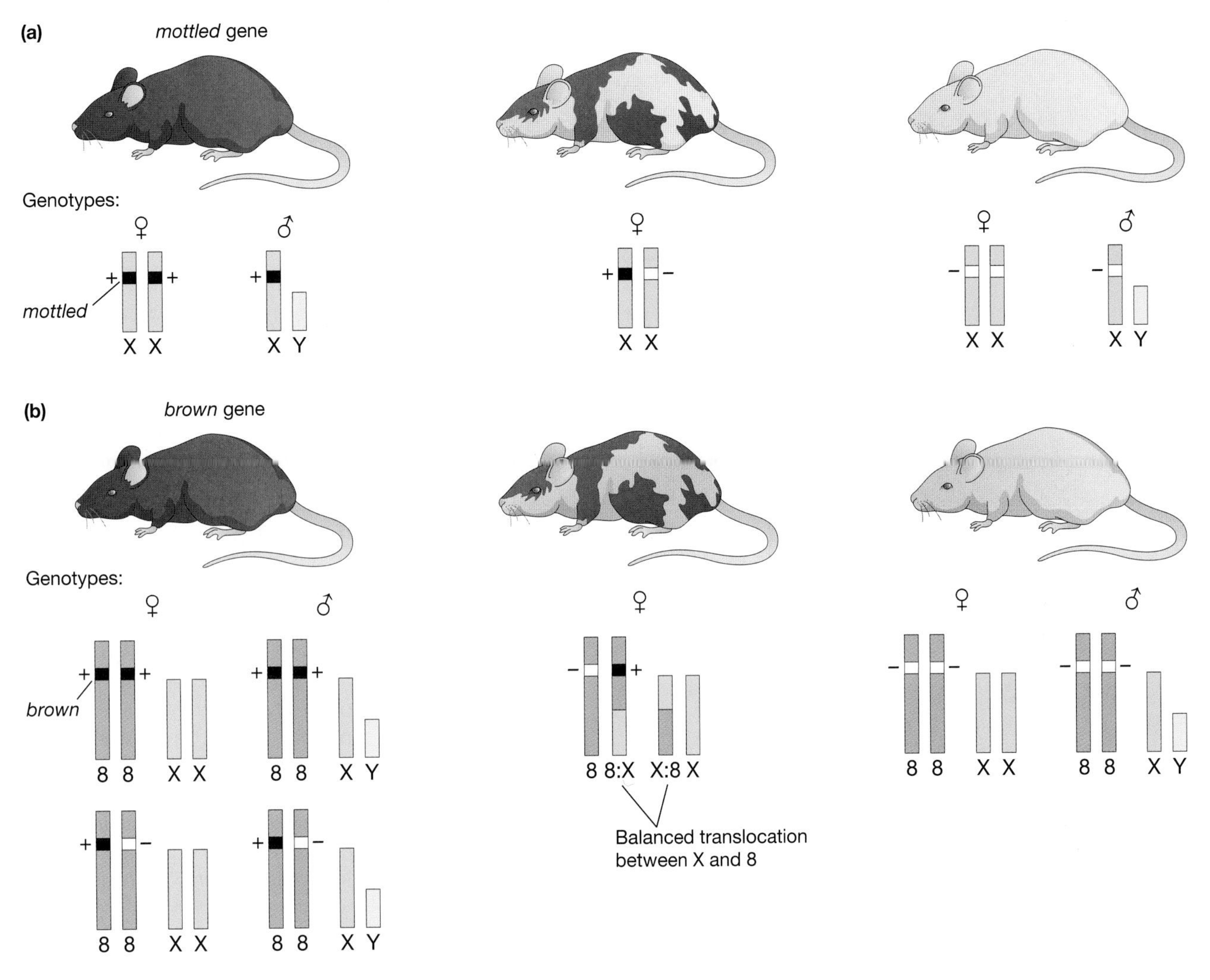

FIGURE 8-3

Variegating coat-color patterns observed in female mice demonstrate random X inactivation in mammals. Mary Lyon analyzed mutations of X-linked coat-color genes. (*a*) The wild-type X-linked pigmentation gene *mottled* produces black mice. Females homozygous for the *mottled* mutation (or mutant males, which carry a single copy of *mottled*) are light gray. Surprisingly, female heterozygotes are variegated, with areas of their coat that are fully black and others that are light gray—apparently a mix of wild type annd mutant! Rare females with single X chromosome (XO) are also light gray. (*b*) Another coat-color gene, *brown*, has also been used to study coat-color variegation. Normally, the *brown* gene is located on the eighth chromosome, and mice that are either homozygous for the wild-type allele—or are heterozygous—are black, regardless of sex, whereas homozygous mutants are brown. However, females having a translocation between chromosomes 8 and X have a variegated coat color. No variegated males were found, even among offspring of the variegated females. These observations suggest that two copies of the X chromosome are required for the variegation effects, and that the X chromosome has the ability to trigger variegation in other genes translocated to it.

as the X chromosome with mutant G6PD was active in the other cells. This finding was corroborated by analysis of G6PD in clonal cell lines produced from skin biopsies taken from six women heterozygous for G6PD isozymes Gd A and Gd B. The clones expressed only one or the other isozyme—none was AB. The experiment also demonstrated that the inactive state of a given X chromosome is maintained through mitosis (Fig. 8-4).

In women heterozygous for X-linked mutations that cause disease, 50% expression of a gene may or may not be sufficient to prevent illness depending on whether the proteins must be made in each cell or can be supplied from a limited population of cells.

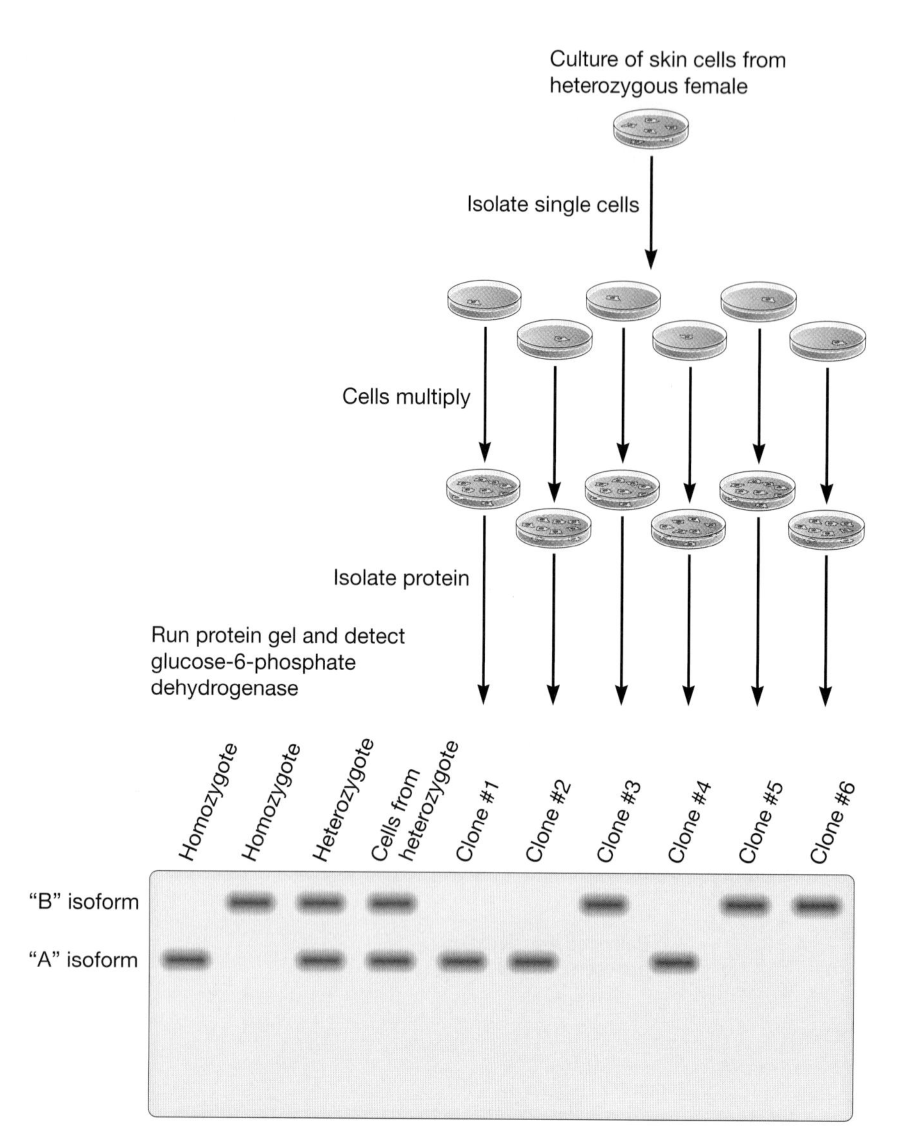

FIGURE 8-4

Glucose-6-phosphate dehydrogenase (G6PD) is used to demonstrate the occurrence and stability of random X inactivation in humans. Skin samples were taken from women heterozygous for two different isoforms of the enzyme, A and B. These two isoforms have slightly different mobility on a gel. The skin samples were disrupted into single cells, and clonal lines were allowed to grow up from each of these cells. Protein was isolated from these lines and analyzed. From each woman, cell lines were recovered expressing either isoform A or isoform B but not both.

For example, females heterozygous for G6PD deficiency suffer bouts of anemia because roughly half of their red blood cells are G6PD deficient. On the other hand, most females carrying a heterozygous mutation in the blood clotting factor VIII do not have hemophilia. These women can pass the mutation on to their sons, but they are rarely themselves affected, because their single good copy of factor VIII supplies adequate levels of the enzyme. This is because the factor VIII protein is secreted from liver cells into the blood plasma, and factor VIII levels that are even just 50% of normal are still sufficient for normal blood clotting. However, about 10% of women heterozygous for factor VIII mutation happen to have a high number of liver cells with an inactivated wild-type X, so that insufficient amounts of the clotting factor are produced. These women have mild symptoms of hemophilia. Below, we will learn more about the mechanisms by which one of the two female X chromosomes is randomly inactivated.

The Inactivated X Chromosome Is Coated with a Noncoding RNA Called *Xist*

Some translocations between an X chromosome and an autosome result in inactivation of translocated autosome genes that are adjacent to X chromosome sequences. Analysis of patients carrying translocations between the X and autosomes demonstrated that there was a segment of the X, termed the X-inactivation center (Xic), that caused the silencing. The silencing influence spreads from the X for a limited distance along the attached autosomal segment. If a translocated autosome is attached to a segment of the X chromosome lacking the Xic, there is no silencing of the autosome. The discovery of the gene responsible for the activity of the Xic had to await the development of methods for molecular cloning.

The analysis of cDNAs made from the X chromosome transcripts led to the discovery of a female-specific cDNA clone. Expression of this transcript, called *Xist*, correlated with the presence of an inactive X chromosome, so that *Xist* RNA was present only in females and not males. Cell lines derived from individuals with X polysomy, such as XXY, XXXXX, and XXXXY, had higher levels of expression of the *Xist* transcript than normal XX cells containing a single inactive X chromosome.

The correlation between the number of inactive X chromosomes and the amount of *Xist* transcript tantalizingly suggested that *Xist* transcription originated from the inactive X chromosome, but definitive proof required study of active and inactive X chromosomes in isolation. Remarkably, it is possible to make hybrid mouse–human cell lines containing single human chromosomes as well as a full complement of mouse chromosomes. Murine and human cells are mixed and treated with Sendai virus or a chemical such as polyethylene glycol, both of which induce fusion of cell membranes. The resulting hybrid cells initially contain both mouse and human chromosomes but tend to spontaneously lose most of the human chromosomes until only a few or even a single human chromosome remains. Fusing cells in this way does not affect X inactivation—an inactive human X chromosome remains inactive. The results of the experiment showed clearly that *Xist* transcription occurred only in cell lines containing an inactive X (Fig. 8-5).

Mouse–human fusion cell lines were also used to map *Xist* to the Xic. Mouse–human hybrid cell lines were established from human females who carried translocations between the X chromosome and autosomes. Each cell contained a different part of the X chromosome. Analysis of *Xist* transcription in each of these cell lines (for which the X-inactivation status was known) revealed that *Xist* mapped to the location of the X-inactivation center.

However, when the full-length *Xist* gene was isolated and sequenced, no protein-coding sequence could be found. Intriguingly, in situ hybridization showed that the *Xist* transcript was present in the nucleus and was localized to the inactive X chromosome. In cells with XXX and XXXXX constitutions, the *Xist* transcript "painted" each of the multiple Barr bodies in these cells (each Barr body corresponds to an inactive X chromosome). It seems that, in ways still not yet understood, *Xist* RNA coats the inactive X chromosome and silences genes.

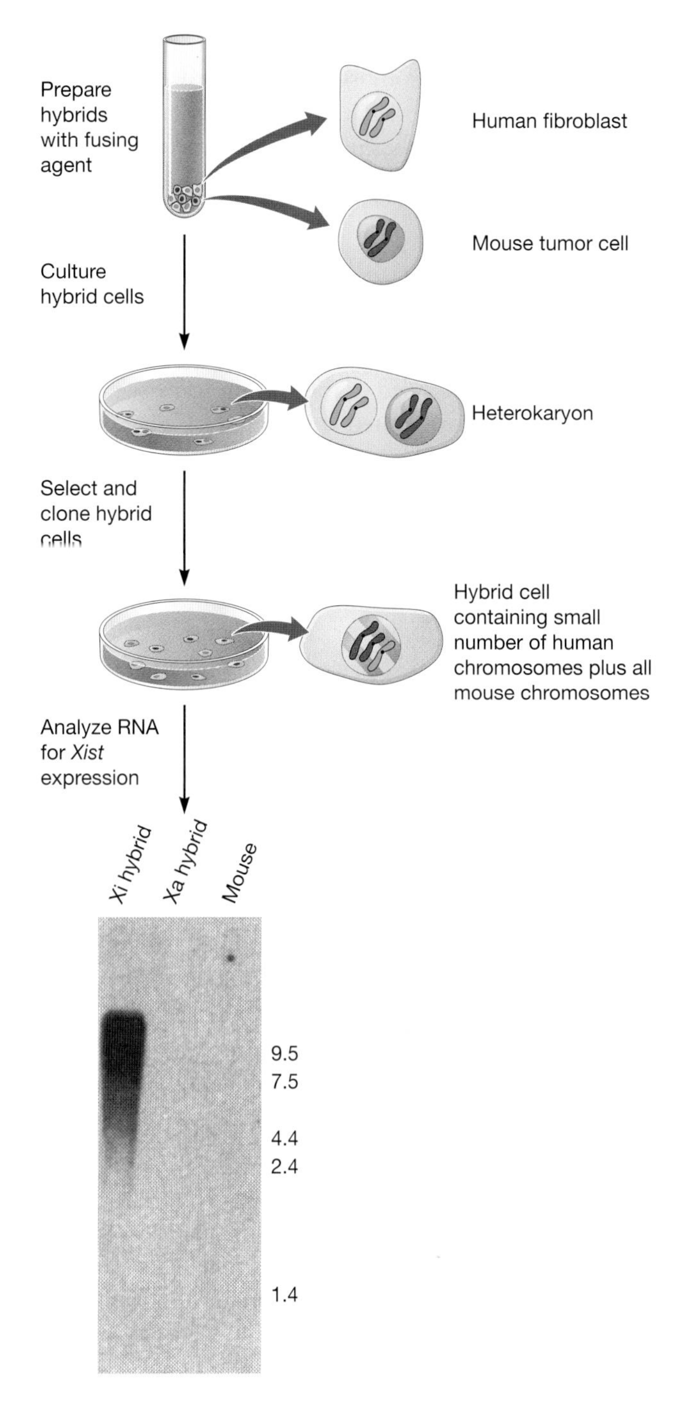

FIGURE 8-5

Somatic cell fusion is used to demonstrate that *Xist* originates from the inactive X. Mouse–human hybrid lines can be prepared by mixing cells and inducing membrane fusion. Most of the human chromosomes are lost at random, but lines can be found with one or more chromosomes remaining. Lines containing active or inactive X chromosomes were prepared. RNA from these lines was analyzed by Northern blot for *Xist* expression. *Xist* could be detected only in lines containing an inactive X (Xi hybrid). Note that the *Xist* transcript is heterogenous in size. Most genes produce one or a few transcripts of defined length—the heterogeneity of *Xist* is atypical.

These experiments correlated *Xist* transcription with the silent X chromosome, but was it required for silencing? Targeted deletions of *Xist* were made in mice. In these animals, an X chromosome with a deletion of *Xist* was never silenced, thereby showing that *Xist* was required for silencing. But was *Xist* itself able to initiate silencing? This was demonstrated using mouse embryonic stem (ES) cell lines containing *Xist* transgenes under the control of an inducible promoter. When expression of the *Xist* transgene was induced, *Xist* RNA was localized to the chromosome where the transgene was integrated and induced silencing of genes along that chromosome. When the transgene was inserted in the X chromosome of male cells and turned on, the cells died because the transgene caused inactivation of the single X chromosome present in male cells, thereby silencing many essential genes.

The silencing ability of *Xist* depends on the differentiation state of the cells. When ES cells were allowed to differentiate before *Xist* expression was induced, the *Xist* transgene no longer could induce silencing. In addition, in undifferentiated ES cells, the *Xist* transgene silenced genes only while it was induced—if the *Xist*-inducing drug was removed, normal gene expression resumed. Differentiation of ES cells triggers the process of X inactivation. In cells undergoing differentiation, there was a brief period in which removal of the *Xist*-inducing drug relieved silencing, but as differentiation progressed, the *Xist*-triggered silencing became irreversible. In fact, *Xist* can be deleted from an inactive X chromosome without disrupting the silencing of the rest of the genes on that chromosome.

The Gene *Xist*, Its Complement, *Tsix*, and Other Nearby Sequences Are Required for X Inactivation

Although an inducible *Xist* transgene can direct the inactivation of a chromosome, subsequent experiments have revealed that other sequences on the X chromosome are necessary to ensure proper X inactivation. There are four separate steps in the process of X inactivation. First, the number of X chromosomes present in the cell is in some way counted, so that all but one will be inactivated. Second, one of the X chromosomes is selected for inactivation. Third, the process of silencing begins on the selected chromosome. Finally, the silencing signal spreads over the entire chromosome. The process of X inactivation has been intensively studied in mice using both targeted deletions on the X chromosome, and insertion of X chromosome segments into autosomal sites. Careful deletion of X-chromosome segments helps to identify regions required for proper silencing, whereas insertion of X-chromosome sequences elsewhere in the genome has helped to identify the minimal sequences needed for silencing.

While characterizing the X-inactivation center, several groups of investigators noticed that *Xist* was not the only transcript found. Surprisingly, long transcripts were found in the antisense orientation to *Xist* and overlapping the *Xist* transcript entirely. These transcripts originate from a gene that was named *Tsix*, in recognition of its antisense orientation to *Xist*. Like *Xist*, *Tsix* has no apparent open reading frame.

The expression patterns of *Tsix* are largely complementary to those of *Xist*. In ES cells, *Tsix* is expressed at high levels from both X chromosomes, whereas *Xist* expression is low. As differentiation begins, one X chromosome is "chosen" for inactivation. *Tsix* expression quickly becomes limited to the future active X, whereas *Xist* expression from that chromosome declines. Simultaneously, *Xist* expression from the future inactive chromosome rises, whereas *Tsix* expression falls. Once X inactivation is established, *Tsix* is completely silent on the inactive X, whereas it continues to be transcribed from the active X for some time.

Female mice that are heterozygous for a *Tsix* deletion preferentially inactivate the X chromosome with the deletion, thereby showing that *Tsix* transcription is required on the chromosome that remains active. As predicted, if *Tsix* helps to inhibit silencing by *Xist*, an X chromosome carrying an overexpressed *Tsix* transgene is always active. However, in male cells, where X inactivation is never initiated, chromosomes with a *Tsix* promoter deletion remain active. Thus, *Tsix* is needed only when an active X must be chosen.

Further deletion experiments with the mouse X chromosome have identified another region, *Xite*, that promotes *Tsix* expression and so fosters the choice of the active X chromosome. The *Xite* region is polymorphic among different strains of mice. Because the activity of transcriptional start sites in the region correlates with the likelihood that an X chromosome will be chosen to remain active, it seems that polymorphism at the *Xite* region may influence which X chromosome outcompetes the other. The *Xite* region and the *Tsix* region can direct pairing of the X chromosomes in cells undergoing X inactivation. This pairing seems to allow cross talk between X chromosomes throughout the process of counting and the eventual choice of the active X.

Not All Genes on the Mammalian X Chromosome Are Silenced

We have seen how one of the two X chromosomes in female mammals is randomly inactivated to achieve proper gene dosage. However, careful examination of gene expression reveals that not all of the genes on the inactivated X are silenced. The *Xist* gene, for example, is actively transcribed from the inactive X. Another such gene is steroid sulfatase, the deficiency of which causes the X-linked skin disease ichthyosis. In contrast to the experiment with G6PD, cell clones deficient in steroid sulfatase could not be isolated from women whose sons were affected by X-linked ichthyosis. (These mothers must be heterozygous for steroid sulfatase deficiency.) Furthermore, all mouse–human hybrid cell lines with single X chromosomes from these women expressed steroid sulfatase, whether the lines contained an inactive X or an active X.

The steroid sulfatase gene is located close to the telomere of the short arm of the X chromosome, near the segment of the X that pairs with the Y chromosome at meiosis. This has been termed the *pseudoautosomal* region, because pairing and crossovers occur there at meiosis in a fashion similar to the meiotic crossovers that occur between autosomes. Other genes in the pseudoautosomal region are expressed from both the X and the Y in males and are expressed from both the active and inactive X in females.

The availability of the complete sequence of the X chromosome has permitted the detailed analysis of gene expression from the inactive X. This has revealed a much more complex pattern of gene expression in females than had been suspected. In mouse–human hybrid cell lines carrying an inactive human X, about 75% of genes are silenced, while some 15% escape inactivation (Fig. 8-6). A few of the latter are expressed at levels nearly equivalent to expression observed from the active X. Surprisingly, another 10% of genes escaping inactivation show differing levels of expression on inactive X chromosomes from different individuals—varying by as much as threefold.

Sex Chromosome Dosage in *Caenorhabditis elegans* and *Drosophila* Is Achieved by Altering Transcriptional Activation

Measurements of phenotypic characteristics and enzymatic activities show that there is dosage compensation of many genes on the *Drosophila* X chromosome. This is not achieved by gene silencing—both X chromosomes in female flies are active. For example, female flies heterozygous for mutations in the X-linked eye color gene *white* (w^+) show an even color rather than patchy pigmentation observed in the coat colors of mice. However, cell-by-cell X-chromosome inactivation might have produced variegation too microscopic to observe. The definitive proof that both X chromosomes are active in flies came from analysis of the enzyme 6-phosphogluconate dehydrogenase, 6-PGD. In *Drosophila* there are two different alleles of 6-PGD, A and B, whose products differ in their electrophoretic mobility. The mature form of 6-PGD is a dimer containing two copies of the polypeptide; the composition of these forms is AA and BB. When flies of the two strains are mated and the 6-PGD of hybrid females analyzed, AA, BB, and a new form intermediate in mobility between A and B are found. The intermediate form is a dimer made up of A and B. Because assembly of dimers can only occur immediately after the protein is produced, the presence of dimers of intermediate size in heterozygous females shows that both X chromosomes are active in each cell (Fig. 8-7).

There are two ways in which gene dosage in males and females could be equalized while keeping all X chromosomes active: Expression of male X chromosome genes could be activated twofold or expression of genes on the two female X chromosomes could be halved. *Drosophila* has adopted the first strategy—transcriptional activity on the male X chromosome is twice that of each female X chromosome (Fig. 8-8). This up-regulation is accomplished in part through the action of noncoding RNAs transcribed from and localized to the male X in a manner similar to that of *Xist*. However, in this case the RNAs direct activation rather than inactivation. Analysis of mutations that were lethal only in the male revealed a number of genes that were essential for the survival of male flies. Two genes, *roX1* and *roX2*, were discovered on the X chromosome and, like *Xist*, these genes express noncoding RNAs that bind to the X from which they are transcribed. However, unlike *Xist*, the *roX1* and *roX2* RNAs exert an activating influence on the chromosome to which they bind.

The nematode worm *C. elegans* has adopted the other solution to dosage compensation—it turns down transcription on each of the X chromosomes of the XX sex so that it equals one-half the transcription from the single male X chromosome. (XX *C. elegans* are *hermaphrodites*—they produce both sperm and eggs. Hermaphrodites can self-fertilize to produce

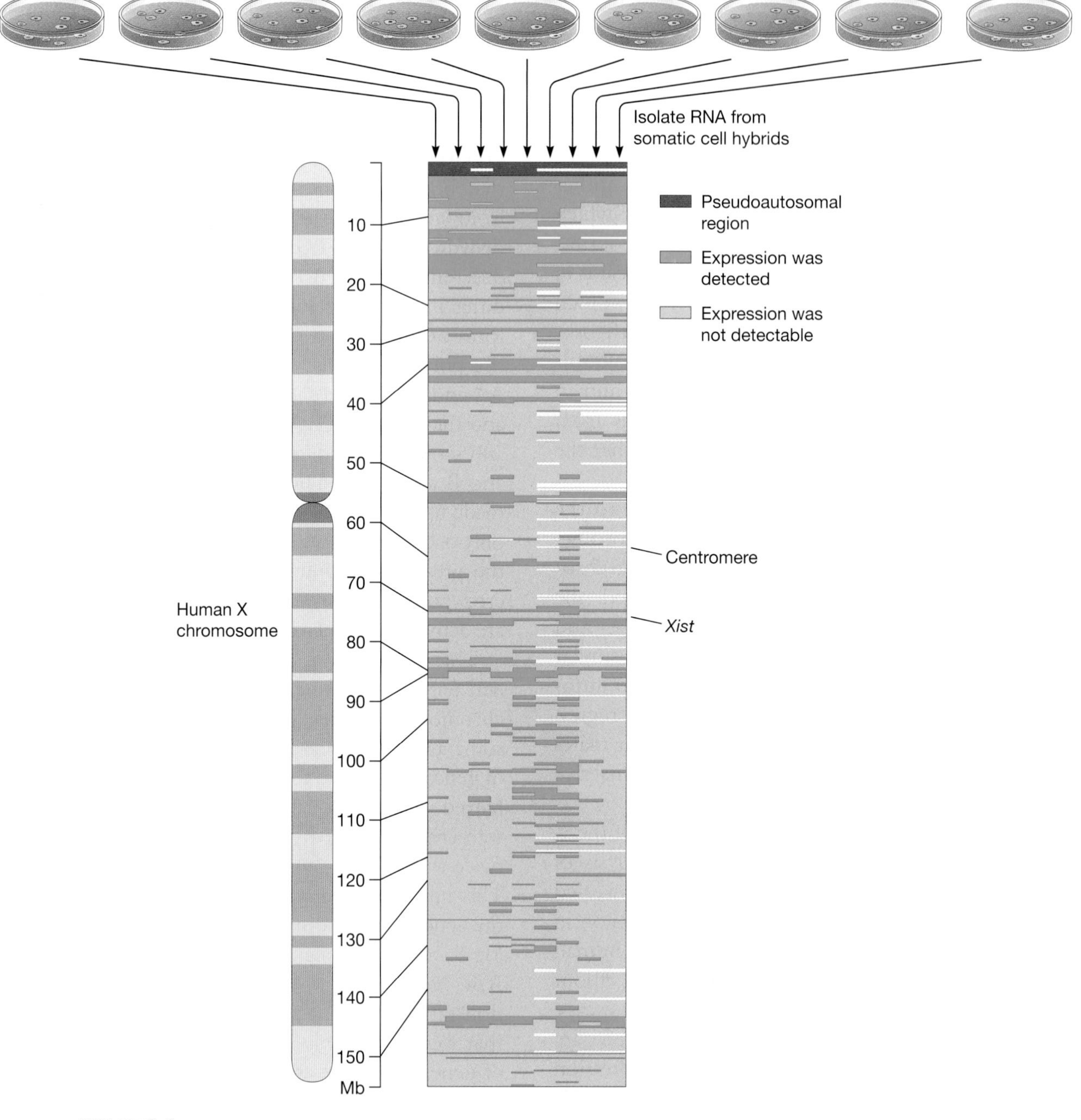

FIGURE 8-6

The inactive X is not completely silent. Mouse–human hybrid cell lines were established containing the inactive X from different women. Gene expression from genes on the X was analyzed in these lines; *yellow* indicates silent genes and *blue* indicates genes that are active. Note that the *Xist* gene is active on each inactive X chromosome, as expected. Also, there are many active genes at the pseudoautosomal region at the tip of the X chromosome. Not all genes are inactive, and the pattern of expression varies between women.

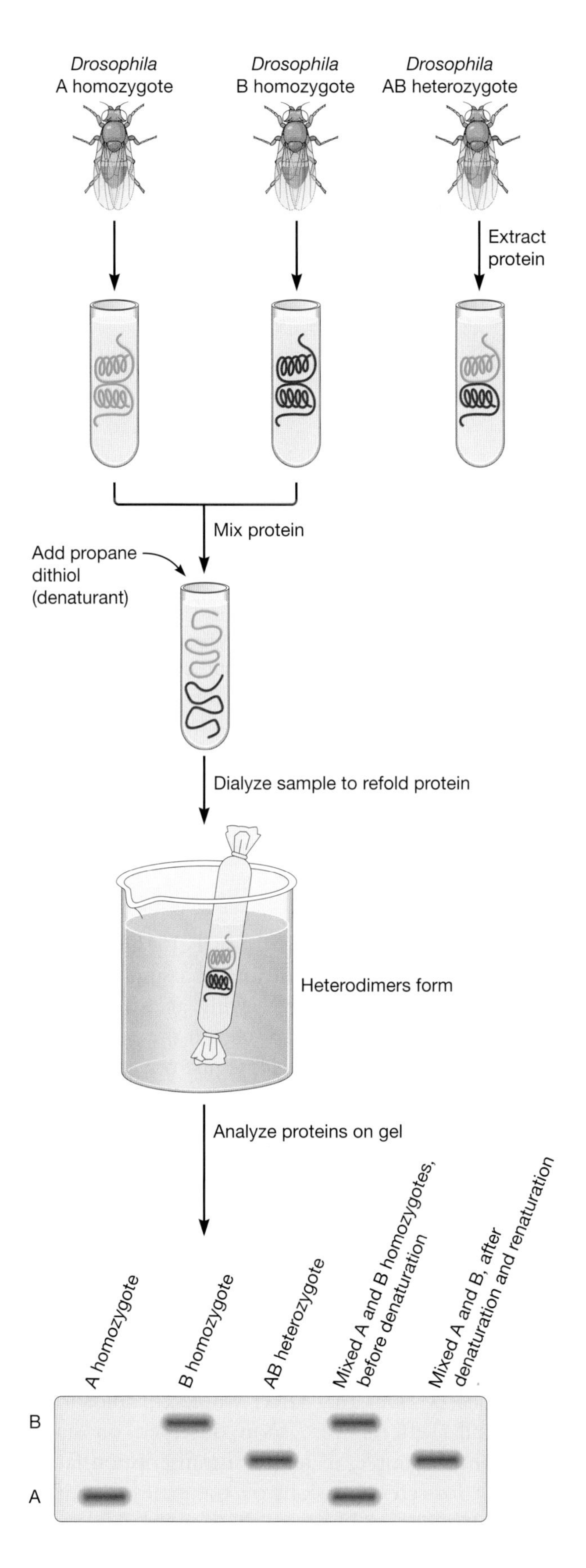

offspring, or they can mate with male worms.) This mechanism of dosage compensation was revealed by careful measurements of the levels of X-linked mRNAs in males and hermaphrodites. Females carrying two copies of a gene had exactly the same level of mRNA as males that had a single copy, and females heterozygous for an X-linked gene carried half the level of wild-type hermaphrodites or males. Furthermore, several mutations were found that blocked dosage compensation—in these mutants, hermaphrodites (XX) had twice the amount of mRNA from X-linked genes compared to males (Table 8-3).

Dosage compensation of sex chromosomes is an epigenetic control imposed during early development and is stably passed on from daughter cell to daughter cell during mitosis. There is also epigenetic control of gene expression, which persists through meiosis but, remarkably, is reset, depending on which parent transmits the epigenetically regulated gene to the offspring.

Parents Do Not Make Equal Genetic Contributions to Their Offspring

Each parent contributes one haploid set of chromosomes to its offspring, and so it was long assumed that the maternal and paternal genomes make an equal genetic contribution of autosomal genes to the offspring. However, experiments transferring nuclei between fertilized mouse eggs showed otherwise. There is a brief time after fertilization when the pronuclei contributed by the parents have not yet

FIGURE 8-7

6-phosphogluconate dehyrogenase reveals that both X chromosomes are active in female *Drosophila*. This enzyme has two isoforms that associate soon after translation to form a dimer. The dimers have differing gel mobility depending on their composition. AA dimers migrate more slowly than BB dimers. The AB heterodimer has an intermediate mobility. Mixing AA and BB dimers after translation does not permit formation of AB heterodimers. Only partial refolding of the mixed proteins will permit heterodimer formation. Therefore, the presence of AB heterodimers in female flies indicates that both X chromosomes are simultaneously active in each cell.

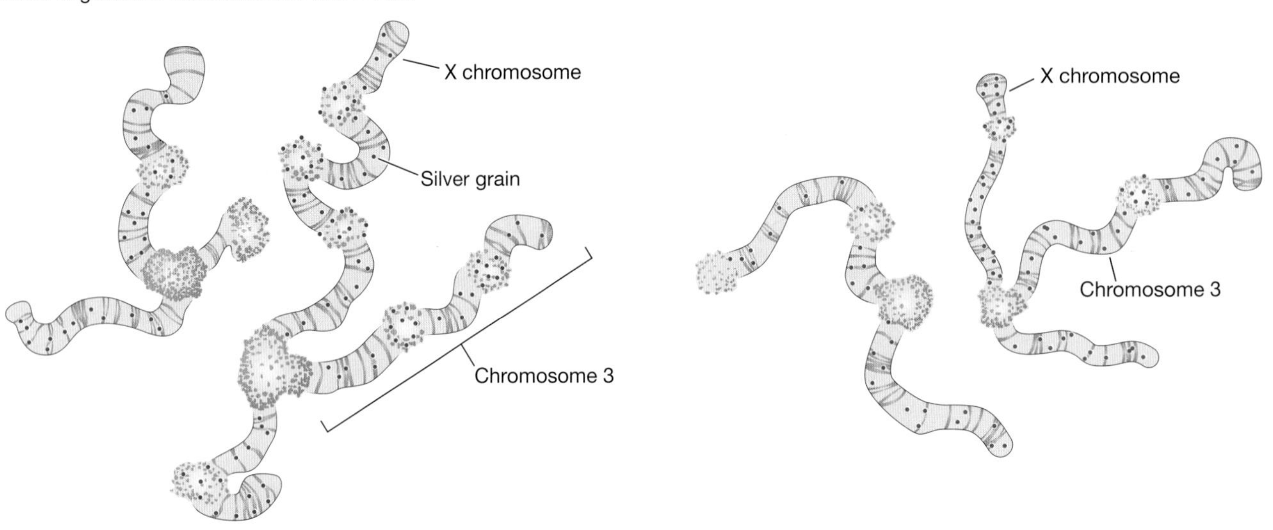

FIGURE 8-8

Transcription on the *Drosophila* male X chromosome is twice as active as on each female X chromosome. Cells were isolated from *Drosophila* salivary glands and incubated with radioactive ribonucleotides to measure transcriptional activity. Regions of high transcription are revealed by autoradiography. The autoradiograph is superimposed over an image of the salivary glands, and the number of silver grains (corresponding to the amount of transcriptional activity) were counted. Note that homologous chromosomes are paired in the *Drosophila* salivary gland, so that the two X chromosomes in the female are paired. In comparing these chromosomes to the single male X, the difference in thickness is apparent.

fused. Microsurgery can remove the male or female pronucleus from a fertilized egg and replace it with the male or female pronucleus from another fertilized egg. In this way, fertilized eggs can be produced readily that contain only maternal (gynogenetic) or paternal (androgenetic) genomes. More than 200 mouse embryos resulted from pronucleus transfer in embryos containing the genomes of both parents; of these, seven embryos developed normally. This is a small number of successes; but no embryo composed of just maternal or paternal genomes survived. Thus, to complete embryogenesis, a fertilized egg must contain one female and one male pronucleus.

If both maternal and paternal haploid genomes are required to complete development, they must make functionally distinct contributions to embryogenesis. This phenomenon of differential genetic contributions of parents to their offspring is called *imprinting*. The first observation of imprinting was in the fungus gnat *Sciara*, where the sex chromosome behaves in an unusual way. All embryos initially receive three copies of the sex chromosome: one from the mother and two from the father. During development of female individuals, a single paternal chromosome is eliminated, leaving an XX constitution. In males, both paternal chromsomes are eliminated. (The male germ cells lose just one paternal X, and in a strange meiosis produce XX sperm.) In sperm, the maternal and paternal sex chromosomes are somehow reset so that they behave paternally in the next generation. Mapping experiments showed a heterochromatic region on the X chromosome that directs the imprinted behavior. The function of this region is determined by the parent; passing the chromosome through the male sets the "imprinting mark," whereas passage through the female resets it.

The concept of imprinting was developed from these observations that affect whole chromosomes, but similar effects were soon observed on single genes. For example, in maize, a transcription factor called *r*, for red color, controls the genes that make pigment in the endosperm, the starchy part of the seed that nourishes the embryo. Only the maternal copy of the *r* gene is expressed. When several other endosperm genes are mutated in the mother, the

TABLE 8-3. Dosage compensation in *Caenorhabditis elegans*

Comparison	Gene studied	Relative dose	Ratio predicted with dosage compensation	Observed ratio of expression
Males XO / Hermaphrodites XX	*myo-2*	1/2	1.0	1.0
Males / Hermaphrodites	*uxt-1*	1/2	1.0	1.0
Males / Hermaphrodites	*uvt-4*	1/2	1.0	0.9
Males / Hermaphrodites	*uxt-2*	1/2	1.0	0.4
Males XO / XX males	*myo-1*	1/2	1.0	0.9
Males / XX males	*uxt-1*	1/2	1.0	0.75
Males / XX males	*uvt-4*	1/2	1.0	0.85
Males / XX males	*uxt-2*	1/2	1.0	0.4
dpy-28 hermaphrodite XX / Hermaphrodite XX	*myo-2*	2/2	1.0	2.2
dpy-28 hermaphrodite / Hermaphrodite	*uxt-1*	2/2	1.0	3.2
dpy-28 hermaphrodite / Hermaphrodite	*uvt-4*	2/2	1.0	2.6
dpy-28 male XO / Normal male XO	*Uxt-1*	1/1	1.0	0.9
dpy-28 male / Normal male	*Uvt-4*	1/1	1.0	1.1

Based on data from Meyer and Casson, *Cell 47:* 871–881 (1986).

Expression of four X chromosome genes in the nematode worm *C. elegans* was measured by Northern analysis and normalized to expression levels of autosomal genes. The gene expression was then compared between males (XO), hermaphrodites (XX), and phenotypic XX males (which are XX but have a mutation that blocks hermaphrodite development). The expression of *myo-2*, *uxt-1*, and *uvt-4* showed evidence of dosage compensation, because expression levels were the same for males, hermaphrodites, and phenotypic males. (The X *uxt-2* did not show evidence for dosage compensation; similar exceptions to dosage compensation are shown by the genes in the pseudoautosomal region of the human X chromosome.)

The expression levels of X-linked genes were then examined in worms carrying the *dpy-28* mutation; *dpy-28* males are phenotypically normal but *dpy-28* hermaphrodites are sick and "dumpy." Although gene expression in *dpy-28* males was unaffected, the mutation appeared to disrupt the process of dosage compensation in hermaphrodites; *dpy-28* hermaphrodites expressed X-chromosome genes at twice the level of wild-type hermaphrodites. This increased expression of X-chromosome genes in *dpy-28* hermaphrodites suggests that dosage compensation in worms is caused by suppression of X chromosomes.

seed is affected. The mutated allele is phenotypically "invisible" when inherited from the father because it has been turned off (i.e., imprinted). This maternal effect observed for imprinted genes differs from the maternal pattern of inheritance of cytoplasmic genes in organelles such as chloroplasts. A mutant chloroplast gene can be inherited only from the mother, but a mutant imprinted gene can be passed through the father.

Imprinting in mammals was discovered through the analysis of a transgene that behaved in an unexpected way. Transgenic mice expressing the growth-promoting transcription factor c-*myc* were produced. However, systematic breeding experiments showed that the transgene was expressed only in the offspring of matings between male carriers of the transgene and noncarrier females. Thus, the transgene was active only when inherited from a male

TABLE 8-4. Expression of the *c-myc* transgene in mice depends on the parental origin of the transgene

Parent		Activity of transgene in transgenic offspring (number of mice)	
Female	Male	Expressed	Not expressed
Transgenic	Not transgenic	0	42
Not transgenic	Transgenic	20	0
Transgenic	Transgenic	6	9

(Table 8-4). However, this unusual pattern of expression could have been related in some way to the fact it was a transgene. As such, it might have been the target of inactivation, albeit in some curious sex-specific manner. It remained to be demonstrated that imprinting occurred under normal circumstances.

Males and Females Compete to Control Fetus Size by Imprinting Key Growth Genes

The insulin-like growth factor II (IGF2, also known as somatomedin A) is a polypeptide growth factor that binds to the cell surface insulin growth factor receptor (IGF1R) to stimulate cell growth. It has an important role in growth and development and is expressed in many tissues. Its function was studied by using homologous recombination to knock out one allele of the *IGF2* gene in embryonic stem cells derived from an agouti strain of mice (Fig. 8-9). The stem cells were injected into blastocysts from albino mice, and mice with chimeric coat colors were obtained. The chimeric coat indicated that the mice contained cells from the albino blastocyst host and from the injected agouti stem cells. These chimeric mice looked normal, presumably because the normal cells from the host blastocyst compensated for the reduction of IGF2 in the cells with the knocked-out gene. When the chimeric males were mated with normal females, all the offspring of some males had agouti hair, showing that their germ lines were derived entirely from the injected ES cells. Surprisingly, half these offspring were small. When the agouti mice were analyzed, the normal-sized progeny were homozygous for wild-type IGF2, whereas the small mice were heterozygous, containing one wild-type and one disrupted allele of *IGF2*. These effects might have been due to gene dosage; the small mice simply had one-half the levels of IGF2 of the normal-sized mice. However, analysis of the small mice showed that the amount of mRNA produced from their one intact allele was about one-tenth, rather than one-half, that found in wild-type embryos. What had happened to the maternally derived allele in these small heterozygous mice? The answer appears to be imprinting. The maternally derived *IGF2* gene is subject to imprinting, so the only *IGF2* alleles active in the offspring are paternally derived, thus producing normal-sized mice (in those inheriting the wild-type paternal allele) or small mice (if they inherited the mutant paternal *IGF2* allele, which means that the mice are depending on a mutant allele for IGF2 production). Because the maternal allele is silent, it has little impact on the development of the mouse.

Other related studies showed that the expression of IGF2R, another IGF2 receptor, located elsewhere in the genome, is also imprinted—but in the opposite way. The receptor is expressed from the maternal allele and silent from the paternal allele. IGF2R is not a typical growth factor receptor; rather than stimulating cell growth when bound by its ligand, IGF2 binding to IGF2R triggers internalization and degradation of IGF2. (The "normal" receptor for IGF2 is IGF1R, which stimulates cell growth when bound by IGF2.) Thus, IGF2 and IGF2R are in competition, and so too are the parents: The mother silences the *IGF2* gene and activates the inhibitory receptor, whereas the father activates *IGF2* and silences the inhibitory receptor.

IGF2 and IGF2R are involved in control of growth, and a bioinformatic analysis of more than 50 genes known to be imprinted shows that a disproportionately large number of imprinted genes play a part in organogenesis, morphogenesis, cell growth, cell proliferation, and regulation of cellular processes. These functions—all associated with growth—lend support to the kinship (or genetic conflict) theory of genomic imprinting, based on the conflicting interests of mammalian mothers and fathers in their offspring. A mother bears the major costs of having offspring by investing very significant resources in nourishing the embryo. She has an equal genetic interest in all of her offspring, by any male, and will want to divide her limited resources equally over multiple pregnancies. A mother, then, to restrict demands made on her by the fetus, will pass on imprinted versions of genes that would otherwise promote fetal growth. On the other hand, a father may conceive only once with a particular

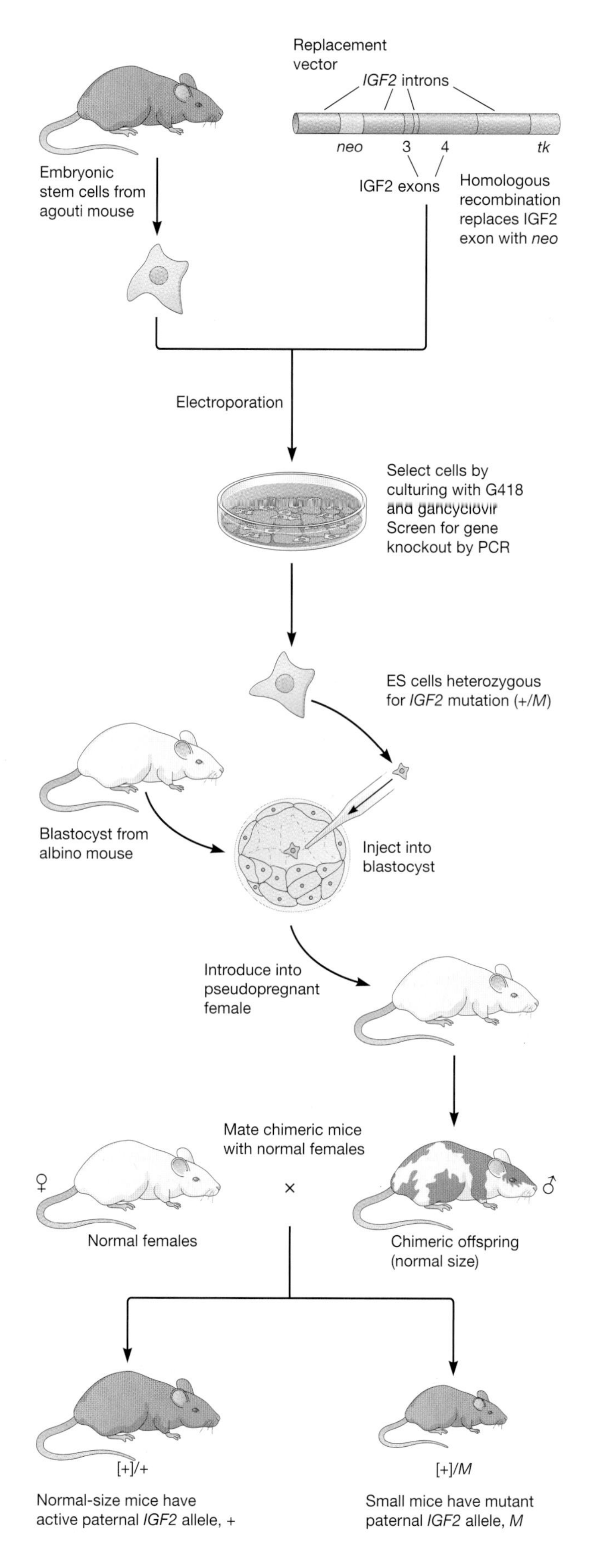

female, and so he is not concerned with her well-being beyond the current pregnancy. It is in the father's genetic interest for his offspring to have the best intrauterine environment possible so that they are most likely to survive in utero and in the perilous perinatal period. He will pass to his offspring silenced versions of those genes that would otherwise inhibit their growth, although doing so will impose a significant cost on the mother. The theory is controversial but it is notable that IGF2 has a placenta-specific promoter. Knockout experiments showed that loss of function of *IGF2*, a maternally imprinted gene, leads to a smaller placenta, which reduces the supply of resources to the fetus. Generally, loss of paternal imprinting leads to a small fetus, whereas loss of maternal imprinting leads to a larger fetus or overgrowth of extraembryonic tissue.

Failures of Imprinting Cause Human Disorders

Several developmental disorders have been linked to disruption of imprinting patterns. A particularly interesting example is provided by two very different

FIGURE 8-9

Imprinting in transgenic mice. The insulin-like growth factor (*IGF2*) gene in embryonic stem cells derived from a black, agouti strain of mice was knocked out by homologous recombination. The replacement vector contained *IGF2* exons 3 and 4, the *neo* gene (conferring resistance to the drug G418), and the *tk* gene (conferring susceptibility to gancyclovir). When the vector undergoes homologous recombination (see Chapter 6) with the endogenous gene, the mice acquire a *neo* gene, which replaces exon II of the mouse gene. At the same time, the *tk* gene is eliminated from the transgene. A double selection eliminates cells that have simply taken up the vector but have not undergone homologous recombination—they contain an active *tk* gene and are killed by the gancyclovir—and cells that do not contain the vector at all—they lack a *neo* gene and are killed by the G418. The resulting cells (+/*M*) were injected into blastocysts from albino mice, which were introduced into pseudopregnant mice. Chimeric mice were recognized by their patches of agouti hair. Males whose germ line was derived entirely from the embryonic stem cells were mated with females. The litters were composed equally of small and normal-size mice. The *IGF2* genes from the females are inactive because of imprinting ([+]); the normal-size offspring have a normal (+) and the small offspring a mutant (*M*) gene from their fathers.

syndromes that affect children, Angelman Syndrome (AS) and Prader–Willi Syndrome (PWS). Children with AS have microcephaly (small skull), hyperactivity, severe mental retardation, and a happy personality—often laughing at inappropriate times. In contrast, children with PWS fail to thrive in the neonatal period and remain short, but begin pathologically overeating in early childhood, which leads to obesity. They frequently have obsessive–compulsive disorder and temper tantrums. Remarkably, a very similar chromosomal alteration—a deletion of the region 15q11 to 15q13 of chromosome 15—was found in patients with these very different syndromes. How could the same genetic defect lead to such dissimilar syndromes? The answer came with the realization that the syndrome depended on whether the maternal or paternal 15q was expressed in the child: PWS results from the loss of function of genes expressed from the paternal chromosome, whereas the converse is true for AS—it results from the loss of genes expressed from the maternal chromosome.

Although PWS and AS can be caused by defects in the imprinting, in most patients other mechanisms prevent the inheritance of correctly imprinted genes from both parents. For instance, three-quarters of cases of PWS are caused by deletions that remove the 15q region from the paternal chromosome, leaving only the maternal allele. Another 20% of PWS patients have uniparental disomy for chromosome 15, in which both copies have been inherited from the mother. This comes about through nondisjunction during meiosis so that the embryo is triploid for chromosome 15. Triploidy of any autosome except chromosome 21 is lethal, but occasionally an extra chromosome 15 is lost so that the embryo is diploid for that chromosome. By chance, the remaining two chromosomes may be the nondisjunction pair from one parent, both carrying the same imprinting pattern. If these are chromosome 15 from the mother, then the child has two copies of the maternal chromosome 15, and these are silenced by imprinting. The 15q11–15q13 region contains many genes and the gene(s) causing PWS have not been identified, although suspicion has settled on a small nuclear ribonucleoprotein that is involved in messenger RNA (mRNA) processing.

Similar mechanisms lead to the loss of maternally expressed 15q genes in AS, although the proportion caused by uniparental disomy is small, perhaps because uniparental disomy of the paternal chromosome is harmful. The gene involved in AS has been identified—it is *UBE3A*, a ubiquitin-protein ligase that marks proteins for destruction by adding a chain of ubiquitin molecules to them. It is expressed only from the maternal allele and so is absent in children with AS. Strikingly, the imprinting of *UBE3A* expression is limited to the brain, which perhaps accounts for the mental retardation in AS children.

DNA Methylation Is a Heritable Mark on the DNA

As we saw earlier, imprinting in mammals was discovered initially not through the analysis of an endogenous gene, but through the analysis of a c-*myc* transgene that was expressed only when inherited from the father. The integration site of the imprinted transgene was carefully analyzed to try to understand the difference between paternal and maternal alleles. Coincidentally, when this unusual transgene was discovered, DNA methylation was beginning to be appreciated as an important hallmark of silenced genes. Consequently, the methylation patterns of the transgenes were tested. A number of restriction enzymes are available that are *isoschizomers* having identical recognition sites but differing in their sensitivity to DNA methylation. By using one enzyme that is insensitive to methylation and another that cannot cut methylated DNA, it is possible to determine the percentage of sites in a sample that are methylated at a given site. One example of such a pair is MspI, which cuts both the sequence methylated and unmethylated CCGG, and HpaII, which cannot cut methylated CCGG. The sensitivity to HpaII digestion differed depending on whether the transgene has been passed through the mother or the father (Fig. 8-10). When the transgene was inherited through the mother, the promoter region of the gene was methylated and transcription of that gene was inactive, whereas the promoters of active transgenes inherited from the father were less methylated.

The discovery that DNA methylation patterns correlate with imprinting status was more than a lucky guess. In fact, DNA methylation had emerged as an important influence on gene expression in many organisms. We will see many more examples of the correlation of DNA methylation with gene expression.

DNA methylation was discovered through chemical analysis of deoxyribonucleotides prepared from organisms. In many organisms, some percentage of the DNA residues are methylated. In bacteria both

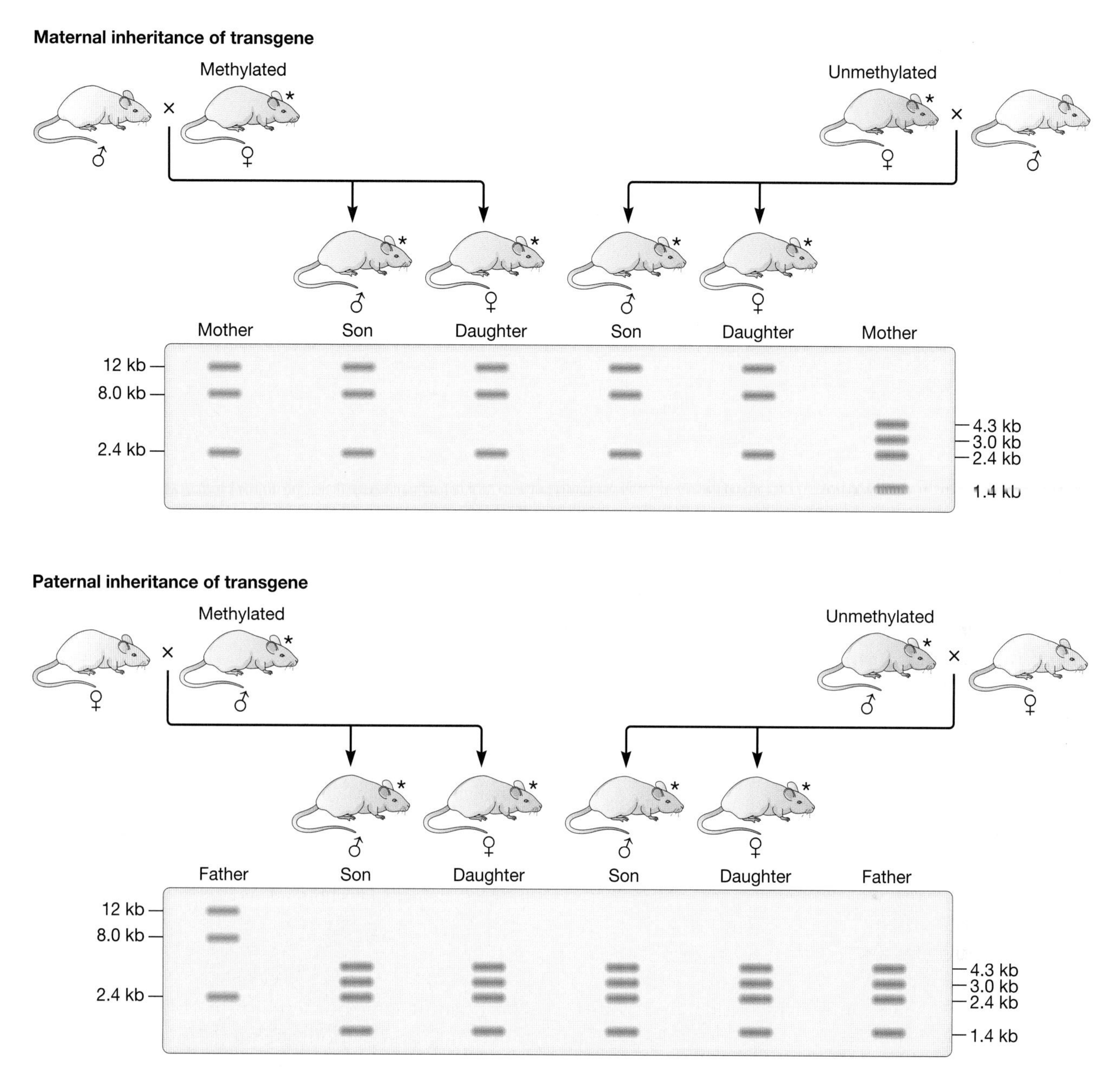

FIGURE 8-10

Gene imprinting depends on methylation of the imprinted gene. A line of mice contains a transgene made up of a fragment of the c-*myc* gene joined to the immunoglobulin locus. By chance, the transgene is subject to imprinting; it is expressed in the offspring of transgenic males and not in the offspring of the transgenic females. This pattern of expression parallels the methylation state of the transgene. The latter can be determined by digesting DNA from the mice with the restriction enzymes HpaII or MspI and probing Southern blots with a DNA fragment from the immunoglobulin locus. The restriction site for both enzymes is CCGG, but HpaII fails to cut when either of the cytosines is methylated. Restriction fragments of 12 kb, 8.0 kb, and 2.4 kb are produced from the methylated transgene (*red*) and fragments of 4.3 kb, 3.0 kb, 2.4 kb, and 1.4 kb are produced from the unmethylated transgene (*blue*), or in MspI digests. (Strictly speaking, the transgene is undermethylated and not totally unmethylated in these animals.) Matings are set up between transgenic (*) and normal mice. The methylation state of the transgene in the offspring depends on the sex of the transgenic parent. A transgenic female *always* passes on a methylated transgene to her offspring, even if her transgene is unmethylated. A transgenic male *always* passes on an unmethylated transgene to his offspring, even if his transgene is methylated.

cytosine and adenine are methylated, whereas in plants and animals only cytosine is methylated. Subsequently, it was observed that the activities of some restriction enzymes are inhibited by DNA methylation. When the structures of such enzymes bound to DNA were solved, it was apparent that the methyl group prevented the restriction enzyme from interacting with the DNA.

Methylated nucleotides are not incorporated into the DNA during replication. Instead, *methyltransferase* enzymes chemically add methyl groups afterward. In bacteria, there are sequence-specific methyltransferases that modify restriction sites to protect the bacterial genome from digestion by its own restriction enzymes (Chapter 4). The methyltransferases of plants and animals do not have this degree of sequence specificity, although there is a strong preference for methylating cytosines in the sequence 5′-CpG-3′. The complementary sequence is also 5′-CpG-3′, and in fact both cytosine residues in such sequences are usually modified. After DNA replication, the template strands retain the methylation mark, whereas the newly synthesized strands lack it. This *hemimethylated* state is quickly repaired by maintenance methyltransferases that recognize the hemimethylated site and methylate the opposite strand so that the CpG pair is fully methylated. Consequently, methylation at CpG sites is maintained as an inherited feature superimposed on the DNA sequence.

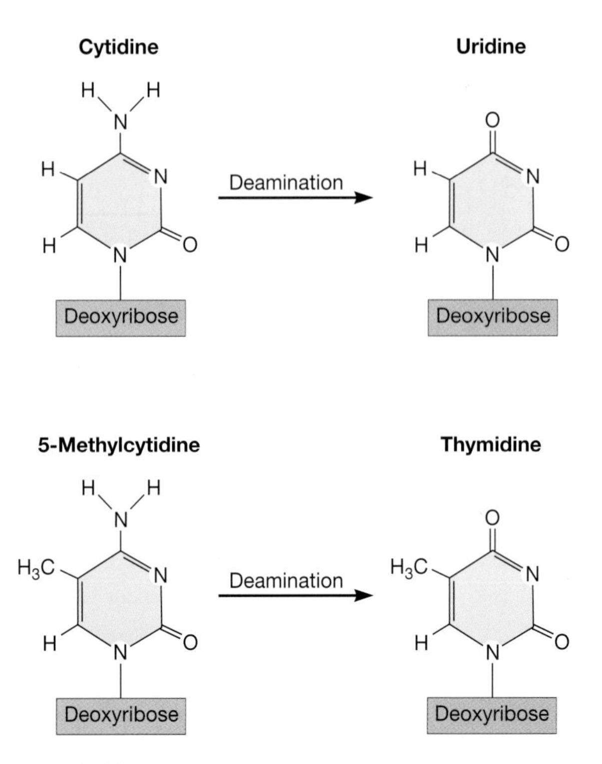

FIGURE 8-11

DNA methylation and the consequence of cytidine deamination. Cytidine deamination occurs spontaneously in DNA. When cytidine is deaminated, the result is the base uracil, but deamination of 5-methylcytidine results in thymidine. Although this mismatch can be recognized by the DNA repair machinery, the cell is unable to resolve whether the thymidine or guanidine residue is the correct base. The mismatched T:G is sometimes repaired to T:A instead of C:G, thus producing a transition mutation.

DNA Methylation Increases the Rate of Mutation

Methylation of the genome has a considerable cost to the organism. Cytidine residues carry the risk of undergoing spontaneous deamination to produce uracil (Fig. 8-11). These unexpected bases are quickly recognized and repaired by DNA repair processes. First, uracil DNA glycosylase cuts off the uracil, an endonuclease cuts the dexoyribose from the DNA backbone, and the gap is repaired by DNA polymerase. However, when 5-methylcytidine deaminates, the result is thymidine, which generates a T-G, a mismatch in the DNA. Faced with this mismatch, the DNA repair machinery cannot distinguish the mutated base from the correct template base, and consequently the result can be a C to T substitution mutation. In fact, a high proportion (often one-quarter to one-half) of spontaneous mutations identified are C to T substitutions at CpG sequences. Such mutations in the p53 tumor suppressor gene have been commonly found in tumors (Fig. 8-12).

The mutagenic burden of 5-methylcytidine is also apparent in the genome sequence as a whole. CpG sequences are five times less abundant than expected, whereas TpG and CpA are accordingly more common than expected. The CpG islands commonly located upstream of eukaryotic genes are an exception. They have a G+C content of greater than 50%. Interestingly, the CpG islands are undermethylated in germ cells. The cause of this undermethylation is not known, but it probably protects the CpG islands from mutation.

Silent Regions of Genomes Often Have Methylated DNA

By the mid-1970s, it was becoming apparent that methylation of new sites occurred infrequently, while the methylation state of existing methylated sites was maintained. Based on those observations,

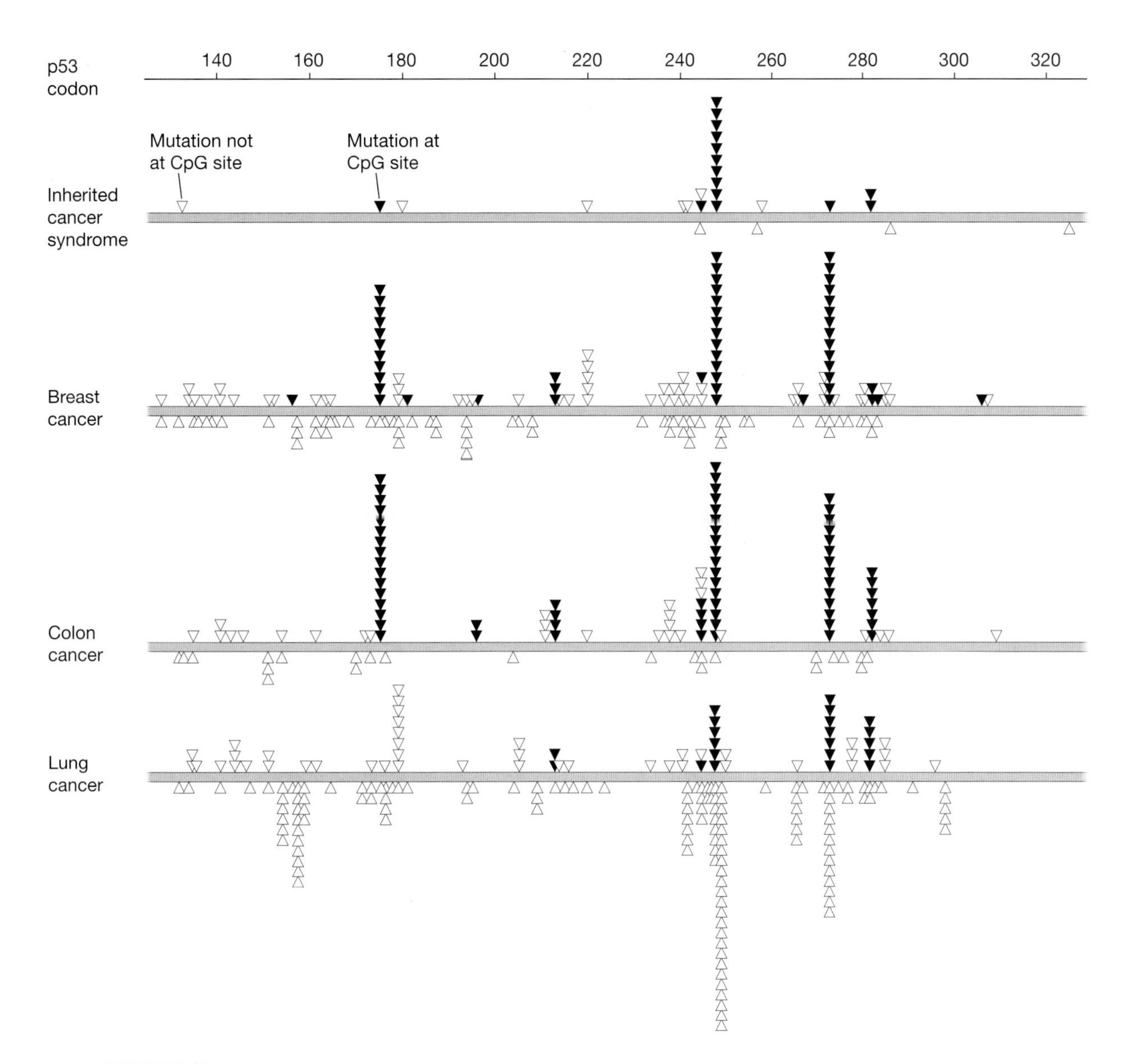

FIGURE 8-12

C to T transitions, possibly produced by 5-methylcytidine deamination, are common in tumors. Each *filled triangle* indicates a CpG transition mutation found in a different individual. *Open triangles* above the lines represent transition mutations at non-CpG sites, and *open triangles* below the lines represent transversion mutations. Transition mutations are from a pyrimidine to a pyrimidine (e.g., cytidine to thymidine) or a purine to a purine (e.g., adenine to guanidine). Transversions are from a purine to a pyrimidine (e.g., from cytidine to adenine). Note the predominance of CpG transition mutations in all of the tissues except lung cancer, in which carcinogens such as radon gas and cigarette smoke play a large role.

and the knowledge that methylation could affect the ability of proteins to bind DNA, in 1975, Robin Holliday and Arthur Riggs proposed independently that methylation and other similar modifications could make "marks" on the DNA while the underlying DNA sequence was preserved. Once made, during the process of cell differentiation, these DNA modifications would be stably inherited from one daughter cell to the next. Soon, a large amount of experimental evidence was collected in support of this hypothesis.

The first link between gene activity and DNA methylation came from analysis of the developmentally regulated genes in the human β-globin cluster (Fig. 5-10). All of these genes code for β-globin, but they are expressed at different times in development: ε is expressed in the embryo; Gγ and Aγ are active in the fetus; and δ and β are expressed in the adult.

(The gene cluster also encodes two nonfunctional pseudogenes: $\psi\beta_1$ and $\psi\beta_2$.) The embryonic β-globin genes were partially demethylated in the embryonic blood island where red blood cells develop, but were fully methylated in all adult tissues. The adult β-globin gene is active in the blood island as well as the adult bone marrow, and in these tissues there was less methylation near the adult β-globin gene than in the tissues such as kidney and brain where β-globin is inactive. The β-globin locus also contains nontranscribed pseudogenes. Methylation near these pseudogenes was at the same level in all tissues. The association of methylation with gene activity was most prominent at regions immediately upstream of transcription start sites. Active genes are often observed to be more susceptible to digestion by nucleases such as DNase I, which is suspected to reflect that the DNA is more accessible to the transcription machinery. Less methylated regions of DNA generally had higher nuclease accessibility, and the correlation between active, open DNA and low methylation strongly suggested that DNA methylation was a marker of inactive genes.

Recombinant DNA techniques were used to directly test this proposed link between DNA methylation and transcriptional activity. Cells were transfected with a single selectable plasmid. Following selection, many fewer stable cell lines were obtained from transfecting a methylated plasmid than when using the same plasmid in an unmethylated state, and the cell lines recovered had often lost much of the methylation on the marker gene. In other experiments, cells were transfected with two different methylated plasmids, each containing a selectable marker (Fig. 8-13). By selecting first for the marker carried by the plasmid, the exper-

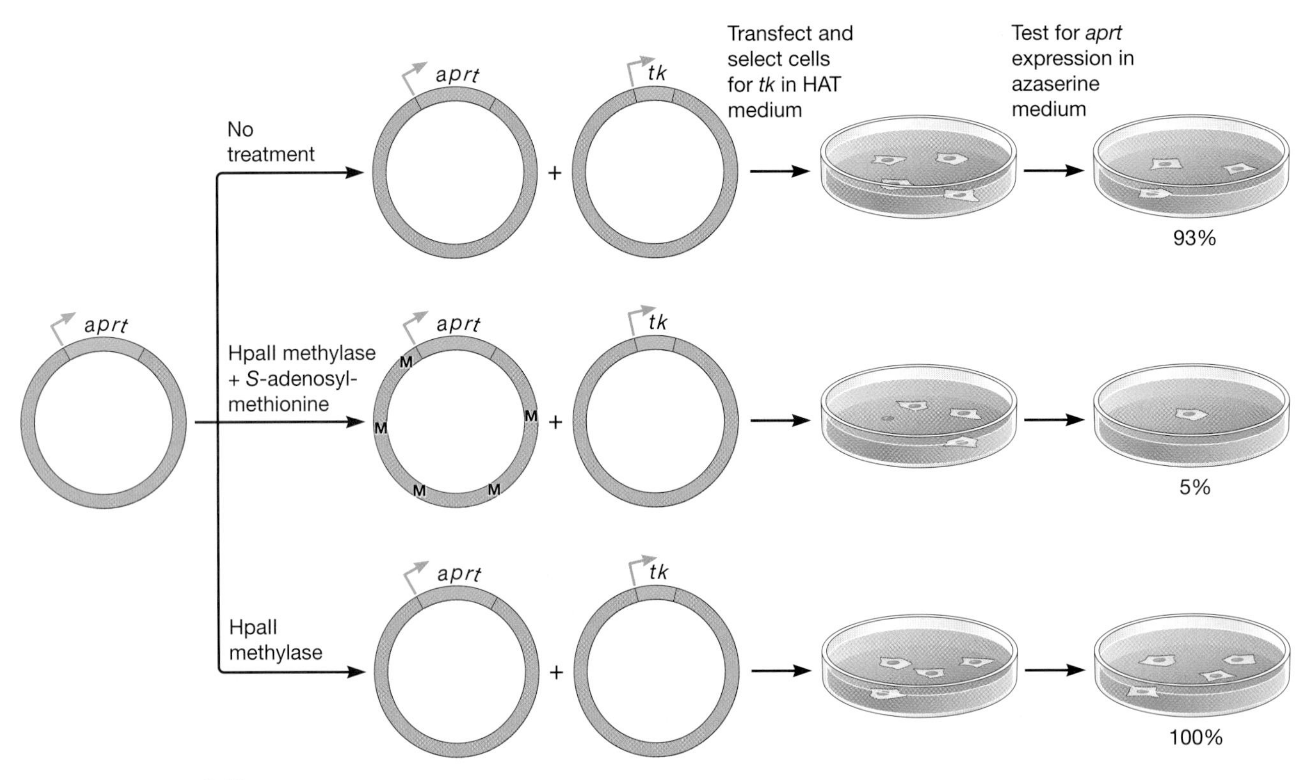

FIGURE 8-13

DNA methylation silences DNA. A plasmid was used containing the selectable gene *aprt*, which confers resistance to the amino acid analog azaserine. The *aprt* plasmid was either untreated, treated with HpaII methylase in the presence of the methyl donor *S*-adenosylmethionine so that all CCGG sites were methylated, or mock-methylated by adding methylase but not *S*-adenosylmethionine. The *aprt* plasmid was then mixed with another plasmid carrying the selectable thymidine kinase gene, and the two plasmids were simultaneously introduced into cells. Clones were selected for *tk* expression, but nearly all clones also carried the *aprt* plasmid because stable transfectants tend to incorporate both DNAs. Then, cells were tested for the ability to grow on HAT medium, which requires the *aprt* gene. Cells that had been transfected with methylated *aprt* plasmid DNA produced fewer resistant colonies. The similar function of the unmethylated and mock-methylated plasmid verified that the methylase enzyme did not damage the DNA in some other way.

imenters could ensure that all of the cells had taken up the second plasmid. However, the cells did not need the second marker for growth, and therefore there was no selective pressure for expression of that gene. Once cell lines containing both markers were established, the susceptibility of these cells to the second drug then was tested. Although the second marker was present in nearly all of the selected cells, cells transfected with an unmethylated version were much more resistant to the second drug than those transfected with methylated DNA, thereby indicating that the methylated marker gene was inactive. Some clones did arise from cells that had received methylated DNA; when the methylation state of these was analyzed, it was found that they tended to have less methylation upstream of the marker gene than the general population of cells before selection.

One question arising from these experiments was whether DNA methylation caused reporter plasmids to integrate into less active regions of the genome. This objection was addressed by two different approaches. First, it was found that methylated DNA was less transcriptionally active than unmethylated DNA when injected into *Xenopus* oocytes, where transcription occurs without DNA integration. Second, when maintenance methylation was inhibited by treating cells with the drug 5-azacytidine, many previously silent genes and retroviruses became active. Their expression (and low levels of methylation) persisted even after the drug was removed. This persistent demethylation is the consequence of the very low rates of de novo DNA methylation.

But how was DNA methylation affecting gene activity? Did DNA methylation directly block the transcriptional machinery or did it promote binding of other, inhibitory proteins? Scientists soon discovered a number of examples in which transcription factor binding was unaffected by DNA methylation, so the search for inhibitory proteins was on. In Edinburgh, Adrian Bird cloned two different proteins, methyl-CpG-binding proteins 1 and 2 (MeCP1 and MeCP2), after searching for proteins that bind to methylated DNA. MeCP1 needs at least 12 methylated cytosine residues to bind DNA, but MeCP2 binds to even a single methyl-CpG pair. In vivo, both proteins localize to the heavily methylated regions of the mouse genome.

An important role of MeCP2 in brain development became apparent when it was linked to Rett syndrome, a progressive neurodevelopmental syndrome observed almost exclusively in females. Children with Rett syndrome appear to develop normally, but around the end of their first year they gradually lose speech and coordinated motion. Very few boys have been described with Rett syndrome. The fact that Rett patients are nearly always female suggested an X-linked dominant trait with male lethality. Eventually, Rett syndrome was mapped to Xq28, and a systematic analysis of candidate genes in that region revealed that affected girls have mutations in *MECP2* that resulted in the partial or total loss of MeCP2 protein. Since the importance of MeCP2 in brain function was realized, physicians have discovered a few mentally retarded boys carrying less severe *MECP2* mutations. Because boys have a single X chromosome, mutations such as those that cause Rett syndrome in girls have a very severe phenotype that leads to an early death. Other boys with Rett syndrome were XXY males or were somatic mosaics in which a mutation had arisen during development so that some cells had wild-type X and others had a mutant *MECP2* allele.

The profound neurological defects in Rett syndrome powerfully illustrate the importance of the interactions of MeCP2 with the genome, but it has been difficult to pinpoint the exact cause of progressive mental decline. It is suspected that MeCP2 affects chromatin structure, because the protein associates with chromatin-modifying enzymes. These enzymes are important regulators of gene activity and mutations in *MECP2* may disrupt gene expression.

Noncoding RNAs Direct Imprinting of Some Genes

We saw above that methylation at the promoter of an imprinted *myc* transgene differed depending on its parental origin. This observation suggested that endogenous imprinted genes like *IGF2* might also vary in their methylation patterns. Surprisingly, in this case, methylation patterns were similar at the maternal and paternal *IGF2* promoters, even though the maternal allele is silent. Subsequently, scientists found a highly methylated region downstream of the *IGF2* paternal allele. Further studies revealed that this methylation permits paternal *IGF2* expression because it blocks an insulator. Insulators are DNA elements that can interrupt the effect of a nearby enhancer. When the insulator downstream of *IGF2* is methylated, it becomes inactive, thereby allowing an upstream enhancer to act on the *IGF2* promoter (Fig. 8-14).

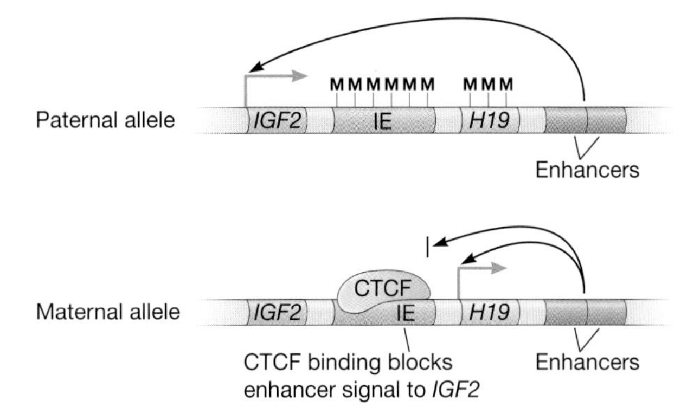

FIGURE 8-14

***H19* controls *IGF2* expression.** This locus contains two genes, *H19* and *IGF2,* separated by an insulator element (IE). On the paternal allele, DNA methylation on the insulator element inactivates it, which permits downstream enhancers to activate the *IGF2* gene. On the maternal allele, the insulator element is unmethylated and binds CTCF, a transcriptional regulator protein. This blocks activation of *IGF2* by the enhancers. Instead, the enhancers activate *H19* expression.

H19, a noncoding imprinted gene, was discovered during studies of the *IGF2*-imprinted region. On the maternal allele, the insulator is not methylated, so it is active and blocks enhancer activation of *IGF2.* Instead, the enhancer activates expression of *H19.* Surprisingly, when *H19* was deleted on the maternal allele, the maternal allele of *IGF2* was no longer silenced, even though the insulator sequence might still have been expected to block enhancer activation. Clearly, *H19* expression was somehow involved in regulating (silencing) *IGF2* expression. Studies using many different targeted deletions of the *H19* locus suggest that the sequence of the *H19* transcript itself is not important. For example, when replaced by another transcriptionally active sequence, *IGF2* is still properly silenced on the maternal allele. However, deleting the promoter of *H19* disrupts imprinting at the locus, which suggests that it is the transcriptional activity of *H19* that helps establish the imprinting pattern at the *IGF2* locus.

Noncoding RNAs have been discovered at other imprinted loci as well. In some cases, a particular noncoding transcript is needed for proper imprinting. In other cases, as with *H19*, the act of transcription seems to be more important. We have already seen examples of dosage-compensation complexes in which RNAs can be either essential positive or negative regulators. In Chapter 9, we will see further examples of the role of RNA in gene regulation. This is currently one of the most exciting areas in biology.

Imprinting Must Be Correct for the Proper Development of Cloned Animals

Two maternal or paternal genomes fail to produce viable embryos, but as the role of imprinting has become increasingly understood, investigators have revisited attempts to produce gynogenetic embryos. One strategy has been to use oocyte pronuclei from immature eggs isolated from 1-day-old mice. However, immature eggs are arrested in a state of meiosis I and they had to be first fused with an enucleated mature oocyte to trigger the completion of meiosis. The resulting haploid nucleus was implanted into a second mature oocyte. This was artificially activated to trigger embryo development, resulting in several relatively normal fetuses, although no animals survived to term. This experiment succeeded where previous attempts had failed because immature eggs do not have fully established imprints, and the transplantation process tricked the nucleus by skipping the maturation period during which imprints are usually set.

In a later series of experiments the transfers were repeated, but the donor of the immature oocyte had a deletion in the *H19* gene so that imprinting of *IGF2* would resemble the paternal pattern. Of 598 oocytes assembled in this manner, two relatively normal pups were born, one of which grew to adulthood and gave birth to normal pups. Apparently, the imprint at the *H19/IGF2* locus is one of the gene expression patterns required for proper development. However, only 28 pups developed and, other than the two live pups, the rest were stillborn or died immediately after birth.

The importance of epigenetic control during development and beyond is shown clearly in attempts to produce cloned animals by transferring nuclei from adult cells into oocytes. Dolly the sheep is the most famous and, indeed, the first example of a cloned mammal—something dismissed as impossible based on the failure to produce parthenogenetic mammals. But the process of cloning is very inefficient; Dolly was the sole survivor of 277 cloning attempts. The success rate varies from species to species—as high as 15% in mice when nuclei from embryonic stem cells are used—but overall very few cloned animals survive.

Animals that survive have various abnormalities—for example, respiratory and circulatory system defects and malformation of the kidneys and brain. Of the embryos that fail to develop, many have overgrowth of the embryo or the placenta. Few cloned embryos show correct imprinting patterns, and abnormalities are apparent even in surviving clones.

It is not surprising that there are epigenetic defects in embryos derived from nuclear transfer. Imprints are modified throughout development and differentiation, so that the pattern of gene expression in an adult cell is very different from that of a gamete. For successful cloning, some process of "nuclear reprogramming" is presumably necessary to reset the imprints of an adult differentiated cell to approximate that of a gamete. One imprint that has been shown to be reset is X chromosome inactivation. This occurs randomly in cloned female embryos, regardless of which X chromosome was inactivated in the donor nucleus. Biologists are still investigating the mysterious processes that lay down patterns of gene expression in early development.

Myriad Histone Modifications Create a "Histone Code" That Guides Gene Expression

DNA methylation is a hallmark of gene silencing in many organisms, but some of biologists' favorite subjects—the yeast *Saccharomyces cerevisae*, the fruit fly *Drosophila*, and the nematode worm *C. elegans*—have little or no DNA methylation. Nevertheless, these organisms display epigenetic inheritance, and it is now apparent that the histones associated with DNA play a key role in determining whether a gene is active or silent.

Careful biochemical analysis during the 1960s revealed that some of the histones isolated from cells were modified by the addition of acetyl and methyl groups. Histones had been thought to be inert molecules whose only function was to act as spools, nucleosomes, around which DNA was wrapped to make chromatin. However, researchers began to appreciate that histone proteins are subject to a wide range of modifications, including phosphorylation, ADP-ribosylation, and ubiquitination (the addition of a small protein called ubiquitin), as well as acetylation and methylation. Aaron Klug and Roger Kornberg elucidated the structure of the nucleosome. Each nucleosome is composed of 147 bp of DNA wound around an octamer of histones—two copies each of H2A, H2B, H3, and H4. Most of the chemical modifications of the histones are made to the flexible "tails" of the molecules, which extend outside the globular core regions where DNA binds (Fig. 8-15). Many different combinations of modifications are possible, and some residues have the potential to be modified in several different ways; for example, lysines in histones can be acetylated, methylated, and ubiquitinated. Many of these modifications are added after the histones have been incorporated into nucleosomes following DNA replication.

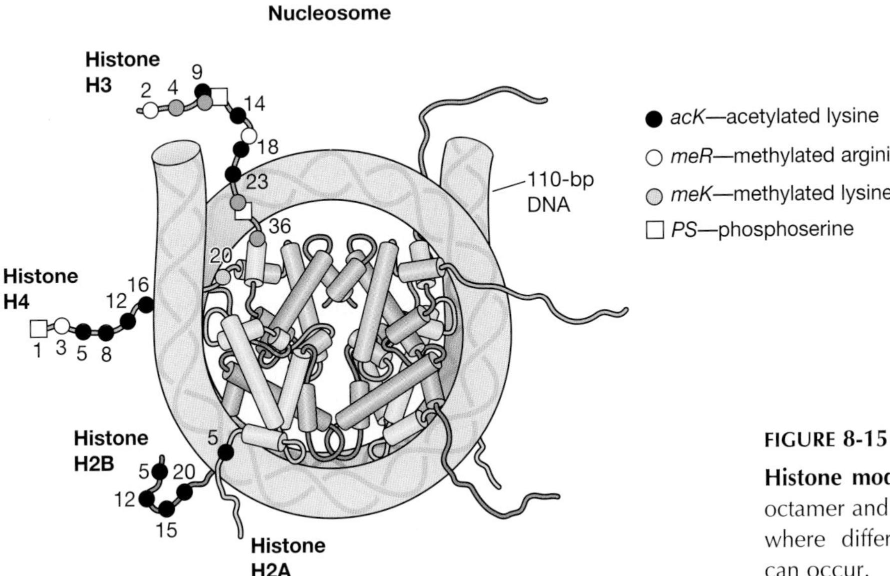

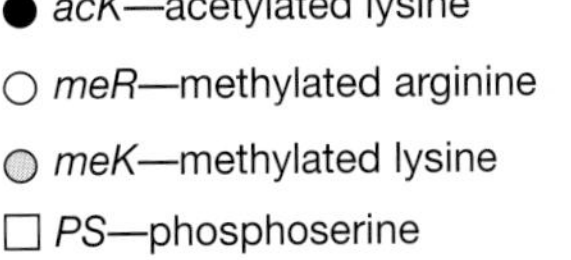

FIGURE 8-15

Histone modifications. A diagram of the histone octamer and the multiple sites on the histone tails where different covalent histone modifications can occur.

Soon after it was recognized that the promoters of inactive genes were methylated, increased levels of histone acetylation were found to be consistently associated with active promoters. For example, actively differentiating white blood cells had increased levels of acetylated histones and nuclease-accessible chromatin (presumably being transcribed) had more acetylated histones than bulk chromatin. The first conclusive experiments were done using antibodies specific for acetylated histones. These antibodies were used to isolate chromatin from chicken blood cells, and although the antibodies precipitated chromatin from the highly active α-globin gene (showing the DNA was enriched for acetylated histones), chromatin from inactive genes could not be detected. In Chapter 13 we will see how this approach has been made more reliable and sensitive through a combination of antibodies, quantitative PCR, and DNA microarrays. Now, chromatin immunoprecipitation (ChIP) experiments can be used to map the distribution of histone modifications and DNA-binding proteins throughout the genome.

As antibodies recognizing other histone modifications were developed and characterized, it was found that some epigenetic phenomena were associated with very specific histone modifications. For instance, the hypertranscribed *Drosophila* male X chromosome contains large amounts of histone H4 acetylated at lysine 16. The inactive mammalian X chromosome is bound by histone H3 methylated at lysine 27, as well as a rare variant histone, macroH2A, which can substitute for the normal H2A. The centromeres in yeast and mammalian chromosomes are bound by another histone H3 variant, CENP-A.

Careful observation of histone modifications associated with many genes in many organisms has led to the proposal that there is a "histone code" that distinguishes active and inactive genes (Table 8-5). For example, methylation on lysine 4 of histone H3 is typically found on active genes, whereas methylation at the nearby lysine 9 is a hallmark of silent genes. As predicted by the early work, several acetylation marks—including those on lysines 9 and 14 of histone H3 and lysine 5 of histone H4—are associated with active genes. However, not all acetylation events are activating; acetylation of lysine 12 on histone H4 is a hallmark of silenced heterochromatin.

How do modifications on the histone tails signal to the gene expression machinery? The modifications change the charge of the histone tails: Acetylation adds a negatively charged acetyl group to a positively charged lysine, whereas methylation neutralizes lysine. These changes in charge may affect the packing of DNA with histones to a small degree and affect access to the DNA strands; the real "reading" of the code is carried out by proteins that bind to the modified histones.

Many of these proteins were found in genetic and biochemical screens for modifiers of epigenetic silencing effects long before it was determined that they bound to modified histones. For example, in *Drosophila* a protein called heterochromatin protein 1 (HP1) was identified because it localized to heterochromatic regions of the genome; HP1 mutants failed to silence genes near heterochromatin. The HP1 protein contains a chromodomain, an amino acid sequence that was soon found in several other chromatin-binding proteins. It turned out that the chromodomain binds to methylated histones; the chromodomain of HP1 binds to histone H3 methylated at lysine 9, which we have seen is a common mark of silent genes.

Histone acetylation is a feature of active genes, and it is not surprising, therefore, that the proteins controlling transcription recognize acetylated histones. Transcriptional activators, as well as the RNA polymerase II initiation complex, include proteins with bromodomains, which recognize and bind to acetylated histones. Consequently, activating complexes are attracted to chromatin with acetylated histones marking active genes.

The chromatin proteins reinforce the histone marks to which they bind. HP1 recruits the histone methyltransferase that methylates histone H3 at lysine 9. Thus H3 lysine 9 methylation tends to induce more methylation of histone H3, which inhibits future recruitment of the transcription machinery. Gene activation is similarly reinforced—transcriptional activators and RNA polymerase II complexes incorporate histone-modifying enzymes, particularly histone acetyltransferases. Once bound to DNA, these transcription complexes acetylate the histones around a gene so that the activation state is maintained. These positive-feedback loops maintain and reinforce histone modifications.

Some Chromatin States Are Inherited through DNA Replication

Although transcriptional activation and repression help reinforce histone modifications, these histone modifications can be easily rewritten, either through

TABLE 8-5. Histone modifications

Histone subunit	Residue	Modification	Consequence
H2A			
	Serine 1	Phosphorylation	Mitosis, transcriptional repression
	Lysine 4	Acetylation	Transcriptional activation
	Lysine 5	Acetylation	Transcriptional activation
	Lysine 7	Acetylation	Transcriptional activation
	Lysine 119	Ubiquitylation	Spermatogenesis
H2B			
	Lysine 5	Acetylation	Transcriptional activation
	Lysine 12	Acetylation	Transcriptional activation
	Serine 14	Phosphorylation	Apoptosis
	Lysine 15	Acetylation	Transcriptional activation
	Lysine 120	Ubiquitylation	Meiosis
H3			
	Threonine 3	Phosphorylation	Mitosis
	Lysine 4	Acetylation	Acetylation: transcriptional activation
		Methylation	Methylation: active euchromatin
	Lysine 9	Acetylation	Acetylation: transcriptional activation
		Methylation	Methylation: transcriptional repression
	Serine 10	Phosphorylation	Transcriptional activation
	Threonine 11	Phosphorylation	Mitosis
	Lysine 14	Acetylation	Transcriptional activation/elongation
	Arginine 17	Methylation	Transcriptional activation
	Lysine 18	Acetylation	Transcriptional activation, DNA repair
	Lysine 23	Acetylation	Transcriptional activation, DNA repair
	Lysine 27	Methylation	Transcriptional silencing
	Serine 28	Phosophorylation	Mitosis
	Lysine 36	Methylation	Transcriptional elongation
	Lysine 79	Methylation	Transcriptional elongation
H4			
	Serine 1	Phosphorylation	Mitosis
	Arginine 3	Methylation	Transcriptional activation
	Lysine 5	Acetylation	Transcriptional activation
	Lysine 8	Acetylation	Transcriptional activation
	Lysine 12	Acetylation	Transcriptional activation, telomeric silencing
	Lysine 16	Acetylation	Transcriptional activation, DNA repair
	Lysine 20	Methylation	Transcriptional silencing
	Lysine 59	Methylation	Transcriptional silencing

Adapted from Peterson and Laniel, *Curr. Biol. 14:* R546–551 (2004).

A listing of known modifications of vertebrate histones is shown. Most of these modifications are also observed in other organisms, although occasionally the histone structure between organisms varies so that the numbering of various modifications changes.

the actions of enzymes that reverse histone modifications or through DNA replication. If DNA replication automatically wrote over histone modifications, the histone code could not heritably affect gene expression and would not qualify for our definition of epigenetics. In fact, chromatin modification states are often very stable and so there must be a way in which the histone code can be faithfully transmitted to newly replicated DNA. In fact, chromatin assembly proteins and histone-modifying enzymes associate with the DNA replication machinery, so that as histones are incorporated into newly synthesized DNA, they can be immediately modified.

Understanding the transmission of the histone code is still an area of very active investigation, but there are several clues as to how this works. First, histone octamers are assembled on DNA by combining "old" with "new" subunits. This was demonstrated by labeling cells with heavy radioisotopes and monitoring the incorporation of newly synthesized histones into newly synthesized chromatin (Fig. 8-16). Existing H2A/H2B

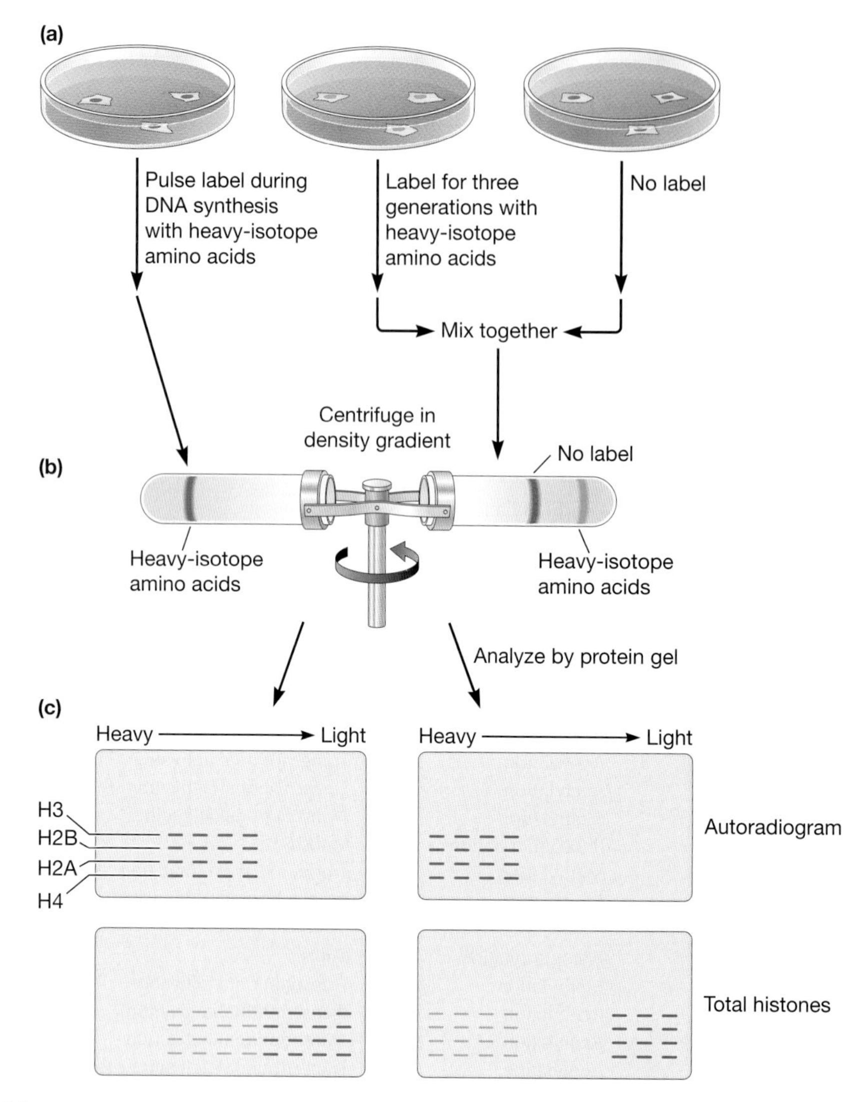

FIGURE 8-16

Newly incorporated histones are mixed with old histones during DNA synthesis. (*a*) A population of cells was pulse labeled with amino acids containing the heavy isotopes ^{15}N, ^{13}C, and ^{2}H. Proteins formed from these amino acids can be separated from proteins containing the normal light isotopes in a density gradient. The cells were also labeled with a small amount of radioactive amino acids to enable the visualization of small amounts of protein. (Heavy amino acids are extremely expensive, and so these experiments were carried out at a very small scale.) A second population of cells was labeled with heavy and radioactive amino acids for many generations, so that their histones were uniformly labeled. The cells were heavily cross-linked with formaldehyde to ensure that the nucleosomes remained intact, and the nucleosomes were then purified from the pulse-labeled and the uniformly labeled cells, as well as from unlabeled cells. (*b*) The histone octamers were fractionated in a density gradient under conditions in which the heavy and light molecules could be separated. The uniformly heavy labeled and unlabeled (light) histones were mixed in a single gradient to act as markers. (*c*) The migration of these nucleosomes was compared to a gradient containing nucleosomes from the pulse-labeled cells (the order of histone proteins in the gel [top to bottom] is H3, H2B, H2A, H4). The nucleosomes in the pulse-labeled cells (visible on the autoradiogram) had an intermediate density, which indicated that they were assembled from old and new subunits. Note that all four histone proteins were present in these nucleosomes of intermediate mass, thus indicating that both old and new copies of each isoform are incorporated into newly synthesized DNA. (Note that the heavy histones are not visible on the protein gels, because these experiments were done at very small scale.)

dimers are assembled together with newly synthesized H3 and H4 proteins, and conversely, old H3 and H4 subunits are brought together with new H2A and H2B. The modifications on the old subunits are recognized by histone-modifying enzymes, which direct the proper modifications to the new histones.

Another way in which histone modifications are controlled seems to be through the timing of DNA replication. Euchromatic (active) regions of the genome are replicated before heterochromatic (inactive) regions, and different histone-modifying enzymes are active early and late during replication. The initial choice between early and late replication is influenced by the histone modifications already present. The result is a self-perpetuating situation in which the modifications that direct late replication result in replication during a time at which enzymes that make heterochromatic modifications are present.

Very recently, biologists have realized that RNA exerts a major influence on the histone code. In the next chapter, we will learn how the RNA interference pathway communicates silencing signals to the genome. Histone modifications are directed to precise locations in the genome based on sequence similarity between small RNAs and the target DNA sequence.

Epigenetic Changes Mean That Identical Twins Are Not Identical

Angelman and Prader–Willi syndromes are examples of circumstances in which epigenetic modifications combined with other changes lead to genetic disorders. But there are also epigenetic changes that contribute to the differences between individuals. Identical, or, more precisely, monozygotic, twins arise by the splitting of an embryo at a very early stage, probably when there are fewer than 500 cells, and they are held up as examples of human beings who are genetic clones of each other. However, those who know monozygotic twins understand that they are not identical—they differ in looks, health, and especially personality—although how these differences arise is far from clear. Because their genome sequences are identical, epigenetic changes are likely candidates to explain the differences. A comparison of epigenetic modifications of DNA between members of 80 monozygotic twin pairs shows clearly that these differ between members of a pair. For example, 35% of the pairs differed in their patterns of DNA methylation and acetylation of histones H3 and H4, both markers of epigenetic modification.

Methylation patterns were also examined using amplification of intermethylated sites (AIMS; Fig. 8-17). One might expect that differences in epigenetic modifications would increase with age and this was the case; members of the youngest twin pairs were more similar than those of the oldest pairs (the age range was from 3 years to 74 years). Remarkably, twins who had lived together for longer periods (i.e., had had more similar life experiences) had more similar epigenetic patterns. Most interestingly, microarray analysis of RNA from twins showed that epigenetic differences are reflected in RNA expression patterns. For example, expression patterns for the 3-year-old twins who were almost identical epigenetically were very similar, whereas there were four times as many more differentially expressed genes in 50-year-old twins who were completely different epigenetically.

Some of the differences found in this study may result from random fluctuations in the transmission of histone modifications and DNA methylation, but the possibility that the environment produces specific changes in epigenetic states is intriguing. How this may happen, we do not yet know, but there is every reason to believe that continuing analysis will reveal many fascinating details of how genomes are modified by epigenetic changes.

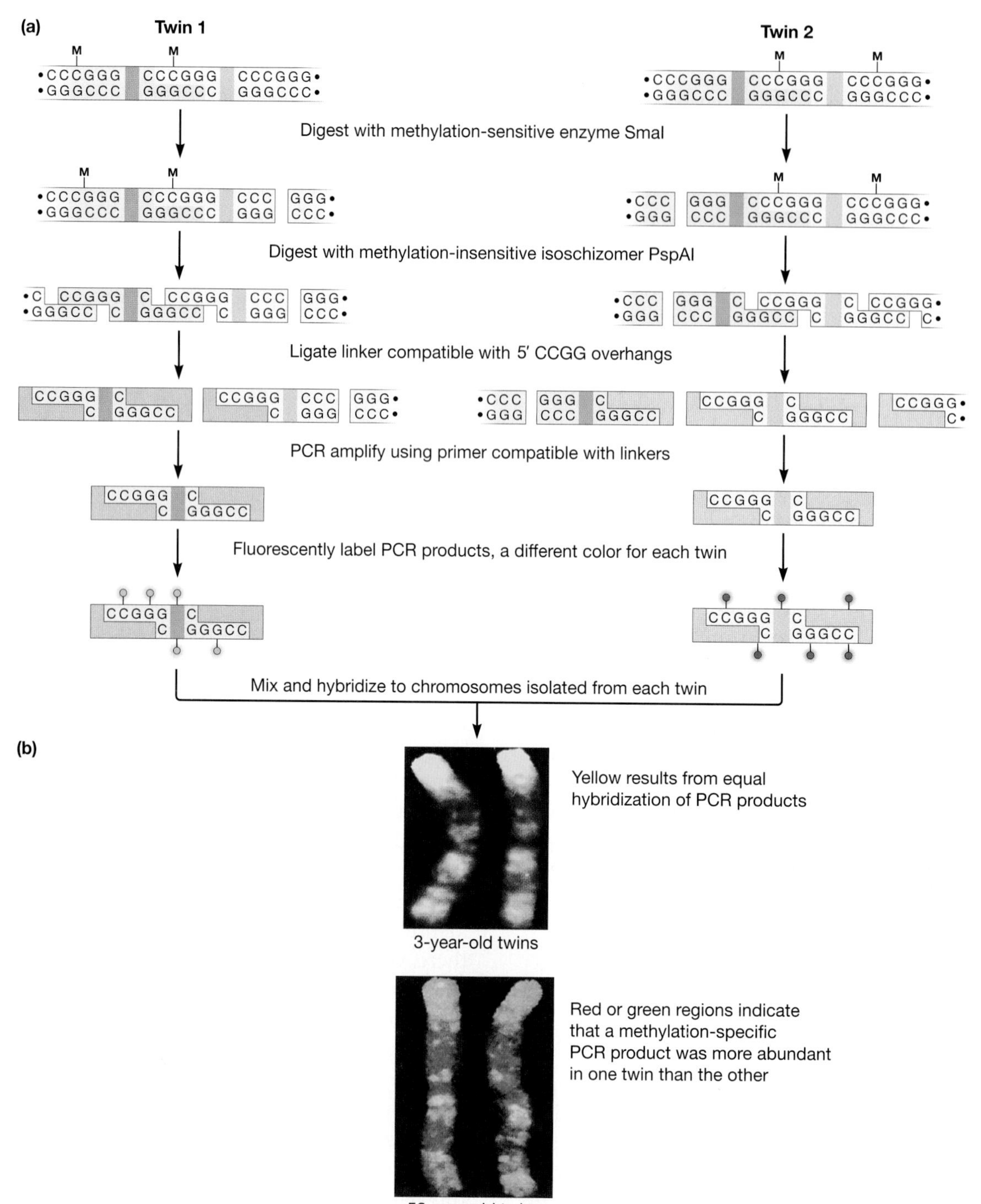

FIGURE 8-17

Epigenetic differences between monozygotic twins. (*a*) Methylation patterns are done using amplification of inter-methylated sites (AIMS). A segment of genomic DNA with three CCCGGG sites is shown. Methylated sites are marked "M." The nonmethylated sites are cut in a first digestion using the methylation-sensitive SmaI restriction endonuclease, which leaves blunt ends. Then, a second digestion is performed using the isoschizomer PspAI, which is capable of cutting methylated CCCGGG sites, and leaves a CCGG overhang. Adaptors are ligated to these sticky ends. (The adapters cannot ligate to the blunt SmaI ends.) Polymerase chain reaction (PCR) is done using specific primers that hybridize to the sequence formed by the adaptor ligation to the restriction site, plus one or more additional nucleotides that are arbitrarily chosen. The addition of this randomly chosen base limits amplification to a smaller number of sequences. (*b*) Comparison of AIMS differences between 3-year-old and 50-year-old twins. The PCR products from each twin were mixed in equimolar quantities with this brother and hybridized to a normal chromosome metaphase spread. Chromosome 1 is shown. The young twins have a similar distribution of methylation changes shown by the yellow color resulting from the even hybridization of red and green AIMS PCR products from each twin. The older twins are very different from one another as shown by the blocks of red and green.

Reading List

Epigenetics

Stillman B. and Stewart D., eds. 2004. *Epigenetics.* Cold Spring Harbor Laboratory Press, Cold Spring Harbor, New York.

Russo V.E.A., Martienssen R.A., and Riggs A.D., eds. 1996. *Epigenetic Mechanisms of Gene Regulation.* Cold Spring Harbor Laboratory Press, Cold Spring Harbor, New York.

X Inactivation

Barr M.L. and Bertram E.G. 1949. A morphological distinction between neurones of the male and female, and the behaviour of the nucleolar satellite during accelerated nucleoprotein synthesis. *Nature* **163:** 676–677.

Carrel L. and Willard H.F. 2005. X-inactivation profile reveals extensive variability in X-linked gene expression in females. *Nature* **434:** 400–404.

Davidson R.G., Nitowsky H.M., and Childs B. 1963. Demonstration of two populations of cells in the human female heterozygous for glucose-6-phosphate dehydrogenase variants. *Proc. Natl. Acad. Sci.* **50:** 481–485.

Lyon M.F. 1961. Gene action in the X-chromosome of the mouse (*Mus musculus* L.). *Nature* **190:** 372–373.

Ohno S. and Hauschka T.S. 1960. Allocycly of the X-chromosome in tumors and normal tissues. *Cancer Res.* **20:** 541–545.

Russell L.B and Bangham J.W. 1961. Variegated-type position effects in the mouse. *Genetics* **46:** 509–525.

Shapiro L.J., Mohandas T., Weiss R., and Romeo G. 1979. Non-inactivation of an X-chromosome locus in man. *Science* **204:** 1224–1226.

Wutz A. and Jaenisch R. 2000. A shift from reversible to irreversible X inactivation is triggered during ES cell differentiation. *Mol. Cell* **5:** 695–705.

Xist

Brown C.J., Ballabio A., Rupert J.L., Lafreniere R.G., Grompe M., Tonlorenzi R., and Willard H.F. 1991. A gene from the region of the human X inactivation centre is expressed exclusively from the inactive X chromosome. *Nature* **349:** 38–44.

Brown C.J., Hendrich B.D., Rupert J.L., Lafreniere R.G., Xing Y., Lawrence J., and Willard H.F. 1992. The human *XIST* gene: Analysis of a 17 kb inactive X-specific RNA that contains conserved repeats and is highly localized within the nucleus. *Cell* **71:** 527–542.

Clemson C.M., McNeil J.A., Huntington H.F., and Lawrence J.B. 1996. XIST RNA paints the inactive X chromosome at interphase: Evidence for a novel RNA involved in nuclear/chromosome structure. *J. Cell Biol.* **132:** 259–275.

Penny G.D., Kay G.F., Sheardown S.A., Rastan S., and Brockdorff N. 1996. Requirement for *Xist* in X chromosome inactivation. *Nature* **379:** 131–137.

Tsix

Debrand E., Chureau C., Arnaud D., Avner P., and Heard E. 1999. Functional analysis of the *DXPas34* locus, a 3′ regulator of *Xist* expression. *Mol. Cell Biol.* **19:** 8513–8525.

Lee J.T., Davidow L.S., and Warshawsky D. 1999. *Tsix*, a gene antisense to *Xist* at the X-inactivation centre. *Nat. Genet.* **21:** 400–404.

Mise N., Goto Y., Nakajima N., and Takagi N. 1999. Molecular cloning of antisense transcripts of the mouse *Xist* gene. *Biochem. Biophys. Res. Commun.* **258:** 537–541.

Xu N., Tsai C.L., and Lee J.T. 2006. Transient homologous chromosome pairing marks the onset of X inactivation. *Science* **311:** 1149–1152.

Ogawa Y. and Lee J.T. 2003. *Xite*, X-inactivation intergenic transcription elements that regulate the probability of choice. *Mol. Cell* **11:** 731–743.

Dosage Compensation

Kazazian H.H., Jr., Young W.J., and Childs B. 1965. X-linked 6-phosphogluconate dehydrogenase in *Drosophila*: Subunit associations. *Science* **150:** 1601–1602.

Meyer B.J. and Casson L.P. 1986. *Caenorhabditis elegans* compensates for the difference in X chromosome dosage between the sexes by regulating transcript levels. *Cell* **47:** 871–881.

Mukherjee A.S. and Beermann W. 1965. Synthesis of ribonucleic acid by the X-chromosomes of *Drosophila melanogaster* and the problem of dosage compensation. *Nature* **207:** 785–786.

Muller H.J. 1948. Evidence of the precision of genetic adaptation. *Harvey Lect.* **43:** 165–229.

Imprinting

Constancia M., Kelsey G., and Reik W. 2004. Resourceful imprinting. *Nature* **432:** 53–57.

Crouse H.V. 1960. The controlling element in sex chromosome behavior in *Sciara. Genetics* **45:** 1429–1443.

DeChiara T.M., Robertson E.J., and Efstratiadis A. 1991. Parental imprinting of the mouse insulin-like growth factor II gene. *Cell* **64:** 849–859.

McGrath J. and Solter D. 1984. Completion of mouse embryogenesis requires both the maternal and paternal genomes. *Cell* **37:** 179–183.

Morison I.M., Ramsay J.P., and Spencer H.G. 2005. A census of mammalian imprinting. *Trends Genet.* **21:** 457–465.

Nicholls R.D. and Knepper J.L. 2001. Genome organization, function, and imprinting in Prader–Willi and Angelman syndromes. *Annu. Rev. Genomics Hum. Genet.* **2:** 153–175.

Pachnis V., Brannan C.I., and Tilghman S.M. 1988. The

structure and expression of a novel gene activated in early mouse embryogenesis. *EMBO J.* **7:** 673–681.

Swain J.L., Stewart T.A., and Leder P. 1987. Parental legacy determines methylation and expression of an autosomal transgene: A molecular mechanism for parental imprinting. *Cell* **50:** 719–727.

Wilkins J.F. and Haig D. 2003. What good is genomic imprinting: The function of parent-specific gene expression. *Nat. Rev. Genet.* **4:** 359–368.

Reik W., Constancia M., Fowden A., Anderson N., Dean W., et al. 2003. Regulation of supply and demand for maternal nutrients in mammals by imprinted genes. *J. Physiol.* **547:** 35–44.

DNA Methylation

Amir R.E., Van den Veyver I.B., Wan M., Tran C.Q., Francke U., and Zoghbi H.Y. 1999. Rett syndrome is caused by mutations in X-linked *MECP2*, encoding methyl-CpG-binding protein 2. *Nat. Genet.* **23:** 185–188.

Holliday R. and Pugh J.E. 1975. DNA modification mechanisms and gene activity during development. *Science* **187:** 226–232.

Lewis J.D., Meehan R.R., Henzel W.J., Maurer-Fogy I., Jeppesen P., Klein F., and Bird A. 1992. Purification, sequence, and cellular localization of a novel chromosomal protein that binds to methylated DNA. *Cell* **69:** 905–914.

Riggs A.D. 1975. X inactivation, differentiation, and DNA methylation. *Cytogenet. Cell Genet.* **14:** 9–25.

Shen C.K. and Maniatis T. 1980. Tissue-specific DNA methylation in a cluster of rabbit β-like globin genes. *Proc. Natl. Acad. Sci.* **77:** 6634–6638.

Stein R., Razin A., and Cedar H. 1982. In vitro methylation of the hamster adenine phosphoribosyltransferase gene inhibits its expression in mouse L cells. *Proc. Natl. Acad. Sci.* **79:** 3418–3422.

Histone Code

Fraga M.F., Ballestar E., Paz M.F., Ropero S., Setien F., et al. 2005. Epigenetic differences arise during the lifetime of monozygotic twins. *Proc. Natl. Acad. Sci.* **102:** 10604–10609.

Firgola J., Ribas M., Risques R.-A., and Peinado M.A. 2002. Methylome profiling of cancer cells by amplification of inter-methylated sites (AIMS). *Nucleic Acids Res.* **30:** e28.

Hebbes T.R., Thorne A.W., Clayton A.L., and Crane-Robinson C. 1992. Histone acetylation and globin gene switching. *Nucleic Acids Res.* **20:** 1017–1022.

Jenuwein T. and Allis C.D. 2001. Translating the histone code. *Science* **293:** 1074–1080.

Turner B.M. 1993. Decoding the nucleosome. *Cell* **75:** 5–8.

Peterson C.L. and Laniel M.A. 2004. Histones and histone modifications (review). *Curr. Biol.* **14:** 546–551.

CHAPTER 9

RNA Interference Regulates Gene Action

Extraordinary advances had been made in understanding the intricate workings of the cell using recombinant DNA techniques, and so biologists were amazed when a completely unsuspected pathway for the control of gene expression was discovered in the 1990s. Researchers studying plants, fungi, worms, flies, and even trypanosomes came to realize that some unexpected results were the consequence of RNA molecules silencing the expression of genes in a process now known by the general term of *RNA interference* (RNAi). New examples of RNAi processes continue to be discovered but all share a common element—small RNAs that pair with other RNAs and possibly DNA, and silence gene expression through a variety of mechanisms, including mRNA degradation, DNA methylation and histone modification, and translational inhibition.

If RNAi were simply an unexpected new mechanism of gene regulation, it would still be a remarkable addition to our understanding of life, but the importance of RNAi extends beyond its biological significance. Just as the polymerase chain reaction (PCR) brought about a revolution in our ability to manipulate and detect nucleic acids, so RNAi is revolutionizing the ways in which we can experimentally control the expression of any gene. RNAi provides a general tool for *reverse genetics*—discovering the function of a gene by interfering with its activity and observing the resulting phenotype, rather than by finding an atypical phenotype and then trying to determine which gene is mutated. As we saw in Chapter 6, biologists have invented clever methods for making specific changes to genes, but many of these are laborious and slow. RNAi is simple and rapid; all that is needed is the sequence of just a portion of the gene to be targeted, the synthesis of the corresponding RNA molecule, and a method to introduce it into cells and organisms.

The technical advance of RNAi could not have come at a more opportune time. Large-scale sequencing is now producing vast amounts of genomic data that are being subjected to computational analysis to predict genes and other genetic elements. However, this is only the first step. As biologists we want to understand the functions of the predicted genes and their roles in the cell. Bioinformaticians are finding thousands—or tens of thousands—of genes. How can the experimentalists begin to make sense of them all? We will review other methods of genome-scale functional analysis in Chapter 13, but here we first examine the fascinating biology of RNAi and then look at how RNAi knockdown of gene expression in plants, yeast, mammals, and fungi provides a powerful tool for studying gene function.

Cosuppression of Transgenes Is Observed in Plants

The introduction of recombinant DNA techniques led to the modern biotechnology industry, promoted by endeavors to create commercially desirable alterations in organisms. Richard Jorgensen and his colleagues undertook the project to produce more richly colored petunia flowers. An additional copy of a petunia pigment gene encoding chalcone synthase was introduced into a petunia plant under the control of a very powerful promoter with the expectation that increased production of the encoded enzyme would produce flowers with a deeper purple color. Remarkably, some of the transgenic plants produced white flowers devoid of any pigment, despite the fact that both the endogenous and transgenic copies of chalcone synthase gene were present (Fig. 9-1). Other transgenic lines had variegated patterns of pigment expression—petals displaying pigmented and unpigmented areas. This surprising phenomenon was named *cosuppression*, because the transgene triggered not only its own silencing but also that of the endogenous chalcone synthase gene, which had the same sequence. This occurred even though the genes were driven by different promoters, thus showing that the effect required only homology between the messenger RNAs (mRNAs).

Interestingly, the cosuppression phenotypes were not stable. Differently patterned flowers appeared on separate branches of a single plant, and seeds grown from these newly patterned flowers sometimes reverted to the original phenotype of the parent plant. Occasionally, progeny of the variant flowers had entirely new patterns different from either the original plant or the variant parent flower. Similar phenotypic instability was also seen in progeny from transgenic plants having fully purple flowers; these could give rise to flowers that lacked pigment. However, the DNA sequence of cosuppressing transgenes was unchanged, even in parts of the plants with altered flowers. This phenotypic instability without change in sequence was similar to many of the unstable epigenetic phenomena we discussed in Chapter 8. Biologists therefore began to regard cosuppression as an epigenetic process.

Further analysis showed that the level of transgene transcription is an essential determinant of cosuppression. Transgenes under the control of strong promoters, transcribed at high levels, are much more likely to cause cosuppression than transgenes transcribed at lower levels. Also, the number of transgene copies affects the likelihood that cosuppression will occur. By crossing transgenic plants with themselves or to wild-type plants, investigators showed that the number of copies of the chalcone synthase transgene determined the cosuppression phenotype. For example, some petunias produced variegated flowers when just a single copy of the chalcone synthase transgene was present, but were pure white when the plants carried two copies of the transgene.

But biologists were using transgenic techniques to do more than produce prettier flowers—they hoped to create disease-resistant plants. Experiments with *Escherichia coli* expressing an altered form of a phage replicase protein had shown that the altered protein could compete with the infecting phage's own replicase and so prevent phage replication. Plant virologists planned to mimic this process in plants by expressing dominant negative forms of key proteins from plant viruses, in the expectation that these proteins would interfere with viral replication. Thus, transgenic plants were created that expressed a mutant form of viral replicase of potato virus X (PVX), and these transgenic plants were indeed

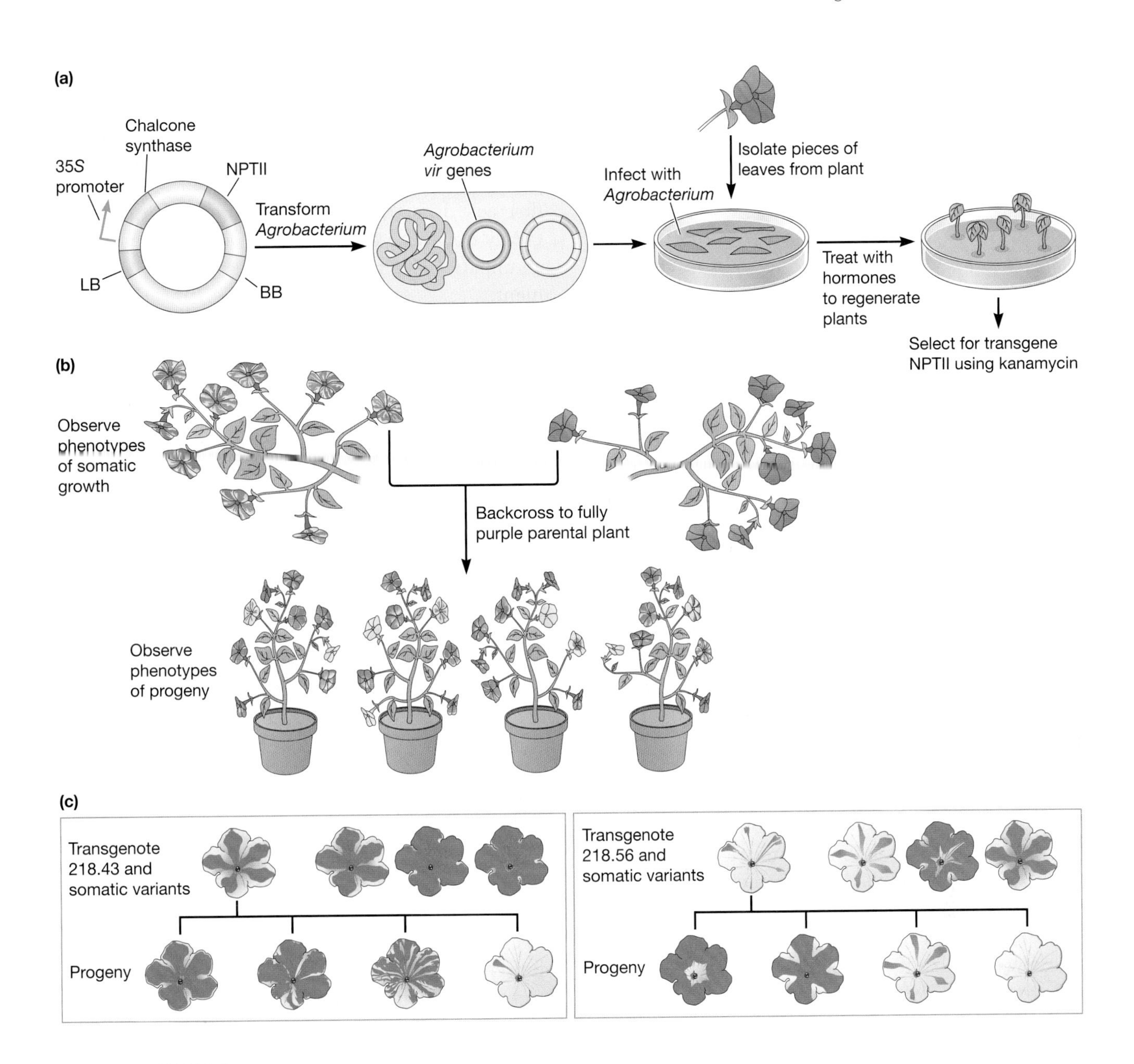

FIGURE 9-1

An overexpressed transgene triggers silencing of the homologous cellular gene. (*a*) In an attempt to create more brightly pigmented flowers, a cDNA encoding the pigment producing enzyme chalcone synthase was placed under the control of the very strong 35*S* promoter from the cauliflower mosaic virus. This construct was placed in a binary vector and transferred into purple petunias using *Agrobacterium*-mediated gene transfer. The binary vector also contained a gene encoding kanamycin resistance to permit selection of transgenic seedlings. (For more details on this method, see Chapter 6.) (*b*) Overexpression of chalcone synthase in petunia resulted surprisingly not in the production of deeply colored flowers but instead in plants bearing white or variegated flowers. To test the pattern of inheritance of the transgene, variegated transgenic progeny were backcrossed to the fully purple parental plant. (*c*) The flower phenotypes are unstable, both within a single plant (somatically) and through multiple generations.

resistant to the virus. The mechanism of viral replication is highly conserved among different strains of viruses, so virologists had hoped that this approach would interfere with the replication of many viral strains. However, these transgenic plants were resistant only to PVX and its very closely related strains. This was not the only surprising result of these experiments—remarkably, the most resistant lines expressed the transgene mRNA at nearly undetectable levels! Nor was the replicase protein needed for inducing resistance—transgenic lines made with an altered replicase gene carrying a frameshift mutation were as resistant as lines expressing full-length protein. It seemed that the infecting virus was being silenced by a mechanism similar to cosuppression.

At what level was cosuppression occurring? The cosuppressed endogenous genes were transcribed at normal levels, as shown in nuclear runoff assays of transcription (Fig. 9-2). Experiments isolating nuclei from cosuppressing plants and radiolabeling the transcripts clearly showed that the promoters of the transgenes and endogenous genes were active. However, the mRNA accumulated to very low levels, which suggest that cosuppression was the result of some kind of posttranscriptional gene silencing in which the mRNA was being destroyed. Other examples of cosuppression, involving a variety of endogenous and viral genes, were found, but cosuppression seemed to be an intriguing peculiarity of only the plant world. Then the unexpected results of an experiment in *Caenorhabditis elegans* revealed another case of unusual gene silencing, and it became clear that a new, perhaps universal, biological process had been discovered.

An Antisense Experiment Gone Awry Points the Way to RNAi in Worms

Embryonic development has fascinated biologists for centuries. How does a single-celled zygote divide and differentiate to produce an adult organism made up of many different cell types arranged in specific patterns? Developmental biologists study simple organisms to tease out the pathways leading from the zygote to the adult animal, and the nematode worm *C. elegans* has proved to be especially valuable.

Specific proteins and RNAs are distributed asymmetrically in the worm zygote. During cell division, different cytoplasmic components are partitioned into specific cells, eventually directing the differentiation of the many specialized cell types that make up the worm. This process can be disrupted, and there are many *C. elegans* mutants with altered or defective embryogenesis. For example, one such mutant, *par-1*, fails to make an asymmetric cleavage in the first cell division; instead, the first division occurs evenly down the center of the zygote, giving rise to an unpatterned embryo that arrests early in development.

The *par-1* mutation was mapped to a specific location in the worm genome, but in 1995, the worm genome had not yet been sequenced and the *par-1* gene had first to be cloned before its analysis could advance. At that time, a standard approach was used to determine which of the several predicted genes in a genetically mapped region was defective in mutants. In this method, genomic DNA or a cDNA corresponding to each candidate gene was first isolated and then expressed in *par-1* mutants. If the DNA was capable of restoring the normal phenotype, the sequence of that candidate gene was analyzed in the mutant. Unfortunately, neither complete genomic clones encompassing the region nor complete cDNA clones of the candidate *par-1* gene could be recovered from cDNA libraries, and so the identity of *par-1* could not be confirmed using the then-standard method.

Instead, Su Guo and Kenneth Kemphues used an antisense RNA to identify the *par-1* gene from among the possible candidates. For each candidate gene, they created a plasmid containing a portion of its cDNA and transcribed that in vitro to produce an antisense copy of the gene. The resulting RNA was then injected into the gonads of wild-type adult hermaphrodites. An antisense RNA is complementary to the mRNA produced from a gene and can decrease levels of the corresponding protein, presumably because the antisense RNA binds the mRNA and inhibits its translation or in some way destabilizes it. In this experiment, the progeny of worms injected with an antisense RNA to one of the candidate genes died with a phenotype very similar to that of the *par-1* mutants. Guo and Kemphues had identified the *par-1* gene!

However, the investigators were surprised at the result of a key control for the experiment. In this control, the worms were injected with the sense *par-1* mRNA, from a partial cDNA clone that could not be translated into a complete protein. Injecting the sense RNA into worms produced a phenotype in their progeny very similar to that seen with the antisense

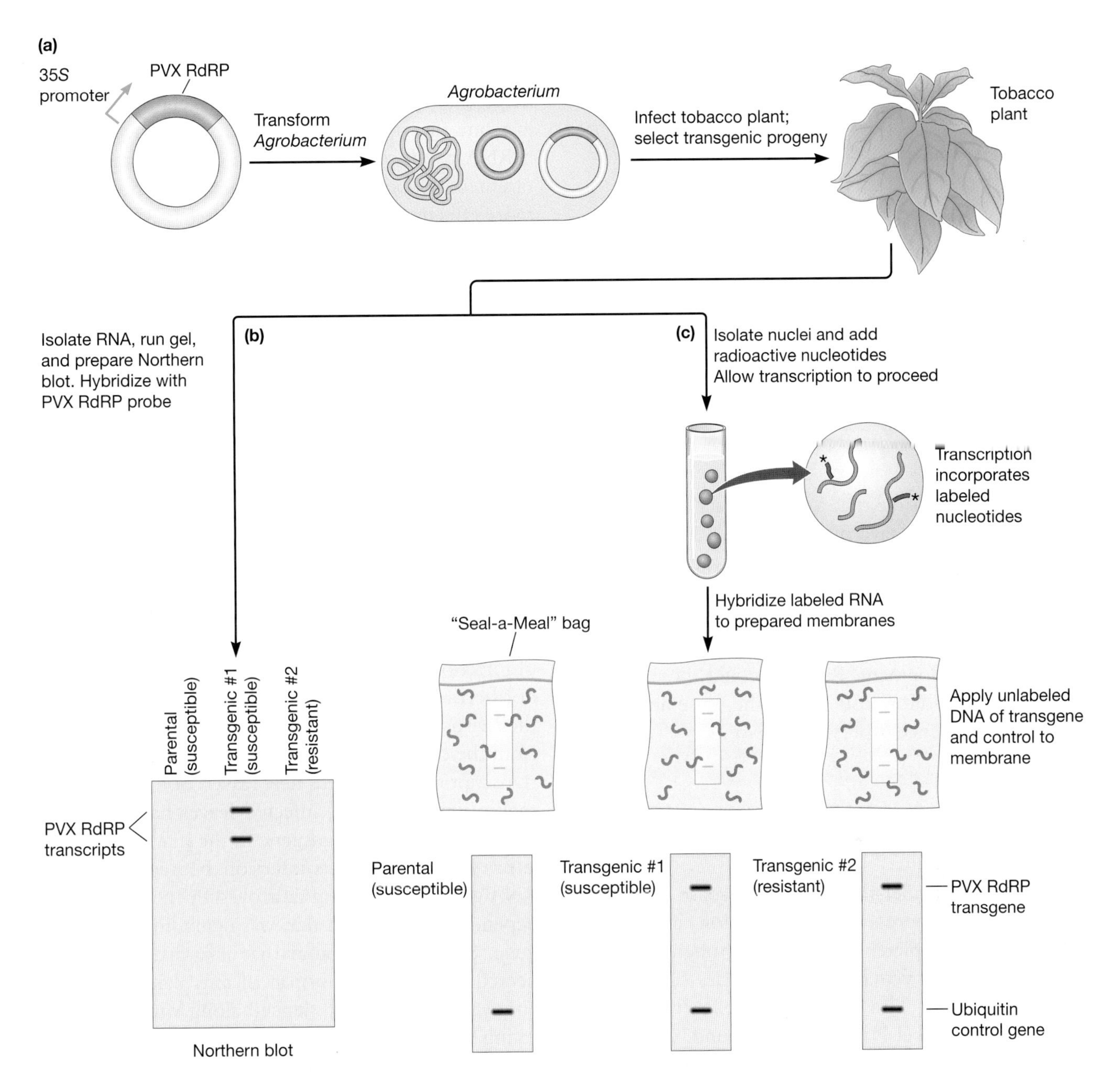

FIGURE 9-2

Nuclear run-on assays show that cosuppression occurs posttranscriptionally. (*a*) The RNA-dependent RNA polymerase (RdRP) from potato virus X (PVX) was placed under the control of a strong promoter, cloned into a binary vector, and transferred into tobacco plants by *Agrobacterium*-mediated gene transfer. Transgenic plants were isolated and tested for their resistance to PVX. (*b*) RNA was isolated from the plants and transgene expression was analyzed by Northern blotting. Surprisingly, those transgenic lines resistant to the virus did not accumulate the transgene transcript. (*c*) However, the transgene was in fact transcribed. Nuclei were isolated from the parental and transgenic plants, and radioactive nucleotides were added so that transcription could proceed. The labeled RNA was then hybridized to RNA immobilized on filters. (The highly expressed gene, ubiquitin, was assayed as a control.) Both susceptible and resistant lines expressed the transgene at equally high levels, thereby demonstrating that the transgene transcript was being eliminated in the resistant lines. This suggested that when the resistant lines were infected with PVX, the viral RNA was quickly degraded, blocking viral replication.

TABLE 9-1. Various sense and antisense RNAs were injected into *Caenorhabditis elegans* and scored for induction of embryonic lethality

Molecule injected	No. of worms injected	Embryonic lethality (%)
par-1 antisense	16	52
par-1 sense	12	54
Drosophila cofilin gene, antisense	8	0
Caenorhabditis elegans zygotic gene, antisense	8	0
Water	4	0

From Guo and Kemphues, *Cell 81:* 611–620 (1995).
Adult hermaphrodite worms were injected, and the number of hatched versus unhatched embryos was counted.

RNA. It was possible that the simple act of injecting the worms might have disturbed their development and, if so, injection of any RNA would cause the same phenotype (Table 9-1). Injection of sense and antisense RNAs from other genes had no effect, however, thus demonstrating that the act of injection did not affect development. But how could these results be interpreted? How could both the sense and antisense RNAs silence expression of *par-1*?

Double-stranded RNA Is the Trigger for RNAi

The search to understand how both sense and antisense RNA suppress gene expression in worms revealed a completely unsuspected phenomenon, one of the most remarkable discoveries of modern biology. Andrew Fire and Craig Mello (who were awarded the Nobel Prize in 2006) found, with their colleagues, neither sense nor antisense RNA alone was responsible, but both were needed. The in vitro transcription method produces a small amount of double-stranded RNA (dsRNA), and it is this product, not the much more abundant single-stranded sense or antisense RNA, that is the active agent in silencing. Used for injection, dsRNA was very effective in reproducing knockout phenotypes of many genes (Table 9-2). Only sequences derived from the mature RNA had this effect—sequences from the promoter or intronic regions of genes did not alter the phenotype of injected worms.

At about the same time, biologists studying post-transcriptional gene silencing of plant viruses realized that plants carrying *both* forward- and reverse-oriented transgenes corresponding to viral sequences were much more resistant to viral challenge than plants carrying a transgene oriented in just one direction. This observation suggested that dsRNA produced in the plants carrying transgenes oriented in both directions might silence RNAs more effectively than sense or antisense RNAs alone. A plasmid was constructed that contained an inverted repeat corresponding to a potato virus Y (PVY) protease gene essential for viral transmission (Fig. 9-3). When such sequences are transcribed, the resulting RNA can fold back on itself to produce a dsRNA. Plants with these inverted repeats were more effectively protected from infection by PVY than even those plants that carried separate sense and antisense transgenes.

Another clue that dsRNA might trigger silencing was the result of a chance observation made in experiments to make the transgenic tobacco plants resistant to the tobacco etch virus (TEV) by expressing the TEV coat protein as a transgene. When the transgenic plants were infected with TEV, there were signs of infection on inoculated leaves, but new leaves did not become infected and they were resistant to further virus infection. It was found that transcripts from the coat protein transgene were degraded post-transcriptionally in these new leaves. Somehow, virus replication in the old, infected leaves had triggered this silencing of the transgene in the new leaves. The latter were resistant to infection because the TEV RNA genome was being destroyed by the homology-dependent mechanism that was degrading the transgene. Soon, it was found that if a virus was engineered to include a portion of any plant gene, the endogenous copy was silenced along with the replicating virus.

One experiment made use of the phytoene desaturase gene, whose expression is easily tracked because leaves deficient in the phytoene desaturase enzyme are white. White leaves were observed whether the phytoene desaturase cDNA was present in the antisense or sense orientation in the virus. When this work was first done, a decade before the observations in worms, silencing was believed to be caused by antisense RNAs produced either directly (in the case of the antisense-oriented clone) or by the process of viral replication. However, the viral replication intermediate is a dsRNA, and once the role of dsRNA in RNAi in worms was apparent, biologists re-interpreted these and other related results from plant viruses. It was now clear that silencing is triggered by the double-

TABLE 9-2. Double-stranded RNA derived from a given gene induces phenotypes resembling a null mutation

Gene	Segment	Size (kb)	Injected RNA	F_1 phenotype
unc-22				*unc-22* null mutants: strong twitchers
unc22A	Exons 21–22	742	Sense	Wild type
			Antisense	Wild type
			Sense + antisense	Strong twitchers (100%)
unc22B	Exon 27	1033	Sense	Wild type
			Antisense	Wild type
			Sense + antisense	Strong twitchers (100%)
unc-22C	Exon 21–22	785	Sense + antisense	Strong twitchers (100%)
fem-1				*fem-1* null mutants: female (no sperm)
fem1A	Exon 10	531	Sense	Hermaphrodite (98%)
			Antisense	Hermaphrodite (98%)
			Sense + antisense	Female (72%)
fem1B	Intron 8	556	Sense + antisense	Hermaphrodite (>98%)
unc-54				*unc-54* null mutants: paralyzed
unc54A	Exon 6	576	Sense	Wild type (100%)
			Antisense	Wild type (100%)
			Sense + antisense	Paralyzed (100%)
unc54B	Exon 6	651	Sense	Wild type (100%)
			Antisense	Wild type (100%)
			Sense + antisense	Paralyzed (100%)
unc54C	Exon 1–5	1015	Sense + antisense	Arrested embryos and larvae (100%)
unc54D	Promoter	567	Sense + antisense	Wild type (100%)
unc54E	Intron 1	369	Sense + antisense	Wild type (100%)
unc54F	Intron 3	386	Sense + antisense	Wild type (100%)
hih-1				*hih-1* null mutants: lumpy-dumpy larvae
hih1A	Exons 1–6	1033	Sense	Wild type (<2% lpy-dpy)
			Antisense	Wild type (<2% lpy-dpy)
			Sense + antisense	Lpy-dpy larvae (>90%)
hih1B	Exons 1–2	438	Sense + antisense	Lpy-dpy larvae (>80%)
hih1C	Exons 4–6	299	Sense + antisense	Lpy-dpy larvae (>80%)
hih1D	Intron 1	697	Sense + antisense	Wild type (<2% lpy-dpy)

From Fire et al., *Nature 391:* 806–811 (1998).

Sense, antisense, and double-stranded mixtures of sense and antisense RNA from several different genes were injected into adult hermaphrodites. The progeny were examined and compared to null phenotypes of the given gene. dsRNAs targeted to exonic sequences generated mutant phenotypes, whereas promoter and intronic sequences did not result in abnormal phenotypes. This suggested that the RNA interference was acting on processed mRNAs and not pre-mRNAs or untranscribed sequences.

stranded replication intermediates rather than just the antisense strand. (There are plant DNA viruses that can trigger similar silencing even though they do not replicate through a dsRNA intermediate. The mechanism of this silencing remains a topic of debate.)

RNAi Is Used to Block Gene Expression in Many Organisms

No general methods have yet been developed for creating targeted mutations or knockouts in *C. elegans*, so finding mutants in a specific gene requires laborious screening of mutagenized populations of worms. In contrast, knocking down gene expression using RNAi is straightforward—all that is needed is the sequence of the target gene so that it can be used to synthesize a dsRNA. Fortuitously, RNAi was discovered just as the worm genome was being sequenced, and soon the sequences of all *C. elegans* genes were available. Very quickly, worm biologists seized on RNAi as a new and precise tool for controlling gene expression.

Initially, RNAi experiments were carried out just as the initial antisense experiments had been—by tedious injection of dsRNA into the gonads of adult hermaphrodites. But investigators soon discovered that RNAi worked equally well if worm larvae were

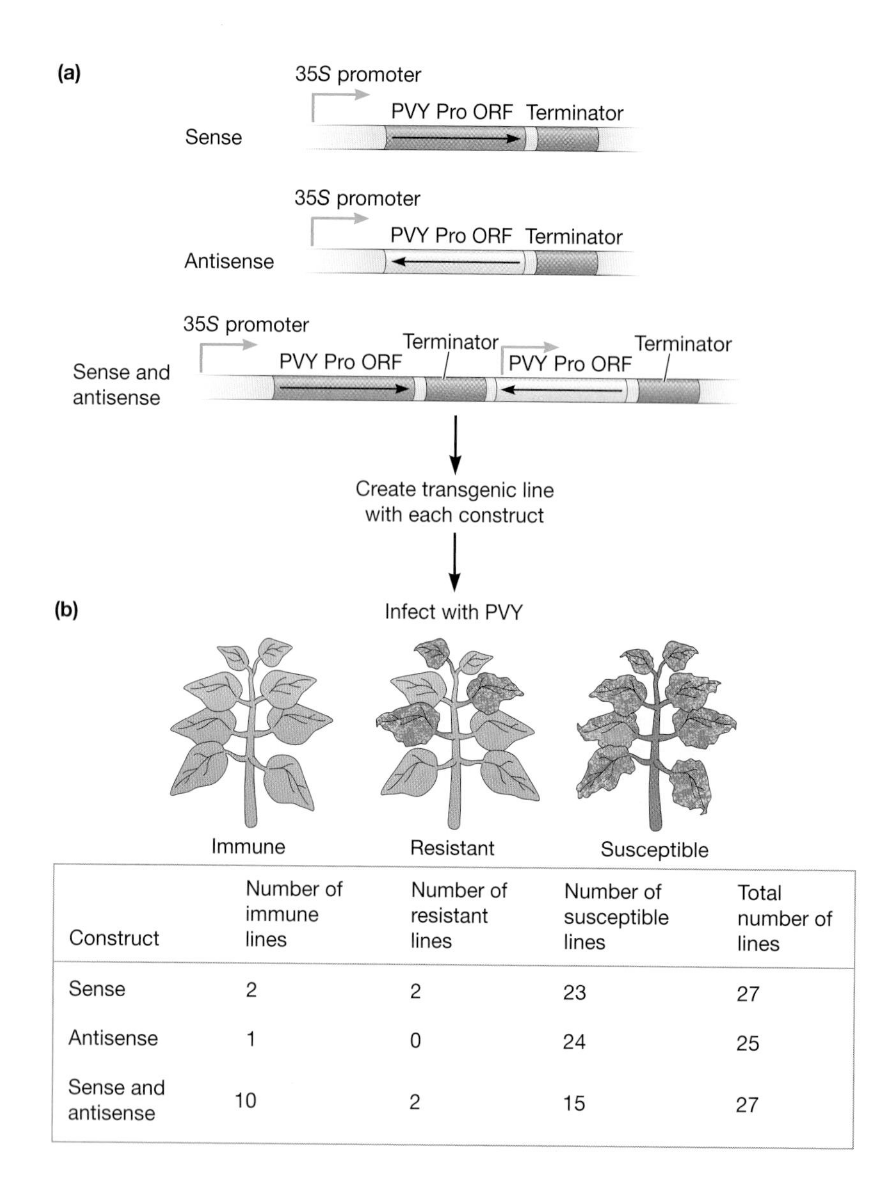

Construct	Number of immune lines	Number of resistant lines	Number of susceptible lines	Total number of lines
Sense	2	2	23	27
Antisense	1	0	24	25
Sense and antisense	10	2	15	27

FIGURE 9-3

Expressing an inverted repeat transgenic construct in plants suggests that double-stranded RNA (dsRNA) can initiate cosuppression. (*a*) Constructs expressing the protease gene of potato virus Y (PVY) were prepared that expressed a sense copy of the gene, an antisense copy, or both sense and antisense copies from the same construct. Each of these PVY protease expression constructs was integrated into a tobacco plant by *Agrobacterium*-mediated gene transfer. (*b*) Transgenic plants were challenged by infecting them with sap from PVY-infected plants and observed for symptoms of leaf mottling and leaf stunting. The construct expressing both sense and antisense of the protease gene was much more effective at producing transgenic plants immune or resistant to viral infection than the sense- or antisense-only constructs. ORF, open reading frame.

mixed directly with dsRNA or, even more remarkably, if worms were fed bacteria expressing dsRNA (Fig. 9-4)! Soon, worm biologists constructed strains of *E. coli* in which each strain expressed a dsRNA homologous to a different *C. elegans* gene. Now *E. coli* strains for feeding to worms are available that express dsRNAs corresponding to every gene in the worm genome.

RNAi has also been enthusiastically embraced by plant biologists. As in worms, there are no general tools for creating specific mutations in plants, but transgenes transcribed as inverted repeats are very effective at suppressing gene expression in plants. RNAi is an especially important tool in plants because many agriculturally important plants are polyploid. When a plant is tetraploid or hexaploid, it can be extremely difficult to isolate a plant with mutation in every copy of a gene. RNA interference silences *all* copies of a gene homologous to the transgene.

Because RNAi requires close sequence similarity between the target gene and the dsRNA, a dsRNA targeted to a particular gene can inhibit related genes. Therefore, the sequences of the dsRNA used for RNAi must be carefully selected so that they are identical only to the gene or genes of interest. This is typically not a problem. There are usually many silent changes in codon usage that make even closely related genes sufficiently different at the mRNA level to permit precise silencing. Even when a great deal of sequence similarity exists in the coding region of related genes, all is not lost. The 5′- and 3′-

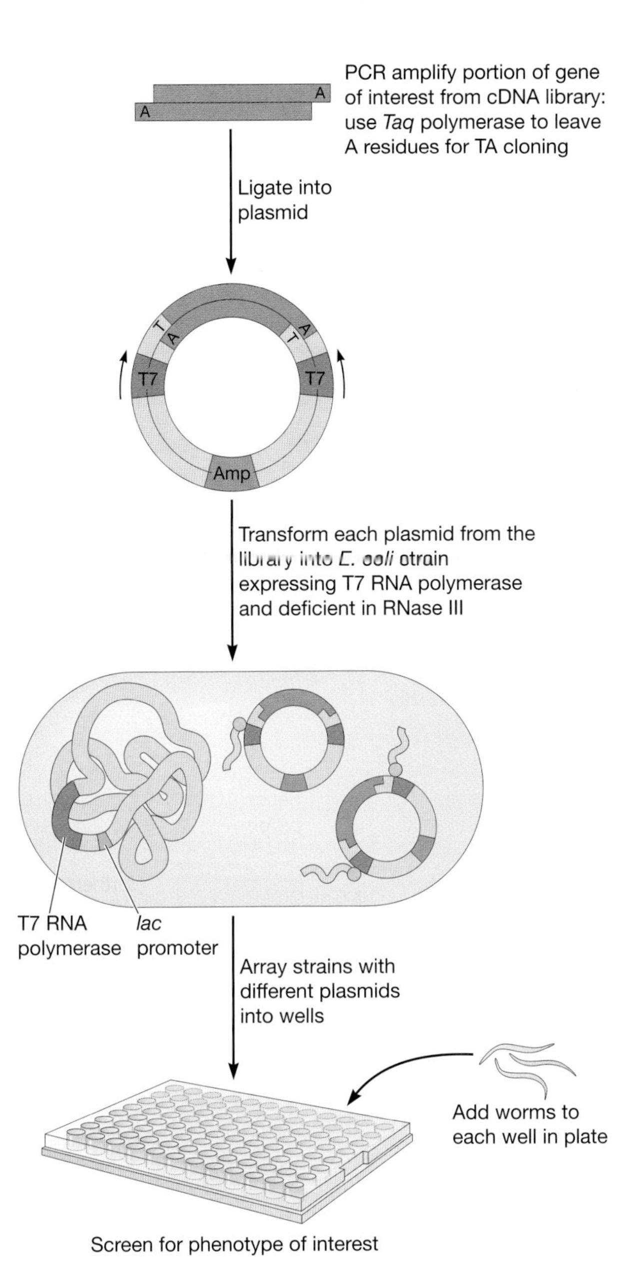

FIGURE 9-4

RNA interference can be induced in worms simply by feeding them *Escherichia coli* expressing dsRNAs. A DNA sequence from a gene is amplified by PCR using *Taq* polymerase to leave A overhangs for TA cloning (Chapter 6). The PCR products were cloned into a vector containing two T7 RNA polymerase promoters in inverted orientations flanking the cloning site. PCR products are ligated into a plasmid and transformed into a special strain of *E. coli* deficient in RNase III, a dsRNase. The strain also contains T7 RNA polymerase under the control of the *lac* promoter, so that it can be induced by addition of isopropyl thiogalactoside (IPTG). Individual strains of bacteria are prepared, each transformed with a plasmid containing a different gene. Each strain is plated on agar in a well of a 96-multiwell microtiter plate. The agar contains IPTG to induce expression of the dsRNA. Then, worms are added and allowed to consume the bacteria. The worms or their progeny are examined for interesting phenotypes.

untranslated regions of mRNAs are typically much less conserved between these genes and so can be used to target RNAi to a specific gene.

A Nuclease Destroys RNAs Homologous to the Silenced Gene

How does dsRNA silence genes? There were four lines of evidence that the silencing dsRNAs were interacting with mRNA. First, transcription of silenced genes typically proceeds at normal rates. Second, mRNA levels were nevertheless decreased in organisms undergoing silencing. Third, dsRNA (or transgenes in the case of plants) could cause silencing only when complementary to the exonic portions of genes, which suggests that the dsRNA was probably interacting in some way with mature mRNA. Finally, it was unlikely that any change had been made in the DNA sequence, because the RNAi effect declined over time.

Biochemical analysis of the mechanism of RNAi began with the use of extracts of *Drosophila* embryos and cells. dsRNA was incubated in these extracts and when RNA homologous to the dsRNA was subsequently added to the extracts, it was quickly degraded. RNAs not homologous to the dsRNA were completely stable. The protein machinery responsible for this sequence-specific nuclease activity was called the RNA-induced silencing complex (RISC). Any RNA could be targeted by RISC if its homologous dsRNA was present in the extract. *Drosophila* cells could even be pretreated with dsRNA. When extracts were prepared from these cells, they contained RISC activity corresponding to the transfected dsRNA.

Small RNAs Are Found in Plants Undergoing Silencing

Because dsRNA triggered the creation of the sequence-specific RISC nuclease, it seemed probable that RISCs would contain an antisense RNA that directed the nuclease to the homologous mRNA. Many investigators used Northern blots to search for these antisense RNAs but without success, as the elusive antisense RNAs were lost on the agarose gels used to separate RNA for standard Northern blot analysis. Agarose gels are excellent for separating RNAs of 500 nucleotides or more in size, but smaller RNAs diffuse rapidly and do not form tight bands, thereby evading detection.

Polyacrylamide gels provide much better resolution of small molecules. Andrew Hamilton and David Baulcombe used these gels to examine RNAs in plants undergoing RNAi and found a population of small RNAs homologous to the gene undergoing silencing (Fig. 9-5). These small RNAs were of uniform length, 21–25 nucleotides long, and were observed in plants where silencing had been triggered by single transgenes, by inverted repeats, and by the replication of RNA viruses. Both antisense and sense RNAs were present, indicating that dsRNA is the common element of all these silencing processes. Surprisingly, antisense RNAs were present even in those plants where cosuppression had been triggered by a single, sense-oriented transgene. We will see below that genetic evidence is beginning to explain how these antisense RNAs are produced.

These small RNAs were not unique to plants; they were soon observed in worms and flies undergoing RNAi. Their origins and functions were investigated by adding radiolabeled dsRNA to extracts from *Drosophila* embryos. The dsRNA was quickly digested into fragments 21–22 nucleotides long, and when RISC was purified from these extracts, the radiolabeled RNAs copurified with the RISC nuclease activity. Presumably, the small RNAs were guiding the RISC nuclease to homologous mRNAs.

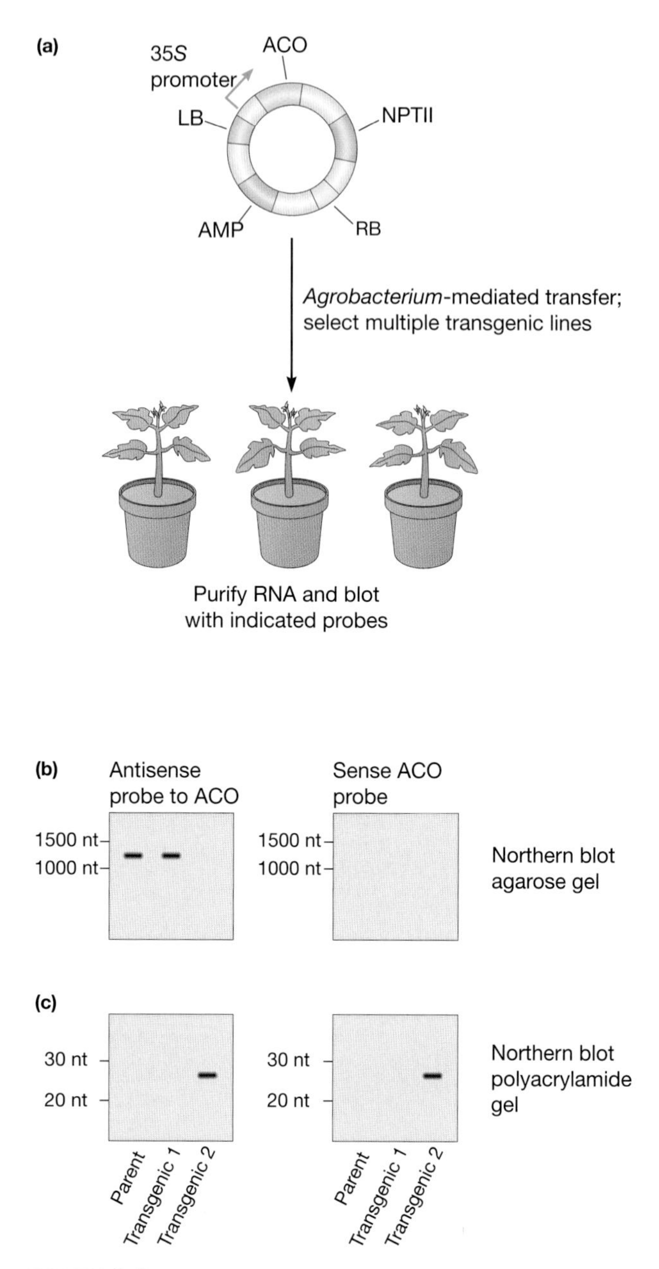

FIGURE 9-5

Small RNAs homologous to the silenced gene are present in plants undergoing silencing. (*a*) Transgenic tomato plants were constructed by *Agrobacterium*-mediated gene transfer to overexpress the enzyme 1-aminocyclopropane-1-carboxylate oxidase (ACO), whose gene is under the control of the very strong 35*S* promoter. (ACO catalyzes a key step in ethylene synthesis, a gas that is important for fruit ripening.) (*b*) Some lines exhibited silencing of the endogenous ACO mRNA, as observed by Northern blotting on agarose gels (the ACO transcript is roughly 1.2 kb). (*c*) However, when Northern blots were prepared from samples electrophoresed on polyacrylamide gels (which are capable of resolving very small RNAs), a population of 22–25-nucleotide RNAs was found. Curiously, both sense and antisense RNAs were found, even though the transgene initiating ACO silencing was present in the sense orientation only.

Dicer Chops Double-stranded RNA to Produce Small RNAs

The discovery of small RNAs in organisms undergoing silencing immediately led to a search for the gene or genes encoding the dsRNA RNases. This research was carried out in *Drosophila*, because the *Drosophila* genome sequence was nearly complete. Computer searches of predicted proteins identified proteins containing RNase III domains, the only type of RNases known to be capable of cutting dsRNA.

One approach to studying these candidate genes would have been to clone the genes, express them in bacteria, purify the proteins, and test their ability to chop dsRNAs into short RNAs. There are limitations to this approach. Not the least is that it presupposes that only a single protein is responsible for the nuclease activity, thus ignoring any cofactors that might be required for assembly of an active complex. A different approach is to add epitope tags to the cDNAs of the candidate proteins and transfect these tagged cDNAs into cells. Extracts prepared from the transfected cells are immunoprecipitated with antibodies recognizing the epitope tags. The expression and purification is performed in *Drosophila* cells that support RNAi. Thus, if some other accessory factor is needed in addition to the protein product of the candidate gene, it is likely to be immunoprecipitated along with the tagged protein. The experiment was performed using the candidate RNase III proteins. Immunoprecipitated proteins were tested, and one of the tagged candidates displayed a nuclease activity that cut long dsRNAs into small RNAs. These were equal in size to those observed in organisms and in purified RISC from in vitro experiments (Fig. 9-6). The candidate gene, a predicted *Drosophila* gene of unknown function, was named *Dicer* for the ability of its protein to chop double-stranded RNA into small, equally sized pieces. The initial identifica-

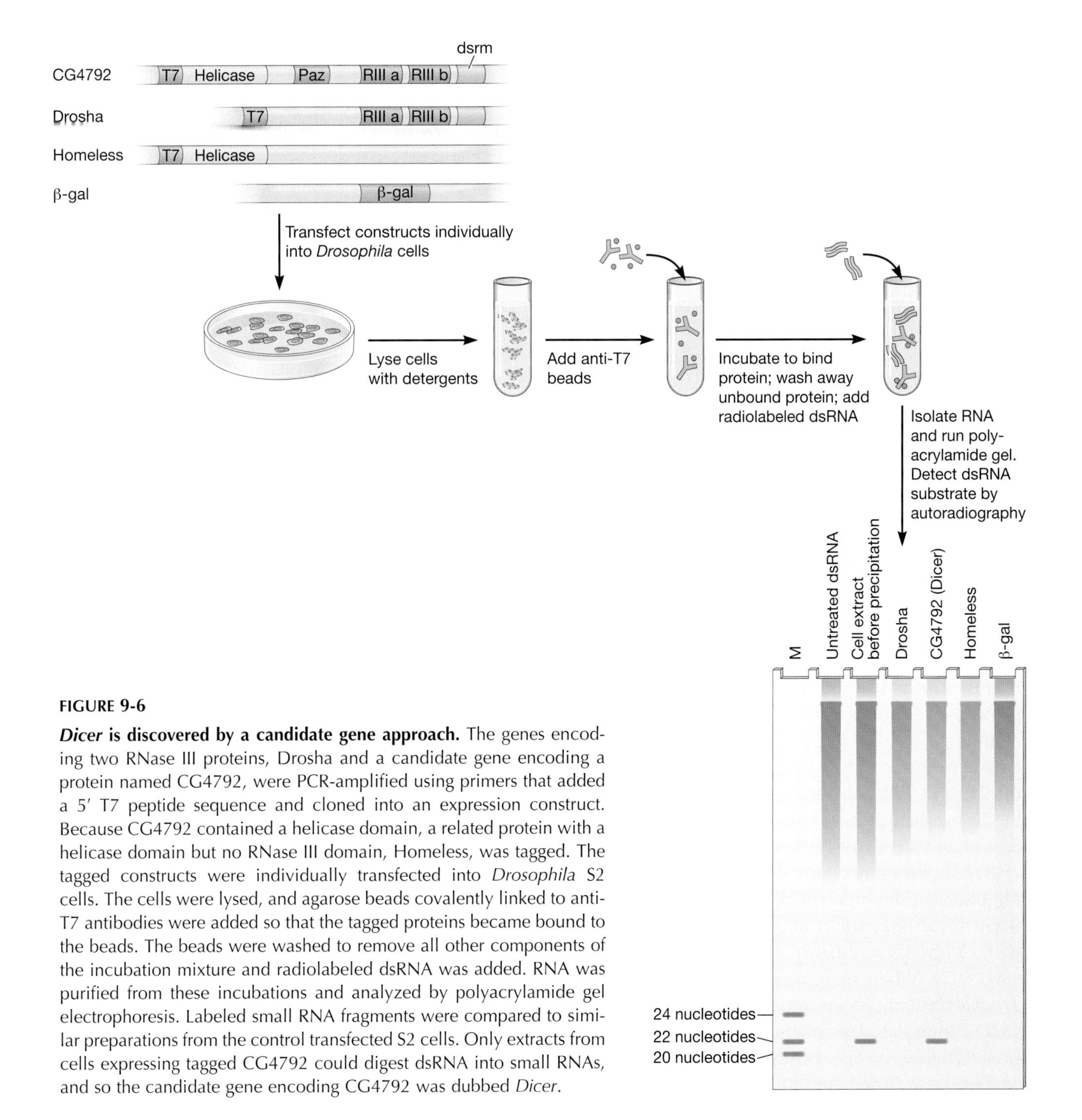

FIGURE 9-6

***Dicer* is discovered by a candidate gene approach.** The genes encoding two RNase III proteins, Drosha and a candidate gene encoding a protein named CG4792, were PCR-amplified using primers that added a 5′ T7 peptide sequence and cloned into an expression construct. Because CG4792 contained a helicase domain, a related protein with a helicase domain but no RNase III domain, Homeless, was tagged. The tagged constructs were individually transfected into *Drosophila* S2 cells. The cells were lysed, and agarose beads covalently linked to anti-T7 antibodies were added so that the tagged proteins became bound to the beads. The beads were washed to remove all other components of the incubation mixture and radiolabeled dsRNA was added. RNA was purified from these incubations and analyzed by polyacrylamide gel electrophoresis. Labeled small RNA fragments were compared to similar preparations from the control transfected S2 cells. Only extracts from cells expressing tagged CG4792 could digest dsRNA into small RNAs, and so the candidate gene encoding CG4792 was dubbed *Dicer*.

tion was carried out using immunoprecipitates that might have contained factors associated with Dicer. But it was later shown, by expressing and purifying the protein from a baculovirus system, that Dicer protein alone digests dsRNA.

With *Drosophila Dicer* in hand, biologists used computers to search for homologous sequences in other organisms. Some organisms have multiple *Dicer* family members—*Drosophila* has two, *Arabidopsis* has four, and *C. elegans* has but one. A search of the not-yet-complete human genome sequence revealed a copy of the *Dicer* gene. This finding suggested that RNAi might well occur in mammalian cells. Indeed, the protein specified by human *Dicer* cDNA was later found to have exactly the same double-stranded RNase activity as that of *Drosophila* Dicer.

Mimicking Dicer Products Induces RNAi in Mammalian Cells

Dicer produces small RNAs from a dsRNA, and these small RNAs are very similar to those observed in organisms undergoing RNAi, which suggests that Dicer is a key player in RNAi. One of the first questions asked was, what exactly is the nature of these small RNAs produced by Dicer? To answer this, dsRNAs of known sequence were digested in extracts of *Drosophila* embryos that contained abundant Dicer activity. The resulting small RNAs were purified, cloned, and sequenced. Because they were so small and their ends were unknown, the RNAs were cloned directly by using *RNA ligase*, an enzyme that ligates single-stranded RNA and DNA, to link primers to the ends of the small RNA. These primers were used for PCR amplification and cloning. These clones were sequenced, revealing that Dicer attacks dsRNAs from the ends, chopping off 21- and 22-nucleotide RNA molecules and leaving 3′ overhangs, 2 bp long, with a phosphate at the 5′ end and a hydroxyl group at the 3′ end.

Based on this information, synthetic RNA oligonucleotides were made to mimic Dicer products—these oligonucleotides were double stranded, with 2-bp 3′ overhangs and 3′-hydroxyl groups (Fig. 9-7). When these small RNAs were added to *Drosophila* embryo extracts, they worked nearly as well as long dsRNAs in triggering mRNA degradation. Therefore, the products of Dicer nuclease activity are sufficient to induce RNAi.

Long dsRNAs were very successful for inducing RNAi in flies and worms, but in most mammalian cells, dsRNA completely stops all protein translation. This is because dsRNA is recognized by antiviral defenses. These defenses likely evolved because many viruses have RNA genomes that replicate through a dsRNA phase. Yet, there were hints that RNAi might work to knock down gene expression in mammalian cells. Mammalian genomes contained homologs of many of the genes required for RNAi in flies, worms, and plants, including *Dicer*. Also, dsRNA could silence genes in embryonic cell types lacking the antiviral responses and in cells in which the dsRNA antiviral response was inhibited. But these seemed exceptional cases, and researchers feared that RNAi would not show the same potential for knocking out gene expression in mammals that had been demonstrated in plants, worms, and flies.

The enzyme protein kinase R (PKR) is a key trigger of the dsRNA translational arrest in mammalian cells. When PKR binds dsRNA, its kinase activity is activated. The activated PKR phosphorylates essential translation initiation factors, thereby blocking translation. The smallest RNA that can be bound by PKR is roughly 30 bp in length, so it seemed possible that synthetic 21-bp small RNAs resembling Dicer products might be small enough to avoid binding by PKR. If this were the case, then it might be possible to induce RNAi in mammalian cells. The reasoning was correct; when 21-nucleotide dsRNAs with 2-bp 3′ overhangs were transfected into mammalian cells, they decreased the level of the complementary mRNA without triggering a general translation arrest. These small RNAs were named short interfering RNAs (siRNAs).

Biologists around the world seized on this new method for directly testing the function of mammalian genes by transfecting the appropriate siRNAs into cells. Other researchers leapt into action to try to find ways to introduce siRNAs into cells to prevent or treat disease. It is difficult to coax normal mammalian cells into taking up DNA and RNA, but there has been recent success in preventing herpesvirus infection by treating the skin of mice with an antiherpes siRNA (Fig. 9-8). These applications of RNAi technology to answer research questions and solve health problems are exciting, but the basic biological questions remains. Why do organisms carry out RNAi? For what purpose did silencing by dsRNA evolve?

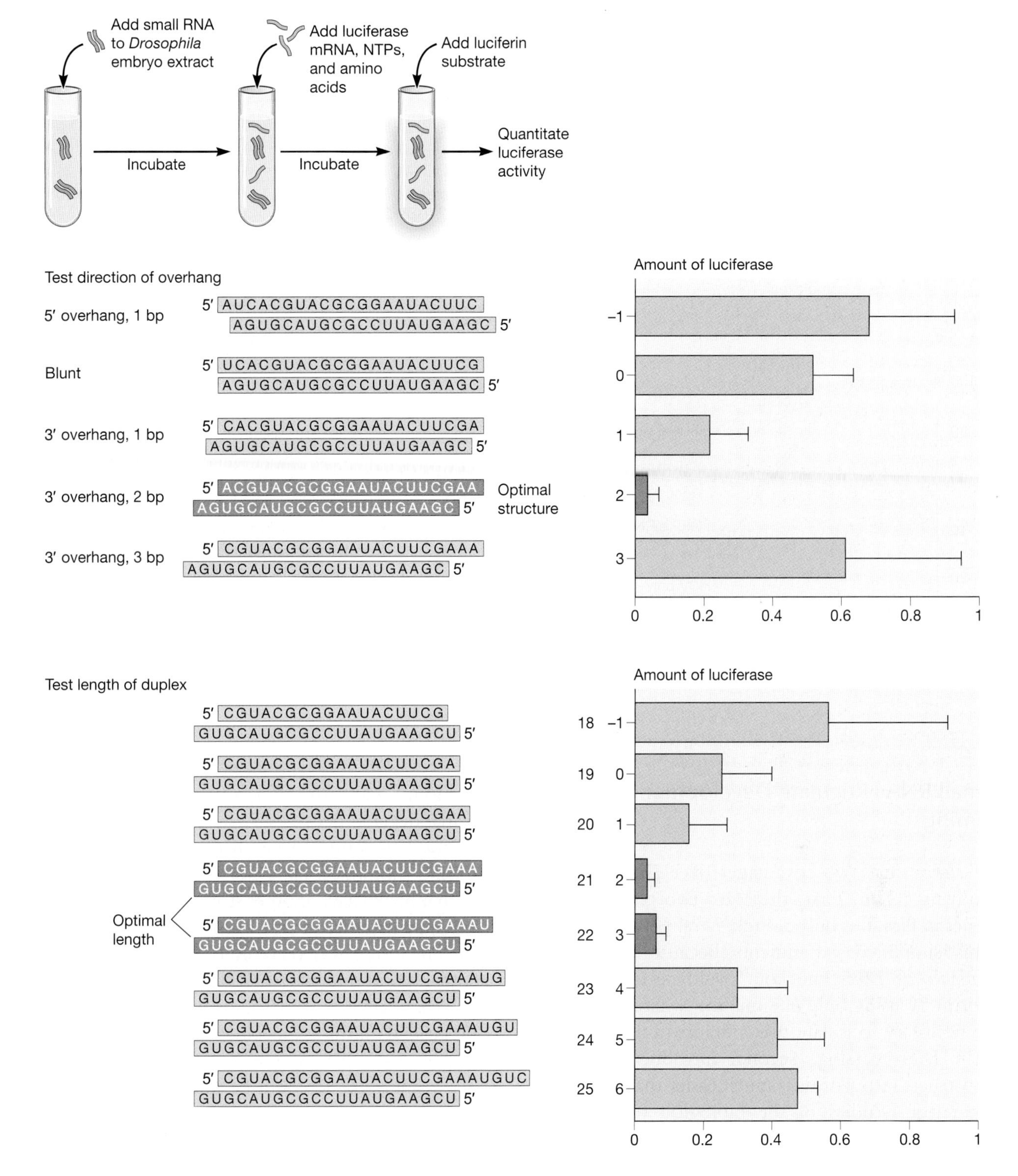

FIGURE 9-7

Synthesized siRNAs mimic Dicer products and trigger RNAi. Small RNA cloning experiments showed that the RNA fragments produced by Dicer were 22 nucleotides in length and, like the products of other RNase III enzymes, had a 2-nucleotide, 3′ overhang. Short siRNA duplexes corresponding to the gene encoding firefly luciferase were synthesized and added to a *Drosophila* embryo extract. After incubation, mRNA for firefly luciferase was added, and the incubation was continued. The extracts contain all of the factors necessary for translation of the luciferase mRNA into luciferase. The amount of luciferase made was assayed by adding luciferin, its substrate, and measuring the luminescence produced as luciferase hydrolyzed luciferin. Extracts with active RISC degraded the luciferase mRNA, and so there was less luciferase activity. Different small RNAs were tested to confirm the importance of both the type of overhang on the duplexes and the length of duplexes on the efficiency of RNAi.

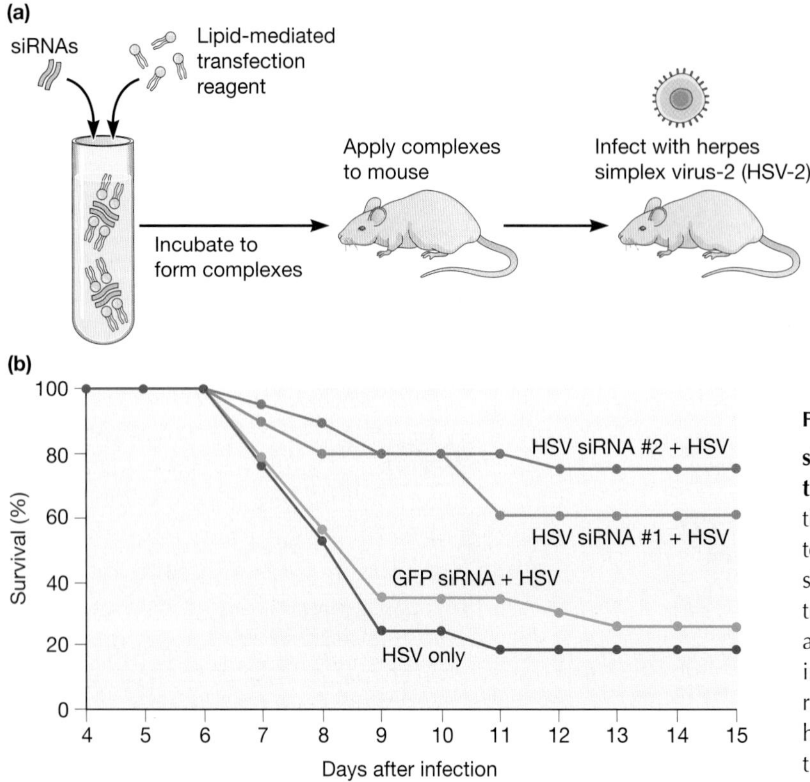

FIGURE 9-8

siRNAs are used to fight herpesvirus infections. (*a*) siRNA complexes corresponding to the mouse herpes simplex virus-2 (HSV-2) and to green fluorescent protein (GFP) were synthesized. The siRNAs were complexed with lipid transfection reagents and applied to the vaginal area of female mice. After 2 hr, the mice were infected with a lethal dose of HSV-2. (*b*) Mice receiving siRNAs targeting HSV-2 had a much higher survival rate than mice receiving a control siRNA or a mock treatment.

Small RNAs Regulate Developmental Timing

C. elegans has two advantages over *Drosophila* for studying RNAi. First, there are two related *Dicer* genes in flies, but only a single *Dicer* gene in worms. This simplified experiments because the consequences of *Dicer* knockout could be examined in worms by using only one mutant, whereas it would be necessary to isolate two mutants in flies, one for each *Dicer* paralog. Second, introducing dsRNA constructs into worms is much easier than into flies. Therefore, studies of the biological functions of *Dicer* were pursued most actively in *C. elegans.*

In these studies, the first question asked was whether *Dicer* mutants could still perform RNAi. To answer it, investigators injected wild-type and *Dicer*-mutant worms with dsRNA directed against a GFP (green fluorescent protein) reporter gene, present in both worms. In wild-type worms, the dsRNA suppressed GFP expression, producing worms that were not fluorescent. However, in the *Dicer* mutants, the dsRNA has no effect on GFP expression. These results confirmed that *Dicer* is essential for RNA interference in *C. elegans*, thus leading to a second question: What is the effect of the *Dicer* deletion on normal development?

Dicer was found to be an essential gene—worms carrying homozygous *Dicer* deletions are sterile. These mutants have another striking phenotype: Adult hermaphrodites frequently burst open from the vulva. This phenotype was very similar to that observed in another worm mutant, *let-7* (*lethal-7*), discovered several years earlier, which fails to produce the correct cells needed for the adult epidermis. The mutation was mapped to a small region of the worm genome, but no open reading frame could be identified at that location. Furthermore, no mRNA could be detected using that region of the genome as a probe in conventional Northern blot analysis. It was only when polyacrylamide gels were used that the *let-7* gene transcript—an RNA 21 nucleotides long—was discovered. The small size of the *let-7* RNA, and also that of another developmental gene *lin-4*, was tantalizingly similar to that of the small RNAs that had been observed just months earlier in plants undergoing silencing.

In *Dicer*-mutant worms, these small 21-nucleotide-long RNAs were much less abundant. Northern blots revealed instead a longer RNA containing the *let-7* and *lin-4* small RNA sequences.

The identity of this longer RNA became apparent upon inspection of the *let-7* and *lin-4* genes. They are encoded in the genome as palindromic sequences so that each transcribed RNA can fold back on itself to form a double-stranded region. This type of folded-back structure is often called a *hairpin* for its resemblance to the old-fashioned pins used to hold hairstyles in place. Dicer processes this hairpin structure by cutting it at the single-stranded loop to produce the small RNAs. It seemed unlikely that *lin-4* and *let-7* would be the only short hairpins in the genome—and so the race was on to find other RNAs that might be processed by Dicer.

Cloning Small RNAs Reveals an Unexpected Universe of microRNAs

RNAi could not have evolved just so biologists could conveniently knock down genes. What might its natural purpose be? Because RNAi requires sequence similarity between small RNAs and their targets, it was a good bet that identifying the small RNAs would reveal genes that are the targets of silencing. These targets, in turn, might help to predict the processes controlled by RNAi. The goal was then to clone genes encoding endogenous Dicer products and to determine which of these other genes were subject to RNA interference. Total RNA from worms, flies, and human beings was isolated and separated on polyacrylamide gels. For each of these organisms, small RNAs, approximately 21 nucleotides long, were cut from the gel, extracted, reverse transcribed, and cloned.

Among the first RNAs identified using this mass isolation strategy were *let-7* and *lin-4*, which showed that this approach was successful and was likely to capture similar RNAs. Unexpectedly, most of 21-nucleotide RNAs expressed in eukaryotes were very similar to *let-7* and *lin-4*. That is, rather than originating from conventional mRNAs, these small RNAs are also encoded in the genome as palindromic sequences. These new genes, with *let-7* as the founding member, were named *microRNAs*, or miRNAs. The palindromic sequences are transcribed as part of a longer precursor from which the 70–80-nucleotide palindromic regions are excised. These then form characteristic double-stranded hairpin structures that are cut by Dicer, and the resulting small RNAs are incorporated into RISCs. In mammals, the miRNAs are recognized in the nucleus by a protein complex called the Microprocessor. This complex contains a dsRNase related to Dicer called Drosha. The Microprocessor trims an miRNA from a long transcript down to its hairpin precursor and presents the hairpin to Dicer for processing. The Microprocessor pathway seems to be unique to animal miRNAs. In plants, miRNA precursors form much longer dsRNAs and are directly processed by Dicer without the help of a Microprocessor complex.

The terminology of small RNAs has become rather confusing (Table 9-3). The term siRNA has been adopted to describe several types of small RNAs, other than miRNAs, affected by RNAi pathways. These siRNAs include both synthetic small RNAs used for gene knockdown and any small RNA other than a miRNA that results from Dicer processing of dsRNA. The dsRNA may have been introduced into

TABLE 9-3. Small RNA nomenclature

Abbreviation	Term	Comments
siRNA	Short interfering RNA	Dicer products of double-stranded RNA. Also used to describe in vitro synthesized RNAs that mimic Dicer products and can be transfected to knock down cellular genes.
miRNA	microRNA	Cellularly encoded hairpin RNAs that are processed by Dicer (and, in animals, Microprocessor) and silence cellular genes.
shRNA	Short hairpin RNA	Synthetic RNA with a short inverted repeat. Designed to act as a Dicer substrate. Short hairpins may be transcribed as RNA or they can be encoded on plasmids.
ta-siRNA	*Trans*-acting siRNA	Short interfering RNAs, produced at one locus in the genome, which may silence other loci. Many ta-siRNAs are dependent on miRNA-directed cleavage of their originating transcript to set the phase of Dicer processing.
ra-siRNA	Repeat-associated siRNA	Short RNAs whose sequence corresponds to repetitive elements, typically those at centromeres or transposable elements.

cells by transfection, produced by replication of an RNA virus, or generated by transcription of inverted repeat structures in the genome. Several subcategories of siRNA have been named to specify their origin or mechanism of action. Typically, use of the term siRNA implies that the small RNAs are completely complementary to their targets, whereas miRNAs often do not match their targets perfectly. We shall see that this difference has important consequences for the mechanism of gene silencing.

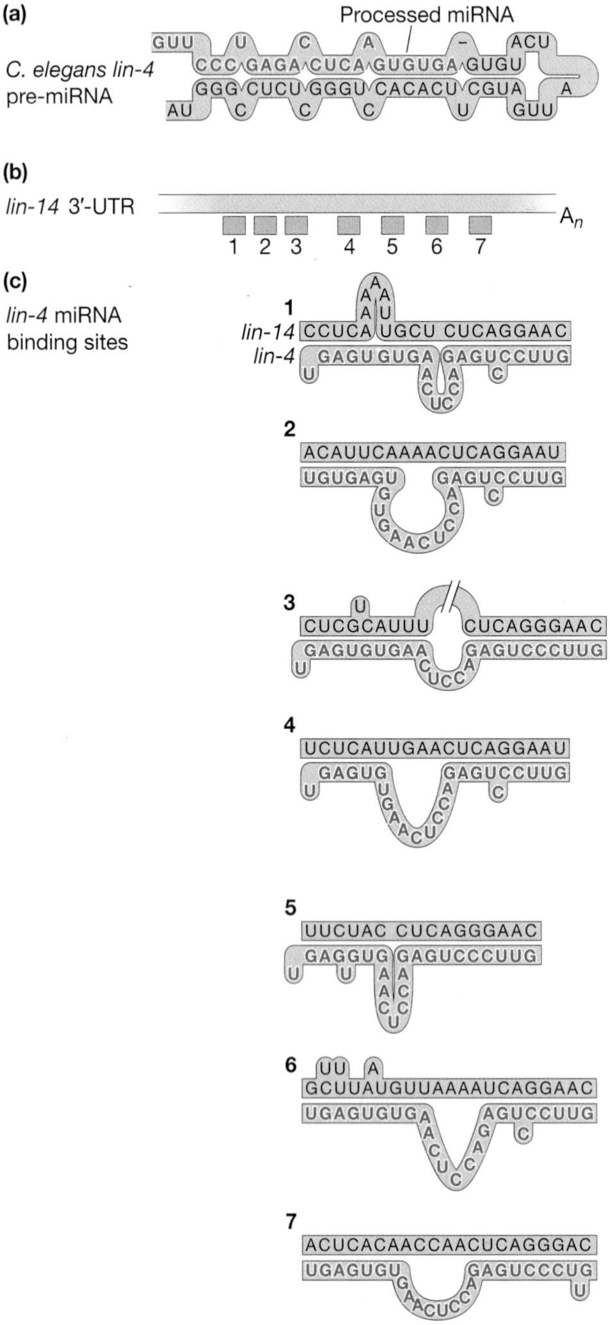

microRNAs Can Suppress Gene Function by Interfering with Translation

Even before the link between miRNAs and RNAi was made, biologists identified other *lineage* (abbreviated *lin*) genes that modified the developmental defects of *let-7* and *lin-4* mutant worms. These interacting genes are normally expressed only at specific phases of development, but in *let-7* and *lin-4* mutants, they are expressed at inappropriate times. When the *let-7* and *lin-4* small RNAs were identified and sequenced, alignments of their miRNAs with the sequences of the interacting *lin* genes revealed that the latter contained several sequences in their 3′-untranslated regions partially complementary to the small RNAs (Fig. 9-9). These potential binding sites immediately suggested that the miRNAs bind to their targets to inactivate them, and indeed *let-7* and *lin-4* are expressed at precisely the same time when targets such as *lin-28* and *lin-14* are silenced.

These binding sites on the target mRNAs are not perfectly complementary to the miRNAs, and mismatches between miRNAs and their targets make it difficult to use bioinformatics to predict target transcripts. In fact, many miRNAs in animals seem to bind primarily through a 6–8-nucleotide "seed" sequence at the 5′ end of the miRNA and have few other base-pairing interactions with their targets. How miRNAs can find such short targets accurately remains a mystery.

At the stage of development when the *lin-4* and *let-7* miRNAs are expressed, their target proteins

FIGURE 9-9

The microRNA *lin-4* has seven potential binding sites in the 3′-untranslated region (UTR) of its target gene, *lin-14*. (*a*) The *lin-4* miRNA is transcribed as a precursor that folds into a hairpin structure. The hairpin is trimmed from a longer precursor by Drosha and then cleaved by Dicer to produce the mature 21-nucleotide *lin-4* miRNA, indicated in *blue*. (The other strand is not stable.) (*b*) After the *lin-4* small RNA was identified, it became apparent that there were complementary sequences in the 3′-UTR of *lin-14* (*numbered boxes*). Geneticists examined *lin-14* because mutations in *lin-14* interact genetically with those in *lin-4*—*lin-14* mutants develop precociously, whereas *lin-4* mutants have slow development. A *lin-14* mutation masks the effect of *lin-4* mutation, which predicts that these genes are in the same pathway and that *lin-4* opposes the development-promoting activity of *lin-14*. (*c*) The *lin-4* binding sites in *lin-14* are not perfectly complementary to the miRNA and instead contain mismatches that create bulges in the base pairing between the miRNA and the target RNA.

disappear. However, the mRNA levels of the target genes are not greatly changed. This was completely different from the action of siRNAs, which direct the degradation of their mRNA targets. To investigate the sequence elements that affect the decision between transcript degradation and translational inhibition, biologists have tested siRNAs and miRNAs with a variety of perfectly matched and imperfectly matched target mRNAs. Perfectly matched siRNAs and miRNAs direct the slicing of their target mRNA opposite the 11th nucleotide of the bound small RNA. If the sequence of an siRNA is mutated so that there is a mismatch at the center of the siRNA, it no longer directs mRNA cleavage but instead suppresses translation (Fig. 9-10). Conversely, if a target is mutated so that it perfectly matches a miRNA, the target mRNA is then cleaved rather than being translationally suppressed. Therefore, near-perfect pairing between small RNA and mRNA is a key determinant of whether silencing comes about through mRNA degradation or translational suppression. Even when pairing is perfect, small RNAs do not always cause degradation of their target. Some perfectly matched miRNAs do not cut but instead silence through translational repression. Biologists are still working to explain what directs the choice of silencing mechanism.

How siRNAs and miRNAs affect translation is also the topic of intense study. The target genes often have multiple miRNA-binding sites, and mutation of even a single site compromises their ability to be properly silenced by the mRNA. miRNAs copurify with actively translating ribosomes, but these messages tend to be bound by relatively few ribosomes. This suggests that the miRNA prevents mRNAs from being translated at a high rate. Other data show that miRNAs are enriched in specialized regions of the cytoplasm (P bodies) from which ribosomes are excluded, and in which many RNases are present. Although most miRNAs do not direct cleavage of their targets, it seems that they may destabilize messages by redirecting them to the portions of the cytoplasm where mRNA is degraded.

Despite these remaining questions in explaining the mechanistic details of miRNA function, it is clear that miRNAs are important gene regulators. More than 2% of animal genes are miRNAs, and one-third of all protein-coding genes have predicted miRNA-binding sites within their messages. The abundance of miRNA molecules in the average cell exceeds the number of U6 RNAs that direct splicing!

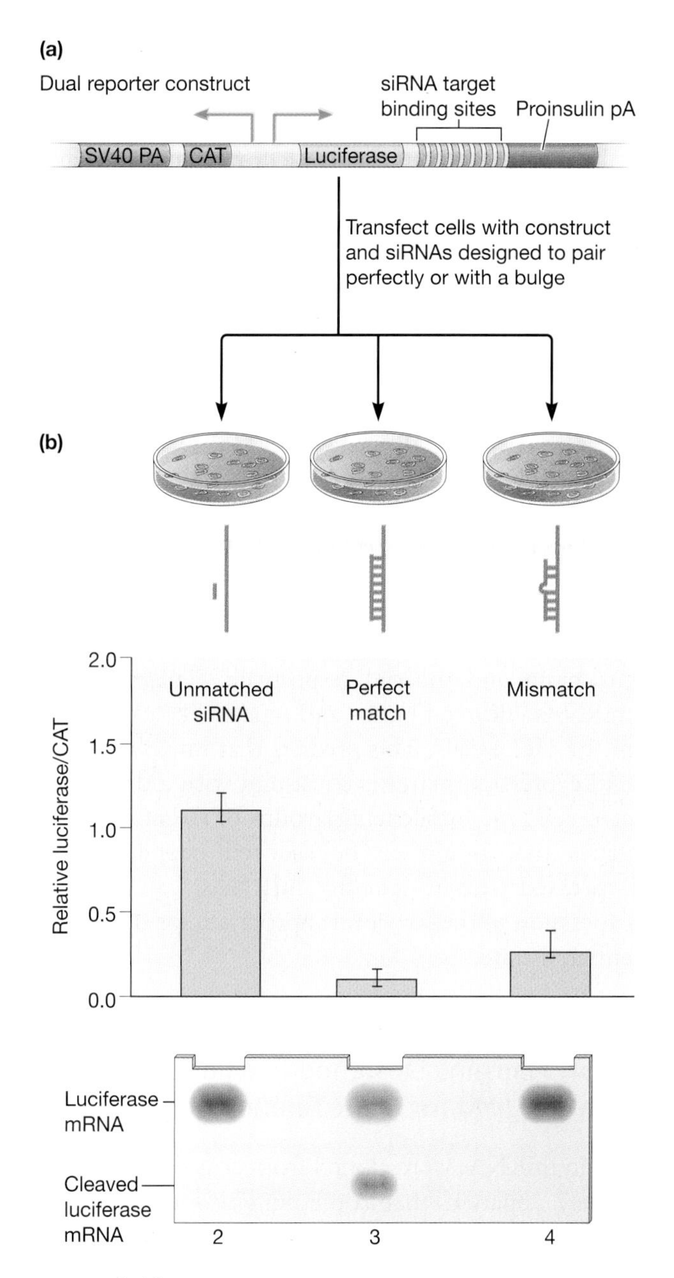

FIGURE 9-10

Mispairing an siRNA causes translational suppression, although mRNA degradation is blocked. (*a*) A dual reporter construct was designed containing the luciferase and chloramphenicol transferase (*CAT*) genes. The 3′-UTR of the luciferase gene contained eight copies of a 22-nucleotide sequence and small RNAs were designed to bind either perfectly or with a central mismatching sequence to the 22-nucleotide sequences in the luciferase-CAT reporter construct. (*b*) These small RNAs were transfected into cells along with the reporter construct. The expression of the reporter genes was measured, and Northern blots were used to determine whether the mRNA of the luciferase-CAT reporter construct had been cleaved at the siRNA-binding sites. Both perfectly matched RNAs and those with a central mismatch suppressed gene expression, but only the perfectly matched siRNA led to the cleavage of its partner mRNA.

In addition to their role controlling worm development, miRNAs regulate cell death in *Drosophila*, spatial patterning in plants, neuron patterning in the worm, patterning of leaves and flowers in plants, and the choice between T-cell and B-cell differentiation in white blood cells. We shall see in Chapter 15 that miRNAs can even affect cell proliferation and survival in cancer.

Detailed surveys of miRNA expression patterns are under way in a variety of organisms and cell types. As was observed with *let-7* and *lin-4*, many miRNAs change expression levels during cell differentiation. Some miRNAs are highly expressed in differentiated cells. When the human brain-specific miR-124 and muscle-specific miR-1 were each overexpressed in the relatively undifferentiated HeLa tumor cell line, most of the genes that were downregulated were those expressed at very low levels in the brain and muscle, respectively. Therefore the miRNAs enforced brain- and muscle-like expression in the HeLa cells. This predicts that miRNAs maintain expression patterns consistent with a differentiated cell type by silencing groups of target genes. If genes that ought to be silenced happen to be expressed, tissue-specific miRNAs can restore expression to the low level appropriate for the differentiated state.

Short Hairpins Designed to Mimic miRNAs Can Be Used for Gene Silencing

When miRNAs were cloned from cells, it was immediately apparent that something was missing—for most miRNAs, only one strand of the starting double-stranded pre-miRNA could be found. In contrast, cloning of siRNAs processed in vitro by Dicer always yielded RNAs from both strands of dsRNAs. How is it that Dicer cleaves the hairpin pre-miRNA so that only one strand survives? When the sequences of cellular miRNAs were examined and compared to their folded-back double-strand precursors, it was apparent that the surviving strand was less stably paired at its 5′ end. This leads to more efficient incorporation into RISC and similar complexes, and so this strand is protected from intracellular nucleases. Meanwhile, the other strand is released (in some cases, after being clipped) after the double-stranded siRNA or miRNA binds the RISC complex. Understanding what determines the choice of which strand of an miRNA is degraded has practical consequences in designing silencing RNAs. Only the siRNA strand antisense to the target mRNA can direct gene silencing, so siRNAs must be designed so that the strand antisense to the target mRNA is efficiently incorporated. As with miRNAs, weaker pairing at the 5′ end of a strand promotes incorporation into RISC. This knowledge has allowed the design of better siRNAs from which the strand antisense to the target message is efficiently assembled into RISC complexes.

Clearly, siRNAs knock down mRNA levels very effectively, but it is difficult to devise general methods by which the tiny, separate strands of an siRNA can be stably expressed within a cell. And, even if they are expressed, the chance of the two strands finding one another so that they can anneal to form a double-stranded siRNA is rather low. Consequently, most siRNA experiments are carried out by transfecting chemically synthesized siRNAs. This silences the target gene, but gene expression returns to normal as siRNA levels decline (siRNAs are diluted with each successive cell division).

A much more efficient way to knock down gene expression is to exploit miRNAs. These are transcribed as single units from genes and form short hairpins before processing by Dicer. Short hairpin RNAs (shRNAs) can be designed to mimic the structure of miRNAs and directly target cellular genes for degradation. Generally, shRNAs are designed to be processed by Dicer to produce an siRNA that perfectly matches its target mRNA, rather than mimicking the mismatched structures typical of miRNA–mRNA pairs. This design helps to ensure that the target mRNA is cleaved and destroyed, which inhibits gene expression far more efficiently than the translational suppression typically triggered by partially base-paired RNA. shRNAs are roughly 75 nucleotides in length and thus short enough to be synthesized as oligonucleotides. Several large-scale efforts have been directed toward constructing libraries of plasmids encoding shRNAs targeting every gene in the human genome. Just as libraries expressing dsRNA have been used for screens in *C. elegans* and *Drosophila*, shRNA libraries are being used for genetic screens in human cells, a feat never before feasible. Libraries of siRNAs have also been prepared, and both shRNA and siRNA reagents are commercially available, so that biologists can buy off-the-shelf reagents ready-made for knocking down their gene(s) of interest.

Slicer Uses siRNA to Cut the Target Message

A standard approach for investigating the steps in a biological process is to search for mutations that disrupt that process. In the case of RNA interference, geneticists began to look for mutations that disrupted dsRNA silencing of genes. One of the first genes they found was a member of the Argonaute family, named for the phenotype of *Arabidopsis* plants carrying mutations in one of the family members. Argonaute is the Greek word for squid, which these tiny, stunted plants resemble. Soon after, scientists studying the process of quelling, a type of transgene-induced gene silencing in the fungus *Neurospora*, mapped a quelling defective mutant to the *qde-2* gene. This gene is also a member of the Argonaute family. The cloning of Argonaute proteins in plants, worms, and fungi was followed by the biochemical identification of Argonaute proteins in RISC complexes. The severe RNAi defects of Argonaute family mutants and the common presence of Argonaute proteins in RISC complexes from multiple organisms led to the conclusion that these proteins were responsible for essential steps in RNAi. Computer programs were used to search genome sequences, and the results revealed that Argonaute proteins were present in many organisms. These proteins contain PIWI and PAZ domains, neither of which had been studied in any detail. The PIWI domain was unique to Argonaute proteins, and PAZ was found only in Argonaute and Dicer. The PIWI and PAZ domains had no obvious similarity to any other known domains.

So, what were the Argonaute proteins doing in the RISC complex, and what function did the mysterious PAZ and PIWI domains have? There was one key step of RNAi that remained unexplained—what protein was the nuclease that destroyed the mRNA? Biochemical studies revealed the activity in the RISC of a nuclease, called Slicer, which cut the mRNA precisely opposite the center of the siRNA. Could the Argonaute protein present in RISC be the Slicer nuclease?

The answer came from X-ray crystallography of the Argonaute protein, which revealed the key roles of the PIWI and PAZ domains. The DNAs for the two domains were cloned, expressed in bacteria, and purified for X-ray crystallography. The PAZ domain, which was solved first, turned out to be ideally suited for binding the 3′ end of the siRNA. This made sense because Dicer, which also interacts with siRNAs, has a PAZ domain. This left the PIWI domain unexplained, but, once crystallized, it was obvious that the PIWI domain is closely related to the well-known RNase H family of nucleases that cut RNA–DNA hybrids. Rather than cutting RNA–DNA hybrids, the PIWI domain cuts RNA–RNA hybrids. Comparison of the crystal structures of RNase H and the PIWI domain suggested which amino acids of the domain were likely to be important for catalyzing RNase activity. When key residues within the PIWI domain of Argonaute were changed using site-directed mutagenesis, the mutant Argonaute proteins lost slicing activity. These results showed definitively that the PIWI domain of these proteins was responsible for slicing. Interestingly, although some members of the Argonaute family of proteins have mutations of their PIWI domains, the overall structure and RNA-binding capacity of these proteins is conserved. This observation suggests that these Argonaute proteins play roles in gene regulation that do not involve slicing mRNAs and would account for the many Argonaute family members—5 in *Drosophila*, 21 in worms, 7 in human beings, and 10 in *Arabidopsis*—each playing a different role.

Our current understanding of RNAi and the molecules involved is illustrated in Figure 9-11. The principal gaps in our knowledge are what role the many different Argonaute proteins play in the cell, the mechanisms by which siRNAs direct histone and DNA methylation, precisely how miRNAs inhibit translation of target messages, and how miRNAs identify their very degenerate target sequences.

Fragile X Mental Retardation Protein Is a Component of RISC

One of the most satisfying circumstances in scientific research occurs when connections are made between apparently unconnected phenomena. Who, for example, would have thought that the esoteric process of RNAi would be associated with a human genetic disorder that has been studied for many years?

Genetic studies revealed genes involved in RNAi, but the biochemical purification of RISC uncovered other players as well. The proteins that copurify with RISC have implicated many other genes that may have roles in RNAi. One of the surprising connections made using this biochemical approach linked

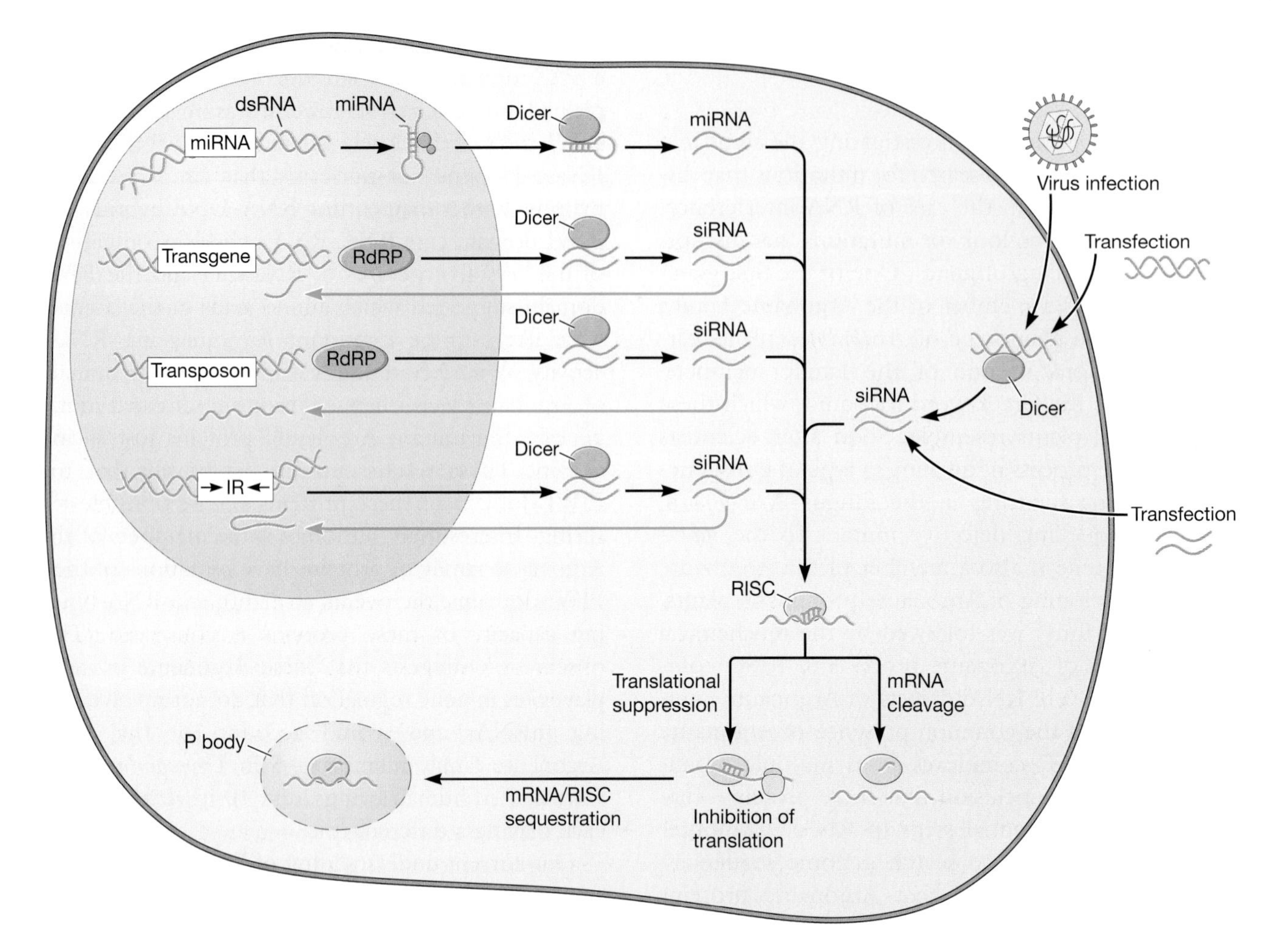

FIGURE 9-11

Overview of RNAi. Silencing can occur at the transcriptional, posttranscriptional, and translational levels. There are several sources of double-stranded RNAs. Some are made in the nucleus and exported to the cytoplasm for processing by Dicer. Hairpin-structured pre-miRNAs are processed from their long pre-miRNA, whereas RNA-dependent RNA polymerase (RdRP) converts single-stranded RNAs to double-stranded RNAs in plants, *Caenorhabditis elegans*, and the yeast *Schizosaccharomyces pombe*. When transcribed, inverted repeats (IRs) produce palindromic RNAs that fold back on themselves, forming dsRNA. Double-stranded RNAs may be introduced directly into the cytoplasm for Dicer processing through infection by an RNA virus, by systemic transport processes in *C. elegans*, or by transfection. The Dicer processing events are shown separately because in some organisms, different Dicer family members specialize in processing particular substrates. The small RNAs produced by Dicer are incorporated into RISC. If the siRNAs find a complementary mRNA, the latter is cleaved if the siRNA is fully complementary to the mRNA sequence. Otherwise, the RISC-mRNA complexes undergo translational inhibition within P bodies in the cytoplasm, sites of RNA turnover. *Blue arrows* indicate situations in which small RNAs can signal heterochromatin formation (which, in many organisms, is accompanied by DNA methylation). This diagram includes RNAi pathways discovered in many different organisms. Not all pathways or particular components are present in all organisms.

the fragile X mental retardation protein (FMRP) to RISC complexes. *FMR1* is an X-linked gene that is the cause of the most common inherited form of mental retardation (Chapter 14). ("Fragile" describes the appearance of the mutant X chromosome when prepared for cytogenetic analysis; the X chromosome is constricted because of sequence amplification at the mutation locus that results in the transcriptional inactivation of *FMR1*.) Patients with fragile X mental retardation do not express FMRP and have a variety of developmental defects as well as mental retardation. The fragile X protein binds to mRNAs and, in cells lacking this protein, many genes are misexpressed. This finding suggests that fragile X is needed to regulate gene expression, perhaps by directly binding mRNAs.

There is a *Drosophila* homolog of the human FMRP gene, and flies with a mutant FMRP have too many synaptic connections between neurons and muscles. In studying the proteins that bind to the *Drosophila* homolog of *FMR1*, one group of researchers discovered that an Argonaute protein was present in the complex. Another group made the same observation using a different approach, by purifying *Drosophila* RISC and finding *Drosophila* FMRP to be one of the proteins associated with the complex. Furthermore, flies heterozygous for both FMRP and one of the Argonaute family members exhibit a much more severe phenotype than that resulting from the *FMR1* mutation alone, as would be expected if the RNAi pathway were involved. Applying these findings to human fragile X syndrome, scientists found that human FMRP protein is bound to Argonaute proteins. The discovery of a link between RNAi and FMRP provides a hint to one mechanism through which FMRP could regulate genes.

Some Single-stranded RNA Becomes dsRNA

Silencing RNAs were first discovered in plants undergoing cosuppression as a consequence of overexpression of single, sense-oriented transgenes. However, in many of these cases, RNAs both sense and antisense to the transgene were present, even though the trigger of cosuppression was a sense transgene. What was the source of these antisense siRNAs?

The answer was revealed in a search for mutants that affected cosuppression. A screen for plants that had lost the ability to cosuppress single-copy transgenes identified one mutant having a disrupted gene homologous to those encoding RNA-directed RNA polymerases (RdRPs). RdRPs use RNA rather than a DNA template to make RNA, but it was not clear why they were needed in eukaryotic cells where RNA is typically transcribed from DNA. These experiments confirmed that RdRP is involved in transcribing or amplifying mRNAs to produce a double-stranded product.

One of the consequences of RdRP amplification of silencing is *spreading*, in which silencing affects nearby sequences (Fig. 9-12). Silencing is triggered by an overexpressed sequence located at one site in the genome, but over time siRNAs appear that correspond to sequences that flank other regions in the genome homologous to the sequence that triggers silencing. In RdRP mutant plants, spreading does not occur. One model for spreading is that the RdRP uses siRNAs as primers to extend along the flanking sequences, thereby producing dsRNA that can then be targeted by Dicer. These new siRNAs enter RISC complexes and recruit RdRP to the original locus or to other similar sequences in the genome. These results helped resolve a long-standing puzzle; the presence of RdRPs in plants was known since the 1970s, but their function remained unknown. Spreading was originally discovered by studying cosuppression, but it was later discovered that certain miRNAs in plants direct the production of siRNA from other target loci. For example, siRNAs produced from loci on *Arabidopsis* chromosomes 1 and 2 are generated under the control of the same miRNA (Fig. 9-13). These siRNAs then target other complementary transcripts for degradation and so are called *trans*-acting siRNAs because they act on other genes. The *trans*-acting siRNAs are apparently produced by Dicer cleavage of a double-stranded transcript initiated from the site at which the perfectly complementary miRNA directs RISC cleavage. RdRPs are required for *trans*-siRNA production, which suggests that they are needed to produce the double strand.

A similar spreading phenomenon occurs in worms and, as might be predicted from the work in plants, RdRP mutants in worms show defects in RNA interference. However, spreading in worms occurs only in the 5′ direction from the silenced gene, whereas it is bidirectional in plants. It is still unclear why this is so. Curiously, no RdRPs are evident in the genomes of flies or mammals. This absence of RdRPs is comforting to those attempting to use RNAi to silence fly or mammalian genes, because silencing cannot spread into adjacent sequences that might trigger silencing of genes other than the desired target. But it is perplexing that RdRPs have been lost in flies and mammals.

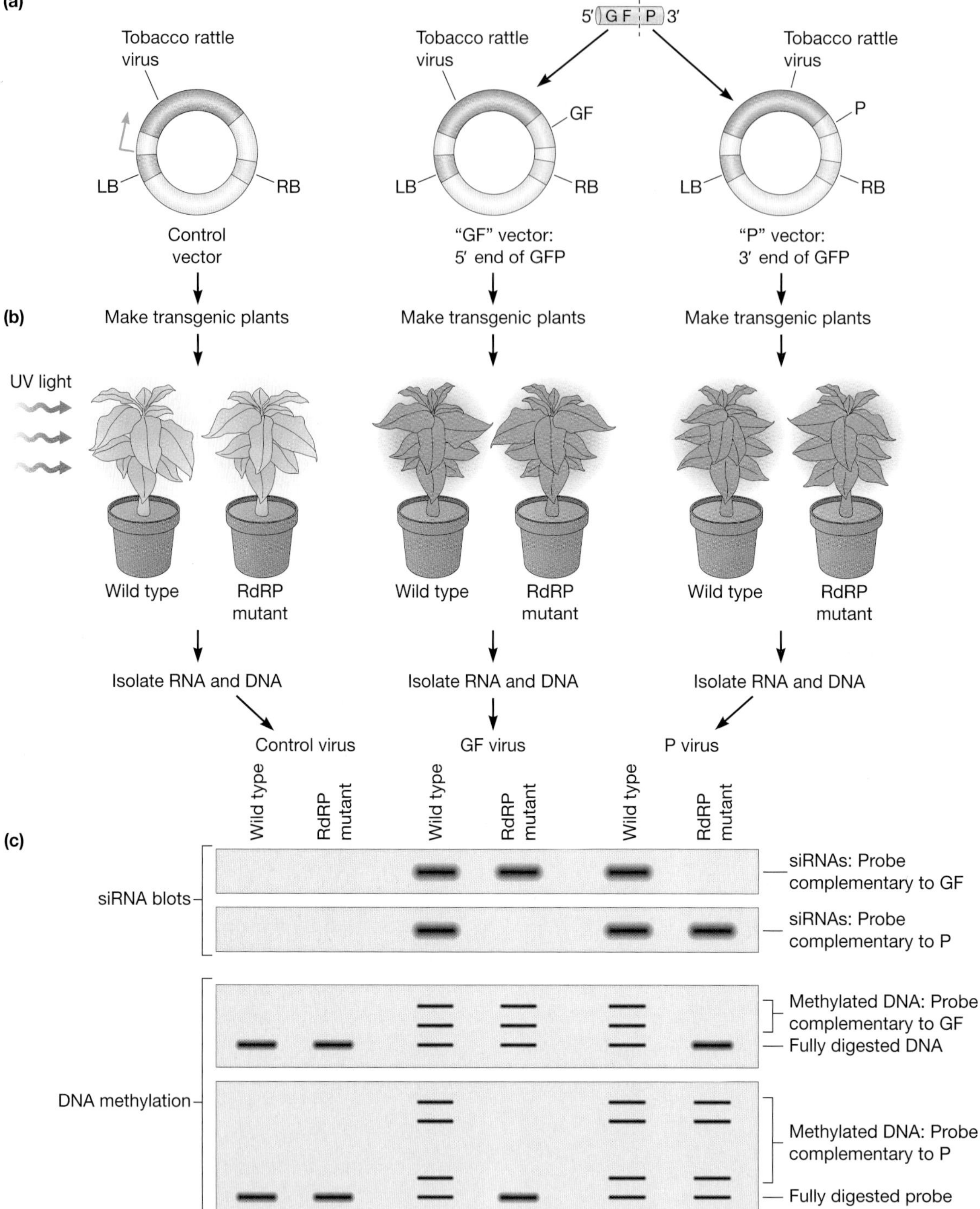

FIGURE 9-12

RNA silencing spreads from a dsRNA-targeted region to adjacent sequences through the action of RNA-directed RNA polymerases (RdRPs). (*a*) Vectors for *Agrobacterium*-mediated gene transfer were prepared that could be transcribed in vivo to produce the tobacco rattle virus. Vectors included a portion of the green fluorescent protein (GFP)—either the 5′ (named GF) or the 3′ (named P) portion. (*b*) The vectors were transformed into *Agrobacterium* and used to infect young tobacco plants that contained a GFP transgene and were either wild-type or RdRP mutants. The virus replicates through a dsRNA intermediate, which can act as an siRNA. Suppression of GFP expression was assayed by illuminating the leaves with UV light; GFP-expressing cells are green, whereas cells without GFP are red from the fluorescence of chlorophyll. Viruses containing either the GF or the P sequence induced GFP silencing. (*c*) In infected wild-type plants, siRNAs corresponding to the entire GFP gene were present, even though only a portion of GFP was included in the replicating virus. Also, DNA methylation appeared throughout the entire GFP transgene in wild-type plants. In RdRP-mutant plants, both siRNA and DNA methylation were limited to the sequences corresponding to the silencing trigger present in the virus.

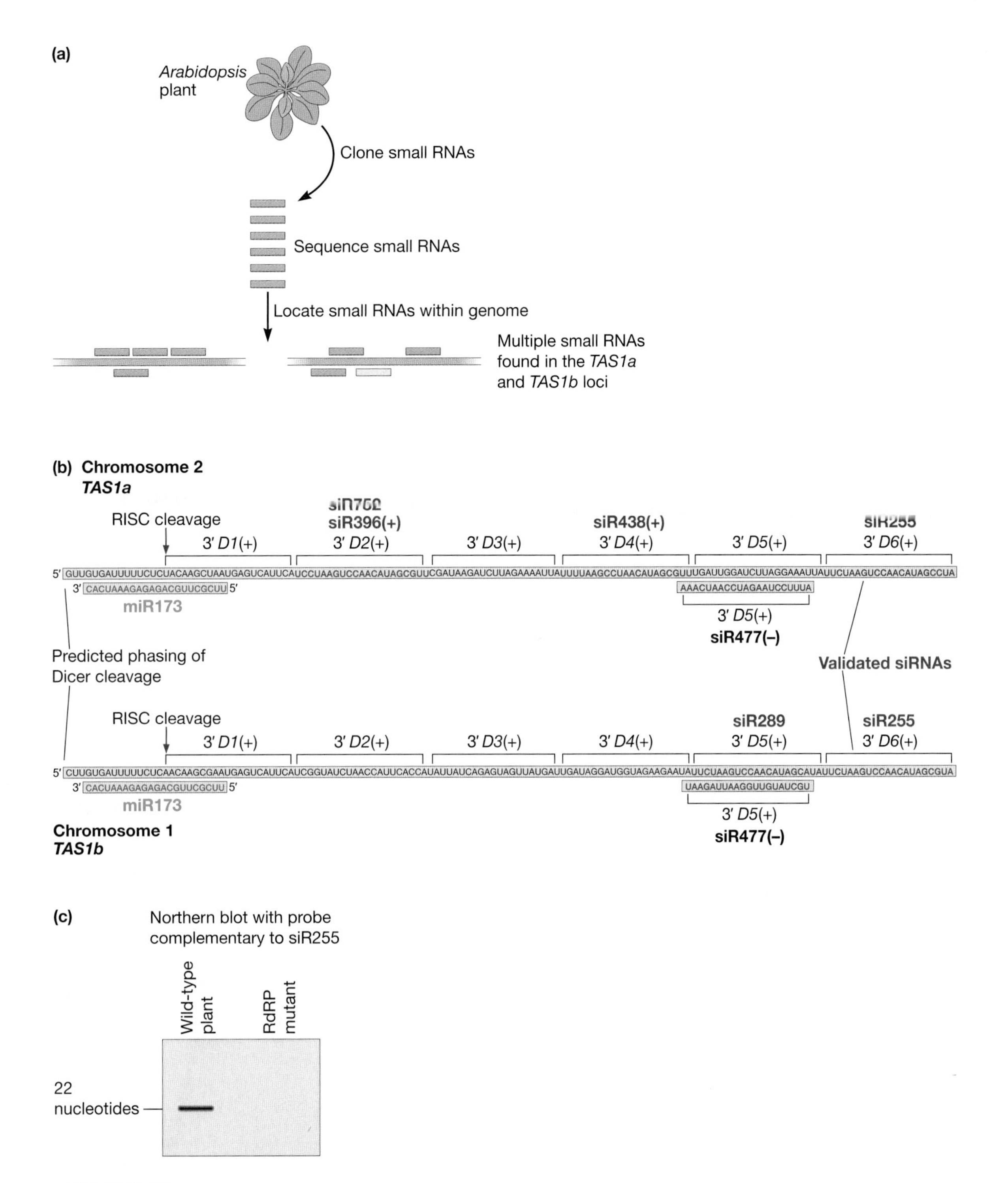

FIGURE 9-13

***Trans*-acting siRNAs can be generated by microRNA-directed cleavage.** (*a*) Thousands of small RNAs were cloned from *Arabidopsis* plants, and their corresponding sequences within the genome were identified. (*b*) Multiple small RNAs were found at two related loci in the genome, *TAS1a* and *TAS1b*, located on chromosomes 2 and 1, respectively. Each transcript (previously identified in a conventional cDNA library) contained multiple small RNAs downstream from a sequence perfectly complementary to *Arabidopsis* microRNA173. *TAS1a* and *TAS1b* transcripts purified from plants were found to have 5′ termini that ended precisely opposite the predicted RISC cleavage location directed by miR173. The siRNAs downstream of this cleavage site exhibited near-perfect 22-nucleotide phasing, as is typical of Dicer-cleavage patterns. This suggests that the miRNA binding and cleavage establishes an end of the RNA for Dicer processing. The siRNAs from the *TAS1a* and *TAS1b* loci are missing in RdRP-mutant plants, showing that RdRPs are necessary to generate dsRNA from these loci for Dicer processing. The siRNAs produced by *TAS1a* and *TAS1b* are predicted to act with other genes. (*c*) As an example, siRNA255 was confirmed to direct cleavage of the mRNA of four previously unstudied genes.

RNAi Can Direct Epigenetic Modifications

As we saw in Chapter 8, extensive research during the 1980s revealed that DNA methylation is a characteristic of epigenetic modifications and is commonly present at or near silent genes. Accordingly, when cosuppression was discovered in plants, it was suspected that silenced endogenous genes and transgenes might also be methylated. This proposed association between DNA methylation and silencing was confirmed first in plants containing a transgene encoding a replication-incompetent genome of a viroid, a tiny plant virus that replicates through a dsRNA intermediate. When these transgenic plants were infected with a related viroid, methylation occurred on transgene regions homologous to the virus. This observation suggests that siRNAs produced by Dicer from the dsRNA replication intermediates direct the DNA methylation process to homologous sequences. Other experiments using viruses and transgenes have confirmed the ability of dsRNA to specify DNA methylation. Surprisingly, methylation occurs at all cytidine residues, not just those located within CpG-rich sequences, the normal targets for methylation.

Another example of RNAi involvement with epigenetic processes comes from studies in the yeast, *Schizosaccharomyces pombe*. *Dicer* and many of the *Argonaute* homologs are lethal for worms, flies, and mice, but, fortunately for genetic analysis, *S. pombe* mutants in *Dicer* and *Argonaute* genes are viable. However, these mutants are not normal; when *Dicer* and *Argonaute*, as well as the gene for RdRP, are knocked out, chromosomes do not segregate correctly to daughter cells. This defect in segregation is caused by a reduced ability of the chromosomes to bind to the mitotic spindle. Histone modifications at the centromeres were studied by chromatin immunoprecipitation, and the results revealed that the RNAi mutants lacked methylation at lysine 9 of histone H3. The absence of this silencing mark meant that the chromosomal protein Swi6 did not bind. Without Swi6, the key centromere protein cohesin could not bind to the centromeric sequences and, as a consequence, the chromosomes failed to attach to the mitotic spindle. Interestingly, the DNA sequence of the outer centromeric repeats of *S. pombe* resembles the sequence of retrotransposons. This link between mobile elements and RNAi has been found in other organisms and hints at the pressures that may have promoted the evolution of RNAi.

RNAi Evolved to Silence Transposons and Viruses

Complex pathways have evolved to carry out RNAi, but to what end? For plants, it has become clear that RNAi plays a vital role in their day-to-day life, a first line of defense against viruses. Plants defective in RNAi genes such as those in the *Dicer* and *Argonaute* families were more susceptible to virus infection; viruses replicated to high levels in these mutant plants. RNAi seems to act as an "immune system" for plants; a virus infecting one leaf of a plant triggers a silencing signal that is transmitted systemically throughout the entire plant, preventing subsequent infection by the original and any other viruses with complementary sequences. Once initiated, the silencing signal does not require further viral replication. In fact, systemic silencing does not require a virus at all; it can even be triggered by introducing a strongly expressed gene by *Agrobacterium*-mediated gene transfer or particle bombardment. The transmission of this signal is strikingly revealed by infection of a GFP-expressing plant with *Agrobacterium* carrying a GFP transgene for

FIGURE 9-14

Gene silencing signals can spread systemically through plants. A single leaf (indicated with arrow) of a tobacco plant expressing a GFP transgene was infiltrated with *Agrobacterium*. The *Agrobacterium* strain contained a T-DNA vector carrying a separate GFP transgene expressed under the control of a strong promoter. The plant is shown 18 days after infiltration. Under UV light, GFP fluorescence appears green. If GFP is absent, the red fluorescence of chlorophyll is visible. The infiltrated leaf has strong GFP fluorescence owing to expression of GFP from the transferred T-DNA, but silencing spreads through the plant away from this infiltrated leaf.

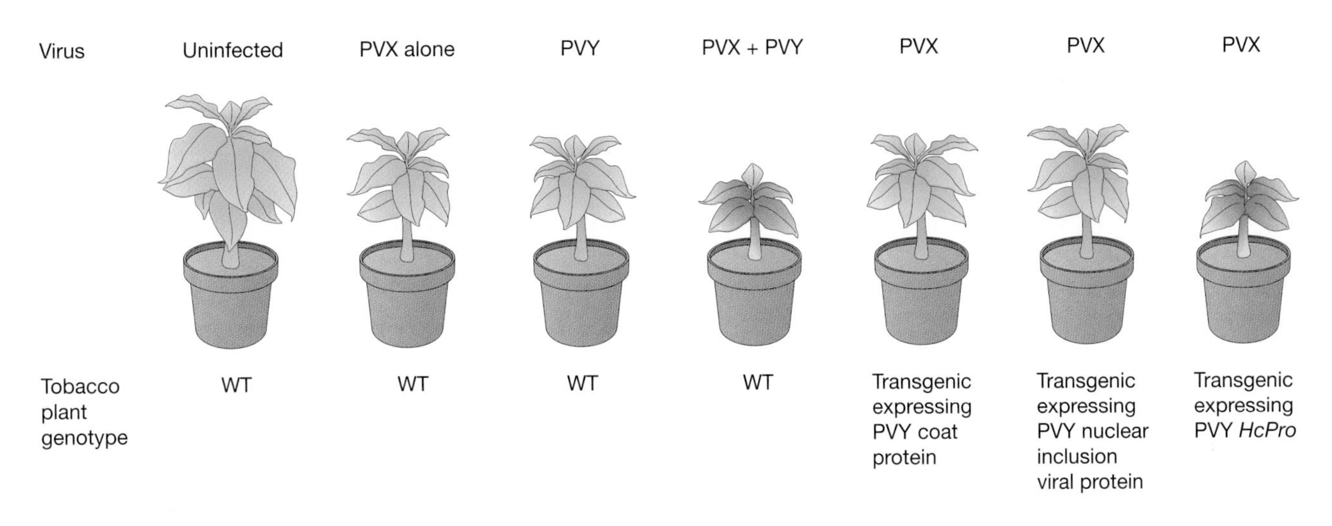

FIGURE 9-15

Viruses have developed strategies to counter the antiviral defenses of plants. Tobacco plants infected simultaneously by the RNA viruses potato virus X (PVX) and potato virus Y (PVY) have much more severe symptoms than those infected by either virus alone. Analysis of RNA from the doubly infected plants revealed that PVX accumulated to high levels in the presence of PVY, which suggests that a PVY gene was responsible. Accordingly, transgenic tobacco plants were constructed that expressed different PVY proteins. Those plants expressing the *HcPro* pathenogenicity gene from PVY had symptoms after PVX infection that were very similar to plants doubly infected with PVX and the complete PVY. WT, wild type.

T-DNA transfer into the plant cell (Fig. 9-14). Even though the construct enters only a few cells, it produces a silencing signal that is transmitted throughout the plant. This is revealed by the disappearance of green fluorescence away from the leaf where the overexpressed gene was introduced.

More evidence of the battle between RNAi and viruses is that viruses have developed defenses against RNAi. The first clues came from the study of *synergism*, in which mixed infection by two viruses produces much more severe symptoms than infection by either virus alone. For example, plants infected with potato virus Y (PVY) and subsequently infected with potato virus X (PVX) support replication of the X virus to higher levels than if it alone was infecting a plant. To determine how these viruses altered replication of other strains, plants were transfected with individual genes from PVY and then tested to see which of these genes was responsible for the increased replication of PVX. The gene responsible for the enhanced effect encodes a pathogenicity factor, HcPro. When *HcPro* was expressed as a transgene, it prevented the plant from mounting an RNAi response to PVX, which allowed PVX to accumulate to high levels and sicken the plant (Fig. 9-15). When these *HcPro* transgenic plants were studied in detail, they were found to have greatly decreased levels of siRNAs. Thus, HcPro interferes with either Dicer activity or the stability of siRNAs. Similar approaches have identified other viral proteins, such as 2b of the cucumoviruses and p19 of tombusviruses, that are suppressors of various steps of RNAi, from siRNA production to systemic spreading of silencing. The interaction between RNAi and viruses is not limited to plants—several genes from insect and worm viruses are now known to help the replicating virus evade RNAi.

RNAi defends against another type of genome invader—transposable elements. As we saw in Chapter 7, transposons are normally transcriptionally silent. However, in many RNAi mutants, the heterochromatic histone modifications typically found on transposons are lost. When RNAi-defective mutants were identified in worms, it was quickly realized that some of the mutant genes had been previously identified through screens for mutator strains in which DNA transposons actively hop in germ cells, causing new mutations. Activation of some RNA transposons occurs in plant RNAi mutants, but it is not yet known whether the same occurs in mammalian RNAi mutants. Although siRNA corresponding to a variety of transposons can be detected in cells, loss of RNAi function activates many fewer types of transposons than does inhibition of DNA methyltransferases and histone deacetylases. This finding suggests that there are several mechanisms other than RNAi that silence transposons. In the silent mating-type genes of *S. pombe*, RNAi is needed only for the initiation of silencing. (The **a** and α mating-type genes combine with the active mating-type locus to switch yeast mating type.) Once these mating-type genes are silent, they are stably inherited in silent form. It is likely that RNAi plays a similar role in establishing silencing in other organisms as well.

One interesting example of RNAi control of transposon expression was discovered in studies of the imprinted *Arabidopsis* gene *flowering locus C* (*FLC*), which encodes a MADS domain transcription factor that represses flowering. *FLC* expression is a major determinant of flowering time—strains that express *FLC* at high levels are generally late flowering, whereas those that express low levels of *FLC* flower early.

One common strain of *Arabidopsis* is early flowering, and the *FLC* gene in this strain was discovered to contain a Mu-like (Mule) transposon insertion within the first intron. The siRNAs corresponding to this element (and others like it in the genome) had already been identified in other sequencing efforts. To test whether RNAi had a role in silencing *FLC* in this early-flowering line, investigators crossed the early- and late-flowering plants with *hen-1* mutant plants, which lack a miRNA methyltransferase required for miRNA and siRNA accumulation. Transposon-generated siRNAs were lost in all strains crossed with the *hen-1* mutant, but flowering time was affected only in the *hen-1*/early-flowering crosses that contained the transposon insertion. This result suggests that the siRNAs from Mule transposable elements silence *FLC* in the early-flowering lines (Fig. 9-16). The *FLC* Mule element in wild-type early-flowering plants was bound by the heterochromatic histone H3-methyl

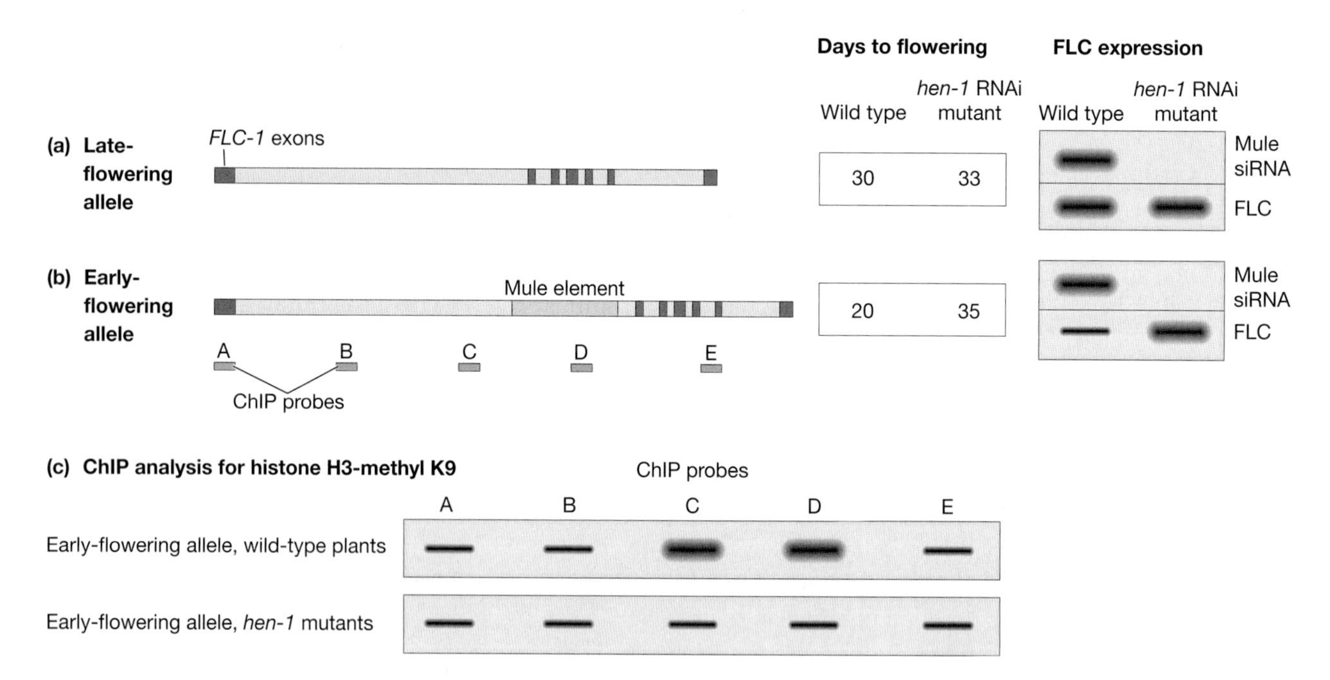

FIGURE 9-16

Plants use RNAi to control the activities of transposons. Expression of the *flowering locus C* (*FLC*) gene in *Arabidopsis* determines when a plant flowers and is regulated by a transposable element inserted near the *FLC* promoter. The sequence of *FLC* was investigated in an early-flowering and a late-flowering strain. (*a*) In a late-flowering strain, *FLC* is expressed at high levels whether the plant is wild type or a *hen-1* RNAi mutant. (The *hen-1*-encoded RNA methyltransferase is required for siRNA accumulation; in mutant plants, siRNAs do not accumulate.) siRNAs produced from the many copies of the Mule element are abundant in wild-type *Arabidopsis* but lost in *hen-1* mutants. (*b*) The early-flowering strain, which expresses *FLC* at low levels and flowers early, contains an insertion of a Mu-like (Mule) transposable element upstream of *FLC-1*. When the early-flowering strain was mated to a *hen-1* mutant, the RNAi-deficient progeny reverted to the usual late-flowering time and expressed high levels of FLC-1. (*c*) Chromatin immunoprecipitation (ChIP) was used to assess levels of the heterochromatic histone H3-methyl K9 across the *FLC* allele in early- and late-flowering plants. There were high levels of H3-K9 at the transposon insertion in early-flowering plants, and this heterochromatic mark was absent in late-flowering plants and lost in the *hen-1* mutants.

K9, but no histone methylation was present in *hen-1* mutant. Thus, siRNAs seem to be necessary for silencing of FLC in early-flowering plants. Flowering time is an important trait for plant survival—plants must flower and set seed well before the killing cold of winter arrives. The Mule transposon insertion in the *FLC* locus demonstrates how a transposon insertion can bring a gene under control of siRNAs from other sites in the genome.

Biologists are making rapid progress in understanding the biological functions of RNAi in the cell, and this understanding has come from, and driven, the continual refinement of RNAi as a simple, direct technique for suppressing expression of any gene. We have always known that no gene acts on its own and now techniques like RNAi, and others that we will discuss in Chapter 13, are beginning to help us understand the complex interactions between and among genes.

Reading List

General

Hannon G.J., ed. 2003. *RNAi: A Guide to Gene Silencing*. Cold Spring Harbor Laboratory Press, Cold Spring Harbor, New York.

Cosuppression

Napoli C., Lemieux C., and Jorgensen R. 1990. Introduction of a chimeric chalcone synthase gene into petunia results in reversible co-suppression of homologous genes in *trans*. *Plant Cell* **2:** 279–289.

van der Krol A.R., Mur L.A., Beld M., Mol J.N., and Stuitje A.R. 1990. Flavonoid genes in petunia: Addition of a limited number of gene copies may lead to a suppression of gene expression. *Plant Cell* **2:** 291–299.

Jorgensen R.A. 1995. Cosuppression, flower color patterns, and metastable gene expression states. *Science* **268:** 686–691.

Muller E., Gilbert J., Davenport G., Brigneti G., and Baulcombe D.C. 1995. Homology-dependent resistance: Transgenic virus resistance in plants related to homology-dependent gene silencing. *Plant J.* **7:** 1001–1013.

Waterhouse P.M., Graham M.W., and Wang M.B. 1998. Virus resistance and gene silencing in plants can be induced by simultaneous expression of sense and antisense RNA. *Proc. Natl. Acad. Sci.* **95:** 13959–13964.

Double-stranded RNA and RNAi

Guo S. and Kemphues K.J. 1995. *par*-1, a gene required for establishing polarity in *C. elegans* embryos, encodes a putative Ser/Thr kinase that is asymmetrically distributed. *Cell* **81:** 611–620.

Fire A., Xu S., Montgomery M.K., Kostas S.A., Driver S.E., and Mello C.C. 1998. Potent and specific genetic interference by double-stranded RNA in *Caenorhabditis elegans*. *Nature* **391:** 806–811.

Kumagai M.H., Donson J., della-Cioppa G., Harvey D., Hanley K., and Grill L.K. 1995. Cytoplasmic inhibition of carotenoid biosynthesis with virus-derived RNA. *Proc. Natl. Acad. Sci.* **92:** 1679–1683.

Timmons L. and Fire A. 1998. Specific interference by ingested dsRNA. *Nature* **395:** 854.

Fragile X Mental Retardation Protein

Garber K., Smith K.T., Reines D., and Warren S.T. 2006. Transcription, translation and fragile X syndrome. *Curr. Opin. Genet. Dev.* **16:** 270–275.

Jin P., Zarnescu D.C., Ceman S., Nakamoto M., Mowrey J., et al. 2004. Biochemical and genetic interaction between the fragile X mental retardation protein and the microRNA pathway. *Nat. Neurosci.* **7:** 113–117.

Caudy A.A., Myers M., Hannon G.J., and Hammond S.M. 2002. Fragile X–related protein and VIG associate with the RNA interference machinery. *Genes Dev.* **16:** 2491–2496.

Ishizuka A., Siomi M.C., and Siomi H. 2002. A *Drosophila* fragile X protein interacts with components of RNAi and ribosomal proteins. *Genes Dev.* **16:** 2497–2508.

RISC

Tuschl T., Zamore P.D., Lehmann R., Bartel D.P., and Sharp P.A. 1999. Targeted mRNA degradation by double-stranded RNA in vitro. *Genes Dev.* **13:** 3191–3197.

Hammond S.M., Bernstein E., Beach D., and Hannon G.J. 2000. An RNA-directed nuclease mediates post-transcriptional gene silencing in *Drosophila* cells. *Nature* **404:** 293–296.

Small RNAs

Hamilton A.J. and Baulcombe D.C. 1999. A species of small antisense RNA in posttranscriptional gene silencing in plants. *Science* **286:** 950–952.

Bernstein E., Caudy A.A., Hammond S.M., and Hannon G.J. 2001. Role for a bidentate ribonuclease in the initiation step of RNA interference. *Nature* **409:** 363–366.

Ketting R.F., Fischer S.E., Bernstein E., Sijen T., Hannon G.J., and Plasterk R.H. 2001. Dicer functions in RNA interference and in synthesis of small RNA involved in developmental timing in *C. elegans*. *Genes Dev.* **15:** 2654–2659.

Grishok A., Pasquinelli A.E., Conte D., Li N., Parrish S., et al. 2001. Genes and mechanisms related to RNA interference regulate expression of the small temporal RNAs that control *C. elegans* developmental timing. *Cell* **106:** 23–34.

Hutvagner G., McLachlan J., Pasquinelli A.E., Balint E., Tuschl T., and Zamore P.D. 2001. A cellular function for the RNA-interference enzyme Dicer in the maturation of the *let-7* small temporal RNA. *Science* **293:** 834–838.

Knight S.W. and Bass B.L. 2001. A role for the RNase III enzyme DCR-1 in RNA interference and germ line development in *Caenorhabditis elegans. Science* **293:** 2269–2271.

Elbashir S.M., Lendeckel W., and Tuschl T. 2001. RNA interference is mediated by 21- and 22-nucleotide RNAs. *Genes Dev.* **15:** 188–200.

Elbashir S.M., Harborth J., Lendeckel W., Yalcin A., Weber K., and Tuschl T. 2001. Duplexes of 21-nucleotide RNAs mediate RNA interference in cultured mammalian cells. *Nature* **411:** 494–498.

Elbashir S.M., Martinez J., Patkaniowska A., Lendeckel W., and Tuschl T. 2001. Functional anatomy of siRNAs for mediating efficient RNAi in *Drosophila melanogaster* embryo lysate. *EMBO J.* **20:** 6877–6888.

Palliser D., Chowdhury D., Wang Q.-Y., Lee S.J., Bronson R.T., Knipe D.M., and Lieberman J. 2006. An siRNA-based microbicide protects mice from lethal herpes simplex virus 2 infection. *Nature* **439:** 89–94.

microRNAs

Bartel D.P. and Chen C.Z. 2004. Micromanagers of gene expression: The potentially widespread influence of metazoan microRNAs (review). *Nat. Rev. Genet.* **5:** 396–400.

Pasquinelli A.E., Hunter S., and Bracht J. 2005. MicroRNAs: A developing story (review). *Curr. Opin. Genet. Dev.* **15:** 200–205.

Lee R.C., Feinbaum R.L., and Ambros V. 1993. The *C. elegans* heterochronic gene *lin-4* encodes small RNAs with antisense complementarity to *lin-14. Cell* **75:** 843–854.

Reinhart B.J., Slack F.J., Basson M., Pasquinelli A.E., Bettinger J.C., et al. 2000. The 21-nucleotide *let-7* RNA regulates developmental timing in *Caenorhabditis elegans. Nature* **403:** 901–906.

Olsen P.H. and Ambros V. 1999. The *lin-4* regulatory RNA controls developmental timing in *Caenorhabditis elegans* by blocking LIN-14 protein synthesis after the initiation of translation. *Dev. Biol.* **216:** 671–680.

Zeng Y., Yi R., and Cullen B.R. 2003. MicroRNAs and small interfering RNAs can inhibit mRNA expression by similar mechanisms. *Proc. Natl. Acad. Sci.* **100:** 9779–9784.

Lim L.P., Lau N.C., Garrett-Engele P., Grimson A., Schelter J.M., et al. 2005. Microarray analysis shows that some microRNAs downregulate large numbers of target mRNAs. *Nature* **433:** 769–773.

Schwarz D.S., Hutvagner G., Du T., Xu Z., Aronin N., and Zamore P.D. 2003. Asymmetry in the assembly of the RNAi enzyme complex. *Cell* **115:** 199–208.

Slicer

Song J.J., Smith S.K., Hannon G.J., and Joshua-Tor L. 2004. Crystal structure of Argonaute and its implications for RISC slicer activity. *Science* **305:** 1434–1437.

Spreading

Voinnet O. and Baulcombe D.C. 1997. Systemic signalling in gene silencing. *Nature* **389:** 553.

Vaistij F.E., Jones L., and Baulcombe D.C. 2002. Spreading of RNA targeting and DNA methylation in RNA silencing requires transcription of the target gene and a putative RNA-dependent RNA polymerase. *Plant Cell* **14:** 857–867.

Allen E., Xie Z., Gustafson A.M., and Carrington J.C. 2005. microRNA-directed phasing during *trans*-acting siRNA biogenesis in plants. *Cell* **121:** 207–221.

Voinnet O., Vain P., Angell S., and Baulcombe D.C. 1998. Systemic spread of sequence-specific transgene RNA degradation in plants is initiated by localized introduction of ectopic promoterless DNA. *Cell* **95:** 177–187.

Viral Suppressors of Silencing

Lecellier C.H. and Voinnet O. 2004. RNA silencing: No mercy for viruses (review)? *Immunol. Rev.* **198:** 285–303.

Vance V.B., Berger P.H., Carrington J.C., Hunt A.G., and Shi X.M. 1995. 5′ proximal potyviral sequences mediate potato virus X/potyviral synergistic disease in transgenic tobacco. *Virology* **206:** 583–590.

Shi X.M., Miller H., Verchot J., Carrington J.C., and Vance V.B. 1997. Mutations in the region encoding the central domain of helper component-proteinase (HC-Pro) eliminate potato virus X/potyviral synergism. *Virology* **231:** 35–42.

Bennasser Y., Le S.Y., Benkirane M., and Jeang K.T. 2005. Evidence that HIV-1 encodes an siRNA and a suppressor of RNA silencing (erratum *Immunity* [2005] **22:** 773). *Immunity* **22:** 607–619.

Transposons and RNAi

Lippman Z. and Martienssen R. 2004. The role of RNA interference in heterochromatic silencing (review). *Nature* **431:** 364–370.

Ketting R.F., Haverkamp T.H., van Luenen H.G., and Plasterk R.H. 1999. *mut-7* of *C. elegans*, required for transposon silencing and RNA interference, is a homolog of Werner syndrome helicase and RNaseD. *Cell* **99:** 133–141.

Sijen T. and Plasterk R.H. 2003. Transposon silencing in the *Caenorhabditis elegans* germ line by natural RNAi. *Nature* **426:** 310–314.

Liu J., He Y., Amasino R., and Chen X. 2004. siRNAs targeting an intronic transposon in the regulation of natural flowering behavior in *Arabidopsis. Genes Dev.* **18:** 2873–2878.

RNA-directed DNA methylation

Matzke M.A., Primig M., Trnovsky J., and Matzke A.J. 1989. Reversible methylation and inactivation of marker genes in sequentially transformed tobacco plants. *EMBO J.* **8:** 643–649.

Wassenegger M., Heimes S., Riedel L., and Sanger H.L. 1994. RNA-directed de novo methylation of genomic sequences in plants. *Cell* **76:** 567–576.

Vaucheret H. 1992. Promoter-dependent *trans*-inactivation in transgenic tobacco plants: Kinetic aspects of gene silencing and gene reactivation. *C.R. Acad. Sci.* **317:** 310–323.

Heterochromatin Formation

Hall I.M., Shankaranarayana G.D., Noma K., Ayoub N., Cohen A., and Grewal S.I. 2002. Establishment and maintenance of a heterochromatin domain. *Science* **297:** 2232–2237.

SECTION 2

FOUNDATIONS OF GENOMICS

The term *genomics* came into use with the development of large-scale sequencing in the 1980s. For the first time it became possible to design research strategies based on knowledge of the complete genetic information of an organism. This genomics-based research differs from other biological research because of its massive scale, whether using powerful computers to search for features in hundreds of millions of nucleotides or to analyze the expression patterns of tens of thousands of genes simultaneously. The genomics era was inaugurated by the sequencing of the first bacterial genome—at the time a remarkable achievement. The sequencing of many other small genomes followed quickly, and we use these examples to introduce some of the fundamental principles of genome-scale sequencing. The seminal event in genomics was the sequencing of the human genome. Orders of magnitude larger than any other previously sequenced genome, the human genome demanded resources and organization never before required in biological research. The story of the Human Genome Project is full of drama and intrigue and remains a pivotal moment in the history of biology. Similarly, this section forms a turning point in this book by providing a foundation for the chapters that follow.

CHAPTER 10

Fundamentals of Whole-Genome Sequencing

The invention of DNA sequencing in the mid-1970s opened a new era of biological investigation, one that led rapidly to an enormous increase in our knowledge of genes—of their structures, functions, regulation, and evolution. Although the techniques were at first somewhat unwieldy, they were quickly adopted by most molecular biology laboratories. Collecting DNA sequence information came to be the norm and the amount of DNA sequence in the GenBank database rose rapidly from 3.3×10^6 base pairs in 1984, to 2.2×10^8 base pairs in 1994, and to more than 10^{11} base pairs in 2006 (Fig. 10-1). The ready availability of DNA sequences drove the design of new experiments and interpretation of gene function and of molecular processes in the cell. Many other studies involving the manipulation of DNA, including site-directed mutagenesis, gene transfer experiments, and analysis of gene regulation and protein expression, have made DNA sequencing an essential component of the molecular biologist's tool kit. DNA sequencing became so much faster, easier, and cheaper than protein sequencing that soon the vast majority of the amino acid sequences available in public databases was deduced from nucleotide sequence data rather than from chemical sequencing of purified peptides.

At first, almost all DNA sequencing was done by individual researchers needing sequence data to test a specific hypothesis or to explore the function of a particular gene. Sequencing, although becoming widespread, was still expensive and labor intensive, so that only the genomes of small plas-

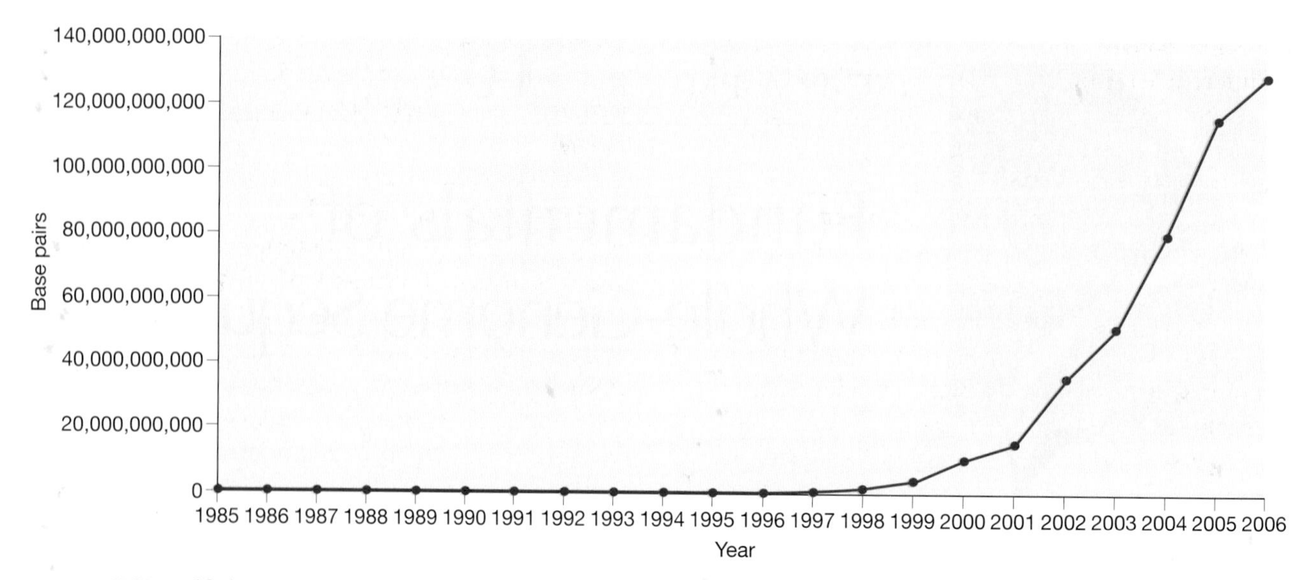

FIGURE 10-1

Growth of DNA sequences deposited into the public databases (GenBank, EBI, DBJL) from 1985 to 2006. GenBank and the other public databases began their operations—storing and disseminating most of the DNA sequences available—in the early 1980s. The increase in sequence data results from the massive increase in sequencing capabilities that came with the Human Genome Project, particularly in the late 1990s when sequencing of the human genome began in earnest. A milestone was reached in 2005 when the 100,000,000,000th (one hundred billionth) base pair was deposited into the databases.

mids and viruses were sequenced completely. By the mid-1980s and early 1990s, a few bold experimenters had produced complete sequences of large bacteriophage and viral genomes—50,000-bp and even 250,000-bp long—but sequencing complete genomes, even as small as those of bacteria, was unimaginable.

However, some scientists recognized that tremendous benefits would come from knowing the complete DNA sequence of an organism. They began discussing whether the newly developed methods of automated DNA sequencing opened the door to sequencing large genomes. Many in the biological research community reacted to these discussions with dismay. They worried that large-scale sequencing projects would be a misuse of scarce research funds and would jeopardize the tried-and-true way of doing biological research—in small groups tackling specific questions. Nevertheless, large-scale international efforts were begun to sequence several genomes, including those of *Escherichia coli*, *Caenorhabditis elegans*, and *Homo sapiens*. The successes of these projects ushered in a whole new way of studying biology. Large-scale, high-throughput, genome-wide experimental and computational analysis is now firmly established as the discipline of *genomics*. This story is one of the most remarkable in the history of biological research, and the following chapters describe the technical and biological foundations of genomics. This chapter introduces the many steps of genome-scale DNA sequencing as illustrated by the sequencing of microbial genomes.

Genomic DNA Sequencing Is Done by Assembling Smaller, Overlapping Sequence "Reads"

Determining the sequence of a long, contiguous stretch of DNA is significantly more complex than sequencing a single piece of DNA a few hundred base pairs long. There are several reasons why this is so. First, the DNA fragments needed for Sanger sequencing (Fig. 4-12) have to be limited in length to a few hundred nucleotides. To be able to read the maximum number of base pairs, there must be a relatively uniform concentration of each of the terminated fragments along the entire length of the DNA being sequenced. However, it is difficult to achieve such uniformity because different concentrations of chain-terminating nucleotides, 2′,3′-dideoxynucleoside triphosphates (ddNTPs), are needed for chains of different lengths. For example, synthesis of fragments of 100 bp long requires a high concentration of ddNTPs so that most chains are terminated soon after synthesis begins. On the other hand, a low concentration of ddNTPs is needed to generate fragments around 1000 bp in length; otherwise the newly synthesized chains are terminated too soon. As a consequence, the concentration of ddNTPs used has to be a compromise and is adjusted to provide a uniform set of terminated fragments ranging from about 50 bp to a few hundred base pairs in length. A second, and more serious, limitation arises from the limited resolution of slab gel electrophoresis. The polyacrylamide gels used for sequencing lose the ability to discriminate between fragments longer than about 800 bases, so that a fragment 800 bases long is hard to distinguish from one 801 nucleotides long. As a consequence, a "sequencing read"—the number of bases that can be determined accurately in a single lane of a gel—is about 750 bases.

Certain refinements, such as carrying out electrophoresis in capillaries, have led to significant improvements. However, determining the sequence of any substantial segment of DNA, and certainly that of a whole genome, involves generating many short sequencing reads from overlapping sections of DNA. (New sequencing technologies that do not use Sanger sequencing produce millions of short reads—20–40 bp—in a single analysis. We discuss these in Chapters 11 and 14.) The complete, contiguous sequence is deduced from reads that overlap one another by performing all possible pairwise comparisons of sequence reads, in a process called "assembling" (Fig. 10-2). One of the major chal-

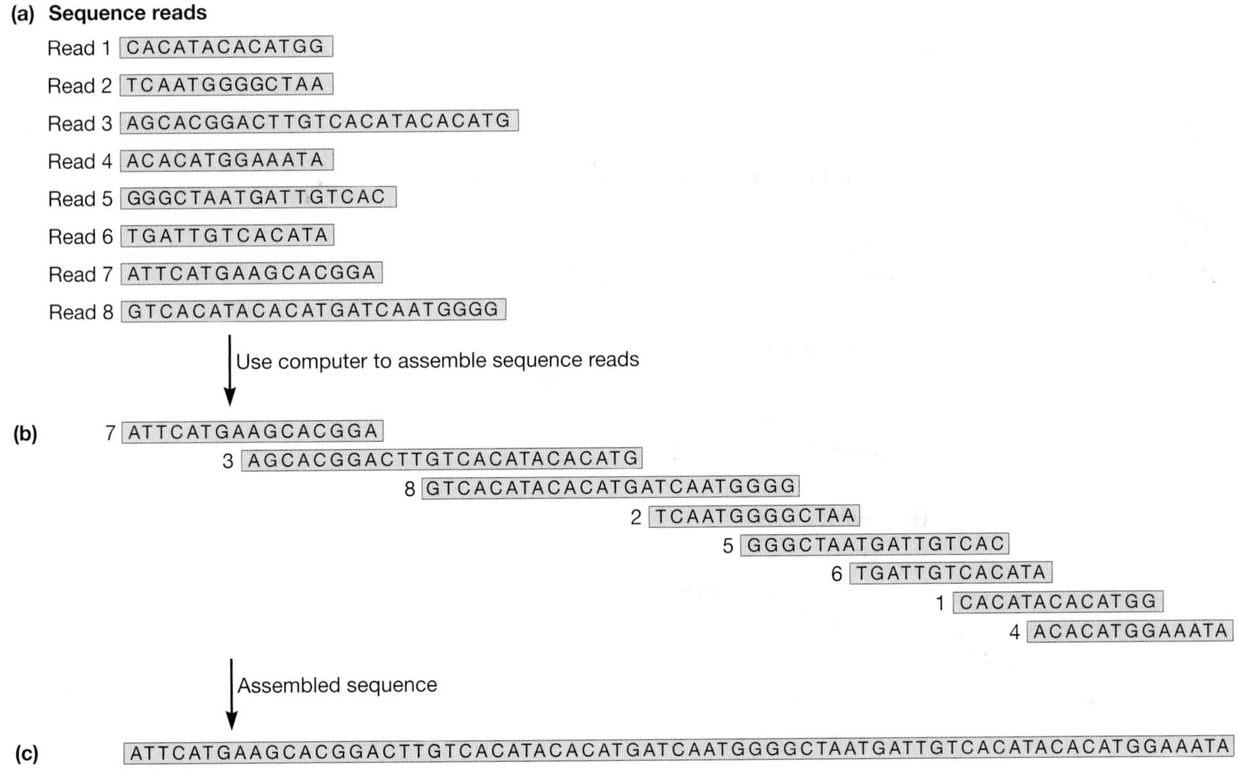

(a) **Sequence reads**

Read 1 CACATACACATGG
Read 2 TCAATGGGGCTAA
Read 3 AGCACGGACTTGTCACATACACATG
Read 4 ACACATGGAAATA
Read 5 GGGCTAATGATTGTCAC
Read 6 TGATTGTCACATA
Read 7 ATTCATGAAGCACGGA
Read 8 GTCACATACACATGATCAATGGGG

Use computer to assemble sequence reads

(b)
7 ATTCATGAAGCACGGA
3 AGCACGGACTTGTCACATACACATG
8 GTCACATACACATGATCAATGGGG
2 TCAATGGGGCTAA
5 GGGCTAATGATTGTCAC
6 TGATTGTCACATA
1 CACATACACATGG
4 ACACATGGAAATA

Assembled sequence

(c) ATTCATGAAGCACGGACTTGTCACATACACATGATCAATGGGGCTAATGATTGTCACATACACATGGAAATA

FIGURE 10-2

The complete sequence of a DNA fragment is determined by assembling overlapping sequencing reads. (*a*) Eight "reads" from a DNA-sequencing project are shown. To simplify the figure, these reads are much shorter than actual reads, but the principles are the same as when longer reads, typically 500–800 bp, are obtained with current technologies. In this example, the reads are ~10–20 bp and each was obtained from a DNA fragment randomly sheared from the initial target segment. (*b*) After the bases are called in the sequencing reactions, the sequences are assembled by the computer, which determines the parts of each read that overlap with another read. (c) The final assembled sequence of the target fragment in the figure is 72 bp, whereas an actual assembly of long sequence reads, counting overlaps, would be a few thousand base pairs.

lenges of a DNA-sequencing project is this process of sequence assembly, which is done not at the laboratory bench, but in the computer.

The Genomes of Bacteriophages, Viruses, and Cell Organelles Were the First to Be Sequenced Completely

The sequences of small RNA molecules were the first to be determined, using cleavage methods similar to those used by Frederick Sanger for protein sequencing. In 1965, after a tour de force of chemical analysis, Robert Holley and his colleagues reported the sequence of an entire nucleic acid molecule, an 80-nucleotide-long alanine transfer RNA (tRNA) from yeast. In 1976, Walter Fiers and his colleagues in Germany described the RNA sequences of two of the genes for the bacteriophage MS2, an *E. coli* virus that uses RNA for its genome. Their sequences, combined with those of other MS2 genes determined by multiple groups around the world, made up the 3569 nucleotides of the MS2 bacteriophage genome, the first complete genomic sequence of a living entity, albeit not a free-living one. These sequences were important landmarks, but the unwieldy methods for sequencing RNA—as well as the fact that almost all genomes are composed of DNA, not RNA—did not signal the revolution that was to come from the DNA-sequencing techniques of Sanger and of Allan Maxam and Walter Gilbert.

In 1977, Sanger and his group made history yet again by reporting the first complete sequence of a DNA genome, that of the *E. coli* bacteriophage ϕX174. Sanger had not quite perfected the dideoxy method by then and instead used the "plus–minus" method (see Chapter 4). There was a special reason for choosing ϕX174. DNA polymerase requires a single-stranded DNA molecule as a template for synthesis of a second strand, and researchers had to use laborious methods to separate the two strands of the DNA to be sequenced. Typically this was done by denaturing the DNA and then separating the strands by ultracentrifugation. ϕX174, on the other hand, had the great advantage of having a single-stranded circular DNA molecule as its genome; the researchers needed only to isolate the bacteriophage and purify its DNA away from the proteins to provide a template for sequencing.

It is hard now to appreciate the heroic efforts needed to sequence DNA prior to the ready availability of reagents from companies. Chemically synthesized oligonucleotides were not yet available for use as primers for initiating DNA synthesis in the sequencing reactions. Instead, Sanger had to prepare primers by identifying and purifying restriction enzyme cleavage fragments of the phage DNA. Large numbers of radioactively labeled plus–minus sequencing reactions had to be carried out, as a single sequencing read was only a few dozen to about 100 bp. The fragments were labeled with ^{32}P and the bands on the autoradiogram images of the sequencing gels had to be recorded manually, a laborious and error-prone procedure.

The data from all of these reads were compared with one another to build a complete sequence assembly of the ϕX174 genome. Although this phage genome is miniscule in length compared to the huge genomes that can be sequenced today, such a sequence assembly required a gargantuan effort in the 1970s. Fortunately, at about the same time, Roger Staden, a computationally minded biologist, recognized that DNA sequencing was the future of biology and began to develop software algorithms and tools to enable computerized management and comparison of DNA sequences.

It took Sanger's laboratory more than 2 years to complete the 5386-nucleotide sequence of the ϕX174 genome, and this pioneering work set the stage for the genomics revolution. It was followed by the sequencing of the complete genomes of other phages and viruses and of organelles like mitochondria, genes from more complex organisms, complete bacterial genomes, and, ultimately, the human and other very large genomes. In retrospect, it is remarkable how few years passed between the first development of DNA-sequencing technology, when a massive effort spread over years was required to sequence a few thousand base pairs, to the present day when a complete bacterial genome can be sequenced in a single day.

Genomes Are Sequenced by Using Map-based and Whole Shotgun Strategies

By the time the idea of sequencing whole genomes was proposed, the principle of sequencing very long stretches of DNA by first cloning and then sequencing small sections, followed by reassembling these to recreate the original sequence, was well understood. However, it was by no means clear whether this strategy would work on whole genomes. There were many uncertainties about how to collect, catalog, and assemble very large numbers of sequencing reads. Assembly was a particular concern for genomes with repetitive sequences. How could these sequences be assigned to their appropriate locations in the genome when they are present at multiple places in the genome? Furthermore, the computer programs for assembling sequences were then not capable of handling the extremely large numbers of sequencing reads that would be necessary for putting together even a small genome.

For these reasons, researchers in the late 1980s and early 1990s decided that assembling genomes would require an additional step—the construction of maps showing the locations of the clones being sequenced. These maps would then provide a framework for ordering the assembled sequence reads. To build maps, bacterial cloning systems were used to make clone libraries with DNA inserts ranging from about 40 kilobases to hundreds of kilobases. Overlapping clones were identified and additional analyses were performed to determine the exact extent of overlap (Fig. 10-3). From these, a set of clones was selected that provided complete coverage of the area but with a minimum of overlap, a "minimum tiling path." The insert DNA in this set of clones was sequenced, one large-insert clone at a time. The final sequence of the complete genome could then be put together based on the order of the clones in the minimum tiling sets. This strategy was adopted for sequencing the genomes of *Saccharomyces cerevisiae* and *C. elegans*, and, as we will see in Chapter 11, it was used by one of the two groups that sequenced the human genome.

Building maps is both time consuming and expensive, and so a different approach to genome-scale sequencing was considered. "Whole-genome shotgun" was proposed in the early 1990s by J. Craig Venter for bacteria and in 1995 by James Weber and Eugene Myers for the human genome. These investigators argued that advances in sequencing technology and computational power meant that it would be possible to sequence a genome by cloning it into many thousands of small plasmids, sequencing these at random, and assembling the reads without knowing the locations of the clones in the genome. This "shotgun" method was used to sequence the genome of the bacterium *Hemophilus influenzae* by Venter and colleagues, most of whom worked at The Institute for Genome Research (TIGR) in Maryland. Because the method worked so well, it has become the standard approach for sequencing bacterial genomes. It has proved successful with much larger genomes, although these require mapping and other additional work to produce a complete sequence (Chapter 11).

Cloned DNA for Sequencing Must Represent the Entire Genome and Be Sequenced Multiple Times

The chain termination method for DNA sequencing requires copying the DNA to be sequenced. Therefore, the first step in sequencing a complete bacterial genome is to obtain a sample of purified genomic DNA from the species. For many bacteria, it is possible to make a pure culture in which all the bacteria in the sample are derived from a single, clonally isolated individual. Because bacteria are haploid, all the DNA molecules in the bacteria of a clonally derived culture are identical, coming from the single DNA molecule in the original bacterial cell used to start the culture. It follows that all sequence reads corresponding to the same region in the genome will have exactly the same sequence. This lack of nucleotide variation makes assembling a bacterial genome from shotgun sequence reads much easier than assembling sequences from outbred, diploid organisms such as human beings. DNA-sequence polymorphisms at many positions in these latter genomes complicate the assembly process.

Because overlapping DNA-sequence reads are required for assembly, the DNA fragments for sequencing must be prepared so that there are overlapping regions. The simplest way to do this is to break the genomic DNA into fragments of a suitable size by mechanical shearing, usually by exposing the DNA to ultrasound, or by forcing a DNA solution through a narrow aperture. The ends of the broken fragments are repaired with DNA-modifying enzymes and then cloned into a bacterial vector, usually a small plasmid, to make subclone libraries.

An essential factor to consider in preparing a subclone library is *coverage*—that is, the number of independent subclones that will be needed to ensure having a complete sequence. The "fold coverage"—or "fold redundancy"—of sequencing data, which is a critical parameter in sequencing projects, is described

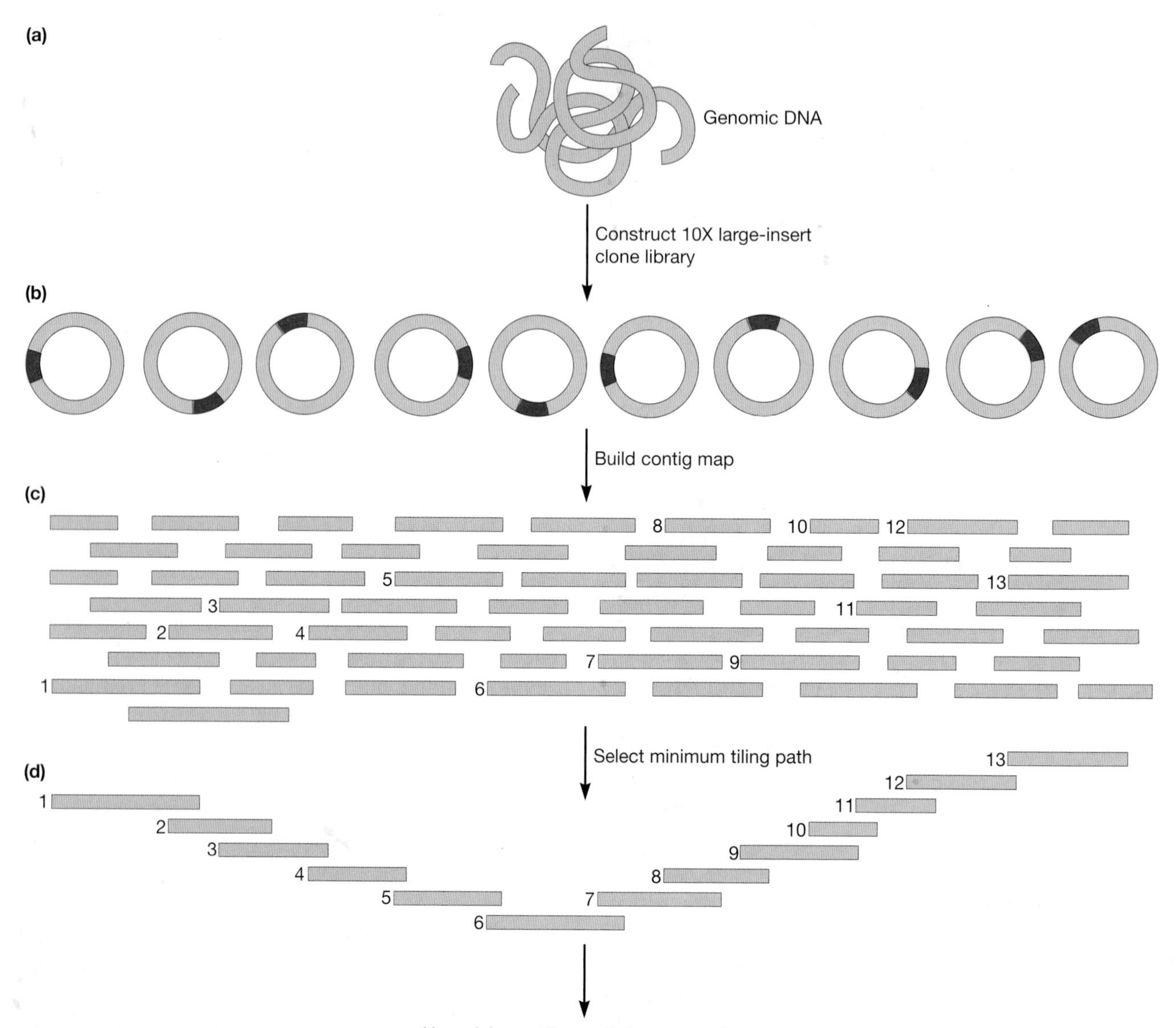

FIGURE 10-3

Sequencing a genome by using an overlapping clone map. (*a*) Genomic DNA is isolated and fragmented, usually by sonication, at random or semirandom positions and the fragments of a desired size range are purified by gel electrophoresis. (*b*) These fragments are then ligated into a cloning vector, such as a cosmid or BAC vector (see Fig. 11-8), and used to transform competent host cells, usually *Escherichia coli*. Sufficient clones are obtained to represent the genome sequence at least ten times; these clones comprise the large-insert clone library. (*c*) The clones in the library are then mapped to determine those that overlap. (*d*) From this contig map, researchers choose a minimal tiling path, which is a set of clones that cover the entire genome but that overlap only enough to confirm that the sequences can be assembled once they are obtained. The minimum path is desirable to minimize the amount of sequencing that must be done, but the choices are restricted to the clones available in the library, which sometimes overlap more or less than desired. Typically, tiling paths chosen have from 15% to 20% overlap at each end.

by a number followed by an "x" (for "times"). A 1x sequence coverage means that the number of base pairs of sequence data determined in a project equals the number of base pairs in the genome of the organism being sequenced. However, a 1x sequence coverage of a genome does *not* mean that every region of the genome has been sequenced. Plasmid subclones for cloning are chosen at random. If only 1x worth of DNA is sequenced from these clones, just by chance some regions of the genome will be sequenced more than once and others not at all. If the plasmid subclones represent a truly random sampling of the genome, a simple equation based on the Poisson distribution gives an estimate of the amount of the genome that is missing for a given amount of sequence coverage. This formula is $P_0 = e^{-m}$, where e is the base of natural logarithms, P_0 is the probability that a base is not sequenced, and m is the amount of sequence coverage (the "x" factor). For example, the amount of a 1.83 million base pair genome (the length of the *H. influenzae* genome) missing after producing 1.83 million bases of sequence (equivalent to 1x coverage), will be $P_0 = 2.73^{-1} = 0.37$, or 37%; that is, 1x coverage produces only about 63% of the genome's sequence.

To ensure that most of a genome is represented in the sequence data, researchers typically sequence to a level of 6x–10x coverage. At 6x coverage and true randomness, 99.75% of the genome should be present in the sequence ($P_0 = 2.73^{-6} = 0.0025$, or 0.25% missing), whereas 99.995% of a genome will be present at 10x coverage. Sequencing to a depth of 6x–10x not only covers almost all of the genome, but also results in sequences that overlap each other to varying degrees, thereby allowing the computational assembly process to work more accurately. However, true randomness is never achieved in practice because, for example, some regions of a genome are more readily cloned than others. This means that a larger fraction of the genome sequence is missing than the calculation suggests. The gaps must be filled by labor-intensive methods directed at each gap, a process referred to as "sequence finishing" (Chapter 11).

How many subclones must be sequenced to obtain 10x coverage of a genome? First, it has been found that the assembly process for sequencing a genome is most efficient if sequence data are obtained from both ends of a subcloned DNA fragment. The advantages of this "paired-end" sequencing strategy will become clear in the next section. Second, although the current generation of sequencing machines produce as many as 1000 bp from a single sequencing reaction, the *average* "read length" is typically 500 bp. This reduction in read length is due to sequencing reaction failures, a decrease in the quality of the sequencing data at greater distances from the sequencing primer, and other vagaries of sequencing. Thus, the 1.83-Mb *H. influenzae* genome would require 3660 sequencing reads (1,830,000 ÷ 500), or 1830 plasmid subclones (each with a 1000-bp insert) sequenced at both ends, for 1x coverage, or 36,600 reads (18,300 subclones) for 10x coverage. The actual number of reads used to sequence the *H. influenzae* genome was larger, however, because the read lengths were shorter than those that can be obtained today.

Sequencing Both Ends of Subclones Assists in Assembling a Complete Sequence

The computation that is needed to assemble a genome from a large number of sequencing reads requires an extremely large number of pairwise comparisons to identify which sequencing reads overlap with which. This identification is not perfect, and sometimes sequence reads are connected incorrectly, placing two segments together that are actually far apart in the genome. *Paired-end sequencing* can help minimize this type of error. Subclone libraries are constructed so that the insert lengths of all the subclones are within 10–20% of each other, and both ends of the inserts of the subclones are sequenced (by starting a sequencing read from each end of the cloning vector). Because these two reads are separated by the length of the subclone, this can be used to determine whether the assembly is correct (Fig. 10-4). For example, if the subclones are 3000 bp long and the sequences from the two ends of a subclone appear 150,000 bp apart in the assembly, it is clear that the computer has made a mistake, and additional analysis, and sometimes experiments, must be done to resolve the discrepancy.

Subclone libraries for paired-end sequencing are made by shearing genomic DNA and then separating the DNA fragments based on their sizes by gel electrophoresis. The area containing the desired

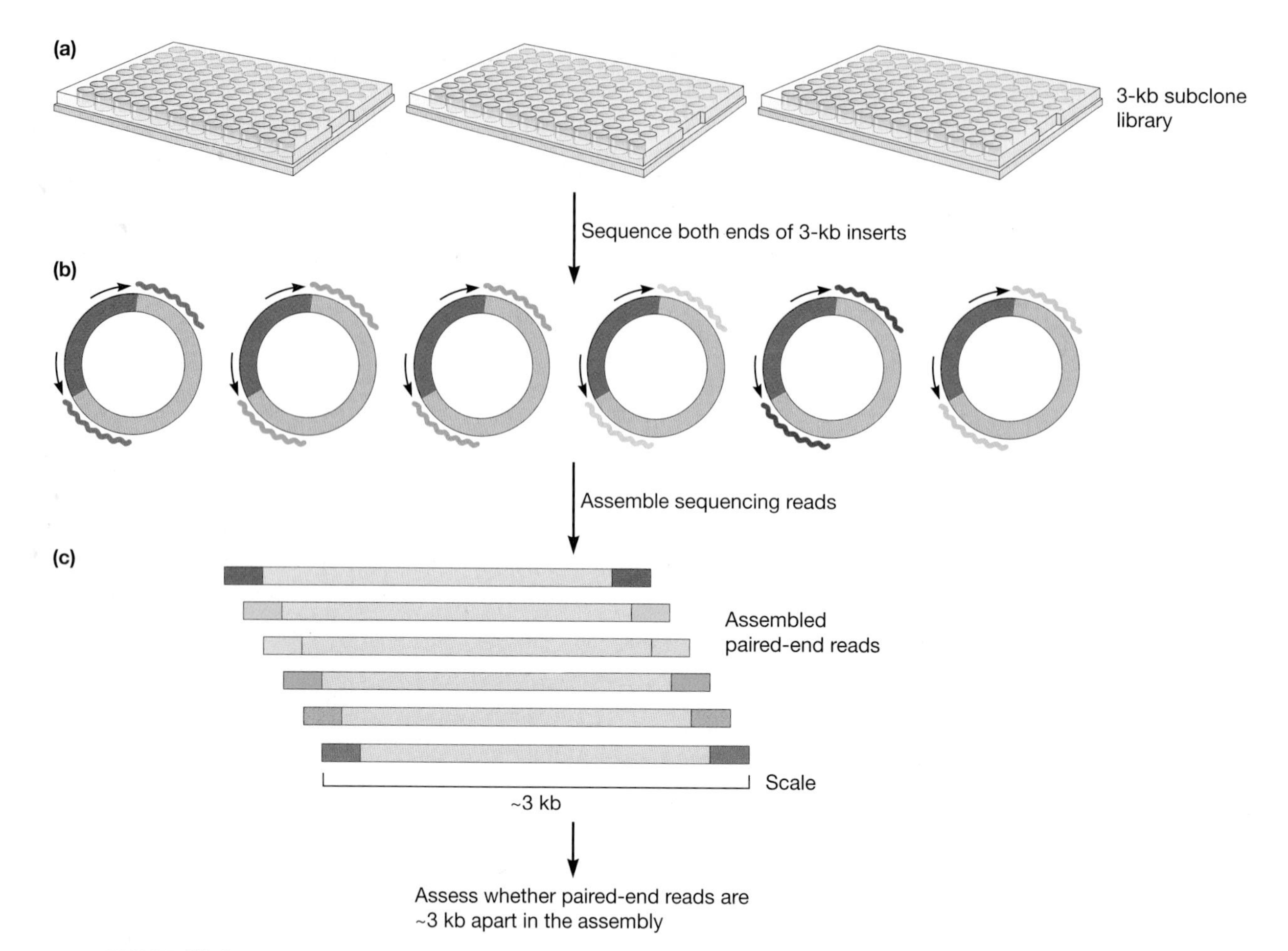

FIGURE 10-4

Paired-end sequencing helps avoid misassembly of sequencing reads. Each DNA-sequencing read provides ~500–800 bp of sequence, usually from subcloned plasmids that carry fragments a few kilobase pairs in length. An advance in sequencing technology occurred when it was realized that by determining the sequences of both ends of the inserts in such subclones, the paired information could be helpful in the later steps of sequencing when the reads are assembled into the final sequence. However, for this trick to be useful, the subclone plasmid library must carry inserts that are approximately the same length, as the length information between the paired-end reads is used to determine whether the assembly is correct. (*a*) Both ends of the cloned insertion of each plasmid in a 3-kb subclone library are sequenced in the shotgun-sequencing stage. (*b*) The sequence derived from each plasmid is represented by a different color. (*c*) When these reads are assembled, the computer checks to see if any pairs of reads from the same plasmid appear in the assembly at places further or closer than about 3000 bp. If so, this is an indication of an assembly error, and additional steps are taken, as part of the "finishing" process, to determine the correct assembly. Most assembly errors are due to the presence of repetitive sequences in one or both ends of the plasmid subclone, which makes it possible to place the reads at multiple places in the assembly.

DNA fragments is cut from the gel and the DNA fragments are eluted for cloning (Fig. 10-5). With this method, it is possible to generate large libraries containing hundreds of thousands of clones, almost all of which are within a very narrow size range. In practice, it has been found that sequencing reads are most readily assembled when two or three separate subclone libraries, each with a different, but narrowly defined, size range, are used. Typically, these libraries might have inserts of 3000 bp, 8000 bp, and 40,000 bp or larger. The majority of the sequence is obtained from the 3000-bp library, whereas the larger libraries are very useful for establishing long-range sequence contiguity in the assemblies, and all the sizes are useful for helping to fill in sequence gaps.

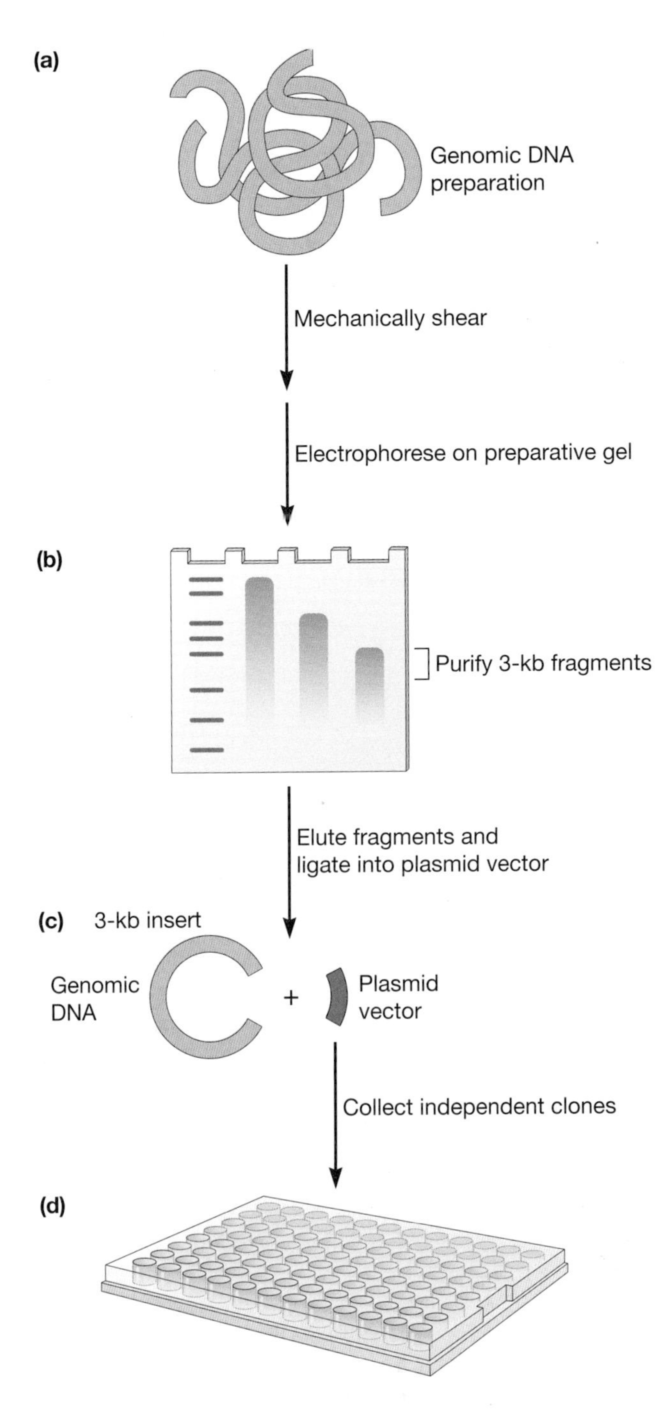

FIGURE 10-5

A plasmid subclone library for paired-end sequences. (*a*) Genomic DNA is sheared by sonication. This is done to different degrees to ensure a wide distribution of fragment sizes, usually from a hundred base pairs to tens of kilobase pairs. (*b*) The sheared DNA is separated by electrophoresis through a preparative agarose gel, and fragments of the desired length are purified from the gel. (*c*) In the gel shown here, the part of the gel containing fragments of ~3000 bp in length, ±300 bp, are chosen. These fragments are eluted and then ligated into a plasmid vector. (*d*) The "3-kb" subclone library is generated following transformation and selection in bacteria.

It is useful to understand the distinction between sequence coverage and physical clone coverage in a sequencing project. The paired-end-sequencing step collects only 1000 bp of sequence (about 500 bp from each end of a clone) from each 3000-bp subclone, leaving about 2000 bp unsequenced. If 10x sequence coverage of a genome is obtained from such sequencing, there is actually an additional 20x of cloned DNA present in the 2000-bp segments in the middle of each of the subclone plasmids. This means that the subclone library actually represents 30x clone coverage of the DNA in the genome. This additional clone coverage is useful for several reasons. It provides purified and identified templates for connecting the gaps in the DNA sequence that inevitably remain, even when sequencing at high fold coverage. And perhaps its greatest value is that it provides a way to overcome the problems posed by repetitive sequences. Consider the example shown in Figure 10-6; here the end sequences from several 3000-bp subclones led to an assembled, contiguous sequence ("contig") that has a 2000-bp repeat. End sequences from other 3-kb clones may lie within this 2-kb segment, but it is not possible to place them there unambiguously because they are repetitive. However, some of the subclones in the assembly span the repetitive region, and the researcher can pick one of these subclones for more directed analysis to determine its entire sequence. As we shall see in Chapter 11, such strategies are essential for sequencing large genomes with many repetitive sequences. Although some repeat sequences are present in microbial genomes, they are less frequent and, on average, much shorter than repeats in the genomes of eukaryotes, which makes it much more straightforward to sequence and assemble a bacterial genome.

Huge Numbers of Chain-terminating Sequencing Reactions Must Be Carried Out

One of the most labor-intensive and costly steps in the sequencing process is synthesizing the dideoxy-terminated DNA fragments that are to be separated by electrophoresis on slab gels or in capillaries. A sequencing reaction requires having the right components in the right amounts: single-stranded DNA template, nucleotides, dideoxynucleotides, an oligonucleotide primer, and a DNA polymerase that

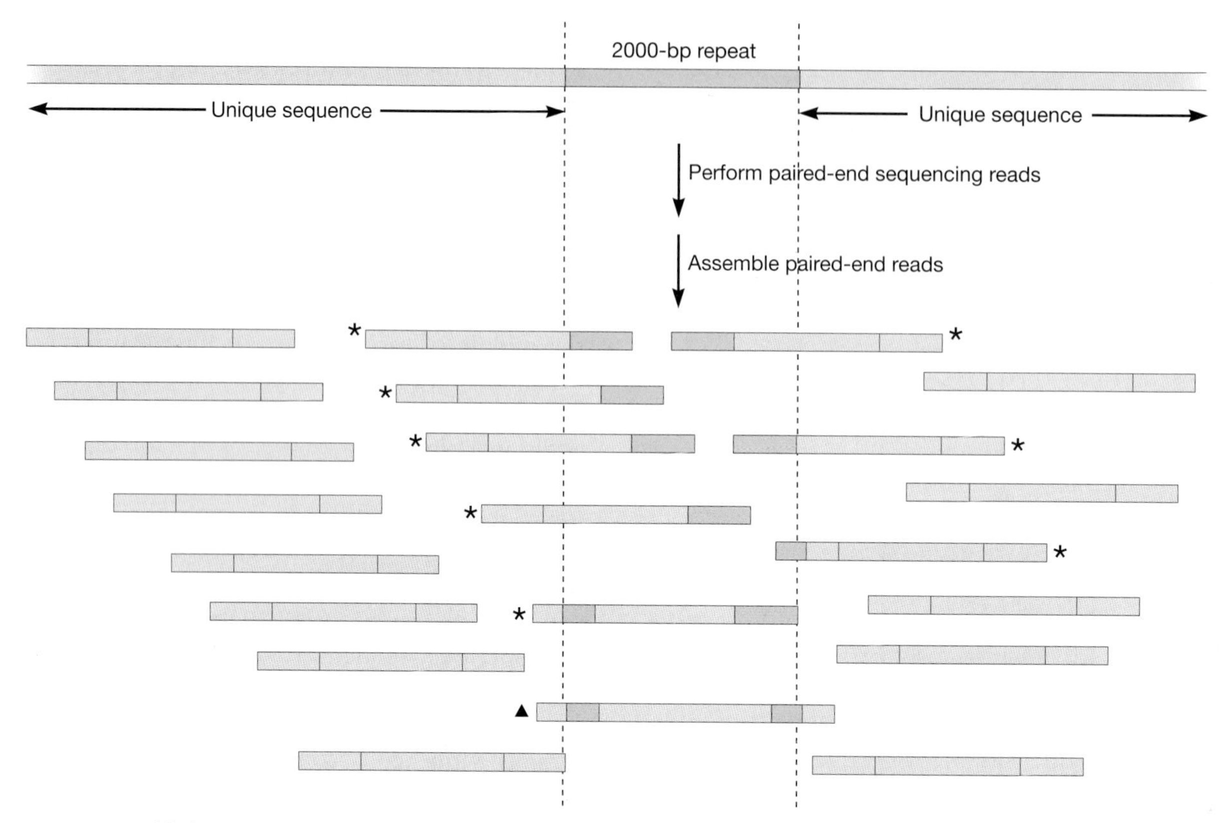

FIGURE 10-6

Clone coverage helps sequence across repetitive regions. In this segment of a genome, most of the 500-bp sequence reads from several 3-kb clones assemble unambiguously with each other because their sequences are unique. However, there is a region of 2000 bp in the middle of this assembly that is found multiple times elsewhere in the genome (shown in *yellow*). Therefore, the assembly algorithm cannot place the end unambiguously, because the sequencing reads fall fully within this repeat segment. However, some reads (marked with *) have unique sequences at one end that allow the repetitive end to be assembled into this contig. Because some of the 3-kb clones span the repeat entirely and contain some unique sequence on either end, it is possible to place the clone reads into the assembly as shown with certainty. One of the spanning 3-kb subclones (marked with a *small triangle*) can be chosen for further sequencing. In this way, the entire sequence of the region of the genome, including the precise sequence of the particular copy of the 2-kb repeat, can be determined.

extends the primer along the DNA template. All of these components have been modified to increase the reliability and efficiency of sequencing reactions; when tens of millions of such reactions are carried out, even the smallest improvement can produce dramatic reductions in time and costs. One of the most interesting improvements is the engineering of a new form of DNA polymerase that synthesizes longer fragments for sequencing and incorporates ddNTPs more evenly.

Stanley Tabor and Charles Richardson had been studying the detailed mechanisms of DNA polymerases for many years. Using their knowledge of the active site of the enzyme and how it interacts with the incoming precursors for DNA synthesis, they decided to alter a DNA polymerase so that it accepted ddNTPs, which are unnatural substrates for the enzyme, more efficiently. Tabor and Richardson changed one amino acid in a DNA polymerase and found that the enzyme now had a preference several thousand times higher for ddNTPs than for dNTPs. This increased preference leads to much more efficient and even incorporation of the chain terminators during the DNA-sequencing reaction. As a consequence, less DNA template is needed, interpreting the sequence is much easier, the sequencing reads are longer, and the accuracy of sequencing reads increases—all critical factors in large genome projects. This is an excellent example of redesigning a protein by combining knowledge of a protein's structure and its

FIGURE 10-7

Robotic devices greatly increase the efficiency of genomic sequencing. The need to perform many millions of DNA-sequencing reads led to the development of robots to assist in most of the steps of the sequencing process. Robots with imaging capabilities pick colonies from Petri plates and transfer them into microtiter plates for growth. Other robots pipette volumes as small as one-millionth of a liter into microtiter plate wells for template preparation, thereby performing the sequencing chemistry and preparing the sequencing reactions for loading into the DNA-sequencing machines. Shown here is a liquid-handling robot that sips and distributes small amounts of liquid from 192 samples at a time.

mechanism of action with molecular techniques of DNA cloning and directed mutagenesis, a field known as bioengineering.

Large-scale sequencing requires large-scale technology, and genome scientists have followed the example of manufacturing industries with the extensive use of robots to increase throughput and to improve reliability and accuracy. These robots are used at all stages of the sequencing assembly line (Fig. 10-7). There are robots that pick bacterial subclones growing on Petri plates and extract the DNA from each clone. There are robotic pipetting devices that deliver miniscule amounts of a solution very accurately, so that the sequencing reactions can be carried out in very small volumes, thus saving expensive reagents. The sequencing machines described below are robots with automatic loading of samples. Reactions are no longer carried out in individual test tubes, but in the wells of plastic vessels called microtiter plates (Fig. 10-8). These plates have been standardized with 96 wells or multiples of 96 wells and the various robots have been designed to handle these plates.

Fluorescent Labels and Electrophoresis in Sequencing Machines Revolutionize DNA Sequencing

In the early days of DNA sequencing, four separate reactions, each containing only one of the four ddNTPs, had to be performed for each DNA template, and each reaction had to have its own lane on

(a)

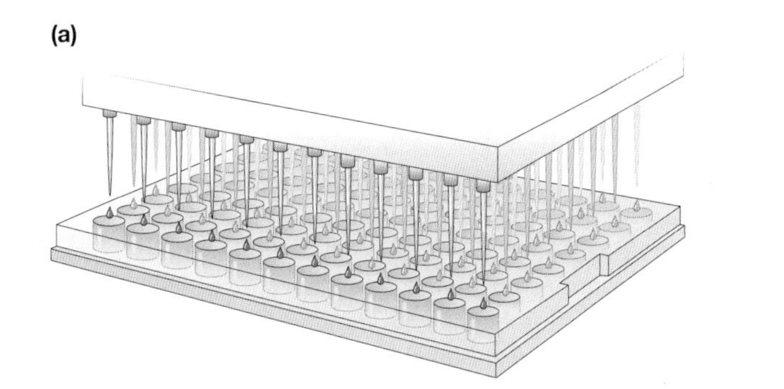

(b)

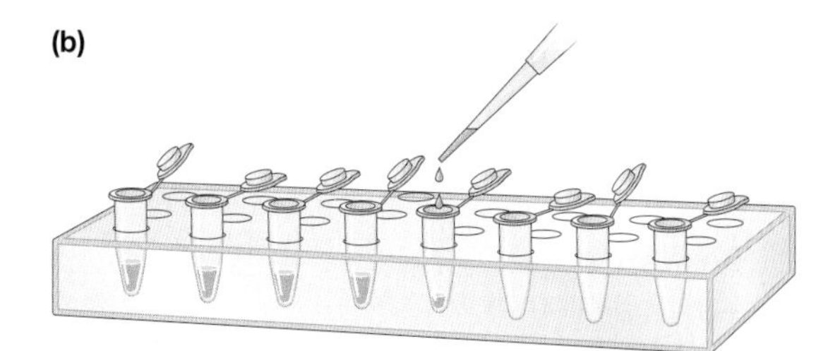

FIGURE 10-8

Microtiter plates. (*a*) Small (3 inches x 5 inches) plastic dishes called microtiter plates were developed in the 1960s by immunologists to perform multiple assays for antibody/antigen measurements. These plates have depressions called "wells" that typically can hold ~150 μl of liquid. The placement of the wells is highly standardized so that their center-to-center distance is always the same. This consistency is important for using multipipette devices, especially robotic ones like those described in Fig. 10-7. The standard microtiter plate contains 12 wells in each of 8 rows, for a total of 96 wells. (*b*) It is easy to see why pipetting into and out of the 96 wells would be easier than doing the same with 96 microcentrifuge tubes or standard test tubes. Unlike test tubes, which can be accidentally switched with one another, the fixed wells in the microtiter plate greatly reduce errors from sample mix-ups.

the electrophoresis gel. The extended DNA fragments were labeled with a radioisotope, typically ^{32}P or ^{35}S, so that the products could be detected by exposure of X-ray film to the gel. The DNA sequence was read from the image of bands on the X-ray film by manually noting which of the A, G, C, or T lanes has a band in each position as the fragments increase in size (Fig. 10-9). This method was tedious and very prone to error. Only about 100–150 bp could be determined in a set of four radioactive sequencing lanes, and the accuracy was so poor that in many cases more than 10% of the bases could not be identified or were incorrectly identified.

At a critical juncture, just as large-scale sequencing was being contemplated, Leroy Hood, Michael Hunkapiller, and Lloyd Smith at the California Institute of Technology began developing an alternative to radioactive-based DNA sequencing. They used fluorescently labeled ddNTPs with a semiautomated sequencing machine that was produced commercially by Applied Biosystems, Inc. In four-color fluorescent sequencing chemistry, each of the ddNTPs is labeled with a different dye (Fig. 10-10). These dyes are attached to the ddNTPs in such a way that they do not interfere with the DNA polymerase as it pairs a ddNTP with the appropriate complementary base in the DNA template. Because the fluorescent signals emitted by each of the dyes can be distinguished one from another, all four chain-terminating reactions can be performed in the same reaction vessel and separated in a single lane of an electrophoresis gel, rather than in the four lanes required with radioactive labeling. The fluorescently labeled fragments are detected automatically in real time during electrophoresis as each successive fluorescently labeled fragment, increasing in size by one base, passes the laser detection system of the sequencing machine (Fig. 10-11). The dye attached to the ddNTP at the end of a DNA fragment is excited by the laser and emits a signal specific for that dye, thus indicating which of the four bases is present at the end of the chain-terminated fragment. Sensitive detectors collect these signals, which are sent to a computer for analysis. The automatic collection of data by the detectors eliminates the most difficult and error-prone step of conventional sequencing, that of reading the image of the radioactive gel, greatly increasing the ease and efficiency of sequencing.

The first versions of the sequencing machines employed slab gels similar to those used for manual sequencing, but with more lanes and producing longer read lengths. These machines still required the time-consuming steps of pouring the gels and loading samples into slots at the top of the gels. Furthermore, because DNA fragments did not move uniformly in the lanes of the gel in these first sequencing machines, researchers had to spend a great deal of

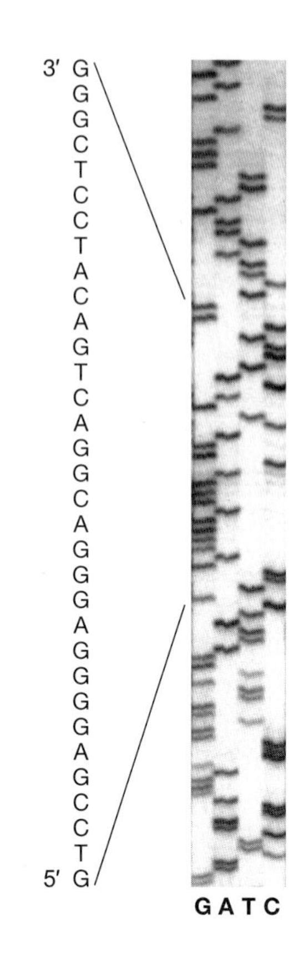

FIGURE 10-9

Interpreting DNA sequence from a radioactive sequencing experiment. An image on X-ray film (called an autoradiogram) shows four lanes of a DNA-sequencing gel in which radioactive phosphate was used to label the terminating DNA-sequencing reactions. The reactions for each of the four bases were performed in separate tubes, one each for the G, A, T, and C dideoxy terminators, and each electrophoresed in a separate lane on the gel. Thus, four lanes were required to sequence each DNA fragment. To determine the DNA sequence from the autoradiogram, beginning at the bottom of the image, bases were "called" at positions in each lane that contain a horizontal "band" of DNA fragments of a precise size. The sequence was read by continuing one base at a time while moving up the image. In this example, a region of the autoradiogram is read from bottom to top, starting with a run of four Gs, followed by an A, then three Gs, an A, etc. Typically only ~100 bp could be determined from one set of four lanes, because of electrophoretic anomalies, problems with the chemistry of the sequencing reaction, and mistakes made by the researchers while manually calling the bases.

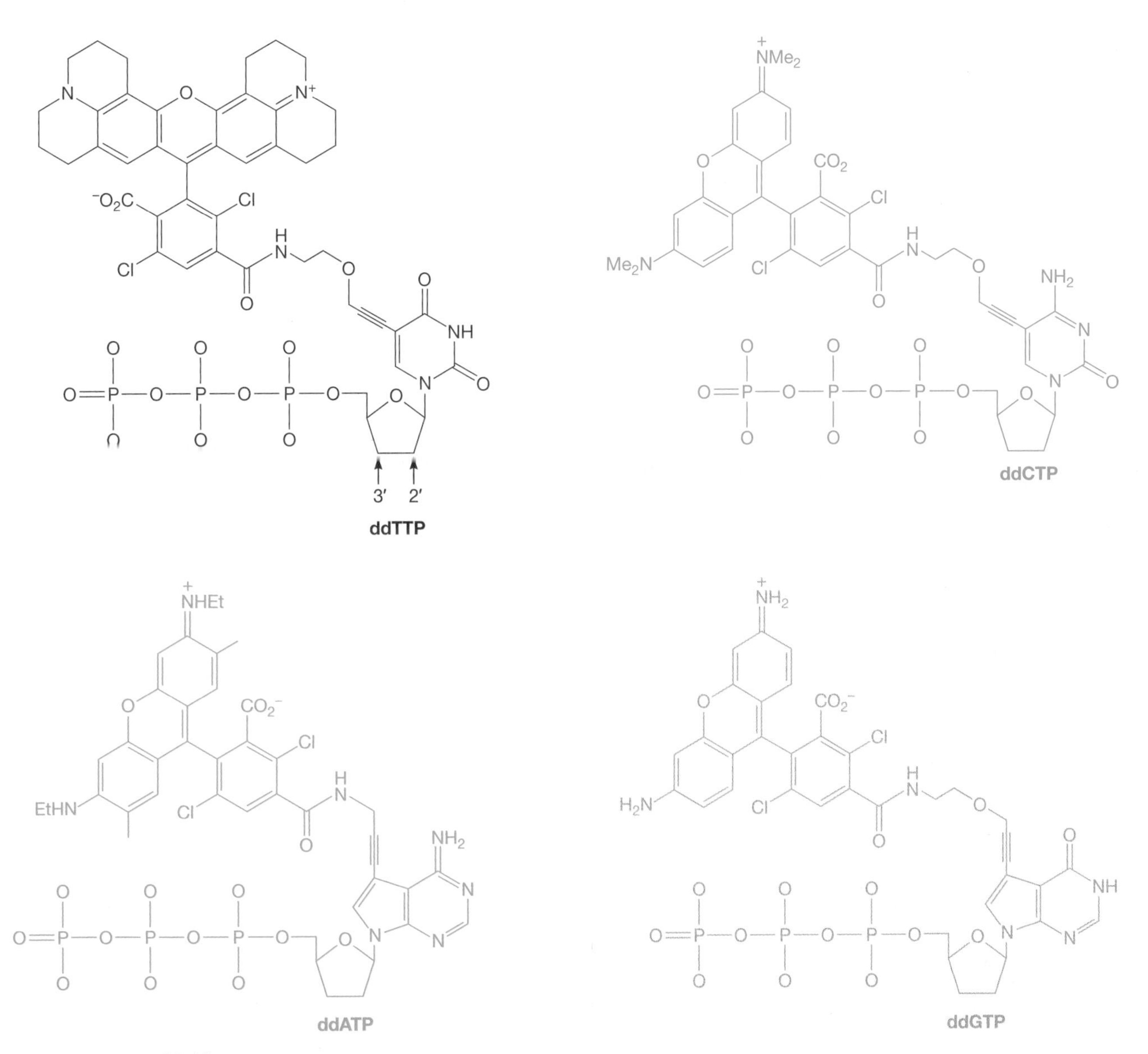

FIGURE 10-10

Structures of fluorescently labeled terminating dNTPs (ddNTPs). Each ddNTP contains a ribose molecule with the triphosphate at the usual 5´ position and the base at the usual 1´ position. The ribose lacks an OH, at the 3´ position, which makes the sugar a dideoxyribose. These precursors are incorporated by DNA polymerase into a growing DNA chain, but terminate synthesis because the next precursor cannot be attached. A different fluorescent dye is attached to each of the four ddNTPs (ddTTP, ddCTP, ddATP, and ddGTP), at positions on the bases that do not interfere with base pairing during DNA synthesis.

time manually aligning, or "tracking," the lanes before the computer could identify or "call" the bases in the lane. It was therefore a great advance when capillary sequencing machines were introduced in the late 1990s. In these machines, the fluorescently labeled DNA-sequencing products are electrophoresed through small-diameter (<0.5-mm) capillary tubes about 48 cm long that contain a material similar to the polyacrylamide used in slab gels. These machines have a much higher degree of automation than the slab gel machines. Instead of manually pipetting the sequencing reactions into wells on the top of a slab gel, the capillary machine automatically "sips" a tiny reaction volume from each well in a microtiter plate and delivers the sample onto the capillary. The same capillary can be used many times, so that with many plates stacked onto a capillary sequencing machine and sequencing reactions auto-

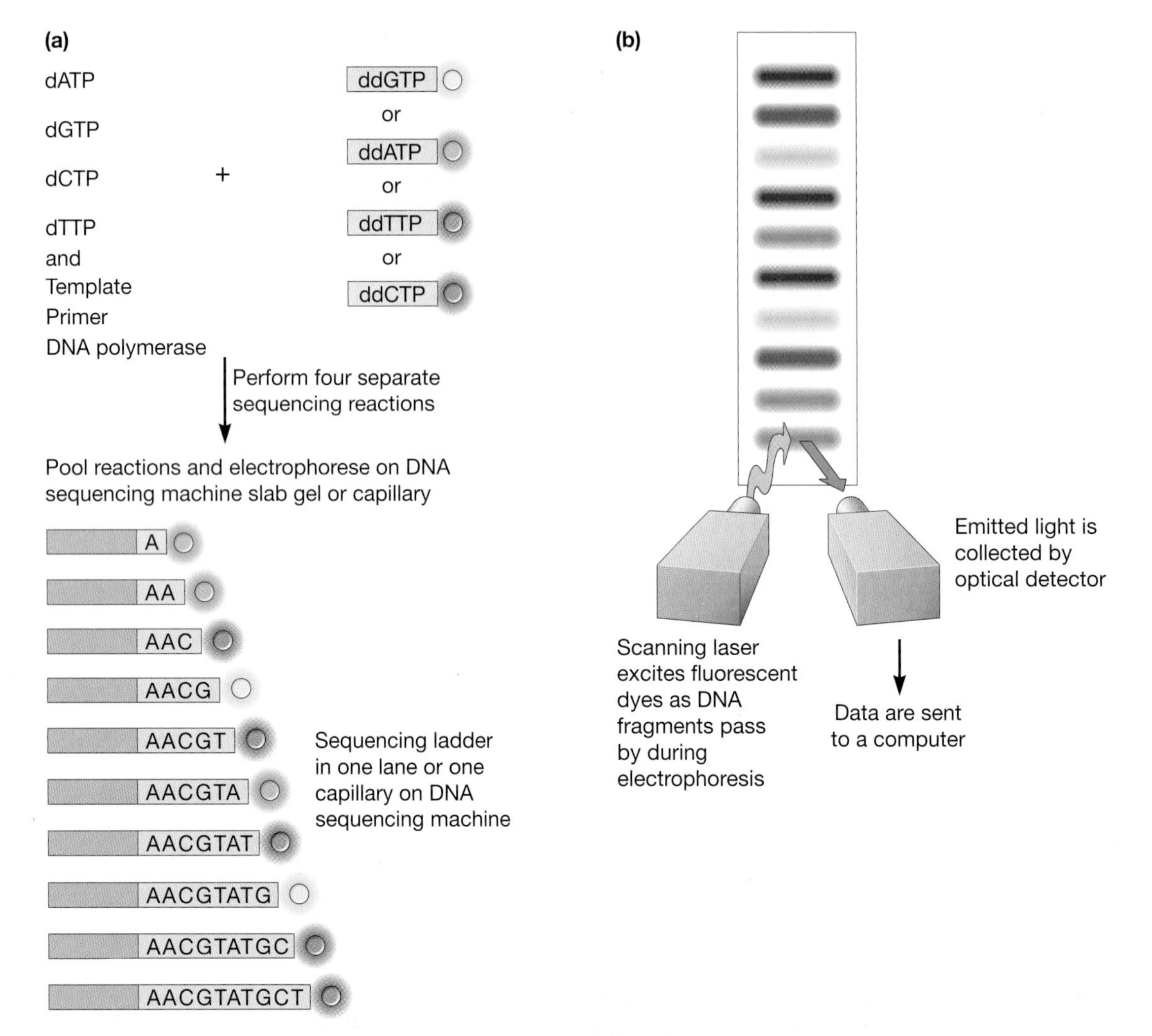

FIGURE 10-11

Automated sequencing. (*a*) Four different termination sequencing reactions are analyzed in a single lane in the automated sequencer. To sequence a DNA template, four separate sequencing reactions are performed, each with three of the standard dNTPs at a high concentration, and a combination of the standard dNTP and a fluorescently labeled ddNTP for each of the four bases found in DNA. After the extension/termination reactions are performed for a template, all four are pooled, and the mixture is electrophoresed on the automated sequencer. One lane separates the ladder of four differently colored DNA fragments as a function of their sizes. The sequence of the subclone, with corresponding colored fragments, is shown. (*b*) The laser detection system in a four-color fluorescence automated DNA sequencer. As the ladder of fluorescently labeled, terminated DNA fragments is electrophoresed in the automated sequencer, each fragment passes a laser and detector. The laser shines a beam of a wavelength that excites the four fluorescent dyes, and each dye emits a beam of visible light of a specific wavelength. The detector collects this light and sends a signal to a computer attached to the sequencing machine.

matically loaded by a robotic device, the machine can be left to analyze a large number of reactions without supervision. The end result of all of these technical advances is that the largest laboratories can sequence tens of millions of nucleotides each day.

Computer Algorithms Identify Which Base Is Which in the Output from a Sequencing Machine

The fluorescent signal emitted by each DNA fragment as it passes the detection laser is sent to a computer that identifies, or "calls," the base at the 3′ end of the fragment. However, it is not a simple matter to assign an A, G, C, or T to a fluorescent signal as it is collected. The behaviors of DNA fragments as they move through a slab gel or capillary, the chemical and fluorescent properties of the dyes, and interactions between bases near each other in the DNA helix all contribute to incomplete separation of the signals as they are collected by the detector. To solve this problem, software algorithms were developed to analyze the signal collected from the detector and assign one of the four bases to the DNA fragment. Base-calling is therefore the process of translating fluorescent signals into a sequence of bases.

These algorithms have been continually refined to improve the accuracy of the base-calling. For example, the program Phred, developed by Phil Green and Brent Ewing, assigns a probability score for the confidence of each assignment (Fig. 10-12). In the first step of the analysis, the program predicts the idealized peak locations in the trace by setting evenly spaced places in a trace where a peak should be. In the second step, the program identifies the actual peaks in the trace. The algorithm matches the observed peaks with those in the idealized trace and, in doing so, omits some of the observed peaks and splits others. Third, the program identifies the base depending on which of the four detectors (one for each of the four bases) has supplied the signal. Finally, the program checks observed peaks that have not been called and tries to fit them into the sequence. Such algorithms can assess the quality of each identified base during a sequencing read, providing "just-in-time" quality control, similar to quality control processes used in manufacturing. This constant assessment of the accuracy of each sequencing read allows researchers to monitor the process and adjust parameters to improve the analysis.

The output from a run on a DNA-sequencing machine is a set of tables that show the order of the bases along the length of the DNA fragment with statistical scores showing the level of confidence that each base has been called accurately. In addition, the computer generates a histogram, called a sequence

Phred scores

9 9 9 9 9 9 10 13 19 14 9 12 10 18 19 27 29 29 29 24 28 28 18 17 15
8 8 27 50 50 50 44 44 47 50 50 50 50 50 56 56 50 50 56 56 18 56 56
51 46 46 42 56 56 42 42 37 32 32 31 31 31 40 50 50 44 44 50 50 43
35 35 35 42 50 50 50 50 50 56 56 46 46 42 42 42 42 50 50 56 56 56
42 42 42 42 36 35 35 35 36 42 50 29 29 28 50 50 47 47 44 40 40 35
42 44 56 56 56 56 56 56 56 56 56 56 .... 56 56 56 56 56 44 42 42
42 42 56 56 56 56 56 56 56 56 56 56 56 42 42 42 42 42 42 44 33 33
31 33 31 36 36 36 36 42 42 42 39 39 42 46 46 46 42 42 42 35 35 35
46 42 15 15 15 15 27 40 20 22 10 10 17 17 24 19 16 9 9 9 15 15 15
15 17 22 19 25 25 22 17 18 14 19 19 21 19 22 25 24 24 27 27 27 27
25 25 9 9 8 11 11 17 16 26 12 17 14 14 10 15 31 27 27 11 12 9 9 9

"Called" DNA sequence

A A G A T G A A G A A T A G T C T T T T G G G
T G T A T T T G G T G A A G G T A T C A A T T
A C C A A G A T A T A A A G C A A T C T T T A
A T T T A T G G G T G C T T T T G A A A T G C
A A G T G G C A C T A T G G G G T A T T T A T
T T T G A C G A A T T .... A C G A C C T T G
C A T T C C G A C T G A T G G T C G T A T T G
A C A G A T T T G C T T G G A T T G A A G T T
G C T G C C A G T T A T A G G C T G G A A T G
C A A C T A G T C T A G C A A A A C T T C T T
T C G G C T T A T G G C G G A A T T A T

FIGURE 10-12

The Phred computer algorithm. The nested DNA fragments labeled with four fluorescent dyes do not produce perfectly spaced, uniformly labeled signals as they electrophorese past the laser and detector in the sequencing machine. A computer program named Phred is used to "call" the bases on each terminated DNA fragment as the signals are collected. The program corrects for electrophoretic and labeling anomalies and, based on behaviors of many DNA fragments whose sequences are known with certainty, defines a probability score for each base at each position in the DNA fragment being sequenced. A Phred score of 20 means that there is a 99% likelihood that the called base is correct. A score of 40 means that there is a 99.99% likelihood that the call is correct, or a 1 in 10,000 chance that it is incorrect. This example shows Phred scores for part of a typical sequencing read. Note the poor scores at the beginning of the read, which occur because it is difficult to collect data from the very short DNA fragments in the nested series. Similarly, the quality of the base calls diminishes at the end of the read, usually at 500–800 bp, because the electrophoresis matrix loses its ability to separate DNA fragments differing by only a single base with longer fragments. The Phred scores for the base calls in the middle of the read are very high, many >40, which means that there is a very high likelihood that the called sequence is correct.

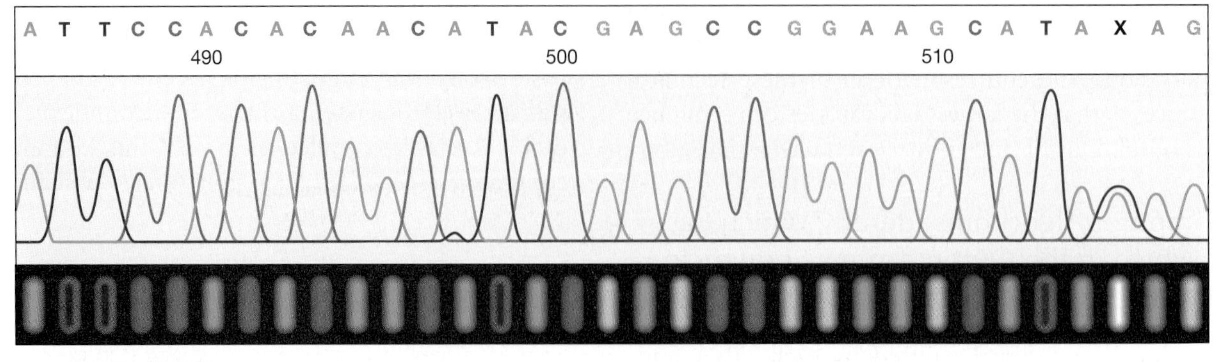

FIGURE 10-13

A sequence trace file showing the readout from an automated DNA sequencer. The Phred-called sequence is shown from *left to right* at the top of the figure. The distribution of the signals from the four colors is shown in the histogram, or trace, for a segment of an electrophoresis run. The trace shows each of the bases in a different color along the *x*-axis, and the intensity of the signal is shown on the *y*-axis. Some of the DNA fragments give a very tight, sharp peak, whereas others are broader and overlap with the signals from the adjacent signals. In this example, there is a segment near the right end of the trace where the sequence quality is particularly poor, as evidenced by low peak heights and overlapping peaks. Phred was unable to call the base at position 515 (marked with an X) because the two peaks overlap.

trace file, that shows the intensity of each of the four fluorescent signals as a function of time of electrophoresis (Fig. 10-13). These histograms are very useful and important for seeing at a glance which regions have poor quality sequence and questionable calls (Fig. 10-13). These regions must be given special attention in the process of finishing a genomic DNA sequence.

Automated sequencing machines and the associated robots for preparing samples are indispensable for genomic sequencing. A major genome-sequencing center resembles nothing as much as an industrial facility, with large rooms filled with machines. In the past, one energetic researcher working very steadily could perform perhaps 50 sequencing reactions in a single workday; now, one person can oversee robots that are carrying out more than 100,000 reactions in a day.

Longer Sequences Are Put Together from Many Sequencing Reads 500–1000-bp Long

Assembly is the computational reconstruction of a long segment of DNA from the overlapping sequence information gathered from the short (500–1000-bp) sequencing reads. The algorithms for assembly were derived from computational methods developed during the 1970s and early 1980s for determining how closely DNA sequences are related to each other. These original sequence alignment algorithms were used to test a new sequence against a database of many other sequences to determine whether the new sequence corresponded to a known gene or a related version of the gene from another organism (Chapter 12). The alignment methods break the sequencing read into small segments or "words" of several base pairs and search the database to identify other sequences that contain those words in the same order. Probability scores are calculated for the likelihood that a match between the sequences is not due to chance.

The same types of algorithms are used in sequence assembly. Consider the relatively simple example of assembling the sequence of a BAC clone containing a 200,000-bp stretch of a bacterial genome that has been sequenced to 8x redundancy (Fig. 10-14). The alignment algorithm considers each sequencing read, one-by-one, and identifies which other reads overlap with it. For a particular read, these overlaps extend the sequence 5′ and 3′ from the starting read, thus generating a contiguous sequence—a "contig"—that is longer than a single read. Each contig is then extended at its ends by additional alignments, and the process is repeated until no additional assembly can be obtained. The end result of this process is a continuous sequence of the 200,000-bp clone. In practice, the assembled sequence will have gaps and a variety of inaccuracies,

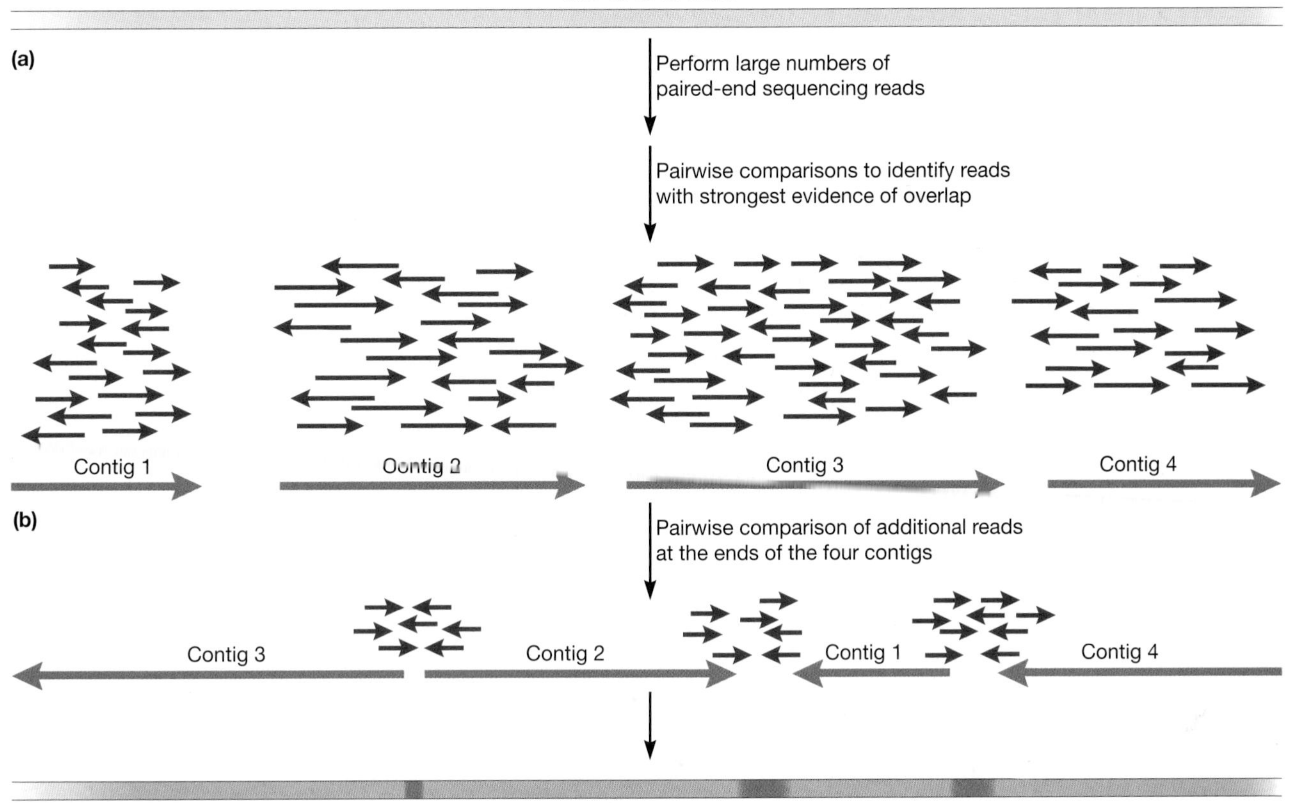

FIGURE 10-14

Assembly of a long sequence from many short sequencing reads. This example depicts a 200,000-bp segment of assembled sequence. (*a*) Large numbers of paired-end sequencing reads are performed on subclones from a 3-kb library constructed from the genomic DNA. The assembly program compares the base-called end read sequences with each other and identifies those reads that have strong evidence that they overlap (shown in *red*). This results in a set of sequence contigs (each shown as a *blue arrow*). (*b*) Another round of assembly is performed that leaves out the reads that provided the strong evidence of overlap and assumes that those contigs are correct. Additional connections, which tie together some of the contigs with each other, are made, although the evidence that these belong together is often not as strong as that used to build the first set of contigs. The result is a fully assembled BAC sequence.

such as unknown base pairs, incorrect base pairs, and misassembled segments (Fig. 10-15). These arise from nonrandomness of the subclone library, poor quality of some reads, chance variations, and, especially, misassembled repetitive sequences.

Special algorithms had to be developed to deal with misassembly caused by repetitive regions. First, because many repeats are shorter than the read lengths that can be obtained with the sequencing machines, each repetitive sequence has some unique sequence on one or both sides. The assembly algorithms can ignore ("mask") the repeats and put the initial sequence together based on only the non-repetitive stretches of sequence. Second, many repeated DNA segments are not identical, but have a range of sequence differences. The assembly algorithms can be trained to recognize these differences and avoid joining two almost identical segments that actually belong in different regions of the genome. Third, the paired-end sequencing strategy described above allows repeat sequences to be assembled into a unique place in the genome when unique sequences are present at one end or in the internal portions of the subclone. These and other strategies have been refined over the years so that a draft sequence of even a very large genome with many

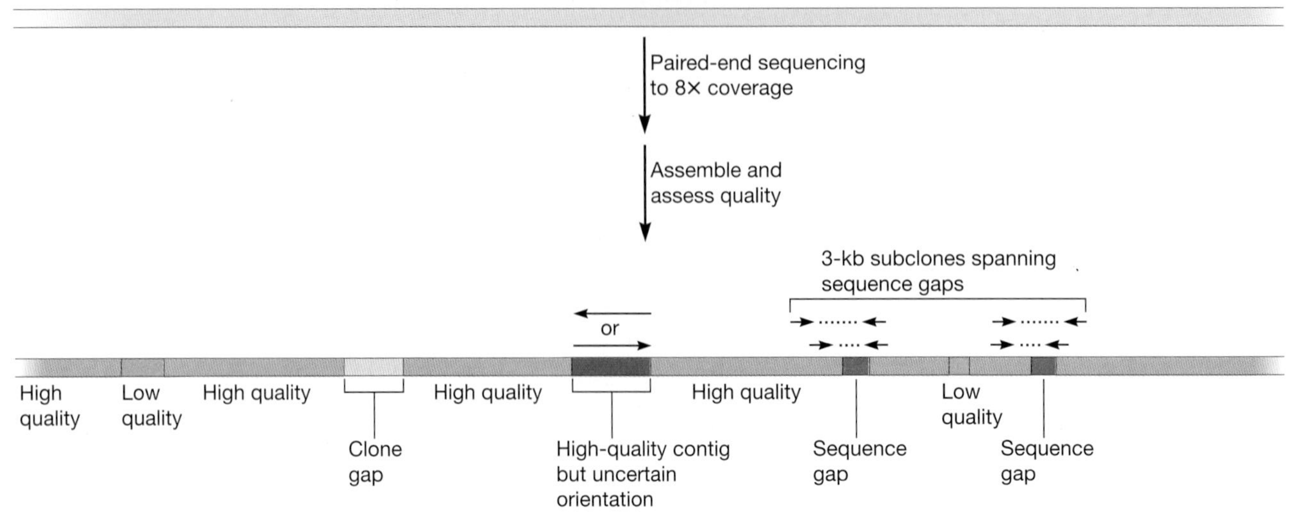

FIGURE 10-15

Several types of problems with assembled sequences. The assembled 175,000-bp segment shown here has several long regions of high-quality sequence (*blue*), but also has regions of poor quality sequence (*orange*) and a region whose orientation in the assembly is uncertain, such that it might be inverted relative to its correct orientation (*purple*). There are two sequence gaps that are spanned by one or more 3-kb subclones (*red*), and a sequence gap that is also a clone gap (*gray*), which means that the missing segment is not present in any of the sequenced 3-kb subclones. The availability of the end sequences of the 3-kb clones that span the two sequence gaps allows the two contigs that contain each of their end segments to be connected to and oriented relative to each other. However, the sequence gap with no clone coverage leaves two large contigs whose order and orientation cannot be determined without additional information, such as genetic or physical mapping data.

repeats can be assembled, albeit with errors, uncertainties, and missing data, from whole-genome shotgun sequence data.

Powerful Computers Are Essential for Genome-Scale DNA Sequencing

Computers and the internet have had a profound impact on all of biology, and genomics is no exception. DNA sequencing presents formidable computational challenges. These include management of the data, sequence assembly, and analysis of the sequence for interesting features. For example, the original approaches for sequence assembly used a computer algorithm to score whether sequence reads matched over part of their lengths. This strategy was computationally intensive, because the algorithms compared each sequence read to every other read, requiring a huge number of pairwise comparisons. For example, sequencing a viral genome that is 5000 bp long with 8x coverage requires 80 sequencing reads of 500 bp ([5000 ÷ 500] x 8). Assembling the data from these reads would require 3160 pairwise comparisons (equal to all combinations of two sequencing reads, $[N^2 - N] \div 2$). This number grows very rapidly as the genome size increases; even a bacterial genome of 5,000,000 bp would require 1.25×10^{11} pairwise comparisons if assembled in this way.

Computers were first used to manage and assemble DNA sequences during the sequencing of ϕX174, which, at 5386 bp, was the longest DNA molecule to be sequenced at that time. Yet this project was by no means the first use of computers in biology. In 1952, John Kendrew used an early electronic computer called EDSAC I in Cambridge for calculating Fourier transforms while determining the structure of myoglobin. In the 1960s, scientists studying protein sequences had made extensive use of computers for analysis and management of data. During this period, computers were relatively rare and very expensive, and biologists had to use large mainframe machines in central computer facilities.

Beginning in the 1970s, smaller, less expensive machines became available and DNA-sequencing laboratories had more ready access to them. Roger Staden, for example, wrote his first programs for the DEC PDP-11/45 minicomputer, a powerful machine for its time with 56 KB of memory and 2.5-MB exchangeable disc drives. Fortunately, an enormous increase in the speed and storage capacity of computers paralleled development of whole-genome sequencing, so that current desktop computers have performances that are thousands of times superior to Staden's PDP-11/45. In addition, new algorithms were developed that counted small "words" of a few base pairs in each sequencing read and matched up reads that had many of the same words. These and other improvements in sequence assembly algorithms make it possible to assemble very large sequences from millions of sequencing reads.

Even the seemingly trivial matter of screening a new sequence against known ones required much more effort in the early days of sequencing than it does now, despite the database of known sequences being miniscule compared to the present versions. This period was, of course, prior to the existence of the internet, and institutions or individual researchers had first to import the entire database of sequences into their computers before they could determine whether a new sequence matched or was similar to one in the database.

Managing and curating the ever-increasing flow of sequence data posed a similar challenge. The data in the early databases were distributed in print. All the world's known protein sequences were published annually in the *Atlas of Protein Sequence and Structure*, founded by Margaret Dayhoff, and Richard Roberts and his colleagues published an annual catalog of restriction enzymes and their recognition sequences in the journal *Nucleic Acids Research*. Later, sequence databases were distributed on compact discs, as well as other databases such as Medline, a catalog of biomedical research papers. All this changed beyond recognition with the development of the internet and, in the mid-1990s, hypertext markup language (HTML) that made the World Wide Web possible. Now terabytes (millions of megabytes) of biological data are instantly accessible and searchable. The databases have become interlinked so that a researcher can move seamlessly between them. Sequencing projects now upload their data to public databases each night, thereby making the data available to laboratories throughout the world.

Hemophilus influenzae Was the First Free-living Organism Whose Genome Was Sequenced

By the early 1990s, the largest complete genome that had been sequenced was that of cytomegalovirus, totaling 229,354 base pairs. Genome researchers were beginning to use map-based strategies to sequence the genomes of multicellular organisms, but automated sequencing methods were still in their infancy. It was expected that much time would pass before any of these very large genome sequences would be completed. Venter at TIGR argued that it should be possible to sequence the entire genome of a bacterium by using only shotgun sequencing, not relying on a set of mapped clones to help with assembly. He and his collaborator, Hamilton Smith, chose to sequence the genome of the human pathogen *H. influenzae*, a small bacterium that causes upper respiratory and middle ear infections, and even meningitis. This bacterium was a good candidate for sequencing—it is medically important, there was no physical clone map, and it has a G + C sequence composition of about 38%, not very different from the genome of human beings. The size of the genome was known to be about 1.8 million base pairs, at the lower end of the size range for bacteria, but still significantly larger than any genome yet sequenced to completion.

Venter and his colleagues obtained a purified sample of genomic DNA from the bacterium, sheared it into fragments of uniform size averaging a little less than 2000 bp, and cloned the fragments into a plasmid library. They also made a second library in a different vector with fewer clones and larger inserts—ranging in size from 15,000 to 20,000 bp. After preparing purified DNA from these subclones, they performed a total of 28,643 sequencing reactions, about two-thirds of which were from one end of a clone, and the remaining one-third from the other end. By obtaining DNA sequence from both ends of some of the subclones, they were then able to use this information to aid in the assembly process. The sequencing reactions were analyzed on 14 of the earlier-generation automated slab gel sequencing machines, at the time an unheard-of number of sequencers for a single laboratory. TIGR used a set of algorithms to assemble the overlapping sequence data into as few contiguous pieces as possible.

The initial assembly, which took advantage of the known distances between the two end reads of many of the sized subclones, resulted in long stretches of contiguous sequences. In theory, given the number of sequencing reactions performed, the read lengths obtained, and the number of failed sequencing reactions, TIGR should have produced a sequence with about 50 gaps. However, the initial assembly resulted in 140 contigs. This was probably because the subclone libraries did not contain truly randomly distributed fragments of the *H. influenzae* genome, or perhaps because of nonrandom shearing and ligation or retention of the fragments in the cloning vector.

This bias was not a serious problem. The TIGR scientists knew that gaps were an inevitable part of the process of shotgun sequencing and that additional work would be necessary to fill them. But, without a long-range physical or genetic map to aid them, there was no way to know the order or orientation of the 140 contigs generated in the initial assembly. So, at this stage in the project, they employed a variety of different strategies to fill the gaps in the sequence. These included making new libraries in cloning vectors different from those used for the initial libraries, and searching for sequences that would extend contigs by screening with probes made from DNA sequences near the ends of the contigs. In addition, TIGR used computational analysis of the sequences of known *H. influenzae* proteins to identify genes split over multiple contigs. This allowed them to place contigs next to each other in the correct orientation.

In the end, the TIGR team produced a single, contiguous sequence of 1,830,137 bp, providing for the first time a glimpse of the complete set of instructions for a free-living organism (Fig. 10-16). The group scrutinized the sequence for its coding content and identified 1743 predicted coding regions, which represented proteins in more than 100 different biological categories. These analyses increased the number of known *H. influenzae* genes by more than tenfold. The bacterial research community was transformed overnight; it suddenly had at its fingertips a massive amount of sequence data containing all the genetic information of the organism. Even more important than the biological messages gleaned from the sequence, this experiment proved that the whole-genome shotgun method worked for organisms with a small, relatively repeat-free genome.

Many Viral and Bacterial Genomes Have Been Sequenced

Methods for whole-genome shotgun sequencing and assembly are now well established for genomes ranging in size from 1 million to 3 billion base pairs, and are especially effective for small bacterial genomes. Many improvements in sequencing strategies and technology have been made, so that now a large sequencing laboratory can produce most of the data for a microbial genome in a single day. These methods have become routine for large sequencing centers, and by 2006, 11 years after the publication of the *H. influenzae* genome sequence, more than 300 prokaryotic genomes have been sequenced.

One of the early motivations for sequencing microbial genomes was medical—microbial pathogens continue to take a terrible toll of human life throughout the world. Tuberculosis, for example, infects one-third of the world's population and kills 2 million people annually. The genome of its causative agent, *Mycobacterium tuberculosis*, has been sequenced, as have those of anthrax, pneumonia, cholera, and many of the other bacterial pathogens. The hope and expectation is that these genome sequences will lead to the identification of genes and their proteins, which can be used to devise new therapies. The genomes of many other bacteria that are important for a wide variety of agricultural, industrial, and environmental problems have also been sequenced, and this trend is likely to continue, given the importance of the millions of species on Earth and the decreasing cost of obtaining complete genomic sequences.

Bacterial genome sequences are particularly valuable for understanding life processes because they are extremely dense with information compared to the much larger genomes of multicellular organisms. The structures of bacterial genes are simple, consisting of continuous open reading frames uninterrupted by introns and with relatively simple regulatory sequences immediately flanking the protein-coding regions. These features make it straightforward to use bioinformatics to identify essentially all the genes in a bacterial genome with a high degree of confidence, based simply on the sequence.

As important and fascinating as microbes are, we are inescapably interested in ourselves. In the next chapter, we will move up three orders of magnitude in scale, from microbial genomes that average about 3,000,000 bp to the human genome at 3,000,000,000 bp.

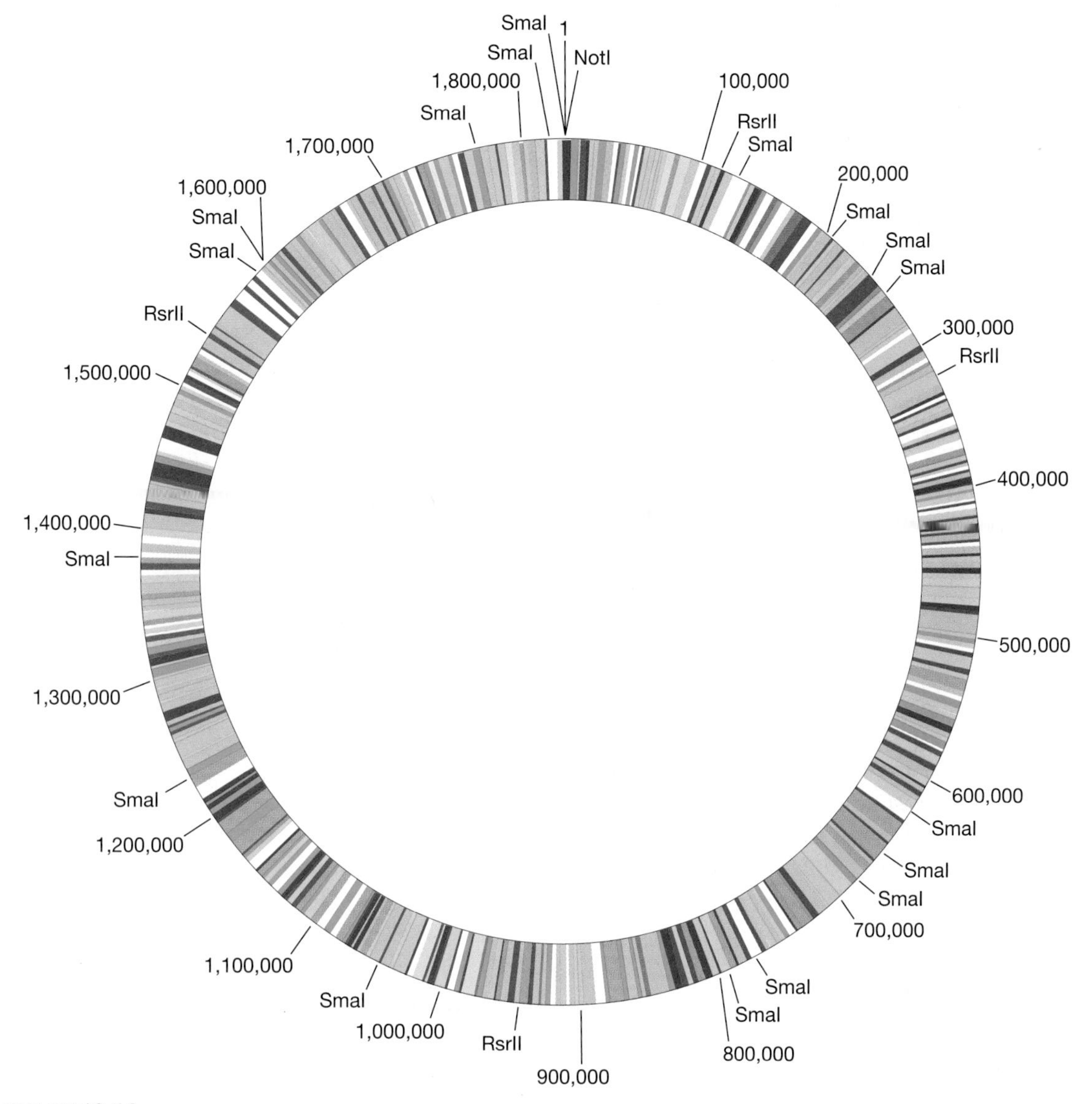

FIGURE 10-16

Sequence map of the *Hemophilus influenzae* genome. The map is circular because the bacterium's genome is a 1.83 million bp (Mb) double-stranded DNA circle. The sequence begins at a unique NotI restriction enzyme cleavage site, designated as base #1, and continues clockwise. The circle shows the locations of the >1700 known or predicted protein-coding segments, as well as other features of the genome such as tRNA and rRNA genes, that were understood at the time the sequence was completed in 1995. Each color represents a functional category (such as transcription factors) for the annotated proteins. The annotation of the *Hemophilus* genome has improved since that time because of additional sequences, research from other laboratories, and comparisons with the sequences of other bacterial species.

Reading List

Early DNA-Sequencing Technology

Sanger F., Nicklen S., and Coulson A.R. 1977. DNA sequencing with chain-terminating inhibitors. *Proc. Natl. Acad. Sci.* **74:** 5463–5467.

Maxam A.M. and Gilbert W. 1977. A new method for sequencing DNA. *Proc. Natl. Acad. Sci.* **74:** 560–564.

Sanger F. 1988. Sequences, sequences, and sequences. *Annu. Rev. Biochem.* **57:** 1–29.

Tabor S. and Richardson C.C. 1990. DNA sequence analysis with a modified bacteriophage T7 DNA polymerase. Effect of pyrophosphorolysis and metal ions. *J. Biol. Chem.* **265:** 8322–8328.

Tabor S. and Richardson C.C. 1995. A single residue in DNA polymerases of the *Escherichia coli* DNA polymerase I family is critical for distinguishing between deoxy- and dideoxyribonucleotides. *Proc. Natl. Acad. Sci.* **92:** 6339–6343.

Sequencing of Phage and Viral Genomes

Holley R.W., Apgar J., Everett G.A., Madison J.T., Marquisee M., et al. 1965. Structure of a ribonucleic acid. *Science* **147:** 1462–1465.

Fiers W., Contreras R., Duerinck F., Haegeman G., Iserentant D., et al. 1976. Complete nucleotide sequence of bacteriophage MS2 RNA: Primary and secondary structure of the replicase gene. *Nature* **260:** 500–507.

Sanger F., Air G.M. Barrell B.G., Brown N.L., Coulson A.R., et al. 1977. Nucleotide sequence of bacteriophage ϕX174 DNA. *Nature* **265:** 687.

Sanger F., Coulson A.R., Hong G.F., Hill D.F., and Petersen G.B. 1982. Nucleotide sequence of bacteriophage λ DNA. *J. Mol. Biol.* **162:** 729–773.

Bankier A.T., Beck S., Bohni R., Brown C.M., Cerny R., et al. 1991. The DNA sequence of the human cytomegalovirus genome. *DNA Seq.* **2:** 1–12.

Computation in DNA Sequencing

Staden R. 1977. Sequence data handling by computer. *Nucleic Acids Res.* **4:** 4037–4051.

Dayhoff M.O., Eck R.V., Chang M.A., and Sochard M.R. 1965. *Atlas of Protein Sequence and Structure.* National Biomedical Research Foundation, Silver Spring, Maryland.

Dayhoff M.O., Schwartz R.M., and Orcutt B.C. 1978. A model of evolutionary change in proteins. In *Atlas of Protein Sequence and Structure* (ed. M.O. Dayhoff), pp. 345–352. National Biomedical Research Foundation, Washington, D.C.

Ewing B., Hillier L., Wendl M.C., and Green P. 1998. Base-calling of automated sequencer traces using Phred. I. Accuracy assessment. *Genome Res.* **8:** 175–185.

Ewing B. and Green P. 1998. Base-calling of automated sequencer traces using Phred. II. Error probabilities. *Genome Res.* **8:** 186–194.

Bonfield J.K., Smith K.F., and Staden R. 1995. A new DNA sequence assembly program. *Nucleic Acids Res.* **23:** 4992–4999.

Dear S. and Staden R. 1991. A sequence assembly and editing program for efficient management of large projects. *Nucleic Acids Res.* **19:** 3907–3911.

Fluorescence-based DNA Sequencing

Hood L.E., Hunkapiller M.W., and Smith L.M. 1987. Automated DNA sequencing and analysis of the human genome. *Genomics* **1:** 201–212.

Horvath S.J., Firca J.R., Hunkapiller T., Hunkapiller M.W., and Hood L. 1987. An automated DNA synthesizer employing deoxynucleoside 3′-phosphoramidites. *Methods Enzymol.* **154:** 314–326.

Hunkapiller T., Kaiser R.J., Koop B.F., and Hood L. 1991. Large-scale and automated DNA sequence determination. *Science* **254:** 59–67.

Rosenblum B.B., Lee L.G., Spurgeon S.L., Khan S.H., Menchen S.M., Heiner C.R., and Chen S.M. 1997. New dye-labeled terminators for improved DNA sequencing patterns. *Nucleic Acids Res.* **25:** 4500–4504.

Heiner C.R., Hunkapiller K.L., Chen S.M., Glass J.I., and Chen E.Y. 1998. Sequencing multimegabase template DNA with BigDye terminator chemistry. *Genome Res.* **8:** 557–561.

Map-based and Whole-Genome Shotgun Sequencing

Chen E.Y., Schlessinger D., and Kere J. 1993. Ordered shotgun sequencing, a strategy for integrated mapping and sequencing of YAC clones. *Genomics* **17:** 651–656.

Weber J.L. and Myers E.W. 1997. Human whole genome shotgun sequencing. *Genome Res.* **7:** 401–409.

Coulson A., Sulston J., Brenner S., and Karn J. 1986. Toward a physical map of the genome of the nematode *Caenorhabditis elegans. Proc. Natl. Acad. Sci.* **83:** 7821–7825.

Olson M.V., Dutchik J.E., Graham M.Y., Brodeur G.M., Helms C., et al. 1986. Random-clone strategy for genomic restriction mapping in yeast. *Proc. Natl. Acad. Sci.* **83:** 7826–7830.

Link A.J. and Olson M.V. 1991. Physical map of the *Saccharomyces cerevisiae* genome at 110-kilobase resolution. *Genetics* **127:** 681–698.

Burke D.T. and Olson M.V. 1991. Preparation of clone libraries in yeast artificial-chromosome vectors. *Methods Enzymol.* **194:** 251–270.

Burke D.T., Carle G.F., and Olson M.V. 1987. Cloning of large segments of exogenous DNA into yeast by means of artificial chromosome vectors. *Science* **236:** 806–812.

Olson M.V., Hood L., Cantor C., and Botstein D. 1989. A common language for physical mapping of the human genome. *Science* **245:** 1434–1435.

Sequencing Bacterial Genomes

Fleischmann R.D., Adams M.D., White O., Clayton R.A., Kirkness E.F., et al. 1995. Whole-genome random sequencing and assembly of *Haemophilus influenzae* Rd. *Science* **269:** 496–512.

Fraser C.M., Gocayne J.D., White O., Adams M.D., Clayton R.A., et al. 1995. The minimal gene complement of *Mycoplasma genitalium*. *Science* **270:** 397–403.

Klenk H.P., Clayton R.A., Tomb J.F., White O., Nelson K.E., et al. 1997. The complete genome sequence of the hyper-thermophilic, sulphate-reducing archaeon *Archaeoglobus fulgidus*. *Nature* **390:** 364–370.

Tomb J.F., White O., Kerlavage A.R., Clayton R.A., Sutton G.G., et al. 1997. The complete genome sequence of the gastric pathogen *Helicobacter pylori*. *Nature* **388:** 539–547.

Blattner F.R., Plunkett G., III, Bloch C.A., Perna N.T., Burland V., et al. 1997. The complete genome sequence of *Escherichia coli* K-12. *Science* **277:** 1453–1457.

Cole S.T., Brosch R., Parkhill J., Garnier T., Churcher C., et al. 1998. Deciphering the biology of *Mycobacterium tuberculosis* from the complete genome sequence. *Nature* **393:** 537–544.

White O., Eisen J.A., Heidelberg J.F., Hickey E.K., Peterson J.D., et al. 1999. Genome sequence of the radioresistant bacterium *Deinococcus radiodurans* R1. *Science* **286:** 1571–1577.

Peterson J.D., Umayam L.A., Dickinson T.M., Hickey E.K., and White O. 2001. The comprehensive microbial resource. *Nucleic Acids Res.* **29:** 123–125.

Fraser C.M., Eisen J.A., Nelson K.E., Paulsen I.T., and Salzberg S.L. 2002. The value of complete microbial genome sequencing (you get what you pay for). *J. Bacteriol.* **184:** 6403–6405.

Liolios K., Tavernarakis N., Hugenholtz P., and Kyrpides N.C. 2006. The Genomes On Line Database (GOLD) v.2: A monitor of genome projects worldwide. *Nucleic Acids Res.* **34:** D332–D334.

CHAPTER 11

How the Human Genome Was Sequenced

No one could have imagined, when DNA sequencing was invented in the mid-1970s, that in less than three decades the complete sequence of the human genome would be known. How scientists advanced from sequencing small gene-sized fragments to taking on whole genomes is a story full of excitement, drama, and intrigue. The Human Genome Project included sequencing and studying the functions of the genomes of other organisms, but its overarching aim was to determine the sequence of the 3 billion base pairs that make up the 24 human chromosomes. A landmark goal was met in 2001, when two groups, one a large international consortium of publicly funded scientists (called the International Human Genome Sequencing Consortium, or IHGSC), and the other a private company called Celera Genomics, separately published draft sequences, each about 90% complete, of the human genome. A second major event occurred in April 2003, when the IHGSC presented a finished reference sequence of the human genome, which contained close to 100% of the entire sequence at a dramatically higher quality than either of the draft sequences. Human genome sequence had been made freely available by the IHGSC as soon as it was generated, but April 2003 was a defining moment. It marked the point in history when we first had access to our entire, albeit not well understood, set of genetic instructions. This chapter recounts some of this history and explains the process that culminated in the finished human genome sequence.

An Astonishing Proposal Is Made to Sequence the Human Genome

Although we may now think of the Human Genome Project (HGP) in terms of its biomedical applications, these were not the initial driving force for the project. In fact, what is now the Keck Telescope on Hawaii played a role in initiating discussions about a human genome project. In the early 1980s, Robert Sinsheimer, then Chancellor of the University of California at Santa Cruz, was trying to raise matching funds for what was planned to be the world's largest telescope. In the end, his efforts failed, but he was reluctant to relinquish the initial gift. Sinsheimer had worked on the bacteriophage ϕX174 and was impressed by Frederick Sanger's sequencing of its genome—at 5400 base pairs, no small feat in 1977. Sinsheimer saw that having the entire human genome sequence would revolutionize human genetics. Although this project would be on a scale far eclipsing anything else in biology, he was familiar with large-scale scientific projects through Santa Cruz's Lick Observatory and the University of California's management of the nuclear weapons laboratories of Los Alamos National Laboratory and Lawrence Livermore National Laboratory. Sinsheimer hoped to persuade the donor of the telescope funds to found a genome institute at Santa Cruz, but he failed. Undaunted, Sinsheimer organized a meeting at Santa Cruz in May 1985, attended by such luminaries as David Botstein, Walter Gilbert, Lee Hood, and John Sulston, to discuss the desirability and feasibility of sequencing the human genome.

At about the same time, Charles DeLisi came to the Department of Energy (DOE) as director of the Office of Health and Environmental Research. The DOE had been studying the effects of radiation on DNA since the end of World War II and was looking for ways to determine whether intense doses of γ irradiation—one product of atomic bomb blasts—caused an increase in mutations in the human germ line. It was hard to design ways to measure mutations in egg or sperm cell DNA following exposure, but it was even more difficult to think about how to measure the background mutation rate in human beings, which was known to be very low. In March 1986, Nobel laureate Renato Dulbecco published in *Science* the text of an address he had given at Cold Spring Harbor in September 1985. Dulbecco had argued that the most effective strategy to understand the genetics of cancer would be to have the entire human genome sequence. The timing could not have been better, for in the same month DeLisi convened a workshop at Santa Fe, again attended by leading lights of the genetics world, which recommended that the DOE proceed with a human genome project. The enthusiasm within DOE enabled DeLisi to set aside DOE funds for the project from his 1987 budget, and in 1988 the DOE officially began its genome project.

The proposal for a human genome project was enthusiastically embraced by several key scientists, notably Sydney Brenner, Gilbert, and James Watson, who believed that such an effort would provide tremendous amounts of extremely valuable data to the scientific community. But there was also opposition from equally notable scientists, including David Baltimore, Robert Weinberg, and Botstein, who then believed that a human genome project might be a terrible waste of money and talent. They worried that a large, coordinated effort would go so much against the grain of single-investigator, hypothesis-driven research that the whole biological research enterprise would be threatened. The debate spread to the wider community, most famously at the 1986 Cold Spring Harbor Laboratory Symposium on "The Molecular Biology of *Homo sapiens*," where Gilbert's prediction that the project would cost $3 billion aroused great interest and consternation.

In 1988, a committee of the National Research Council chaired by Bruce Alberts and a report issued by the Office of Technology Assessment, both in the United States, recommended that an international effort should be mounted to sequence the human genome. The National Institutes of Health (NIH) had been conspicuously absent from these early discussions, but, based on these reports and advocacy by Watson and others, James Wyngaarden, who was then the Director of NIH, actively pushed for NIH involvement. An additional motivation for NIH was the decision to include other organisms—*Escherichia coli*, *Drosophila*, and *Caenorhabditis elegans*—under the umbrella of the HGP, which greatly extended the number of scientists who would benefit from the project. Wyngaarden persuaded Watson to become director of the Office for Human Genome Research (which later became the National Center for Human Genome Research and subsequently the National Human Genome Research Institute, or NHGRI), a move that gave the NIH effort immediate credibility and influenced the U.S. Congress to provide additional funds for the project. The official start of the U.S. program as a joint DOE-NIH effort was October 1990, with the goal of finishing in 2005. The NIH initially established genome centers at four universities, later expanded to 16 centers, and finally contracted back down to three centers. The DOE set up genome centers at three of the National Laboratories—large government laboratories involved in nuclear weapons and radiation biology research. The project immediately became international in scope. The Sanger Centre in Cambridge, England, funded by the Wellcome Trust, greatly expanded its mapping and sequencing efforts and ultimately sequenced about one-third of the human genome. Research groups in France, Germany, and Japan, all supported by their governments, joined in and the international, publicly funded effort was dubbed "the Human Genome Project."

In an unprecedented move, Watson, recognizing that the data of the HGP would have long-lasting implications for our health and our understanding of ourselves, recommended that 3% of the NIH funds (later raised to 5% by Congress) be used for research and other activities to address the ethical, legal, and social issues arising from the HGP. The DOE followed suit and a joint NIH–DOE Ethical, Legal, and Social Issues (ELSI) Working Group was established. It was a prescient step—many ELSI-related issues, including paternity testing, forensics, ethnicity, and privacy, have indeed arisen as a consequence of the new information and capabilities made possible by the HGP. We will later discuss the benefits as well as some of the problems that have arisen from applications of human genome data in medicine and forensics (Chapters 14 and 16).

In the first few years of the HGP, some scientists continued to oppose the idea of such a large project. By the mid-1990s, however, large amounts of physical mapping data, human and mouse genetic maps, cDNA sequences, and sequences of complete bacterial genomes, including *Hemophilus* (Chapter 10) and *E. coli*, as well as parts of the yeast genome, became available. Therefore, most biologists, including most of the former skeptics, began to see the great value that the project could provide. Soon large amounts of physical and genetic mapping data, as well as DNA sequence data, began pouring into the public databases. Researchers were able to use these data immediately, and they began to change the way they conducted their research. The massive increases in mapping and sequencing efficiency, and the analysis of extremely large sequence datasets, led to the development of general approaches and principles that are now used in almost every field of biological research. These general approaches and principles constitute a new and extremely powerful scientific field referred to as genomics.

Repeat Sequences in the Human Genome Pose a Great Technical Challenge

From the very beginning it was clear that sequencing the human genome posed formidable technical and logistic problems on a scale never before tackled in biological research. One problem was precisely to do with scale. When discussion of the human genome began, a genome as small as that of a typical bacterium (about 3 million base pairs) had not yet been sequenced completely. The human genome would require more than 1000 times as many sequencing reactions and electrophoresis runs, and vastly more pairwise comparisons for sequence assembly, than a bacterial genome.

A second major problem concerned the nature of the human sequence. Hybridization kinetics experiments in the 1970s, and the sequences of single genes determined by many researchers, had shown that the human genome contains a very large amount of repetitive DNA, which causes difficulties in DNA sequencing. For example, DNA polymerases have difficulty moving along repeating A and T bases or repeating G and C bases if these continue for a hundred or more base pairs. In the test tube, the polymerase tends to stall at or skip repeating sequences, which leads to missing base pairs. And the human genome has hundreds of thousands of such sequences.

In addition, there are longer repeat sequences (e.g., SINEs [short interspersed elements] and LINEs [long interspersed elements]) dispersed throughout the genome on all chromosomes (Chapter 7). As we discussed in Chapter 10, these repeats cause difficulties when assembling sequencing reads into a long contiguous segment with computational techniques. The assembly process involves performing many pairwise comparisons of the DNA sequences from a large collection of 500-bp reads and identifying reads that overlap with one another because they share a contiguous stretch of identical sequence (Chapter 10, Fig. 10-2). If two copies of a repeat segment, each from a different place in the genome, are identical or almost identical in sequence over a long stretch, the computer often incorrectly assembles those segments. This problem is easily circumvented for repeat segments, such as the 300-bp *Alu* repeat family, that are shorter than the length of a sequencing read, because there is unique sequence elsewhere in the read. The problem becomes more complicated when repeats, such as those of the 7-kb LINE family, are longer than a sequencing read. Sequencing reads contained completely within these repeats have no unique sequence at one end and are harder to place uniquely into an assembly. Pairwise comparisons of such reads typically lead to their assembly with each other even though they may come from different parts of the genome (Fig. 11-1).

Another class of repeats, called "segmental duplications," ranges in size from about 1000 base pairs to more than 400,000 base pairs. These are present in more than one copy in the genome, sometimes in tandem (i.e., repeated next to each other) and some-

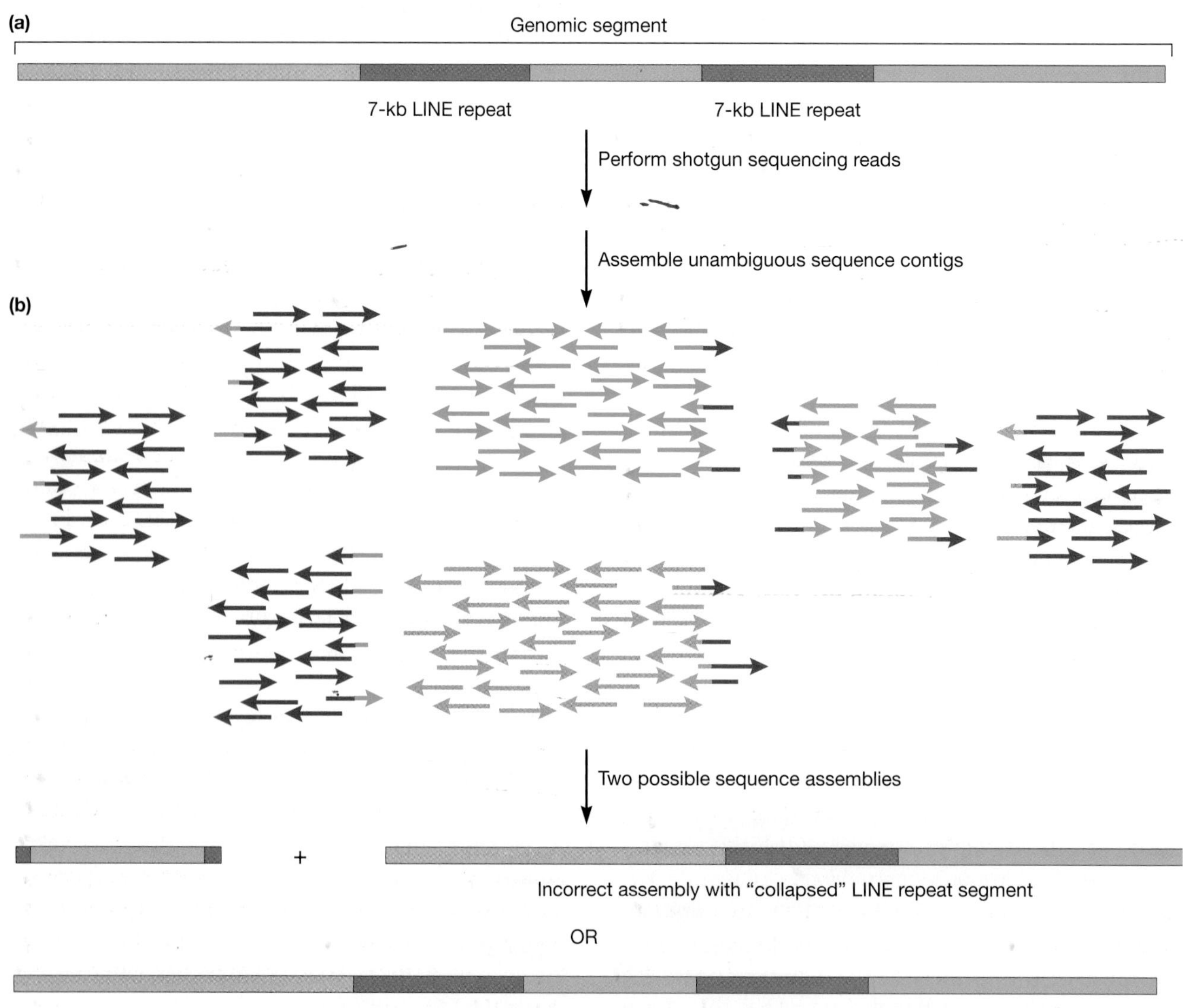

FIGURE 11-1

Difficulties in assembling sequence when long repeat segments are present. (*a*) The genomic structure of a region of the human genome that contains two LINE (long interspersed element, *red*) repeats, each about 7 kb in length and almost identical in sequence. (*b*) Paired-end sequencing reads from 3-kb clones are obtained and assembled. Many of the sequence reads fall into one or the other of the two 7-kb repeats, and the assembly algorithm has trouble determining where these reads belong in the assembly, particularly those that lie completely within the LINE repeats. This confusion can lead to an incorrect "collapsed" assembly, in which only a single LINE repeat is present, and the unique DNA sequences that lie between the two repeats are treated as a separate contig.

times dispersed. Segmental duplications cause so much trouble in sequencing because of their large size, pervasiveness (they make up ~5% of the total sequence of the human genome), and sequence similarity. (These duplications are often more than 90% identical to one another.) This similarity, and sometimes absolute identity, over large stretches of sequence is due to the evolutionary history of segmental duplications. Most occurred very recently in our ancestry (at least on an evolutionary timescale) so that not enough time has elapsed to allow accumulation of DNA sequence differences between the two or more copies. The near-identity of the repeats makes it impossible to assign the location of the sequence to only one of the regions in the genome from which it might have been derived. The result is that the assembly "collapses," so that the final genome sequence appears to have only one copy when, in fact, there are two or more. For repeats that are located in tandem relatively close to one another in the genome, this misassembly may exclude any unique sequence lying between the repeats. So, not only does the assembled genome have only one copy of a segment that is actually present more than once, it may also be missing sections of unique sequence.

In view of the problems posed by repetitive sequences, why bother to sequence them at all? Why not concentrate on the information-containing unique sequences and not worry about getting the repeats correct? Indeed, during the early planning of the HGP, there were discussions about directing the sequencing effort only to unique segments. This could be done, for example, by using hybridization methods to select out those subclones that contain repeat sequences and remove them from the sequencing pipeline. But this approach would cause serious errors in the final sequence. The sizes of the many different kinds of repeats and their distribution in the human genome are such that most subclones that contain repeat segments also contain unique sequences. Eliminating these clones from sequencing would produce a genomic sequence missing many gene segments and other functional regions of DNA. Furthermore, the hybridization methods for identifying repeat-containing subclones are effective only if one knows the sequences of the repeats or if the repeats are extremely abundant in the genome (as are LINE and *Alu* repeats). Thus, using hybridization to eliminate repetitive regions would not work for the large number of low-copy repeats in the human genome.

Concentrating the sequencing effort only on unique sequences was rejected for an important conceptual reason. The final product would be a collection of unconnected short sequence contigs that would not represent the whole genome but only a fraction (albeit an important fraction). The assemblage of unconnected contigs would not provide information on the order and orientation of the contigs in the genome. It was feared also that regions containing repetitive sequences might have important functions not yet discovered and that would never be discovered if they were not included in the HGP. It was clear that the problems of repetitive sequences could not be ignored.

Large Efforts Were Mounted to Sequence cDNAs

In 1990, Sydney Brenner suggested sequencing cDNAs so as to capture just the protein coding regions of the human genome. Because cDNAs are produced by reverse transcription of messenger RNAs, they do not contain introns or any other intergenic, noncoding sequences. Therefore, their use would reduce the amount of sequencing by as much as 100-fold. Brenner was unable to obtain funding to carry out this strategy, but, in 1991, Craig Venter, then at the NIH, reported sequencing a large number of cDNAs prepared from brain tissue. These were partial sequences of the cDNAs, derived from a single sequence read, called expressed sequence tags (ESTs). Of the 609 ESTs obtained in this study, 337 were not homologous to sequences then in the databases and were assumed to represent new genes. These EST sequences, and later the sequences of full-length cDNAs, were very helpful in confirming and validating the assembly of the human genome sequence. In addition, plasmid clones that carry the ESTs and full-length cDNAs have been extremely valuable for functional experiments, including gene expression microarrays (Chapter 13).

NIH decided to patent Venter's ESTs and this provoked great controversy as almost everyone believed that all DNA sequences should be freely available (eventually, NIH dropped the patent application). Venter left NIH and established The Institute of Genome Research (TIGR), which was devoted to using high-throughput methods to sequence (and patent) very large numbers of human

ESTs. In response, a group of pharmaceutical companies led by Merck and Company funded the Genome Sequencing Center at Washington University to sequence ESTs. These results were immediately downloaded to public databases so that they were free for use by all investigators. Other sequencing centers, especially in Japan and France, followed suit, and ultimately many millions of sequences generated from mRNAs from a very wide range of human cell types were available through the public databases. Beginning about 1999, attention turned to making "full-length" cDNA libraries and sequencing these, thereby making available the complete mRNAs for as many genes as possible.

Certainly ESTs have been invaluable; however, cDNA sequencing was rejected as an alternative to genomic sequencing for the same reasons that repetitive sequences were included. It was argued that sequencing only cDNAs would miss the many noncoding regions of the genome that are not present in mature mRNAs but have important functions (e.g., transcriptional regulatory sequences and splice junctions).

Genetic Maps of Human Beings Were Difficult to Make

Aware of the problems that large amounts of repetitive DNA would cause in sequencing the full human genome, most investigators agreed it would be necessary to generate maps, both genetic and physical, to make it possible to assemble the sequences.

Genetic maps are built by measuring the frequencies of recombination between traits or polymorphic DNA markers. A low recombination frequency between the two traits means that their genes must lie close to each other on the same chromosome; they are tightly "linked." Higher recombination frequencies indicate that the genes are far apart—loosely linked. (Genes on different chromosomes are unlinked.) Therefore, careful measurements of the recombination frequencies between several traits allow the order and the "genetic distances" between the genes to be determined. The earliest genetic linkages in human beings were found for genes on the X chromosome. In 1911, the cytologist Edmund B. Wilson argued that the inheritance patterns of male color blindness indicated that this trait is recessive and is due to a mutation in a gene on the X chromosome. Genes for other traits (e.g., hemophilia and Duchenne muscular dystrophy) were assigned to the X chromosome by using similar reasoning. Assigning genes to autosomes, the non-sex chromosomes, was much more difficult. It was not until 1968 that the Duffy blood group gene, which encodes a protein on the surface of red blood cells, was assigned chromosome 1 through linkage to a structural change in the chromosome that could be observed microscopically.

A significant advance came in the 1960s with the development of *somatic cell genetics* (Chapter 8). It was found that cells growing in tissue culture could be made to fuse with each other by treating the culture with inactivated Sendai virus or with a chemical such as polyethylene glycol. Hybrid cell combinations—human–mouse, human–hamster, and mouse–hamster—could be made by using this approach. When human and rodent cells are fused and then grown in culture, there is a progressive loss of the human chromosomes until only one or a few are left. Single hybrid cells are isolated and grown to make lines of mouse cells containing just a small amount of human DNA, which can be characterized by using probes of known chromosomal locations. DNA of unknown origin can be localized by using a panel of hybrid cell lines, each with a different human chromosome or part of human chromosome. In 1971, geneticists using this approach localized the human thymidine kinase (TK) gene to chromosome 17 (Fig. 11-2). A mutant mouse cell line that lacked TK enzyme, because of a mutation in the *tk* gene (at that time not yet mapped or isolated), was fused with nonmutant human cells, and after selection for the presence of TK function in special media, several hybrid cell lines containing one or a few human chromosomes were isolated. These hybrid cells were able to grow because they had retained the human chromosome with an intact *tk* gene. By using cytology to determine which human chromosomes were present in the collection of cell lines, investigators found that the presence of human chromosome 17 was sufficient to allow the mutant mouse cells to grow in selective media. They therefore concluded that the human *tk* gene is located on human chromosome 17. By 1979, some 200 genes had been assigned, more or less confidently, to human chromosomes by a combination of family studies, somatic cell hybrids, and other techniques. A more sophisticated variation of this approach—radiation hybrid mapping—played a major role in the HGP (see below).

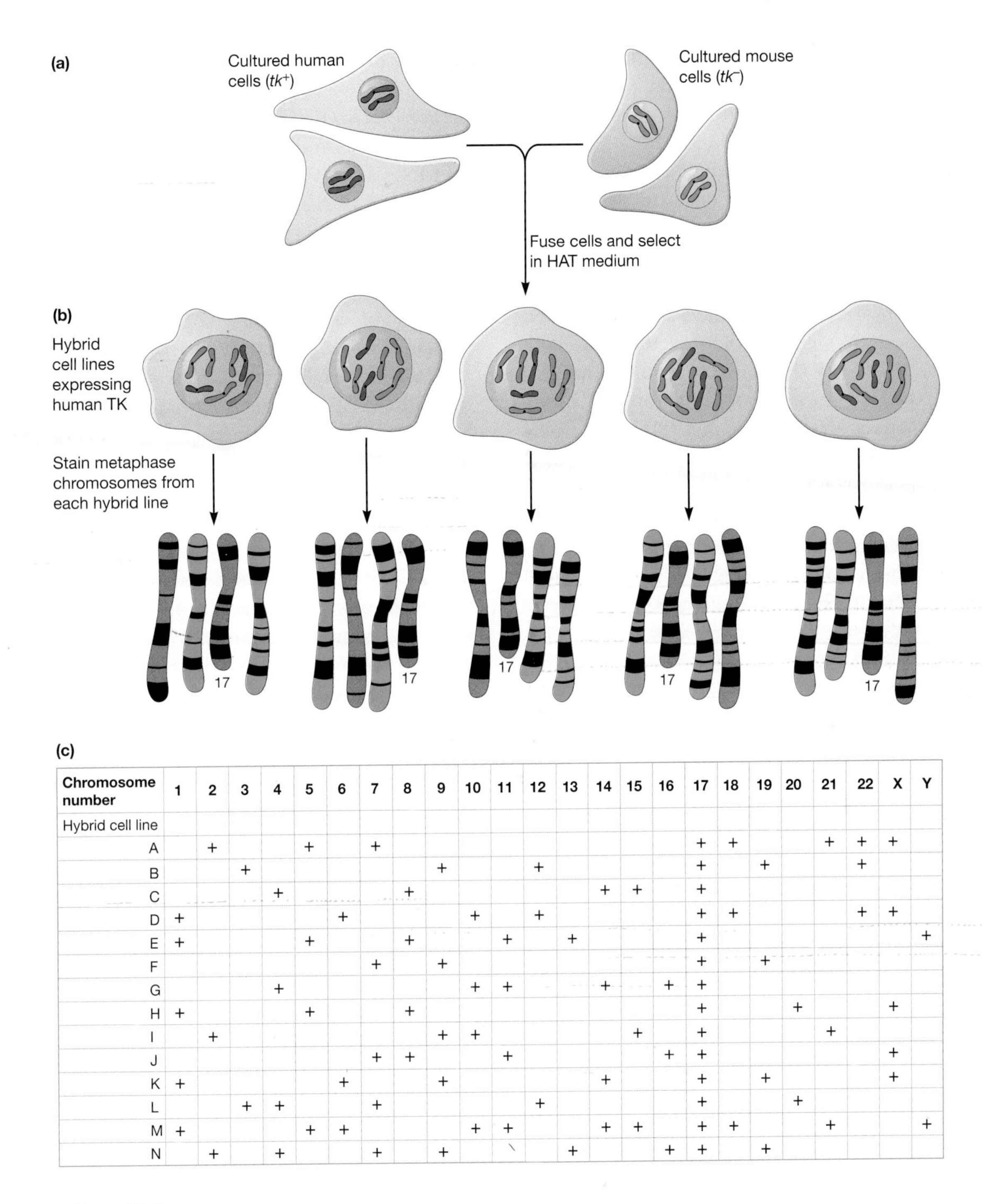

Chromosome number	1	2	3	4	5	6	7	8	9	10	11	12	13	14	15	16	17	18	19	20	21	22	X	Y
Hybrid cell line																								
A		+			+		+										+	+			+	+	+	
B			+						+			+					+		+			+		
C				+				+						+	+		+							
D	+					+				+		+					+	+				+	+	
E	+				+			+			+		+				+							+
F							+		+								+		+					
G				+						+	+			+		+	+							
H	+				+			+									+			+			+	
I		+							+	+					+		+				+			
J							+	+			+					+	+						+	
K	+					+			+					+			+		+				+	
L			+	+			+					+					+			+				
M	+				+	+				+	+			+	+		+	+			+			+
N		+		+			+		+				+			+	+		+					

FIGURE 11-2

Localization of the *thymidine kinase* gene to human chromosome 17 by using somatic cell hybrids. (*a*) Cultured human cells that contain a functional copy of the *tk* gene are fused with mutant mouse cells that lack the *tk* gene (tk^-). The fused cells are put into HAT medium, which contains chemicals that allow cells to grow only if they have a functional *tk* gene, and a set of mouse–human hybrid cell lines are selected. (*b*) For reasons that are not understood, the mouse chromosomes are retained in the nuclei of the hybrid cells, but most of the human chromosomes are discarded. However, because of the selection in HAT medium, the human chromosome, or in some cases, portions of the chromosome, that contains the *tk* gene is retained in all of the hybrid cells (otherwise the cells would not grow), along with a few other, nonselected human chromosomes. The hybrid cells are screened by staining their metaphase chromosomes to identify, in the microscope, which human chromosomes are present in each cell line. (*c*) By determining that all of the hybrids contained human chromosome 17, the location of the gene was identified.

Restriction Fragment Length Polymorphisms Serve as Markers for Linkage Analysis

By 1968, 68 traits had been assigned to the X chromosome, but determining the order of these genes along the length of the X chromosome was much more challenging. It was difficult to find individuals with two or more X-linked traits, and because the families with affected individuals were so rare, recombination rates could not be determined with any accuracy. This situation, for mapping genes not only on the X chromosome but for the entire genome, changed dramatically in 1980 when David Botstein, Ronald Davis, Mark Skolnick, and Ray White proposed that genetic maps of the human genome could be generated much more efficiently and at much higher resolution by using polymorphic DNA markers. These are segments of DNA present at specific sites in the genome, in different, distinguishable variants, just like the alleles of a gene. These DNA markers need not necessarily be associated with a gene and could be, and often would be, in noncoding DNA. Botstein and his colleagues recognized that the sites cut by restriction endonucleases could be polymorphic and these could be used as genetic markers. Single base-pair differences occur approximately every 1000 base pairs between any two people, and some of these differences alter a restriction enzyme site so that an enzyme no longer cuts DNA at that site. For example, the restriction endonuclease EcoRI cuts DNA at a specific sequence:

```
...G|A A T T C...          ...G        A A T T C...
                  ──────→
...C T T A A|G...          ...C T T A A        G...
```

When one or more nucleotides in the endonuclease recognition sequence are altered, EcoRI fails to cut the DNA strands at the altered site, and a longer DNA fragment is produced. When a given restriction enzyme site is present in the DNA molecule of one chromosome, but the other homologous chromosome lacks the site, treatment with the enzyme produces a shorter fragment from the chromosome having the site, and a longer fragment from the chromosome without it. Longer and shorter fragments would also be produced in those cases in which both of the restriction sites are intact, but there is an inserted or deleted stretch of DNA between them. We can now distinguish the two chromosomes in an individual and between individuals on the basis of this *restriction fragment length polymorphism* (RFLP). An individual with two different versions of the locus is heterozygous for this RFLP and is said to be *informative* at that locus. Because the RFLP is inherited just like a gene, the individual chromosomes can be followed as they pass from generation to generation by tracing the inheritance of the marker fragments (Fig. 11-3). The process of scoring individuals for their DNA sequence variants is called *genotyping*.

How does having a polymorphic DNA marker like an RFLP help map a trait in the genome? Suppose the EcoRI fragment described above is located near a gene that, when mutated, results in a disease. As shown in Figure 11-4, when this DNA marker is genotyped in families with the disease, one of the two variants is coinherited with the disease. That is, in any given family, individuals with the disease always (or almost always) have the same variant, and unaffected individuals have the other variant. With enough family studies, one can do a simple statistical test to determine whether the association of the DNA marker with the disease in each family occurs at a higher frequency than would be expected by chance. If so, the DNA marker is said to be linked to the disease locus and the genetic distance between marker and gene can be calculated, just as Alfred Sturtevant did 100 years ago (Chapter 1). It is very important to note that the RFLP is not the cause of the genetic disorder but merely a marker for it. (Occasionally, the RFLP may be within a gene and be the mutation itself; for example, sickle-cell anemia results from the change of an A to a T in the β-globin gene. This changes a glutamic acid residue for a valine and eliminates an MstII restriction site.)

Linkage analysis with RFLPs propelled human genetics into the molecular age. Geneticists quickly took up this new approach and used RFLP linkage mapping to identify many disease loci. This proved to be much faster than previous methods. However, it was still a relatively slow process, because polymorphic DNA markers had to be identified and each one had to be tested in families by Southern blotting. As the chance that any randomly chosen DNA marker lies close to a particular disease locus of interest is low, many markers had to be identified and tested before a "hit" was obtained. Once a marker showing linkage to the disease has been found, its chromosomal location must be deter-

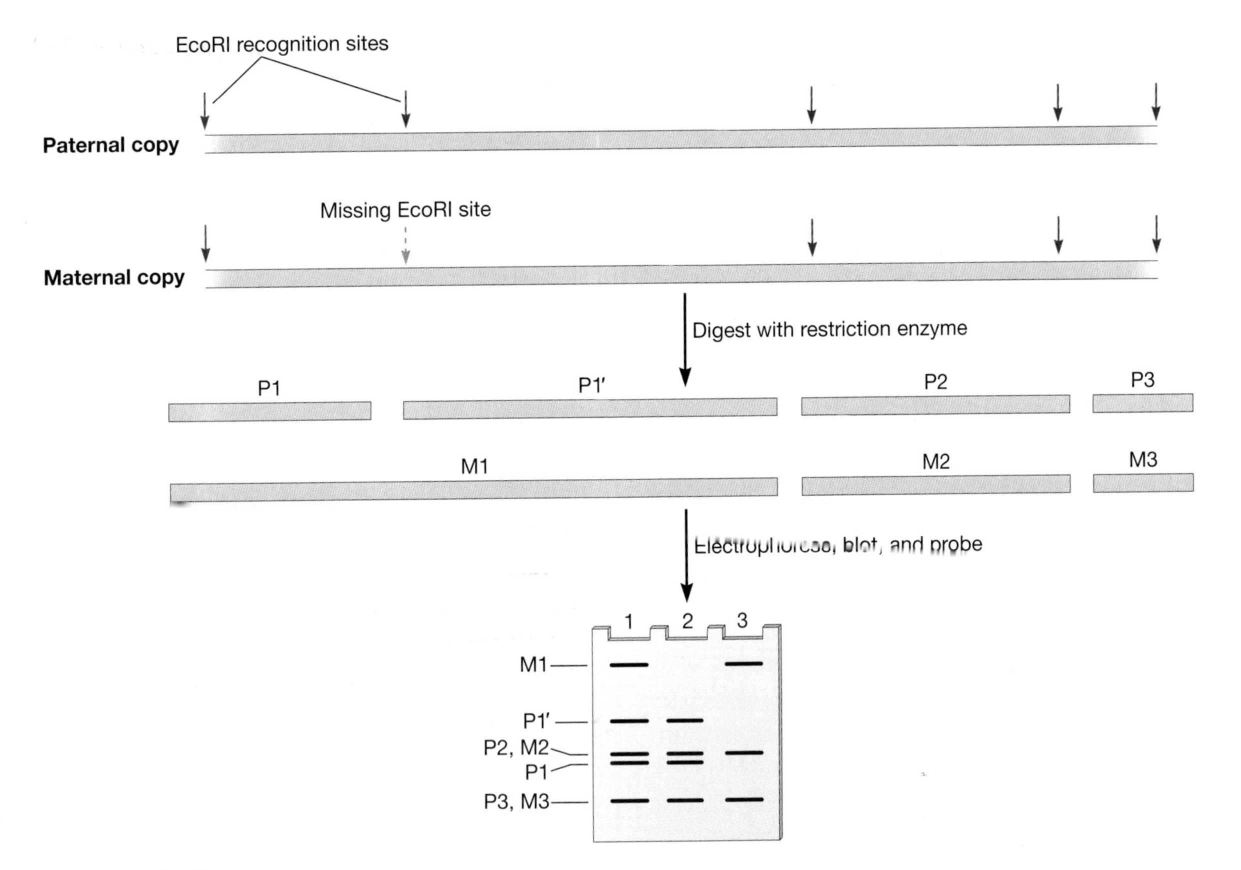

FIGURE 11-3

Restriction fragment length polymorphism (RFLP). Each bar represents the double-stranded DNA molecules of two homologous chromosomes that an individual inherited from her parents, one from her father and the other from her mother. The paternal copy has five recognition sites for the restriction enzyme EcoRI (*red arrows*). However, the maternal copy has only four EcoRI sites because of a single-base-pair change at one site. Genomic DNA was cut with EcoRI, electrophoresed on an agarose gel, transferred to a nylon filter by Southern blotting, and probed with a labeled DNA fragment corresponding to this region of the genome. Five bands are observed (lane *1*): P2, M2 and P3, M3, which are the same sizes on both chromosomes; two fragments P1 and P1′ from the paternal chromosome; and a larger fragment M1 from the maternal chromosome because of the missing EcoRI site in her mother's copy of the genomic region. Lanes *2* and *3* show the patterns that result from individuals who are homozygous for the allele seen in the father (lane *2*) and the mother (lane *3*).

mined. Before the onset of the HGP, this process was very slow and tedious. Furthermore, there was little effort among investigators to cooperate in this early phase of disease gene hunting. Rather, there was fierce competition to be the first to find important genes like those involved in cystic fibrosis or Duchenne muscular dystrophy, and so many redundant, inefficient efforts proceeded simultaneously. It was clear that a much more efficient and coordinated approach was needed. What was needed were genetic maps of the entire human genome constructed with polymorphic DNA markers and made available to all researchers interested in genetic diseases. It was also clear that mapped markers would be extremely helpful in sequencing the human genome by providing a framework for assembling contigs.

Instead of mapping a disease trait relative to a DNA marker, the new genetic maps were made by mapping polymorphic markers relative to each other. The same principles of genetic linkage apply. Markers are genotyped in genomic DNA from multigenerational families and the genetic distances between the markers are calculated from the number of times they recombine during meiosis. White and colleagues at the University of Utah and, independently, Helen Donis-Keller and her colleagues at a biotechnology company, Collaborative Research,

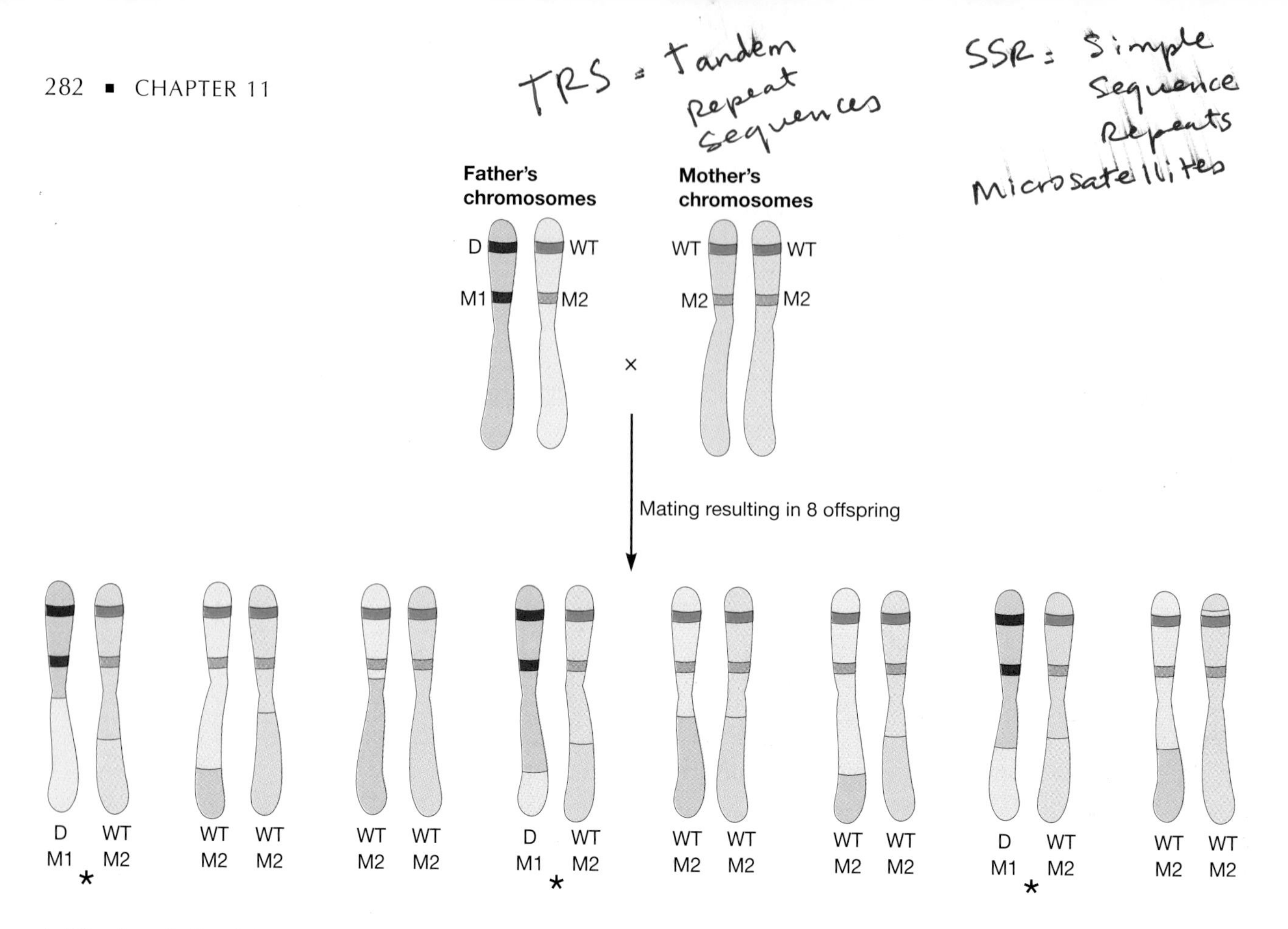

FIGURE 11-4

Using RFLPs in genetic linkage to localize a disease gene to a region of a human chromosome. In this mating, the father carries a dominant mutation in a single gene that results in a genetic disease (D). A few million base pairs away, which is very close considering the length of the entire genome, is a polymorphic DNA marker, M. The father is heterozygous (M1, M2) for this polymorphism. The mother is homozygous for the wild-type version of the disease gene (WT) and for the M2 allele at the DNA marker. The five children who are homozygous for M2 are not affected by the disease. As one of their M2 chromosomes must have come from their father, we can deduce that the disease gene cannot be on his M2 chromosome and must be on his M1 chromosome. This is confirmed by the three children with the disease; they are heterozygous M1 and M2. The M1 chromosome must have come from their father and must be carrying the dominant mutation. The association between M and this mutation would have to be confirmed by testing other families.

TRS

Inc., built maps of the human genome with RFLPs. These maps had about 400 DNA markers on them, but because of their nonrandom spacing, the markers did not tie the entire genome together. Where the markers were very far apart, there were gaps in the maps that limited their usefulness for finding genes. Nevertheless, much of the genome was covered, and human geneticists were much better off than they had ever been before. Progress was rapid building on these early efforts, and by 1990, almost 2000 genes and 5000 DNA fragments had been mapped to human chromosomes.

Tandem Repeat Sequences Make Better DNA Markers

The next significant improvement came in 1989 when James Weber and, separately, Michael Litt described a new type of polymorphic DNA marker called a simple sequence repeat (SSR) marker or a "microsatellite" marker. These are loci in the human genome where very short DNA sequences, 2- to 5-bp long (and sometimes longer), are repeated many times next to each other (in "tandem"). For example, the dinucleotide CA (and its complement GT on the opposite strand) might occur about 30 times in a row at a specific place in the genome. These simple repeats can be polymorphic, because the number of repeats at any locus may vary. For example, the number of CA repeats at a specific locus on different chromosomes might be 15, or 21, or 28, or 32, and so on (Fig. 11-5). These are valuable as markers not only because they are often extremely polymorphic, but also because there are tens of thousands of them in the human genome. Such markers can be scored by determining their lengths, typically by using polymerase chain reaction (PCR) to amplify a specific locus with oligonucleotide primers that hybridize to the unique sequences flanking the SSR, and separating the fragments by electrophoresis in a denaturing polyacrylamide gel. One of the oligonucleotide primers is labeled with a fluorescent dye attached to its 5′ end, and the electrophoresis and detection are carried out in an automated DNA sequencer where a laser "reads" the fluorescent signal as the DNA fragments move through the gel (Chapter 10). By using four different dyes to label the oligonucleotide primers, four different genetic markers can be run in the same lane of a gel. And if the PCR products are designed so that they can be distinguished by size as well as color, as many as four sets of markers, each with four fluorescent labels, can be electrophoresed in one lane (Fig. 11-6).

Several groups began building microsatellite marker maps of the human genome. At first, this was done on a chromosome-by-chromosome basis, but it was quickly determined that it was more efficient to build a map of the entire genome all at once. In 1993, Jean Weissenbach, Daniel Cohen, and their colleagues at Généthon, a large genome laboratory near Paris, published a microsatellite map of the human genome, primarily composed of markers with polymorphic CA repeats. The Généthon map contained more than 2000 ordered microsatellite markers and was the first genetic map that covered almost the entire genome (90%). Soon afterward, the Cooperative Human Linkage Center, a consortium of several laboratories in the United

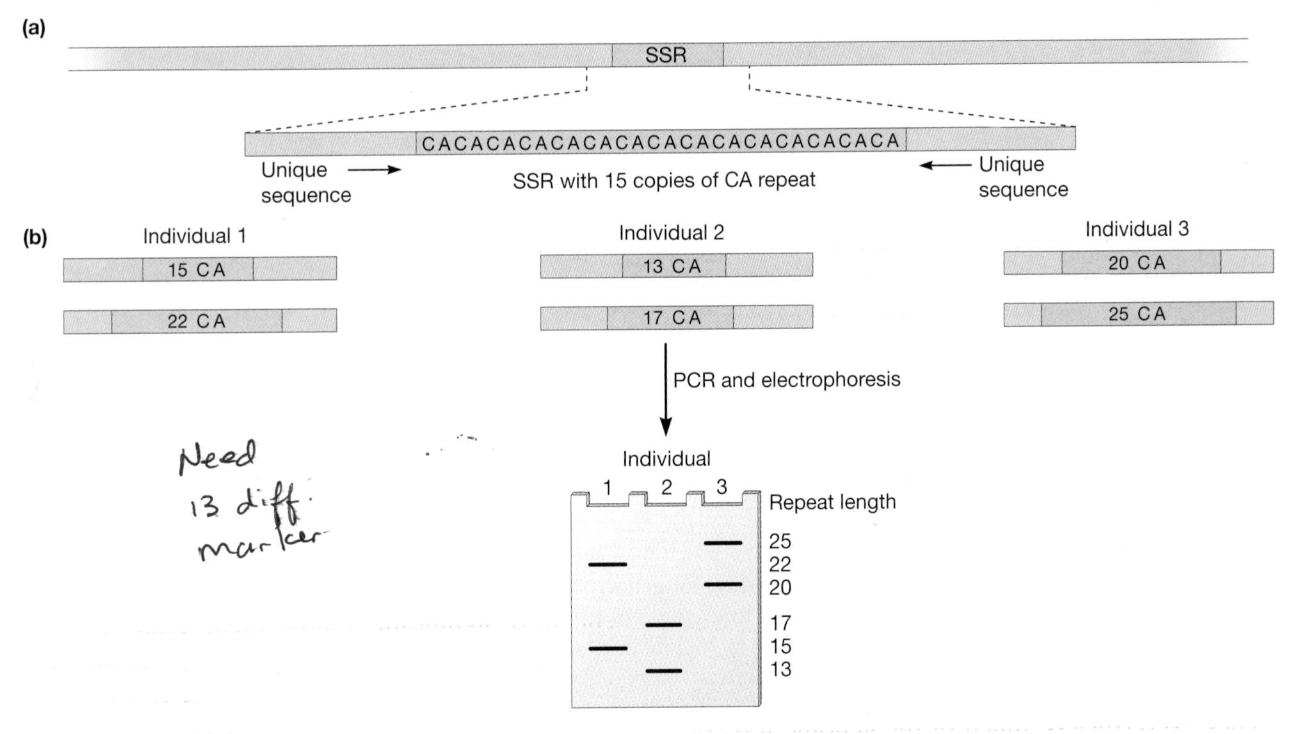

Need 13 diff. marker

FIGURE 11-5

Microsatellite sequences can be used as genetic markers. (*a*) A simple sequence repeat (SSR) is made up of the dinucleotide CA repeated 15 times. Unique sequences on either side of such "microsatellite" markers can be used to amplify a specific locus from genomic DNA by PCR. (*b*) The alleles are scored by measuring the lengths of the PCR products from each person by gel electrophoresis. In this example, Individual 1 has 15 and 22 copies of the dinucleotide CA. Individual 2 is "13/17" and Individual 3 is "20/25."

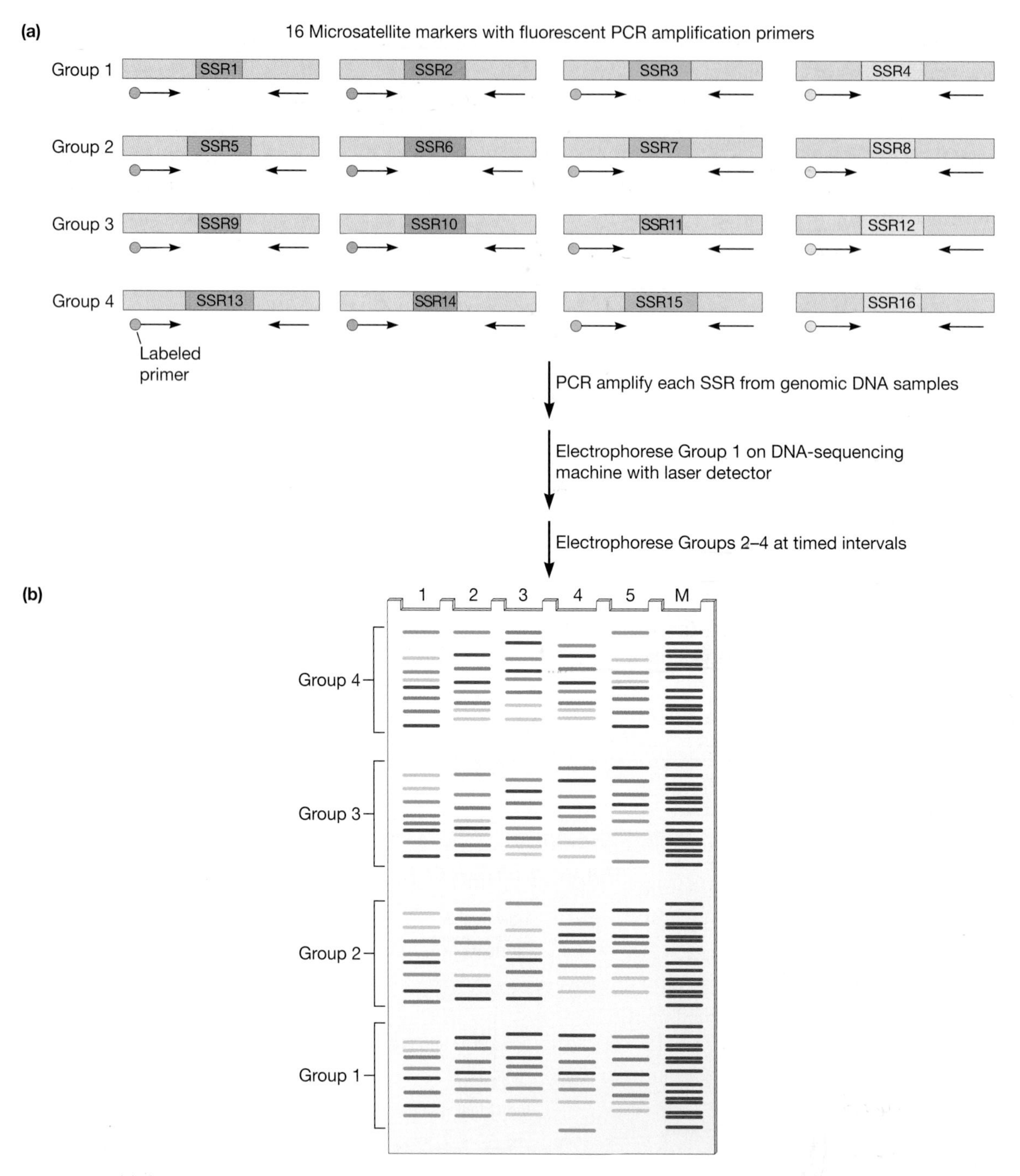

FIGURE 11-6

Multiplex measurement of microsatellite markers for genotyping. Because electrophoresis is time consuming and expensive, researchers developed ways to examine many markers in each lane of a gel using modified fluorescent DNA sequencers. (*a*) PCR primers for four different microsatellite markers are labeled with four different fluorescent dyes. Multiple sets of reactions are carried out (groups 1–4). The four amplification reactions are loaded into a single gel lane (or capillary) in the sequencing machine. (*b*) Each of lanes *1–5* contains the analysis for 16 markers for one individual. (M contains a set of size markers.) Group 1 contains markers SSR1 through SSR4, and two PCR products, corresponding to the two alleles of each marker can be seen. SSR1, *red*; SSR2, *blue*; SSR3, *green*; and SSR4, *yellow*. An additional level of multiplexing can be obtained by loading the gel with three or four additional sets of four labeled microsatellite markers in succession (groups 2, 3, and 4). Each set of four labeled markers is not confused with the set in front of it as all the fragments from the previous run have moved past the laser by the time the next set arrives. This degree of multiplexing reduces the number of gels that would have to be run by almost 20-fold.

States, published a genetic map of tri- and tetranucleotide repeat markers, which are easier to score because the differences in size between variants are more readily distinguished by gel electrophoresis than are those of CA repeats. Finally, in 1994, these groups and others coordinated their efforts and built a single comprehensive map containing 5870 polymorphic DNA markers, with an average spacing of one marker every 0.7 cM (centiMorgan) corresponding to roughly 700,000 bp (Fig. 11-7). This comprehensive map was a major achievement because it covered the complete human genome with markers at very high density, it was easy to use, and it was freely available. The map achieved one of the first goals, and became one of the first successes, of the HGP.

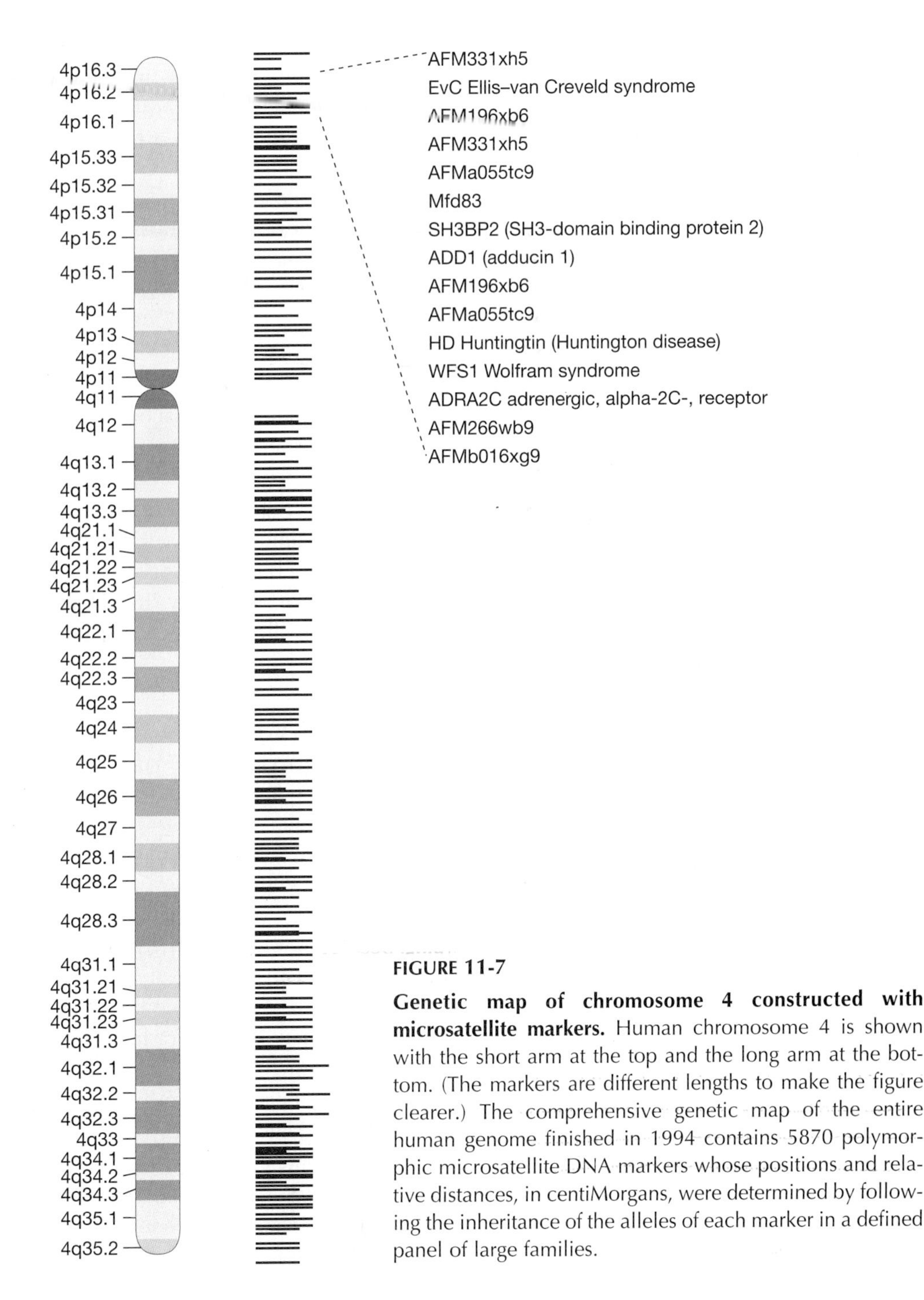

FIGURE 11-7

Genetic map of chromosome 4 constructed with microsatellite markers. Human chromosome 4 is shown with the short arm at the top and the long arm at the bottom. (The markers are different lengths to make the figure clearer.) The comprehensive genetic map of the entire human genome finished in 1994 contains 5870 polymorphic microsatellite DNA markers whose positions and relative distances, in centiMorgans, were determined by following the inheritance of the alleles of each marker in a defined panel of large families.

Cloning Vectors Are Developed for Very Large Fragments of Human DNA

It was clear that maps of overlapping clones were proving indispensable for the yeast and *C. elegans* sequencing projects. It was believed that a similarly ordered set of overlapping clones of human DNA would be essential for providing a physical map for assembling the very much larger human sequence. Furthermore, the clones could be used as a framework for organizing the sequencing in an orderly way, thereby ensuring that the entire human genome would be covered.

The initial goal was to produce a set of overlapping cloned human DNA fragments that extended from one end of a chromosome to the other for each of the human chromosomes. As we discussed in Chapter 10 (see Fig. 10-3), these sets of overlapping clones were called *clone contigs* (in the same way that overlapping sequencing reads were called sequence contigs), and ideally the number of clone contigs for a genome would be equal to the number of chromosomes. In the case of the human genome and other complex genomes, the DNA of the centromeres is very long and composed of repetitive sequences, which are very difficult, if not impossible, to clone. This creates a "gap" in each chromosome so the minimum number of clone contigs for the human genome is 43. (There are two contigs for each of the 17 autosomes that have a short arm and a long arm; one for each of the five acrocentric chromosomes where the centromere lies very close to one end of the chromosome; and two for each of the X and the Y chromosomes.)

The number of clones needed to cover the human genome depends on the sizes of the genomic DNA inserts in the clones; the larger the inserts, the fewer clones are needed. At the time the HGP began, cosmids were the cloning vehicle of choice. These are circular plasmid vectors with elements of bacteriophage λ, which can carry inserts up to about 40,000-bp long (Fig. 11-8). Other cloning systems that can carry much larger DNA segments were developed, including yeast artificial chromosomes (YACs) and bacterial artificial chromosomes (BACs). Because YAC clones can accommodate much larger inserts than the other vector systems, many fewer clones are required to cover the human genome at 1× coverage. However, even though much effort was spent building YAC clone maps of the human genome, these maps were not as useful as had been hoped. The yeast strains in which YACs were cloned and grown frequently deleted, inverted, or otherwise rearranged segments of the cloned human DNA, probably because of the repetitive segments present throughout our genomes. So, although YACs would have been the most economical means of producing a physical map of the human genome, such a map would not have been an accurate representation of the sequence. Nevertheless, because much effort was put into building YAC clone maps of the human genome early in the HGP, a significant amount of information from these maps was used in the initial sequencing and genetic mapping efforts. In the end, most of the sequencing was done by using other types of clone maps, as we shall describe below.

By about 1995, it became apparent that BACs and a similar cloning system called PACs (for P1 artificial chromosomes) had several advantages over other vector systems. The circular BAC and PAC vectors are simple and small (~7 kb in length) and yet they can incorporate relatively large inserts, up to about 200,000 bp long. The vectors are designed to allow only a single copy of each clone to replicate in a bacterial cell, thereby preventing the inserts from rearranging. And on a practical level, BACs and PACs are easier to prepare and maintain than YACs. Several groups built BAC and PAC libraries of the human genome, but most of the sequence of the human genome produced by the IHGSC was determined from BAC libraries made by Pieter deJong.

Physical Maps Were Made by Restriction Enzyme "Fingerprinting" of Large-Insert Clones

Having prepared a library of large-insert DNA covering the entire genome, how does one construct a physical map? One of the key methods used to build maps was restriction site fingerprinting. First developed for producing maps of *C. elegans* by Alan Coulson and his colleagues at Cambridge, it was applied to human BAC libraries by Marco Marra, John McPherson, and Robert Waterston. BAC libraries were generated from human DNA that had been partially digested with HindIII or MboI, restriction enzymes that recognize a 6-bp and 4-bp site, respectively. Partial digestion with these enzymes (the reaction was not allowed to go to completion) produced DNA fragments averaging

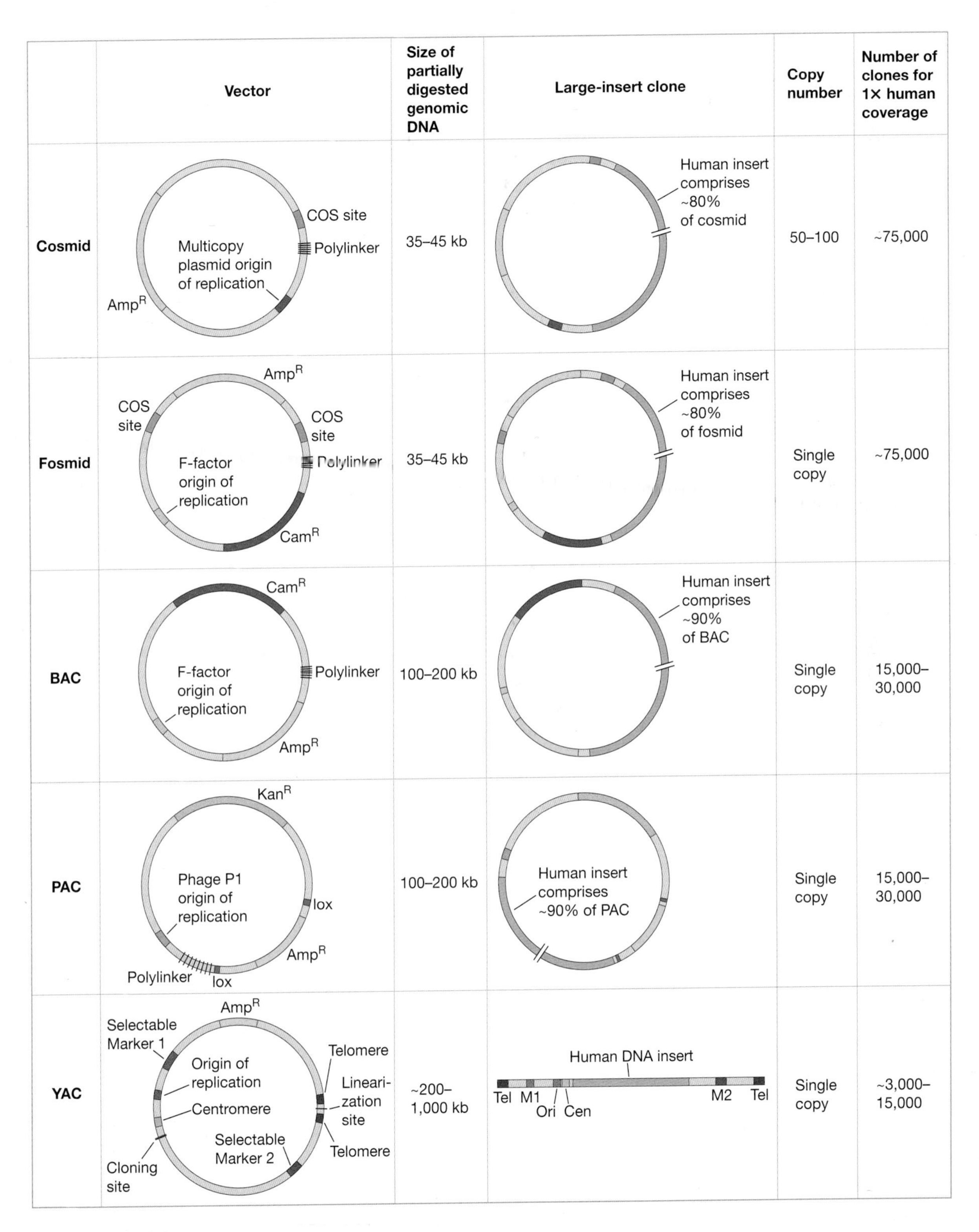

FIGURE 11-8

Large-insert cloning systems are essential for genome studies. Five of the large-insert cloning systems that were used by the IHGSC to build maps for sequencing the human genome are shown here, with their range of insert sizes, copy number, and host used for cloning and propagating the clones. Cosmids were the first of the vector systems to be developed, and they were used almost exclusively at the beginning of the Human Genome Project. YACs (yeast artificial chromosomes), BACs (bacterial artificial chromosomes), and PACs (P1 artificial chromosomes) were developed in the early 1990s, and by the time most of the sequencing was being done by the IHGSC, BACs were the major source of mapped large-insert clones. Fosmids, which are similar to cosmids but are typically more stable because they are present at single copy in the host *Escherichia coli* cells, were used to aid in finishing and checking the assembly.

STS - seq tagged sites

about 170,000 base pairs long. These fragments were cloned in BACs so that there were enough clones produced in each experiment (a "library") to ensure that the human genome was present many times over. DNA from each individual BAC clone was cut with a restriction enzyme, the products separated by size on an agarose gel, the gels imaged, the sizes of the restriction fragments calculated automatically, and the data stored in a database. The set of DNA fragments from a particular BAC is that clone's fingerprint (Fig. 11-9). The complete set of BACs was fingerprinted in this way. The data were compared to determine which clones had restriction fragments in common, and those that did were thought highly likely to overlap. Although fingerprinting could, in theory, build a clone map covering an entire genome, in practice it fails for a large and complex genome like that of human beings. Nonrandomness of libraries and restriction sites, sizing errors on the fingerprint gels, rearranged clones, incomplete data, and the presence of a wide variety of repeat sequences in the human genome all conspire to produce a map with many, rather than fewer, contigs of BAC clones. At this stage, the map requires additional work to order and orient all the clone contigs.

Sequence-tagged Sites Help Develop Large-Scale Contigs

Genome scientists needed other markers and therefore turned to *sequence-tagged sites* (STSs). These are short unique sequences—200- to 500-bp long—used to orient and order genomic clones (BAC and PAC) to generate physical maps of the genome. Maynard Olson, Leroy Hood, Charles Cantor, and David Botstein argued that the official "currency" of genome maps and sequences should not be segments of cloned DNA, but instead should be PCR assays that can be used to amplify a unique segment of the genome. These segments need not be polymorphic; the presence or absence of an STS in two clones would tell whether or not the clones overlapped. There were two major advantages in using STSs. First, whereas an STS represents a physical piece of DNA, unique in the genome, it is a "virtual" DNA marker. Researchers do not need a DNA clone of an STS in order to use it; databases would contain the sequence data for PCR primers flanking the STS, so that the primer sequences could be downloaded by other investigators and their DNA clones tested for the presence of the STS with a PCR assay. This possibility had a dramatic, liberating effect, making the use of genomic resources much cheaper and fairer. Second, the fact that STSs are detected by PCR greatly increased their utility. PCR is a much more rapid and robust technique than Southern blotting; it requires much less DNA, and it can be performed rapidly on thousands of samples.

Large numbers of STS markers for the human genome were developed. These included PCR versions of many of the earlier hybridization-based microsatellite markers that had been mapped by blotting and other means, and STSs generated from a large number of sequences determined at random from the human genome. When a sufficient number of STSs were used to screen BAC libraries, each BAC clone was found to contain multiple STSs and its "STS content" was scored. Comparing the STS content of each BAC made it possible to determine which BACs shared STSs and therefore contained overlapping sequences. This information was then used to build contigs of BAC clones (Fig. 11-10). These contigs also provided data on the order of the STSs along the genome. Because these clones were constructed with DNA fragments produced by partial restriction enzyme digestion, the ends of the BAC clones extend to different places in the genome. Thus, overlapping BACs share some but not all of their STSs. The patterns of the presence and absence of STSs allow one to deduce not only where the end points of the BAC clones lie relative to each other, but also the order of the STSs (Fig. 11-10).

Radiation Hybrids Provided the High Resolution Needed for Physical Mapping

On occasion, a technique is developed before its time, in the sense that the ancillary tools are not yet available for it to be used to its full potential. The technique of genome mapping by using *radiation hybrids* is an example. First developed in 1975 by Stephen Goss and Henry Harris, it found little application until 1990 when David Cox and Richard Myers refined and modernized the technique by exploiting the new molecular tools that had been developed in the intervening years.

Hybrid cells between human cells and rodent cells progressively eliminate the human chromosomes until only one is left (or a portion of one if a human chromosome with a deletion is used). Goss and

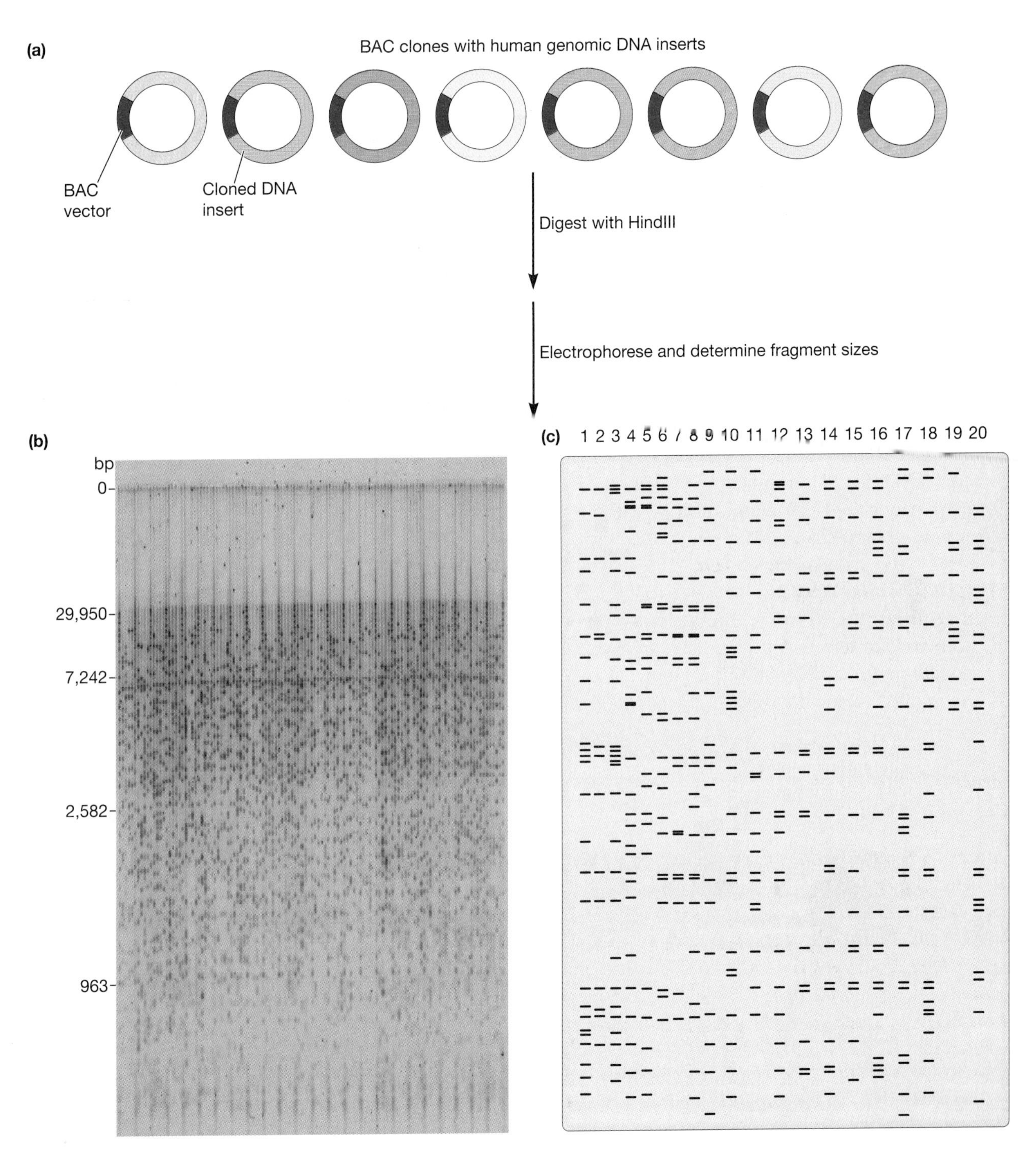

FIGURE 11-9

Bacterial artificial chromosome (BAC) fingerprinting is used to build contigs of BAC clones. (*a*) DNA preparations of each clone in a BAC library are digested with a restriction enzyme such as HindIII, which cuts approximately every 5000 bp, generating a unique set of an average of 40 DNA fragments for each BAC. (*b*) The digested DNAs are separated by size on agarose gel electrophoresis, and the lengths of the restriction fragments for each BAC are measured and recorded as that BAC's fingerprint. A computer algorithm then compares the fingerprints of all the BACs to one another, identifying those that share a significant number of fragments. By comparing a large number of BACs, multiple overlapping clones can be identified, and a physical map that extends for many BACs can be built. (*c*) In this example, several overlapping BACs sharing common restriction fragments can be identified in adjacent lanes on the drawing of a gel to demonstrate the sharing of restriction fragments. (The original gel has 96 lanes of BAC DNA and 24 lanes with size standards.)

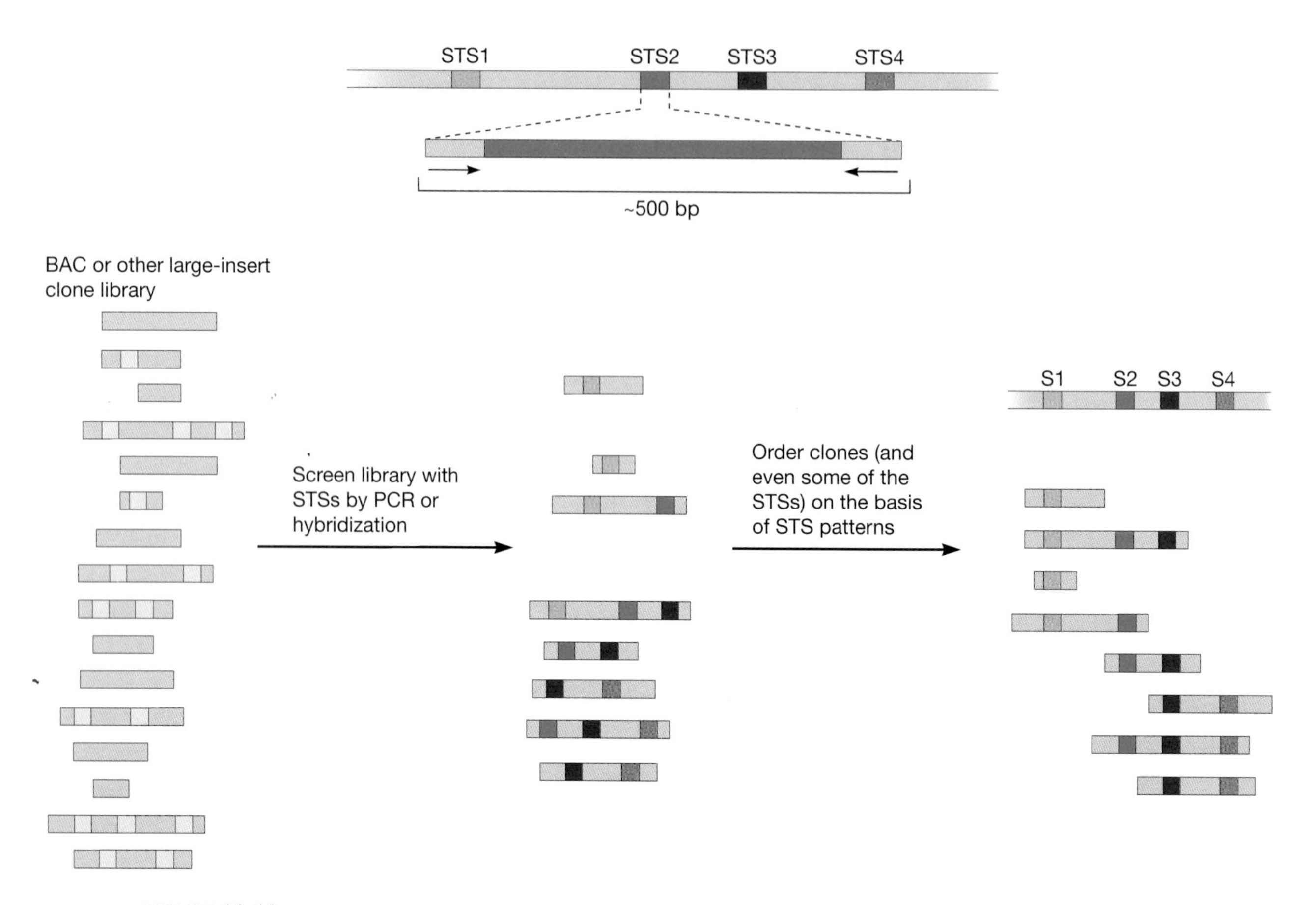

FIGURE 11-10

STS-content mapping. A large number of sequence-tagged sites (STSs), which are short (200–500-bp) fragments that can be amplified by PCR (polymerase chain reaction) from a unique locus in the genome, are generated. A library of BAC (bacterial artificial chromosome) clones that cover the human genome many times over is generated, and each BAC is scored for the presence or absence of each of the STSs, either by PCR or hybridization. BACs that contain some of the same STSs must come from the same region of the genome and overlap one another. With enough STSs and deep enough BAC coverage, two maps can be built simultaneously—a BAC contig map, just like the ones built by fingerprinting, and a map of the STSs in the genome, which can be determined by their presence or absence in the overlapping BACs.

Harris created hybrid cells with very small fragments of human chromosomes by first breaking the chromosomes of the human cells with very strong doses of γ rays. These cells were fused to rodent cells and sets of hybrid cell lines were isolated, each cell line retaining only a fraction of the human genome. Goss and Harris then carried out linkage studies on the hybrid cells by determining the frequency with which two genes or markers occurred together in the set of hybrids.

Cox and Myers developed this method further so that it could be used to map large numbers of DNA markers. First, they generated a set of 90 radiation hybrids, each of which retained large numbers of broken DNA fragments comprising about 25% of the human genome (Fig. 11-11). The 90 cell lines were scored for thousands of STSs, many of which were used in the genetic mapping and BAC mapping studies described above, by using PCR and agarose gel electrophoresis. The patterns of the presence or absence of each of the STSs in each of the hybrids were compared, and statistical algorithms were developed to predict the likely order of the STSs relative to each other and the distances between them (Fig. 11-12). The resolution—the average minimum distance that can be distinguished between markers—of a radiation hybrid map can be adjusted by varying the dose of radiation used to fragment the human chromosomes. The higher the dose, the more frequently breaks will occur and the higher the

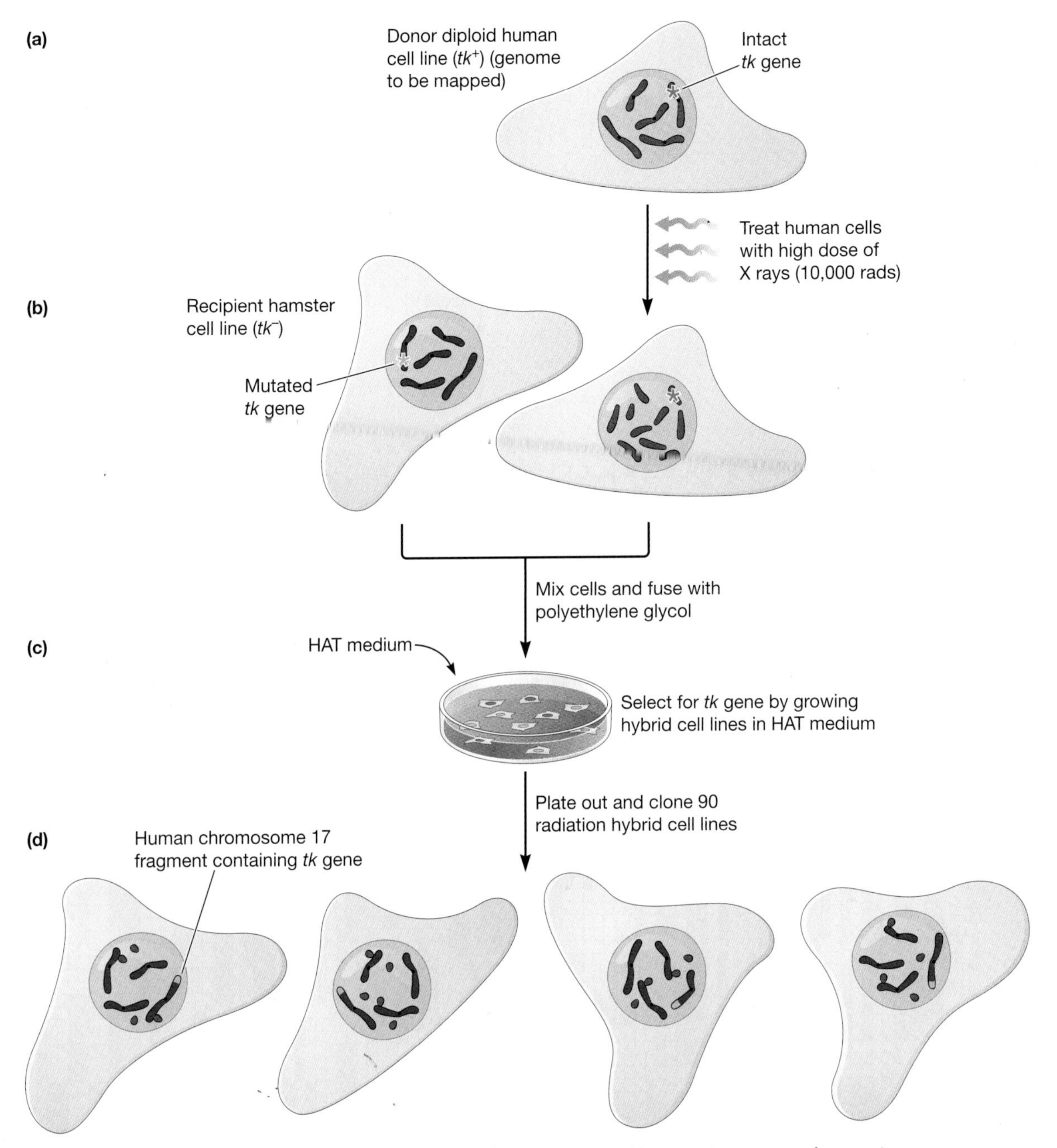

FIGURE 11-11

Generation of hybrid cell lines for use in radiation hybrid mapping. (*a*) The diploid human cell line contains a functional *thymidine kinase* (tk^+) gene on each of its copies of chromosome 17, shown by a *turquoise asterisk*. This cell line is irradiated with very high doses of X rays, which causes double-stranded breaks at thousands of places in the genome. (*b*) The irradiated cells are mixed with a hamster cell line that lacks the *tk* gene. Polyethylene glycol is added, and this causes the membranes of the cells to fuse with one another. (*c*) The fused cells are then placed in tissue culture plates and HAT medium is added; this growth medium contains a mixture of chemicals that causes cells that lack the *tk* gene to die. (*d*) During this selection process, stable cell lines form, all of which contain the hamster chromosomes and a fragment of human chromosome 17 that contains the *tk* gene. Many other human chromosomal fragments are also retained, probably because they fuse and form new chromosomes or integrate into the hamster chromosomes by the repair mechanisms in the cell nuclei. Each radiation hybrid cell line retains about 25% of the human genome, and, in theory, each is a random collection of thousands of radiation-produced fragments from all over the genome. About 90 such cell lines are constructed to allow a radiation hybrid map of the genome to be built.

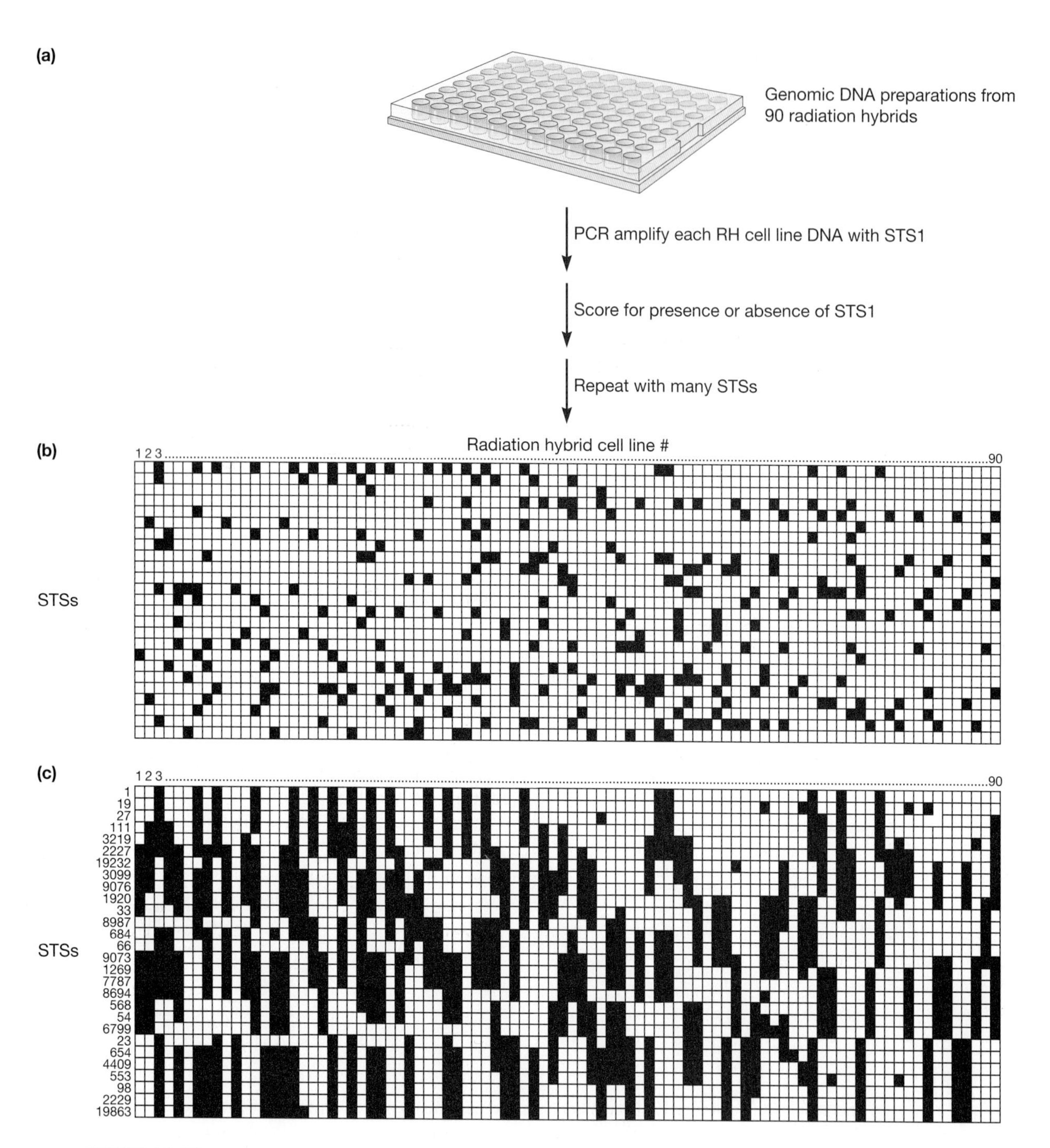

FIGURE 11-12

Using radiation hybrid (RH) mapping data to determine the order of sequence-tagged sites (STSs). (*a*) 90 RH lines are produced in large quantities and genomic DNA is prepared from each. The genomic DNAs are then scored for the presence or absence of thousands of STSs from all over the human genome. (*b*) Each row is data obtained from one STS. *Filled boxes* mean that the STS is present in that cell line, and *open boxes* mean that it is absent. (*c*) The panel shows a reordering of the rows from the experiment such that STSs with similar (but not identical) retention patterns are next to each other. For eample, STS1 and STS19 in the top two rows are both present in several of the cell lines, but each is present in additional cell lines that are not the same for each STS. The co-occurrence of these STSs in multiple cell lines is significantly more frequent than would be expected by chance, which indicates that the two markers are located near one another in the genome. With large numbers of STSs, a complex computer program determines the orders of the STSs in the genome. In addition, this analysis provides an estimate of the distances between the STSs, which are described in units of centiRays (cR), analogous to centiMorgans in genetic mapping.

resolution of the map that can be generated. Although the resolution of a radiation hybrid map is much lower than a BAC fingerprint, it provides an independent way to order segments of DNA. Two radiation hybrid maps were used to help build the physical map of the human genome, one with a resolution of about 1.5 million base pairs and the other with about 400,000 base pairs.

Completing the Physical Map of the Human Genome Required Further Cloning and Integration of Map Data

Considering the size of the human genome and the complexity of the analyses, it is perhaps not surprising that the physical map was incomplete even after this extensive analysis. There remained many gaps—places where contigs did not connect with other contigs. A primary cause appeared to be that some regions of the human genome simply could not be cloned in BAC vectors, possibly because of unusual sequences or because the cloned human DNA had some biological activity that harmed the host *E. coli* cells. This problem was dealt with in two ways. First, some of the missing regions were cloned in other vectors, such as YACs. Second, some of these missing regions were isolated in *E. coli* vectors, but only as much shorter pieces rather than as long, 100- to 200-kb segments in BACs. These shorter pieces could be cloned, perhaps because they contained only a portion of whatever element had been biologically active when intact in a large clone or because some DNA structural oddity was disrupted in these smaller pieces. Although this approach led to an improvement, the BAC map at this stage still consisted of thousands of contigs, far more than the 48 expected if the contigs were completely connected. It is likely that this result is the best that can be expected using only physical clone mapping and no other technique. For this reason, another step was added that greatly improved the BAC clone map.

The strategy followed was to compare the BAC physical map with other genomic maps that had been obtained by independent means. These maps were genetic maps with polymorphic markers, STS maps derived from radiation hybrids, and maps produced by using fluorescence in situ hybridization (FISH), which showed the positions of BACs and markers along chromosomes. Because the orders of these markers had been determined by means other than BAC contig building, they served as independent landmarks in the genome that allowed the BAC contigs to be ordered. In fact, similar to the principle of STS-content mapping, there were enough ordered genetic and radiation hybrid markers not only to allow the orders of BAC contigs to be determined, but also to work out how the contigs were oriented in the genome. In addition, in the same way that paired-end sequencing of 3-kb subclones provides valuable mapping information on a very fine scale of resolution, the ends of most of the BAC clones, as well as the other large-insert clones such as YACs and fosmids, were sequenced to provide additional, long-range mapping information that helped to sew the sequences together. The final physical map of the human genome was an integrated map containing as many different kinds of mapping information as were available. It contained almost 400,000 BAC and other large-insert clones and had fewer than 1000 gaps. And because so many varied sources of data had been used, the map had a very high level of contiguity—the orders, orientations, and even distances between the BAC contigs were known.

This BAC clone map was invaluable in producing the draft human genome sequence that was reported in 2001 and in producing the finished sequence 2 years later. In addition, the clones themselves have provided an extremely valuable resource for researchers who want to work with a particular section of the human genome. Instead of having to screen a library to find the DNA segment of interest, researchers can simply inspect the physical map, locate the clones that they need, and order them from a repository. Even though today most organisms are sequenced by using a whole-genome shotgun strategy, BAC clones and at least partial maps are still generated for many of these projects because they greatly improve the quality of the assembly of the sequence and provide useful reagents for studying the genome.

Sequencing of the Human Genome Began Slowly while Improved Technologies Were Developed

The HGP began formally in 1990, but little human sequence was generated during the first 8 years of the project. There were many reasons for this delay, some deliberate and others unexpected. First, generating physical and genetic maps of the human genome was a first priority and took precedence over sequencing. Second, sequencing on this scale had never been attempted and much needed to be done to upgrade all steps of the process. The first sequencing machines were in use and were a great improvement over manual sequencing, but as samples still had to be loaded by hand, the process was slow. Third, sequencing had been up to that point primarily a cottage industry, but the HGP was going to require large multidisciplinary teams, a high level of coordination from large numbers of laboratories spread around the world, and implementation of industrial principles and production schedules. Biologists had little or no experience in building and operating such an infrastructure, and there was much trial and error during the first few years as these skills were learned.

By 1995, improvements in technology and infrastructure encouraged the genome community to begin a large-scale pilot human sequencing phase. The aim was to generate enough data to determine whether it was feasible to scale up to sequence the whole human genome. Researchers had almost no experience in sequencing the difficult, repeat-rich regions known to be present in the human genome. Also, the genome landscape was so unexplored at the time that it was possible that other, as-yet-unknown, problems would be encountered. It was also unknown whether the goal of producing sequence of very high quality (≥ 99.99% accuracy) was achievable at a reasonable cost. Several large genome centers scaled up their equipment and their research teams and, together with smaller groups, toiled for the next 3.5 years to generate about 15% of the human genome as high-quality finished sequence. These centers gained additional experience by contributing to the sequencing of various smaller genomes, including those of the yeast *Saccharomyces cerevisiae* and the nematode worm *C. elegans.*

Much was learned during this pilot phase. The experience showed that sequencing large-insert BAC clones to high levels of accuracy was possible and that many of the newly developed tools and procedures were likely to work when applied on a large scale. The funding agencies and large sequencing groups now felt that the time was right to tackle the whole genome and set the target date of 2005 for completing the genome, 15 years after the official start of the HGP. However, two quite different developments—new technology and a competition—caused a major shake-up in the HGP, which ultimately led to the completion of the sequence earlier than was originally planned.

The "Bermuda Principles" Made the Human Genome Sequence Freely Available

A critical part of scientific research is communicating—results, technological breakthroughs, new theories, and creative thinking—to other scientists. Clearly, there would be no progress if researchers kept their findings, data, and ideas to themselves. Research in the public sector, and to a large extent in the private sector, has been reported by a variety of means, including publishing in science journals, presenting at scientific conferences, and, increasingly, posting on the internet. Typically, moving from analysis of results to publication is a long process—the time for a paper to be written, reviewed, and appear in the press is usually many months. Research done in companies typically is treated differently because commercial research is concerned with making products. Here, results are often kept as trade secrets and are not published in journals or, if so, only after patent protection is obtained. Without such protections, there would be little motivation for private investment. (Academic laboratories also patent their research findings, but the requirements for funding, promotion, and recognition encourage rapid publication.)

The scientists who planned the HGP in the late 1980s argued early on that data from the mapping and sequencing projects would have to be treated differently. First, if standard academic practice were followed, very large amounts of data would not be available to other researchers for years. However, in contrast to most research, even incomplete genomic data, such as maps or sequences of only parts of genomes, are extremely useful to researchers. Much time would be wasted and many opportunities lost if the data were not released until the genome was completely sequenced. Second, it was argued that it would be unfair to allow the scientists running the genome centers to have exclusive access to the unpublished but invaluable data. Finally, it was forcefully maintained that this publicly funded resource should not be patented and that the sequence of the human genome should not be controlled or owned by any individual, university, company, or government.

From the beginning of the HGP in 1990, a policy was introduced requiring that every group participating in publicly funded sequencing projects release their data rapidly and with no strings attached. This policy was formalized at an international conference in Bermuda in 1996, funded by the Wellcome Trust, and attended by sequencing groups from around the world. The so-called "Bermuda Principles" called for the automatic release of sequence data into public databases within 24 hours of their assembly. This extraordinarily rapid release ensured that the large-scale sequence data were available to all. Furthermore, it prevented blanket patenting of the sequences, not only because they were in the public domain, but also because there would be no known utility (a requirement of the U.S. and European patent offices) for any particular small segment of sequence. The sentiments of the Bermuda Principles—similar to the "open source" model adopted by many in software development—became the standard for almost all publicly funded, large-scale sequence data. Biology researchers the world over have come to rely on access to this freely available DNA sequence data.

Capillary Sequencing Machines Make Sequencing Much More Efficient

Fluorescent sequencing was a major improvement over manual sequencing, which uses radioactively labeled nucleotides and X-ray films to record images of the bands. However, it became clear during the pilot phase of the HGP that it was going to be an almost impossible task to complete the sequencing by using slab gel electrophoresis. With 85% of the human genome still to be sequenced, at least a million slab gel runs, each containing 48 sequencing reactions, would be needed. Fortunately, another advance in sequencing technology was implemented at about this time. New machines that performed electrophoresis in capillaries rather than in slab gels were much easier to use than the slab gel machines and resulted in at least a tenfold increase in the efficiency and rate at which DNA could be sequenced (Chapter 10). In 2001, a second-generation capillary machine became available that gave an additional boost in throughput of about fivefold. Although there may be further improvements, it is likely that sequencing based on the dideoxy method is approaching its peak of efficiency and that new methods based on different principles will be required for further improvements. At the start of the HGP, exotic methods of sequencing, including hybridization, single molecule analyses, and several others, were explored but none came to fruition in time to sequence the human genome. Recently, however, some new strategies for sequencing have begun to show promise, thus offering the possibility of a "$1000 genome sequence" (Chapter 14).

Two Independent Groups, One Public and One Private, Raced to Sequence the Human Genome

At a meeting at Cold Spring Harbor Laboratory in May 1998, J. Craig Venter, who led the effort to sequence the *Haemophilus influenzae* genome (Chapter 10), dropped a bombshell on his colleagues in the public human genome sequencing effort. He announced that he was starting a company, Celera Genomics, that would sequence the entire human genome in just 2 or 3 years. The company would use the new capillary machines and a whole-genome shotgun method, thereby omitting the expensive and time-consuming mapping step. This announcement and events over the next few years produced great unease, heated debates, a tremendous amount of generally unhelpful publicity, and recriminations from both Celera and the publicly funded project, the IHGSC. The latter argued that the whole-genome shotgun approach would produce an incomplete genome sequence of poor quality, that Celera would have access to the IHGSC data but not vice versa, and that no private company should own the human genome sequence. Venter depicted the IHGSC as a bloated and inefficient academic enterprise and attempted to convince the U.S. Congress that it was a waste of public money to fund the IHGSC, arguing that Celera would sequence the human genome and the public effort should sequence the mouse or some other genome. The Wellcome Trust in the United Kingdom responded immediately by doubling its financial support of the public project and, together with other agencies around the world, undertook to fund the entire public effort if the U.S. government backed out. The U.S. Congress, however, did not accept Venter's argument and continued to fund the large U.S. contribution to the IHGSC.

Soon after Celera's announcement of its strategy, the company and the IHGSC centers acquired large numbers of the capillary sequencing machines. Celera quickly built an efficient automated sequencing pipeline and developed improved methods for sample preparation and sequencing chemistry. As a "proof of principle," Celera applied the whole-genome shotgun approach to sequencing the *Drosophila melanogaster* genome, at 180 million base pairs very much larger than any previously shotgun-sequenced genome. The Celera group sheared *Drosophila* genomic DNA, cloned the fragments into plasmids to obtain three different size classes, and sequenced both ends of more than 1.5 million subclones. To attack the daunting problem of assembling the genome, Celera recruited a computer scientist named Eugene Myers, who built a computational team and developed new algorithms and strategies to assemble the *Drosophila* sequence. The assembly contained only the two-thirds of the *Drosophila* genome that makes up the euchromatin and lacked the third that is heterochromatin—long segments of highly repetitive sequences. There were a significant number of gaps, misassemblies, and missing sequence, but this first attempt at sequencing a eukaryotic genome by a whole-genome shotgun approach was successful in providing a very useful draft sequence of *Drosophila*. More importantly, it demonstrated to Celera scientists that whole-genome shotgun sequencing and assembly was feasible when applied to a genome 20 times larger than had yet been attempted.

Turning to the human genome, Celera used as starting templates DNAs that were short-insert subclones of total human genomic DNA, not mapped and fingerprinted large-insert clones. The company produced more than 27 million high-quality sequencing reads in less than 1 year. The challenge remained of assembling the millions of unmapped sequencing reads into an ordered sequence. Myers refined and further developed the methods used to assemble the *Drosophila* genome by inventing ways to deal with the complex sets of repetitive sequences that make up half of our genome. This was a major tour de force, and the Celera team produced an assembled human genome sequence within 2 years by using a combination of Celera and IHGSC sequence data, as we shall discuss below.

For the IHGSC, Francis Collins (National Human Genome Research Institute, NIH), Michael Morgan (The Wellcome Trust, United Kingdom), and Aristides Patrinos (U.S. Department of Energy), worked together to coordinate the 20 genome centers around the world that generated the public sequence. The vast majority of sequence was produced by five very large centers (Table 11-1). Based on the nearly finished physical map, the IHGSC

TABLE 11-1. The finished sequence of the human genome produced by the International Human Genome Sequencing Consortium included contributions from 20 different centers from six countries

Center	Finished sequence totals (kb)
SC	919,388
WUGSC	645,062
WIBR	562,096
JGI/SHGC	485,085
BCM	320,735
RIKEN	155,769
UWGC	145,745
GS	99,970
GTC	45,710
UWMSC	39,227
Keio	44,905
IMB	73,677
Beijing	38,079
MPIMG	9,838
GBF	8,325
UOKNOR	18,657
TIGR	10,390
CGM	2,768
SDSTDC	7,792
UTSW	8,555

From International Human Genome Sequencing Consortium, *Nature 431:* 931–945 (2004).

The left column lists abbreviations for the 20 centers. The right column contains the total number of base pairs finished by each center. The first five centers on the list produced more than 90% of the total sequence. Abbreviations: SC, Wellcome Trust Sanger Institute (Hinxton, United Kingdom); WUGSC, Washington University Genome Sequencing Center (St. Louis, Missouri); WIBR, Whitehead Institute of Biomedical Research (Cambridge, Massachusetts); JGI/SHGC, U.S. DOE Joint Genome Institute and Stanford Human Genome Center (Walnut Creek and Stanford, California); BCM, Baylor College of Medicine Human Genome Sequencing Center (Houston, Texas); RIKEN, RIKEN Genomic Sciences Center (Yokahama, Japan); UWGC, University of Washington Genome Center (Seattle, Washington); GS, Genoscope (Evry, France); GTC, Genome Therapeutics Corporation (Waltham, Massachusetts); UWMSC, Institute for Systems Biology Multimegabase Sequencing Center (Seattle, Washington); Keio, Keio University School of Medicine (Tokyo, Japan); IMB, Institute for Molecular Biology (Jena, Germany); Beijing, Beijing Genomics Institute (Beijing, China); MPIMG, Max Planck Institute for Molecular Genetics (Berlin, Germany); GBF, German Research Center for Biotechnology (Braunschweig, Germany); UOKNOR, University of Oklahoma Advanced Center for Genome Technology (Norman, Oklahoma); TIGR, The Institute for Genome Research (Rockville, Maryland); CGM, Center for Genetics in Medicine (Perkin Elmer/Washington University); SDSTDC, Stanford DNA Sequencing and Technology Development Center (Stanford, California), UTSW, University of Texas Southwestern Medical Center (Dallas, Texas).

groups chose a minimum tiling path of BAC clones and used shotgun methods together with capillary sequencing machines to sequence them to a depth of 6x–8x coverage. The massive increase in data generation identified many problem regions in the genome and the methods were improved to help deal with these. However, even when the collection of the sequencing data was nearly complete, the IHGSC had not yet solved the problem of piecing the individual assembled BAC sequences into an assembly of the whole genome. They therefore recruited computer scientists David Haussler and James Kent to work on this crucial step of the process. Kent worked around the clock writing computer code and in a very short time developed an effective method for assembling the entire sequence. At about the same time that Celera announced that it had produced an assembly of the human genome sequence, the IHGSC also had an assembled sequence, called at this draft stage "the Golden Path" sequence.

Draft Sequences of the Human Genome Were Announced in 2000

By June 2000, Celera and the IHGSC had each produced an assembly of the human genome, and a joint press conference, called by U.S. President William Clinton and U.K. Prime Minister Tony Blair, was held to announce that both groups had sequenced the human genome. The achievement was heralded as a historic contribution to knowledge and was described by the politicians and much of the media as the resolution of a fierce competition, which had ended in a tie. In reality, the human genome sequence was nowhere near "done." The two assembled sequences were *drafts* of the human genome sequence, with many base-pair errors, small and large misassemblies, and a significant fraction of the genome missing. The draft sequences from the IHGSC and from Celera were published simultaneously in February 2001, but in different journals. *Nature*, which published the IHGSC sequence, had refused to accede to Celera's demand that their sequence be kept secret, and so the Celera sequence was published in *Science*, which did not require Celera to release all of its data. There were various commentaries and other papers related to the drafts, but the key issues were how the two groups had produced their drafts, how the drafts compared, and what sorts of biological messages were present in the sequences.

The IHGSC draft sequence assembly was generated from the shotgun sequencing of 29,298 mapped large-insert clones, mostly BACs, that added up to 7.5x coverage of the human genome. Celera generated whole-genome shotgun sequence data that totaled 5.1x coverage of the genome. The Celera sequence was combined with the IHGSC's 7.5x reads (which were freely available in the public databases) for an overall 12.6x coverage. With this 12.6x data, Celera generated two different types of assemblies. In one, they used no mapping data, thereby producing what they called a "pure whole-genome assembly" (WGA). In their second assembly, called a "compartmentalized shotgun assembly" (CSA), Celera included map information that had been generated by the IHGSC. Unfortunately, detailed comparisons of the different assemblies were hampered by restrictions that Celera placed on the use of their sequence data and their assemblies. Effectively, no large-scale analysis was allowed without payment of substantial fees that only a few large pharmaceutical companies could afford. Nevertheless, academic researchers did gain limited access to Celera's sequences by purchasing a less expensive license that greatly restricted the use and types of analyses that could be done. Several comparisons, some more scientifically rigorous than others, were performed and reported at the time of and following the publications of the two draft sequences.

The most relevant comparisons were of Celera's CSA (their 12.6x assembly that included mapping information) and the IHGSC's 7.4x mapped-based assembly. The most important conclusion was that, as expected, neither assembly produced anything close to a complete sequence of the human genome. About 90% of the euchromatic portion of the human genome was contained in the Celera CSA and IHGSC assemblies, but these had incorrect base calls, unassembled regions, and, most seriously, a large number of incorrectly assembled segments. The Celera sequence had slightly more segments of the genome in assembled contigs greater than 100 kb in length than did the IHGSC sequence (49% vs. 46%). Yet the longest contig in the IHGSC sequence (28.5 Mb) was significantly longer than Celera's longest contig (2 Mb). Both assemblies had a huge number of gaps in the assembled sequence—170,033 for Celera and 149,821 for the IHGSC. It was difficult to compare their accuracy, but it is likely that the two assemblies were very similar, averaging one error every 100 base pairs. One difference was that about 35% of the genome had been subjected to "sequence finishing" (which we describe in the following section) by the IHGSC, thus producing sequence having a much higher degree of accuracy than the draft regions, with less than one error in 10,000 base pairs.

Although both draft sequences of the human genome were incomplete, they contained far more sequence information for a complex genome than had ever been reported. Much was learned from the assemblies about the architecture of the human genome, such as its GC content and the distributions of genes and repeat sequences. A remarkable and surprising finding was that the estimated number of human genes was much smaller than originally thought. Based on the analysis of the sizes of a few genes, biologists had assumed for two decades that human beings had about 100,000 genes, but the draft sequences suggested that there are fewer than 30,000 genes, not dissimilar from the number of genes in the fruit fly genome! The draft sequences found many enthusiastic users. Hundreds of Mendelian disease genes were identified much more quickly than before, as the sequence and maps from the HGP provided much of the data needed to locate and clone genes (Chapter 14). Investigators who were studying human genes and cells used these maps, clones, cDNA sequences, and draft genome sequences as they became available throughout the course of the HGP. Many considerable advances were made that would otherwise have been significantly delayed; thus the IHGSC policy of rapid release of map and sequence data was validated.

Finishing a Genome Sequence Requires Special Attention to the Problem Regions

A great deal of additional work was required to fill the gaps and correct the errors in the draft sequence to finish the sequence. Finishing is a complex, iterative process that involves using a variety of experimental approaches in combination with computation and data analysis. Consider, for example, the assembly of a human BAC clone that is about 200,000 base pairs in length (see Fig. 10-15). The BAC sequences, when subcloned into 3-kb plasmids and shotgun sequenced to 6x–8x sequence coverage, will typically produce a small number (usually 2–10) of contigs that together add up to about 190,000 base pairs. There will be a large number of sequence reads that do not fit into the assembled contigs—some because they are repetitive, some because they simply did not find an overlapping mate, and others because the sequence quality was poor.

A first step in determining the order of the contigs is electronic screening for unique DNA markers for which other data, such as meiotic, physical, or radiation hybrid mapping information, are available. This process usually establishes the order and orientation of some of the contigs. Second, plasmids that lie near the ends of the sequence contigs are resequenced, as are the ends of plasmid subclones that were of low sequence quality (Fig. 11-13). After this round of finishing, the assembly process is run again. The quality of the sequence will have improved, and the additional reads from the subclones near the ends of contigs, and some of the subclones that were not originally assembled, will have filled some of the sequence gaps. In addition, whereas some of the newly assembled subclones may not yet fill the gaps, they do extend further into the missing regions at the ends of the contigs. Custom oligonucleotide primers are synthesized for sequencing the internal portions of these subclones to provide additional sequence coverage. The assembly is repeated and reassessed, and the whole process repeated until only one or two gaps remain. It is probable that the starting subclone library does not have any clones with these missing sequences and DNA corresponding to the gaps must be cloned from total human genomic DNA by using PCR primers from the ends of the two contigs flanking the gap. These clones (or sometimes the direct PCR products) are sequenced and the assembly is repeated. If the BAC clone is still not complete, further cloning, sequencing, and assembly must be carried out until it is completely contiguous and there is less than one error in 10,000 base pairs. The BAC clone is now considered finished.

The IHGSC applied this laborious process to the 29,298 mapped large-insert clones that had been the source of the subclones used for the shotgun sequencing. Some large-insert clones were especially difficult because they contained long or almost identical repeat segments or both that were impossible to assemble, regions of low-complexity sequence (such as high-GC or high-AT content), regions that did not clone well in *E. coli*, and even regions that were

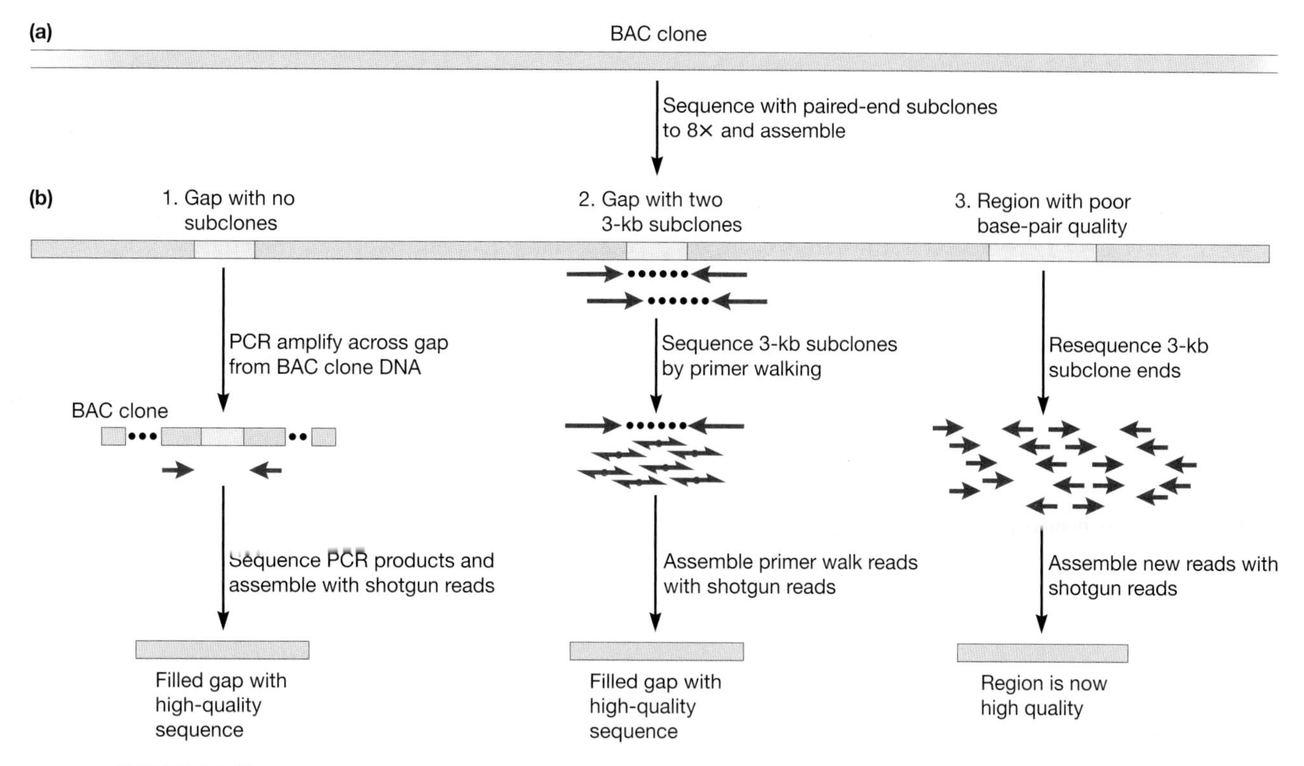

FIGURE 11-13

Sequence finishing. In theory, deep shotgun sequencing to 8x sequence coverage should result in a completely assembled, unambiguous high-quality sequence of a BAC clone or even a whole genome. In practice, this rarely happens, for a variety of reasons, but particularly because of problems caused by repetitive and difficult-to-traverse sequences. (*a*) A BAC clone has been shotgun sequenced with 8x coverage. (*b*) The reads have been assembled and the *light blue* regions are where the assembled sequence is accurate and the base-pair quality exceeds the minimum requirements (usually set at a maximum of 1 error in 10,000 bp). There are three problem regions: (*1*) a gap in the sequence that is not covered by any of the subclone templates, (*2*) a gap in the sequence that is covered by several 3-kb subclones, and (*3*) a region that is assembled properly but where the base-pair quality is poor. To fix these regions, the process of sequence finishing is performed.

1. The clone gap can be solved by several approaches, including isolating the region in another type of subclone library or, more commonly, by using PCR. In this case, primers for amplification are designed based on the sequences on either side of the gap, and PCR is used to amplify the region from the BAC clone. The amplified product is either then sequenced directly or cloned into a plasmid vector and sequenced. To achieve the highest standards of completeness, at least two sequencing reactions, preferably one in each direction, are performed on the PCR product.
2. The sequence gap is straightforward to fix because two 3-kb subclones are available that span the gap. These subclones were identified in the shotgun sequencing of the paired-end phase of the sequencing project. To fill in the gap, both subclones are subjected to complete sequencing, usually by designing primers based on the known sequence that extend further into the 3-kb insert. If the gap is longer than a sequencing read, typically "primer walking" is performed, in which a second set of primers, now based on new sequence information obtained from the first set of primers, is used.
3. The region of poor sequence quality is usually fixed simply by choosing the subclones whose end sequences were used to assemble the region and performing the sequencing reactions a second time. In cases where this does not work, perhaps because the region is very GC-rich and the DNA polymerase used in sequencing stalls in the region, additional sequencing reactions with an alternative DNA polymerase or a variety of chemical additives are performed.

not amenable to PCR amplification. These required many more iterations and additional steps to obtain accurate and contiguous sequence. The end result was that there were only 281 locations in the genome where, despite applying every known method to finish these sections, it was not possible to close the gaps. The IHGSC annotated each of these gaps and determined their approximate sizes by measuring clone or PCR fragment sizes or by using other mapping information.

The IHGSC had argued from the beginning that generating a finished product from a complex genome would require mapped large-insert clones, and this was affirmed by experience. It was clear that the number and complexity of finishing steps needed to assemble a whole shotgun sequence of the human genome without mapping data would be astronomical, and still would leave many gaps, assembly errors, and incorrect base pairs. Subsequently, almost all projects sequencing large genomes incorporate the advantages of whole-genome shotgun sequencing with a variety of mapping and other steps.

The Public Sequence of the Human Genome Was Finished to a High Degree of Accuracy

The accuracy of the final public version of the human genome was assessed by using base-calling and assembly algorithms to calculate probability scores for the accuracy of the base pair call at each position in the genome. Stringency for correct assembly was set by requiring a minimum of two sequencing reads (although the vast majority had many more) for independent subclones of every assembled segment, together with confirmation by mapping information from genetic, physical, radiation hybrid, and other mapping methods. An additional quality assessment (QA) exercise involved resequencing a sample of randomly chosen BACs representing about 1% of the human genome to provide an independent analysis of the finished sequence.

The finished sequence covered more than 99% of the euchromatic regions of the human genome. The 281 sequence gaps were all in regions of extremely complicated repetitive sequences. Of note, 50% of the base pairs were in contigs of at least 27.5 Mb. The finished sequence had less than one error in 100,000 base pairs, which exceeded the original goal for accuracy by more than tenfold. These improvements over the draft IHGSC sequence published 2.5 years earlier were dramatic (Table 11-2). Not only was the overall base-pair accuracy much higher in the finished sequence, but also most striking was the reduction of gaps from almost 150,000 in the draft sequence to 281 in the finished sequence, a reduction of more than 400-fold. The vast number of assembly errors in the draft was reduced to close to zero in the finished sequence. The final sequence is a very close representation of a reference human genome sequence.

Was the cost of finishing the human genome worth it? At the beginning of the HGP, it was argued that because understanding our genome is so important to us, we should put special emphasis on making the human reference sequence as close to perfect as possible. There were good scientific arguments for this. First, a jumbled, inaccurately assembled genome sequence would make it extremely difficult to identify genes and determine their structures. Many genes in the draft sequences were not assembled or were incorrectly assembled, even though cDNA sequences were used to annotate gene content and structure in the drafts. Second, if the error rate of identifying bases was significantly higher than the rate of sequence variation between individuals (which was known to be about one in 1000 base pairs), the sequence would not be

TABLE 11-2. The IHGSC finished sequence of the human genome is drastically improved compared to the draft sequence

	No. of base pairs	Euchromatin (%)	No. of gaps	Base pairs in gaps	No. of scaffolds
Draft (February 2001)	2,692,000,000	92	145,514	152,000,000	87,757
Finished (Build 34 July 2003)	2,843,433,602	99.3	281	20,335,408	80

The table lists the number of base pairs covered, the percentage of the euchromatic regions covered, the number of gaps, and the number of base pairs in the gaps in the 2001 draft sequence compared to the 2003 finished sequence of the human genome.

a good reference for identifying and studying human sequence variation. Third, an inaccurate sequence would reduce its utility for comparing the human sequence to the genome sequences of other organisms, which is essential for identifying functionally significant sequences (Chapter 12). Fourth, a highly accurate sequence is needed to identify duplicated genes and members of large gene families, which constitute a significant fraction of our genome.

The initial date for the completion of the HGP was 2005. In the late 1990s, partly as a result of the competition with Celera, the IHGSC announced that it believed that the sequence could be completed 2 years earlier than originally planned, and set April 7, 2003 as the target date. They chose this date for its historical significance; this would be the 50th anniversary of the date that Watson and Crick submitted their manuscript describing the structure double helix to the journal *Nature*. Eighteen minutes before midnight on April 7 on the west coast of the United States—the westernmost site of IHGSC scientists—the finished sequence for the last human chromosome was submitted to GenBank. The announcement was made public on April 25, the anniversary of the date that Watson and Crick's paper appeared in *Nature*.

Whose Genome Was Sequenced Anyway?

During the early days of the project, there was much discussion concerning whose DNA would be sequenced. Surely some scientists, and perhaps others who anticipated the excitement of the upcoming HGP, imagined the attention and perhaps the glory that would come with being known as the person whose genome was the first sequenced. There was even a lighthearted proposal that the IHGSC, as a way of raising funds to support the project, should hold an auction and award to the highest bidder the right to have their genome chosen. In fact, there were serious scientific and ethical issues to be considered, and as it turned out, the IHGSC and Celera took very different approaches to selecting their starting DNAs.

Before the beginning of the HGP, human genome sequences had come from a wide variety of different genomic libraries prepared in different laboratories. No particular attention had been paid to the sources of DNA, and no efforts were made to conceal the identities of those who donated their DNA to make these libraries. By the mid-1990s, with strong urging from the funding agencies, scientists and ethicists carefully considered how the sources of genomic DNA for sequencing the human genome should be chosen. A genome sequence contains unique information about an individual—information that could cause harm if misused. The IHGSC adopted the policy that donors who provided DNA for sequencing would be anonymous and, as far as possible, untraceable. Volunteers were recruited by an independent contractor, and blood samples were taken only after the volunteers had a session with a genetic counselor and provided written informed consent. All identifiers were removed, and the samples were assigned random numbers. From these, about one in ten samples was randomly selected, and all labels were removed prior to shipping the samples to the IHGSC scientists who made BAC libraries from them. Because all records linking the donors to DNA samples were destroyed, no one, not even the donors themselves, knew which samples were used to sequence the genome by the IHGSC. About 70% of the final sequence from the IHGSC was obtained from a single sample and the remaining 30% came from a number of different anonymous individuals.

Although it first appeared that the company took a similar approach to the IHGSC, the situation at Celera turned out to be very different. Their plan was to obtain anonymously a set of 21 DNA samples from individuals who were considered to belong to five different ethnic groups and then to select several samples from this set for sequencing. The company hoped to gain information about sequence variation among ethnic groups while maintaining anonymity of the donors. In the publication of their draft sequence in 2001, Celera described how this process had produced a composite genome sequence that could not be attributed to any individual or group of individuals. However, in 2002, after he had left Celera, Venter declared that this was not the case. He had secretly overridden the process for donor selection and anonymity and had used his own DNA for the vast majority of the sequencing at Celera. Venter announced that he carries a genetic marker indicating he is at increased risk for developing heart disease and Alzheimer disease and that he had begun taking fat-lowering drugs in hopes of staving off the effects of these sequence variants.

Celera's bioethics board, which included a number of prominent biologists and bioethicists, believed that Venter had committed a serious ethical infraction. Whereas he might have the right to reveal his own DNA sequence to the world, he was also revealing information about his genetic relatives. Venter justified his actions by claiming that he wanted to set an example of how genome information could be used to promote lifesaving measures.

Resequencing Parts of the Human Genome from Many People Identifies DNA Sequence Variants

One significant application of the information and technology developed by the HGP was to identify DNA sequence variations. These variations are then analyzed to help us understand how genetics contributes to important traits like diseases, variable responses to drugs, and differences in susceptibility to infectious agents. Sequence variants are also essential tools for understanding our genetic and cultural history (Chapter 14) and for developing forensic applications (Chapter 16). Many of these variants are single base substitutions; for example, one chromosome may have a G at a particular position, whereas the homologous chromosome in the same or other individuals may have an A at that same position (Fig. 11-14). These variants have been called single-nucleotide polymorphisms (SNPs, pronounced "snips").

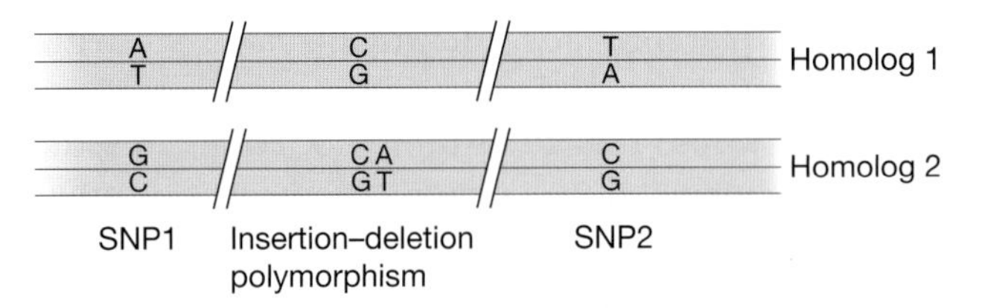

FIGURE 11-14
Single-nucleotide polymorphisms (SNPs). Many DNA sequence variants are single-base substitutions, for example, a G at a particular position in the genome versus an A at that same position. Variants such as this that are present at a frequency of 1% or more in human populations are called single-nucleotide polymorphisms (SNPs, pronounced "snips"). Any two humans, or for that matter, the two copies of the genome that an individual inherited from his parents, have a SNP approximately once every 1000 base pairs on average. The vast majority of SNPs fall in regions of the genome that are not important for function and are therefore likely to be neutral. The two alleles of a SNP at a locus can be measured in people by using PCR and various genotyping methods. More than 9 million SNPs have been identified in human populations.

Many small, and several very large, studies have identified a huge amount of human sequence variation in a very short time. Even before the draft sequences were complete, The SNP Consortium, an alliance of academic and industry groups, identified more than a million SNPs by shotgun sequencing genomic DNA from individuals from a variety of geographic locations. Subsequently, Perlegen Sciences, a private company founded by David Cox, and a public effort called the International HapMap Project used oligonucleotide microarray hybridization methods and DNA sequencing to identify a very large fraction of the common DNA sequence variation that is present in the world's populations, as well as a very large number of rarer SNPs. As of 2006, more than 9 million SNPs have been identified and are available in public databases from Perlegen and the HapMap Consortium. This information has already proven extremely valuable in human genetics and population studies. Discovering SNPs is likely to become much more efficient in the near future as new technologies for ultrarapid and inexpensive DNA sequencing, perhaps several orders of magnitude cheaper than even the most efficient capillary methods, come on line. These methods will enable resequencing of the entire genomes of very large numbers of people, so that all of the sequence variants in each person, both common and rare, will be determined. As we shall discuss in Chapter 14, this will radically transform the way that researchers and clinical practitioners use genetics to understand human traits.

In addition to SNPs, the human genome has a significant amount of DNA sequence variation resulting from insertions and deletions ("indels"), inversions, and other rearrangements, which range in size from a few to millions of base pairs. Such variation is difficult to detect but it has become clear that it is much more common than had previously been thought. It is important to identify these large changes because they are more likely to alter the function and expression of genes, with serious biological consequences. For example, as we shall see in Chapter 15, large-scale rearrangements are common in cancers. New methods can detect these changes on a genome-wide scale, for example, by hybridizing genomic DNA to arrays of oligonucleotides representing the entire genome.

Many Other Genomes Are Being Sequenced

The power of whole-genome sequencing was further confirmed by the information flowing from the hundreds of bacterial and archaeal genomes sequenced in their entirety. The very first bacterial genome sequence had a transforming effect on the prokaryotic research community; suddenly those studying *H. influenzae* had a detailed picture of the entire coding potential, genomic architecture, biochemical pathways, and regulatory networks of the organism. Researchers no longer spent time working to isolate and sequence a single gene, but could use the internet to download the information from the database. As additional bacterial genome sequences became available, comprehensive sequence comparisons could be made, evolutionary histories determined, and gene functions and pathways assigned with much greater confidence.

The Human Genome Project included the mapping and sequencing of the genomes of a set of important "model" organisms—those that are used by experimentalists to study biological processes—including the mouse, *E. coli*, the nematode worm *C. elegans*, the yeast *S. cerevisiae*, the fruit fly *Drosophila*, and a mustard weed, *Arabidopsis thaliana*. Early sequence data from these eukaryotic genomes revolutionized their research communities, and by the mid-1990s, it was clear to many biologists that progress in their own field would be advanced immeasurably by having the genome sequence of the organism under study. As a consequence, efforts were made to sequence the genomes of every organism important in health, agriculture, economics, and the environment. Now the genomes of the rat, dog, cow, chimpanzee, macaque monkey, and several other mammals have been sequenced, and the sequencing of many more mammals and other vertebrates is under way. All of the major human pathogens, many fungi, several important crop plants like rice, and plant pathogens have been or are being sequenced. The genomes of organisms that are located at key nodes in the phylogenetic tree of life (e.g., noneutherian mammals like the platypus) are being sequenced in the expectation that they will help us establish more reliable evolutionary relationships between organisms (Chapter 12). Indeed, some have argued that we should sequence the genomes of every species on the planet.

Although many genomes have been sequenced and many others will be sequenced in the future, there is a very wide range in the degree of completeness of the various genome sequences. For economic reasons, most of the large genomes are not being taken through finishing, and their levels of completeness vary with the depth of shotgun data obtained. Each new genome seems to bring a unique set of problems, which complicate its sequencing. Repetitive sequences continue to be the major culprit; the distribution, copy number, type, age (and consequently degree of sequence similarity), and length of the repeats all present challenges to the assembly algorithms. A second variable is the degree of polymorphic sequence variation present in the two chromosomes of a diploid individual or between individuals. The more variable the two copies of the genome in an individual, the harder it is to assemble the sequence accurately, because it is difficult to determine whether two different, but similar, sequencing reads are from different alleles of the same locus, or from different loci of a repeat sequence. Table 11-3 shows the status of various genome sequences for a sample of both prokaryotic and eukaryotic organisms as of 2006.

The sequencing of the human genome was the first large-scale project in biology and it is not surprising that it did not go entirely according to plan. The "genome war" between the publicly and commercially funded projects attracted a great deal of popular attention, but in the end two very similar, incomplete drafts of the genome were produced, and much of the hyperbole died down quickly. The competition had some positive effects. The IHGSC was galvanized and probably reached its goals faster, and Celera eventually released its sequence with relatively few restrictions. Perhaps the most important contribution by Celera was their demonstration that whole-genome shotgun sequencing is feasible for sequencing genomes, even very large ones. Clearly, the demand for sequence and other genomic data seems to be insatiable, and genomic sequencing will continue for years to come.

There is no doubt that biology has been, and medicine will be, transformed by access to complete sequences of genomes. In much the same way as the periodic table is the touchstone for chemists, so the genomes of yeast, *Drosophila*, *Homo sapiens*, and any other organism are the reference points for the biologists studying those organisms.

TABLE 11-3. Some bacterial and eukaryotic organisms whose genomes have been sequenced

Common name (scientific name)	Description	Genome size (Mb)	Approximate no. of protein-coding genes
Ulcer bacterium (*Helicobacter pylori*)	Causative agent in human stomach ulcers	1.66	1,491
Plague bacterium (*Yersinia pestis*)	Rod-shaped bacterium that causes systemic invasive infectious disease in humans classically referred to as "the plague"	4.82	4,012
Baker's yeast (*Saccharomyces cerevisiae*)	Key agent in making wine, beer, and bread. One of the most useful model organisms for studying basic biological functions, with well-established genetics, biochemistry, cell biology, and genomics.	12.1	6,600
Marine diatom (*Thalassiosira pseudonana*)	Ocean-dwelling photosynthetic eukaryotic microbe producing about one-fourth of the oxygen in the atmosphere. This species is useful for studying diatom physiology.	34	11,200
Nematode worm (*Caenorhabditis elegans*)	A small (1-mm) soil worm that serves as an excellent model organism for studying development and many other multicellular processes. Complete lineage of all 959 cells is known.	97	19,000
Mustard weed (*Arabidopsis thaliana*)	A small flowering plant that is widely used as a model organism. The first plant genome to be sequenced.	125	25,500
Fruit fly (*Drosophila melanogaster*)	Classic model organism, where first principles of genetics were developed. Particularly useful for studying development.	180	14,000
Rice (*Oryza sativa*)	Major food source for more than half the world's population. Smallest genome of the cereal crop plants.	430	37,500
Sea squirt (*Ciona savignyi* and *Ciona intestinalis*)	Marine organisms with the smallest known genomes of chordates. Useful in comparative sequence analysis for studying fundamental properties of vertebrates. Has an unusually high (>1.5%) rate of polymorphism.	180	14,000
Mouse (*Mus musculus*)	Common laboratory mouse. The ability to target genes for loss- and gain-of-function mutations makes it the most useful mammalian model organism.	2,500	25,000
Dog (*Canis familiaris*)	"Man's best friend." Its unique breeding history makes it ideal for studying rapid evolution of mammalian traits. Hundreds of diseases similar to diseases in humans.	2,450	19,300
Common chimpanzee (*Pan troglodytes*)	Great ape that is the closest relative to human beings. Its DNA sequence is almost 98% identical to ours.	2,800	23,000
Human (*Homo sapiens*)	Obtaining the sequence of our own genome was one of the primary motivations for the Human Genome Project. The genome sequence has greatly increased our understanding of human biology and disease.	2,850	23,000

By the end of 2006, the genomes of more than 300 bacterial species and 100 eukaryotic organisms were either completely sequenced or had substantial amounts of genomic sequence data available in public databases. The table demonstrates the wide range of types of organisms and the rationale for choosing some of them. Genome sizes are shown in millions of base pairs (Mb). The numbers of protein-coding genes in the genomes are estimated by annotation efforts based on cDNA sequences, comparison with other genome and gene sequences, and computation.

Reading List

Early Discussions about Sequencing the Human Genome

Dulbecco R. 1986. A turning point in cancer research: Sequencing the human genome. *Science* **231:** 1055–1056.

Sinsheimer R.L. 1989. The Santa Cruz Workshop—May 1985. *Genomics* **5:** 954–956.

U.S. Department of Energy, Office of Health and Environmental Research. 1986. *Sequencing the Human Genome: Summary Report of the Santa Fe Workshop, Santa Fe, New Mexico.* Los Alamos National Laboratory, Los Alamos, New Mexico.

U.S. Congress, Office of Technology Assessment. 1988. *Mapping Our Genes—Genome Projects: How Big, How Fast?* OTA BA 373, U.S. Government Printing Office, Washington, D.C.

National Academy of Science. 1988. *Report of the Committee on Mapping and Sequencing the Human Genome.* National Academy Press, Washington, D.C.

Cook-Deegan R. 1989. The Alta summit, December 1984. *Genomics* **5:** 661–663.

Davis B.D. 1990. The human genome and other initiatives. *Science* **249:** 342–343.

Sinsheimer R.L. 1990. Human genome initiative. *Science* **249:** 1359.

Olson M.V. 1995. A time to sequence. *Science* **270:** 394–396.

Cook-Deegan R.M. 1994. *The Gene Wars: Science, Politics, and the Human Genome.* W.W. Norton, New York.

Arguments for and against Whole-Genome Shotgun Sequencing

Weber J.L. and Myers E.W. 1997. Human whole genome shotgun sequencing. *Genome Res.* **7:** 401–409.

Green P. 1997. Against a whole-genome shotgun. *Genome Res.* **7:** 410–417.

Venter J.C., Adams M.D., Sutton G.G., Kerlavage A.R., Smith H.O., and Hunkapiller M. 1998. Shotgun sequencing of the human genome. *Science* **280:** 1540–1542.

STSs and cDNAs

Olson M.V., Hood L., Cantor C., and Botstein D. 1989. A common language for physical mapping of the human genome. *Science* **245:** 1434–1435.

Adams M.D., Kelley J.M., Gocayne J.D., Dubnick M., Polymeropoulos M.H., et al. 1991. Complementary DNA sequencing: Expressed sequence tags and human genome project. *Science* **252:** 1651–1656.

Adams M.D, Dubnick M., Kerlavage A.R., Moreno R., Kelley J.M., et al. 1992. Sequence identification of 2,375 human brain genes. *Nature* **355:** 632–634.

Boguski M.S., Lowe T.M., and Tolstoshev C.M. 1993. dbEST—Database for "expressed sequence tags." *Nat. Genet.* **4:** 332–333.

Schuler G.D., Boguski M.S., Stewart E.A., Stein L.D., Gyapay G., et al. 1996. A gene map of the human genome. *Science* **274:** 540–546.

Deloukas P., Schuler G.D., Gyapay G., Beasley E.M., Soderlund C., et al. 1998. A physical map of 30,000 human genes. *Science* **282:** 744–746.

Strausberg R.L., Feingold E.A., Grouse L.H., Derge J.G., Klausner R.D., et al. (The MGC Program Team). 2002. Generation and initial analysis of more than 15,000 full-length human and mouse cDNA sequences. *Proc. Natl. Acad. Sci.* **99:** 16899–16903.

Gerhard D.S., Wagner L., Feingold E.A., Shenmen C.M., Grouse L.H., et al. (The MGC Project Team). 2004. The status, quality and expansion of the NIH full-length cDNA project: The Mammalian Gene Collection (MGC). *Genome Res.* **14:** 2121–2127.

Carninci P., Kasukawa T., Katayama S., Gough J., Frith M.C., et al. 2005. The transcriptional landscape of the mammalian genome. *Science* **309:** 1559–1563.

Genetic Mapping

Botstein D., White R.L., Skolnick M., and Davis R.W. 1980. Construction of a genetic linkage map in man using restriction fragment polymorphisms. *Am. J. Hum. Genet.* **32:** 314–331.

Weber J.L. and May P.E. 1989. Abundant class of human DNA polymorphisms which can be typed using the polymerase chain reaction. *Am. J. Hum. Genet.* **44:** 388–396.

Litt M. and Luty J.A. 1989. A hypervariable microsatellite revealed by in vitro amplification of a dinucleotide repeat within the cardiac muscle actin gene. *Am. J. Hum. Genet.* **44:** 397–401.

NIH/CEPH Collaborative Mapping Group. 1992. A comprehensive genetic linkage map of the human genome. *Science* **258:** 67–86.

Gyapay G., Morissette J., Vignal A., Dib C., Fizames C., et al. 1994. The 1993–94 Généthon human genetic linkage map. *Nat. Genet.* **7:** 246–339.

Murray J.C., Buetow K.H, Weber J.L., Ludwigsen S., Scherpbier-Heddema T., et al. 1994. A comprehensive human linkage map with centimorgan density. *Science* **265:** 2049–2054.

Dib C., Faure S., Fizames C., Samson D., Drouot N., et al. 1996. A comprehensive genetic map of the human genome based on 5,264 microsatellites. *Nature* **380:** 152–154.

Broman K.W., Murray J.C., Sheffield V.C., White R.L., and Weber J.L. 1998. Comprehensive human genetic maps: Individual and sex-specific variation in recombination. *Am. J. Hum. Genet.* **63:** 861–869.

Dietrich W.F., Miller J., Steen R., Merchant M.A., Damron-Boles D., et al. 1996. A comprehensive genetic map of the mouse genome. *Nature* **380:** 149–152.

Physical Mapping

Olson M.V., Hood L., Cantor C., and Botstein D. 1989. A common language for physical mapping of the human genome. *Science* **245:** 1434–1435.

Burke D.T., Carle G.F., and Olson M.V. 1987. Cloning of large segments of exogenous DNA into yeast by means of artificial chromosome vectors. *Science* **236:** 806–812.

Ioannou P.A., Amemiya C.T., Garnes J., Kroisel P.M., Shizuya H., et al. 1994. A new bacteriophage P1-derived vector for the propagation of large human DNA fragments. *Nat. Genet.* **6:** 84–89.

Shizuya H., Birren B., Kim U.J., Mancino V., Slepak T., Tachiiri Y., and Simon M. 1992. Cloning and stable maintenance of 300-kilobase-pair fragments of human DNA in *Escherichia coli* using an F-factor-based vector. *Proc. Natl. Acad. Sci.* **89:** 8794–8797.

Osoegawa K. 2001. A bacterial artificial chromosome library for sequencing the complete human genome. *Genome Res.* **11:** 483–496.

The BAC Resource Consortium. 2001. Integration of autogenetic landmarks into the draft sequence of the human genome. *Nature* **409:** 953–958.

Marra M.A., Kucaba T.A., Dietrich N.L., Green E.D., Brownstein B., et al. 1997. High throughput fingerprint analysis of large-insert clones. *Genome Res.* **7:** 1072–1084.

Hudson T.J., Stein L.D., Gerety S.S., Ma J., Castle A.B., et al. 1995. An STS-based map of the human genome. *Science* **270:** 1945–1954.

McPherson J.D., Marra M., Hillier L., Waterston R.H., Chinwalla A., et al. (International Human Genome Mapping Consortium). 2001. A physical map of the human genome. *Nature* **409:** 934–941.

Lai Z., Jing J., Aston C., Clarke V., Apodaca J., et al. 1999.A shotgun optical map of the entire *Plasmodium falciparum* genome. *Nat. Genet.* **23:** 309–313.

Radiation Hybrid Mapping

Goss S.J. and Harris H. 1975. New method for mapping genes in human chromosomes. *Nature* **255:** 680–684.

Goss S.J. and Harris H. 1977. Gene transfer by means of cell fusion I. Statistical mapping of the human X-chromosome by analysis of radiation-induced gene segregation. *J. Cell Sci.* **25:** 17–37.

Cox D.R., Burmeister M., Price E.R., Kim S., and Myers R.M. 1990. Radiation hybrid mapping: A somatic cell genetic method for constructing high-resolution maps of mammalian chromosomes. *Science* **250:** 245–250.

Stewart E.A., McKusick K.B., Aggarwal A., Bajorek E., Brady S., et al. 1997. An STS-based radiation hybrid map of the human genome. *Genome Res.* **7:** 422–433.

Boehnke M., Lange K., and Cox D.R. 1991. Statistical methods for multipoint radiation hybrid mapping. *Am. J. Hum. Genet.* **49:** 1174–1188.

Olivier M., Aggarwal A., Allen J., Almendras A.A., Bajorek E.S., et al. 2001. A high resolution radiation hybrid map of the human genome. *Science* **298:** 1298–1302.

Sequence Assembly, Finishing, and Quality Assessment

Gordon D., Abajian C., and Green P. 1998. Consed: A graphical tool for sequence finishing. *Genome Res.* **8:** 195–202.

Gordon D., Desmarais C., and Green P. 2001. Automated finishing with autofinish. *Genome Res.* **11:** 614–625.

Felsenfeld A., Peterson J., Schloss J., and Guyer M. 1999. Assessing the quality of the DNA sequence from the Human Genome Project. *Genome Res.* **9:** 1–4.

Semple C.A., Evans K.L., and Porteous D.J. 2001. Twin peaks: The draft human genome sequence. *Genome Biol.* **2:** comment2003.1–2003.5.

Schmutz J., Wheeler J., Grimwood J., Dickson M., Yang J., et al. 2004. Quality assessment of the human genome sequence. *Nature* **429:** 365–368.

Eichler E.E., Clark R.A., and She X. 2004. An assessment of the sequence gaps: Unfinished business in a finished human genome. *Nat. Rev. Genet.* **5:** 345–354.

The Draft and Complete Sequences of the Human Genome

Bentley D.R. 1996. Genomic sequence information should be released immediately and freely in the public domain. *Science* **274:** 533–534.

Guyer M. 1998. Statement on the rapid release of genomic DNA sequence. *Genome Res.* **8:** 413.

Lander E.S., Linton L.M., Birren B., Nusbaum C., Zody M.C., et al. (International Human Genome Sequencing Consortium). 2001. Initial sequencing and analysis of the human genome. *Nature* **409:** 860–921.

Venter J.C., Adams M.D., Myers E.W., Li P.W., Mural R.J., et al. 2001. The sequence of the human genome. *Science* **291:** 1304–1351.

Wolfsberg T.G., Wetterstrand K.A., Guyer M.S., Collins F.S., and Baxevanis A.D. 2003. A user's guide to the human genome. *Nat. Genet.* (suppl. 1) **35:** 4.

International Human Genome Sequencing Consortium. 2004. Finishing the euchromatic sequence of the human genome. *Nature* **431:** 931–945.

Finished Sequences of the Human Chromosomes

Gregory S.G., Barlow K.F., McLay K.E., Kaul R., Swarbreck D., et al. 2006. The DNA sequence and biological annotation of human chromosome 1. *Nature* **441:** 315–321.

Hillier L.W., Graves T.A., Fulton R.S., Fulton L.A., Pepin K.H., et al. 2005. Generation and annotation of the DNA sequences of human chromosomes 2 and 4. *Nature* **434:** 724–731.

Muzny D.M., Scherer S.E., Kaul R., Wang J., Yu J., et al. 2006. The DNA sequence, annotation and analysis of human chromosome 3. *Nature* **440:** 1194–1198.

Schmutz J., Martin J., Terry A., Couronne O., Grimwood J., et al. 2004. The DNA sequence and comparative analysis of human chromosome 5. *Nature* **431:** 268–274.

Mungall A.J., Palmer S.A., Sims S.K., Edwards C.A., Ashurst J.L., et al. 2003. The DNA sequence and analysis of human chromosome 6. *Nature* **425:** 805–811.

Hillier L.W., Fulton R.S, Fulton L.A, Graves T.A, Pepin K.H, et al. 2003.The DNA sequence of human chromosome 7. *Nature* **424:** 157–164.

Nusbaum C., Mikkelsen T.S., Zody M.C., Asakawa S.,

Taudien S., et al. 2006. DNA sequence and analysis of human chromosome 8. *Nature* **439:** 331–335.

Humphray S.J., Oliver K., Hunt A.R., Plumb R.W., Loveland J.E., et al. 2004. DNA sequence and analysis of human chromosome 9. *Nature* **429:** 369–374.

Deloukas P., Earthrowl M.E., Grafham D.V., Rubenfield M., French L., et al. 2004. The DNA sequence and comparative analysis of human chromosome 10. *Nature* **429:** 375–381

Taylor T.D., Noguchi H., Totoki Y., Toyoda A., Kuroki Y., et al. 2006. Human chromosome 11 DNA sequence and analysis including novel gene identification. *Nature* **440:** 497–500.

Scherer S.E., Muzny D.M., Buhay C.J., Chen R., Cree A., et al. 2006. The finished DNA sequence of human chromosome 12. *Nature* **440:** 346–351.

Dunham A., Matthews L.H., Burton J., Ashurst J.L., Howe K.L., et al. 2004. The DNA sequence and analysis of human chromosome 13. *Nature* **428:** 522–528.

Heilig R., Eckenberg R., Petit J.L., Fonknechten N., Da Silva C., et al. 2003. The DNA sequence and analysis of human chromosome 14. *Nature* **421:** 601–607.

Zody M.C., Garber M., Sharpe T., Young S.K., Rowen L., et al. 2006. Analysis of the DNA sequence and duplication history of human chromosome 15. *Nature* **440:** 671–675.

Martin J., Han C., Gordon L.A., Terry A., Prabhakar S., et al. 2004. The sequence and analysis of duplication-rich human chromosome 16. *Nature* **432:** 988–994.

Zody M.C., Garber M., Adams D.J., Sharpe T., Harrow J., et al. 2006. DNA sequence of human chromosome 17 and analysis of rearrangement in the human lineage. *Nature* **440:** 1045–1049.

Nusbaum C., Zody M.C., Borowsky M.L., Kamal M., Kodira C.D., et al. 2005. DNA sequence and analysis of human chromosome 18. *Nature* **437:** 551–555.

Grimwood J., Gordon L.A., Olsen A., Terry A., Schmutz J., et al. 2004. The DNA sequence and biology of human chromosome 19. *Nature* **428:** 529–535.

Deloukas P., Matthews L.H., Ashurst J., Burton J., Gilbert J.G., et al. 2001. The DNA sequence and comparative analysis of human chromosome 20. *Nature* **414:** 865–871.

Hattori M., Fujiyama A., Taylor T.D., Watanabe H., Yada T., et al. 2000. The DNA sequence of human chromosome 21. *Nature* **405:** 311–319.

Dunham I., Shimizu N., Roe B.A., Chissoe S., Hunt A.R., et al. 1999. The DNA sequence of human chromosome 22. *Nature* **402:** 489–495.

Ross M.T., Grafham D.V., Coffey A.J., Scherer S., McLay K., et al. 2005. The DNA sequence of the human X chromosome. *Nature* **434:** 325–337.

Skaletsky H., Kuroda-Kawaguchi T., Minx P.J., Cordum H.S., Hillier L., et al. 2003. The male-specific region of the human Y chromosome is a mosaic of discrete sequence classes. *Nature* **423:** 825–837.

Genome Architecture

Bailey J.A., Church D.M., Ventura M., Rocchi M., and Eichler E.E. 2004. Analysis of segmental duplications and genome assembly in the mouse. *Genome Res.* **14:** 789–801.

Horvath J.E., Bailey J.A., Locke D.P., and Eichler E.E. 2001. Lessons from the human genome: Transitions between euchromatin and heterochromatin. *Hum. Mol. Genet.* **10:** 2215–2223.

Torrents D., Suyama M., Zdobnov E., and Bork P. 2003. A genome-wide survey of human pseudogenes. *Genome Res.* **13:** 2559–2567.

Roest Crollius H., Jaillon O., Bernot A., Dasilva C., Bouneau L., et al. 2000. Estimate of human gene number provided by genome-wide analysis using *Tetraodon nigroviridis* DNA sequence. *Nat. Genet.* **25:** 235–238.

Genome Sequences of Other Eukaryotes

C. elegans Sequencing Consortium. 1998. Genome sequence of the nematode *C. elegans*: A platform for investigating biology. *Science* **282:** 2012–2018.

Adams M.D., Celniker S.E., Holt R.A., Evans C.A., Gocayne J.D., et al. 2000. The genome sequence of *Drosophila melanogaster*. *Science* **287:** 2185–2195.

Celniker S.E., Wheeler D.A., Kronmiller B., Carlson J.W., Halpern A., et al. 2002. Finishing a whole-genome shotgun: Release 3 of the *Drosophila melanogaster* euchromatic genome sequence. *Genome Biol.* **3:** RESEARCH0079.

Arabidopsis Genome Initiative. 2000. Analysis of the genome sequence of the flowering plant *Arabidopsis thaliana*. *Nature* **408:** 796–815.

Waterston R.H., Lindblad-Toh K., Birney E., Rogers J., Abril J.F., et al. (Mouse Genome Sequencing Consortium). 2002. Initial sequencing and comparative analysis of the mouse genome. *Nature* **420:** 520–562.

El-Sayed N.M., Myler P.J., Bartholomeu D.C., Nilsson D., Aggarwal G., et al. 2005. The genome sequence of *Trypanosoma cruzi*, etiologic agent of Chagas disease. *Science* **309:** 409–415.

International Rice Genome Sequencing Project. 2005. The map-based sequence of the rice genome. *Nature* **436:** 793–800.

Lindblad-Toh K., Wade C.M., Mikkelsen T.S., Karlsson E.K., Jaffe D.B., et al. 2005. Genome sequence, comparative analysis and haplotype structure of the domestic dog. *Nature* **438:** 803–819.

Chimpanzee Sequencing and Analysis Consortium. 2005. Initial sequence of the chimpanzee genome and comparison with the human genome. *Nature* **437:** 69–87.

SECTION 3

ANALYZING GENOMES

Acquiring DNA sequence has now become routine, and, in fact, centers equipped for large-scale sequencing can sequence a bacterial genome in a single day. But DNA sequence coming from sequencing machines is nothing more than a list of As, Ts, Gs, and Cs. Understanding the meaning of those billions of letters stored in computer databases is a daunting challenge. Deciphering the genetic information encoded in these DNA sequences requires knowledge of what to look for, algorithms capable of detecting interesting features in sequences, and computers powerful enough to perform complex analyses efficiently and rapidly. Fortunately, advances in all three areas have kept pace with the generation of sequence, and the computational analysis of DNA constitutes the new field of *bioinformatics.* Although bioinformatics provides biologists with very powerful tools for interpreting DNA sequences, the algorithms for detecting genes and other functional units are not perfect. We cannot rely on them exclusively, and so new techniques for large-scale experimental analysis have developed in parallel with genome-scale sequencing. Bioinformatic and experimental analysis of the ever-increasing amount of sequencing data continues to provide unprecedented knowledge about life.

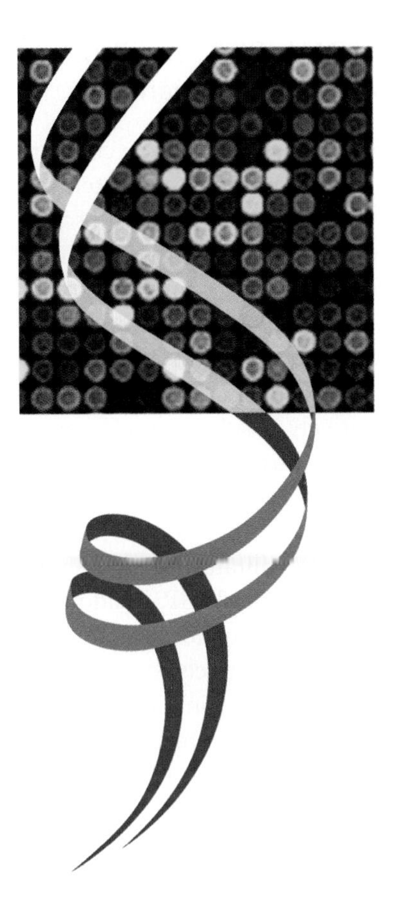

CHAPTER 12

Comparing and Analyzing Genomes

Analyzing DNA and protein sequences is a powerful tool for understanding the functions of genes and their encoded products. Biologists began using computers to manage sequence data and to analyze protein and DNA sequences soon after techniques for sequencing were developed. But when large-scale sequencing began in earnest in the late 1980s, leading to the Human Genome Project (HGP) in the 1990s, computational analysis became ever more important and a new field of *bioinformatics* emerged. Bioinformatics is a field of study that seeks to characterize functional features in genes and genomes by exploiting computational models of the biological and evolutionary processes that underlie the data. Bioinformatics, however, goes far beyond interpreting data. The findings of primary data analyses must be managed—stored in a form that is useful to the experimental biologist and is readily accessible. This has led to the development of many databases dedicated to specific topics. There are, for example, databases with information on genome-wide gene expression, transcription factor binding, chromatin, and DNA modifications, as well as many dedicated to specific organisms. A key element is that the databases communicate with each other seamlessly via the internet so that a researcher is unaware that she has moved from a database in the United Kingdom, to one in Italy, and then to one in the United States.

Evolution underpins most aspects of bioinformatics analysis and the analysis of large amounts of DNA sequence data generated by genome researchers has provided unequivocal evidence for, and amazing insights into, the realities of evolution. In this chapter, we discuss how understanding the evolutionary relationships among species and their genomes is crucial to interpreting genome sequence and review some of the strategies of DNA sequence analysis. The fruits of bioinformatics analysis will be evident throughout the book.

Comparative Sequence Analysis Shows That an Oncogene Is a Growth Factor

The knowledge of biological functions that can come from making amino acid and nucleotide sequence comparisons was demonstrated long before any of the genome projects were under way. In 1980, for example, a spectacular and unexpected discovery was made in cancer when two quite different lines of research came together. The first was the work of Russell Doolittle, who was making a database of protein amino acid sequences using the *Atlas of Protein Sequence and Structure* and culling sequences from recently published papers. His database contained 1700 partial and full-length protein sequences, a total of 250,000 amino acids, and he made it available to other researchers.

The second was the discovery by Michael Waterfield that cancer cells growing in tissue culture had a lower requirement for platelet-derived growth factor (PDGF) than did normal cells. His group determined a partial amino acid sequence of PDGF, and Waterfield searched Doolittle's database to discover that the PDGF sequence was similar to that of the *sis* oncogene, which encodes the transforming protein of simian sarcoma virus. At the same time, Michael Hunkapiller (of later DNA sequencer fame, Chapter 10) sequenced part of the PDGF protein supplied by Harry Antoniades. Working independently of Waterfield, Doolittle searched his database with this sequence and found the same similarity between *sis* and PDGF.

This was the first time that an oncogenic protein was found to be the same as a cellular protein of known function, and the discovery illustrates how experimental research and bioinformatics are synergistic. These and many other important connections between genes of similar sequence were made prior to the development of large-scale sequencing projects. Now, making sequence comparisons is crucial for gleaning knowledge of gene and protein functions from the ever-increasing number of sequences in the databases.

Public Databases of DNA, RNA, and Protein Sequences Are Essential for Research

Initially, the information in data banks was circulated in print in journals with yearly updates. Later, the contents of the databases were provided by subscription on magnetic tape or on compact discs, so that large institutions could copy them onto mainframe computers and individual laboratories could load them onto personal computers. Even though the amount of DNA sequence information available in the mid-1980s was relatively small, it soon became too large to be distributed through these channels. The development of the internet in the early 1990s and the wide availability of ever more powerful personal computers helped to facilitate the analysis of sequence data that came from the HGP, in addition to allowing scientists across the world to share and analyze these data easily and effectively.

Three tightly linked repositories, in England, the United States, and Japan, are the primary resources for sequence storage and dissemination for biologists the world over. In the early 1980s, the European Molecular Biology Laboratory in Heidelberg, Germany, founded the EMBL-Bank, now hosted by the European Bioinformatics Institute (EBI) at Hinxton, United Kingdom, near the Sanger Centre. In 1982, the Los Alamos National Laboratory established the database GenBank, now hosted by the National Center for Biotechnology Information (NCBI) in Bethesda, Maryland. Soon afterward, in 1986, the National Institute of Genetics in Mishima, Japan, formed the DNA Data Bank of Japan (DDBJ). In 1987, a formal alliance among the three groups (the International Nucleotide Sequence Database Collaboration) allowed DNA and RNA sequence data submitted to any single database to be mirrored immediately and in the same format on the other databases. Today, when a researcher in the United States refers to finding a sequence "in GenBank," in fact, the sequence is from the combined "GenBank-EBI-DDBJ" and can be thought of as coming from a single repository. In March 2005, GenBank-EBI-DDBJ reached an impressive milestone when it reported that the database contained 100,000,000,000 base pairs (or 100 gigabases) of DNA sequence derived from more than 165,000 different organisms. Sequence is now submitted at the rate of many millions of base pairs per month and is queried in some form more than 100,000 times in a single day.

There are now several hundred databases storing and providing data and analytical tools for researchers (see Table 12-1 for some examples). Two prominent resources for researchers interested in the human and other vertebrate genomes are the Human Genome Browser, hosted by the University of California at Santa Cruz (UCSC), and Ensembl, hosted at the EBI in Hinxton, United Kingdom. These databases contain not only the assembled sequence of the human genome, but gene annotations, cDNA (complementary DNA) and EST (expressed sequence tag) sequences, comparative analyses to other genomes (like that of the mouse, as discussed below), and a long list of other related resources. The Online Mendelian Inheritance in Man (OMIM) and GeneCards databases contain a wealth of information about human diseases, genes, proteins, sequences, mutations, and biochemical pathways.

There are also other organism-specific databases, such as FlyBase, WormBase, and the *Saccharomyces* Genome Database, that are central supports to the communities focused on these model organisms. These databases are much more than collections of sequence data; they act as central clearinghouses for researchers wanting any kind of information about the organism. For example, the *Arabidopsis* Information Resource provides information on stocks, protocols, and naturally occurring variants, as well as news about meetings and jobs in *Arabidopsis* biology.

In addition, there are databases dedicated to particular biological molecules or processes. Some are highly specialized; for example, ARABI-COIL is for *Arabidopsis* proteins with coiled-coil domains (which indicate sites of protein–protein interactions), whereas RKD is the rice kinase database. Others have ambitious goals of integrating genetic and metabolic data of all kinds to develop a systems description of the organism. EcoCyc is such a database for *Escherichia coli*. This database has annotated metabolic pathways cross-linked to information in other databases, as well as to the scientific literature on *E. coli*. Such a resource is an immensely valuable tool for gaining insights into fundamental metabolic pathways.

For many years, the release of sequence data was not problematic, but times have changed and there has been much soul searching on what should be entered into public databases. Since September 11, 2001, we cannot assume that what the research biologists do, intended to improve the welfare of humanity, will necessarily be used by others for benevolent purposes. Should the sequences of pathogenic organisms, of plants and animals, as well as human beings, most of which have been determined by publicly funded laboratories, be kept secret or made freely available on public databases?

A panel of experts convened by the U.S. National Academy of Sciences concluded that it would be better to have open access to the data rather than to restrict it. Among the many reasons given by the panel was the argument that any restrictions effective enough to prevent misuse would also impede any efforts to develop new vaccines, preventative measures, or treatments of infections by these agents. Furthermore, given the ease of sequencing microorganisms, a terrorist group could itself sequence a pathogen's genome, negating any benefit arising from restricting database submissions.

If restrictions had been in place in the early 1980s, we might not have had the sequence of the human immunodeficiency virus to study its biology and search for therapeutic targets. More recently, within six weeks of the outbreak of severe acute respiratory syndrome (SARS), the virus genome had been sequenced and posted on public databases. Scientists around the world therefore had immediate access to the information and were able to begin to develop diagnostic tests, vaccines, and treatments. Clearly, it is essential that scientific and public health efforts are coordinated and quickly mobilized to tackle newly emerged infectious diseases like AIDS, SARS, and avian flu. These efforts require the free exchange of data unhampered by bureaucratic restrictions.

The primary mission of these databases is to provide unlimited access, free of charge and without constraints on use of the data, to anyone interested in studying genomic and cDNA sequences of any organism. It is no exaggeration to say that these databases, and the immediate access to them through the internet, have changed the way that nearly all biological research is done. Almost every topic and example we discuss in the following chapters has made use of these databases, and it is anticipated that they will remain central to the life sciences community for the indefinite future.

Finding Genes by Sequence Similarity Works Because of Evolution

One of the most powerful tools for determining whether a DNA (or protein) sequence serves a biological function is to compare it to other sequences whose functions are known. As organisms diverged, giving rise to new species, key proteins such as enzymes that catalyze fundamental processes, including energy metabolism, DNA replication, and

TABLE 12-1. Some of the databases that contain genomic information

Resource and URL	Comments
National Center for Biotechnology Information (NCBI): http://www.ncbi.nlm.nih.gov/	NCBI is a central resource for a wide variety of existing and newer information, including literature, sequences, and much more.
GenBank: Access through NCBI site	The NCBI database that contains all freely available sequences for genes, proteins, RNAs, genomes, etc.; this database is coordinated with similar centers in Japan and Europe.
PubMed: Access through NCBI site	A central repository (within NCBI) and search engine for the vast majority of scientific publications.
Online Mendelian Inheritance in Man (OMIM): Access through NCBI site	A database in NCBI that is an on-line extension of a catalog originally developed by Victor McKusick, OMIM is a central resource for phenotypes (largely diseases) that are known to have a genetic basis in humans, literature relevant to the subject, and links to other important sources of biological information.
National Human Genome Research Institute (NHGRI): http://www.genome.gov/	The institute within the National Institutes of Health (NIH) that is responsible for many of the major sequencing and genomics science efforts in the United States and abroad.
U.S. Department of Energy (DOE): http://www.doegenomes.org/	This site describes DOE genomics research programs and has useful materials for education.
University of California at Santa Cruz (UCSC) Human Genome Browser: http://genome.ucsc.edu/	The UCSC browser contains the human and many other genomes, providing access and analysis to tools to learn about, compare, and annotate genome sequences.
ENSEMBL: http://www.ensembl.org	This database, at the European Bioinformatics Institute in England, is similar to the UCSC browser, but offers a different set of tools, annotations, etc., to analyze the human and other genomes.
Berkeley *Drosophila* Genome Project: http://www.fruitfly.org/	One of the primary resources for the *Drosophila* research community, and a central resource for data about *Drosophila* genes, genomes, mutations, etc.
Saccharomyces Genome Database: http://www.yeastgenome.org/	This curated database is hosted at Stanford University and provides access to genes, genomes, mutations, etc.
EcoCyc: http://ecocyc.org/	A curated database, hosted by SRI International, of the genes, transcriptional regulation, and metabolic pathways of *Escherichia coli*.
SwissProt/TrEMBL: http://ca.expasy.org/sprot/	A curated protein sequence database that provides high-quality protein annotations for many organisms, with data about alternative splicing isoforms, posttranslational modifications, protein domains, etc.
The Institute for Genome Research (TIGR) Comprehensive Microbial Resource: http://cmr.tigr.org/	A database with many bacterial (>250 complete) and archaeal (>20) genomes, with related information and comparative genomic resources.
Tree of Life: http://tolweb.org/tree/	A resource for information about many organisms on Earth and how they are all evolutionarily related to one another.
p53 Mutation Database: http://www-p53.iarc.fr/	A database of mutations that have been identified in the p53 gene, a critical cell cycle regulator, in many human cancers and cell lines.
Protein Data Bank (PDB): http://www.rcsb.org/pdb/	A database of protein sequences and structure information, including detailed 3D information about proteins and their interactions.
Nucleic Acids Research Database Issue: http://nar.oxfordjournals.org/	A database of databases, so to speak. The journal *Nucleic Acids Research* has a yearly issue dedicated to describing new biological databases and significant updates to existing biological databases.

This list contains the names and URLs for some of the major databases available on-line for genomic data, including data on DNA sequences, gene expression, genetics, sequence variation, and phenotypes. Each site has many connections to additional sites. Also included are several more specialized databases, with detailed information on particular organisms or even specific genes.

transcription, changed little. Drastic changes that would affect, for example, the active site of an enzyme were eliminated because the individual carrying that mutation would be at a selective disadvantage. In general, nucleotide and amino acid sequences that are similar across species are conserved because they perform the same important functions. For example, despite more than a billion years of evolution since the last common ancestor of human beings and the bacterium *E. coli*, the amino acid sequences of their glyceraldehyde 3-phosphate dehydrogenase (GAPDH) proteins are more than 65% identical. This level of similarity is far greater than expected by chance, which reflects the fundamentally important function of this gene in cellular metabolism. This is why searching out sequence similarities is an essential tool for the bioinformatician. Conservation of sequence hints at conservation of function, even over vast periods of time.

However, similarities among sequences between different organisms may occur simply by chance rather than a shared evolutionary history. For example, consider the sequence GATTACA present immediately 5′ to the start site of a gene. Given its position, it seems possible that it could play some important regulatory role. How unusual is this sequence? The human genome is composed of approximately 60% A:T base pairs, so that the probability of observing an A (or T) at any given location is 0.3 (looking at only one strand), and G or C is 0.2. Thus, drawing nucleotides at random means that the chance of observing GATTACA is $0.3^5 \times 0.2^2$ (0.3 is raised to the power of 5 because there are five As and Ts in the sequence and there are two Gs and Cs). This probability (~0.01%) may seem small, but because the human genome has about 3 billion base pairs, this sequence would occur 300,000 times just by chance alone. Even if the sequence GATTACA is used at some positions in the genome to perform a critical function, most such copies of the sequence have no function. Thus, sequence similarity alone may not be sufficient for our assumption of functional importance. Additional information and, in many cases, experimental evidence are required to understand the meaning of sequence similarity.

On the other hand, many similar sequences do have similar functions. Sequence motifs, ranging in length from a handful to a few dozen nucleotides in DNA and RNA, and amino acids in proteins, have functional roles in a wide variety of biological processes. For example, the DNA sequence TATAA is a motif that is recognized and bound by the protein TATA-binding factor (TBF), which is important in initiating transcription at the promoters of many human genes. Protein domains are somewhat larger regions that confer particular functional properties on a protein (Fig. 12-1). For example, "Src homology" (SH) domains are 60–100 amino acids in length and are used to mediate physical interactions between proteins. Zinc fingers are protein domains that bind to DNA and RNA sequences and are often seen in transcription factor proteins. The ATP-binding cassette is a protein domain that recognizes ATP and is present in proteins involved in metabolism, cell signaling, and other basic cellular processes. Domains and motifs are often used by organisms in a modular fashion. These small regions can be inserted into or deleted from a larger molecule to add or remove a function without grossly affecting the global structure of the larger molecule.

Sequences that are similar because they are evolutionarily related are said to be *homologous*. Two sequences cannot be "highly homologous" or have "low homology"; they are either homologous or they are not. (Two sequences can, however, have differing degrees of *similarity*.) A good example is that of the FOXP2 protein, which is involved in human speech. Human FOXP2 is homologous to the FOXP2 protein in chimpanzees because both are related to an ancient FOXP2 protein that existed in the last common ancestor of humans and chimpanzees. Such closely related proteins are often referred to as "close homologs." Homology can also extend across far larger evolutionary distances. For example, DNA and RNA polymerases found in distantly related species such as humans and bacteria are homologous proteins that have been retained across vast evolutionary times because they perform essential functions required by all cells.

We distinguish between two different types of homology. Two related genes in different species are said to be *orthologs* if they are derived from the same gene in a common ancestral species (Fig. 12-2). *Paralogs* are two related genes that have arisen by a duplication event within a species. Members of the *patched* gene family provide a good example of both types of homology (Fig. 12-3). The Patched protein is the receptor for Hedgehog, a secreted protein that plays a central role in development. There are two vertebrate *patched* genes, *patched1* and *patched2*. The human and chimpanzee *patched1* genes are

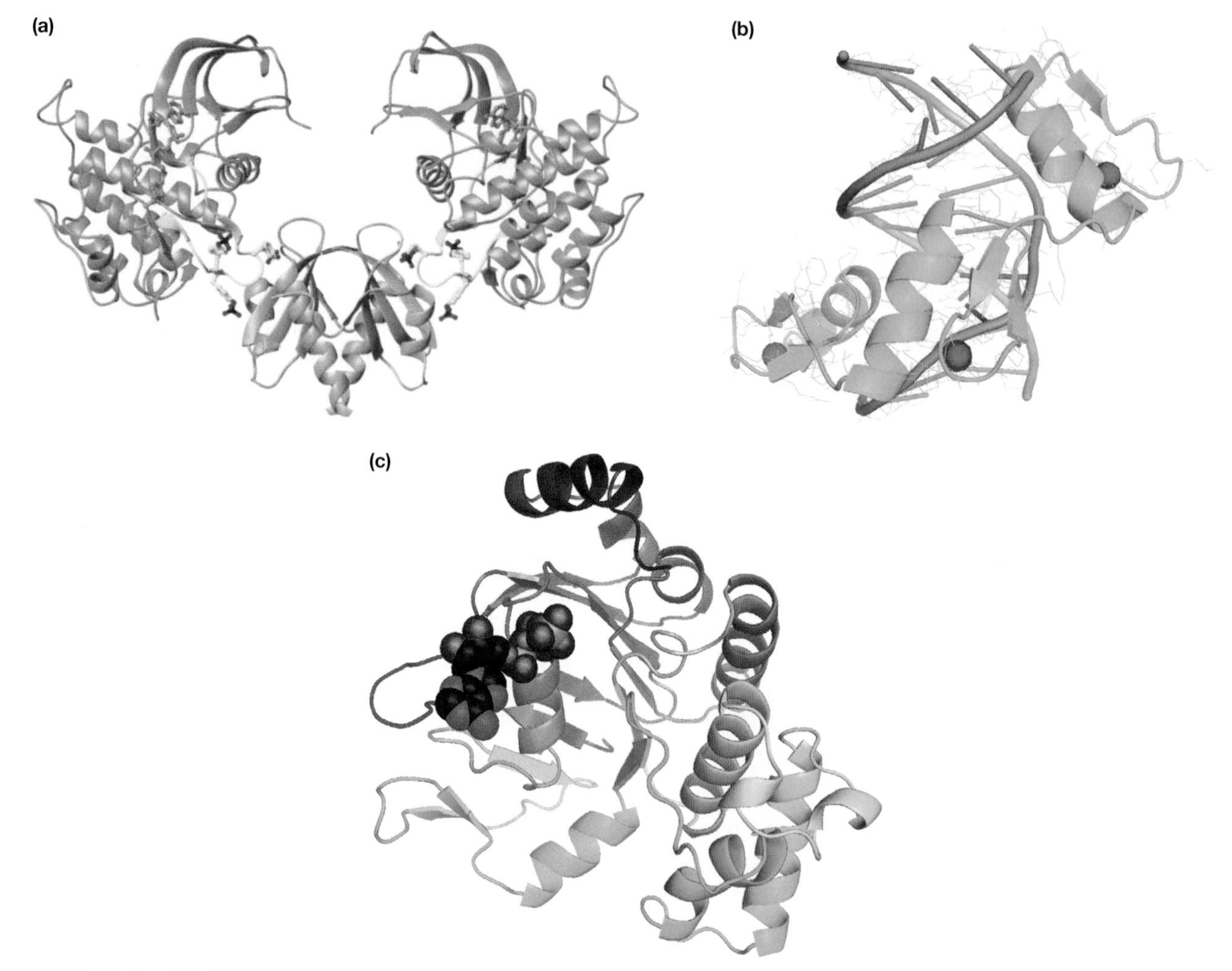

FIGURE 12-1

Some of the known protein domains. (*a*) A structural depiction of an SH2, or Src Homology Region 2 (so named because it is related to the heavily studied oncogene *src*) domain. SH2 domains are regions in proteins involved in binding to other proteins. In general, protein–protein-binding domains allow two proteins to bind in a specific spatial or temporal manner. SH2 domains bind phosphotyrosine-containing proteins and are commonly seen in many cell biological proteins. The image shows the APS SH2 dimer bound to *tris*-phosphorylated IRK. The two IRK molecules are colored *cyan*, except for the activation loop, which is colored *yellow*. The two APS(SH2) protomers are colored *green* and *purple*, and the bisubstrate inhibitor is colored *orange*. (*b*) Zinc finger domain. The image shows the protein–DNA complex of the zinc finger domain binding to the TATA box sequence. The zinc fingers are drawn in *green*, the zinc ions in *purple*, and the DNA in *gray*. (*c*) ATP-binding domain. Protein domains are often crucial to the use of biologically important small molecules, such as ATP. Shown here is ATP (cluster of spheres toward the *top left*) bound to the ATP-binding subunit of the histidine permease from *Salmonella typhimurium*.

orthologs—they are derived from the common ancestor of human beings and chimpanzees. On the other hand, *patched1* and *patched2* in any one species are paralogs—they arose from a gene duplication very early in vertebrate evolution. Orthologs and paralogs need not be protein-coding genes. For example, each human *Alu* repeat sequence (Chapters 7 and 11) is paralogous to all other *Alu* sequences in the human genome, as they are all the result of duplications of previous *Alu* elements.

Usually, orthologs behave in functionally similar ways within their respective organisms and are typically very similar at the amino acid level. For example, histone proteins, critical components of chromatin structure in all eukaryotes, are identical or nearly identical between humans and other mammals. Knowing that two proteins in human and mouse are orthologs suggests that what we learn about the mouse protein is likely to apply to the human protein. This concept is the basis for doing

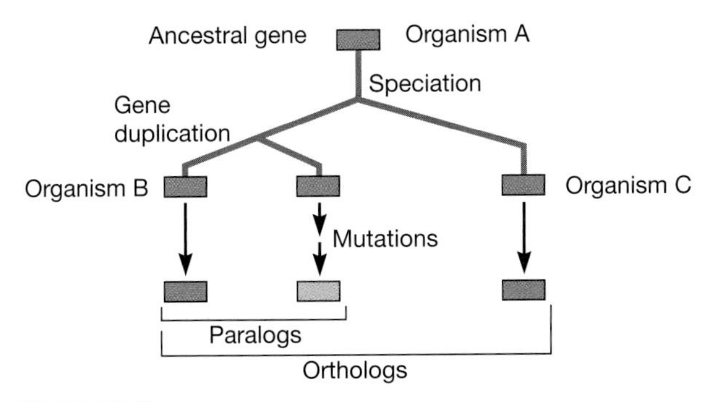

FIGURE 12-2

Paralogs and orthologs are two distinct types of homologs. Organism A undergoes speciation and gives rise to Organism B and Organism C. Subsequently, a gene undergoes duplication in B. One copy of the gene is under strong negative selection, so it remains unchanged and maintains its original function. This gene and the same gene in C are *homologs*, related through speciation. The second copy of the gene is free to undergo changes that may alter the function of the protein so that the two genes come to have different roles in the cell. (It may follow a different path and become a nonfunctional pseudogene.) The two copies of the duplicated gene in B are *paralogs*.

experiments in "model" organisms—organisms that are particularly amenable to experimental manipulations—as a means to learn about human biology. Paralogs, on the other hand, often perform different functions within an organism. The olfactory receptor gene family, for example, is composed of hundreds of distinct genes in the human genome. Each paralogous receptor encoded by this gene family recognizes a different molecule, but all the receptors have a similar overall structure. Because they arise by gene duplication, paralogs are initially identical, but diverge over time. Often one copy of the gene is free to change over time because the other copy continues to perform the original function (Fig. 12-4). The former may acquire a new function or may be expressed in different tissues or at different times during development compared to its paralogous sibling or it may simply become nonfunctional. The human β-globin gene cluster provides examples of these possibilities. Here the ψβ-globin pseudogene, a "dead" copy of a duplicated gene, lies upstream from

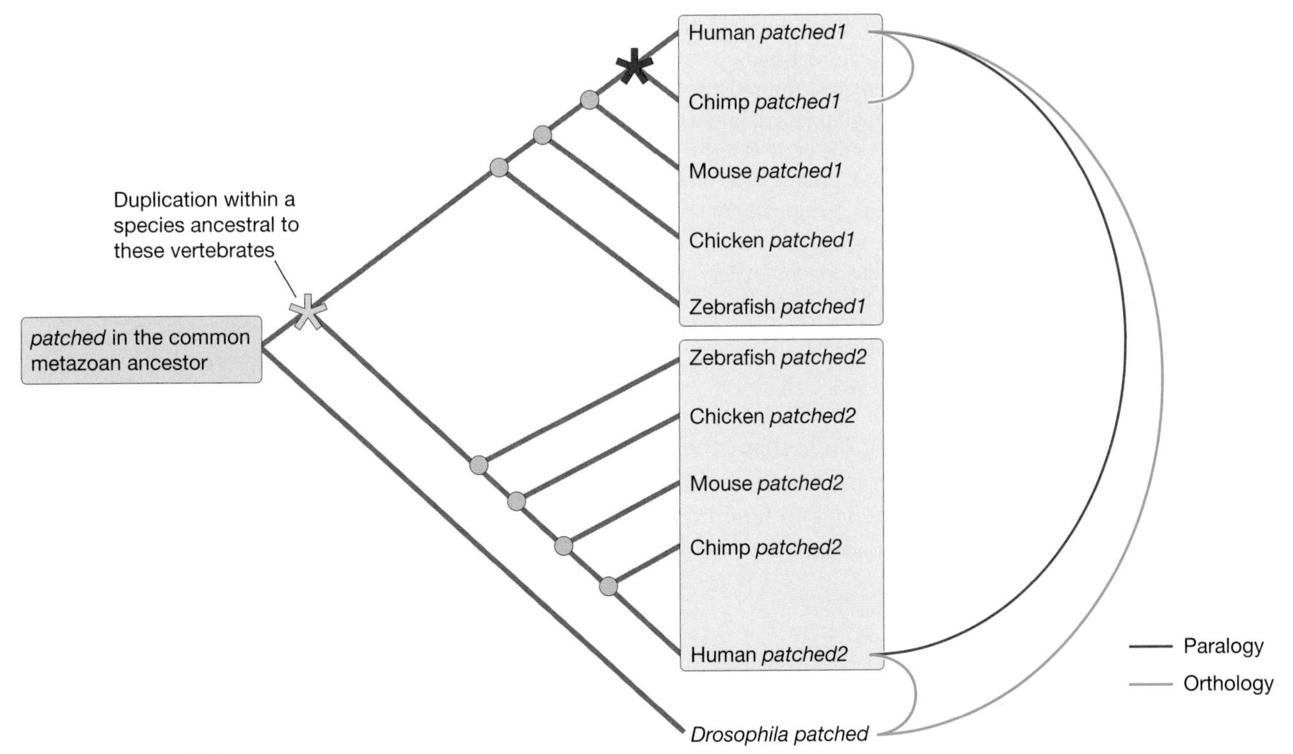

FIGURE 12-3

A tree showing the relationships between orthologs and paralogs of the patched proteins. The *patched* gene family in humans has two members, *patched1* and *patched2*, as do the genomes of other vertebrates. Human *patched1* and chimpanzee *patched1* are orthologs—they are related by the speciation event that separated humans from chimpanzees (*red star*). On the other hand, human *patched1* and *patched2* are paralogs—they are the result of a gene duplication event (*yellow star*). Human *patched1* is much more closely related to chicken *patched1* (their encoded proteins are ~80% identical) than it is to human *patched2* (human *patched1* and *patched2* proteins are ~52% identical). All members of the *patched* family in vertebrates are orthologous to a single *patched* gene in *Drosophila*.

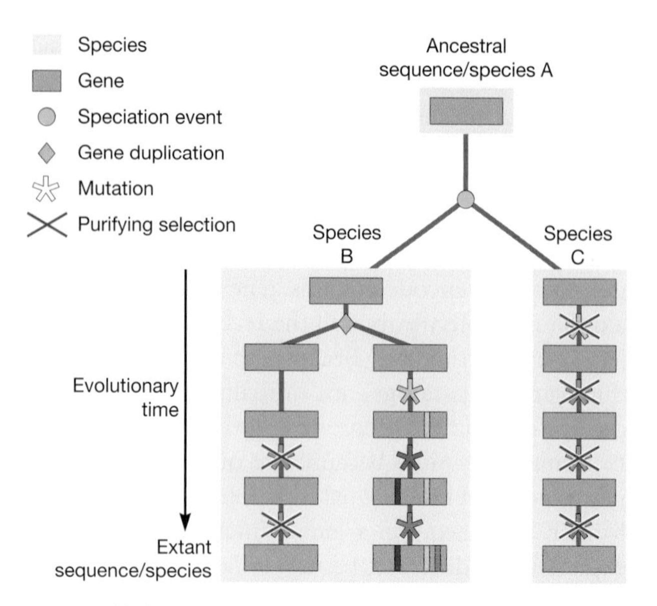

FIGURE 12-4

Gene duplications allow new gene functions to arise. A gene (*brown box*) in an ancestral species A (*green*) is absolutely required for survival and reproduction. A speciation event occurs (*blue circle*), which produces two new species B and C. A gene duplication event subsequently occurs in species B (*orange diamond*) that produces two identical copies of the same gene. A random mutation arises that changes the sequence of one of the two genes in species B (*yellow star*). Because one good copy of the gene remains, the mutation remains in the population. As more mutations accumulate (*red* and *purple stars*), one of the copies of the gene is further changed, which eventually produces a new protein with new functions. The original copy of the gene is continually maintained by purifying selection, as its function is still required. In contrast, in species C, purifying selection eliminates all changes arising in the gene, because one good copy is required for the organism.

the adult β-globin gene (Fig. 12-5). The other genes in the cluster are expressed in different combinations and at different times in development: ε in fetal life, γ in embryonic life, and δ and β in adult life.

Alignment of Sequences Is an Important Challenge Underlying Many Aspects of Sequence Analysis

Before functional or evolutionary hypotheses about pairs or groups of sequences can be tested, the sequences must first be matched up to one another, or *aligned*. In a pairwise alignment, the residues of two sequences are arranged with respect to one another to identify similarities and differences between them, which allows gaps and mismatches to occur. Aligning more than two sequences together simultaneously, known as a *multiple sequence alignment*, is computationally more difficult, but provides more information than does a simple pairwise alignment. For example, Figure 12-6 shows a comparison of base pairs from the 3′ end of a gene from 21 different mammalian species, produced by a multiple sequence alignment of genomic DNA. Mismatches are present between the sequences, and these are the result of substitution events (such as the "A" or "G" at position 44). Gaps also occur in the alignment; these are the result of insertion and deletion events that occurred during and after the separation of the organisms into different species (e.g., the gap from position 52 to 55). Interpreting these events can provide information about the evolutionary history of an organism. For example, the fact that mouse and rat have a G at position 44 compared to the A in all the other species is due to a substitution event that occurred after the

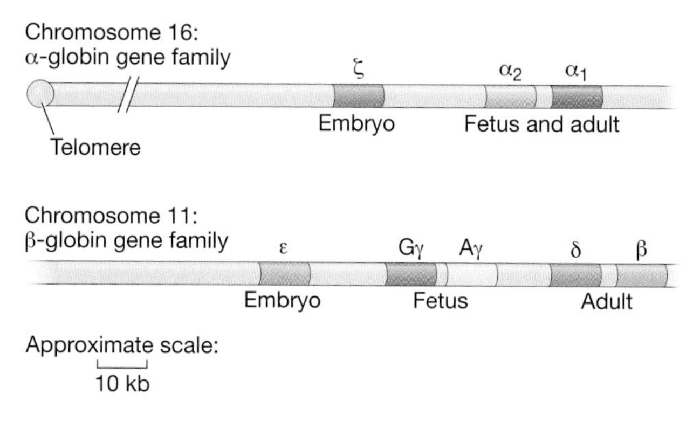

FIGURE 12-5

Gene duplication produces family of related genes. Two regions of the human genome—one on chromosome 11 and the other from chromosome 16—each contain members of the hemoglobin gene family. The order of the genes in each cluster occurs in the order of their expression during development. In the α-globin cluster on chromosome 16, the ζ gene is expressed in embryonic stages, whereas the α_2- and α_1-globin genes are expressed in fetal and postbirth stages. In the β-globin cluster on chromosome 11, the ε gene is expressed in the embryo, Aγ- and Gγ-globin genes are expressed in the fetus, and the δ- and β-globin genes are expressed after birth. Both regions contain pseudogenes that do not produce functional proteins. Two copies of α-like chains and two copies of β-like chains are expressed at each stage, thereby producing a tetramer protein that is the functional hemoglobin molecule. The different types of tetramers produced from the family members at each stage in development have different oxygen-binding properties that are needed at each stage.

split of the mouse and rat ancestor from all other mammals during evolution. The presence of this change in the two organisms furthermore supports the hypothesis that mouse and rat are more closely related than are, for example, mouse and human.

There are two primary types of sequence alignments: local and global. Selection of the appropriate type of alignment depends on the goals of the researcher. Global alignments produce only one best alignment and are used only on homologous sequences. They require that the entirety of each sequence be considered and that the result should be the best alignment that correlates the sequences from beginning to end. A group of homologous proteins, for example, is usually subjected to global alignment. The genomic sequence aligned in Figure 12-6 was generated using a global alignment strategy. Local alignments, in contrast, are used to identify all regions of similarity between sequences, including all subsequences of the full sequence. Local alignments are useful for identifying many of the similarities that we have described, including homology, but local alignments are also useful for identifying short matches of high similarity between sequences, such as motifs and small domains, which are not necessarily evolutionarily related.

The primary tool used for generating global sequence alignments, called the Needleman–Wunsch algorithm, was developed by Saul Needleman and Christian Wunsch in 1970. A wide array of modern software tools exist for computing global alignments, including programs such as ClustalW (for both DNA and protein sequences), ProbCons, and TCoffee (for protein alignments), and tools such as MLAGAN and MAVID (for alignment of large segments of genomic DNA).

One of the primary tools for generating local alignments, called the Smith–Waterman algorithm,

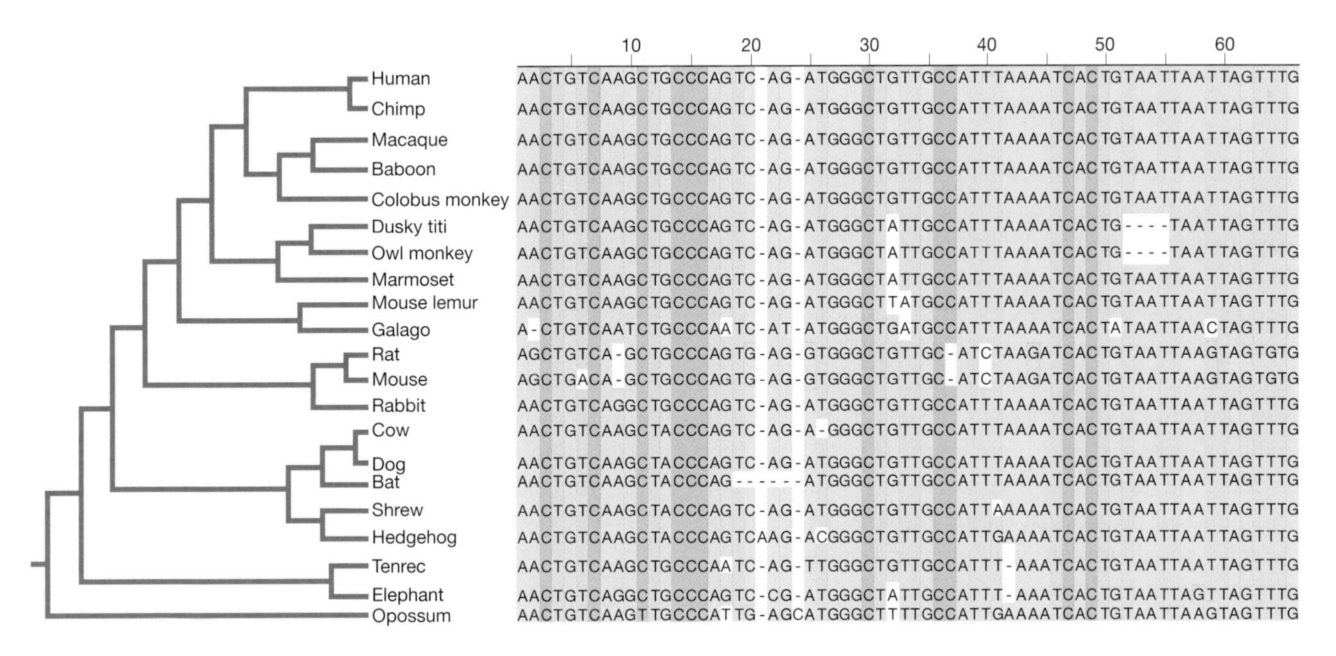

FIGURE 12-6

Alignments can be built for many genomic sequences and provide information about the evolution and function of genes and genomes. The sequences in this alignment are from the 3′-UTR of the *CAPZA2* gene of human beings and the orthologous DNA from 20 other mammalian species. These are listed in the tree, which shows the evolutionary relationships between them. Species in the same subbranches are more closely related to one another than are species further apart. For example, dogs are more closely related to cows than they are to hedgehogs. The *CAPZA2* gene encodes for a protein involved in the processing of actin and the sequences used in this alignment are highly conserved. The alignment reveals evolutionary changes. For example, a 4-bp deletion must have occurred in the common ancestor of the dusky titi and owl monkey species as these bases are present in all other species. The fact that the dusky titi and owl monkey share this deletion suggests they are closely related. Similarly, the mouse and rat have many changes in common, for example, at positions 9 and 37, as well as substitutions of A to G in position 44, T to C (position 40), and T to G (position 59), showing that they are close relatives.

was developed by Temple Smith and Michael Waterman in 1981. Today, there is a wide array of programs designed to generate local alignments, including the FASTA program developed in 1985 by David Lipman and William Pearson, and BLAST, which we describe below, and all use the Smith–Waterman algorithm. In some cases, the distinction between local and global alignments is biologically artificial, as both techniques may be required to identify accurately all biologically relevant alignments. This is especially true in generating whole-genome alignments and maps of conserved synteny (those genomic regions for which both the order and orientation of genes are preserved). Here, a local alignment step is required to identify gene or region-level relationships, followed by a global alignment step that identifies all the nucleotide-level homologous relationships within that particular gene or region.

A quantitative scoring system is applied in sequence alignments to take into account sequence gaps and base-pair mismatches that occur between the sequences. Significant penalties in the scores are given to gaps. These penalties must be stringent enough to provide realistic alignments, but lenient enough to allow real similarities to be identified. If the gap penalty were set to 0, the alignment would produce a good score, but the alignment itself would contain so many gaps as to be biologically meaningless. On the other hand, if gap penalties are too stringent, then it would be easy to miss regions of high similarity in cases where insertions or deletions have occurred. An additional important aspect of gap penalties is that of gap size. Insertions and deletions can affect thousands or millions of base pairs in one event. If large gaps such as these are incorrectly considered to be due to many independent mutational events and are penalized too heavily, an alignment will not be detected. Thus, alignment algorithms often penalize the first gapped base heavily (a "gap open penalty") and penalize subsequent gaps more lightly (a "gap extension").

In addition to gaps, scoring systems for sequence alignments take into account the fact that not all base-pair and amino acid substitutions are equal in their biological effects. For example, isoleucine and leucine are physically and chemically quite similar to one another, and therefore can be exchanged for one another without producing major effects on the protein's structure or function. Indeed, these two amino acids are found to substitute for one another frequently throughout the evolution of protein sequences. Therefore, such changes are not heavily penalized in alignment scoring schemes. Proline, however, is physicochemically quite distinct from other amino acids. Thus the loss or gain of proline is much more likely to disrupt a protein's structure and function. These changes occur much more rarely than do other types of amino acid changes and alignment scoring schemes therefore place a higher penalty level on these changes.

Crucial to all alignment techniques is the use of *scoring matrices*, which list the scores for each aligned position in the nucleotide or amino acid sequences. For nucleotide sequence alignments, the scoring scheme can be very simple, +5 is scored for a match and –4 is scored for a mismatch. In contrast, amino acid comparison scores in the matrices are much more complex for reasons we have just discussed. Here the matches of rare or unique amino acids in an alignment are assigned a higher score than is given for matches of more common amino acids. Moreover, conservative substitutions (e.g., lysine for arginine) are given a positive score, whereas substitutions of structurally or chemically different types of amino acids (e.g., tryptophan for alanine) are given negative scores. Thus the number and sign of the score for each position reflects the physicochemical properties and similarity (or difference) between the amino acids (Fig. 12-7).

The massive increase in DNA sequence data has driven the development and constant improvement of sequence alignment tools, and the ability to align and compare sequences has turned out to be one of the most powerful tools for the biologist. Interestingly, the alignment methods have led to a greatly improved understanding of evolution—how the molecules have come to be what they are—but, at the same time, understanding how molecules evolve has led to better sequence alignments.

The BLAST Programs Are the Most Effective and Popular Alignment Tools

Researchers want to know many things about a gene or a protein once it has been identified. Does the protein contain any known domains or motifs? Does the gene have any homologs in the public databases? Is there a mouse ortholog and, if so, has the ortholog been studied or tested experimentally? Does the protein, or any other protein related to it, contribute to a human disease? To answer these questions, researchers must align the sequence and compare it to other sequences, a task that requires an enormous amount of computational power given the huge quantity of sequence data in the public databases.

BLOSUM62 matrix values

Amino acids, Protein 2	Amino acids, Protein 1																			
	A	**R**	**N**	**D**	**C**	**Q**	**E**	**G**	**H**	**I**	**L**	**K**	**M**	**F**	**P**	**S**	**T**	**W**	**Y**	**V**
A	4	−1	−2	−2	0	−1	−1	0	−2	−1	−1	−1	−1	−2	−1	1	0	−3	−2	0
R		5	0	−2	−3	1	0	−2	0	−3	−2	2	−1	−3	−2	−1	−1	−3	−2	−3
N			6	1	−3	0	0	0	1	−3	−3	0	−2	−3	−2	1	0	−4	−2	−3
D				6	−3	0	2	−1	−1	−3	−4	−1	−3	−3	−1	0	−1	−4	−3	−3
C					9	−3	−4	−3	−3	−1	−1	−3	−1	−2	−3	−1	−1	−2	−2	−1
Q						5	2	−2	0	−3	−2	1	0	−3	−1	0	−1	−2	−1	−2
E							5	−2	0	−3	−3	1	−2	−3	−1	0	−1	−3	−2	−2
G								6	−2	−4	−4	−2	−3	−3	−2	0	−2	−2	−3	−3
H									8	−3	−3	−1	−2	−1	−2	−1	−2	−2	2	−3
I										4	2	−3	1	0	−3	−2	−1	−3	−1	3
L											4	−2	2	0	−3	−2	−1	−2	−1	1
K												5	−1	−3	−1	0	−1	−3	−2	−2
M													5	0	−2	−1	−1	−1	−1	1
F														6	−4	−2	−2	1	3	−1
P															7	−1	−1	−4	−3	−2
S																4	1	−3	−2	−2
T																	5	−2	−2	0
W																		11	2	−3
Y																			7	−1
V																				4

FIGURE 12-7

A substitution matrix used to score protein sequence alignments. This matrix, originally developed by Steven Henikoff and Joria G. Henikoff, is known as a BLOSUM62 matrix and is the standard substitution matrix used to assess protein alignment in BLAST. The 20 possible amino acids are listed along the *x*- and *y*-axes. The cell at the intersection of each row and column contains the alignment score that would be given if the amino acids in the given row and column were to be aligned to one another. For example, an alanine (A) to alanine match scores a +4 (*pink box*), whereas an arginine (R) to alanine mismatch scores a −1 (*blue box*). This scoring matrix was generated by using protein sequences that were known to be homologous to one another. When two different amino acids are seen to replace (or substitute) for one another very often throughout evolution, the more likely it is that those two amino acids in a query alignment are truly homologous residues. That pairing receives a high score. Conversely, when two amino acids are rarely seen to substitute for one another, such an amino acid pair gets a low score because such a mismatch is less likely to indicate homology. Other factors are taken into account. Amino acids that are physicochemically similar to one another have high scores because similar amino acid replacements are likely to conserve the function of the protein. Conversely, some amino acids are very dissimilar from most other amino acids; they are given very negative scores when aligned to other amino acids and very positive scores when aligned to themselves. Examples of these include cysteine (C), tyrosine (Y), and tryptophan (W). The latter scores +11 when paired with itself (*yellow box*).

GenBank contains more than 100 billion base pairs, so that searching a single query sequence against this database involves millions of computations.

The Basic Local Alignment Search Tool, or BLAST algorithm, first introduced in 1990, provides a powerful method for biologists to rapidly compare and align protein and nucleotide sequences. A key concept that makes BLAST effective is that of "word" matching, which significantly reduces the complexity of sequence alignment (generating complete Smith–Waterman alignments would require far too much computing time). Rather than comparing each letter in a sequence to each of the letters from another sequence, BLAST first identifies short strings ("words")—four amino acids or 11 nucleotides, for example—that match between the two sequences (Fig. 12-8). These word matches are used as a starting point to compute longer alignments in which the word match serves as a seed that is extended both to the left and to the right. Those word matches that are surrounded by significant levels of similarity become alignments, whereas those word matches not flanked by such similarity (many word matches will occur by chance) are deleted. This concept and variations on it, originally utilized in computer science, are extremely important not only in BLAST, but also in all forms of sequence alignment, including that done for whole genomes.

There are now several BLAST programs provided by NCBI that allow several different types of sequence comparisons to be made. These include *blastp* for querying a protein sequence against the protein database, *blastn* for querying a nucleic acid sequence against the nucleic acid database, and *blastx* for translating a nucleic acid sequence into protein, and then comparing it against the protein database. The choice of which type of comparison depends on the type of biological question being asked. For example, it is very difficult to identify similarity between *E. coli* and human genomic DNA sequences because they are so distantly related. However, the use of blastp allows comparison of the amino acid sequences encoded by *E. coli* with human protein-coding genes. These observations reveal important similarities between bacterial and human DNA polymerases, many of their metabolic enzymes, and other proteins involved in universal functions. An important feature of BLAST is that it can identify sequence similarities arising for any of the many reasons we have described, including homology (both orthology and paralogy), simple sequence repeats, and motifs, and therefore can be used as a general tool for many applications.

A critical component of BLAST is its use of statistical methods to distinguish alignments that are biologically significant from those that arise due to chance. These methods ask how good an alignment is compared to how good it would be if nucleotides (or amino acids) were drawn at random. As we described earlier, comparing the 7-mer GATTACA to the human genome would be expected to yield 300,000 perfect matches even if nucleotides were drawn completely randomly from the human genome. Therefore, it is dangerous to draw biological meaning from an exact 7-mer match if you are analyzing three billion nucleotides. Furthermore, as described above, alignments use scoring schemes that take into account matches and mismatches of different types (such as a proline to valine, or isoleucine to leucine in proteins) based on their different biological and evolutionary properties. Samuel Karlin and Stephen Altschul developed a statistical approach used in BLAST alignments that helps identify meaningful from nonmeaningful sequence matches. A BLAST analysis of an alignment provides an "Expect value," or E-value, which is an estimate of how many random alignments would be found that score at least as high as the given alignment. The E-value

FIGURE 12-8

How BLAST compares sequences with one another. BLAST works by analyzing "words," or stretches of amino acids (or nucleotides) in a sequence alignment. The algorithm first identifies all perfect matches of at least four amino acids (or seven nucleotides) between two sets of sequences being compared and lines them up, or "maps" them relative to one another (shown by *vertical dotted lines* in the *middle panel* of *a* and *b*). In the next step, the algorithm attempts to align residues to the left and right of each word match (shown in the *third panel* of *a* and *b*). (*a*) In this case, a good, high-scoring alignment was found, where several words were aligned and it was possible to extend to the right and left to additional identical or similar amino acids (the new aligned amino acids are shown by the *vertical dotted lines* connecting the two sequences in the *third panel*). Because the alignment was high scoring, it is retained in the sequence comparison search. (*b*) This case shows an example where a word was identified between two sequences, but no extension to additional neighboring amino acids could be found. In this case, the alignment algorithm quits and wastes no more time comparing these two sequences. (*c*) Here, the output of an alignment search of the *green* query protein sequence from *a* and *b* against an entire database of sequences is shown. In the initial search, BLAST retains only those sequences in the database with at least one word match (these are shown as the *darker boxes* in each of the proteins). This means that most sequences in the database are immediately eliminated, so the algorithm spends no more additional time on them (the *yellow*, *beige*, and *blue* protein sequences are examples). For those sequences that have matching words, a second step is done that searches for additional amino acid matches to the left and right of the word matches. The *white*, *gray*, and *lavender* proteins score poorly in this second alignment step (i.e., the words cannot be extended, shown by *X marks*), so they are then eliminated in the search. The *green* and *orange* proteins, however, score high because the words match well and are extended (shown by *check marks*). In this manner, only 5 of 8 sequences even have an alignment attempted, whereas only two complete alignments are generated (the *orange* and *green* proteins). Thus, matching words within sequences provides a way to reduce the number of sequences that need to be analyzed and to restrict the subsequent alignment to only the residues near a word match and not the whole of each sequence. This is extremely important considering the more than 100 billion base pairs in GenBank and the other public databases that are queried every day by thousands of researchers.

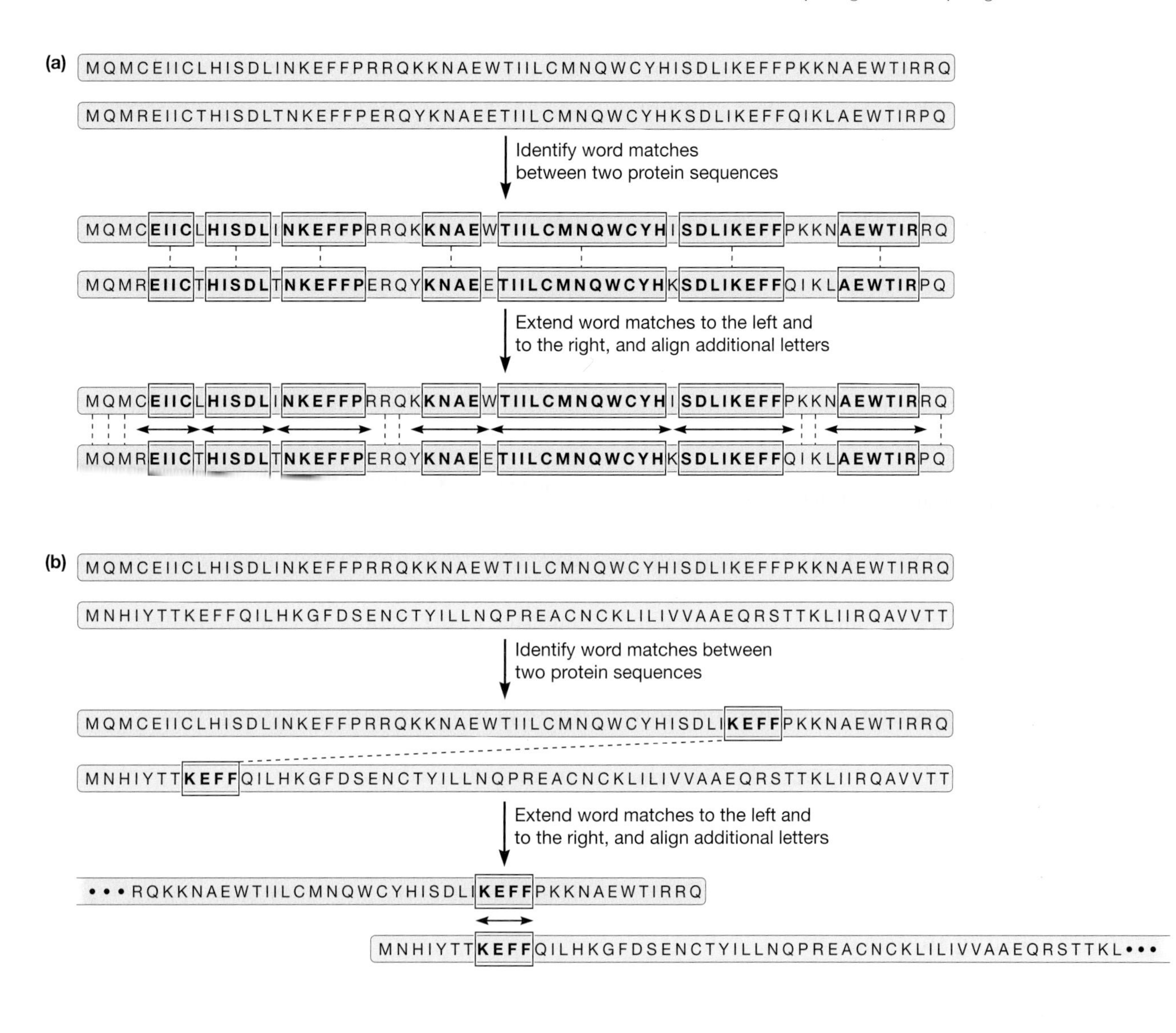

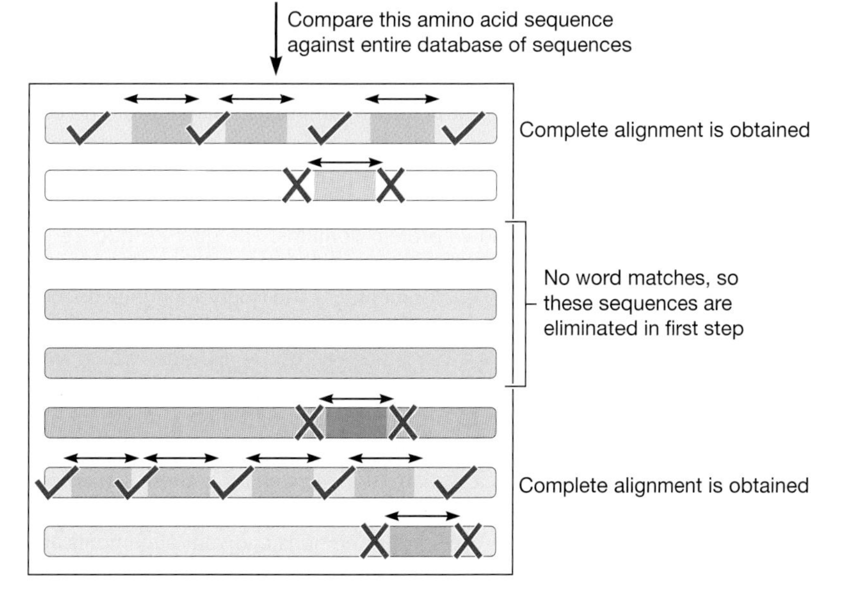

FIGURE 12-8 *(See facing page for legend.)*

depends not only on the score of the alignment, but also on the size and complexity of the query sequence and on the size of the database. Biologically meaningful alignments often have very tiny E-values, lower even than 10^{-10} (1 in 10 billion), which indicates that such an alignment is very unlikely to occur by chance and therefore very likely to be biologically significant. In the example in Figure 12-9, the results show that a query sequence contains a hedgehog (HH) signal domain and has dozens of strong BLAST "hits" (alignments) in organisms ranging from mouse and rat to chicken, frog, and even *Drosophila*. It also is clear that the sequence is related to the protein Sonic hedgehog (Shh), an important developmental signaling protein in all animals, which we discuss in more detail below.

BLAST is available through GenBank and is widely used by many other internet-based and computational resources. With a simple copy-and-paste action and a few mouse clicks, a sequence can be entered into the BLAST query box, after which BLAST quickly provides a list of E-values and other data. Every day, thousands of scientists run analyses comparing sequences they are interested in with the entries in GenBank, hoping to glean useful data about their sequence.

Linked Databases Provide Immediate Access to Further Information

Because of the ways in which diverse databases are integrated, a great deal more information relevant to the new sequence can be found with just a few additional clicks of your mouse. For example, from the *Sonic hedgehog* section of the UCSC Web page, there are links to the Mouse Genome Informatics database at Jackson Laboratory and to the Mutant Mouse Database at Oak Ridge National Laboratory. One learns that mice lacking *Shh* do not survive to birth and that humans with mutated Shh protein have a condition known as holoprosencephaly (HPE). From there, another click leads to OMIM, which provides extensive information on this human genetic disorder. There we find that severe cases are characterized by brain defects and cyclopia (one-eye) and that this disorder occurs with a frequency of 1 in 16,000 live births and 1 in 200 spontaneous abortions. There are eight genetic loci associated with HPE; one of these is HPE4, which maps to chromosome 7q36. These database sites in turn provide many links to the hundreds of papers on the hedgehog family listed in PubMed of the National Library of Medicine. These can be searched by keyword, so that the researcher can narrow down the search for published articles to those most relevant to the study at hand.

Since the late 1990s, there has been a movement in the biological community to make publications freely available on-line, which is similar in concept to previous standards established by the HGP to make all sequence data freely available. The most prominent example is the journal *PLoS* (for *Public Library of Science*) started by Patrick Brown, Michael Eisen, and Harold Varmus, which is freely available immediately on publication. Open access has become recognized as critical to the scientific endeavor. Now nearly all commercial and scientific society publish-

FIGURE 12-9

BLAST is a powerful tool for annotating and characterizing biological molecules. The output from a Web-based BLAST query can be used to quickly learn much about the mystery protein sequence. (*a*) The sequence of a human protein 462 amino acids in length is shown. Without further analysis, the sequence itself is not particularly informative and so it is queried against the GenBank database. (*b*) The Web tool first compares the query sequence with the sequences of known protein domains. The query sequence is found to contain two domains, one of which is an "HH_signal" domain (*green*). HH is an abbreviation for "hedgehog," a protein that plays a key role in signaling during development. (*c*) The query identifies dozens of BLAST alignments, or "hits," other proteins with similarities to the unknown protein. The degree of similarity is given by the bit score and the statistical significance of a score is given by the E-value, which indicates how many alignments would be expected to score at least as high as the given hit, just by chance. The smaller the E-value, the more signficant is the alignment. Sequences from other mammals are at the top of the list, and the top-scoring match has an E-value of 0, showing that the query sequence is identical to the human protein Sonic hedgehog. Nineteen hits have E-values less than 10^{-77}, so small that these sequences must be homologous. These include hedgehog proteins from more distantly related vertebrates like the African Clawed Toad (XENLA), and even for the fruit fly (DROME). There are some nonsignificant alignments at the bottom of the list; those with an E-value of 8.9, for example, are expected to occur approximately nine times by chance. Each hit on the Web page displaying the results is a hyperlink to details about the proteins and to related resources, including scientific literature, and other relevant databases.

(a) Query sequence

>NM_000193 (SHH)
MLLLARCLLLVLVSSLLVCSGLACGPGRGFGKRRHPKKLTPLAYKQFIPNVAEKTLGASG
RYEGKISRNSERFKELTPNYNPDIIFKDEENTGADRLMTQRCKDKLNALAISVMNQWPGV
KLRVTEGWDEDGHHSEESLHYEGRAVDITTSDRDRSKYGMLARLAVEAGFDWVYYESKAH
IHCSVKAENSVAAKSGGCFPGSATVHLEQGGTKLVKDLSPGDRVLAADDQGRLLYSDFLT
FLDRDDGAKKVFYVIETREPRERLLLTAAHLLFVAPHNDSATGEPEASSGSGPPSGGALG
PRALFASRVRPGQRVYVVAERDGDRRLLPAAVHSVTLSEEAAGAYAPLTAQGTILINRVL
ASCYAVIEEHSWAHRAFAPFRLAHALLAALAPARTDRGGDSGGGDRGGGGGRVALTAPGA
ADAPGAGATAGIHWYSQLLYQIGTWLLDSEALHPLGMAVKSS

(b)

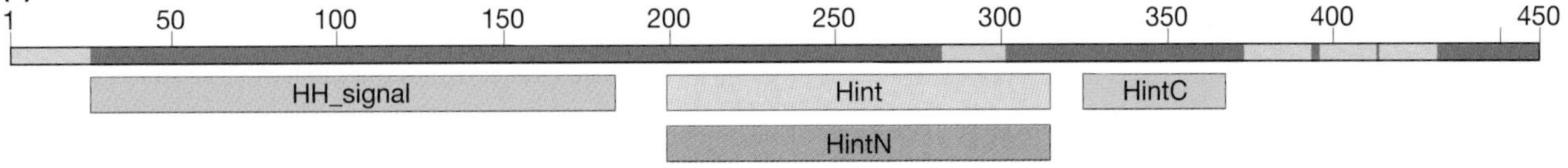

(c)

Sequences producing significant alignments:		Score (bits)	E-value
gi 6094283 sp Q15465 SHH_HUMAN	Sonic hedgehog protein precurs...	711	0.0
gi 6094284 sp Q62226 SHH_MOUSE	Sonic hedgehog protein precurs...	674	0.0
gi 6094286 sp Q63673 SHH_RAT	Sonic hedgehog protein precurs...	670	0.0
gi 6094281 sp Q91035 SHH_CHICK	Sonic hedgehog protein precurs...	618	1e-176
gi 6175032 sp Q92000 SHH_XENLA	Sonic hedgehog protein precurs...	529	8e-150
gi 6174983 sp Q92008 SHH_BRARE	Sonic hedgehog protein precurs...	523	3e-148
gi 6094282 sp Q90385 SHH_CYNPY	Sonic hedgehog protein precurs...	500	3e-141
gi 6136068 sp Q90419 TWHH_BRARE	Tiggy-winkle hedgehog protein...	490	4e-138
gi 6016342 sp Q98938 IHH_CHICK	Indian hedgehog protein precur...	437	4e-122
gi 6016351 sp Q91612 IHH_XENLA	Indian hedgehog protein precur...	437	4e-122
gi 6166227 sp P97812 IHH_MOUSE	Indian hedgehog protein precur...	434	3e-121
gi 33112634 sp Q14623 IHH_HUMAN	Indian hedgehog protein precur...	433	5e-121
gi 6016340 sp Q98862 IHH_BRARE	Indian hedgehog B protein precur...	411	2e-114
gi 6014965 sp Q61488 DHH_MOUSE	Desert hedgehog protein precur...	394	4e-109
gi 61666118 sp Q43323 DHH_HUMAN	Desert hedgehog protein precur...	392	9e-109
gi 6014961 sp Q91610 DHH1_XENLA	Desert hedgehog protein 1 precur...	377	5e-104
gi 6014962 sp Q91611 DHH2_XENLA	Desert hedgehog protein 2 precur...	376	9e-104
gi 37999912 sp Q02936 HH_DROME	Hedgehog protein precursor [Co...	305	2e-82
gi 38258879 sp P56674 HH_DROHY	Hedgehog protein precursor [Co...	289	1e-77
•••			
gi 21431732 sp 077013 AMYR_DROKI	Alpha-amylase-related protein p	34.3	0.80
gi 21431733 sp 077019 AMYR_DROBA	Alpha-amylase-related protein p	32.7	2.3
gi 18202173 sp 077022 AMYR_DROPN	Alpha-amylase-related protein p	32.3	3.1
gi 18202159 sp 076459 AMYR_DROSR	Alpha-amylase-related protein p	32.0	4.0
gi 17865446 sp Q9GQV3 AMYR_DROJA	Alpha-amylase-related protein p	31.6	5.2
gi 62287605 sp Q5WCL2 TAGH_BACSK	Teichoic acids export ATP-bindi	31.2	6.8
gi 32130356 sp Q8TVI1 YE08_METKA	Hypothetical UPF0245 protein MK	30.8	8.9
gi 13432238 sp P46391_RECF_MYCLE	DNA replication and repair prot	30.8	8.9
gi 27734302 sp P59175_GPMI_SHEON	2,3-bisphosphoglycerate-inde...	30.8	8.9
gi 2497140 sp Q04264_PDS5_YEAST	Sister chromatid cohesion pro...	30.8	8.9

FIGURE 12-9 *(See facing page for legend.)*

ers provide free access to papers six months after publication, and many do so immediately upon publication. Most institutions have on-line subscriptions so that a researcher, having found citations to relevant papers, can go to the appropriate Web sites and immediately download high-resolution PDFs of the papers. This network of data and information resources, accessible through the internet, has transformed the ways in which scientists perform their work by greatly speeding up the process of collecting and analyzing information.

Identifying Genes in Genomes Requires Both Computation and Experiments

Having sequences of entire genomes has completely changed the way that researchers identify genes. Until the genome revolution, genes were identified by researchers with specific interests in a particular protein or cellular process. Once identified, these genes were isolated, typically by cloning and sequencing cDNAs, usually followed by targeted (and labor-intensive) sequencing of the longer genomic segments that code for the cDNAs. Once an organism's entire genome sequence becomes available, there is strong motivation for finding all the genes encoded by a genome at once rather than in a piecemeal approach. Such a catalog is immensely valuable to researchers, as they can learn much more from the whole picture than from a much more limited set of genes. For example, genes of similar sequence can be identified, evolutionary and functional relationships can be elucidated, and a global picture of how many and what types of genes are present in a genome can be seen. A significant portion of the effort in genome sequencing is devoted to the process of *annotation*, in which genes, regulatory elements, and other features of the sequence are identified as thoroughly as possible and catalogued in a standard format in public databases so that researchers can easily use the information.

Identifying all the genes in a complex genome like that of humans has turned out to be a major challenge and is perhaps a more difficult problem than geneticists had originally envisioned at the outset of the genome sequencing projects. The problem is much easier to solve with the compact genomes of bacteria, where simply finding open reading frames and using the wealth of biological data collected over the decades has identified most of the protein-coding genes in many species. Even in the eukaryotic microbe *Saccharomyces cerevisiae*, because it has few introns, similar signals can be used effectively to identify protein-coding genes in raw genomic sequence. In fact, the first set of protein-coding genes for yeast was identified largely on the basis of a simple rule: Any open reading frame in the genome larger than 100 codons was considered to be a gene. Such a rule is founded on a simple statistical principle: There are three stop codons out of a total of 64 different codons, and thus you would expect to see a stop roughly once every 21 codons (3/64) at random throughout the genome. Conversely, the odds of seeing 100 codons without a stop codon at random (i.e., the region does not actually code for a protein) is approximately 0.8%, and thus fairly rare.

In the much larger and more complex genomes of vertebrates, however, the issue of splicing poses a major obstacle to the identification of genes. The vast majority of proteins encoded in the human genome are larger than 100 amino acids and result from the splicing of two or more exons separated by substantial amounts of intronic DNA. The typical protein-coding exon in the human genome is about 150 bp long, and the typical protein-coding gene is composed of seven to ten distinct exons (including untranslated sequence at both the 5′ and 3′ ends) often spanning tens to hundreds of kilobases in the genome.

This fragmentation of human genes, the large size of the human genome, and the fact that only a small portion of the genome codes for proteins make it very difficult to identify genes in the human genome. In fact, the catalog of genes is not yet complete for any eukaryotic genome sequence. Even in yeast, the precise number of protein-coding genes is still uncertain. Each year, the *Saccharomyces* Genome Database updates and refines the catalog of yeast genes; in some years hundreds of genes are changed (there are between 5000 and 6000 genes in yeast). In animals and plants, this uncertainty is much worse. For example, even when we know the identity of a particular protein-coding gene, alternative splicing may produce additional distinct versions of the protein. Furthermore, recent evidence suggests that there are potentially many more "noncoding" RNA genes—those that code for RNAs that serve regulatory and other functions—than was previously recognized (Chapter 9). Although the number, types, and functions of these RNAs are not fully understood, it remains possible that there are thousands, or even tens or hundreds of thousands, of RNA genes still to be identified.

Despite this complexity, significant strides have been made in recent years to identify and catalog all the genes in our genome. These methods make use of the sequence patterns that are seen in known genes. For example, a program to identify genes in the genome first looks for the signals at the start of a gene, such as a TATAA sequence in a promoter. After identifying a potential start of a gene, the program searches for patterns that look like internal parts of genes, such as open reading frames longer than those expected by chance, and splice donor and acceptor sequences. Finally, the program looks for a signal to end a gene prediction, such as the end of a long open reading frame or a transcription termination signal. This type of analysis is then combined with sequence comparison with other species. For example, most human genes have an ortholog in the mouse genome, so if signals associated with genes are seen in the same locations in both the human and mouse genome, it becomes much more likely that such a prediction is real.

In practice, these computational predictions require refinement with substantial amounts of experimental evidence, largely through the alignment of cDNA sequences that provide concrete evidence that a particular gene is transcribed and spliced. The integration of all these data is a challenge, and these efforts have been compiled by a few major bioinformatics centers (such as Ensembl in the United Kingdom and the Joint Genome Institute in Walnut Creek, California) into gene annotation "pipelines." These pipelines require significant amounts of time and effort to compile and maintain and are continually undergoing improvement through the efforts of many biological and bioinformatics researchers.

Uncertainty aside, it is clear that catalogs of genes, especially of protein-coding genes, are more comprehensive in all sequenced genomes than they have ever been, especially for the human. Based on a very large number of cDNA sequences, de novo predictions, and other computational and experimental work, the current best estimates are that there are about 23,000–25,000 protein-coding genes in the human genome. Surprising to many, this number is similar to the number of genes seen in *Drosophila* (14,000–18,000) and *Caenorhabditis elegans* (~19,000) and is also similar to the number of genes found in all other mammals whose genomes have been sequenced (Chapter 11). Rodents, for example, are thought to have 20,000–25,000 genes. Interestingly, there is no good correlation between genome size and numbers of genes; for example, the pufferfish *Fugu* is predicted to have more genes than humans but has a genome about one-ninth the size of that of human. In addition, gene count and genome size also do not correlate with organismal complexity; grasshoppers have genomes many times larger than that of humans, despite being simpler than human beings. The number of human protein-coding genes is slightly less than in *Fugu* or zebrafish, and substantially smaller than the number of genes in some plants. In any case, it is certain that considerable resources and research will continue to be dedicated to the identification of genes in all sequenced species.

Syntenic Maps Reveal Large-Scale Relationships between Genomes

Mammalian genomes are surprisingly malleable in terms of their overall architecture—whole segments of chromosomes have often been shuffled around throughout evolution by inversions, translocations, segmental duplications, and chromosomal fusions. Nevertheless, mammalian genomes have very large regions in which the order and orientation of the genes is preserved among species. This phenomenon is sometimes referred to as *conservation of synteny* (synteny literally means "same thread"), and these regions are called *syntenic blocks* (Fig. 12-10). For complicated genomes containing a large amount of duplicated sequence, like those of mammals, identifying regions and genes that are syntenic with respect to one another is one of the strongest arguments that those regions and genes are truly orthologs and not paralogs. For example, there are hundreds of olfactory receptor genes located throughout the human genome, many of which appear to be highly similar in sequence. Knowing that a particular olfactory receptor gene is embedded between two other known genes on the X chromosome in both humans and chimps prevents a false alignment between that gene and a paralogous olfactory receptor gene on another chromosome, despite the fact that the paralogs might be highly similar.

Syntenic maps have been generated between the human, mouse, rat, chimpanzee, and dog genomes, thus making it possible to identify how each human chromosome is related to the chromosomes of other species (Fig. 12-10). Based on these maps, evolutionary biologists have hypothesized what the mammalian

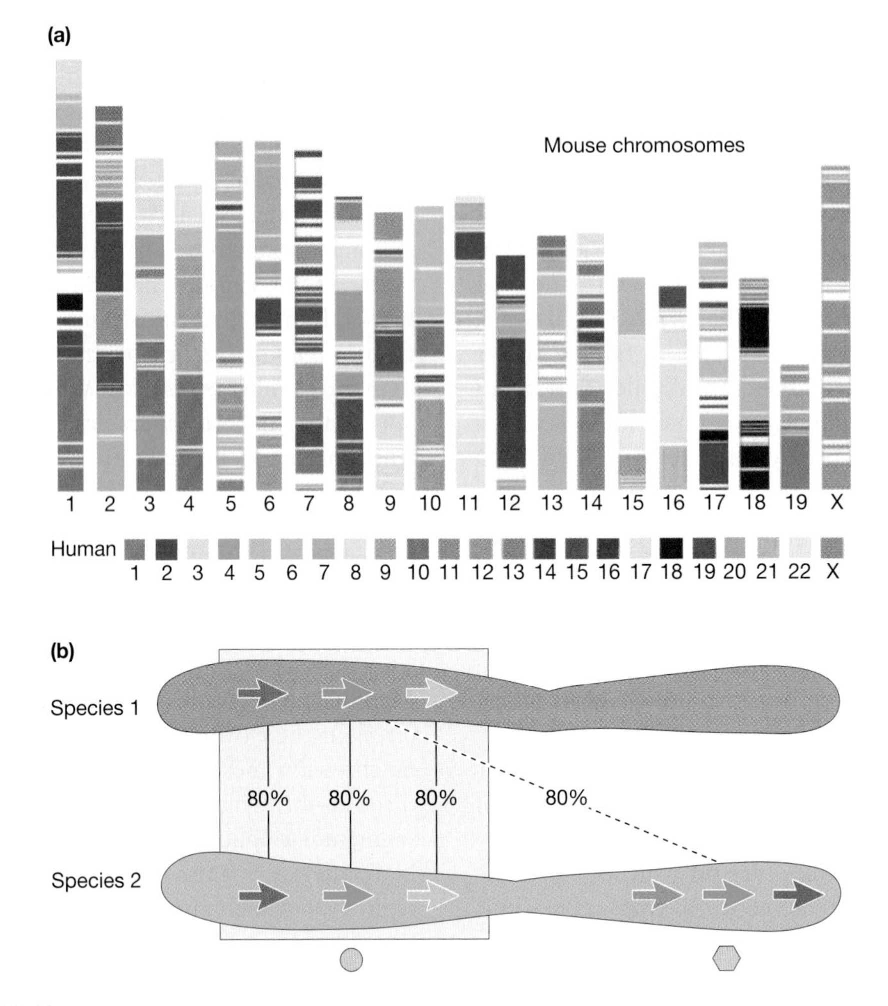

FIGURE 12-10

Syntenic relationships between genes in different species. Mammalian genomes are characterized by large chromosomal segments in which a few up to several thousand genes, spanning hundreds of thousands to millions of base pairs, are in the same order and orientation in distinct species. (*a*) Each of the 20 mouse chromosomes (autosomes 1–19 and X) is colored according to the chromosome(s) in the human genome with which it shares syntenic blocks. The key to the colors of the human chromosomes (1–22 and the X) is shown at the *bottom*. For example, there is a large block at the tip of mouse chromosome 1 that is orthologous to a segment from human chromosome 8 (*yellow*). These regions were part of the same chromosome in the last common ancestor of humans and mice. The complete synteny between the X chromosomes of mice and human beings is striking. This is characteristic of the X chromosomes of placental mammals—all are derived from the same ancestral chromosome and there is great evolutionary pressure to keep X-chromosome genes together and not mixed in with autosomal genes. (*b*) Synteny can be used to identify the likely ortholog of a protein. The *arrows* represent genes. When the sequence of *blue* gene in Species 1 is compared to the sequence of the entire genome of Species 2, two genes in Species 2 are found to match it at the level of 80% sequence similarity (*gray circle* and *hexagon*). A similar comparison is done for the neighboring genes. One of the *blue* genes in Species 2 is found to be flanked by the same neighboring genes (*brown* and *yellow*) as is the *blue* gene in Species 1. In contrast, the second *blue* gene in Species 2 (*hexagon*) is flanked by two genes (*orange* and *red*) that have no similarity to the flanking genes in Species 1. The location of the *blue* gene between two similar genes in Species 2 makes it a very strong candidate for the ortholog of the *blue* gene in Species 1.

ancestral chromosomes probably looked like, and also how many and what types of changes must have occurred along the branches of the mammalian tree to produce the modern chromosomes. Such an analysis of the human and mouse genomes suggests that about 200 major rearrangements—translocations, fusions, or inversions—have occurred in the time since these two species shared a common ancestor (Fig. 12-10). In addition to understanding chromosomal evolution within mammals, syntenic maps are critical to the generation of whole-genome alignments, which have many important utilities that we describe below.

Comparison of the Mouse and Human Genomes Reveals That Most Sequences Are Not Functionally Important

One of the most commonly used concepts in comparative sequence analysis is that of sequence conservation. When a mutation arises in a genome, it may have functional consequences; for example, when a stop codon or frameshift deletion occurs in a protein-coding gene, the result is a truncated or otherwise greatly altered protein that is likely to be nonfunctional or unstable. If the function of the protein is important to the organism, then individuals with the mutation will be less likely to reproduce and pass it along to their offspring. The degree to which mutations, compared to the original (or "wild-type") allele, are more or less likely to be passed on to future generations is the phenomenon Charles Darwin referred to as *natural selection*. Mutations that are deleterious, such as the protein truncation described above, reduce the chance that the individual will reproduce and are removed from populations. This type of selection is often called *purifying* or *negative selection*. Regions of a genome that are under negative selection are said to be under *evolutionary constraint*. On the other hand, mutations that improve an organism's chances of survival and reproduction—disease resistance is a good example—are beneficial and become more prevalent in a population. This type of selection is often referred to as *positive* or *adaptive selection*. There are also neutral mutations that are neither deleterious nor advantageous and are not under selective pressure; these mutations, which make up many of the genetic differences between species, are said to "drift" within a population. Because many of the mutations that affect functional regions of the genome are under negative selection and are weeded out, functional regions change (evolve) more slowly. By identifying the regions of the genome that are evolutionarily conserved, we can identify the regions of genomes that are likely to encode functional elements.

With the sequencing of the mouse genome in 2002, biologists were able to compare the human genome with that of another mammal for the first time. By doing a pairwise comparison, it was possible to determine the fraction of the human genome that has been under evolutionary constraint since the time of the last common ancestor of humans and mice some 80 million years ago. Researchers first estimated the "baseline" neutral rate of evolution between the two species by determining the percent sequence identity between transposon-derived segments, regions of the genome that are believed to have neutral mutations. They then determined the percent identity between all the aligned sequences of the entire genome and compared it to the baseline neutral rate. The surprising finding was that only a tiny portion of the human genome appears to be under evolutionary constraint; the vast majority (>95%) appears to have drifted, which suggests that these segments are not functionally important. Conversely, the remaining 5% is under evolutionary constraint, which suggests that only a small portion of our genome codes for important functions.

All of the protein-coding exons of our genes comprise only about one-fourth (1.2%) of the 5% of the human genome under evolutionary constraint. The remaining 3.8% must be important for other functions—almost surely for transcriptional regulation, DNA replication, and other regulatory processes. Some of the constrained regions encode microRNAs and likely other regulatory transcripts (Chapter 9).

Comparative Genomics Helps to Identify Protein-coding Genes

Comparative genomics refers to the use of evolutionary sequence comparisons, in combination with comparison of functions of genes by experimental tests in different organisms, to understand biological processes. Comparative genomics can be exploited to verify and characterize protein-coding genes. A remarkable early example was the use of human and mouse genomic sequence comparisons to identify a gene known as *ApoAV*. This gene codes for an apolipoprotein that has sequence similarity to several other proteins that were already known to play roles in lipid metabolism. After using computational comparisons to identify this previously unknown gene (Fig. 12-11), Len Pennacchio and Edward Rubin at the Lawrence Berkeley National Laboratory conducted experiments in mice that showed that the ApoAV protein is an important regulator of triglyceride levels in the bloodstream. Triglycerides are fatty molecules circulating in the blood that, like cholesterol, influence susceptibility to coronary artery disease. Pennacchio and Rubin found that the presence of excess ApoAV protein is sufficient to reduce triglyceride levels severalfold. Conversely, by genetically engineering mice to disrupt the *ApoAV* gene, they showed that loss of the protein results in triglyceride levels severalfold higher than in normal mice. Furthermore, in a human clinical study, polymorphisms in the *ApoAV* gene are sig-

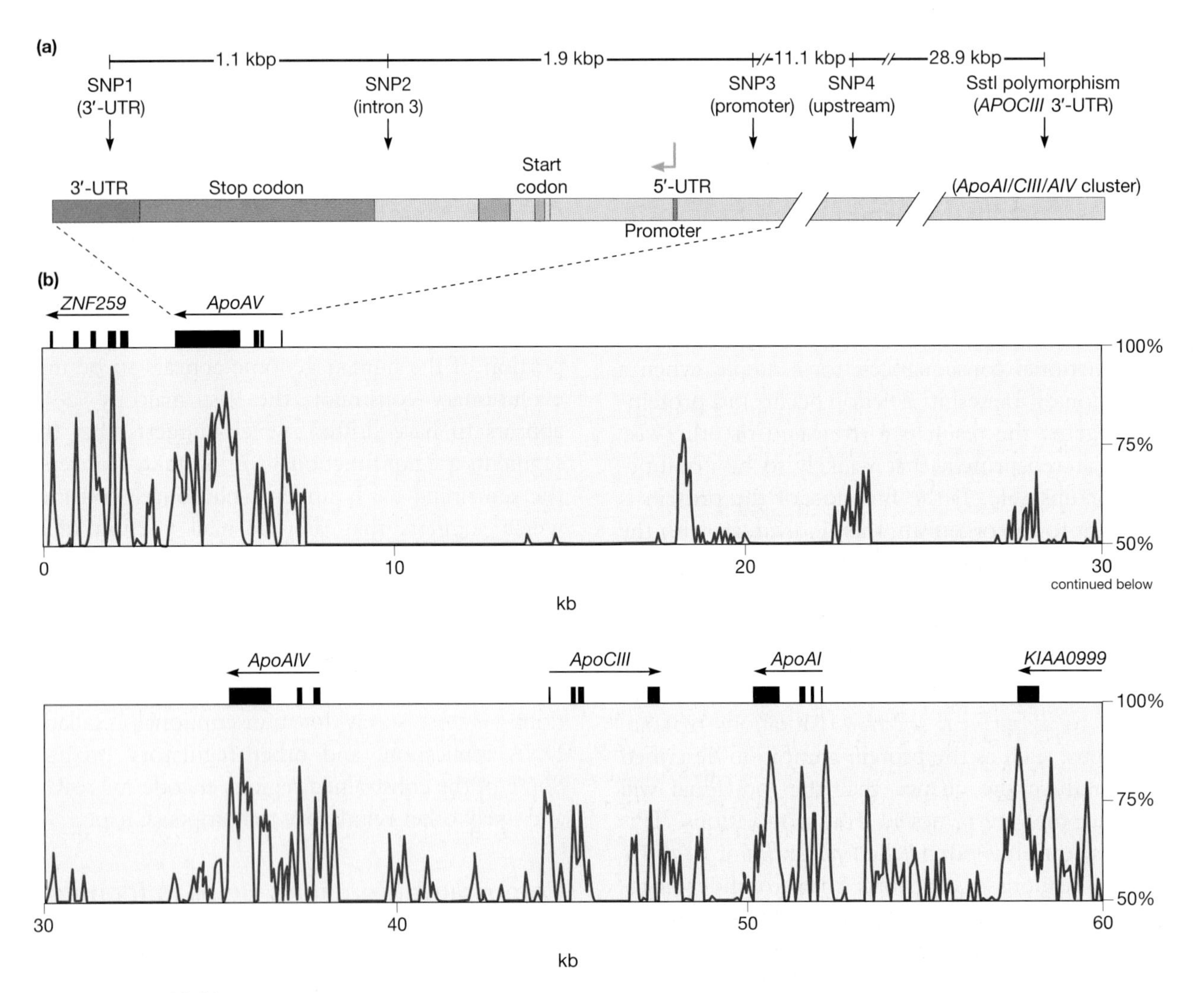

FIGURE 12-11

Comparative genomics identifies the *ApoAV* gene. (*a*) The structure of the human *ApoAV* gene on chromosome 11 is shown. It contains four exons and is transcribed from *right to left* in this depiction. (*b*) Comparison of the human and mouse genomes was a valuable tool in identifying the *ApoAV* gene. The plot shows the percent identity between the human and mouse DNA sequences in the *ApoAV* region. The exons are shown as *filled black boxes*. Dramatic peaks of conservation are seen in the exons, particularly compared to the neighboring introns. In the comparative analysis used to generate the plots, sequence conservation is indicative of evolutionary constraint and, therefore, function. In this case it is indicative of the function of the *ApoAI* gene. Similarly, an analysis of the region marked *ApoAV* also shows high levels of sequence similarity between mouse and human. Prior to this study, the *ApoAV* gene had not been identified or annotated. This sequence conservation in the region led researchers to expect that it is a new apolipoprotein gene family member. Their experiments then showed that this is indeed the case, and *ApoAV* is now known as a important regulator of fatty acids in the blood.

nificantly correlated with triglyceride levels, thus confirming that the gene also functions in human lipid metabolism. This striking example shows how computational comparative analyses, grounded in basic principles of molecular evolution, can have a profound impact on our understanding of the human genome and how it influences human health.

Comparative Sequence Analysis of Protein Sequences Reveals Key Amino Acids

Another very powerful use of phylogenetics and sequence comparison is to identify which amino acids in a protein are most important for function. As these amino acids are under strong negative selection, they evolve very slowly and can be identified by comparing the sequences of the protein in several species. An excellent example of this is the DNA binding protein p53, which is a tumor suppressor and is mutated in many human cancers. Because of its clinical and biological importance, the p53 protein has been sequenced and studied in a large number of distinct mammalian and vertebrate species. These sequences can all be aligned together, and, by determining how fast particular sites and regions of p53 have changed throughout evolution, those parts of the p53 protein that have evolved very slowly can be identified. Such an analysis reveals that the most important amino acids within p53 are those that make contacts with specific base pairs in DNA (Fig. 12-12). Furthermore, mutations in these amino acids are more likely to contribute to cancer development than are mutations elsewhere in the protein. In this manner, comparative analyses not only can help us to understand the basic biology of how a protein functions, but can also provide insights into how mutations in proteins can affect human phenotypes such as cancer. Importantly, this approach can be used to identify functional amino acids and domains in any protein for which multiple sequences are available for comparison (there are thousands of proteins that can be analyzed in this way, with more sequences becoming available each day), including the many proteins that are not yet functionally characterized at any level.

Transcriptional Regulatory Elements Can Be Identified by Using Sequence Comparisons

Comparative sequence analysis can also be used to identify elements in the genome that regulate transcription, such as promoters and enhancers (Chapter 3). Although these elements play critical roles in virtually all biological processes, they can be very difficult to identify, much more so than protein-coding sequences, particularly in genomes of complex, multicellular organisms. Experiments to identify and verify them are laborious, although, as we shall see in Chapter 13, genome-scale functional approaches are starting to be applied to identify regulatory elements. So far, it has not been possible to define a specific sequence vocabulary for regulatory elements in the same way that the genetic code is used to translate nucleotide sequence to amino acid sequence; therefore, purely computational approaches are not likely to be the solution to the problem. However, because these elements are so important to the organism, they are often under evolutionary constraint and therefore amenable to discovery by comparative sequence analysis (Fig. 12-13). In fact, a number of human transcriptional enhancers have now been

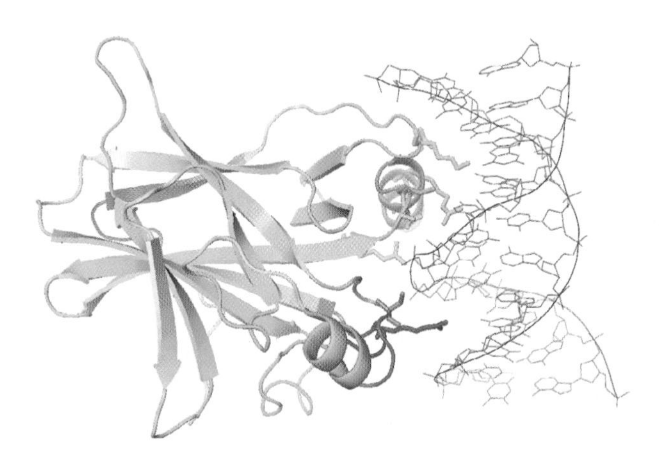

FIGURE 12-12

Comparative sequence analysis can be used to identify the most important amino acids of proteins. A model of the p53 protein (to the *left*) interacting with DNA (on the *right*) provides clues about the important regions of the protein. The human p53 sequence was compared with that of many other homologous vertebrate proteins, and here the degree of evolutionary constraint at each amino acid is plotted on the model of the structure. The regions of the p53 protein that evolved very slowly (indicative of especially strong evolutionary constraint) are colored *blue* or *green* (with *blue* being the slowest evolving). The regions of the protein that evolved more swiftly are painted *red* or *yellow* (*red* being the fastest). As can be seen, the region of p53 that interacts with DNA is mostly *blue* (and a little *green*), whereas structural parts of the protein farther from the DNA-binding region are *reddish* or *yellow*, thus indicating that they evolve more swiftly and are therefore less functionally important. In fact, mutations affecting the DNA-binding domain of p53 are much more likely to result in cancer than are mutations in other regions.

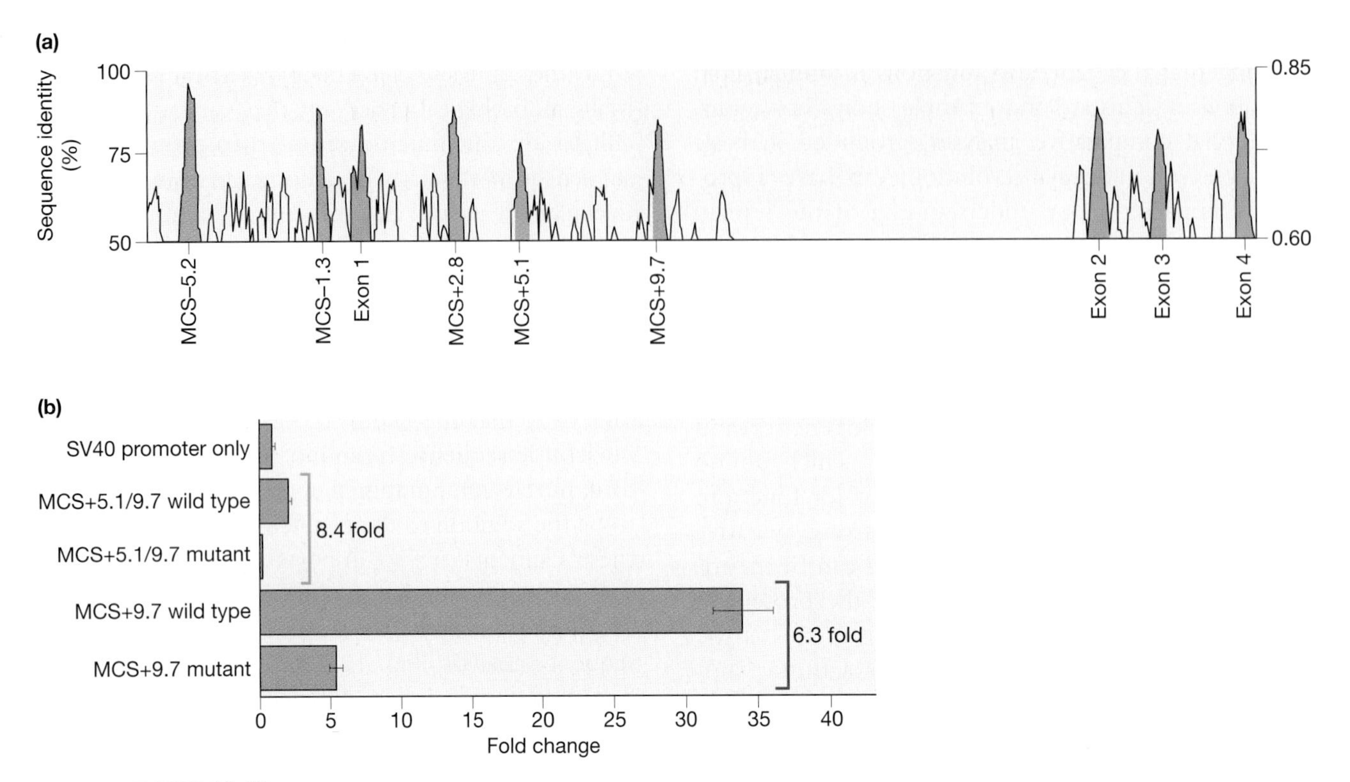

FIGURE 12-13

Comparative sequence analysis can be used to identify regions of genomes that are *cis*-acting transcriptional regulatory elements. (*a*) Alignment of the human and mouse genomes was used to highlight regions of high sequence conservation near the human *RET* gene, a major susceptibility factor for Hirschsprung disease. The peaks of sequence conservation are in *pink* and *blue*. The *pink* peaks correspond to noncoding conserved segments, and the *blue* peaks mark highly conserved exons. MCS, multiple conserved sequence. (*b*) Several of the regions corresponding to peaks in *a* were cloned from people with and without Hirschsprung disease and tested for their ability to act as transcriptional enhancers. In the top row, the transcriptional activity of a promoter without any other DNA is measured (this serves as a background measurement), and very low levels of transcripts are seen. The nondisease allele of conserved fragment 5 (*second row from the bottom*) was found to be a transcriptional enhancer that increases transcription ~35 times higher than background levels. The allele of fragment 5 from a patient with Hirschsprung disease, however, activates transcription only about fivefold, six times less than the normal allele. Thus, the failure of this enhancer to produce sufficient RET protein is a likely contributor to the development of Hirschsprung disease.

computationally annotated in this way and then verified by experiments in cultured cells or transgenic mice where the putative enhancer is shown to increase transcription of an attached reporter gene (see Chapter 13). These discoveries have exploited comparisons of the genomes of fish, frog, chicken, and many mammals, thereby underscoring the value of having many genome sequences spanning large evolutionary intervals.

The ready availability of DNA and amino acid sequences has already changed the way biologists design experiments and interpret the results and will continue to do so. Genome-scale sequencing is becoming less expensive and in the near future it will not be the limiting step in producing sequence data. The power of comparative genomic analyses will grow as the sequences of an ever more diverse set of organisms become available. Furthermore, although we have taken a largely human-centric view in this book, sequence analyses have applications well beyond the human genome and are being exploited to better understand organisms in all phyla, including bacteria, archaea, yeast, fruit flies and other insects, and plants (including the small weed *Arabadopsis*, the mighty redwood tree, and agricultural species like rice, wheat, sorghum, and corn).

Many bioinformatics experts, particularly in the early days of the HGP, were computer scientists who formed partnerships with biologists. With the growth of the field of genomics, it is not unusual today for a student to be trained in a truly interdisciplinary way by developing deep expertise in both biology and computational science. Although we have learned much from sequence analysis, the knowledge we have gained so far represents only the tip of the iceberg. The field of bioinformatics will continue to be indispensable, and those trained in it will remain in great demand for a long time.

Reading List

Databases

Benson D.A., Karsch-Mizrachi I., Lipman D.J., Ostell J., and Wheeler D.L. 2006. GenBank. *Nucleic Acids Res.* **34:** D16–D20.

Diehn M., Sherlock G., Binkley G., Jin H., Matese J.C., et al. 2003. SOURCE: A unified genomic resource of functional annotations, ontologies, and gene expression data. *Nucleic Acids Res.* **31:** 219–223.

Issel-Tarver L., Christie K.R., Dolinski K., Andrada R., Balakrishnan R., et al. 2002. *Saccharomyces* Genome Database. *Methods Enzymol.* **350:** 329–346.

Pruitt K.D. and Maglott D.R. 2001. RefSeq and LocusLink: NCBI gene-centered resources. *Nucleic Acids Res.* **29:** 137–140.

Stover N.A., Krieger C.J., Binkley G., Dong Q., Fisk D.G., et al. 2006. *Tetrahymena* Genome Database (TGD): A new genomic resource for *Tetrahymena thermophila* research. *Nucleic Acids Res.* **34:** D500–D503.

Sequence Analysis Tools

Altschul S.F., Gish W., Miller W., Myers E.W., and Lipman D.J. 1990. Basic local alignment search tool. *J. Mol. Biol.* **215:** 403–410.

Altschul S.F., Madden T.L., Schaffer A.A., Zhang J., Zhang Z., Miller W., and Lipman D.J. 1997. Gapped BLAST and PSI-BLAST: A new generation of protein database search programs. *Nucleic Acids Res.* **25:** 3389–3402.

Blanchette M., Kent W.J., Riemer C., Elnitski L., Smit A.F., et al. 2004. Aligning multiple genomic sequences with the threaded blockset aligner. *Genome Res.* **14:** 708–715.

Brudno M., Do C.B., Cooper G.M., Kim M.F., Davydov E., et al. 2003. LAGAN and Multi-LAGAN: Efficient tools for large-scale multiple alignment of genomic DNA. *Genome Res.* **13:** 721–731.

Karlin S. and Altschul S.F. 1990. Methods for assessing the statistical significance of molecular sequence features by using general scoring schemes. *Proc. Natl. Acad. Sci.* **87:** 2264–2268.

Thompson J.D., Higgins D.G., and Gibson T.J. 1994. CLUSTAL W: Improving the sensitivity of progressive multiple sequence alignment through sequence weighting, position-specific gap penalties and weight matrix choice. *Nucleic Acids Res.* **22:** 4673–4680.

Notredame C., Higgins D.G., and Heringa J. 2000. T-Coffee: A novel method for fast and accurate multiple sequence alignment. *J. Mol. Biol.* **302:** 205–217.

Analysis of Sequenced Genomes

Aparicio S., Chapman J., Stupka E., Putnam N., Chia J.M., et al. 2002. Whole-genome shotgun assembly and analysis of the genome of *Fugu rubripes*. *Science* **297:** 1301–1310.

Chimpanzee Sequencing and Analysis Consortium. 2005. Initial sequence of the chimpanzee genome and comparison with the human genome. *Nature* **437:** 69–87.

Hillier L.W., Miller W., Birney E., Warren W., Hardison R.C., et al. (International Chicken Genome Sequencing Consortium). 2004. Sequence and comparative analysis of the chicken genome provide unique perspectives on vertebrate evolution. *Nature* **432:** 695–716.

Lander E.S., Linton L.M., Birren B., Nusbaum C., Zody M.C., et al. (International Human Genome Sequencing Consortium). 2001. Initial sequencing and analysis of the human genome. *Nature* **409:** 860–921.

International Human Genome Sequencing Consortium. 2004. Finishing the euchromatic sequence of the human genome. *Nature* **431:** 931–945.

Lindblad-Toh K., Wade C.M., Mikkelsen T.S., Karlsson E.K., Jaffe D.B., et al. 2005. Genome sequence, comparative analysis and haplotype structure of the domestic dog. *Nature* **438:** 803–819.

Marra M.A., Jones S.J., Astell C.R., Holt R.A., Brooks-Wilson A., et al. 2003. The genome sequence of the SARS-associated coronavirus. *Science* **300:** 1399–1404.

Waterston R.H., Lindblad-Toh K., Birney E., Rogers J., Abril J.F., et al. (Mouse Genome Sequencing Consortium). 2002. Initial sequencing and comparative analysis of the mouse genome. *Nature* **420:** 520–562.

Gibbs R.A., Weinstock G.M., Metzker M.L., Muzny D.M., Sodergren E.J., et al. (Rat Genome Sequencing Project Consortium). 2004. Genome sequence of the Brown Norway rat yields insights into mammalian evolution. *Nature* **428:** 493–521.

Identifying Genes

Burge C. and Karlin S. 1997. Prediction of complete gene structures in human genomic DNA. *J. Mol. Biol.* **268:** 78–94.

Cliften P., Sudarsanam P., Desikan A., Fulton L., Fulton B., et al. 2003. Finding functional features in *Saccharomyces* genomes by phylogenetic footprinting. *Science* **301:** 71–76.

Kellis M., Patterson N., Endrizzi M., Birren B., and Lander E.S. 2003. Sequencing and comparison of yeast species to identify genes and regulatory elements. *Nature* **423:** 241–254.

Lim L.P., Glasner M.E., Yekta S., Burge C.B., and Bartel D.P. 2003. Vertebrate microRNA genes. *Science* **299:** 1540.

Roest Crollius H., Jaillon O., Bernot A., Dasilva C., Bouneau L., et al. 2000. Estimate of human gene number provided by genome-wide analysis using *Tetraodon nigroviridis* DNA sequence. *Nat. Genet.* **25:** 235–238.

Pennacchio L.A., Olivier M., Hubacek J.A., Cohen J.C., Cox D.R., et al. 2001. An apolipoprotein influencing triglycerides in humans and mice revealed by comparative sequencing. *Science* **294:** 169–173.

Comparative Sequence Analysis and Evolution

Dermitzakis E.T., Reymond A., Lyle R., Scamuffa N., Ucla C., et al. 2002. Numerous potentially functional but nongenic conserved sequences on human chromosome 21. *Nature* **420:** 578–582.

Jukes T.H. and Cantor C.R. 1969. Evolution of protein molecules. In *Mammalian Protein Metabolism* (ed. H.N. Munro), pp. 21–132. Academic Press, New York.

Kimura M. 1983. *The Neutral Theory of Molecular Evolution.* Cambridge University Press, Cambridge.

Li W.-H. 1997. *Molecular Evolution.* Sinauer Associates, Sunderland, Massachusetts.

Mayor C., Brudno M., Schwartz J.R., Poliakov A., Rubin E.M., et al. 2000. VISTA: Visualizing global DNA sequence alignments of arbitrary length. *Bioinformatics* **16:** 1046–1047.

Nobrega M.A., Ovcharenko I., Afzal V., and Rubin E.M. 2003. Scanning human gene deserts for long-range enhancers. *Science* **302:** 413.

Simon A.L., Stone E.A., and Sidow A. 2002. Inference of functional regions in proteins by quantification of evolutionary constraints. *Proc. Natl. Acad. Sci.* **99:** 2912–2917.

Yang Z. 1997. PAML: A program package for phylogenetic analysis by maximum likelihood. *CABIOS* **13:** 555–556.

Identifying and Characterizing Regulatory Sequences

Cooper G.M. and Sidow A. 2003. Genomic regulatory regions: Insights from comparative sequence analysis. *Curr. Opin. Genet. Dev.* **6:** 604–610.

Göttgens B., Barton L.M., Chapman M.A., Sinclair A.M., Knudsen B., et al. 2002. Transcriptional regulation of the stem cell leukemia gene (*SCL*)–Comparative analysis of five vertebrate SCL loci. *Genome Res.* **12:** 749–759.

Emison E.S., McCallion A.S., Kashuk C.S., Bush R.T., Grice E., et al. 2005. A common sex-dependent mutation in a *RET* enhancer underlies Hirschsprung disease risk. *Nature* **434:** 857–863.

Hardison R.C. 2000. Conserved noncoding sequences are reliable guides to regulatory elements. *Trends Genet.* **16:** 369–372.

Loots G.G., Locksley R.M., Blankespoor C.M., Wang Z.E., Miller W., Rubin E.M., and Frazer K.A. 2000. Identification of a coordinate regulator of interleukins 4, 13, and 5 by cross-species sequence comparisons. *Science* **288:** 136–140.

Pennacchio L.A. and Rubin E.M. 2001. Genomic strategies to identify mammalian regulatory sequences. *Nat. Rev. Genet.* **2:** 100–109.

Sumiyama K., Kim C.B., and Ruddle F.H. 2001. An efficient *cis*-element discovery method using multiple sequence comparisons based on evolutionary relationships. *Genomics* **71:** 260–262.

Uchikawa M., Ishida Y., Takemoto T., Kamachi Y., and Kondoh H. 2003. Functional analysis of chicken Sox2 enhancers highlights an array of diverse regulatory elements that are conserved in mammals. *Dev. Cell* **4:** 509–519.

Woolfe A., Goodson M., Goode D.K., Snell P., McEwen G.K., et al. 2004. Highly conserved non-coding sequences are associated with vertebrate development. *PLoS Biol.* **3:** e7.

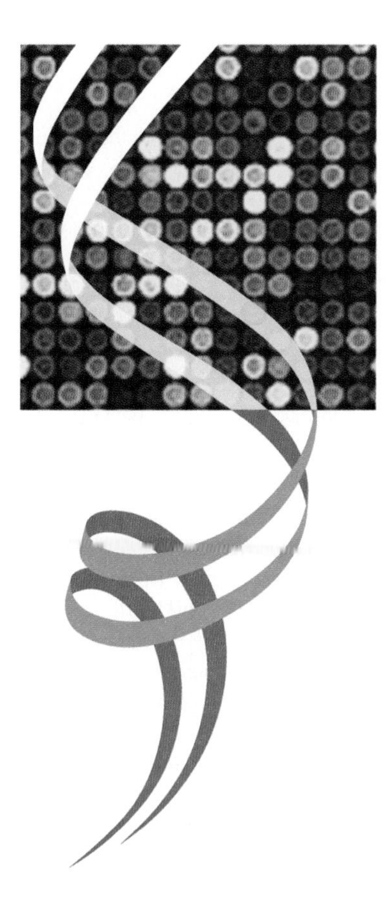

CHAPTER 13

From Genome Sequence to Gene Function

When the Human Genome Project began, many scientists assumed that the project would require at least 15 years of hard work to obtain the sequences of the human genome and those of a few other organisms—yeast, *Escherichia coli*, and *Drosophila*. However, as we saw in Chapters 10 and 11, genomic technologies developed rapidly and improved so much that many genomes were sequenced before the 2005 deadline originally planned for the human genome expired. Very quickly, the complete genomic sequences of many organisms, chiefly bacteria, began to make their way into public databases. As they did so, scientists began to develop a variety of methods for understanding the functions of genomes on a large scale by emulating some of the strategies used in the mapping and sequencing projects—most notably by taking advantage of the efficiencies that can be gained from economies of scale. For example, microarray hybridization methods were invented for measuring the transcript abundance of thousands of genes simultaneously. Similarly, very efficient large-scale methods were developed for genetic analysis. These included generating knockouts for many, or even all, of the genes in a genome and developing methods for identifying and assaying huge numbers of DNA sequence variants. And it was not only nucleic acid–based techniques that changed. Experimental approaches that typically had been performed on one or a few genes or proteins at a time were modified and new methods developed to study many of these biological processes more efficiently and on a large scale. These evolving methods advanced the studies of protein–DNA and protein–protein interactions, *cis*-acting transcriptional regulatory sequences, the locations of mRNAs in cells and tissues, and protein expression levels. This approach of "high-throughput biology" set in motion a decade of creative technology development, which led to new experimental tools that biologists have added to their recombinant DNA tool kits. This chapter describes some of the approaches that have been developed so far in this exciting new field of *functional genomics.*

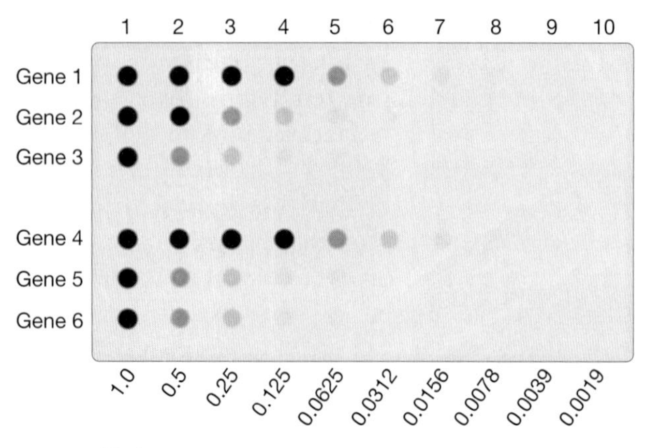

FIGURE 13-1

Measuring steady-state levels of mRNAs by "dot blot" hybridization. Serial dilutions of plasmids containing cDNAs for six genes (rows 1–6) have been applied in dots to the same type of nylon filter used for Southern and Northern blotting. The highest amount of DNA is on the *left* (1) and decreases by twofold at each position to the *right* (2–10). The DNA on the filter is denatured and hybridized with labeled mRNA isolated from the biological sample (cultured cells or tissues). After washing, the amount of radioactivity is determined by autoradiography. It is difficult to use these radioactive signals to determine the exact amounts of mRNAs that were present in the cells, but it is possible to estimate the *relative* amounts of the different mRNAs compared to each other. Gene 1, for example, gives a saturating amount of hybridization in dilution 4, whereas gene 2 shows saturation in dilution 2, indicating that the transcript from gene 1 is present at levels about four times higher than that of gene 2.

Microarrays Allow Simultaneous Analysis of Tens of Thousands of Nucleic Acid Sequences

Many of the fundamental measurements used by molecular biologists involve determining the presence, absence, or relative amounts of a gene or its messenger RNA in a cell. In some cases, it is necessary only to determine whether a whole gene or mRNA is present or not. In other cases, it may be desirable to know which particular variants of a gene—a single base pair or perhaps a larger segment of DNA—are present. Southern and Northern blots and the polymerase chain reaction (PCR) are used to measure the quantity and often the specific quality (size, exact DNA sequence) of a specific nucleic acid in a complex biological sample, such as an extract from a collection of cells or a tissue. As powerful as these techniques are, in most cases they are used to measure one or only a few DNA or RNA sequences at a time. To study multiple genes or mRNAs, the experimenter performs additional experiments. A simple and effective advance was the development of the so-called "dot blots" or "slot blots," in which DNA fragments are denatured and then attached to a filter in a simple array either by manual spotting or by using an inexpensive apparatus. DNA or mRNA from a biological sample is labeled and then hybridized to the filter, just as in Southern blotting, to determine whether the sequence is present (Fig. 13-1). Dot blots offered a significant improvement in efficiency of hybridization measurements, as the same effort used to analyze one sample provided data for dozens of genes. Dot blotting improved efficiency severalfold; however, it did not provide the scale required to cope with the flood of gene sequences coming from genome projects. Now the challenge was to analyze the expression levels of thousands of genes in many different cell types and under a variety of experimental conditions, and to identify DNA sequence variations, whether single-base changes or larger deletions, insertions, and rearrangements.

In a major breakthrough in the early 1990s, scientists developed microarrays for hybridizing tens or even hundreds of thousands of nucleic acid fragments in a single experiment. The principle is similar to that of a dot blot, but instead of just a few dozen dots of DNA, as many as several million different DNA fragments are deposited in an ordered array on a glass slide using automated methods.

Two different ways of constructing and using microarrays were developed. One approach synthesizes oligonucleotides at each spot on the microscope slide, whereas the second deposits, or "spots," DNA clones, PCR products, or oligonucleotides on a slide. The preparation processes, resulting microarrays, and how the experiments are performed on the two platforms are quite different. But no matter which method is used, many nucleic acid, protein, and small molecule measurements and interactions can be determined by using a single microarray. These applications include, for example, analyzing the transcript levels as well as the copy number of essentially all the genes from an organism, genotyping the millions of DNA sequence variants, and identifying the genomic DNA segments bound by proteins in the cell. Microarray technology has produced a dramatic increase in efficiency and has changed the way that researchers study almost every aspect of biology.

Photolithography Is Used to Make Very High Density Oligonucleotides with Microarrays

In the late 1980s, Stephen Fodor and his colleagues adopted the process of photolithography, which is used to manufacture microprocessor chips, to make microarrays with chemically synthesized oligonucleotides. The microarrays, produced by Affymetrix, are often referred to as oligonucleotide "chips" because their synthesis is similar to that of a computer chip. In the first step, a quartz wafer is coated with a photosensitive chemical that prevents nucleotides from attaching to the wafer surface. A mask is used to expose only the sites on the surface where the oligonucleotide will have, for example, an A as the first base. The blocking agent at these sites is destroyed by illumination with a powerful light source. The surface is bathed with an activated version of dATP, and this first base becomes covalently linked to the surface. The activated nucleotides carry a blocking group so, once again, all positions on the surface are blocked. The unbound dATP is washed away, and a new mask is used that exposes only the sites where the first oligonucleotide will be a G. This time, an activated dGTP is used, and the cycle repeated for C and T so that the first base for the entire set of oligonucleotides is now present on the microarray. The synthesis continues with the four bases for the second position, and these and all subsequent bases are joined to the growing oligonucleotide chain by a phosphodiester bond. After the requisite number of bases have been added in the proper sequence, the chip has more than one million regularly spaced sites, called "features," each of which contains many copies of a specified single-stranded oligonucleotide (Fig. 13-2). Hundreds of individual microarray chips can be synthesized on a single wafer. The wafer is cut up and the microarray chips mounted in special holders. They are now ready for use in experiments.

The gene expression microarrays developed by Affymetrix typically contain a set of 22 different oligonucleotides that average 25 nucleotides in length (called "25-mers") for each gene to be tested. Eleven of the 25-mers have the exact sequence of different segments of the mRNA transcribed from the gene, and the other 11 are identical except for an incorrect base at the middle of the oligonucleotide, usually at nucleotide number 13 (see Fig. 13-3). There are several reasons for this design. The oligonucleotides on the chip are short, and they sometimes hybridize poorly or give incorrect signals; having 11 perfectly matched oligonucleotides for each gene provides multiple independent measurements, which results in a much more accurate estimate of how much of the transcript is present. Poorly performing oligonucleotides—outliers—are removed from the analysis, and the average signal from the remaining oligonucleotides for a particular gene is used to estimate the amount of transcript present. The 11 mismatched oligonucleotides help to establish the background signal for each correct oligonucleotide in the hybridization step. The single mismatched base in a sequence of 25 causes these oligonucleotides to hybridize extremely poorly (or not at all) to their near-complement in the mRNA test sample. These signals are subtracted from the signals of each of the perfectly paired oligonucleotides during the data analysis step. As we shall discuss in Chapter 14, using hybridization to distinguish matched versus mismatched oligonucleotides is also exploited to detect single-nucleotide polymorphisms (SNPs).

Oligonucleotide microarrays can be used to measure the levels of a large number of different nucleic acid fragments in a complex solution, such as steady state levels of transcripts in a study of gene expression. The nucleic acid mixture is isolated from the cells of interest, labeled with a fluorescent tag, and then incubated with the microarray so that the test fragments find their complementary segments among the single-stranded oligonucleotides present on each feature of the chip (Fig. 13-4). The hybridized microarray is analyzed by a special scanning machine that detects the quantity of fluorescence at each feature. These results are used to estimate the amount of each fragment present on the chip, and the intensity scores are collected by computer for analysis. Background signals are subtracted for each matched and mismatched pair, and an average value of intensity is calculated from the multiple oligonucleotides representing each gene. Researchers then interpret the large resulting data set for its biological meaning. We shall describe some examples in the sections to follow.

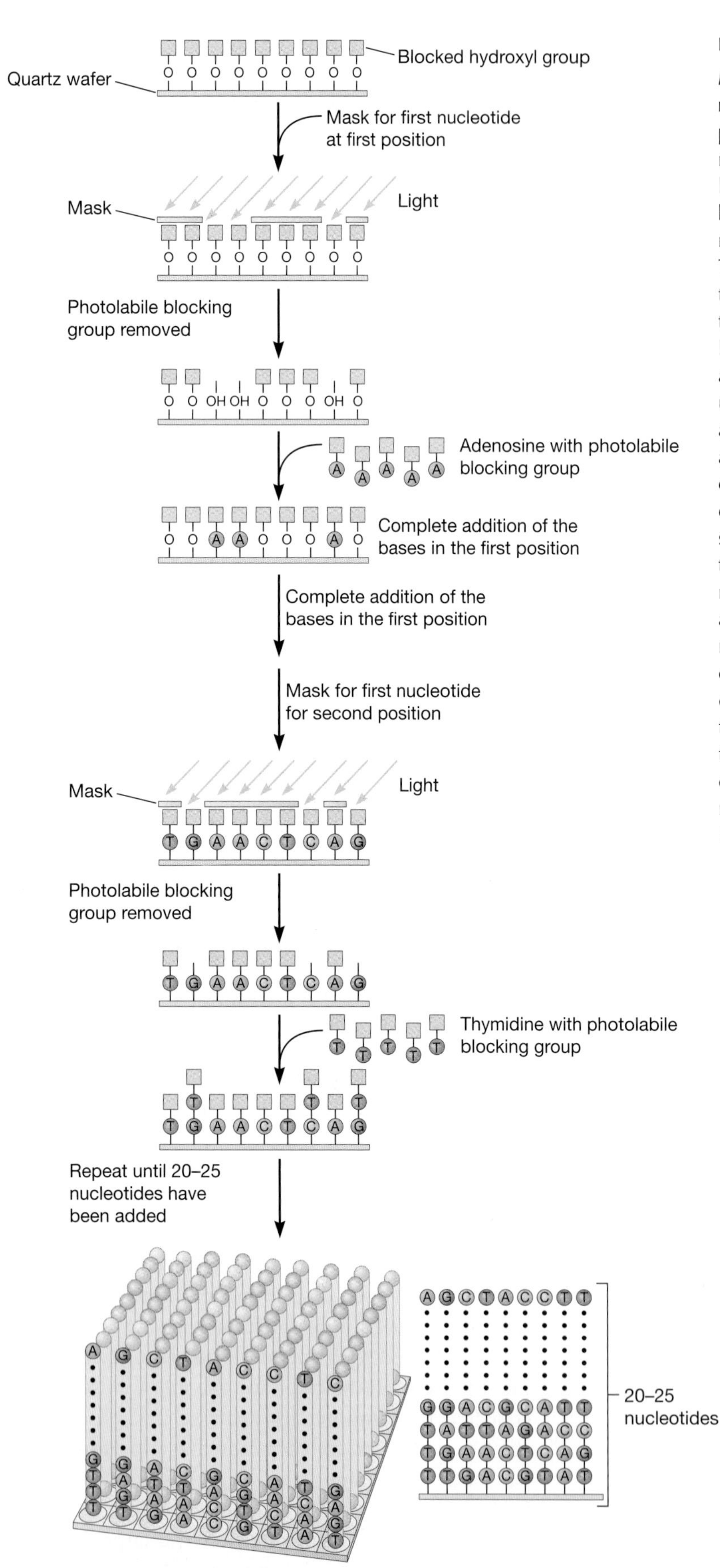

FIGURE 13-2

Making very high density oligonucleotide microarrays using a photolithographic process. A quartz wafer of the kind used to make computer chips is treated with a photolabile chemical that prevents nucleotides binding to the surface. A photolithographic mask is overlaid on the chip and illuminated. The holes in the mask correspond to the positions on the chip where the first nucleotide is to be added (in this case, adenosine). The light destroys the blocking chemical and adenosine is added so that it binds to the unprotected locations. The adenosine carries a blocking group so all these positions are not available until they are once again deprotected. Three different masks are used in turn to deprotect the positions for thymidine, guanosine, and cytidine. At this stage, the first position of every oligonucleotide—about one million—on the chip has been put in place and all carry a blocking group. The process is repeated for the four nucleotides that will occupy the second place in the oligonucleotide. By the end of the second cycle, there are dinucleotide chains at every location. This cycle of reproduction and addition of a nucleotide is repeated until the array carries oligonucleotides 20–25 nucleotides in length.

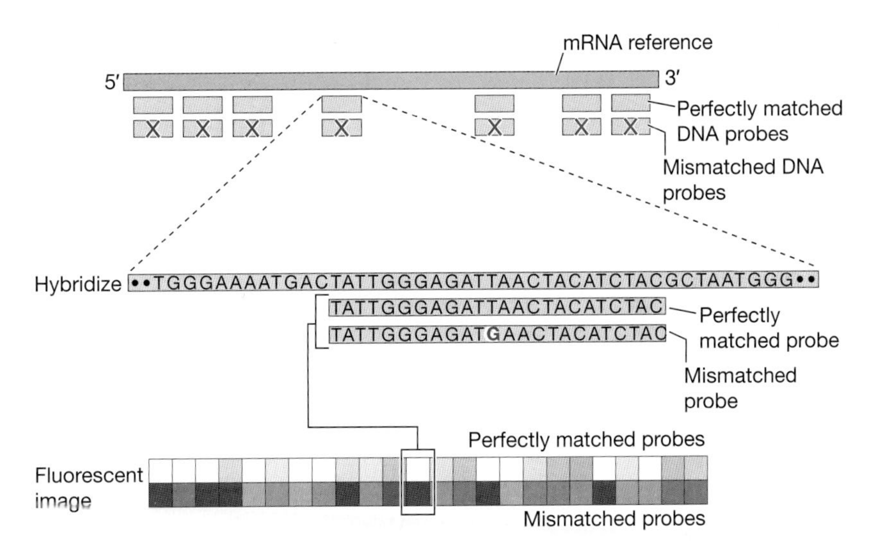

FIGURE 13-3

Multiple oligonucleotides are used to detect each locus on microarrays constructed with photolithography. Oligonucleotide microarrays are designed to provide several independent tests of hybridization for each mRNA or DNA fragment being analyzed. Typically, 21 segments are chosen for a region, such as an mRNA or a genomic segment. For each one of these segments, a 22- to 25-nucleotide-long oligonucleotide that is perfectly matched with its complement is synthesized on the microarray. For each perfectly matched oligonucleotide, a second oligonucleotide is synthesized at a different site on the microarray that is exactly like it, except that the base in the middle position of the oligonucleotide is mismatched. The mismatched oligonucleotide serves as a measure of the background hybridization for its perfectly matched oligonucleotide; the mismatched hybridization signal is subtracted from the perfectly matched signal to give the level of hybridization for the DNA segment. The subtracted signals from all the perfectly matched oligonucleotides are averaged to provide an estimate of the amounts of the mRNA or genomic region that were present in the test sample. (The lighter the square, the more hybridization signal.)

In recent years investigators have developed several additional approaches to constructing microarrays by synthesizing oligonucleotides in situ. One method, commercially provided by Agilent Technologies, uses technology similar to that of an ink-jet printer to deliver nucleotides for each DNA synthesis step to tiny spots on a glass slide. Another, developed by NimbleGen Systems, uses a "maskless synthesis" photolithography that employs digital light processing instead of masks to generate ultra-high-density microarrays on small glass slides. These produce oligonucleotides that are longer—typically 50–70 nucleotides long but as long as 100 nucleotides—than those made by photolithography.

Spotted Microarrays Use Large DNA Segments

In the mid-1990s a different approach for generating and using microarrays—a miniaturized but greatly expanded version of the dot blot technique—was developed by Pat Brown and his colleagues. First, the genes to be assayed are obtained as a set of plasmid cDNA clones (Chapter 4); large sets of these clones became available from the expressed sequence tag (EST) and cDNA sequencing projects begun early in the Human Genome Project (HGP). Second, PCR is used to amplify the cDNA inserts, and the double-stranded DNA fragments are denatured and then "spotted" by a robotic device onto a glass microscope slide. The amount of liquid deposited for each gene is on the order of a few nanoliters, so that each spot dries quickly and the DNA becomes attached to the glass. Although the density of features can be high (as many as 20,000 separate features per centimeter), the process does not allow the ultrahigh densities that can be achieved with photolithography. However, because the DNA fragments on the array are large (from several hundred to several thousand base pairs), hybridization is more specific so that only one feature per gene is needed. Because the largest genomes typically have 20,000 or so genes, a single spotted cDNA microarray can assay all the genes in an organism.

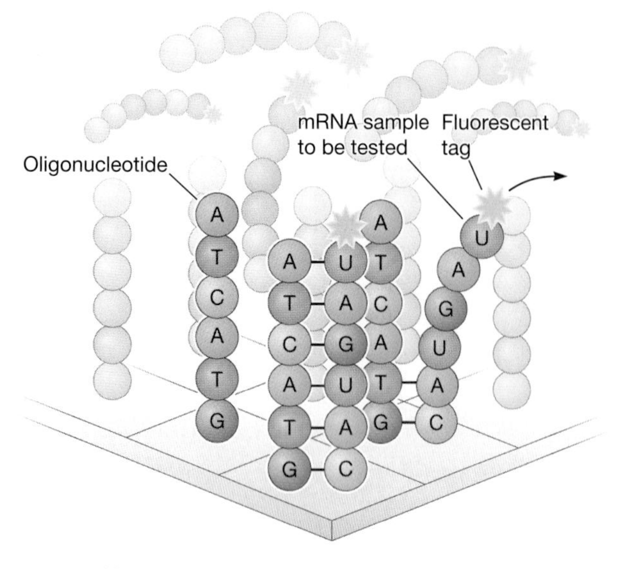

FIGURE 13-4

Hybridization of mRNA fragments to oligonucleotide microarrays. The mRNAs to be measured are fragmented and the fragments are labeled with a fluorescent dye, depicted by the *yellow star* at the end of each molecule. In some experiments, the mRNA fragments themselves are directly labeled; in others, the mRNA is made into cDNA and the cDNA fragments are labeled. The microarray is bathed with the labeled mixture and each mRNA fragment finds its complementary mate at a feature on the microarray. Each square feature has millions of copies of a specific oligonucleotide. After hybridization and washing, the microarray is placed in a special detector, which scans and measures the amount of fluorescent signal at each feature. The amount of fluorescence at each set of probes corresponding to a particular gene is used as a measure of the amount of each mRNA in the biological sample being studied.

Hybridization experiments that use spotted cDNA microarrays are performed in a different manner than those that use oligonucleotide arrays. For spotted arrays, two different sets of complex mRNA samples are used, each labeled separately with a different fluorescent label, usually Cy3 (green) or Cy5 (red). One is a control mRNA mixture from one particular cell type or from a large number of different cell types; an aliquot of this labeled sample is used in every experiment. The other sample is the test sample for which the gene expression pattern is to be determined. The two labeled samples are mixed together and hybridized to the microarray. A detector determines the *ratio* of the hybridization signals of the two fluorescent labels present at each spot on the microarray. If the test sample has a higher level of mRNA for a particular gene than the control sample, the signal will be higher for the label used to tag the test sample, whereas the reverse is true for an mRNA present at a higher level in the control sample (Fig. 13-5). The strategy of two-color hybridization provides an internal standard for every gene that is tested on the microarray by controlling for differences in concentrations and hybridization conditions in the large set of genes being tested in one experiment.

mRNA Profiling with Microarrays Reveals New Relationships between Biochemical and Cellular Pathways

Although all of the genes in the genome are present in almost every cell of an organism, all genes are not expressed in every cell type. The pattern of genes that is turned on or off determines whether a cell is, for example, a liver cell or a neuron. Much control of gene expression occurs at the level of transcription; that is, whether or not to make an mRNA, and how much to make. Nuclease protection assays, Northern blots, and quantitative PCR have been used to measure mRNA levels, but these cannot follow the time course of expression of more than a few genes at a time. However, microarrays can examine the expression of thousands of genes following different treatments at multiple time points.

Such experiments produce very large quantities of data; an experiment that tests 20,000 human cDNAs for relative levels of transcripts in just two different samples generates hundreds of thousands of data points. It is impossible just to look at the data and make sense of it. Instead, extensive statistical analysis and innovative ways of displaying the analysis are needed. One of the most useful ways to analyze such complex data is to use a method called hierarchical clustering. Genes are ranked in a table according to how similar their gene expression levels are in a single experiment or across different experiments in different cell types or under different physiological conditions. To make it easy to visualize these rankings, a graphical representation called a "heat map" is produced, in which each data point is shown as a shade of green or red or other pairs of colors.

cDNA microarrays were used to study the gene expression changes that occur when cultured human fibroblast cells that have been starved of growth factors are suddenly shifted into serum, a medium that is a rich source of proliferative factors. Release from serum starvation has been analyzed for years to study how cells

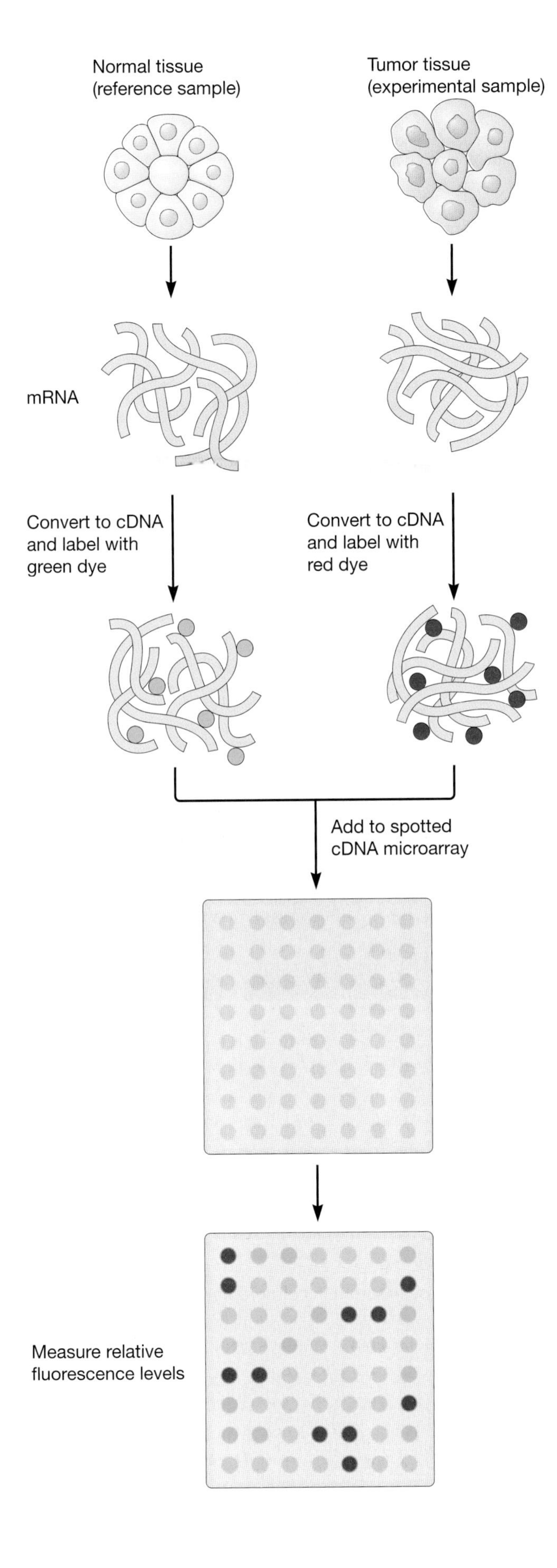

proliferate in response to growth factors, and previous experiments with fibroblasts had identified several genes that are involved in the proliferative response. To obtain a more complete picture of the process, researchers used microarrays to measure the expression profiles of human fibroblasts during starvation and at 12 time points after addition of serum to the growth medium (Fig. 13-6). Hierarchical clustering was used to order the genes for each time point from those expressed at the highest levels during starvation to those with the highest levels after serum addition. About 500 genes were found to vary in their expression levels during the course of the treatment. Many genes not expressed during starvation were very rapidly induced as soon as growth medium was added, whereas several classes of genes showed a gradual increase in expression as the proliferation began. The increased expression of some of the induced genes, including known growth response genes, transcription factors, and various signaling pathways, was expected, although this experiment provided an unprecedented level of detail on the kinetics of expression of these genes.

What was striking and unexpected, however, is that at later time points, after the proliferative response was well established, numerous additional genes known to be important in the process of

FIGURE 13-5

Spotted cDNA microarrays are more accurate with a "two-color" experimental design. A two-color experimental design strategy achieves greater accuracy as it controls for differences in concentrations and hybridization conditions across experiments and large sets of genes in a single experiment. In this example, mRNA (converted to cDNA) from a reference or control sample from normal tissues is labeled with one fluorescent tag, whereas the experimental sample from tumor tissue is labeled with a second tag. (Both samples come from the same individual.) The ratio of the amounts of each mRNA in the experimental to control samples is determined by assessing relative fluorescence levels at each spot. If the signal at one of the spotted cDNAs on the microarray has a higher ratio of green fluorescence to red fluorescence, it means that the mRNA for that gene is expressed at a higher level in the normal tissue than in the tumor. If the signal at a gene's feature on the microarray has a higher ratio of red signal to green, then the mRNA is expressed at a higher level in the tumor. Gene features that have equal levels of the two colors (depicted as *yellow* on this array) are expressed at equal levels. By measuring the two labeled samples on the same array, problems caused by variation in labeling, hybridization, and detection are minimized.

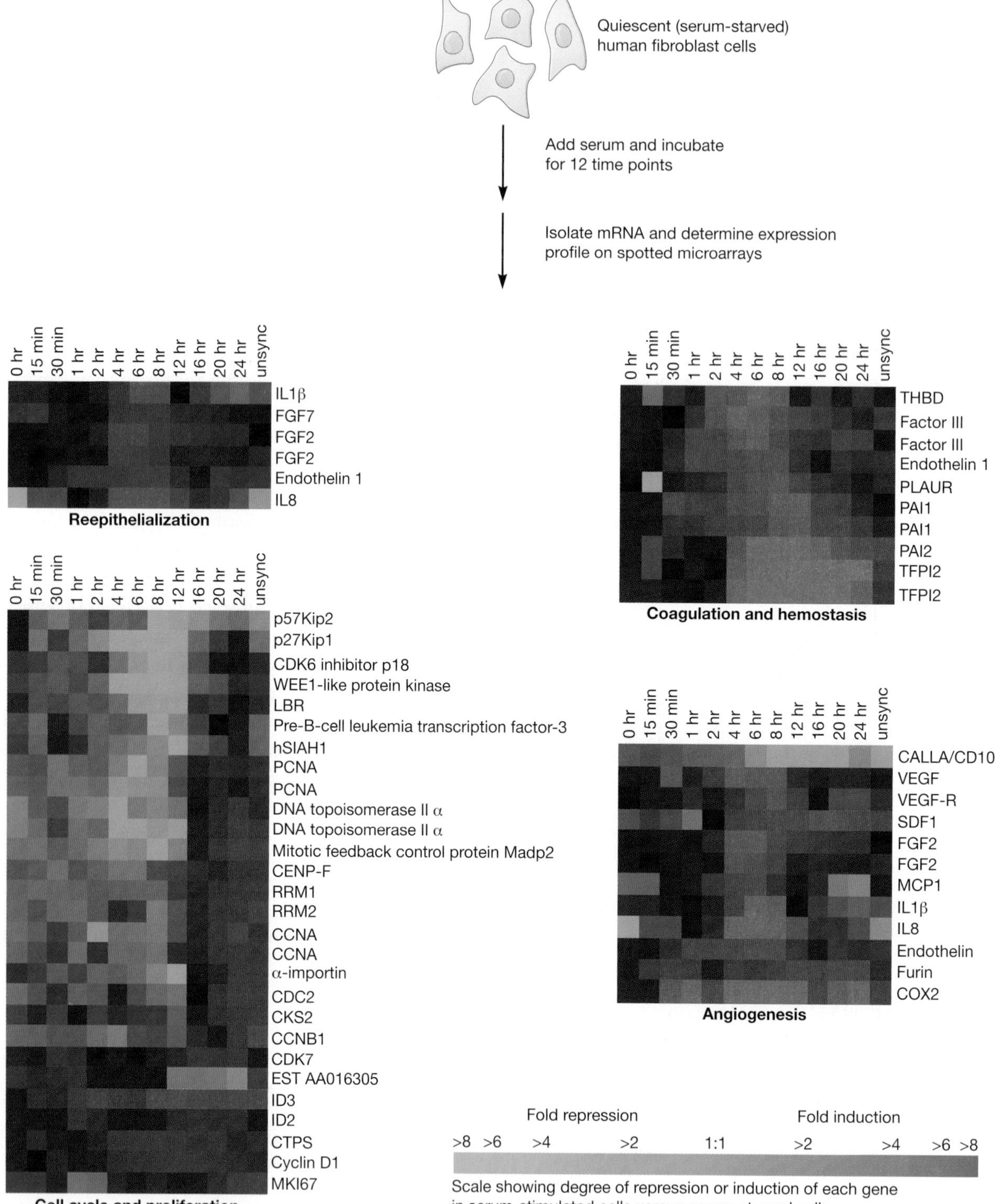

FIGURE 13-6

Microarrays reveal that genes whose expression changes during response to serum in cultured fibroblast cells are involved in wound healing. Microarrays are used to track expression changes that occur at 12 time points after the addition of nutrient-rich serum to quiescent cultured human cells. Four heat maps are shown of the clusters of genes whose expression patterns changed with addition of serum. For each gene or data point, the more intense the red color, the greater the increase in expression in serum-stimulated cells compared to the level of expression in serum-starved cells. In contrast, an increase in intensity of the green color reflects decreasing, or repressed, levels of gene expression in the presence of serum compared to those in serum-starved cells. Each of the clusters are genes that are known to play important roles in healing of wounds.

wound healing were induced. These included genes that the fibroblast uses in the remodeling of blood clots and the extracellular matrix, in promoting the activation of immune cells, in regenerating and repairing blood vessels, and in signaling between cells at the wound. Because so many genes in the wound healing pathway show increased expression in a highly coordinated manner during the proliferative process, there can be no doubt that the two processes are physiologically connected in an intimate way. Thus, a new, important biological relationship between two different pathways was discovered, because a large number of human genes were tested and analyzed for one process without a preconceived notion of how it might work.

mRNA Levels Are Also Measured by High-Throughput Sequencing of cDNAs

Microarrays are powerful tools for examining the expression of thousands of mRNAs, but they rely on hybridization, a complex process subject to experimental variation. One factor complicating the accurate measurement of hybridization is that oligonucleotides or spotted DNA fragments at each feature on the microarray must be in great excess over the amount of mRNA in the test sample. This is particularly difficult if a microarray contains a large number of genes, as the expression levels of different genes in a cell can range over several orders of magnitude. Many genes are expressed at high enough levels in some cells that it is difficult, if not impossible, to have excess probe on the microarray. A second source of error with microarrays, particularly spotted cDNA microarrays, is cross-hybridization. Because many genes have segments with very similar sequences, a feature or set of features on a microarray that is meant to detect a single gene may, in fact, also hybridize with related mRNAs, thereby giving imprecise estimates of specific mRNA levels.

DNA sequencing avoids these problems. Plasmid libraries containing cDNAs (or small fragments of cDNAs) are made from the mRNA pool from a cell or tissue sample, and large numbers of clones are sequenced. The sequences are analyzed, and the number of times a sequence is present for a particular gene is used as an indication of its abundance in the original mRNA pool. This approach is essentially a digital method, one in which *counting* the number of discrete sequencing reads corresponding to a particular gene, not continuous quantitative measurement of a fluorescent signal (as in hybridization), is used to estimate the relative levels of transcripts.

Whether this is feasible depends on how much DNA sequencing the researcher is willing to do (or can pay for). For the counting to be accurate, it is important that multiple "hits"—several instances where at least part of the sequence of a gene is observed—are obtained for each gene. Typically a subset of genes is expressed at very high levels in any given cell type, and these tend to dominate the data set. (This is a serious issue in making cDNA libraries where normalization techniques are used to try to ensure that rare mRNAs are represented in the libraries.) Therefore, a very large number of reads is needed for transcripts expressed at modest or low levels. Two different strategies have been used for this type of sequencing.

The first is derived from the projects to clone and sequence ESTs. In the early days of the HGP, it was recognized that obtaining the sequences of mRNAs would be an invaluable resource for identifying genes. Today, many millions of EST sequencing reads have been obtained from cDNA clones generated from different tissues. Because these data are usually available in databases, many researchers have simply used the frequency of ESTs for each gene as an estimate of the expression level of that gene. This approach is limited, however, to analysis of the tissues and cell types that have been chosen in the EST projects. Nevertheless, valuable data are available, especially for genes that are expressed in a highly tissue-specific manner.

A second strategy is much more applicable to studying gene expression patterns in a variety of cells under different conditions. Serial analysis of gene expression (SAGE) infers the abundance of an mRNA by counting the number of times it is represented in a sample. Instead of devoting a single sequencing read to the analysis of a single gene, SAGE and its derivatives determine the presence of as many as 50 gene segments in a single sequencing read (Fig. 13-7). A plasmid library is produced for each cell type to be tested. Each plasmid insert is made up of multiple, concatenated short cDNA segments, called tags, that are derived from the cell mRNAs and separated from one another by short segments of DNA containing the recognition sequences for two restriction enzymes. In one version of SAGE, a tag and its attached restriction sites are 17-bp long, so a plasmid insert of 700 bp will contain tags for about 40 different mRNAs. Because the sequencing reads are about 700-bp long, the sequence for 40–50 SAGE

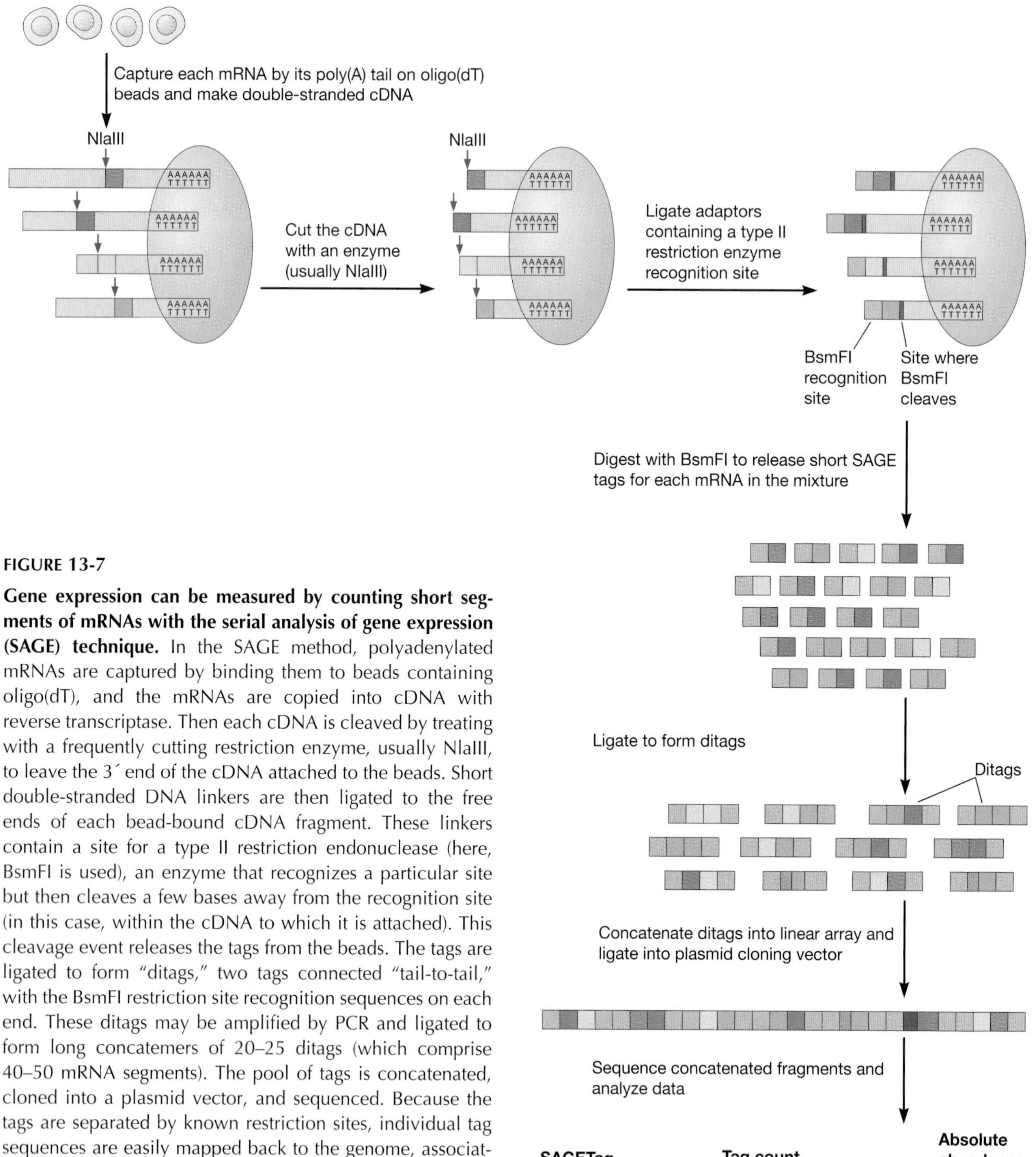

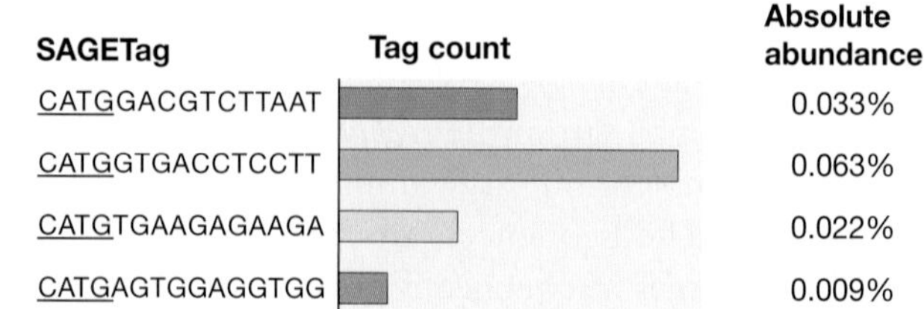

FIGURE 13-7

Gene expression can be measured by counting short segments of mRNAs with the serial analysis of gene expression (SAGE) technique. In the SAGE method, polyadenylated mRNAs are captured by binding them to beads containing oligo(dT), and the mRNAs are copied into cDNA with reverse transcriptase. Then each cDNA is cleaved by treating with a frequently cutting restriction enzyme, usually NlaIII, to leave the 3′ end of the cDNA attached to the beads. Short double-stranded DNA linkers are then ligated to the free ends of each bead-bound cDNA fragment. These linkers contain a site for a type II restriction endonuclease (here, BsmFI is used), an enzyme that recognizes a particular site but then cleaves a few bases away from the recognition site (in this case, within the cDNA to which it is attached). This cleavage event releases the tags from the beads. The tags are ligated to form "ditags," two tags connected "tail-to-tail," with the BsmFI restriction site recognition sequences on each end. These ditags may be amplified by PCR and ligated to form long concatemers of 20–25 ditags (which comprise 40–50 mRNA segments). The pool of tags is concatenated, cloned into a plasmid vector, and sequenced. Because the tags are separated by known restriction sites, individual tag sequences are easily mapped back to the genome, associated with a transcript, and added to a tally. The abundance of each tag from such a sequencing experiment is calculated, and then the absolute abundance of each associated mRNA is calculated by dividing the abundance of each tag by the total number of tags obtained in the experiment. (The NlaIII sites of the tags are underlined.)

tags can be obtained in one read, thus greatly increasing efficiency. Although some effort is required to produce plasmid libraries containing the concatenated tags, SAGE has been developed as an alternative to microarrays for analyzing differences in gene expression. In one study, SAGE libraries were generated from normal colon tissue and from a colon tumor sample. DNA sequencing was performed on about 2000 SAGE clones to generate more than 60,000 independent tag sequences from each library. Bioinformatics analysis was performed to assign each tag to a specific gene in the human genome, and the number of tags per gene was counted and compared in the tumor and normal samples, which provided striking results. More than 500 genes were found to differ significantly in their expression levels in normal compared to cancerous cells. In many cases, the differences in levels were greater than tenfold, providing many new leads for investigating the causes of colon cancer. The quantity of data generated in this experiment was actually quite modest. As sequencing has become easier and less expensive, SAGE experiments typically produce hundreds of thousands of tags per sample and ten times the number of sequencing reads.

SAGE and other similar sequence-based techniques are likely to become more widely used with the development of radically new DNA sequencing technologies that are potentially vastly faster and cheaper than capillary sequencing. As the cost of sequencing falls, greater numbers of tags and a larger variety of samples can be analyzed. It is conceivable that fast sequencing methods will eventually supplant hybridization-based techniques.

Chromatin Immunoprecipitation Is Used to Measure Transcription Factor Binding in a Genome

The techniques we have discussed so far were designed to measure the steady-state levels of mRNA from a large number of genes. None of these methods, however, directly analyzes the mechanisms of control of gene expression, that is, *how* the levels of transcripts change with time or differ between one cell and the next. As we have seen, a key event in turning transcription on or off is the binding of transcription factors to specific DNA sequences within the promoters and other *cis*-acting DNA elements near or within genes (Chapter 3). Biochemical methods have provided a great deal of detailed information concerning regulation by transcription factors, including their preferred recognition sequences in the DNA and the consequences of their binding to these sequences. Most of these studies have been performed in vitro by using techniques such as nuclease protection and gel mobility shift assays with purified transcription factors and cloned DNA fragments. These techniques are extremely useful; however, they do have some shortcomings. First, they can be used to study a single or, at the most, a few DNA-binding sites at a time. Therefore, much of what we know about most transcription factors is limited to detailed studies of the regulation of only a few genes. Second, studying the binding of a transcription factor to a site in a purified DNA fragment in the test tube may not necessarily tell us much about binding that occurs in the cell. Supporting evidence that the interaction occurs in vivo can come from functional experiments, such as mutational analysis of the *cis*-acting site, but these experiments are slow and do not provide a complete picture of how a transcription factor behaves in the cell.

Chromatin immunoprecipitation (ChIP) provides a partial solution to some of these problems. In ChIP, living cells, either grown in tissue culture or isolated from dissociated tissues, are briefly treated with formaldehyde to form chemical cross-links between proteins and between proteins and nucleic acids (Fig. 13-8). At these concentrations of formaldehyde, cross-links occur only between proteins and DNA that are close to one another, as is the case for transcription factors bound to their sites in the genome. Cross-linking "freezes" transcription factors (as well as other proteins associated with chromatin) to the sites that they occupy in the cell at the moment formaldehyde is added. The cells are then disrupted, and the genomic DNA is sheared by sonication into fragments about 500 bp in length. An antibody that recognizes a particular transcription factor is then added to the sheared DNA, and those fragments that are cross-linked to that factor are immunoprecipitated. The mixture is centrifuged, which brings the specifically bound fragments into the pellet. Immunoprecipitation also brings down chromatin fragments from random positions in the genome, but a good antibody can provide several-hundredfold enrichment for fragments cross-linked to the desired factor. These fragments can be identified by hybridizing the "ChIPed" DNA to a microarray that tiles across the genome, or across selected parts of the genome, including only selected sequences such as transcriptional promoters. This

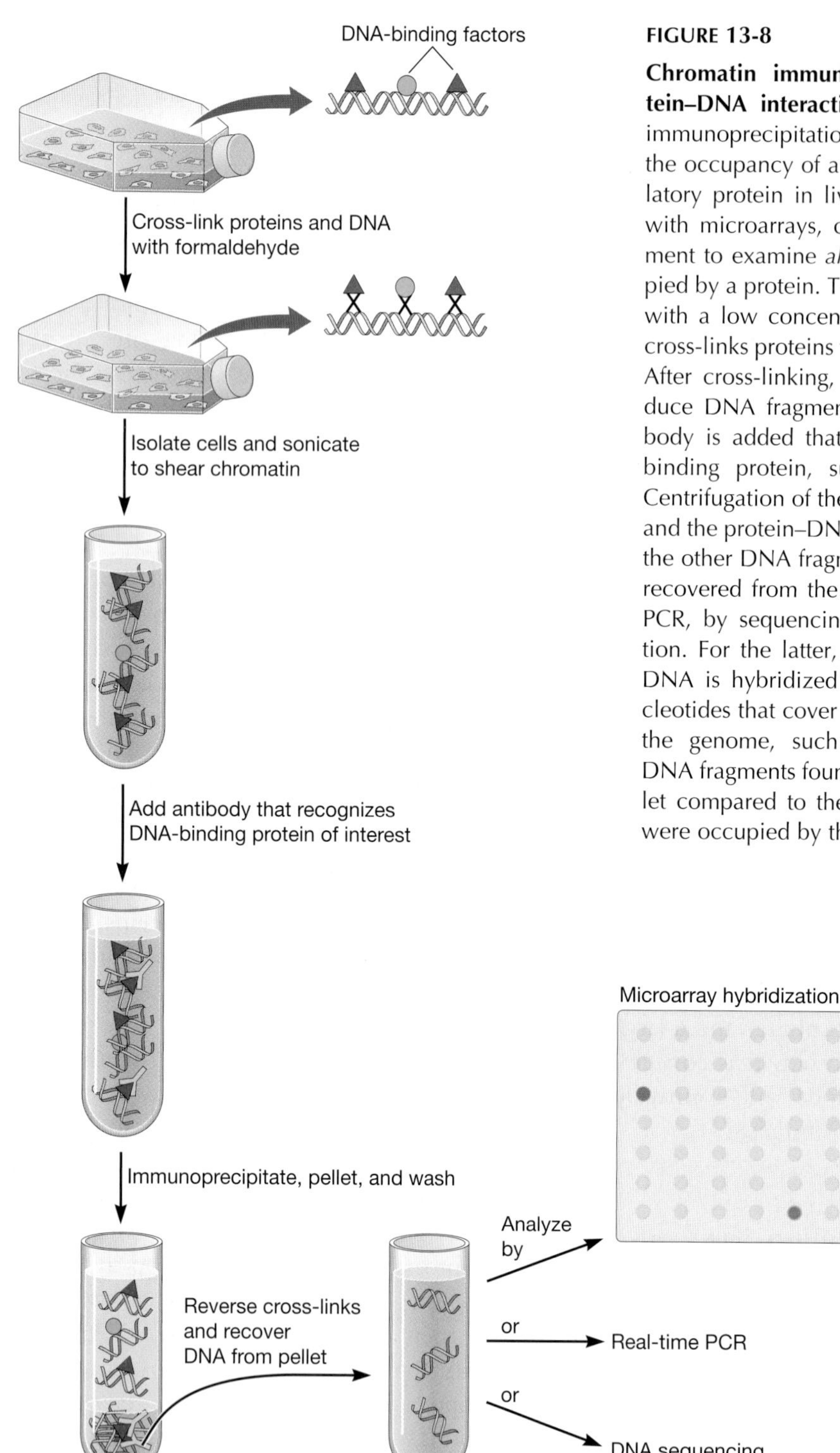

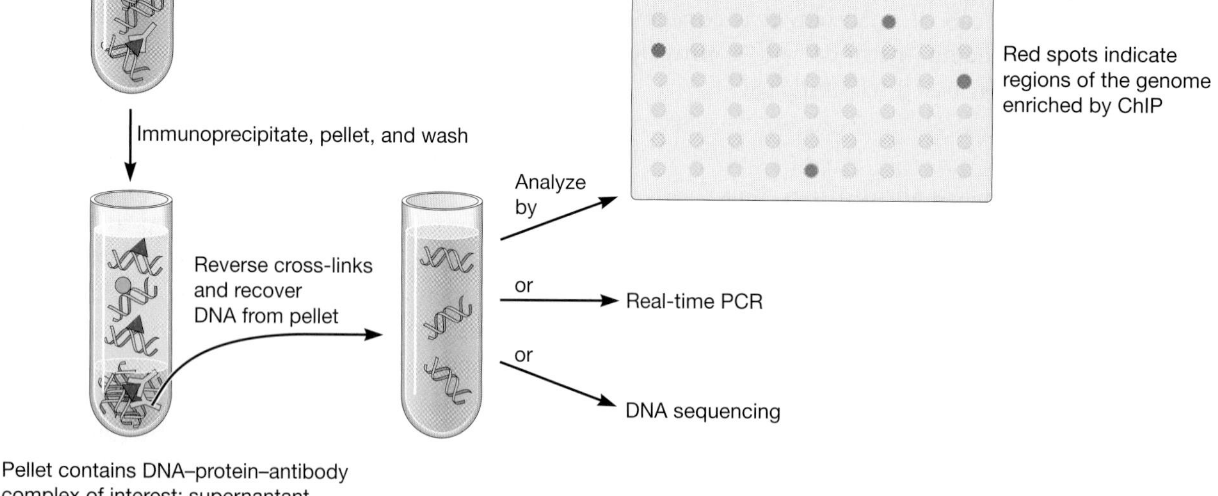

FIGURE 13-8

Chromatin immunoprecipitation identifies protein–DNA interactions in living cells. Chromatin immunoprecipitation (ChIP) experiments measure the occupancy of a *cis*-acting DNA site by its regulatory protein in living cells, and, in combination with microarrays, can be used in a single experiment to examine *all* the sites, genome-wide, occupied by a protein. To perform ChIP, cells are treated with a low concentration of formaldehyde, which cross-links proteins to proteins and proteins to DNA. After cross-linking, the mixture is sheared to produce DNA fragments ~500-bp long, and an antibody is added that recognizes a particular DNA-binding protein, such as a transcription factor. Centrifugation of the mixture separates the antibody and the protein–DNA complexes (in the pellet) from the other DNA fragments (in the supernatant). DNA recovered from the pellet is analyzed by real-time PCR, by sequencing, or by microarray hybridization. For the latter, a "ChIP-chip" experiment, the DNA is hybridized to a microarray with oligonucleotides that cover the genome or selected parts of the genome, such as transcriptional promoters. DNA fragments found in high abundance in the pellet compared to the supernatant identify sites that were occupied by the factor in the cell.

strategy is called "ChIP-chip," but other methods, such as selected quantitative PCR or high-throughput DNA sequencing, can also be used to determine which regions of the genome are occupied in the cell.

How does one identify which DNA fragments were bound by the protein in a ChIP experiment? If the goal of the experiment is to confirm or further study a few specific sites in the genome that are known or suspected to bind a protein, then quantitative PCR (QPCR) assays can be used to measure the enrichment of an immunoprecipitated fragment compared to unbound DNA fragments. However, a much more efficient approach is to hybridize the DNA fragments enriched by immunoprecipitation to a microarray that contains DNA fragments representing all or a part of the genome of the organism being studied. This ChIP-chip technique is particularly powerful—it tests every fragment in an entire genome for its binding of a particular protein—and provides a complete picture of where a transcription factor acts under a specified set of conditions. The microarrays need not contain the entire genome sequence; the set of oligonucleotides or DNA fragments that represent a short sequence from every few hundred base pairs is sufficient to allow detection of the sheared immunoprecipitated fragments. A small genome such as that of yeast can easily be analyzed on a single microarray in a ChIP-chip experiment.

In yeast, a well-studied DNA-binding protein called Rap1 is known to regulate the expression of genes involved in a variety of important cellular processes. These include the genes that determine the mating type, ribosomal genes, genes involved in metabolism, and sequences involved in controlling transcription at telomeres. The factor has both activating and repressing activities depending on the gene it is regulating. ChIP experiments were performed with anti-Rap1 antibody and microarrays containing PCR products of DNA fragments spaced on average every 2000 base pairs throughout the entire yeast genome. The experiments identified about 300 sites in the yeast genome where Rap1 binds, the majority of which were not known from previous biochemical or genetic studies (Fig. 13-9). Thus, in a simple set of experiments, the researchers were able to scan the entire yeast genome to identify the locations of many binding sites for this transcription factor.

In a very short time, the methods for analyzing immunoprecipitated DNA fragments and for using ChIP technology have become more sophisticated and are now in wide use. New, cheaper, and faster sequencing methods are being applied to ChIP-enriched DNA fragments, and it is possible in the not-distant future that analysis will shift from microarrays to sequencing. Regardless of how the immunoprecipitated fragments are analyzed, it is already possible to screen a large genome, such as that of humans, throughout most of its length for binding of specific factors. An almost unlimited number of ChIP-type experiments can be envisioned, given the large numbers of DNA-binding proteins, the hundreds of cell types that make up complex organisms, and the enormous range of physiological conditions that can be tested.

ChIP and Other Genome-wide Methods Can Be Used to Assay Modifications in the Structure of Chromatin in Living Cells

Sequence-specific DNA-binding proteins are only one element of transcriptional control. Modifications of chromatin structure, including chemical changes of specific amino acids of histones and methylation of DNA, also play a significant role in gene expression. Determining when and where these modifications occur throughout the genome is essential if we are to have a thorough understanding of gene control. Early experiments examining chromatin modifications were typically performed one gene at a time, and progress toward a global analysis was slow. Now, ChIP is used to analyze the state of chromatin proteins and DNA methylation at all positions in a genome.

Three classes of chromatin modifications are known to be biologically important (Fig. 13-10). The first are chemical changes, such as acetylation and methylation of amino acids in the carboxy-terminal tails of histones, that lead to activation or repression of transcription. A second type of modification is methylation of DNA at cytosines in CpG dinucleotides, which are enriched near the transcriptional start sites of a large number of genes in vertebrates. Methylation of the cytosines in many of these CpG dinucleotides is associated with repression of transcription. A third class of chromatin modification is detected by the degree of sensitivity of DNA to various nucleases. It has been known for many years that when gently isolated nuclei are subjected to nuclease treatment, specific patterns of DNA digestion occur. Genes that are active are sensitive to modest concentrations of DNase throughout their transcribed length and are cut into small fragments. Regions where key regulatory events occur are cleaved at specific sites when treated with very low concentrations

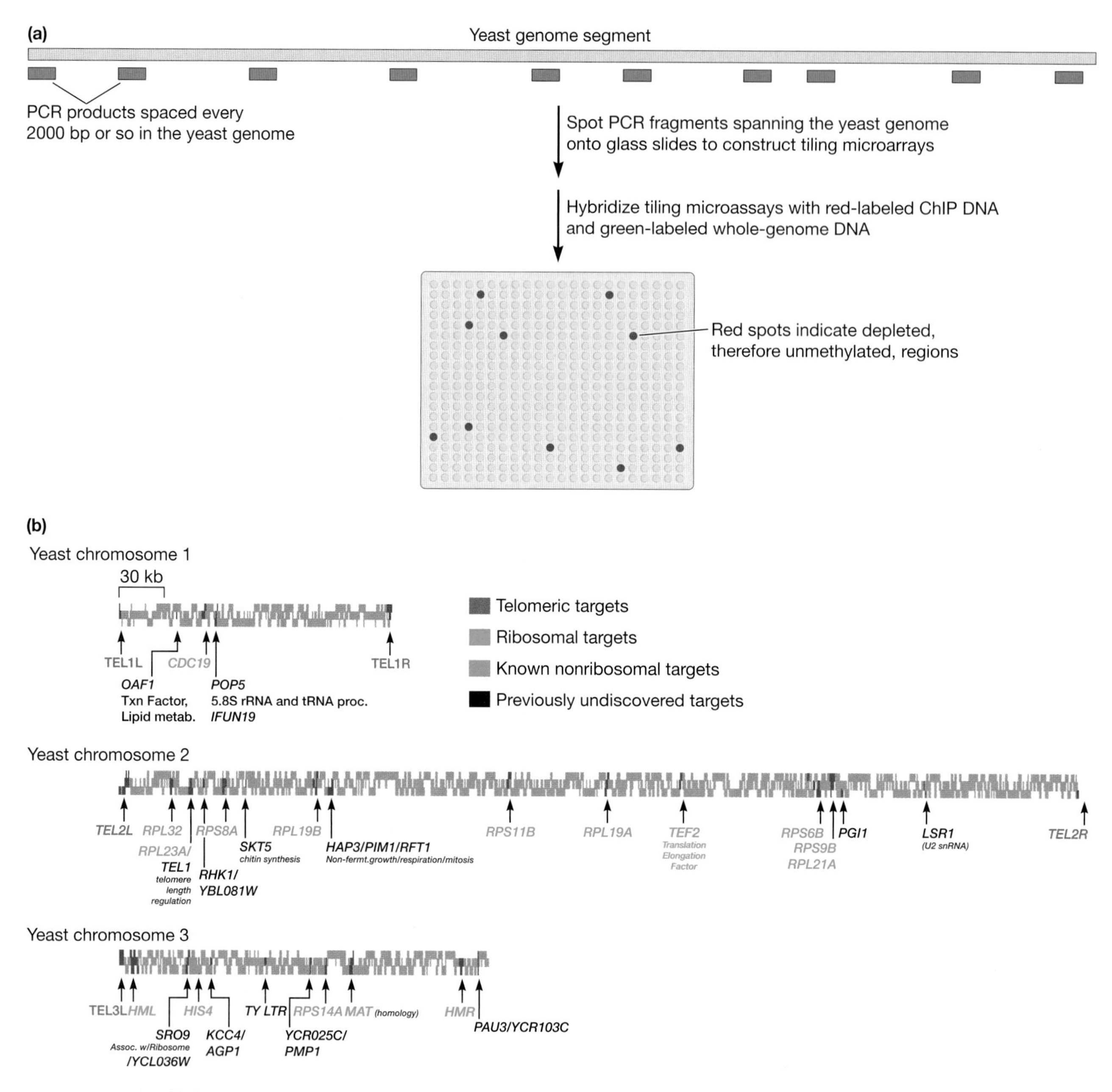

FIGURE 13-9

ChIP-chip experiment identifies new binding sites for a transcription factor. Chromatin immunoprecipitation followed by hybridization to a microarray (ChIP-chip) identifies nearly every binding site for Rap1, an important transcription factor in the yeast *Saccharomyces cerevisiae*. (*a*) A "tiling" microarray is made that contains DNA fragments spaced every 2000 base pairs throughout the yeast genome. These microarrays are hybridized to DNA fragments that were immunoprecipitated from chromatin with an anti-Rap1 antibody. Microarray fragments that produced higher signals in the ChIP samples compared to a sample containing the whole genome identified those regions of the genome enriched by virtue of Rap1 occupying the site in the living cells. (*b*) Rap1 occupancy map in the yeast genome determined by ChIP-chip. By analyzing hybridization data from the tiling microarrays, a genome-wide occupancy map for Rap1 was produced. The figure shows the results for yeast chromosomes 1, 2, and 3. The color coding of the gene names shows the types of targets that were occupied. (*Red blocks* indicate that this was identified on the microarray.)

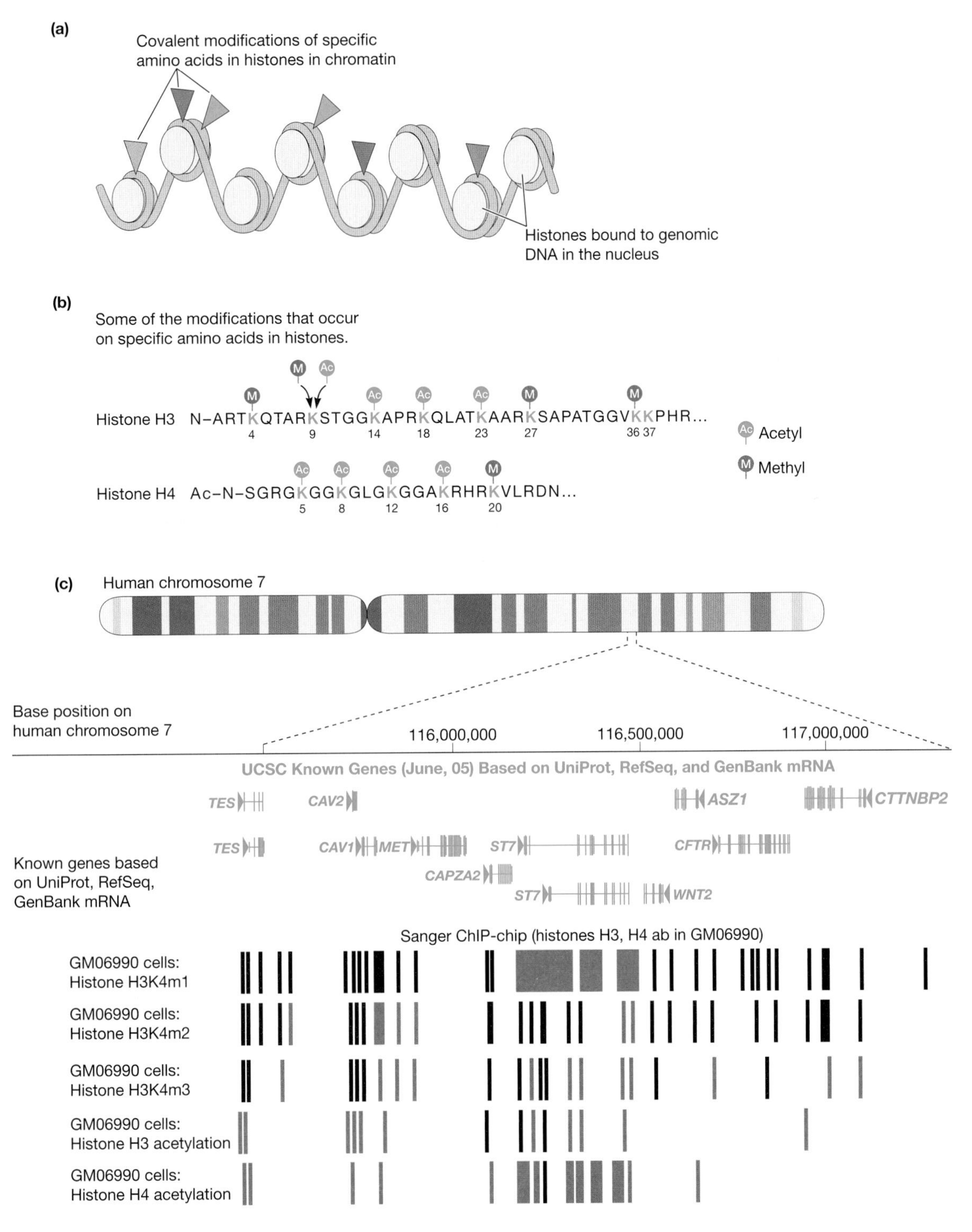

FIGURE 13-10

The ChIP technique is used to measure modifications of histones in chromatin in living cells. (*a*) Histone octamers bound to genomic DNA in the classical "beads on a string" model of chromatin. Each complex of a histone core and the DNA wrapped around it constitutes a nucleosome. The *colored arrowheads* indicate covalent modifications that can occur on specific amino acid residues. (*b*) Acetylation and methylation occur at specific amino acids in human histones H3 and H4. The *purple circles* show methylation sites, and the *blue circles* show acetylation sites. (*c*) A ChIP-chip experiment examined chromatin modifications in the human lymphoblastoid cell line GM06900. A 2-Mb region of human chromosome 7 with nine genes is pictured in detail, with the *vertical blue lines* representing exons and *arrowheads* representing transcription start sites. The data are shown for experiments measuring five different histone modifications in this region: mono-, di-, and trimethylation on lysine 4 (K4) of histone H3, and acetylation of histone H3 and H4. The bands are signals showing where the modifications occurred; the darker the band, the more those DNA fragments were enriched by ChIP. Most of these chromatin modifications are localized to genes.

of DNase; these are said to be *hypersensitive* sites. The reasons for DNase sensitivity and hypersensitivity are not completely understood, although they are thought to be related to the relaxation of chromatin that must occur before and during transcription.

ChIP techniques can be used to study all three classes of chromatin modification. With ChIP-chip or ChIP-QPCR experiments, antibodies that recognize each of the specific histone amino acids in their modified and unmodified state are used to determine the status of the chromatin at each position. The measurement of DNA methylation throughout the genome is even easier, as no antibodies are needed. Genomic DNA is isolated and cut to completion with a mixture of six restriction enzymes that cleave only unmethylated CpG bases. The unmethylated CpG-rich regions are cut into pieces so tiny that they can be eliminated by a simple column chromatography step. The remaining DNA is hybridized to a microarray containing DNA fragments spanning the entire genome, and the regions of unmethylated CpGs, depleted by the restriction enzyme digestion, are easily identified (Fig. 13-11). Similar approaches can be used to profile sites of DNase hypersensitivi-

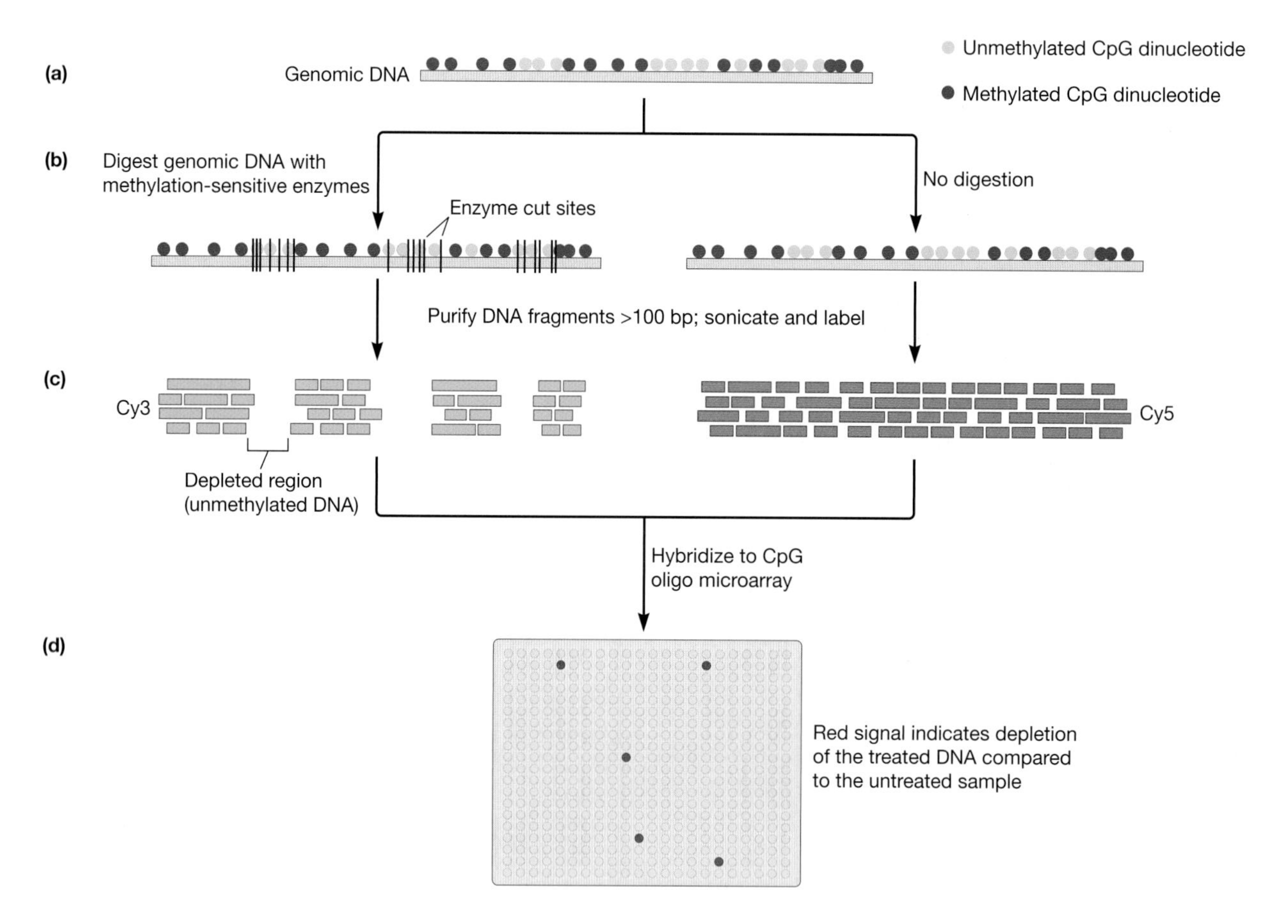

FIGURE 13-11

DNA methylation can be measured on a genome-wide scale. The approach shown here for measuring methylation at CpG dinucleotides is carried out using a depletion method, in which sites that are unmethylated are destroyed and the remaining sequences are measured. (*a*) Genomic DNA. *Purple circles* represent methylated CpG dinucleotides, and the *light blue circles* represent unmethylated CpGs. (*b*) Genomic DNA is separated into two samples. One is treated with a cocktail of restriction enzymes that specifically digest unmethylated CpG-rich regions into very small fragments that are lost upon purification of the DNA sample. (*c*) Both nuclease-digested and untreated DNA samples are sonicated and aliquots of the resulting fragments in each sample are differentially labeled. An aliquot of digested DNA is labeled with Cy3, an untreated genomic DNA aliquot is labeled with Cy5, and when the two are competitively hybridized to a microarray, the unmethylated CpG-rich regions can be identified. (*d*) The signals corresponding to these sequences are red on the microarray, because the undigested DNA sample is overrepresented as a result of the depletion in the digested sample. If the microarray includes all CpG-rich regions from a genome, nearly all unmethylated CpG-rich regions (associated with transcriptional activation) may be identified for the experimental condition under study.

ty throughout the genome. These experiments have provided highly detailed pictures of the wide variety of changes in chromatin on a genome-wide scale.

Reporter Plasmids Are Used to Measure Regulatory Elements on a Large Scale

Since the 1970s, transient DNA transfection has been widely used for studying both the expression and functions of genes (Chapter 6). Typically plasmid DNA containing a human gene and its transcriptional promoter is delivered into cells growing in tissue culture, with DNA uptake facilitated by one of a variety of chemical agents. A small fraction of the DNA molecules are taken up and make their way into the nucleus where the cloned gene is transcribed. Much has been learned about promoter and enhancer function from these transient transfection experiments; however, the results were limited in scope as the experiments were typically performed with one gene at a time with cells grown in large tissue-culture plates. As with so many other techniques, transient transfections have become genomics based and are now performed on hundreds or thousands of DNA samples at a time. In one example, cultured cells were grown in the wells of microtiter plates, and each well was transfected with a plasmid containing a promoter from a human gene driving the expression of a reporter gene, such as that for luciferase (Fig. 13-12). The cells were grown for two days after transfection to allow expression to occur, and the level of chemiluminescence—an indicator of transcription of the reporter gene—was measured in each well simultaneously. With appropriate controls, this method can be used to test unknown promoters, confirm and quantitate known promoters, and dissect the functional elements within promoters. One such experiment showed that about 10% of human promoters are "bidirectional," that is, they transcribe divergent genes facing in opposite directions.

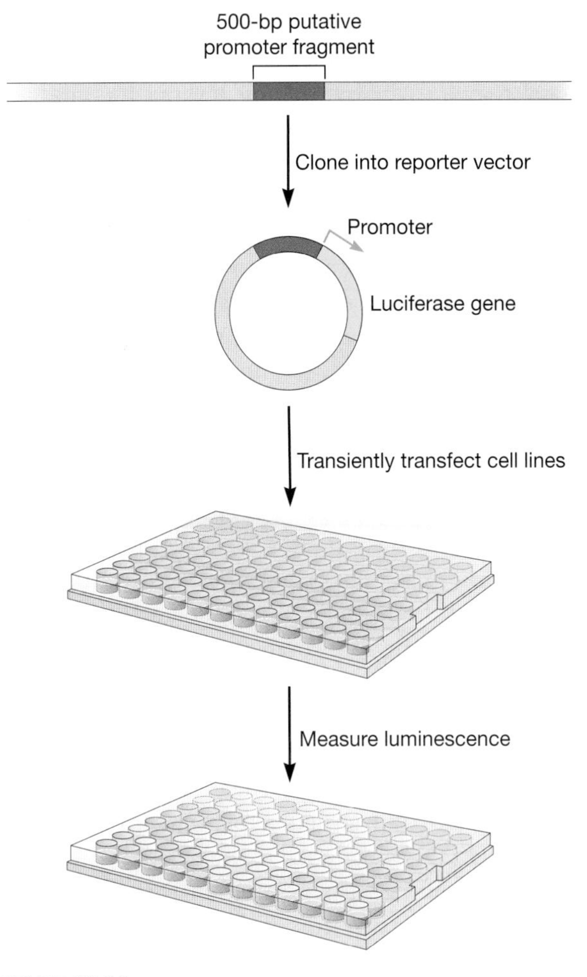

FIGURE 13-12

Reporter assays in living cells measure promoter activity. Scientists have modified the techniques involving the transfection of reporter constructs so as to measure the activity of hundreds to thousands of promoters simultaneously. Each promoter is cloned upstream of a luciferase reporter cassette on a plasmid. Each plasmid is then transfected into a pool of living cells and incubated to allow expression. After addition of a luciferase substrate, the amount of visible light coming from the cells is proportional to the transcriptional activity of the promoter. By conducting cloning, transfection, and luminescence assays in a microwell format, this process can be used to study many promoters in a single experiment.

Knocking Out All the Genes in the Yeast Genome

Mutational analysis continues to be the fundamental experimental method for determining the importance of proteins and DNA sequences. In the 1980s a major advance occurred with the invention of methods for targeting mutations to specific genes. During the following decades, mutational analysis of genes in yeast, mice, and cultured cells led to a more complete understanding of the functions and workings of many genes.

Having complete genomic sequences and knowledge of all or most of the genes of an organism makes possible the extraordinary goal of generating knockout mutations in every gene, a goal already achieved for the yeast *S. cerevisiae*. Yeast has a small genome, a modest number of genes, and is particularly amenable to efficient gene targeting. A consortium of yeast biologists used gene targeting to generate a collection of yeast strains, each strain being deleted for one gene and replaced by a selectable marker. In each case, the marker is flanked by TAGS (also called molecular bar codes), short unique segments of DNA, detectable by hybridization or PCR.

With such a resource in hand, it has become possible to undertake a systematic analysis of gene function. For example, determining which strains grow as homozygous diploids when no functional copy of the gene is present identifies which genes are crucial for normal function. This type of experiment showed that about one-third of the yeast genes are essential for survival. By testing growth under a wide variety of growth conditions, it was possible to determine which genes are important for various metabolic functions. In one example, 4706 strains carrying homozygous deletions were mixed—in equal numbers—and grown under conditions that do not allow fermentation (i.e., they are required to utilize glycerol, lactate, or ethanol). Mutants that have defects in respiration grow poorly or die under these conditions. Because the targeting plasmid for each gene integrated a DNA bar code at the deleted gene site in the yeast genome, it is possible in this type of "fitness" experiment to determine quickly—by hybridization to a microarray that contains sequences complementary to each of the bar codes—which yeast strains are depleted after the competitive growth step (Fig. 13-13). In this single experiment, 400 of the yeast deletion strains were found to be depleted when grown under conditions requiring respiration, which more than doubled the number of yeast proteins known to participate in respiration. This approach has been used to identify new genes important for biological processes at a much higher rate than can be done with nongenomic approaches.

In a remarkably ambitious effort, experiments are under way to produce strains that have a disruption in every gene. The generation of targeted mutations is much more difficult in mouse than in yeast; the targeting must be done in embryonic stem (ES) cells, and mice have to be generated from these ES cells. Furthermore, experiments performed with these mice require much more work than do those with yeast; for instance, the competitive growth strategy used in yeast is not applicable to complex organisms like the mouse. Nevertheless, having a catalog and repository of mouse strains with disruptions in all or almost all of the genes will be an indispensable tool for researchers studying gene function in mammals.

The Yeast Two-Hybrid System Is Used to Identify Protein–Protein Interactions on a Genome-wide Scale

The yeast two-hybrid system is a powerful assay for identifying protein–protein interactions (Chapter 3). Early versions of the method tested one protein (the "bait") against all or most of the other proteins in a cell for interactions. The availability of complete genome sequences, particularly those with well-annotated coding sequences, makes it possible to use the yeast two-hybrid system to screen for *all* possible interactions between the proteins in an organism rather than to test one bait at a time. The open reading frames (ORFs) from an organism's genome are cloned into two plasmid vectors—one that expresses the bait and another that produces the prey—and each is introduced into yeast cells by transfection. A high-throughput mating method is then used to introduce each bait plasmid into yeast cells with each prey plasmid, and the hybrids are screened for expression of the reporter gene. This is a tremendous number of assays; a comprehensive screen of all 6000 yeast genes would require more than 36 million tests for interactions. Although no study of this scale has yet been carried out, "interactome" maps have been produced for thousands of proteins in yeast, *Caenorhabditis elegans*, mouse, and human beings. These have led to many new insights about proteins whose functions are not known, as well as new interactions for even well-studied proteins. When an unknown protein is found to interact with other proteins that are part of a biochemical pathway, the unknown protein is then implicated in that process—"function by association" (Fig. 13-14).

Protein Arrays Also Reveal Interactions between Proteins

Even though the yeast two-hybrid system has been extremely useful for identifying which proteins interact with each other, it has some shortcomings. For example, the assay is based on transcription, so the bait and prey proteins must enter the nucleus and interact in a cellular location very different from their normal environment. An alternative approach is to use microarrays to screen for protein–protein interactions. In this method, all the ORFs from the yeast genome are used to express each yeast protein tagged with a glutathione-*S*-transferase (GST) epi-

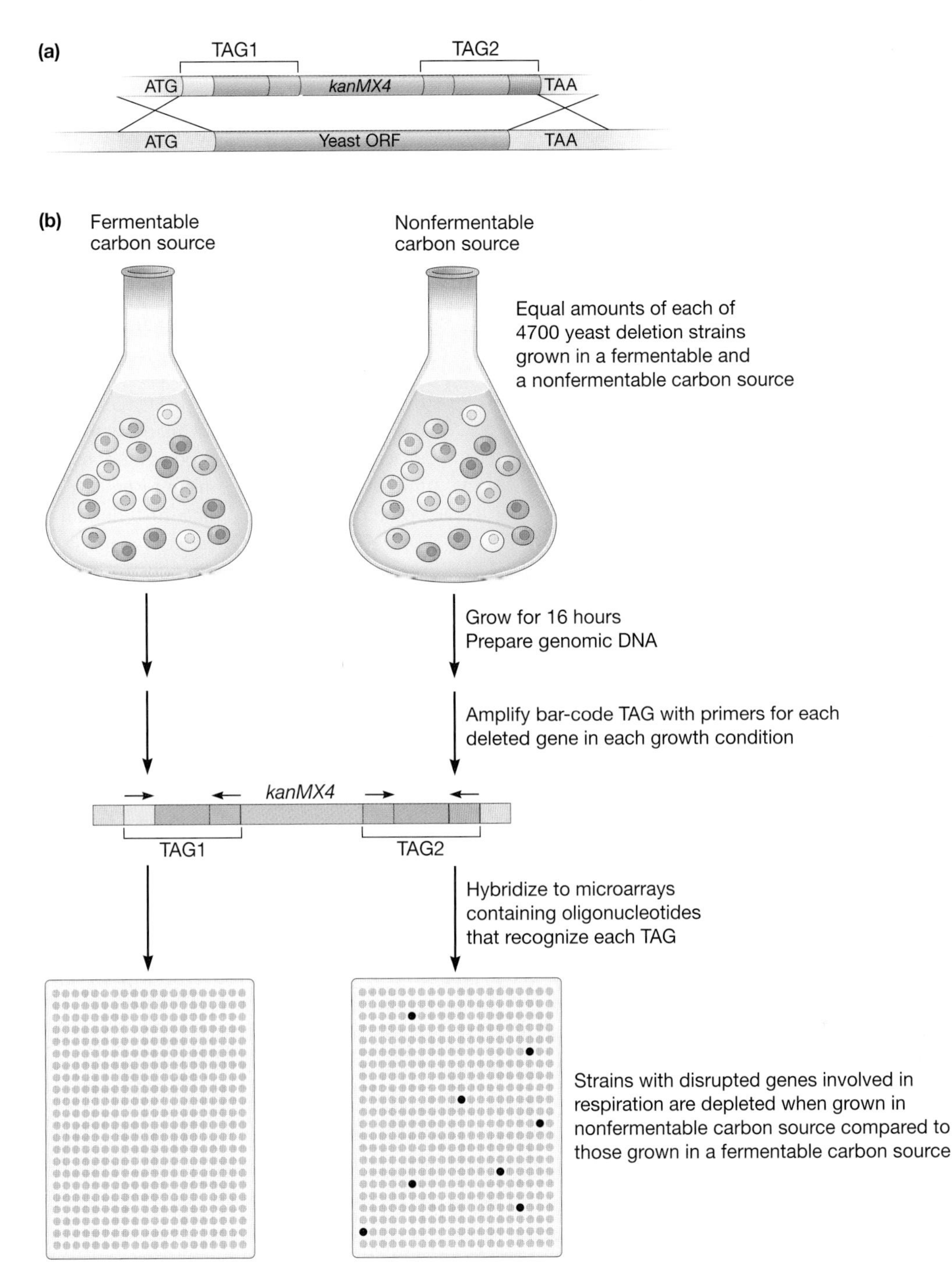

FIGURE 13-13

A functional genomic experiment identifies the genes that are important for aerobic respiration in yeast. (*a*) A collection of yeast strains, each representing a targeted knockout of a single gene, was generated for 4700 of ~6000 genes in the yeast genome. Each strain was generated by homologous recombination of a segment of DNA that removes the coding region of the gene and replaces it with a cassette containing a selectable marker (*kanMX4*) and flanked by two different TAGs. Each gene has its own pair of TAGs; each TAG is a unique sequence that will hybridize to a specific oligonucleotide on a microarray and is flanked by PCR primer sites. (*b*) An experiment to determine which genes among the 4700 in the deletion strain set play an important role in respiration. Equal amounts of each of the deletion yeast strains are grown together in a fermentable carbon source and a nonfermentable carbon source. After 16 hr of growth, genomic DNA is prepared from each pool of cells, and each is amplified with 4700 sets of primers that amplify the two TAGs at each deleted gene. One of the PCR primers is labeled with a fluorescent dye. The mixture is then hybridized to a microarray that has oligonucleotides complementary to each TAG. The amount of signal at each spot on the microarray is an indication of the amount of each yeast deletion strain that is remaining after the 16 hr of growth. Strains showing depletion, as seen on the *right-hand* microarray for the NF growth condition, indicate that these genes are required for respiration. This experiment identified 466 yeast genes with a role in respiration, of which 265 had not been previously shown to be involved in the process.

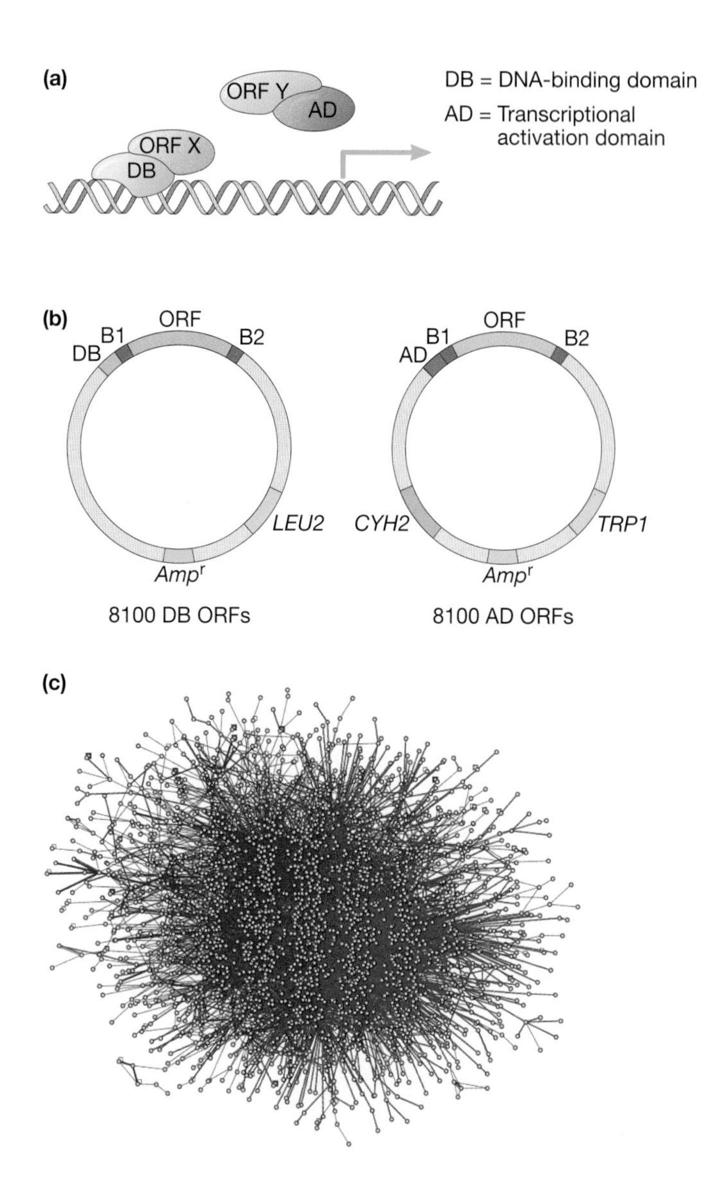

FIGURE 13-14

High-throughput yeast two-hybrid screens can be used to build a human "interactome" map. (*a*) A library is made containing all the open reading frames (ORFs) from the organism (the human genome, in this case) in a "prey" vector that makes fusion proteins between the ORFs and a transcriptional activating domain (AD). Unlike the standard method, the "bait" is not made with a single fusion protein, but instead is made as a library containing all the ORFs from the organism fused to a DNA binding domain (DB). (*b*) For the first human interactome map, 8100 human ORFs were made in both the AD and the DB fusion protein constructs. To perform the high-throughput screen, automated methods were used to mate pools of 188 yeast strains carrying the AD ORFs with each of the 8100 DB ORF yeast strains. Each plamid carries a nutrient selectable marker, *LEU2* or *TRP1*. Positive pools were deconvoluted and individual pairwise interactions were confirmed by additional methods. This experiment tested 66 million pairwise combinations. (*c*) A human protein interaction map based on the high-throughput yeast two-hybrid screen combined with large numbers of interactions from the literature. Proteins are shown as *yellow dots*, and interactions are shown as *red* (from yeast two-hybrid screening) and *purple* (from literature information) lines. Maps from such experiments contain extremely large numbers of data points, and the databases containing the data allow researchers to zoom in on small regions of the network map to examine particular protein interactions in detail.

tope. As GST is a bacterial enzyme, antibodies directed against it are unlikely to cross-react with proteins in the host cell, thereby allowing purification of the GST-tagged proteins by high-throughput chromatography. The purified proteins are spotted onto glass slides to generate protein microarrays, and the protein under investigation is labeled and added to the array under gentle conditions that allow the proteins to interact. The spots on the microarrays are then analyzed for the intensity of the signal from the labeled interacting test protein (Fig. 13-15). It is possible to screen many different test proteins for thousands of possible interactions in just a few experiments. In addition to identifying interactions and providing new insights into the functions of a test protein, this type of study can sometimes reveal details about which amino acids in the proteins are responsible for the interactions. For example, the set of proteins on the array that interact with the labeled test protein may have a stretch of amino acids in common, thus indicating a candidate motif for the interaction domain (Fig. 13-15).

Mass Spectrophotometry Is Used to Identify Proteins in Complex Samples

Each "hit" in a yeast two-hybrid or protein microarray experiment detects an interaction between two proteins, and several pairwise interactions can be measured. However, in the cell, most proteins interact with each other in multiprotein complexes; the best known of these are the ribosome and the spliceosome. The study of these complexes requires a combination of molecular and biochemical experiments,

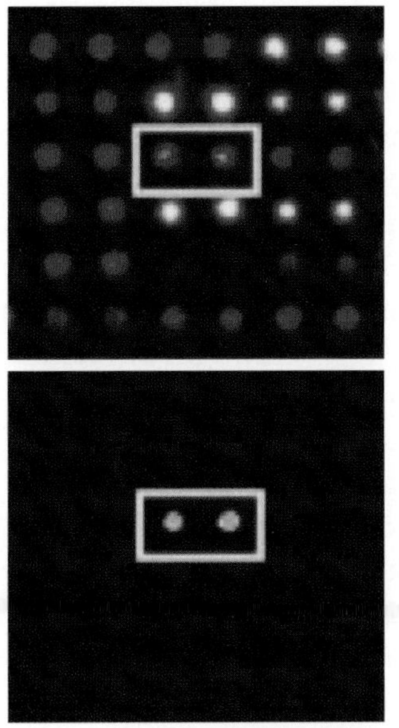

(c)

Protein	Sequence	Position
MSH4	LKET LQSVKSLKD ALND	390
OPI3	HSVD LQSSKFQLA IVCT	28
REG2	DEHF IQRLPSTRL NSTD	196
SPS19	AKIP LQRLGSTRD IAES	246
CMK2	DDLR LQSQKKGGE LTEE	395
IPP1	LNPI IQDTKKGKL RFVR	80
RPL26B	RKAL IQR.KGGKL E...	129
RPN11	VVDP IQSVKGKVV IDAF	154
RPB3	GHPI IQD.KEGNG VLIC	143
PUS2	RVWG IQPVNKKFN ARSA	103
SRP101	LLRE IQSKRSKKD EEGK	388
CMK1	LNMK IQKLRDLYL EQTE	346
MET8	DLFG IQHCHN.ID VKRL	242
MYO4	NGLL IQSSKFISK VLLT	1167

I/LQXK(K/X)GB

14
7
0

Motif consensus sequence (X=any residue; B=basic residue)

FIGURE 13-15

Protein microarrays can be used to identify which proteins interact with a labeled query protein or small molecule. (*a*) A portion of a microarray containing 6566 spotted proteins from yeast cells is shown. In the top part of the panel, a portion of the array is stained with anti-GST (glutathione-*S*-transferase) antibody, which detects every protein on the array and is used as a control to indicate the relative mass of each protein on the array. (*b*) In the *bottom* panel, the array has been probed with labeled calmodulin, a protein that also binds calcium and is involved in modulating multiple cellular processes. The two green signals are from duplicates of the protein spotted at those positions. The arrays can also be probed with labeled small molecules, such as signaling molecules or cofactors, to determine which proteins bind them. (*c*) Because the yeast genome sequence and almost all of protein-coding genes have been identified, the amino acid sequences of each protein on the microarray are known. By comparing the sequences of all the calmodulin-binding proteins with each other, a segment, or motif, of amino acids of similar sequence was identified. Fourteen proteins with this motif are shown. The panel shows the consensus amino acid sequence with the size of the letter indicating the relative frequency with which it appears in the motif. The *red Q* (glutamine) is present in all the proteins, and the *dark blue* and *purple* amino acids are present in many of them. Note that the spacing of the amino acids in the consensus is the same in all the proteins.

mass spectrometry, and genomic sequence analysis.

An expression plasmid is constructed that contains the gene for a "bait" protein connected to an epitope tag such as a small portion of the myc protein. This fusion protein is expressed in the cell of interest and affinity chromatography is used to purify it, along with the other proteins that interact with it (Fig. 13-16). The purified complex, which may contain dozens of proteins, is subjected to electrophoresis through a denaturing protein gel, and the separated protein bands are extracted from the gel and cleaved into small peptide fragments with a protease. The fragments are analyzed by high-throughput mass spectrometry, a technique that determines with absolute accuracy the molecular weights of all the fragments in the mixture. If the complete genome sequence for the organism is available, and particularly if the genome is well annotated for its entire protein-coding component, it is possible to assign many of the molecular weight peaks in the mass spectrometer readout to specific proteins encoded by the genome. Because multiple peptide fragments are analyzed for each protein, the ambiguous data—molecular weight peaks that cannot be uniquely assigned to a protein—can be discarded, thereby leaving enough information to determine with high confidence which proteins are present in the complex.

An example showing the power of this approach analyzed the polyadenylation complex in yeast. This is a protein machine that cleaves pre-mRNAs and adds a stretch of adenosine bases to the 3′ end of messages in eukaryotic cells. By applying the mass spectrometry experiment, as described above, researchers identified several new proteins that had not been previously implicated in polyadenylation (Fig. 13-16).

The application of large-scale, high-throughput techniques to proteins has greatly increased our knowledge of the network of interactions that occur in many complex cellular processes. This approach represents the emerging field of proteomics, which has the goal of understanding the entire protein complement (their amounts, locations, interactions, and even activities) of an organism's cells.

Arrayed Antibodies Are Used to Measure Protein Levels in Cells

Comparing the relative levels of stable mRNAs in different cell types or under different conditions is very informative, but mRNA levels do not necessarily correspond to the abundance of the proteins they encode. There are posttranscriptional mechanisms that regulate the translation of an mRNA into protein, as well as the splicing, stability, and location of the mRNA in the cell. Although still in development, microarray-based techniques can be used to compare the relative levels of a large number of proteins in a cell or tissue. In one approach, high-throughput methods are used to produce antibodies that recognize specific regions of a large number of different proteins. These antibodies are spotted onto microarrays in such a way that they are able to recognize and bind their target proteins. Protein extracts from two cells being compared are labeled, each with a different fluorescent dye. Equal amounts of the two labeled extracts are mixed and the mixture is added to the antibody microarray. The comparative levels of each protein from the two cell samples are determined by measuring the relative amounts of the two different fluorescent signals at each of the antibody positions on the microarray, in the same way that mRNA levels are measured in two-color microarrays. Many questions can be answered with this type of analysis. For example, a comparison of stable mRNA levels determined using cDNA microarrays, with the protein levels determined with antibody microarrays in the same cell, can help establish relationships between synthesis, processing, and stability of mRNAs and proteins.

Determining the Locations of Proteins in Cells and Tissues

Knowing all the proteins of the cell, the quantities of each, and which protein interacts with which is not sufficient for a complete description of the role of proteins in the cell. What is also needed is knowledge of the distribution of proteins in the cell and of which cells in a tissue express each protein. The most commonly used method for protein localization is immunocytochemistry—using a fluorescently labeled antibody that recognizes a single protein to locate that protein in cells in tissue culture or in sections of tissue. The labeled antibody is detected by fluorescence microscopy. This method is slow, even for the analysis of one protein, because making and testing antibodies, the experiments themselves, and the data analysis all require much effort.

Two approaches have been developed to speed up this process by using the high-throughput data gener-

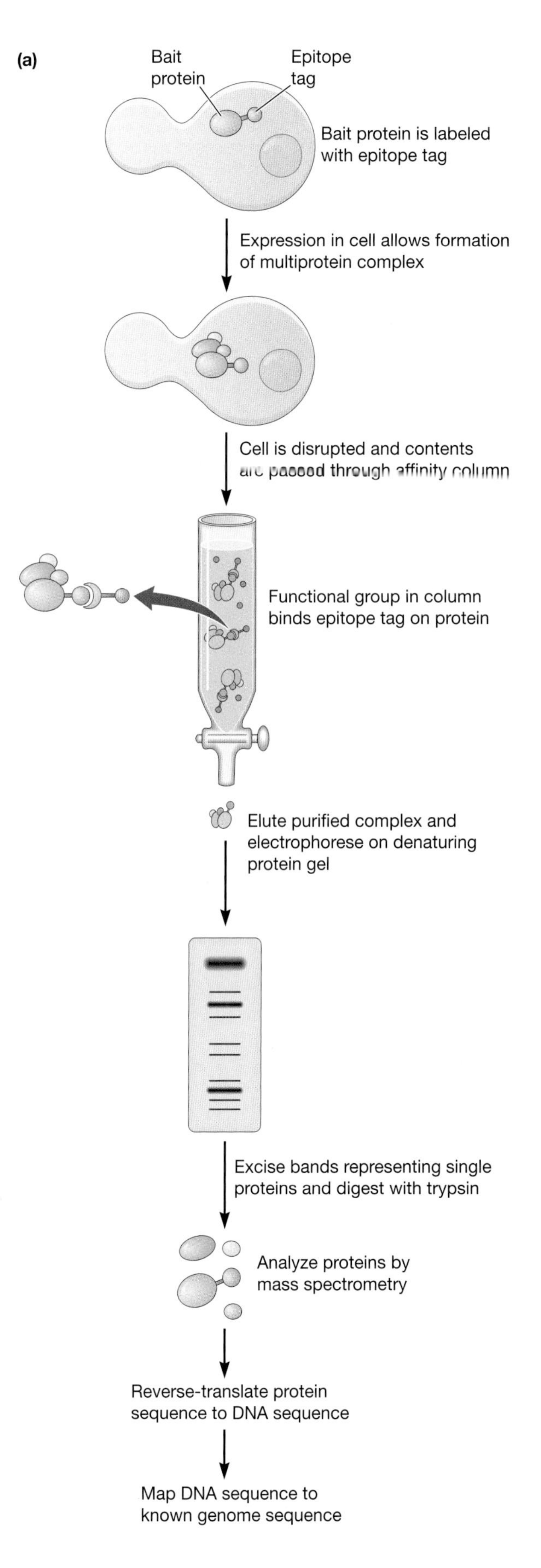

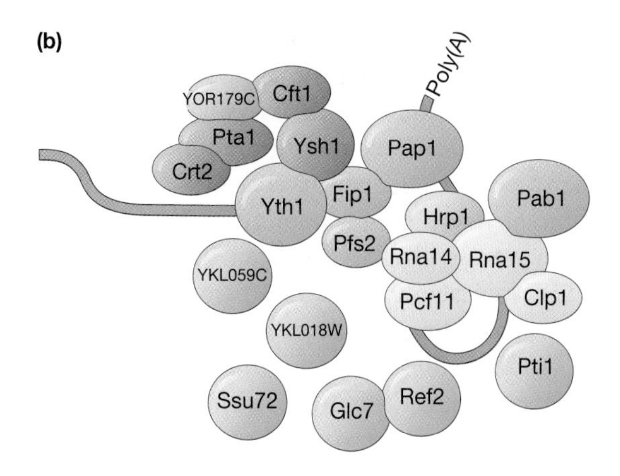

FIGURE 13-16

Affinity chromatography is used to purify protein complexes for analysis by mass spectrometry. Many important proteins function as multiprotein complexes in the cell. To isolate and analyze the contents of these complexes, affinity chromatography and mass spectrometry can be used. (*a*) One member of a protein complex, called the "bait," is labeled with an epitope tag and expressed in a living cell, which allows the multiprotein complex to form. The cell is disrupted, and the contents are passed over an affinity column that contains a functional group that binds the epitope tag on the bait protein. The tagged protein and the other proteins in the complex bind to the affinity column, whereas the other cell contents pass through. The purified complex is subjected to electrophoresis on a denaturing protein gel, and bands representing single proteins are excised, digested into small fragments, and analyzed by mass spectrometry. The segments of the amino acid sequences of the protein fragments are reverse-translated to DNA sequence, and these sequences are mapped to the known genome sequence. (*b*) The use of the method in a identified multiple components of the yeast poly(A) addition machinery. This protein complex binds to pre-mRNAs near the 3´ end at a specific RNA sequence and conducts the cleavage reaction of the mRNA followed by polymerization of poly(A) onto the cleaved end. This proteomic experiment confirmed several proteins known to be involved in the process, but also identified many unknown proteins (*blue*), thereby increasing the knowledge of the components important in the mechanism of poly(A) addition.

ation and analysis that typifies the methods described in this chapter. In one, the antibodies generated to measure protein levels with arrays have been used for immunocytochemistry of large numbers of cells and tissue types in "tissue microarrays" (Fig. 13-17). Pathology tissue samples are typically embedded in paraffin wax so that sections can be cut for staining and histological examination. A tissue array is made by taking small (0.6-mm-diameter) cores from selected places in the embedded tissue and arraying them in a new block. Thin sections are cut from this block and each section is probed with an antibody. As each block can contain as many as 1000 cores, and a single block can provide 200 tissue arrays, the distribution of multiple proteins can be examined in many samples from, for example, a single tumor, or compared between different tumor types. Hundreds of slices are probed with antibodies in a single test, and the systematic analysis of the subcellular and tissue locations of the proteins can provide inferences about functions of proteins.

A method can be used for protein localization that does not rely on antibodies. In this approach, each gene is cloned into an expression plasmid so that the encoded protein is synthesized as a fusion protein with the green fluorescent protein (GFP) at its carboxyl or amino terminus (Fig. 13-18). In yeast, the fusion proteins were expressed under the control of their natural transcriptional elements, and the subcellular localization of each was determined by rapid fluorescence microscopy, an automated process to detect GFP fluorescence with systematic data collection. These experiments can be done quickly because the time taken to make, apply, and detect antibodies is eliminated. Application of this approach to 1500 yeast genes increased our information about localization of proteins by more than 30% (Fig. 13-18).

Although the field of genomics is relatively new, many of its innovations—large-scale experiments, use of automated techniques, and the application of intensive computational analyses—have become commonplace. Similar approaches are beginning to be applied to characterize all the proteins of the cell and their interactions with other cell components. These large-scale approaches would not be possible without the sequences of genomes and mRNAs, and because biological systems are so complex, it is likely that many additional clever approaches will be developed in the coming years to study functions of genes and proteins on a genome-wide scale.

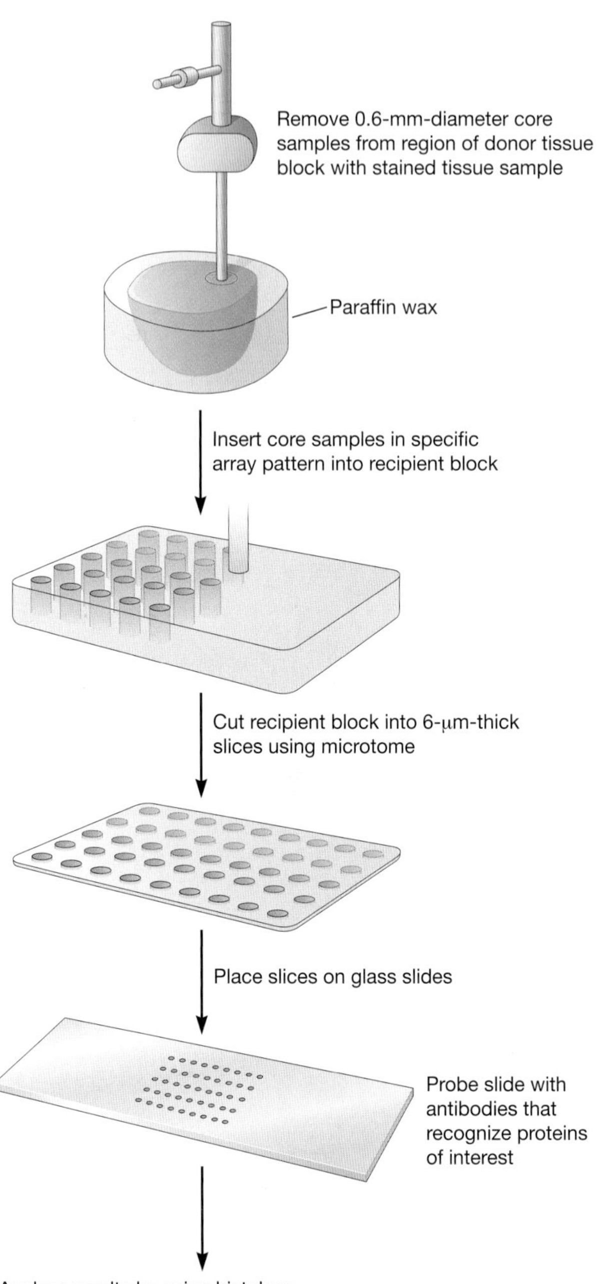

FIGURE 13-17

Probing "tissue microarrays" with antibodies identifies the location of proteins in tissues and cell types. Tissue microarrays are constructed by taking "core samples," typically a few hundred micrometers in diameter from a tissue sample. The cores are then inserted in a wax block at defined, arrayed positions. The array is then cut into very thin slices with a microtome, placed onto glass slides, and fixed. The slides are then probed with antibodies that recognize specific proteins of interest. The locations of many proteins throughout a tissue can be analyzed much more quickly and precisely than with conventional methods.

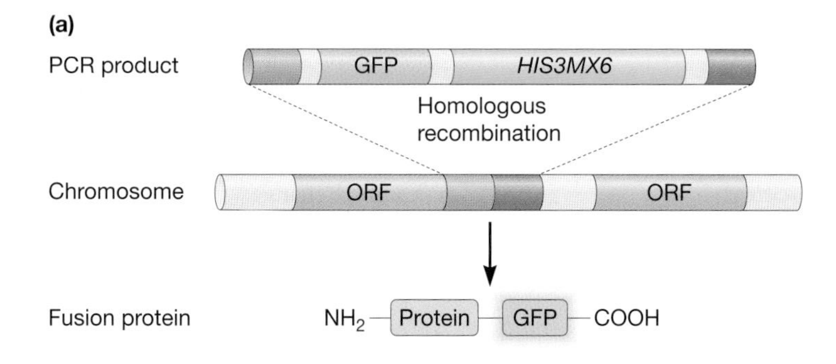

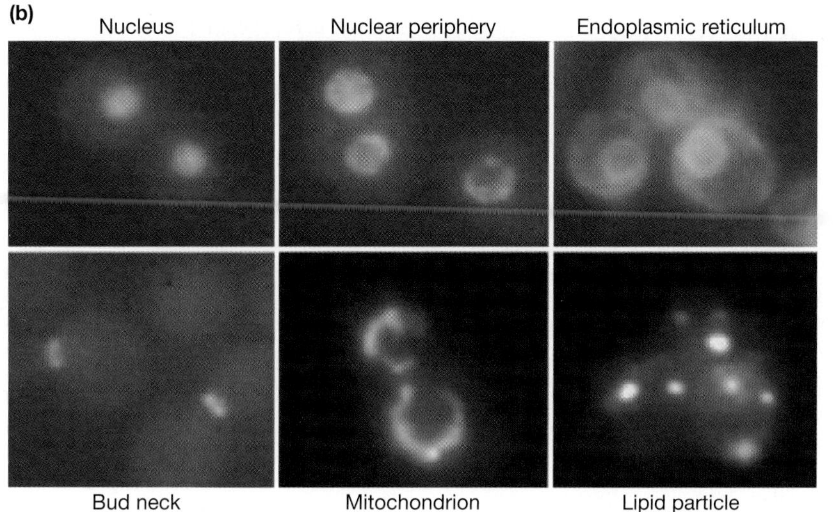

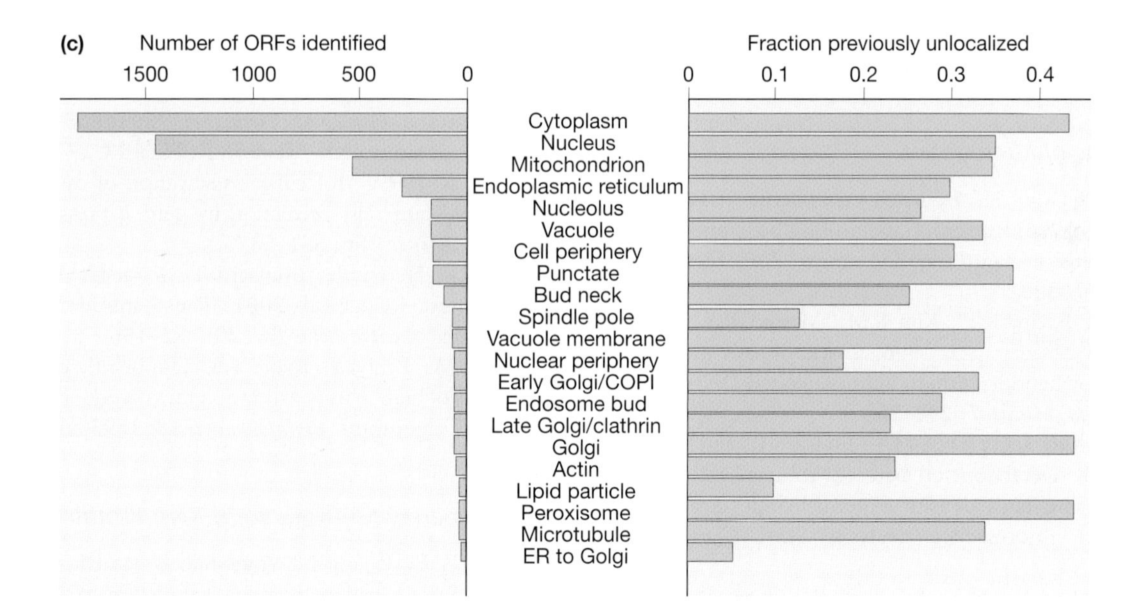

FIGURE 13-18

Tagging open reading frames (ORFs) with green fluorescent protein (GFP) allows the subcellular localization of all yeast proteins in living cells. (*a*) Every yeast protein is tagged with GFP by using homologous recombination to engineer each gene. (*b*) Each fusion line of yeast is analyzed by an automated fluorescence microscopy approach to determine the location of the GFP signal in the cells. Examples of proteins localized to various compartments in living yeast cells by GFP tagging. (*c*) The histogram shows the number of proteins that localize to various compartments in the cell (*left*), and the fraction of the localized proteins whose locations were previously unknown (*right*). Thus, the subcellular localization of each protein can be determined on a large scale without producing unique antibodies for each gene.

Reading List

Construction of Oligonucleotide and Spotted DNA Microarrays

Fodor S.P., Read J.L., Pirrung M.C., Stryer L., Lu A.T., and Solas D. 1991. Light-directed, spatially addressable parallel chemical synthesis. *Science* **251:** 767–773.

Fodor S.P., Rava R.P., Huang X.C., Pease A.C., Holmes C.P., and Adams C.L. 1993. Multiplexed biochemical assays with biological chips. *Nature* **364:** 555–556.

Pease A.C., Solas D., Sullivan E.J., Cronin M.T., Holmes C.P., and Fodor S.P. 1994. Light-generated oligonucleotide arrays for rapid DNA sequence analysis. *Proc. Natl. Acad. Sci.* **91:** 5022–5026.

Schena M., Shalon D., Davis R.W., and Brown P.O. 1995. Quantitative monitoring of gene expression patterns with a complementary DNA microarray. *Science* **270:** 467–470.

Chee M., Yang R., Hubbell E., Berno A., Huang X.C., et al. 1996. Accessing genetic information with high-density DNA arrays. *Science* **274:** 610–614.

Schena M., Shalon D., Heller R., Chai A., Brown P.O., and Davis R.W. 1996. Parallel human genome analysis: Microarray-based expression monitoring of 1000 genes. *Proc. Natl. Acad. Sci.* **93:** 10614–10619.

Shalon D., Smith S.J., and Brown P.O. 1996. A DNA microarray system for analyzing complex DNA samples using two-color fluorescent probe hybridization. *Genome Res.* **6:** 639–645.

Applications of Microarrays

Lockhart D.J., Dong H., Byrne M.C., Follettie M.T., Gallo M.V., et al. 1996. Expression monitoring by hybridization to high-density oligonucleotide arrays. *Nat. Biotechnol.* **14:** 1675–1680.

Cronin M.T., Fucini R.V., Kim S.M., Masino R.S., Wespi R.M., and Miyada C.G. 1996. Cystic fibrosis mutation detection by hybridization to light-generated DNA probe arrays. *Hum. Mutat.* **7:** 244–255.

Hacia J.G., Brody L.C., Chee M.S., Fodor S.P., and Collins F.S. 1996. Detection of heterozygous mutations in *BRCA1* using high density oligonucleotide arrays and two-colour fluorescence analysis. *Nat. Genet.* **14:** 441–447.

Kozal M.J., Shah N., Shen N., Yang R., Fucini R., et al. 1996. Extensive polymorphisms observed in HIV-1 clade B protease gene using high-density oligonucleotide arrays. *Nat. Med.* **2:** 753–759.

Zhang L., Zhou W., Velculescu V.E., Kern S.E., Hruban R.H., et al. 1997. Gene expression profiles in normal and cancer cells. *Science* **276:** 1268–1272.

Iyer V.R., Eisen M.B., Ross D.T., Schuler G., Moore T., et al. 1999. The transcriptional program in the response of human fibroblasts to serum. *Science* **283:** 83–87.

Pollack J.R., Perou C.M., Alizadeh A.A., Eisen M.B., Pergamenschikov A., et al. 1999. Genome-wide analysis of DNA copy-number changes using cDNA microarrays. *Nat. Genet.* **23:** 41–46.

Perou C.M., Sorlie T., Eisen M.B., van de Rijn M., Jeffrey S.S., et al. 2000. Molecular portraits of human breast tumours. *Nature* **406:** 747–752.

Kapranov P., Cawley S.E., Drenkow J., Bekiranov S., Strausberg R.L., et al. 2002. Large-scale transcriptional activity in chromosomes 21 and 22. *Science* **296:** 916–919.

Sorlie T., Tibshirani R., Parker J., Hastie T., Marron J.S., et al. 2003. Repeated observation of breast tumor subtypes in independent gene expression data sets. *Proc. Natl. Acad. Sci.* **100:** 8418–8423.

Chang H.Y., Sneddon J.B., Alizadeh A.A., Sood R., West R.B., et al. 2004. Gene expression signature of fibroblast serum response predicts human cancer progression: Similarities between tumors and wounds. *PLoS. Biol.* **2:** E7.

Frigessi A., van de Wiel M.A., Holden M., Svendsrud D.H., Glad I.K., and Lyng H. 2005. Genome-wide estimation of transcript concentrations from spotted cDNA microarray data. *Nucleic Acids Res.* **33:** e143.

Shyamsundar R., Kim Y.H., Higgins J.P., Montgomery K., Jorden M., et al. 2005. A DNA microarray survey of gene expression in normal human tissues. *Genome Biol.* **6:** R22.

Analysis of Microarray Data

Eisen M.B., Spellman P.T., Brown P.O., and Botstein D. 1998. Cluster analysis and display of genome-wide expression patterns. *Proc. Natl. Acad. Sci.* **95:** 14863–14868.

Golub T.R., Slonim D.K., Tamayo P., Huard C., Gaasenbeek M., et al. 1999. Molecular classification of cancer: Class discovery and class prediction by gene expression monitoring. *Science* **286:** 531–537.

Sherlock G., Hernandez-Boussard T., Kasarskis A., Binkley G., Matese J.C., et al. 2001. The Stanford Microarray Database. *Nucleic Acids Res.* **29:** 152–155.

Liu W.M., Mei R., Di X., Ryder T.B., Hubbell E., et al. 2002. Analysis of high density expression microarrays with signed-rank call algorithms. *Bioinformatics* **18:** 1593–1599.

Nonmicroarray Methods for mRNA Profiling

Velculescu V.E., Zhang L., Vogelstein B., and Kinzler K.W. 1995. Serial analysis of gene expression. *Science* **270:** 484–487.

Zhang L., Zhou W., Velculescu V.E., Kern S.E., Hruban R.H., et al. 1997. Gene expression profiles in normal and cancer cells. *Science* **276:** 1268–1272.

Velculescu V.E., Vogelstein B., and Kinzler K.W. 2000. Analysing uncharted transcriptomes with SAGE. *Trends Genet.* **16:** 423–425.

Chen J. and Sadowski I. 2005. Identification of the mismatch repair genes *PMS2* and *MLH1* as p53 target genes by using serial analysis of binding elements. *Proc. Natl. Acad. Sci.* **102:** 4813–4818.

Uses of the Yeast Deletion Collection

Winzeler E.A., Shoemaker D.D., Astromoff A., Liang H., Anderson K., et al. 1999. Functional characterization of the *S. cerevisiae* genome by gene deletion and parallel analysis. *Science* **285:** 901–906.

Steinmetz L.M., Scharfe C., Deutschbauer A.M., Mokranjac D., Herman Z.S., et al. 2002. Systematic screen for human disease genes in yeast. *Nat. Genet.* **31:** 400–404.

Giaever G., Chu A.M., Ni L., Connelly C., Riles L., et al. 2002. Functional profiling of the *Saccharomyces cerevisiae* genome. *Nature* **418:** 387–391.

Deutschbauer A.M., Williams R.M., Chu A.M., and Davis R.W. 2002. Parallel phenotypic analysis of sporulation and postgermination growth in *Saccharomyces cerevisiae. Proc. Natl. Acad. Sci.* **99:** 15530–15535.

Lum P.Y., Armour C.D., Stepaniants S.B., Cavet G., Wolf M.K., et al. 2004. Discovering modes of action for therapeutic compounds using a genome-wide screen of yeast heterozygotes. *Cell* **116:** 121–137.

Functional Analysis of Transcriptional Regulatory Elements

Trinklein N.D., Force Aldred S., Saldanha A., and Myers R.M. 2003. Identification and functional analysis of human transcriptional promoters. *Genome Res.* **13:** 308–312.

Trinklein N.D., Force Aldred S., Hartman S.J., Schroeder D.I., Otillar R., and Myers R.M. 2004. An abundance of bidirectional promoters in the human genome. *Genome Res.* **14:** 62–66.

van Steensel B. 2005. Mapping of genetic and epigenetic regulatory networks using microarrays. *Nat. Genet.* (suppl.) **37:** S18–S24.

Cooper S.J., Trinklein N.D., Anton E.D., Nguyen L., and Myers R.M. 2006. Comprehensive analysis of transcriptional promoter structure and function in 1% of the human genome. *Genome Res.* **16:** 1–10.

Measuring Protein–DNA Interactions

Fields S. and Song O. 1989. A novel genetic system to detect protein–protein interactions. *Nature* **340:** 245–246.

Lieb J.D., Liu X., Botstein D., and Brown P.O. 2001. Promoter-specific binding of Rap1 revealed by genome-wide maps of protein-DNA association (erratum *Nat. Genet.* [2001] **19:** 100). *Nat. Genet.* **28:** 327–334.

Ito T, Chiba T., Ozawa R., Yoshida M., Hattori M., and Sakaki Y. 2001. A comprehensive two-hybrid analysis to explore the yeast protein interactome. *Proc. Natl. Acad. Sci.* **98:** 4569–4574.

Rice J.C. and Allis C.D. 2001. Histone methylation versus histone acetylation: New insights into epigenetic regulation. *Curr. Opin. Cell Biol.* **13:** 263–273.

Oberley M.J., Tsao J., Yau P., and Farnham P.J. 2004. High-throughput screening of chromatin immunoprecipitates using CpG-island microarrays. *Methods Enzymol.* **376:** 315–334.

Kim J., Bhinge A.A., Morgan X.C., and Iyer V.R. 2005. Mapping DNA–protein interactions in large genomes by sequence tag analysis of genomic enrichment. *Nat. Methods* **2:** 47–53.

Fields S. 2005. High-throughput two-hybrid analysis. The promise and the peril. *FEBS J.* **272:** 5391–5399.

Measuring Protein–Protein Interactions

Uetz P., Giot L., Cagney G., Mansfield T.A., Judson R.S., et al. 2000. A comprehensive analysis of protein–protein interactions in *Saccharomyces cerevisiae. Nature* **403:** 623–627.

Gavin A.C., Bosche M., Krause R., Grandi P., Marzioch M., et al. 2002. Functional organization of the yeast proteome by systematic analysis of protein complexes. *Nature* **415:** 141–147.

Ho Y., Gruhler A., Heilbut A., Bader G.D., Moore L., et al. 2002. Systematic identification of protein complexes in *Saccharomyces cerevisiae* by mass spectrometry. *Nature* **415:** 180–183.

Kumar A. and Snyder M. 2002. Protein complexes take the bait. *Nature* **415:** 123–124.

Li S., Armstrong C.M., Bertin N., Ge H., Milstein S., et al. 2004. A map of the interactome network of the metazoan *C. elegans. Science* **303:** 540–543.

Rual J.F., Venkatesan K., Hao T., Hirozane-Kishikawa T., Dricot A., et al. 2005. Towards a proteome-scale map of the human protein–protein interaction network. *Nature* **437:** 1173–1178.

Protein Levels and Localization

Kononen J., Bubendorf L., Kallioniemi A., Barlund M., Schraml P., et al. 1998. Tissue microarrays for high-throughput molecular profiling of tumor specimens. *Nat. Med.* **4:** 844–847.

Zhu H., Bilgin M., Bangham R., Hall D., Casamayor A., et al. 2001. Global analysis of protein activities using proteome chips. *Science* **293:** 2101–2105.

Ghaemmaghami S., Huh W.K., Bower K., Howson R.W., Belle A., et al. 2003. Global analysis of protein expression in yeast. *Nature* **425:** 737–741.

Huh W.K., Falvo J.V., Gerke L.C., Carroll A.S., Howson R.W., et al. 2003. Global analysis of protein localization in budding yeast. *Nature* **425:** 686–691.

Nilsson P., Paavilainen L., Larsson K., Odling J., Sundberg M., et al. 2005. Towards a human proteome atlas: High-throughput generation of mono-specific antibodies for tissue profiling. *Proteomics* **5:** 4327–4337.

SECTION 4

HUMAN GENOMICS

The study of human genetics began in the years just after the discovery of Mendel's studies, but rapid advances had to await two developments—the development of recombinant DNA techniques, so that human genes could be isolated and manipulated, and the development of mapping techniques, so that genes involved in human inherited disorders could be located. We are now entering a third era of human genetics, which emerged from the Human Genome Project. The chapters in this final section provide a sampling of the many and extraordinary ways in which the human genome sequence is advancing our understanding of human biology. The first chapter discusses clinical genetics, the study of how our genes and their errors lead to disorders. Here genomic strategies are likely to help dissect those complex disorders involving several genes and interactions with the environment. The genetic aspects of cancer, which typifies complex disorders, are covered in the second chapter. Here we see, again, that the human genome sequence and genomic tools like microarrays promise to reveal all the mutations that lead to cancer. We close with DNA fingerprinting, which has transformed the practice of forensics with striking advances in methods for individual identification, whether in forensic cases, paternity suits, missing persons cases, or mass catastrophes. Now, for example, suspects can be unequivocally associated (or excluded) from a crime scene and the wrongly convicted freed, even years after a crime was committed. We end this final chapter by considering some of the problems that may arise as a consequence of the very rapid penetration of DNA technology into human society.

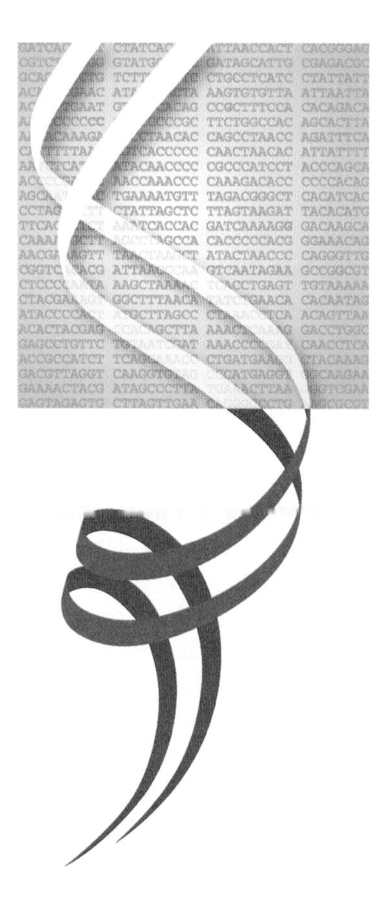

CHAPTER 14

Finding Human Disease Genes

Genetic factors play a role in all illnesses—disorders arise from missing or extra chromosomes, or when pieces of a chromosome are deleted, duplicated, or swapped between chromosomes, or when there are mutations altering codons or regulatory sequences. There are disorders with a clear pattern of inheritance such as sickle-cell anemia, Huntington disease, and cystic fibrosis. There are also many very common disorders that have a strong, although less well-defined, genetic component. These include heart problems, diabetes, and mental disorders such as schizophrenia and bipolar disorder. Genes also modulate our interactions with environmental factors, whether bacterial, viral, or chemical, whether noxious or beneficial.

It is the hope and expectation of those who sequenced, and continue to analyze, the human genome that the data of the Human Genome Project (HGP) will very soon lead to a much greater understanding of the genetic basis of human disorders. The HGP also led to significant technical developments in instrumentation for automated sequencing, microarrays, and other functional genomics techniques. These new technologies are also being harnessed for use in diagnosis and in searching for many mutations and polymorphisms simultaneously. We begin this chapter with a brief description of the cloning and analysis of human genes that was achieved by using recombinant DNA techniques available in the period before the HGP. The remainder of the chapter shows how the HGP and associated new genomic techniques are revolutionizing our understanding of human genetics and inherited disorders.

The First 40 Years of Human Genetics Proceeded Slowly

Almost immediately following their discovery, Mendel's laws were applied to human beings. Archibald Garrod, a physician at St. Bartholomew's Hospital in London, England, observed a patient whose urine turned black after exposure to air (a condition now known as alkaptonuria). This fascinated Garrod, who found five more families with this odd affliction and noticed that the parents of affected children were first cousins. Garrod consulted William Bateson, who pointed out that this was exactly the circumstance in which a rare, recessive character would appear. Garrod went on to determine the biochemical change that produced the black urine and coined his famous phrase by describing these as cases of "inborn errors of metabolism." Other traits, less spectacular than alkaptonuria, were studied (e.g., eye and hair color), but these generally did not have clear patterns of Mendelian inheritance.

Some early geneticists turned their attention to the genes that they believed underlie intelligence, mental illnesses, and antisocial and psychopathic behaviors. They established a movement, eugenics, that sought to improve the human genetic stock by better mate selection and, failing that, sterilization of undesirables. Among the leading eugenicists were Charles Davenport and Harry Laughlin at Cold Spring Harbor in the United States, and Leonard Darwin in the United Kingdom. Eugenics attracted widespread support with enthusiastic proponents in many countries. It was particularly influential in the United States where Congress passed severely restrictive immigration legislation based in part on data provided by Laughlin. However, by about 1930, it was recognized that the traits that interested the eugenicists were likely not due to single genes and that the scientific basis for eugenics actions had little or no merit. The Nazi regime, however, continued to use genetic and eugenic arguments to dress up its political and economic goal of ridding the Third Reich of "undesirables." First, the mentally ill and incurable individuals, and then what the Nazis considered to be the socially undesirable—such as gypsies, homosexuals, and Jews—were labeled as inferior beings that should be exterminated. Tarnished forever by its association with the Nazi regime, the eugenics movement faded away.

Important discoveries in human genetics had been made prior to 1940—Asbjørn Følling discovered phenylketonuria and J.B.S. Haldane and Julia Bell demonstrated linkage between color blindness and hemophilia, but after World War II, human genetics became a neglected field. Human clinical genetics dealt with rare disorders such as alkaptonuria and did not attract physicians, while geneticists could do much better with *Drosophila*, bacteria, and bacteriophage. It was not until the late 1940s that there was a significant advance.

James Neel then showed that sickle-cell anemia and sickle-cell trait (a milder version) were caused by mutations in a gene that, when homozygous, produced anemia and, when heterozygous, caused sickle-cell trait. And in 1949, Linus Pauling and Harvey Itano compared hemoglobins prepared from normal individuals with those from patients with sickle-cell anemia and showed a difference in their electrophoretic mobility. Their paper was titled provocatively "Sickle Cell Anemia, a Molecular Disease" at a time when it was not yet clear that all molecules of a particular protein had the same arrangement of amino acids. Seven years later, Vernon Ingram, working with Max Perutz in Cambridge, England, showed the precise nature of the molecular defect. Ingram treated normal and sickle-cell hemoglobin with trypsin and separated the peptides by paper using electrophoresis in one dimension followed by chromatography in the second. This treatment revealed a single-peptide difference between the two trypsin digests. Ingram then sequenced the peptide and found a single-amino-acid change, from glutamic acid to valine. That a change of one amino acid out of 146 could produce a severe disease was a sensational finding.

Cytogenetics Marks a Major Advance in Human Genetics

Few of these findings contributed to clinical genetics, but in 1956 Albert Levan and Joe Hin Tjio showed that human beings have 46 chromosomes, not 48 as had been believed for more than 30 years. And, just as significant, their new technique for preparing chromosomes led to studies of developmental disorders. In quick succession, it was shown that individuals with Down syndrome had 47 chromosomes, as did patients with Klinefelter syndrome, whereas Turner syndrome patients had 45. However, at this time, the individual chromosomes could not be distinguished. All that could be done was to assign chromosomes to one of seven groups, depending on sizes of the chromosomes and the positions of their centromeres. Thus, the extra chromosome in Down syndrome was in Group G, a set of the smallest chromosomes. It was not until 1969 that a new technique was developed that distinguished each chromosome.

Torbörn Caspersson reasoned that if base ratios varied along a chromosome, then the variation might be detectable using mutagens that bind to DNA. A chemist colleague synthesized quinacrine mustard and Caspersson found that it stained human chromosomes with banding patterns that were unique for each chromosome. These Q-band patterns revealed dark staining heterochromatic regions and lighter staining euchromatic regions of chromosomes. Later, simpler and more effective methods such as pretreating a chromosome spread with trypsin and then staining with Giemsa stain became standard. Now it was possible to recognize many subtle chromosomal abnormalities and the discipline of cytogenetics flourished. However, very little could be said about human genes as opposed to chromosomes. As we discussed in Chapter 11, the discovery that cells could be fused to make hybrids led to the development of somatic cell genetics and the assignment of the thymidine kinase gene to chromosome 17. But these techniques for locating genes were very laborious and limited in their application. It was the development of recombinant DNA techniques in the 1970s that changed human genetics by providing the means to identify and clone human genes so that they could be studied in detail.

Recombinant DNA Techniques Enable Cloning of Human Disease Genes

Recombinant DNA techniques provided two indispensable sets of tools to scientists interested in human genetics. The first included all the enzymes, vectors, and other materials for cloning genes as described in Chapters 4 and 6. The second set of tools enabled geneticists to locate human genes, and included linkage analysis using restriction fragment length polymorphisms (RFLPs) and the linkage maps produced with them (Chapter 11).

Duchenne muscular dystrophy (DMD) is an X-linked disorder causing progressive muscle degeneration in young boys. There is no cure, and patients die in their late teens or early twenties. It is a relatively common disorder, occurring in about one in 3500 live male births. Despite intensive research, very little progress was made in understanding the pathogenesis of the disorder. Yet, within the span of a few years, *dystrophin*, the gene mutated in DMD, had been cloned. Women who are heterozygous for an X-linked disorder have one X chromosome carrying a normal copy of the mutated gene and are not usually affected by the disease. However, women have been described who appear to have DMD, and it was found that some of these women had a chromosomal translocation with the breakpoint on their X chromosomes always located in band Xp21. This suggested that a gene involved in DMD was located at Xp21. Subsequently, RFLPs detected by probes that mapped to Xp21 showed linkage to DMD. Cytogenetic analysis of a young boy with a bizarre group of disorders including DMD revealed a very large deletion of the Xp21 locus—so large that it affected several genes. Lou Kunkel and his colleagues used this DNA in a clever subtractive hybridization experiment to clone normal DNA from the deleted region (Fig. 14-1).

It was a different matter to search for a gene relying only on RFLP linkage analysis. There were many naysayers, then, when a group of geneticists inspired by Nancy Wexler decided to search for the gene involved in Huntington disease (HD). This is a devastating neurodegenerative disorder in which neurons in the striatum, a region of the brain involved in the control of movement, die. Initial symptoms often appear in midlife, between the ages of 30 and 45, and the disease inexorably progresses over the course of 10–20 years, leading to death. HD has an autosomal-dominant pattern of inheritance—each child of an affected parent has a 50% chance of inheriting the fatal gene.

The hunt for the HD gene began in the 1980s, when James Gusella and his colleagues began to test RFLPs in a huge Venezuelan family afflicted by HD that had been studied by Wexler. In the hope of identifying a polymorphic marker associated with the disease, DNA samples collected from family members were digested with restriction enzymes. The samples were then subjected to Southern blotting (Chapter 5) and probed with DNA fragments that served as anonymous probes to detect the polymorphism.

In 1983, HD was found to be linked to a polymorphic marker, G8, on chromosome 4p. This was

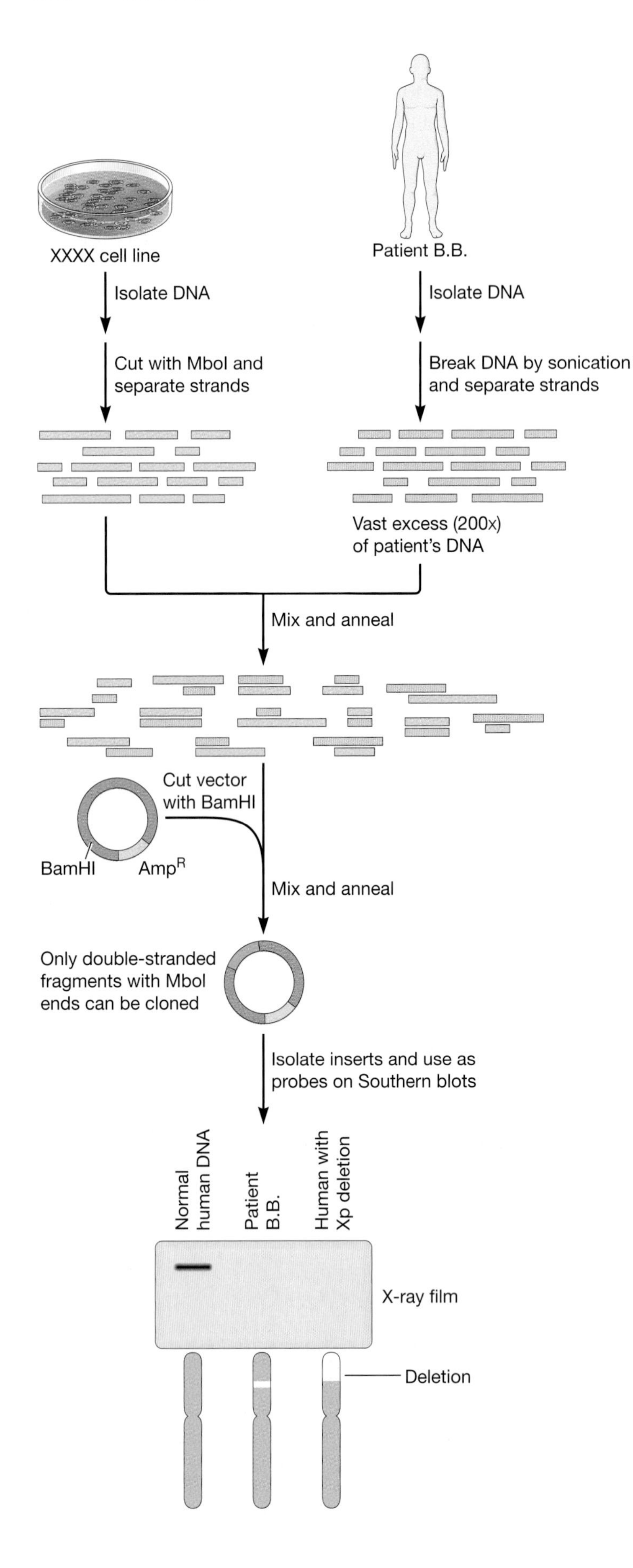

FIGURE 14-1

Cloning DMD sequences by subtraction. DNA from a human cell line containing four X chromosomes was cut with the restriction enzyme MboI, while DNA from a DMD patient with a very large deletion of the DMD region was broken into pieces at random sites by sonication. A 200-fold excess of the patient's DNA fragments was added to the normal X chromosome MboI fragments, and the DNA fragments were annealed for 37 hr in a phenol emulsion to enhance reassociation. Under these conditions, three types of double-stranded molecules were produced. Most molecules were formed by reassociation of strands of the patient's DNA. Because these strands came from sonicated DNA, the ends of these double-stranded molecules were random sequences and thus could not be cloned by ligation into a vector. A smaller fraction were hybrid molecules between patient and cell line DNA. These double-stranded molecules could not be cloned because their ends were also incompatible with the cloning sites in the vector. The third class of double-stranded molecules were formed by reassociation of DNA strands from the cell line. These molecules were enriched for sequences present in the cell line and absent from the patient's DNA, because the vast excess of patient DNA had "mopped up" sequences common to both sources of DNA. Double-stranded molecules derived from the normal X chromosomes of the cell line were the only molecules in the annealing reaction that had ends that could be cloned into a vector, in this case a cloning site with BamHI sticky ends that are compatible with the MboI sticky ends of the cell line–derived DNA. The origin of cloned inserts was determined by using them as probes in Southern blotting of DNA from a normal individual, from the patient, and from two rodent–human hybrid cell lines containing either a whole human X chromosome or an X chromosome with the tip deleted. A probe called pERT87 failed to hybridize to DNA from the patient or from the hybrid cell with the partially deleted X chromosome, showing that pERT87 comes from the deleted region.

extraordinarily lucky; G8 was only the 12th marker that Gusella and his colleagues tested! Identification of the disease gene within this linked region, however, required another 10 long years. The linkage studies suggested the gene lay between G8 and the telomere, a region four million base pairs long, and probably close to the telomere. Researchers concentrated on this region but eventually new data indicated that the gene lay at the other end of the linked region. Linkage analysis narrowed the region to 500,000 bp and candidate genes were cloned from the region and screened for potential mutations. One of these mutations, known initially as IT15, was found within the first exon of an anonymous gene and was shown to consist of a stretch of DNA containing multiple copies of the repeat CAG, which codes for glutamine. Upon closer inspection, geneticists discovered that the length of the repeated region was polymorphic in normal chromosomes, but was significantly longer in HD chromosomes. This was the first occasion on which such a triplet expansion mutation had ever been observed.

The work of the HD researchers led the way for others to rely on RFLP linkage analysis alone, and another disease-associated gene that attracted early attention was that involved in cystic fibrosis (CF). The hunt for the CF gene began with screening a large number of CF families with a large number of RFLPs, looking for RFLPs that could be used to track the CF gene. An RFLP on chromosome 7 was found to be loosely linked to CF and the region containing the gene was defined by flanking RFLPs. The gene itself was cloned using a novel strategy called chromosome "jumping" to move rapidly through this region (Fig. 14-2). A 500-kb of stretch of DNA encompassing the CF gene was analyzed by restriction mapping but no deletions or other abnormalities were detected in patients with CF. Instead, mutations had to be searched for by sequencing. Just three base pairs resulting in the loss of a phenylalanine residue were found to be deleted in the cystic fibrosis transmembrane conductance regulator (CFTR) gene (Fig. 14-3). These successes, the finding of the genes involved in three such different disorders—Duchenne muscular dystrophy, Huntington disease, and cystic fibrosis—helped pave the way for the HGP.

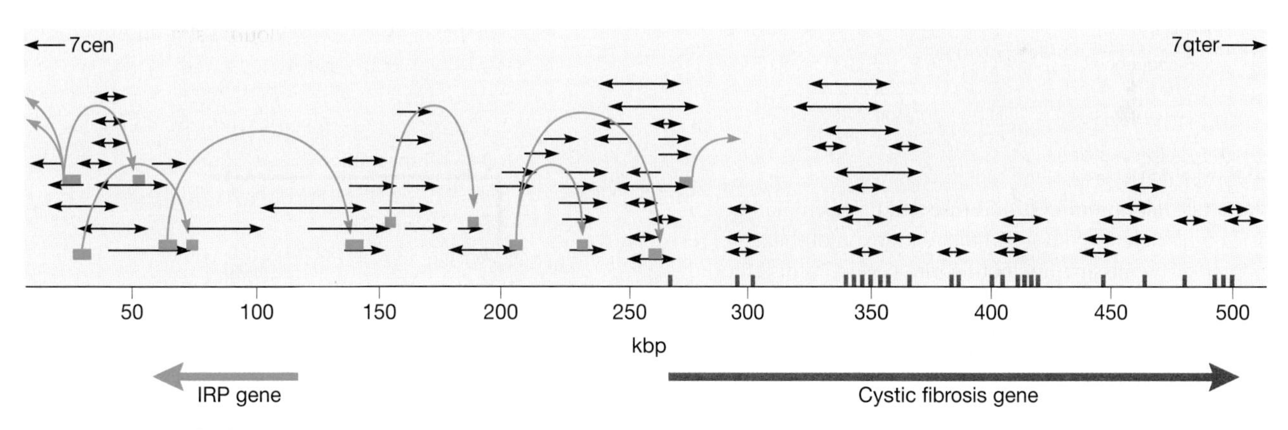

FIGURE 14-2

Cloning the CF gene by jumping and walking. The long continuous line represents a DNA strand more than 500 kb long (the length markers are at 50-kb intervals from the region of chromosome 7 containing the CF gene). The centromere (7cen) is to the *left* and the tip of chromosome 7 (7qter) to the *right* of the diagram. *Curved arrows* show the length and direction of each jump, and the *blue boxes* are the clones that were the starting and end points of the jumps. The *horizontal arrows* are overlapping phage and cosmid clones containing DNA isolated from the regions at the ends of each jump. The directions of these arrows show the directions of cloning; *double-headed arrows* indicate that cloning proceeded in both directions from the starting clone. The location of the CF gene is indicated by the *large arrow* below the DNA strand, and the positions of its 24 exons are shown by the *small vertical bars* on the horizontal axis. The position of the IRP gene (for *int*-related protein), a "landmark" for CF, is shown by the *large leftward-pointing arrow*.

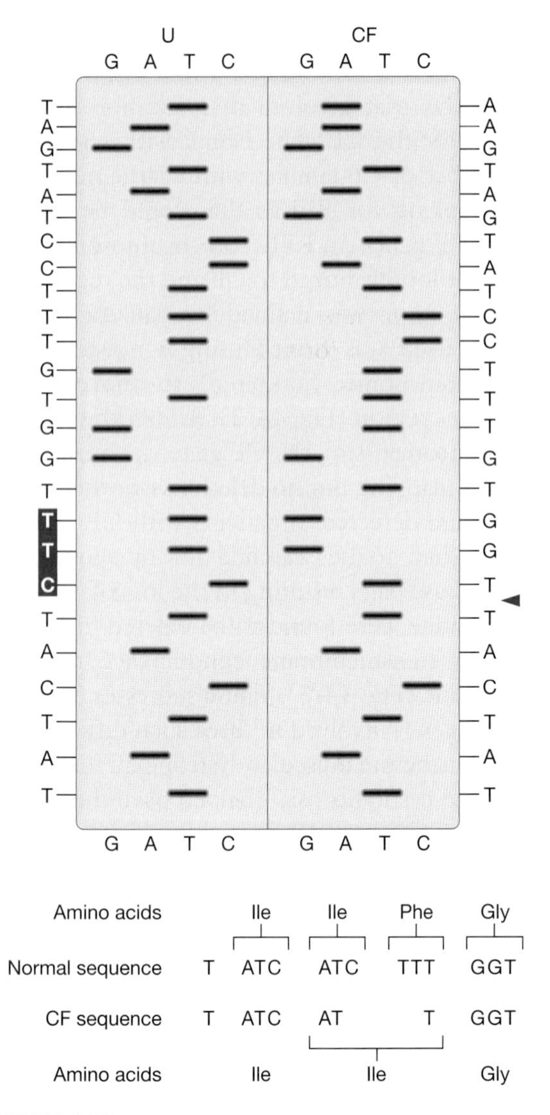

FIGURE 14-3

The 3-bp deletion in cystic fibrosis (CF). This mutation consists of the deletion of CTT in the tenth exon of the gene (*red triangle*). The missing nucleotides are boxed in this autoradiograph showing this sequence in an unaffected individual (U) and a patient with cystic fibrosis (CF). Loss of 3 bp maintains the reading frame so that in this case only a single phenylalanine is lost from the protein. The frequency of this mutation in CF patients ranges from as low as 30% in parts of Southern Europe to 95% in Denmark, and in the U.S. population it is about 70%. More than 1500 other mutations have been found in the CFTR gene.

Cloning of the Fragile X Gene Reveals a Mutation with Unexpected Properties

Physicians had noticed that in some genetic disorders—fragile X syndrome, Huntington disease, myotonic dystrophy—the severity of symptoms increased, and the age of onset decreased, in successive generations of affected families. This phenomenon, known as *anticipation*, was initially ascribed to increased vigilance on the part of family members and clinicians once an affected family was identified. In fact, the explanation proved to be molecular.

Fragile X syndrome is the most common cause of inherited mental impairment; it occurs primarily in males (Chapter 9). The syndrome gets its name from the appearance of X chromosomes in cytogenetic preparations of patients' cells. For reasons still not clear, reducing folate in the culture medium creates a "fragile site," seen as a constricted site or gap, in the X chromosome. The gene involved in fragile X syndrome, *FMR1*, was cloned in 1991 and found to have a CGG repeat in the 5′-untranslated region (Fig. 14-4). There are between 6 and 40 copies of this repeat in unaffected individuals and this number of copies of the repeat is stably inherited. However, when the repeat number expands to 41–60 repeats (known as the intermediate state), the CGG region becomes unstable. As the repeat expands to more than 200 copies, the *FMR1* gene begins to be transcriptionally silenced, which leads to loss of the fragile X mental retardation protein (FMRP) and produces the fragile X phenotype. The molecular basis of anticipation, then, is the increase in length of the repeat region from generation to generation, primarily during sperm and egg development. Multiple

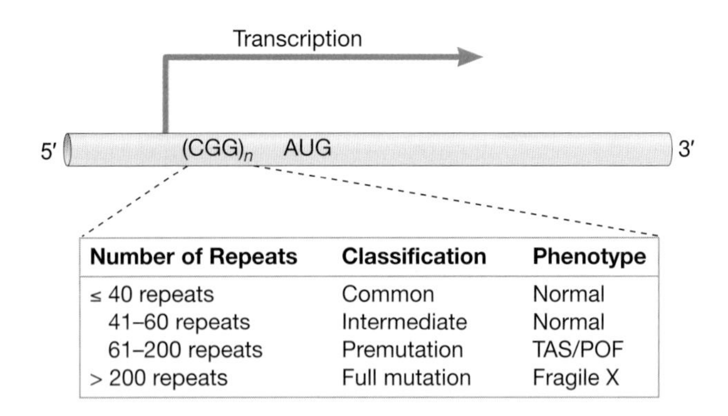

Number of Repeats	Classification	Phenotype
≤ 40 repeats	Common	Normal
41–60 repeats	Intermediate	Normal
61–200 repeats	Premutation	TAS/POF
> 200 repeats	Full mutation	Fragile X

FIGURE 14-4

CGG repeat structure in the 5′-untranslated region of *FMR1*, the gene responsible for fragile X. The majority of individuals have repeat sizes of 40 alleles or less. Individuals with 41–60 CGG repeats have no physical symptoms, but their alleles can become unstable, expanding during meiosis. Among individuals with premutation-sized repeats, older males often exhibit tremor and ataxis syndrome (TAS), whereas younger women undergo premature ovarian failure (POF). Individuals with more than 200 repeats have fragile X syndrome. AUG marks the start of translation.

processes, including faulty DNA replication and improper genome maintenance and repair, are believed to contribute to the repeat instability and the expansion.

Surprisingly, individuals carrying a premutation allele produce significantly more mRNA and FMRP than normal. This leads to a curious clinical phenotype. Many premutation males develop a tremor and ataxia syndrome in their 60s and 70s, whereas premutation females often struggle with premature ovarian failure in their 20s and 30s. How this comes about is not well understood, but there are some intriguing studies that implicate FMRP in the process of RNA interference (RNAi). In fact, FMRP binds to mRNA and has been shown to associate with the RNA induced silencing complex (RISC; Chapter 9).

Since this form of mutation was first detected, nearly 20 triplet repeat diseases have been identified. Despite sharing a common pattern, these mutations do not affect the gene expression in the same way. Some cause loss of gene function, whereas others lead to a gain of function or to altered RNA function. The central nervous system appears to be particularly vulnerable to the effects of repeat expansion. Why this should be so is not known.

The HGP Revolutionizes Human Genetics

Cloning human disease genes demanded extraordinary effort. In the 1980s, Huntington disease required more than 10 years from locating the gene to cloning it, whereas DMD and CF each took a relatively short 4 years. The data now available to us from the HGP have dramatically decreased the time and effort needed to locate, identify, and clone genes mutated in single-gene disorders. Take, for example, the discovery of the gene responsible for one form of hereditary spastic paraplegia (HSP), reported in 2006.

The HSPs are a group of some 30 inherited, untreatable disorders characterized by progressive frailty and stiffness of the leg muscles, caused by degeneration of the upper motor neurons in the brain and spinal cord. After the human genome sequence and dense genetic maps became available, a group of geneticists embarked upon a study of two large families, with a total of 46 individuals, 18 of whom were affected by an autosomal-dominant form of HSP. In a genome-wide search for new HSP genes, these geneticists, using 341 microsatellite markers, identified a region on chromosome 2p12 tightly linked with HSP. An additional ten microsatellite markers were genotyped, and the region was narrowed down to nine million bases.

At this point the researchers turned from mapping to searching the human genome sequence and very rapidly identified more than 60 genes, any one of which could have been the culprit gene. These genes were then examined for features that suggested they might be involved in a neurodegenerative disorder. For example, failure of axonal transport contributes to neurodegeneration, and four of the genes in the region coded for proteins involved in membrane transport. These candidates were given the highest priority for analysis, but none showed any variation associated with the disease. These four candidates and an additional five candidates from the region were selected for detailed mutation analysis by sequencing (Fig. 14-5). Knowing the sequences of these genes enabled the selection of primers for polymerase chain reactions (PCRs) and the rapid sequencing of the exons and exon–intron boundaries for all nine genes in all members of both families. The sequencing results revealed two simple changes—a single-base deletion and a splice site mutation—in the gene encoding receptor expression–enhancing protein 1 (REEP1). Each of these mutations resulted in a frameshift leading to a premature stop codon. Additional loss-of-function mutations in *REEP1* were identified in another group of HSP-affected families. These results provided very strong evidence that the *REEP1* gene, when mutated, is responsible for HSP in these families.

The HSP project took approximately 2 months, from the beginning of the genome-wide linkage to the discovery of the *REEP1* mutations. At each stage, these investigators relied on information gathered by the HGP. Rather than using random genetic markers, the researchers used finely mapped, well-characterized markers in both the genome-wide and the targeted linkage scan. Their selection of the candidate genes was guided by comparisons of the structures and functions of proteins encoded by genes known to be involved in other neurodegenerative diseases. And, finally, the availability of the sequences for these candidate genes led to their rapid sequencing and the identification of mutations in affected families. These new strategies, all arising from the HGP, have led to the identification of the genes involved in more than 2000 genetic disorders.

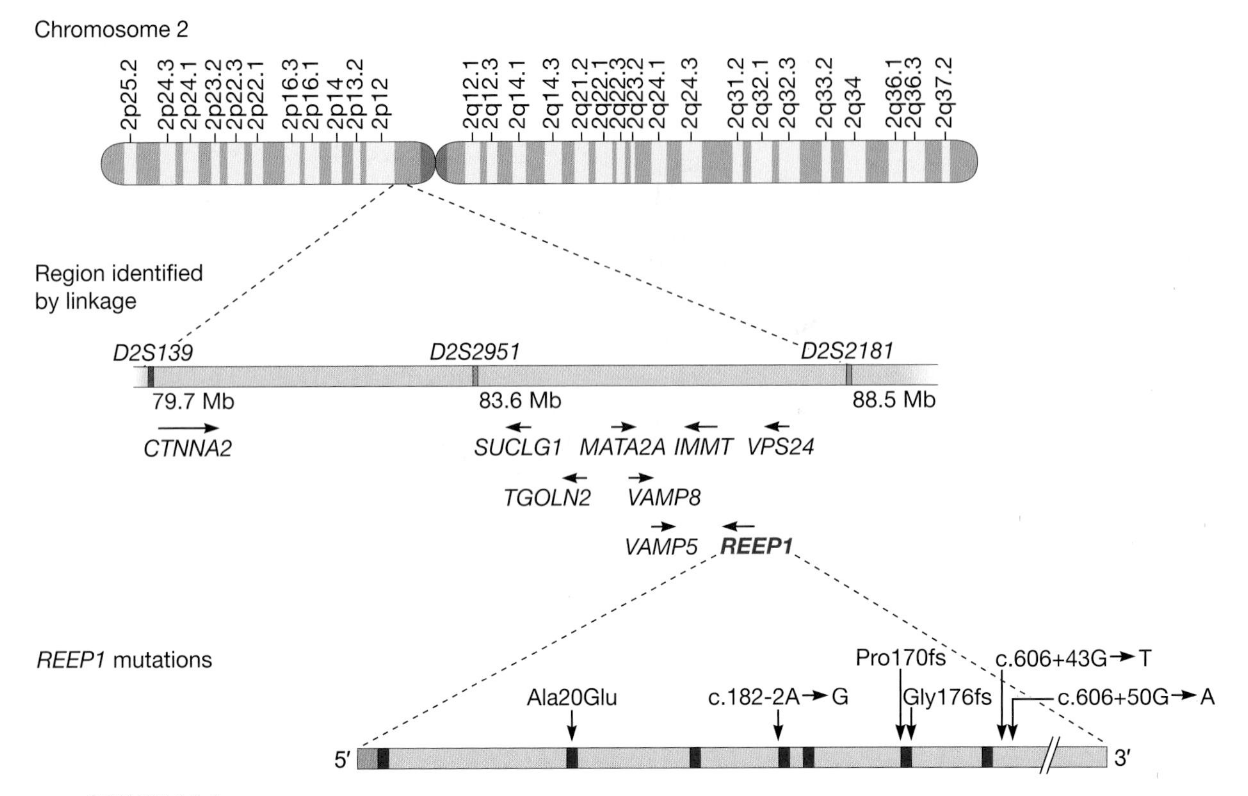

FIGURE 14-5

Identification of the gene for a form of hereditary spastic paraplegia (HSP). Linkage analysis identified a region of nine megabases on the short arm of chromosome 2 as a possible location for the gene involved in HSP. Inspection of the human genome sequence showed that the region contained more than 60 genes. Nine of these were selected as candidate genes (*black arrows* indicate the direction of transcription). All nine were screened for mutations by sequencing genomic DNA from individuals in six families with HSP. *REEP1*, a gene with seven exons (*red boxes*), was found to contain mutations in individuals with HSP. These included two single-base-pair deletions leading to frameshifts, a splice site mutation, and a missense mutation in the coding sequences. There were two mutations in the 3′-untranslated region that would lead to increased miRNA-mediated repression of translation, and hence to less REEP1 protein.

Modifiers of "Single"-Gene Disorders Lead to Complex Phenotypes

We are accustomed to think of disorders such as sickle-cell disease (SCD), the most common inherited blood disorder in the United States, as a prototypical single-gene, Mendelian disorder. SCD results from a mutation in the β-globin gene, which leads to the substitution of a valine for a glutamic acid at amino acid number 6 in the β-globin protein. The active hemoglobin protein is a tetramer with two α and two β chains. Under deoxygenated conditions, hemoglobin molecules containing sickle mutant β chains (HbS) polymerize, thus forming aggregates that distort the shape of red blood cells, damaging the cell wall and producing brittle cells that frequently block capillary beds (Fig. 14-6). These blockages prevent oxygen from reaching tissues, which results in many complications, including stroke, periods of intense pain, and heart and kidney failure. Individuals who are heterozygous for the mutation are largely asymptomatic and are said to have sickle-cell trait. Those individuals with mutations in both copies of the gene suffer from SCD. Sickle-cell anemia, then, appears to be a classic textbook example of an autosomal-recessive disorder, yet two SCD patients can have very widely differing symptoms—ranging from death during childhood to a relatively pain-free existence through adulthood. Until recently, this variation has posed a mystery and was ascribed to "variable expressivity" with no understanding of how the variability arose.

We now know that much of this variation is due to genetic modifiers, allelic variants of genes elsewhere

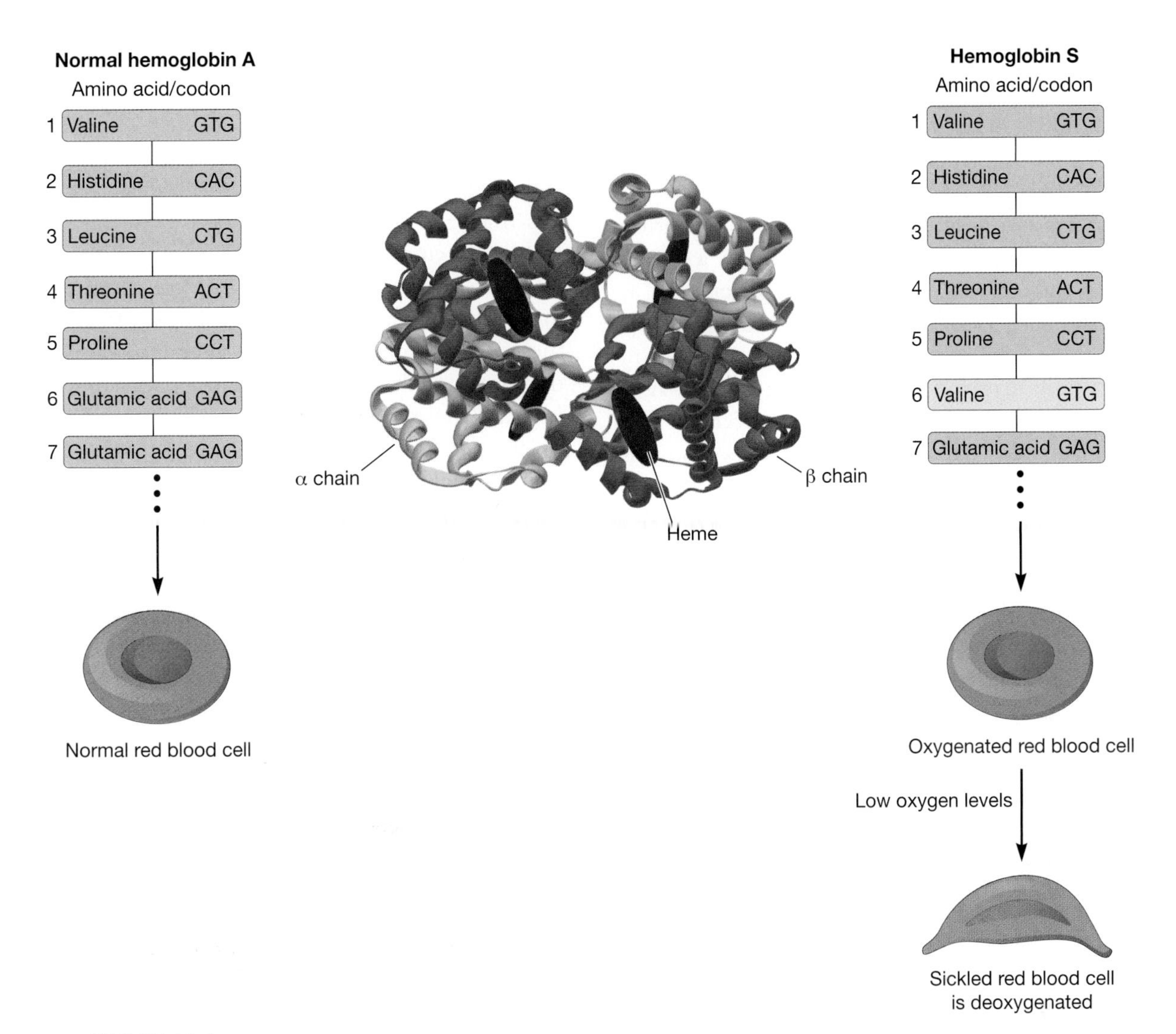

FIGURE 14-6

Normal hemoglobin is composed of two α and two β chains. A mutant form of hemoglobin (HbS) is present among individuals with sickle-cell disease (SCD). The first seven amino acids of normal and sickle-cell hemoglobin are shown. Under deoxygenated conditions, HbS forms aggregates that distort the shape of the red blood cell. These sickled cells can block small vessels, which leads to the clinical features associated with SCD. SCD, an autosomal-recessive disease, is primarily caused by a mutation in the β-globin gene that substitutes valine for glutamic acid at the sixth amino acid position. However, the clinical features can be modified by alleles at other genes that serve to slow the incidence or rate of HbS aggregation.

in the genome that modulate the effects of the primary β-globin mutation. One factor that influences sickling is the intracellular concentration of the HbS protein. Some people have two copies of the β-globin gene with the sickle-cell mutation together with a second mutation in the α-globin locus that deletes one of the four α hemoglobin genes (causing a mild form of α-thalassemia). This combination of mutations results in lower numbers of α-globin subunits, which leads to a subsequent reduction in the total number of fully assembled hemoglobin complexes. This, in turn, lowers the overall intracellular concentration of the mutant HbS, which results in a milder disease than in someone with two sickle mutant genes and normal levels of α-globin.

Individuals who are homozygous for the sickle-cell mutation, but who also carry a mutation in the gene that produces fetal hemoglobin, may also have a milder form of sickle disease. Fetal hemoglobin, or HbF, is the version of β-globin expressed during the development of the fetus. HbF is usually turned off at birth but there are HbF mutations that cause fetal β-globin to continue to be expressed after birth. Fetal hemoglobin is a more effective oxygen carrier than is adult hemoglobin, so its expression in individuals with sickle hemoglobin results in a higher intracellular oxygen saturation. Thus, HbF blocks polymerization of deoxygenated HbS, interfering with the sickling process, which leads to a better clinical prognosis for the sickle-cell patients. It is likely that many other genes carry DNA sequence variants that modify the sickle phenotype, including those that are involved in cell–cell adhesion, coagulation, platelet function, and other aspects of red blood cell activity. Understanding these mechanisms will provide opportunities for new, more effective therapeutic approaches.

SCD is, therefore, a complex disease with multiple players. However, even allowing for the effects of modifier genes, the β-globin mutation, which leads to changes in the hemoglobin molecule, is the cause of the disease. In contrast, in the truly complex disorders, any one of the several genes involved provides only a modest contribution to overall disease risk, perhaps increasing it by 5–10% (Table 14-1). These susceptibility alleles should not be considered "disease genes." Many individuals inherit a susceptibility allele without developing the disease. The correct view, therefore, considers these genes as predisposing, or risk-increasing, genes. It is the specific combination of multiple predisposing alleles and environments that leads to clinical symptoms. Indeed, most diseases that afflict humans are like this—complex combinations of alleles of multiple genes interacting with the environment.

TABLE 14-1. Common multifactorial diseases

Congenital disorders	Disorders of childhood and adult life
Cleft lip and palate	Alzheimer disease
Congenital heart defects	Asthma
Neural tube defects	Atherosclerosis
Pyloric stenosis	Autism
Talipes	Bipolar disorder
	Diabetes mellitus
	Epilepsy
	Glaucoma
	Hypertension
	Inflammatory bowel disease
	Ischemic heart disease
	Ischemic stroke
	Macular degeneration
	Multiple sclerosis
	Parkinson disease
	Psoriasis
	Rheumatoid arthritis
	Schizophrenia

Association Studies Are Used to Locate Genes Involved in Complex Genetic Traits

Even though there are thousands of chromosomal and single-gene disorders, they are rare, affecting less than 3% of the population. By contrast, other diseases, including cardiovascular disease, psychiatric disorders, autoimmune diseases, and cancer, affect much of the world's population. However, although these diseases all have a strong genetic component, they are not inherited in a simple Mendelian fashion. Instead, these complex diseases, sometimes referred to as *multifactorial* diseases, arise from combinations of alleles in multiple genes and are modified by environmental factors (Fig. 14-7).

Genetic linkage methods, applied so successfully to mapping Mendelian disease genes, do not work for complex diseases. Instead, geneticists use an approach called *genetic association.* In a genetic association study, scientists measure the frequency of specific DNA sequence variants (usually single-nucleotide polymorphisms [SNPs]) between two groups. One group is the *case group* (a population of unrelated individuals affected with the disease), and the other is the *control group* (a population of ethnically similar, unrelated persons without the disease). Genetic association either tests a specific gene (a candidate gene association study) or examines polymorphic markers across the entire genome (a whole-genome association study). In both cases, a gene is suspected to be associated with the disease when particular alleles of the gene are present at a statistically higher frequency in cases than in controls. The strength of an association is usually reported as an *odds ratio*, calculated from the frequency of the specific allele in the disease and control groups. For example, the hypothetical study in Figure 14-8 shows an association between allele 3 of a candidate gene and the disease. Allele 3 is present at a much higher frequency among the case group than among the controls.

Genetic association studies are conducted by collecting clinical data and DNA samples from affected individuals, regardless of whether they come from a family with a history of the disease. Association studies require large numbers of cases and controls, often in the hundreds or thousands, to achieve statistical significance for the typically small increased risk—1.2- to 1.5-fold—that a particular allele contributes to the disease. Members of the case and control groups must be drawn from the same population because the frequency of alleles in the controls can vary significantly from one group to another. If

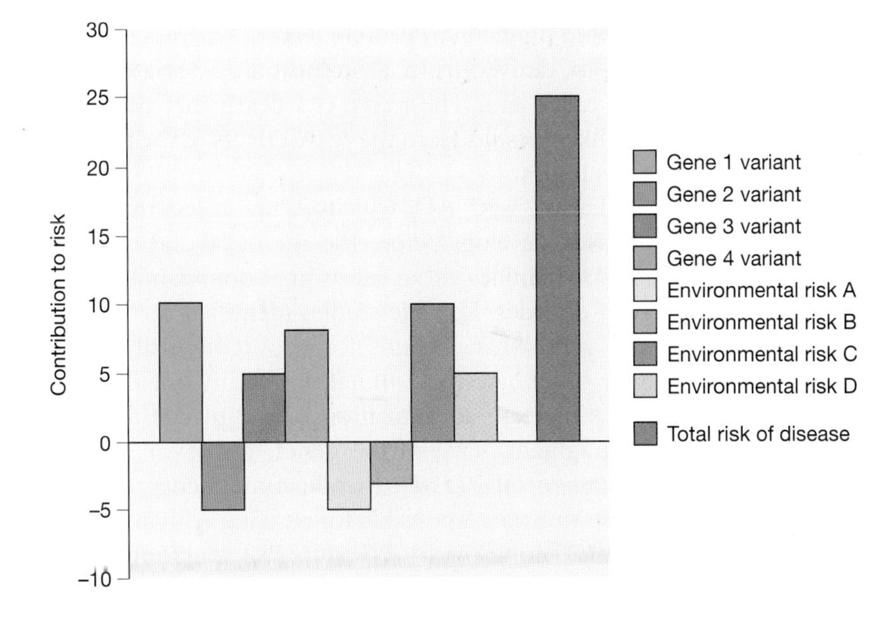

FIGURE 14-7

Genetic and environmental contributors for a hypothetical multifactorial disease. Some factors increase the likelihood of developing the disease whereas others protect against disease occurrence. The overall risk for the individual is the composite contribution by all factors.

the cases and controls are not drawn from the same ancestral groups, observed differences in allele frequencies may have little to do with underlying disease etiology.

Cardiovascular disease is a classic example of a complex disorder. An important indicator of a healthy (or unhealthy) heart is the length of time that the muscle takes to repolarize after an electrical pulse, the QT interval. A QT interval that is either too long or too short is a significant cause of severe heart disease. For example, a short QT interval is the cause of sudden cardiac death, a tragic disorder in which the heart of a seemingly healthy person, such as a star athlete, abruptly stops. The QT interval is variable in the population, and twin and family studies indicate that it has a strong genetic component.

A study that found a gene affecting the QT interval illustrates the complexities and scale of using

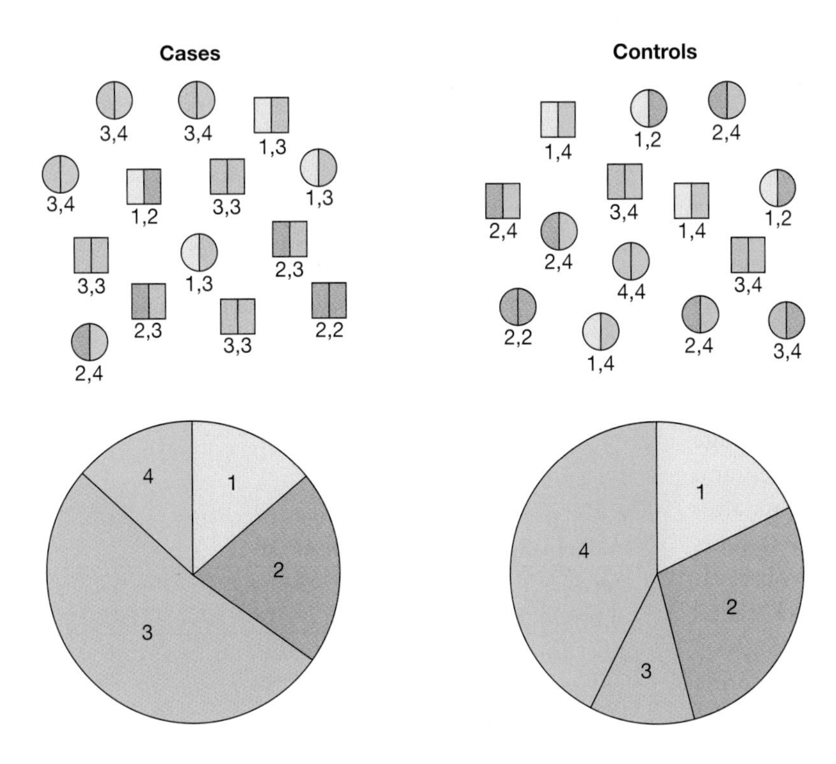

FIGURE 14-8

In an association study, individuals with the disease (cases) and closely matched individuals lacking the disease (controls) are collected and genotyped at various genetic markers. The allele frequencies at any given marker are compared between the cases and controls. Association occurs when the allele frequencies are significantly different between the two groups. In this illustration, allele 3 is more frequently found among the cases than the controls, which suggests its possible involvement in the disease under study.

genetic association analysis. The study began with a sample of 3966 individuals, recruited in 1994, and was carried out in five stages. First, the researchers studied the QT intervals in the 2001 women in the sample, choosing to study only women to avoid complications of any possible gender effect on QT interval. Second, 100 women were selected from each extreme of the QT interval distribution—100 having the longest and 100 the shortest QT intervals—and their genomic DNA was genotyped for 115,000 SNPs using oligonucleotide microarrays. Ten loci were found having significant association with QT interval. Third, the researchers added another 45 SNPs for candidate genes, based on what is known of the biology of cardiac repolarization. All these were tested in the original 200 women plus 400 additional women. From this analysis, they identified eight loci with significant association. Fourth, these eight loci were analyzed in the remaining 3366 males and females who had yet to be tested. This analysis identified a highly significant association between QT interval variation and a SNP in the *NOS1AP* gene (also called the *CAPON* gene), which encodes a regulator of neuronal nitric acid synthase that likely affects calcium flux through an ion channel. Fifth, and most importantly, follow-up studies in two populations of 2646 and 1805 individuals confirmed the association. This study, determining more than 23 million genotypes, thus identified a gene, not previously suspected to be associated with cardiac repolarization, that is a strong genetic contributor to an important clinical phenotype. The finding points to new areas of research and drug development.

Genetic Basis of Alzheimer Disease Is Untangled Using Association Studies

Alzheimer disease (AD) is a dreaded neurodegenerative disorder that is characterized by a progressive and irreversible decline in intellect, memory, and social skills. The disease is frighteningly common; it affects up to 50% of people who live beyond the age of 80. AD advances at widely different rates and death may occur anywhere from 3 to 20 years after the onset of symptoms. Massive neuronal loss occurs in some regions of the brain, and autopsy observation reveals deposits in the brain that are called extracellular "plaques" and intracellular "tangles." The plaques are mainly made of a short Aβ peptide (42 amino acids long), a proteolytic fragment of the amyloid β precursor protein (APP), whereas the tangles are derived from a microtubule-associated protein called tau. The nerve cell death that occurs in AD likely results from the toxic effects of the plaques and tangles.

"Early-onset" AD, which occurs in less than 10% of cases, develops before the age of 60 and typically runs in families as an autosomal-dominant, single-gene disorder. Linkage studies identified three separate genes that account for about one-half of the early-onset cases, and mutations in any one of these genes increase accumulation of the plaque-forming Aβ fragment. Two of the genes, *presenilin 1* (*PS1*) and *presenilin 2* (*PS2*), encode components of a secretase enzyme responsible for cleaving APP into fragments. People with *PS1* and *PS2* mutations have abnormal cleavage of APP, which leads to an increase in the concentration of the Aβ peptide. In addition, in some rare families, mutations have been identified in the gene that encodes APP itself, and these mutations also increase the production of the Aβ fragment. These findings suggest that the initiating event in AD development is an increase in Aβ, thus beginning a cascade of neuronal injury that ultimately triggers the formation of tau-based tangles.

For the majority of AD cases, where the disease occurs in individuals over the age of 60, mutations are not found in the genes encoding APP or presenilin. Therefore investigators have examined genes encoding proteins that interact with APP to increase Aβ accumulation as possible candidate genes for the more common forms of AD. Genetic linkage experiments identified a region on the long arm of chromosome 19, which includes the gene encoding apolipoprotein E (APOE). Subsequent genetic analysis has shown that it is indeed a significant genetic contributor to AD. *APOE* encodes a protein component of the low-density lipoprotein particle that binds the Aβ fragment and has been identified within amyloid plaques found in AD patients. There are three major alleles at the *APOE* gene—ε2, ε3, and ε4; their frequencies in the population and amino acid variations are shown in Table 14-2.

TABLE 14-2. Allele variation in *APOE*

APOE allele	Position 112	Position 168	Allele frequency (%)
ε2	Cys	Cys	8
ε3	Cys	Arg	75
ε4	Arg	Arg	15

TABLE 14-3. ***APOE* allele frequencies**

		APOE genotype frequency (%)						*APOE* allele frequency (%)		
Ethnic group	**No.**	**ε2/ε2**	**ε2/ε3**	**ε2/ε4**	**ε3/ε3**	**ε3/ε4**	**ε4/ε4**	**ε2**	**ε3**	**ε4**
European–American										
Case patients	5107	0.2	4.8	2.6	36.4	41.1	14.8	3.9	59.4	36.7
Controls	6262	0.8	12.7	2.6	60.9	21.3	1.8	8.4	77.9	13.7
African–American										
Case patients	235	1.7	9.8	2.1	36.2	37.9	12.3	7.7	59.1	32.2
Controls	240	0.8	12.9	2.1	50.4	31.8	2.1	8.3	72.7	19.0
Hispanic										
Case patients	261	0.4	9.6	2.3	54.4	30.7	2.7	6.3	74.5	19.2
Controls	267	0.4	12.0	0.8	67.4	17.6	1.9	6.7	82.3	11.0
Japanese										
Case patients	336	0.3	3.9	0.9	49.1	36.9	8.9	2.7	69.5	27.8
Controls	1977	0.4	6.9	0.8	75.7	15.5	0.8	4.2	96.9	8.9

From Farrer et al. 1997. *JAMA 278:* 1349–1356, Table 2.

To determine whether these different forms of *APOE* influence the risk of developing AD, researchers conducted a genetic association study. In a typical European–American population, the frequencies of occurrence of the *APOE* alleles ε2, ε3, and ε4 are 8%, 75%, and 15%, respectively (Table 14-2). Data gathered from more than 40 research teams were used to compare the frequencies of the different *APOE* genotypes between AD individuals and a group of unaffected controls matched for age, gender, and ethnicity (Table 14-3). Among European–Americans, the presence of at least one ε4 allele was observed to occur nearly three times more frequently in individuals with AD than in the control group. In addition, people heterozygous for the ε4 allele, those who have the ε2/ε4 or ε3/ε4 genotype, were, respectively, 1.2 or 2.7 times more likely to have early-onset AD than those homozygous for the ε3 allele, the most common genotype. Strikingly, the ε4/ε4 genotype resulted in an odds ratio of 12.5. That is, an individual with two copies of ε4 is 12 times more likely to develop AD than an individual with two copies of the ε3 allele. Similar increases in risk were observed among African–American, Hispanic, and Japanese populations. Studies have also shown that *APOE* genotypes influence the age at which AD appears. Typically, the age of onset of AD is 80–85 years, but *APOE* ε4 heterozygotes who develop AD begin to show symptoms in their mid-70s. Onset occurs even earlier in ε4/ε4 homozygotes—in their mid-to-late 60s. Although the association between AD and APOE has been confirmed in many studies, the underlying mechanism linking APOE and AD is unclear.

The *APOE* genotype is believed to account for approximately 50% of the susceptibility to late-onset AD, but it is neither necessary nor sufficient for the disorder. An estimated one-half of all individuals with ε4 will never develop AD, and many patients who have AD do not have an ε4 allele. For these reasons, many physicians recommend against offering APOE testing to patients concerned about developing AD. As an unfortunate consequence, many "direct to consumer" DNA diagnostic companies are marketing an APOE test. The availability of these tests places consumers in the grim position of having information about their risk of developing this dreadful and untreatable disease, but not knowing how to interpret it.

Cytogenetic Abnormalities Reveal Genes Involved in Schizophrenia

We have seen that even with refined association studies, millions of markers, and high-throughput array techniques, it is still very difficult to find the key players among the many loci that come under suspicion. Large chromosomal abnormalities visible by light microscopy proved invaluable in the early days of hunting for genes involved in human genetic disorders. For example, researchers trying to clone the gene associated with DMD exploited a rare translocation to pinpoint and clone the dystrophin gene. Similar translocations have recently fingered four genes that may be involved in schizophrenia. Using such translocations is a powerful gene-finding strategy; it is a genome-wide approach that pinpoints a candidate locus in a family or even a single case.

A set of families with cytogenetic abnormalities was screened for associated mental illness. The screening identified one large family in which there was an association between the occurrence of a serious mental disorder and a balanced translocation between the long arms of chromosomes 1 and 11, at 1q42;11q14.3. Eighteen out of 29 family members with the translocation, and none of 38 without it, had a diagnosis of major mental illness. The regions around the breakpoints were isolated by microdissection and cloned. Few if any genes were found in the region of the breakpoint on chromosome 11, but there were two genes, *Disrupted-in-Schizophrenia 1* and *2* (*DISC1* and *DISC2*), at the 1q42 breakpoint. Subsequently, linkage studies in more than 200 Finnish families with schizophrenia (but without the translocation) showed convincing linkage to 1q42. *DISC1* is a good candidate for a gene involved in a psychiatric disorder. It is expressed in many tissues, including the brain, and the DISC1 protein interacts with proteins known to be involved in neurodevelopment, neurotransmission, and cell signaling. RNAi knockdown of *DISC1* in mouse embryonic brains reduces neuronal migration and neurite extension. Most interestingly, transgenic mice with homozygous mutations in *Disc1*, the mouse homolog of the human gene, have working memory deficits.

A second translocation has been found in an individual with psychosis, involving breakpoints at 1p31 and 16q21. The former disrupts the gene encoding phosphodiesterase 4B (*PDE4B*), a gene of special interest in the context of nervous system disorders. The PDE4B phosphodiesterase inactivates cyclic AMP, and flies having mutations in the *Drosophila* gene homologous to human *PDE4B* have learning and memory deficits, whereas corresponding mutant mice exhibit depressive-like behavior. Remarkably, DISC1 and PDE4B, independently identified as possibly contributing to schizophrenia, were shown to interact in a yeast two-hybrid screen (Chapter 13).

A third translocation between chromosomes 9 and 14 is associated with schizophrenia, and again the disrupted gene is a promising candidate. Molecular analysis revealed that the breakpoint at 14q13 disrupted the gene *NPAS3*, which encodes a helix–loop–helix transcription factor. Studies of transgenic mice with mutations in *Npas3* show various behavioral abnormalities, including impaired social interactions and altered motor behavior. The mutant mice also have reduced neurogenesis, which leads to developmental deficiencies in the hippocampus.

In all these cases, cytogenetic abnormalities have unambiguously defined genes that contribute to some cases of schizophrenia. In addition, the family with (1;11)(q42;q14.3) translocation has shown that the same mutation can be found in individuals with quite different diagnoses. Of the 29 family members with the translocation, seven were diagnosed with schizophrenia, one with bipolar affective disorder, and ten with recurrent major depression. The differing expression of the mutation presumably reflects the effects of modifying genes elsewhere in the genome and shows that diagnostic heterogeneity may mask a common underlying mutation. Cytogenetic abnormalities have also been found in other complex disorders including autism, Tourette syndrome, dyslexia, and bipolar affective disorder. If these cytogenetic changes are the underlying cause of the disorders in these individuals, then the genes flagged by these chromosomal changes will provide a well-defined starting point for exploring the pathways in which they operate.

The HapMap Project Has Collected Many Millions of SNPs

One challenging goal to come out of the HGP is to catalog the vast amount of DNA sequence variation present in our genome. The International HapMap Project and Perlegen Sciences have identified about ten million human SNPs in four populations, two Asian, one European–American, and one African. Many of these polymorphisms are believed to contribute to common, polygenic traits, including diseases. It would be a daunting and practically impossible task to examine each SNP for its association with a disease. Fortunately, *haplotypes* can help. SNPs that are adjacent to each other on a chromosome are often inherited as a unit, where many, closely spaced SNPs do not separate from each other during meiosis when chromosomes cross over and exchange DNA. Thus, the presence of a particular variant in one region of the genome often can predict the occurrence of specific variants at other nearby sites. The set of alleles that remain associated is called a haplotype and the stretch of DNA sequence containing them is a *haplotype block*. Haplotypes make it possible to use only one or a few SNPs to determine the genotype of a large region of the genome, which greatly reduces the number of SNPs needed to scan the whole genome. Indeed, a few hundred thousand SNPs are sufficient for a genome-wide scan in a genetic association study.

HapMap researchers search the genome for regions of *linkage disequilibrium*, the preferential association between specific alleles at neighboring markers. For example, two adjacent markers, A and B, each have two alleles, A1 and A2, and B1 and B2 (Fig. 14-9). There are four possible haplotypes: (A1, B1), (A1, B2), (A2, B1), and (A2, B2). If all four are observed with equal frequency in a population, markers A and B are considered to be in linkage equilibrium. If however, a particular haplotype, such as (A1, B2), occurs more frequently than expected, the markers are said to be in linkage disequilibrium with each other. The presence of allele A1 indicates that the allele at marker B is B2, and conversely.

Linkage disequilibrium is believed to result from distant founder effects, in which chromosomes carrying susceptibility or disease-causing alleles have descended from a common ancestor. Meiotic recombination did not shuffle the susceptibility allele onto other haplotypes and, as a result, the ancestral haplotype is observed at a high frequency for the region surrounding the disease gene (Fig. 14-9). Linkage disequilibrium occurs at varying levels throughout the genome. Some parts of the genome have very little linkage disequilibrium and, by inference, very high levels of recombination or mutation that prevent the inheritance of large haplotype blocks. And there are also long stretches where specific haplotypes have been passed from parent to child through many generations. Although most of the common haplotypes occur across all human populations, some variation does occur. To take account of this variation, researchers in the

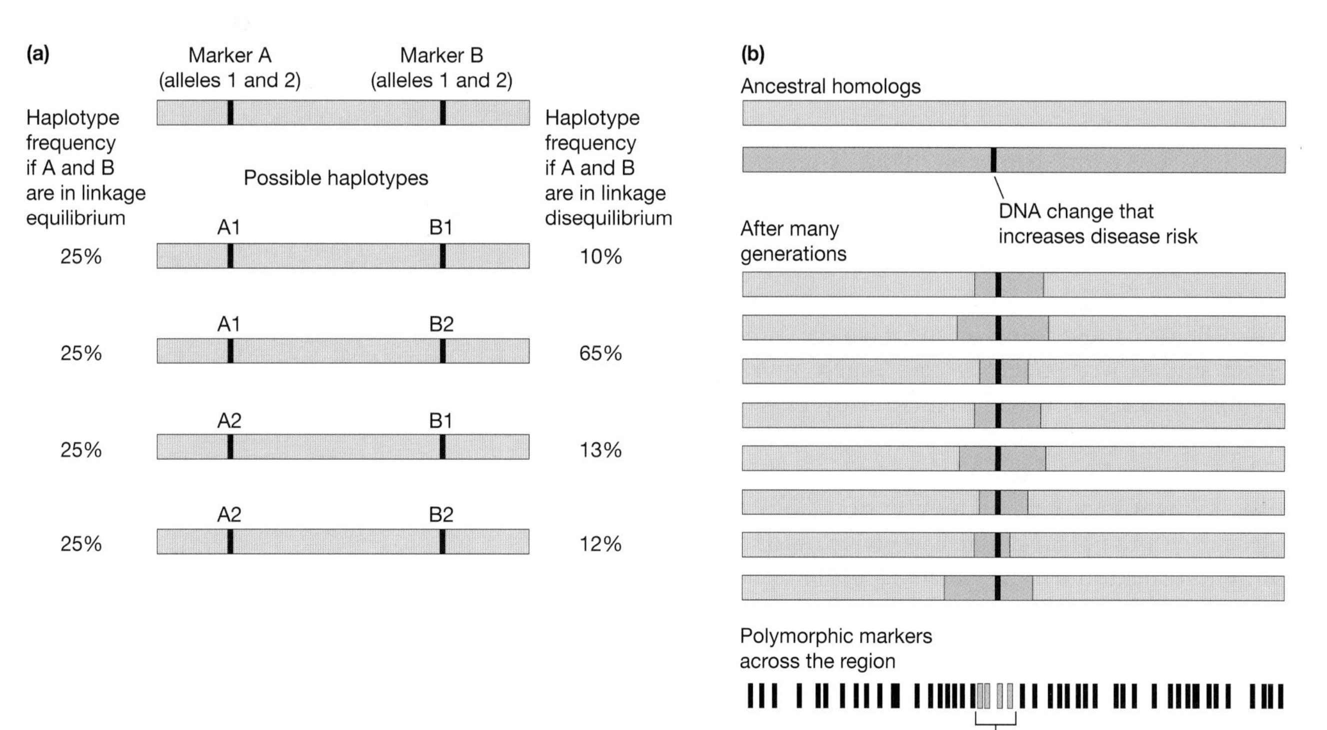

FIGURE 14-9

Linkage disequilibrium (LD). (*a*) LD is the coinheritance of specific combination of alleles at neighboring markers (haplotypes) more often than expected by chance. (*b*) LD is found in regions of the genome where specific allele combinations have not been separated by meiotic recombination over many generations. The *green* represents the set of alleles in the ancestral chromosome. Over generations, recombination leads to new alleles except in the region immediately surrounding the disease locus. It can be used to explain positive genetic association studies, if the associated marker is in LD with the causative risk allele.

HapMap Project gathered samples from a range of different populations, including Nigeria, Japan, China, and the United States (primarily populations representing Northern and Western European ancestry). New populations continue to be studied and added to the databases.

Geneticists exploit the information from the HapMap Project when selecting a panel of SNP markers for use in candidate gene or genome-wide association studies. Most of the genetic variation present in humans can be identified by genotyping approximately 400,000 so-called "tag" SNPs, which represent about 4% of the total number of human SNPs. Each tag SNP serves as a signpost for the information contained within its haplotype block. By examining the genotype of the tag SNP, researchers can infer the genetic sequence of the other SNPs in the haplotype block, thereby reducing the number of genotyping tests to be performed. The HapMap tag SNPs identify most, but not all, of the genetic variation present in human beings. Rare nucleotide changes that alter protein function, resulting in some of the susceptible risk alleles, are not detected in HapMap studies. Therefore, if a positive association is found with a tag SNP, the entire haplotype block is searched in hope of discovering a candidate gene. Each candidate gene must be sequenced in patients and in normal individuals to determine whether or not there are mutations that could be responsible for the disease. Before the advent of the HapMap Project, these studies were onerous and hugely time-consuming. Investigators had to identify SNPs by resequencing the region of interest in a group of individuals that represented the study population, a process that often required months of tedious work. Today, with the HapMap data available on-line, the same information can be obtained within a matter of minutes.

Macular Degeneration Gene Is Located and Cloned by Using HapMap SNPs

Age-related macular degeneration (AMD) is characterized by progressive degeneration of the macula, the central region of the retina, and is one of the major cause of blindness. Individuals with AMD may first notice a blurring of central vision, which progresses to blind spots (Fig. 14-10). Inheritance studies indicate that there are genetic contributors to AMD, but the inheritance is complex, involving multiple genes of modest effect interacting together. In view of this complexity, researchers have used genetic association studies to identify three genes that increase the risk of AMD. These studies were greatly aided by data from the HapMap Project and are an indication of the value of having haplotype information on very large numbers of freely available, mapped SNPs.

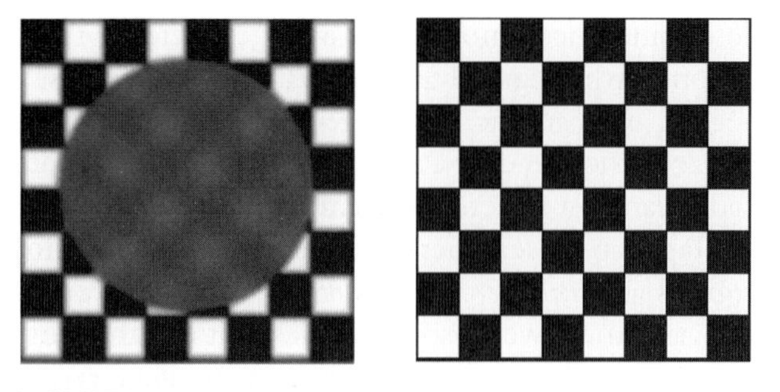

FIGURE 14-10

Loss of central vision due to age-related macular degeneration (AMD). Peripheral vision is often spared, but objects in the center of vision are blurred or completely blocked. The image at *left* has been altered to show the vision lost due to AMD.

One study involved 96 individuals with AMD and 50 control subjects. The population variation in SNP frequencies was controlled by examining individuals who self-identified as "white, not of Hispanic origin," and great care was taken to ensure that the diagnosis was correct. Genotyping analysis was carried out using 103,611 autosomal SNPs. Of these, only two showed association with the disease. Both SNPs are located within an intron of the gene encoding complement factor H (CFH) on chromosome 1q31. This protein regulates the complement system of innate immunity, which recognizes, and targets for destruction, viruses, bacteria, and damaged cells. As the two SNPs do not change amino acids in the protein and are not thought to be in a regulatory element, it is unlikely that either is the primary causative mutation. More likely, the SNPs are in linkage disequilibrium with a mutation that results in a functional change in the *CFH* gene.

The investigators defined a haplotype block of 500 kb using 19 SNPs from their own data set, but when they turned to the International HapMap Project, they found a further 152 SNPs from the region, which they used to narrow the linkage-disequilibrium region to just 41 kb. They subsequently sequenced the *CFH* gene in 96 individuals and found another SNP in exon 9 of the *CFH* gene that leads to the substitution of histidine for tyrosine in the protein. The same association was identified in a second, independent study of 404 patients and 131 controls. This change is likely to alter the binding of the complement H factor to downstream regulators of the complement system, thereby increasing inflammation and its consequences. Inheriting a single copy of the allele with histidine carries an almost fivefold increase in risk for macular degeneration, and inheriting two copies increases risk by a factor of more than 7. That the complement system and inflammation play a role in AMD is further supported by a third study that found association between AMD and two other components of the complement system, complement factors B and 2. In this case, the data from the HapMap project effectively reduced the search for the cause of AMD from years to months.

Many Human Genetic Disorders Occur More Frequently in Specific Populations or Ancestral Groups

The frequency of some disorders varies among different populations. For example, women of Ashkenazi Jewish descent are more likely than women from other ethnic backgrounds to have a certain form of breast cancer. This is because a particular mutation in the *BRCA1* gene—inherited in an autosomal-dominant manner—is present at higher frequency among individuals of Ashkenazi Jewish ancestry. *Positive selection* and *founder effect* provide two alternative explanations for the increased frequency of a mutation in a particular population.

In the 1940s, investigators noted that the sickle-cell trait (present in individuals heterozygous for the sickle-cell mutation HbS) is curiously prevalent in certain parts of the world. Among African populations from the coastal regions of Kenya and the area surrounding Lake Victoria, the frequency of sickle-cell trait can be as high as 20–30%. As individuals with sickle-cell anemia (homozygous for HbS) rarely survive into adulthood, it was puzzling that the frequency of the single allele should be so high. Anthony Allison realized that the high frequency of the sickle-cell trait occurred in regions where the transmission of the malarial parasite *Plasmodium falciparum* was also high. He proposed that the sickle-cell trait conferred some resistance against malaria. And, indeed, this hypothesis has been confirmed—a recent study in Kenya has reported that sickle-cell trait is associated with 50% protection against mild clinical malaria, 75% protection against hospitalization due to malaria, and a 90% reduction in severe malarial episodes. Thus, although sickle-cell disease is a negative selective force against the HbS allele, the *positive selection* afforded by resistance to severe malaria in HbS carriers maintains a relatively high frequency of the allele in populations where malaria is endemic. It is thought that the high frequencies of cystic fibrosis in Northern Europeans and of Tay–Sachs disease in the Ashkenazi Jewish population may be explained by a similar logic—that these diseases may confer some protection against cholera and tuberculosis, respectively.

A second factor that affects the distribution of allele frequencies is the *founder effect*, in which certain alleles in a limited population are inherited from a common ancestor. Suppose that a group from a large and diverse population moves from its original location to a second location where it is reproductively isolated from other populations. The genetic composition of this subgroup then represents only a fraction of the overall genetic variation present in the original population (Fig. 14-11). If one (or a few) of the families in the subgroup carries a particular recessive mutation associated with a genetic disease, the mutation will be passed down through generations and gradually become more frequent. In a small population, the likelihood is increased that two people (each carrying one copy of the mutation) will mate and, among their children, one-quarter will be homozygous for the mutation. This pattern is called a founder effect and the initial mutation is called a founder mutation (Fig. 14-11). The same effect can occur if a population undergoes what is called a bottleneck—a drastic reduction in size, followed by expansion from the members of the surviving subgroup. It is for this reason that clinical geneticists and genetic counselors will discuss a client's racial and ethnic background when taking a family history. This information may help identify mutations or diseases for which a family may be at higher risk.

DNA Microarrays Detect Copy Number Polymorphisms in Chromosomes

Abnormalities on the order of millions of base pairs can be detected by light microscopy using stains that produce banding patterns on chromosomes. Smaller genetic alterations escape detection with these methods, but fluorescence in situ hybridization (FISH)

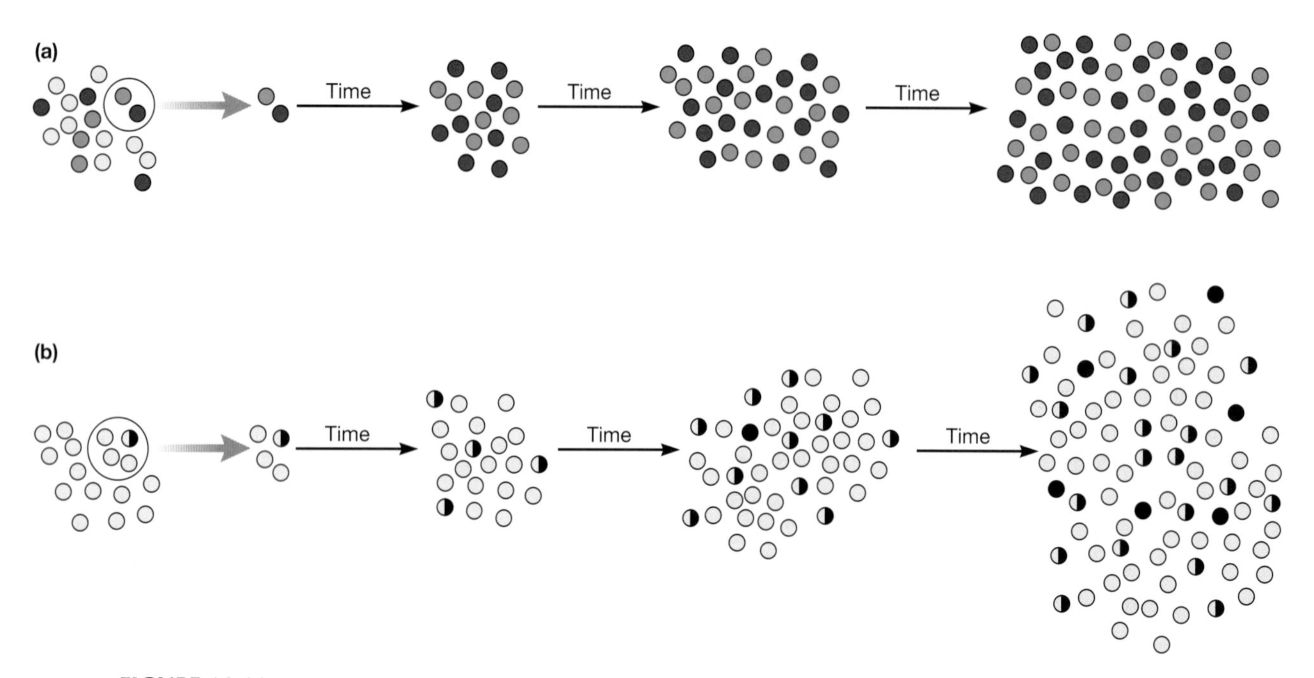

FIGURE 14-11

Founder effect on distribution of allele frequencies. (*a*) As indicated by the different shades of color, the original population encompasses a great deal of variation. When a small subpopulation moves away from the original population and reproduces, the resulting population does not represent the full diversity of the original population and does not include all the variants in the population. This is known as genetic drift. (*b*) If a mutation carrier for an autosomal-recessive disorder (*half-black circle*) is present in the migrating subpopulation, the carrier frequency will increase over time as the subpopulation reproduces. This increases the likelihood that two carriers will reproduce and have a child homozygous for the mutation, so the frequency of the disorder is higher in the subpopulation than in the original population from which it was derived. The mutation is known as a founder mutation.

can identify changes involving tens of thousands of nucleotides. A FISH probe for an autosomal locus will normally highlight two chromosomal segments, one for each locus of the homologous autosomes. If one copy of the sequence is deleted, the signal is observed on only the normal chromosome; a duplication of the sequence, in contrast, will produce three distinct signals. FISH can be performed on both metaphase cells and interphase cells. The former takes several days because the cells have to be grown and then arrested in metaphase, whereas FISH analysis of interphase cells can be completed in as little as 24 hours.

A recent adaptation of microarray analysis, array comparative genomic hybridization (array CGH) provides an even higher resolution than FISH. It is capable of detecting and measuring changes in DNA copy number (the extent of DNA amplification or deletion) at the level of the gene on a genome-wide scale (Fig. 14-12). These changes are commonly recognized as the underlying basis for many genetic disorders (monosomy, trisomy, and submicroscopic deletions and translocation) as well as complex diseases such as cancer. The microarray is composed of mapped genomic clones, typically genomic libraries made in high-capacity vectors such as BACs (bacterial artificial chromosomes). The array is thus the equivalent of performing hundreds or thousands of FISH assays simultaneously, at a vastly higher resolution than that provided by cytogenetic staining, and allows identification of tiny genomic changes that have previously gone undetected.

SNP-based Diagnosis Makes Use of Large-Scale, High-Throughput Techniques

Once the gene for an inherited disorder has been identified and characterized, it becomes possible to develop molecular tests for the disorder. In the second edition of this book, we commented on the large number of DNA-based tests then available—in 1992 more than 200 disorders could be diagnosed using DNA-based methods. Today, molecular tests

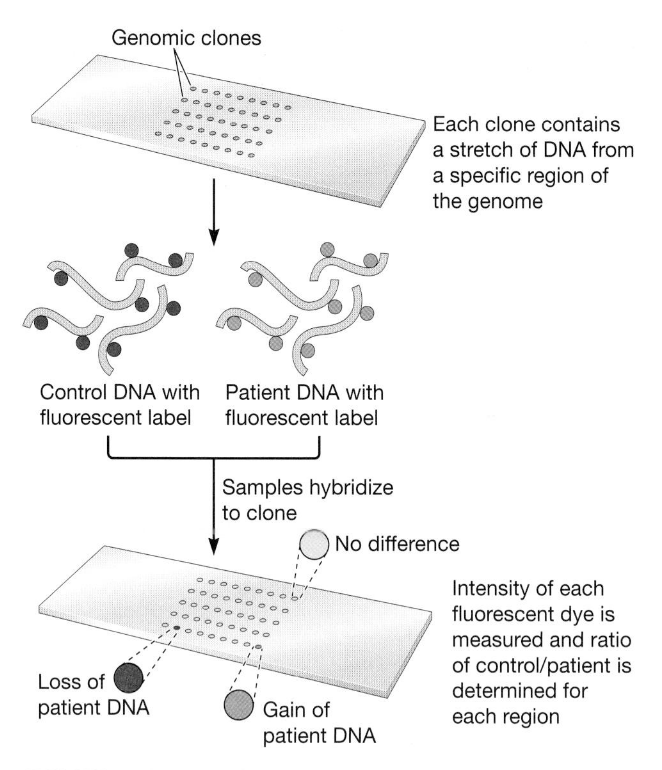

FIGURE 14-12

Array comparative genome hybridization (array CGH). Array CGH is performed with a panel of clones containing DNA fragments from specific regions across the genome. Depending on the type of study, hundreds or thousands of these clones are fixed to a small glass slide. Two sets of DNA are prepared: DNA from a wild-type or reference set and the DNA from the patient under study. Each preparation is labeled with a different fluorescent dye to distinguish wild type from patient, and both sets of DNA are hybridized to the same array. Hundreds of copies of each clone are present at each site of interest, so both reference and patient DNA can hybridize the same position on the array. The relative intensity of the fluorescent dyes at each position reflects how much DNA hybridized to each clone, and a ratio of control to patient DNA is developed at each of the probes on the array. An equal amount of the two dyes represents equivalent amounts of DNA between control and patient. Higher levels of the control dye suggest a loss (deletion) of patient DNA at the probe, whereas higher levels of patient dye suggest a duplication of patient DNA at the probe region.

based on a variety of techniques exist for more than 2000 genetic disorders. The particular method chosen depends on characteristics of the disease—the range of mutations involved, their frequency in the patient's ethnic group, and their effects on the encoded protein. These approaches are, however, time consuming and expensive. To exploit the information of the HGP, investigators need techniques to examine the sites of mutations quickly and on a much larger scale, hundreds or thousands of nucleotide positions at the same time. Three techniques—mismatch detection, single-base extension, and allele-specific primer extension—may prove invaluable in diagnostic work.

Mismatch detection on microarrays can be used for the simultaneous genotyping of up to 100,000 mutations (Chapter 13). Not all mutations can be detected because the stability of hybridization between the patient DNA and oligonucleotide bound to the array is highly dependent on the sequence surrounding the polymorphic site. For some SNPs, flanking sequences prohibit their inclusion on the array platform, a serious drawback for this technology. For single-base extension analysis, the SNP genotyping reaction occurs directly on nonamplified genomic DNA. PCR amplification then follows this genotyping step to increase the detection of the genotyped molecule. In contrast to microarray methods, this approach can amplify nearly any SNP and provides a much clearer distinction between alleles than does hybridization to an oligonucleotide. Molecular inversion technology is a clever approach to single-base extension (Fig. 14-13). Allele-specific primer extension is similar to single-base extension methods. The SNP-genotyping reaction is carried out directly on genomic DNA. This step is followed by PCR amplification to increase the number of genotyped molecules (Fig. 14-14). Tens or even hundreds of thousands of SNPs can be genotyped in a single experiment. As with the single-base extension methods, this technology is capable of assaying nearly any SNP using a standard set of reaction conditions—a marked advantage compared to standard microarray analysis.

DNA Sequencing Provides Definitive Diagnosis of Mutations in Breast Cancer

The most direct way to determine whether a mutation is present in a gene is to sequence the gene in genomic DNA isolated from a patient. This approach is, of course, expensive but has been used for diagnostic testing for mutations in the *BRCA1* and *BRCA2* genes. Between 5% and 10% of the nearly 200,000 women in the United States diagnosed with breast cancer have an inherited form of the disease, and many of these heritable cases are due to mutations in either *BRCA1* or *BRCA2*. Women with a mutation in one of these genes are three to seven times more likely to develop breast cancer

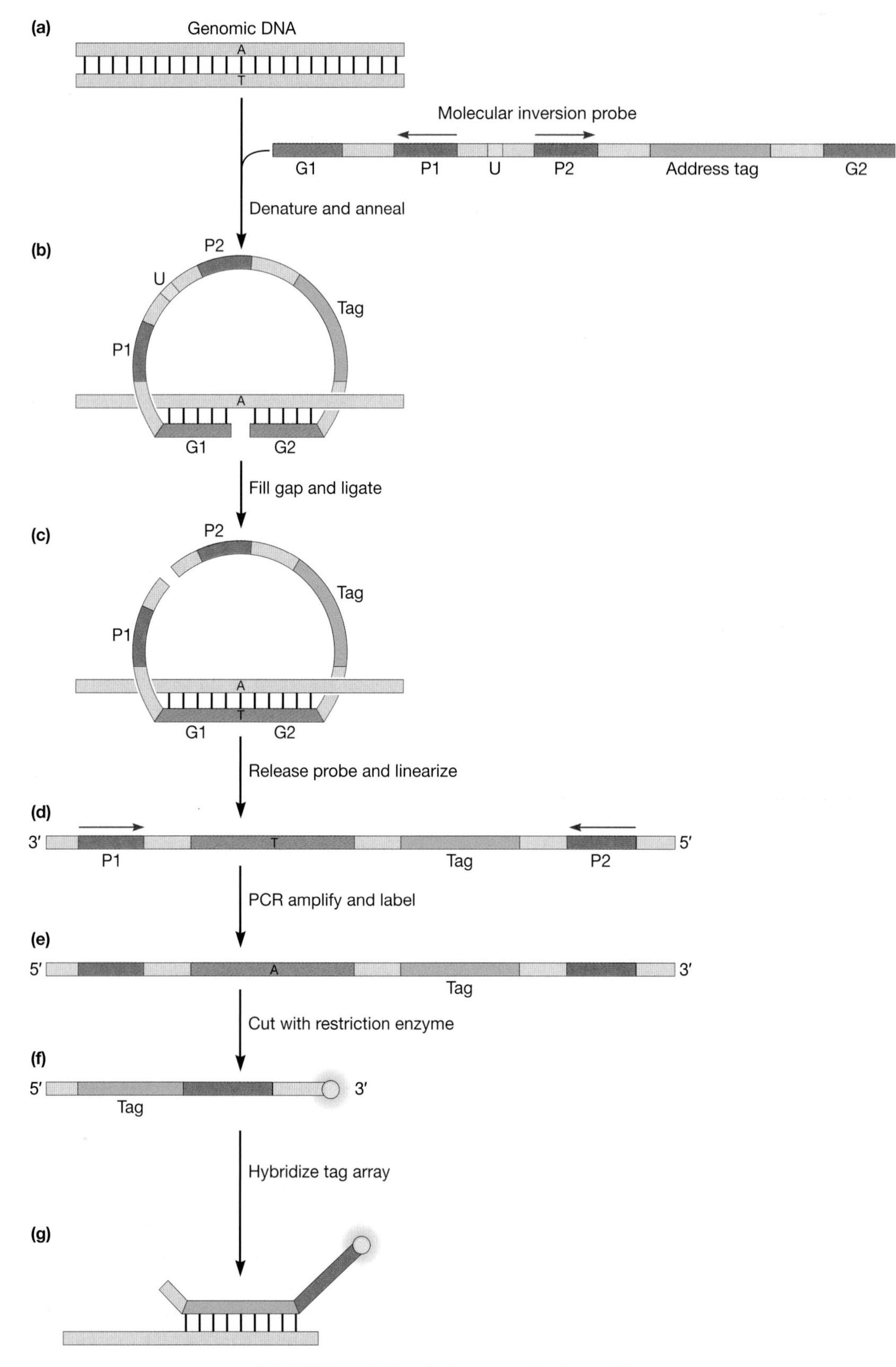

FIGURE 14-13 (See facing page for legend.)

than are women with normal *BRCA1* and *BRCA2* sequences, and onset often occurs at a young age (before menopause). Mutations in these genes are also associated with an increased risk of ovarian cancer as well as prostate cancer (*BRCA1*) and breast cancer (*BRCA2*) in men.

More than 600 different mutations have been identified in the *BRCA1* gene and more than 450 have been detected in *BRCA2*. The majority of the mutations are unique to each family and so testing cannot be done for just a few common mutations, which is a complicating factor for widespread diagnostic analysis. Instead, DNA sequencing detects *all* changes in *BRCA1* and *BRCA2* in women at risk for developing breast cancer. Typically, the family analysis begins with a woman who has already been affected with breast or ovarian cancer and is most likely to reveal what, if any, mutation is present in the family. It is important to appreciate that the failure to identify a *BRCA1* or *BRCA2* mutation does not eliminate the possibility that a mutation is present in the family—not all heritable cases of breast cancer are due to mutations in *BRCA1* or *BRCA2*. Furthermore, the clinical significance is unclear for approximately one-third of the sequence changes found in these genes. It is likely that a subset of these changes will ultimately prove to be normal variants that do not lead to an increased cancer risk.

New Technologies Offer Large-Scale Sequencing-based Diagnosis

Current methods of DNA sequencing are based on the dideoxy method developed 30 years ago by Frederick Sanger (Chapter 4). Early in the HGP, it was thought that the Sanger method and DNA sequencing machines would not be able to sequence a genome of 3 billion base pairs, and several projects were begun to devise new sequencing strategies. None of these came to fruition, and, in the event, conventional sequencing was simply scaled up to provide the needed capacity. Industrial-scale sequencing has led to a significant drop in cost, but the standard method of sequencing remains too expensive to be widely applied for diagnosis. As a consequence, there is a renewed interest in developing new sequencing techniques.

One method, pyrosequencing, does not use Sanger chemistry at all. Instead of reading the dideoxy nucleotide that terminates the fragments

FIGURE 14-13

Molecular inversion probes are used for large-scale detection of SNPs. (*a*) An individual is heterozygous for a SNP, having an A:T allele on one chromosome and a G:C allele on the other. (To simplify the diagram, only the A:T allele is shown.) The DNA is denatured and a molecular inversion probe (MIP) is annealed to it. Molecular inversion probes contain two regions (G1 and G2), which are complementary to sequences that flank the SNP. There are two primer-binding sites, orientated away from each other (P1 and P2) and, between them, a segment of DNA containing several residues of the unusual deoxynucleotide, dUTP, which is not normally found in DNA (although the nucleotide uracil is a component of RNA). There is also a tag or address sequence (*green*), designed as a unique stretch of DNA that is specific to this particular locus. This segment is used, in the last step of the procedure, to capture the oligonucleotide by hybridizing to its complementary oligonucleotide attached to an array. (*b*) The MIP anneals to the DNA, leaving a single-nucleotide gap aligned at the SNP site. As the probe anneals to the genomic DNA, it becomes entwined. (*c*) When the probe is circularized by filling the gap with an allele-specific deoxynucleotide, here an A, and stitched together with DNA ligase, the circular probe is "padlocked" to the DNA (shown here by a single interlocking). Four separate reactions are carried out, one for each base. If the added deoxynucleotide is not complementary to the allele present at the SNP site, the gap is not filled, ligation will not occur, and the probe remains linear. The reaction mixture is treated with exonuclease to destroy all linear DNA molecules and then heated to inactivate the exonucleases. The circular probes are then released from the genomic DNA in two steps. First, the reaction mixture is treated with uracil-*N*-glycosylase, which removes the uracil from the deoxyribose in the DNA. Second, on heating, the probe DNA backbone is broken at the points where the uracils were removed. (*d*) The probe is melted from the genomic DNA and released. The linear probe now has the primers pointing in the opposite direction from the starting probe (i.e., it is "inverted"). (*e*) Amplification by PCR with fluorescently labeled primers leads to a labeled product of only those molecules that had been circular. (*f*) After trimming with a restriction enzyme, the molecules are hybridized to an array with address tags complementary to those of many different probes. (*g*) Detection of a specific fluorescent marker associated with a given SNP tag genotypes the SNP.

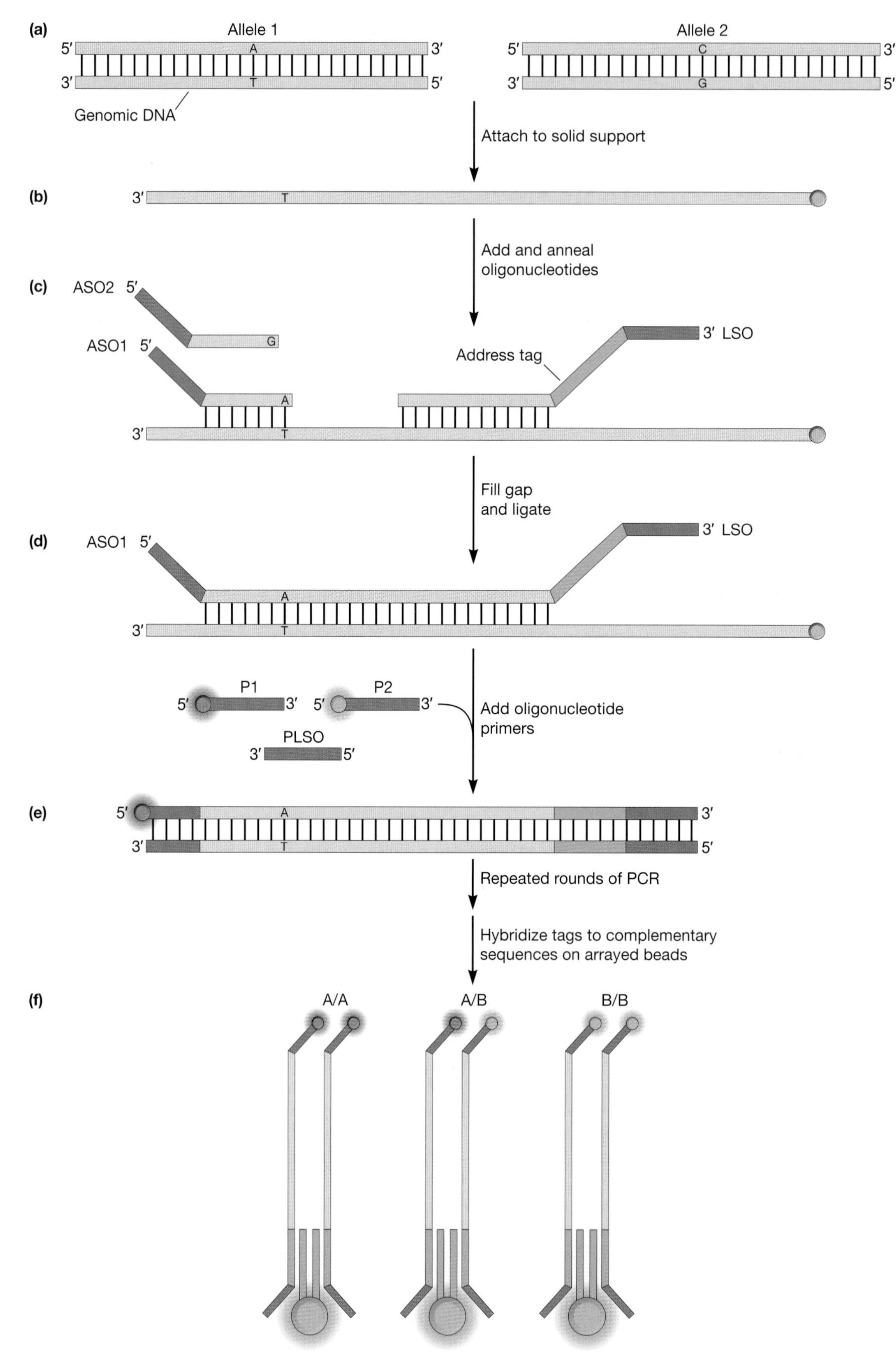

FIGURE 14-14 (See facing page for legend.)

synthesized in the sequencing reaction, the DNA to be sequenced is denatured and the single strands attached to beads. These beads are spread over a surface and immobilized to the surface. DNA polymerase and bioluminescent proteins (including luciferin, the firefly enzyme luciferase) are added to the wells, together with one nucleotide triphosphate, for example, dGTP. This base pairs to the next position on the template strand where the base is a C. Two of the phosphate groups are released as pyrophosphate (PP_i), which is converted to ATP by the enzymes in the reaction mixture. The ATP is used by luciferase to cleave luciferin, producing a flash of light wherever the G was added. The solutions are removed and replaced with another nucleotide triphosphate and the process is repeated. In this way, the sequence of the template strand can be built up for each of the beads. The reads obtained with this method are shorter than for Sanger sequencing, but so many reactions are performed simultaneously that a very large amount of data can be collected in a single experiment.

A second strategy uses an approach similar to dideoxy sequencing but with terminators that are reversible. DNA fragments to be sequenced are immobilized at their 5′ ends to a glass surface and amplified to produce a small (1-μm-diameter) island of identical fragments. All four fluorophore-labeled reversible terminators, together with primers and DNA polymerase, are added and one nucleotide is added to the growing chain. The unincorporated chemicals are washed out and the terminator present at the end of all the newly synthesized chains is determined. Next, the blocking groups are removed, the chemical synthesis repeated, and the second terminator is read. The sequencing reads are shorter than those obtained with Sanger sequencing, but, once again, the huge numbers of fragments that are sequenced in parallel makes this an extremely fast method, with potentially tens of millions of bases sequenced in a single analysis.

Pharmacogenomic Strategy Leads to Improved Treatment for Epilepsy

Our genes influence everything about us, including our responses to drugs. Two patients can respond differently to the same medication: One receives clear benefit whereas the other experiences no improvement or even adverse effects. Current approaches to identifying the correct medication and dosage follow a trial-and-error format, often with tragic results. In the United States alone, adverse drug reactions have been estimated to cause more than two million hospitalizations and 100,000 deaths annually. Variability in efficacy and toxicity of medication is likely due to

FIGURE 14-14

High-throughput detection of SNPs using allele-specific probes and tag addresses (Illumina, Inc.). (*a*) An individual is heterozygous for a SNP, having an A:T allele on one chromosome and a G:C allele on the other. (To simplify the following diagram, only the A:T allele is shown.) (*b*) Following denaturation, the DNA is biotinylated and attached to an avidin-coated bead. (*c*) Three oligonucleotides are added. Two allele-specific oligonucleotides (ASOs), ASO1 and ASO2, are complementary to the sequence 3′ to the SNP (*blue*) and end in a base complementary to the SNP, in this case an A (ASO1) and a G (ASO2). They each have a unique sequence (gray) that is complementary to PCR primers used later in the procedure. The third, locus-specific oligonucleotide (LSO), has three regions. At the 5′ end is a sequence complementary to the sequence a few bases 5′ to the SNP (*blue*). The second region is a tag or address sequence (*green*), designed as a unique stretch of sequence that is specific to this particular locus. This sequence is used to capture the oligonucleotide by hybridizing to its complement attached to a bead in the last step of the procedure. There are 1152 address tags so that 1152 SNPs can be tested in one multiplex reaction. The third region is a sequence that will be used for PCR (*gray*). (*d*) After annealing the ASO (in this case ASO1 carrying an A) and the LSO to the single-stranded genomic DNA template, the gap is filled and then ligated to form a continuous second strand. (*e*) PCR is then carried out with primers P1 and P2 complementary to the sequences in either ASO1 or ASO2 (here P1 for ASO1) and primer PLSO, complementary to the sequence in the LSO. P1 and P2 are fluorescently labeled at their 5′ ends, each with a different dye, so that a different color is associated with either A or G for the other allele. The products of the PCR are P1-A-LSO for an AA homozygote, P2-G-LSO for a GG homozygote, and a mixture of both for an A/G heterozygote. (*f*) The PCR products are captured by the unique address tag on the LSO to a bead carrying the complementary sequence. In this diagram, an AA homozygote produces a red signal, a GG homozygote a green signal, and a heterozygote a mixed color.

multiple factors, including genetic and environmental influences on drug metabolism as well as the drug's mode of action. One of the most interesting and potentially most important outcomes of the HGP will be to use genetic information to guide decisions on the medications an individual will receive. This is the field of *pharmacogenomics*, sometimes referred to as "personalized medicine."

Children with a type of progressive myoclonic epilepsy, called EPM1, have seizures and a progressive neurological deterioration that results in motor problems and dementia. The symptoms in these children are similar to a more generalized form of epilepsy, and often even well-trained neurologists do not recognize that a patient has EPM1. Indeed, EPM1 is one of five main causes of progressive myoclonic epilepsy, and the optimal treatment is different in each subtype. Misdiagnosis of EPM1 can be harmful for these children if they are prescribed dilantin, the most common antiepileptic medication, as this drug can significantly worsen many of the symptoms of EPM1 by precipitating neurological deterioration. EPM1, which is inherited in a clear, autosomal-dominant manner, was found to be due to loss-of-function mutations in the gene encoding cystatin B, a protease inhibitor found in all cells in the body. Because we know the genetic cause of EPM1, doctors can perform a DNA test to determine whether a child with the symptoms of progressive myoclonic epilepsy symptoms carries mutations in this gene and hence has EPM1. In this case, the doctor can treat the child with another drug (valproate) appropriate for the condition, rather than prescribing dilantin and causing harm to the patient.

The case of EPM1 is an example of the growing field of "personalized medicine," in which genetic analysis is being used to improve the safety and efficacy of the treatment of diseases with drugs. Sequence variants that alter the function of the protein target of a drug can affect the drug's action by either enhancing or reducing its efficacy. One broad class of molecules that are particularly important in drug response and toxicity are the proteins that transport and metabolize drugs. Some of the best studied drug-metabolizing enzymes include members of the cytochrome P450 (CYP) gene family. These proteins function in the liver and code for oxidases that modify the chemical structures of many drugs when they have been absorbed from the gut. These modifications cause the drugs to be excreted, thereby reducing their concentration and therapeutic contribution. Remarkably, six of these CYP enzymes, accounting for more than 90% of total oxidative activity, clear almost 60% of all drugs from the body. Polymorphisms in the genes encoding these enzymes modulate their effects, thus leading to a range of enzyme activities. Patients can be screened for genetic variations in two of the CYP enzymes by microarray analysis. One microarray identifies variations in *CYP2D6*, which encodes an enzyme that metabolizes antidepressants, antipsychotics, antiarrhythmics, and β-blockers. The second microarray examines variations in *CYP2C19*, encoding the enzyme that metabolizes anticoagulants, anticonvulsants, proton pump inhibitors, benzodiazepines, antimalarials, and other medications. These tests can identify the range of drug metabolism in patients ranging from poor metabolizers (who may develop serious side effects from high levels of unmetabolized drugs) to ultrarapid metabolizers (who may modify the drugs so quickly that they receive little therapeutic effect at standard doses). This screening therefore enables drug doses to be tailored to an individual's metabolic activity. As we come to understand better the genes that control drug metabolism, pharmacogenetics will become increasingly common, thereby reducing the frequency of adverse side effects and identifying the most effective drug for a patient.

RNAi Is Used to Inhibit Overgrowth of Blood Vessels

Our ability to treat genetic diseases has been generally disappointing. A few therapies have been effective for treating inherited disorders of metabolism. For example, a strict phenylalanine-free diet can diminish the effects of phenylketonuria, and enzyme-replacement therapy has been used successfully to treat certain metabolic defects—for example, mucopolysaccharidoses and lipidoses. But in general our lack of understanding of the underlying genetic basis for most inherited disorders accounts for our relative lack of success. Fortunately, as the genetic basis of these disorders is determined, new strategies for therapy become available for investigation.

For example, remarkable and unexpected progress has been made in correcting genetic disorders through suppressing and correcting the relevant mRNAs in patients. Early mRNA-based therapies using antisense mRNAs showed some promise, but proved ultimately disappointing. Recent discoveries in the field of RNAi (Chapter 9), however, have renewed interest in RNA-based strategies and have led to some quite encouraging results. Short double-stranded RNA molecules (siRNAs) complementary to the gene of interest combine with the RNA-induced silencing complex (RISC) and direct it to bind the mRNA of the targeted gene. Once bound, the mRNA is selectively degraded, which effectively silences the gene from which it was transcribed.

The first RNAi therapy to enter clinical trials was for the "wet" form of age-related macular degeneration (AMD), a disorder that destroys central vision. AMD is hastened by high levels of vascular endothelial growth factor (VEGF), a protein that promotes blood vessel development behind the retina. In AMD, these vessels often leak, thereby clouding and degrading central vision. In a series of remarkable clinical trials, double-stranded RNA complementary to VEGF was injected directly into the eyes of individuals with AMD. Although the trials were intended primarily to assess safety issues, the initial therapeutic results were promising. Two months after injection with the drug, a quarter of the patients experienced significantly clearer vision, and the vision of the other patients had stabilized.

The effective application of RNAi therapies still faces daunting hurdles. These include identifying the best methods for delivery of RNAi molecules to the right targets, avoiding siRNAs veering "off-target," silencing other genes or cellular processes, and determining how to ensure that the drugs remain active long enough to provide clinical benefit.

Severe Combined Immune Deficiency Is Treated by Retroviral Gene Therapy

Gene therapy—replacing a mutant gene with a normal copy—has a long and controversial history. Despite the successful "treatment" of numerous animal models using gene therapy protocols, correcting human hereditary disorders has proven much more difficult. Once the target disease gene has been identified, a way must be found to deliver the normal gene so that it can compensate for the mutant gene; great care must be taken to assure that the gene is expressed at appropriate levels (not too much and not too little) and at the right time. The normal gene must be inserted in a sufficient number of the affected cell type so that the disease phenotype is corrected or alleviated, without causing new and adverse clinical symptoms. The HGP has helped tremendously with the first step—identifying disease genes in more than 2000 inherited diseases—but the difficulties of the subsequent steps of gene delivery and expression have continued to plague researchers.

Recently, scientists in France reported the treatment of several boys affected with an X-linked form of severe combined immunodeficiency (X-SCID). Children with X-SCID lack white blood cells necessary to fight infection and die within 1 year without a bone marrow transplant. The mutation lies in the γc gene, which encodes one subunit of the receptor for several interleukins, proteins that help stimulate the immune response. This receptor, found on the surface of immature blood cells in the bone marrow, helps direct the growth and maturation of a number of different cells in the immune system, including T cells, B cells, and natural killer cells. These cells, which kill invading viruses and bacteria and produce antibodies, are severely depleted in SCID children. Retroviruses are a well-established method for introducing genes into mammalian cells (Chapter 6). The researchers developed a vector in which the normal gene for the γc interleukin receptor subunit replaced the genes required by the retrovirus to replicate. Stem cells were isolated from the bone marrow of each patient and infected with the modified retrovirus. These cells, now containing an integrated copy of the normal γc gene, were returned to each donor and allowed to repopulate the marrow.

The initial results were striking and encouraging—nine of the ten boys treated showed dramatic immune system improvement. They have normal levels of T cells, live at home, and have been able to receive childhood immunizations and generate antibodies against infectious agents. However, it has been recognized for many years that retroviral vectors are potentially very dangerous and in fact may cause disease. One reason retroviruses were attractive as gene therapy vectors is that, even when modified, they integrate their DNA at random sites in the DNA of the host cell. As we will see in Chapter 15, however, such integrations may disrupt or alter the expression pattern of an important host gene and lead to cancer. This has happened in two of the children in the SCID gene therapy trial, who have as a result developed a leukemia-like disorder. Analysis of their DNA shows that in both children, the vector has integrated adjacent to, and activated, the *LMO2* gene, a zinc-finger proto-oncogene on chromosome 11. This is a frequent site of chromosomal translocations in childhood T-cell leukemia, the translocations leading to overexpression of *LMO2*, resulting in cancer.

All treatments have some side effects and gene therapy is no exception. In this particular case, an assessment of the risks and benefits lead to the conclusion that it is justifiable to use the gene therapy for treatment of SCID; in a majority of cases, leukemia in children is treatable, whereas SCID is invariably fatal.

The HGP and the HapMap Project, together with the development of high-throughput sequencing, genotyping, and cytogenetic tools, are transforming the task of identifying the genetic and environmental influences on simple and complex genetic disorders. With these tools, investigators have already identified hundreds of genes underlying single-gene disorders and continue to uncover the genetic and environmental risks for common complex diseases. These advances in our genetic knowledge of human beings are being put to good use in clinical genetics, but what is needed to make optimal use of the HGP is "genetic medicine," an approach to medicine that uses genetic information to help understand all aspects of illness. This approach will help provide accurate diagnostic techniques, more informative risk assessments, and effective preventative measures and treatments, all benefiting the patient.

Reading List

General

Strachan T. and Read A.P. 2004. *Human Molecular Genetics*, 3rd ed. Garland Press, New York.

Nussbaum R.L., McInnes R.R., and Willard H.F. 2004. *Thompson & Thompson Genetics in Medicine*, 6th ed. Saunders, Philadelphia.

Pasternak J.J. 2005. *An Introduction to Human Molecular Genetics: Mechanisms of Inherited Diseases.* Wiley-Liss, Wilmington, Delaware.

The First 40 Years of Human Genetics Proceeded Slowly

Scriver C. E. and Childs B. 1989. *Garrod's Inborn Factors in Disease.* Oxford University Press, Oxford.

Carlson E.A. 2001. *The Unfit: A History of a Bad Idea.* Cold Spring Harbor Laboratory Press, Cold Spring Harbor, New York.

Muller-Hill B. 1997. *Murderous Science: Elimination by Scientific Selection of Jews, Gypsies, and Others in Germany, 1933–1945.* Cold Spring Harbor Laboratory Press, Cold Spring Harbor, New York.

Neel J.V. 1949. The inheritance of sickle cell anemia. *Science* **110:** 64–66.

Pauling L., Itano H.A., Singer S.J., and Wells I.C. 1949. Sickle cell anemia, a molecular disease. *Science* **110:** 543–548.

Ingram V.M. 1956. A specific chemical difference between globins of normal and sickle-cell anæmia hæmoglobins. *Nature* **178:** 792–794.

Cytogenetics Marks a Major Advance in Human Genetics

Hsu T.C. 1979. *Human and Mammalian Cytogenetics: An Historical Perspective.* Springer-Verlag, New York.

Harris H. 1995. *The Cells of the Body: A History of Somatic Cell Genetics.* Cold Spring Harbor Laboratory Press, Cold Spring Harbor, New York.

Tjio J. and Levan A. 1956. The chromosome number of man. *Hereditas* **42:** 1–6.

Lejeune J., Gautier M., and Turpin R. 1959. Etude des chromosomes somatique de neuf enfant mongoliens. *Compt. Rend.* **248:** 1721–1722.

Caspersson T., Zech L., and Johansson C. 1970. Analysis of human metaphase chromosome set by aid of DNA-binding fluorescent agents. *Exp. Cell Res.* **62:** 490–492.

Seabright M. 1971. A rapid banding technique for human chromosomes. *Lancet* **2:** 971–972.

Recombinant DNA Techniques Enable Cloning of Human Disease Genes

Verellen-Dumoulin C., Freund M., De Meyer R., Laterre C., Frederic J., et al. 1984. Expression of an X-linked muscular dystrophy in a female due to translocation involving Xp21 and non-random inactivation of the normal X chromosome. *Hum. Genet.* **67:** 115–119.

Kunkel L.M., Monaco A.P., Middlesworth W., Ochs H.D., and Latt S.A. 1985. Specific cloning of DNA fragments absent from the DNA of a male patient with an X-chromosome deletion. *Proc. Natl. Acad. Sci.* **82:** 4778–4782.

Ray P.N., Belfall B., Duff C., Logan C., Kean V., et al. 1985. Cloning of the breakpoint of an X;21 translocation associated with Duchenne muscular dystrophy. *Nature* **318:** 672–675.

Botstein D., White R., Skolnick M., and Davis R. 1980. Construction of a genetic linkage map in man using restriction fragment length polymorphisms. *Am. J. Hum. Gen.* **32:** 314–331.

Wexler A. 1995. *Mapping Fate: A Memoir of Family, Risk, and Genetic Research.* Times Books, New York.

Gusella J.F., Wexler N.S., Conneally P.M., Naylor S.L., Anderson M.A., et al. 1983. A polymorphic DNA marker genetically linked to Huntington's disease. *Nature* **306:** 234–238.

The Huntington's Disease Collaborative Research Group. 1993. A novel gene containing a trinucleotide repeat that is expanded and unstable on Huntington's disease chromosomes. *Cell* **72:** 971–983.

Rommens J.M., Iannuzzi M.C., Kerem B., Drumm M.L. Melmer G., et al. 1989. Identification of the cystic fibrosis gene: Chromosome walking and jumping. *Science* **245:** 1059–1065.

Riordan J.R., Rommens J.M., Kerem B.-S., Alon N., Rozmahel R., et al. 1989. Identification of the cystic fibrosis gene: Cloning and characterization of complementary DNA. *Science* **245:** 1066–1073.

Cloning of the Fragile X Gene Reveals a Mutation with Unexpected Properties

Gatchel J.R. and Zoghbi H.Y. 2005. Diseases of unstable repeat expansion: Mechanisms and common principles. *Nat. Rev. Genet.* **6:** 743–755.

Verkerk A.J., Pieretti M., Sutcliffe J.S., Fu Y.H., Kuhl D.P., et al. 1991. Identification of a gene (*FMR-1*) containing a CGG repeat coincident with a breakpoint cluster region exhibiting length variation in fragile X syndrome. *Cell* **65:** 905–914.

Oberlé I., Rousseau F., Heitz D., Kretz C., Devys D., et al. 1991. Instability of a 550-base pair DNA segment and abnormal methylation in fragile X syndrome. *Science* **252:** 1097–1102.

Yu S., Pritchard M., Kremer E., Lynch M., Nancarrow J., et al. 1991. Fragile X genotype characterized by an unstable region of DNA. *Science* **252:** 1179–1181.

Hagerman P.J. and Hagerman R.J. 2004. The fragile-X premutation: A maturing perspective. *Am. J. Hum. Genet.* **74:** 805–816.

The HGP Revolutionizes Human Genetics

Antonarakis S.E. and Beckmann J.S. 2006. Mendelian disorders deserve more attention. *Nat. Rev. Genet.* **7:** 277–282.

Zuchner S., Wang G., Tran-Viet K.N., Nance M.A., Gaskell P.C., et al. 2006. Mutations in the novel mitochondrial protein REEP1 cause hereditary spastic paraplegia type 31. *Am. J. Hum. Genet.* **79:** 365–369.

Zuchner S., Kail M.E., Nance M.A., Gaskell P.C., Svenson I.K., et al. 2006. A new locus for dominant hereditary spastic paraplegia maps to chromosome 2p12. *Neurogenetics* **7:** 127–129.

Modifiers of "Single"-Gene Disorders Lead to Complex Phenotypes

Steinburg M.H. 1996. Modulation of the phenotypic diversity of sickle cell anemia. *Hemoglobin* **20:** 1–19.

Chui D.H.K. and Dover G.J. 2001. Sickle cell disease: No longer a single gene disorder. *Curr. Opin. Pediatr.* **13:** 22–27.

Chang Y.P., Maier-Redelsperger M., Smith K.D., Contu L., Ducroco R., et al. 1997. The relative importance of the X-linked FCP locus and β-globin haplotypes in determining haemoglobin F levels: A study of SS patients homozygous for β S haplotypes. *Br. J. Haematol.* **96:** 806–814.

Association Studies Are Used to Locate Genes Involved in Complex Traits

Lander E.S. and Schork N.J. 1994. Genetic dissection of complex traits (review). *Science 265:* 2037–2048.

Arking D.E., Pfeufer A., Post W., Kao W.H., Newton-Cheh C., et al. 2006. A common genetic variant in the NOS1 regulator NOS1AP modulates cardiac repolarization. *Nat. Genet.* **38:** 644–651.

Genetic Basis of Alzheimer Disease Is Untangled Using Association

Corder E.H., Saunders A.M., Strittmatter W.J., Schmechel D.E., Gaskell P.C., et al. 1993. Gene dose of apolipoprotein E type 4 allele and the risk of Alzheimer's disease in late onset families. *Science* **261:** 921–923.

Farrer L.A., Cupples L.A., Haines J.L., Hyman B., Kukull W.A., et al. 1997. Effects of age, sex, and ethnicity on the association between apolipoprotein E genotype and Alzheimer disease. A meta-analysis. APOE and Alzheimer Disease Meta Analysis Consortium. *J. Am. Med. Assoc.* **278:** 1349–1356

Mahley R.W., Weisgraber K.H., and Huang Y. 2006. Apolipoprotein E4: A causative factor and therapeutic target in neuropathology, including Alzheimer's disease. *Proc. Natl. Acad. Sci.* **103:** 5644–5651.

Cytogenetic Abnormalities Reveal Genes Involved in Schizophrenia

Pickard B.S., Millar J.K., Porteous D.J., Muir W.J., and Blackwood D.H. 2005. Cytogenetics and gene discovery in psychiatric disorders. *Pharmacogenomics J.* **5:** 81–88.

Blackwood D.H.R., Fordyce A., Walker M.T., St. Clair D.M., Porteous D.J., and Muir W.J. 2001. Schizophrenia and affective disorders—Cosegregation with a translocation at chromosome 1q42 that directly disrupts brain-expressed genes: Clinical and P300 findings in a family. *Am. J. Hum. Genet.* **69:** 428–433.

Ekelund J., Hennah W., Hiekkalinna T., Parker A., Meyer J., et al. 2004. Replication of 1q42 linkage in Finnish schizophrenia pedigrees. *Mol. Psych.* **9:** 1037–1041.

Pieper A.W., Wu X., Han T.W., Estill S.J., Dang Q., et al. 2005. The neuronal PAS domain protein 3 transcription factor controls FGF-mediated adult hippocampal neurogenesis in mice. *Proc. Natl. Acad. Sci.* **102:** 14052–14057.

Millar J.K., Pickard B.S., Mackie S., James R., Christie S., et al. 2005. DISC1 and PDE4B are interacting genetic factors in schizophrenia that regulate cAMP signaling. *Science* **310:** 1187–1191.

The HapMap Project Has Collected Many Millions of SNPs

International HapMap Consortium. 2005. A haplotype map of the human genome. *Nature* **437:** 1299–1320. www.hapmap.org

Crawford D.C. and Nickerson D.A. 2005. Definition and clinical importance of haplotypes. *Annu. Rev. Med.* **56:** 303–320.

Goldstein D.B. and Cavalleri G.L. 2005. Understanding human diversity. *Nature* **437:** 27.

Wall J.D. and Pritchard J.K. 2003. Haplotype blocks and linkage disequilibrium in the human genome. *Nat. Rev. Genet.* **4:** 587–597.

Montpetit A., Nelis M., Laflamme P., Magi R., Ke X., et al.

2006. An evaluation of the performance of tag SNPs derived from HapMap in a Caucasian population. *PLoS Genet.* **2:** pe27.

Macular Degeneration Gene Is Located and Cloned by Using HapMap SNPs

Klein R.J., Zeiss C., Chew E.Y., Tsai J.Y., Sackler R.S., et al. 2005. Complement factor H polymorphism in age-related macular degeneration. *Science* **308:** 385–389.

Edwards A.O., Ritter R., Abel K.J., Manning A., Panhuysen C., et al. 2005. Complement factor H polymorphism and age-related macular degeneration. *Science* **308:** 421–424.

Haines J.L., Hauser M.A., Schmidt S., Olson L.M., Gallins P., et al. 2005. Complement factor H variant increases the risk of age-related macular degeneration. *Science* **308:** 419–421.

Many Human Genetic Disorders Occur More Frequently in Specific Populations or Ancestral Groups

Allison A.C. 1954. Protection afforded by the sickle-cell trait against subtertian malaria infection. *Br. Med. J.* **1:** 290–294.

Doo M., Terlouw D.J., Kolczak M.S., McElroy P.D., ter Kuile F.O., et al. 2002. Protective effects of the sickle cell gene against malaria morbidity and mortality. *Lancet* **359:** 1311–1312.

Diamond J.M. 1988. Tay-Sachs carriers and tuberculosis resistance. *Nature* **331:** 666.

Gabriel S.E., Brigman K.N., Koller B.H., Boucher R.C., and Stutts M.J. 1994. Cystic fibrosis heterozygote resistance to cholera toxin in the cystic fibrosis mouse model. *Science* **266:** 107–109.

Risch N., Tang H., Katzenstein H., and Ekstein J. 2003. Geographic distribution of disease mutations in the Ashkenazi Jewish population supports genetic drift over selection. *Am. J. Hum. Genet.* **72:** 812–822.

Slatkin M. 2004. A population-genetic test of founder effects and implications for Ashkenazi Jewish diseases. *Am. J. Hum. Genet.* **75:** 282–293.

DNA Microarrays Detect Copy Number Polymorphisms in Chromosomes

Rosenberg C., Knijnenburg J., Bakker E., Vianna-Morgante A.M., Sloos W., et al. 2006. Array-CGH detection of micro rearrangements in mentally retarded individuals: Clinical significance of imbalances present both in affected children and normal parents. *J. Med. Genet.* **43:** 180–186.

Lugtenberg D., de Brouwer A.P.M., Kleefstra T., Oudakker A.R., Frints S.G.M., et al. 2006. Chromosomal copy number changes in patients with non-syndromic X linked mental retardation detected by array CGH. *J. Med. Genet.* **43:** 362–370.

Rickman L., Fiegler H., Shaw-Smith C., Nash R., Cirigliano V., et al. 2006. Prenatal detection of unbalanced chromosomal rearrangements by array CGH. *J. Med. Genet.* **43:** 353–361.

Genome-based Diagnosis Makes Use of Large-Scale, High-Throughput Techniques

Syvane A.C. 2005. Toward genome-wide SNP genotyping. *Nat. Genet.* (suppl.) **37:** 5–10.

Steemers F.J., Chang W., Lee G., Baker D.L., Shen R., and Gunderson K.L. 2006. Whole-genome genotyping with the single-base extension assay. *Nat. Methods* **3:** 31–33.

Shen R., Fan J.B., Campbell D., Chang W., Chen J., et al. 2005. High-throughput SNP genotyping on universal bead arrays. *Mutat. Res.* **573:** 70–82.

Fan J.-B., Oliphant A., Sen R., Kermani B.G., Garcia F., et al. 2003. Highly parallel SNP genotyping. *Cold Spring Harbor Symp. Quant. Biol.* **68:** 69–78.

DNA Sequencing Provides Definitive Diagnosis of Mutations in Breast Cancer

Palma M., Ristori E., Ricevuto E., Giannini G., and Gulino A. 2006. *BRCA1* and *BRCA2*: The genetic testing and the current management options for mutation carriers (review). *Crit. Rev. Oncol. Hematol.* **57:** 1–23.

New Technologies Offer Large-Scale Sequencing-based Diagnosis

Margulies M., Egholm M., Altman W.E., Attiya S., Bader J., et al. 2005. Genome sequencing in microfabricated high-density picolitre reactors. *Nature* **437:** 376–380.

Church G.M. 2006. Genomes for all. *Sci. Am.* **294:** 46–54.

Pharmacogenomic Strategy Leads to Improved Treatment for Epilepsy

Pennacchio L.A., Lehesjoki A.-E., Stone N.E., Willour V.L., Virtaneva K., et al. 1996. Mutations in the gene encoding cystatin B in Progressive Myoclonus Epilepsy (EPM1). *Science* **271:** 1731–1734.

Eldridge R., Iivanainen M., Stern R., Koerber T., and Wilder B.J. 1983. "Baltic" myoclonus epilepsy: Hereditary disorder of childhood made worse by phenytoin. *Lancet* **2:** 838–842.

Shahwan A., Farrell M., and Delanty N. 2005. Progressive myoclonic epilepsies: A review of genetic and therapeutic aspects. *Lancet Neurol.* **4:** 239–248.

RNAi Is Used to Inhibit Overgrowth of Blood Vessels

Dykxhoorn D.M., Palliser D., and Lieberman J. 2006. The silent treatment: siRNAs as small molecule drugs. *Gene Ther.* **13:** 541–552.

Shen J., Samul R., Silva R. L., Akiyama H., Liu H., et al. 2006. Suppression of ocular neovascularization with siRNA targeting VEGF receptor 1. *Gene Therapy* **13:** 225–234.

Tong A.W., Zhang Y.A., and Nemunaitis J. 2005. Small interfering RNA for experimental cancer therapy. *Curr. Opin. Mol. Ther.* **7:** 114–124.

Tan F.L. and Yin J.Q. 2004. RNAi, a new therapeutic strategy against viral infection. *Cell Res.* **14:** 460–466.

Severe Combined Immune Deficiency Is Treated by Retroviral Gene Therapy

O'Connor T.P. and Crystal R.G. 2006. Genetic medicines: Treatment strategies for hereditary disorders. *Nat. Rev. Genet.* **7:** 261–276.

Hacein-Bey-Abina S., Le Deist F., Carlier F., Bouneaud C., Hue C., et al. 2002. Sustained correction of X-linked severe combined immunodeficiency by ex-vivo gene therapy. *N. Engl. J. Med.* **346:** 1185–1193.

Hacein-Bey-Abina S., Von Kalle C., Schmidt M., McCormack M.P., Wulffraat N., et al. 2003. LMO2-associated clonal T cell proliferation in two patients after gene therapy for SCID-X1. *Science* **302:** 415–419.

Gaspar H.B., Parsley K.L., Howe S., King D., Gilmour K.C., et al. 2004. Gene therapy of X-linked severe combined immunodeficiency by use of a pseudotyped γ retroviral vector. *Lancet* **364:** 2181–2187.

McCormack M.P. and Rabbitts T.H. 2004. Activation of the T-cell oncogene *LMO2* after gene therapy for X-linked severe combined immunodeficiency. *N. Engl. J. Med.* **350:** 913–922.

Fischer A. and Cavazzana-Calvo M. 2005. Integration of retroviruses: A fine balance between efficiency and danger. *PLoS Med.* **2:** e10.

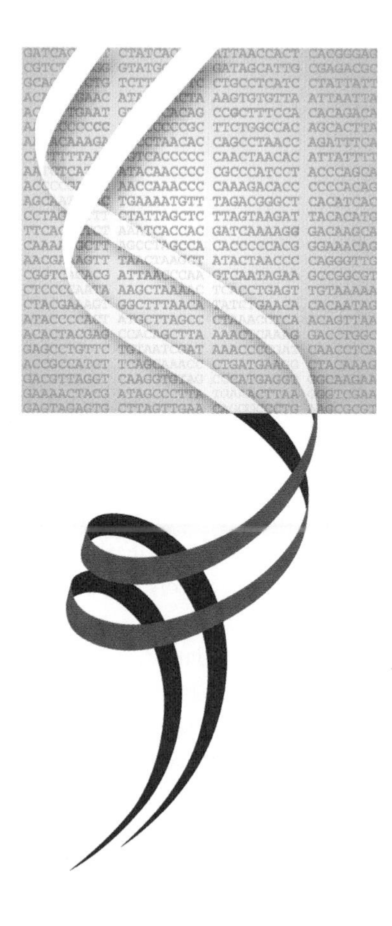

CHAPTER 15

Understanding the Genetic Basis of Cancer

There is no diagnosis that strikes such great fear as that of cancer. Cancer is the second leading cause of death in Western countries; in 2003, it accounted for 23% of deaths in the United States and 27% of deaths in the United Kingdom. However, earlier diagnosis and treatment, and a fall in the numbers of men who are smokers, have led for the first time to a small decline in cancer rates. Indeed, the most effective tool against developing cancer is minimizing exposure to environmental carcinogens. On the other hand, treatments for cancers have remained the same for many years, and many cancers ultimately overwhelm those afflicted. Although chemotherapy kills cancer cells, it is not specific, and its side effects cause the death of normal cells and tissues. Radiation therapy can be delivered more accurately to a tumor, but it, too, is not selective for cancer cells. Surgery can be very effective when used to excise solid tumors in the early stages of their growth, but is of limited benefit once a tumor has spread. There is a long way to go in the treatment of cancer. This is made clear from comparisons with other major killers; in the past 50 years, the death rate for heart diseases has fallen by one-half and that for stroke by two-thirds.

The sooner any treatment begins, the better the prognosis. For example, 15-year survival rates for melanoma are closely correlated with the stage of development of the cancer when the diagnosis is first made. So improved diagnostic tools for early diagnosis are urgently needed. Not only are more sensitive methods for detecting and diagnosing cancers needed, we need new tests that will enable us to characterize on a molecular level the same type of tumor in different individuals. We must be able to distinguish, for example, a breast cancer tumor that will metastasize rapidly from one that will remain in its primary site. The former should be treated aggressively, whereas treatments with fewer side effects could be used in the latter case. New, rapid, large-scale techniques for examining the molecular details of tens of thousands of samples should allow us to determine the general features of cancer as well as those specific to particular types of cancer.

Cancer has been subject to intensive scientific investigation since the end of the 19th century. Mouse breeding experiments carried out 100 years ago showed that cancer has a genetic component, but it was not until the tools of recombinant DNA became available that it became possible to locate, clone, and examine the genes that when mutated lead to cancer. We have learned that many of the genes that control cell growth have the potential to cause cancer, and that there are other genes whose role is to block cancer. We have discovered the cellular pathways in which these genes work and the properties of the cancer cell. However, cancer is a most complex process and it is only recently that we have begun to make progress developing effective therapies by exploiting what we know of the differences between normal and cancer cells.

The comprehensive knowledge we now have of human genes through the Human Genome Project is enabling us to understand better the networks of genes and their mutations that transform normal cells into cancer cells. Genomic approaches are already providing clues for diagnoses and better therapies. We begin this chapter with a general discussion of the causes of cancers, and then go on to review the work that led to the discovery of cancer-causing genes—oncogenes. Finally, we look at some examples of how genomics-based research is being used to identify new targets for therapy and to help refine diagnoses.

Environmental Factors Are the Primary Initiators of Cancer

Environmental factors are of paramount importance in cancer. This was recognized as long ago as 1775, when Percivall Pott, a surgeon in London, found that chimney sweeps had a high incidence of scrotal cancer. He ascribed this to "...a lodgement of soot in the rugae (skin folds) of the scrotum," the first description of an occupational cancer. The powerful effects of chemical carcinogens are demonstrated most convincingly by considering the relationship between smoking tobacco and the incidence of lung cancer (Fig. 15-1). The association of lung cancer and smoking was put on a firm footing by Richard Doll and A. Bradford Hill in the United Kingdom and Ernst Wynder and Evarts Graham in the United States in 1950. Their epidemiological studies, and every study since, show unequivocally that smoking is responsible for about 80% of lung cancer deaths, and thus for 30% of all cancer deaths, in the United States. Furthermore, the decline in lung cancer death rates for men of about 1.9% per year between 1991 and 2002 parallels the decline in smoking.

The classic example of the association between radiation and cancer comes from the early 20th century, when watch dials and hands were painted with radium-containing paint so that they would glow in the dark. The women employed in this work licked the paint brushes to make the tips pointed and, as a consequence of ingesting the radium, developed bone cancers 8–40 years later. Workers in uranium mines breathed an atmosphere high in the radioactive gas radon^{222}, which comes from radium, a decay product of uranium^{238}. Radon emits α particles and it was these that caused a much-higher than expected level of lung cancer in the miners. Melanoma provides another clear example of cancer arising from environmental exposure to radiation, in this case ultraviolet light from the sun. Australia has one of the highest rates of melanoma in the world, but as a result of a vigorous public health campaign to persuade people, especially children, to reduce their exposure to sun, these rates have fallen.

A third environmental exposure that can affect us is infection by viruses. As we shall discuss later in this chapter, the first clear case of a virus causing cancer was discovered by Peyton Rous in 1911. He was able to transmit a sarcoma from chicken to chicken using a cell-free tumor extract. This result led to extensive research exploring the possibility that human cancers, too, were caused by an infectious agent. In fact, although tumor viruses have been key contributors to our understanding of the biology of cancers, they appear not to be a major cause of human cancer. A most important exception is human papilloma virus (HPV), which is found in about 50% of cervical cancers. In the United States,

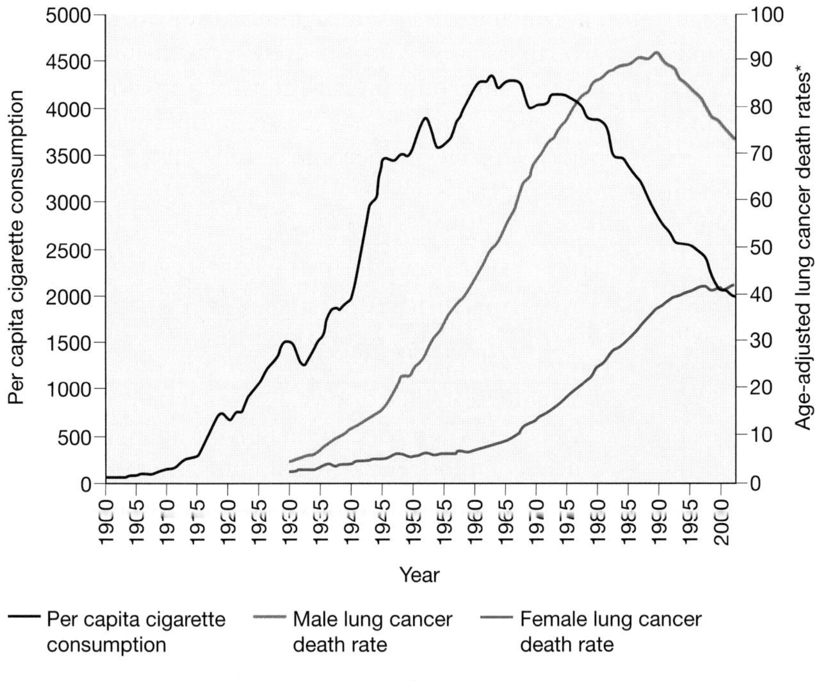

FIGURE 15-1

Tobacco use and rates of death in the United States due to lung cancer from 1900 to 2002. The graph illustrates and compares the incidence of per capita cigarette consumption over a 102-year span and the rates of death resulting from lung cancer for both men and women over this time period. *Age-adjusted rates* take account of the fact that cancers occur at different rates in populations of different ages. The absolute cancer rate in a particular age group is multiplied by the proportion of the general population in that age group.

there are some 10,000 new cases of cervical cancer with almost 4,000 deaths reported each year. Worldwide, this is the second most common cancer in women, causing an estimated 233,000 deaths each year. The U.S. Federal Drug Administration (FDA) has recently (2006) approved a vaccine that will provide protection against HPV types 6, 11, 16, and 18. In trials involving 21,000 women, the vaccine provided almost 100% protection against the development of precancerous lesions.

It is clear that reducing exposure to environmental insults that lead to cancer is the most effective means to prevent cancer. But not all cancers have readily identifiable causes and preventative measures cannot help those already afflicted. To do so, we must understand the biology of cancers and identify those cellular pathways that offer targets for therapies.

Cancer Is a Genetic Disease

Cancer is a genetic disease; this was known long before we had tools to examine the genes in cancer cells and compare them with their counterparts in normal cells. For example, different inbred strains of mice differ not only in their susceptibility to developing spontaneous cancers, but also in the cell types affected. And cancer cells, when they divide, give rise to cells with the same cancer phenotype. Indeed, it can be shown that many tumors are clonal—that is, they are derived from a single common progenitor cancer cell that divides incessantly to generate a tumor of identical sibling cells. For example, myelomas, cancers of the B cells of the immune system, each produce only one antibody unique to that tumor, which is a characteristic exploited in the making of monoclonal antibodies. This is best explained by assuming that each myeloma arises from a single cell. Cells within a tumor have similar genetic changes, which suggests that the disease phenotype is genetically determined—that is, it is encoded within tumor cell DNA.

If cancers arise from mutations, then one would expect the incidence of cancers to increase steadily with advancing age, as genetic changes accumulate in the DNA of an individual. This is, in fact, the case. Population-based estimates show steadily rising age-specific incidence rates for epithelial tumors such as colorectal and prostate carcinomas. For example, prostate cancer incidence rises from 24 cases per 100,000 men in the age group 20–54, to 454 cases in age group 55–64, to 991 cases in the group 65–74, a 40-fold increase. However, not all cancers follow this pattern. Breast cancer rates (overall) increase rapidly until age 50 years, then continue to rise more slowly, which perhaps reflects the contribution of early-onset inherited forms of breast cancer. (As we have come to live longer through better health care, so the proportion of deaths due to cancer has risen—there is simply more time for a cancer to appear.) The environmental factors that we discussed in the previous section—chemical carcinogens and radiation—are mutagens that damage DNA. This common property of carcinogens and radiation suggests that their ability to inflict genetic damage is the basis for their carcinogenic properties. As we shall see in this chapter, the foundation for modern research on the nature of cancer is that cancers arise from genetic changes in our cells.

Tumor Cells Grow Abnormally in Culture

When normal diploid human fetal lung fibroblasts are grown in tissue culture, they form a monolayer on the Petri dish surface and grow steadily for about 8 months, during which time they divide some 60 times. Although cell division stops, the cells remain alive and change in shape from the characteristic elongated fibroblast shape to a more spread-out form. Such cells are referred to as *senescent.* The limit on division of normal cells is called the *Hayflick limit* after Leonard Hayflick, who first investigated the phenomenon in the mid-1960s. In contrast to human fibroblasts, mouse embryonic fibroblasts readily undergo changes that make them immortal; that is, they become cell lines and can be grown indefinitely. These cells often have chromosomal changes and a distinctive shape and are less dependent on specific factors in their growth medium. Such cells are typified by the 3T3 mouse embryonic cell line developed by George Todaro and Howard Green. The difference in behavior is due in part to the differences between mouse and human telomeres (Chapter 7), which we will discuss later in this chapter.

However, cells that can grow indefinitely do not necessarily form tumors on injection into animals. To do so, further changes are needed, for example, those induced by carcinogenic chemicals and radiation, by infection with tumor viruses, or by the introduction of cancer-causing genes (oncogenes). Once these changes occur, the cells continue growing; they pile atop one another and form tumors when injected into animals. It is also possible to grow human cancer cells in culture, and they continue to behave as cancer cells. The classic HeLa cell line was derived from a cervical adenocarcinoma in 1951 and has been growing in culture ever since. As we shall see later, these cells have undergone genetic changes that have disabled the cellular mechanisms that keep cell growth in check. It has proved very difficult to transform normal human cells growing in tissue culture, and it seems that these cells require as many as five genetic changes to become cancer cells.

The First Cancer Genes Were Found by Studying Tumor Viruses

The discovery that certain viruses produced tumors when inoculated into animals considerably simplified our ideas about cancer. On a practical level, perhaps cancer could be regarded as an infectious disease and the public health measures that had helped reduce mortality from bacterial diseases would work for cancer. For the researcher seeking to understand the biology of cancer, these tumor viruses seemed to be much more tractable subjects for analysis. Somewhere within the tiny genomes of these viruses—nearly a millionth the size of the genome of an animal cell—lurked the genetic instructions that enabled the virus to subvert the metabolism of the cell to serve its own ends.

Tumor viruses come in two types: those whose genome is encoded in RNA and those whose genome is encoded in DNA. As discussed earlier in the chapter, the first tumor virus to be discovered was an RNA virus, the Rous sarcoma virus (RSV), named for Rous who isolated it from a chicken in 1911. Avian sarcoma virus, feline leukemia virus, and mouse mammary tumor virus are other RNA tumor viruses. DNA tumor viruses include the mouse polyoma virus, simian virus 40 (SV40), HPV, and human adenovirus. Tumor viruses carry oncogenes that are responsible for the viruses' ability to transform cells. Oncogenes encode proteins, often termed *oncoproteins*, that play a number of important roles in the viral life cycle, such as initiation of DNA replication and transcriptional control of viral genes. The ability of oncogenes to transform cells is a consequence of these activities.

Viruses normally infect nongrowing cells, which represent the vast majority of cells in an animal. Because they require host enzymes to replicate their DNA, one of the critical things a virus must do after infecting a cell is to prod the cell into synthesizing the cellular DNA replication machinery. The chief role of the oncoproteins of the DNA tumor viruses is to activate the growth of dormant cells so that viral DNA can be replicated. In some cells, these tumor viruses grow lytically; they enter the cell, multiply rapidly, and kill the host cell by rupturing its cell membrane. Certain cells, however, resist lytic infection and, at a low frequency, become transformed into cancer cells. Invariably, transformation is associated with integration of the viral genome (or portions thereof) into host cell DNA and its stable passage to daughter cells. Thus, tumor viruses create genetic alterations that transform cells.

A major advance came when Marguerite Vogt and Renato Dulbecco showed that mouse embryonic cells growing in tissue culture became transformed when infected with polyoma virus. Such cells were much more amenable than animals for experimental investigation of cancer and this discovery initiated a new field of research. For example, temperature sensitive RSV mutants showed that the continued activity of a viral gene (or genes) was needed to maintain the transformed state. Cells infected with these mutants were transformed at the lower nonpermissive temperature (36°C) but were normal at the higher temperature (41°C). Thus, the cells could be switched between the two states simply by changing the temperature. Similar mutants were found for DNA tumor viruses. Other RSV mutants carrying a deletion were isolated that could replicate in cells but not transform them. At the same time, biochemical analyses of DNA tumor viruses were revealing virus-encoded proteins that interacted in a complicated manner with the host cell proteins to regulate viral replication and cell transformation. These proteins included the large T antigen of SV40, the large and middle T antigens of polyoma virus, the E1A and E1B proteins of adenovirus, and E6 and E7 of papillomaviruses.

RSV and other RNA tumor viruses provided the breakthrough to understanding the genetics of cancer. In one of the most startling discoveries in cancer research, J. Michael Bishop and Harold Varmus showed that the RSV oncogene was derived from a cellular gene. These investigators made use of a defective virus that had extensive deletions, which presumably included the viral oncogene. They purified RNA from large quantities of normal virus and used it to make radiolabeled complementary DNA (cDNA), which was hybridized to RNA from the transformation-defective virus. Those labeled cDNA sequences that did not hybridize contained the sequences deleted from the defective virus, which included the RSV oncogene named *src* (Fig. 15-2). When this cDNA was hybridized to DNA from uninfected cells, the cDNA hybridized stably, thus showing that *normal* chicken cell DNA contained sequences very closely related to *src*, a potent oncogene! These *src*-related sequences were found not only in chicken, but also in all vertebrates, including human beings. These normal cellular equivalents of retroviral oncogenes were called *proto-oncogenes.*

With the ability to generate molecular clones of retroviral genomes, it soon became apparent that each of the acutely transforming retroviruses, those that cause rapidly growing tumors within a few weeks of inoculation, carried oncogenes derived from cellular DNA. During their evolution, these viruses captured cellular genes that provided the viruses with their dramatic growth-transforming properties. Although retroviruses do not appear to cause human cancer, the study of these viruses has revealed human genes that probably play a very direct role in the disease.

Acutely transforming retroviruses like RSV are quite rare; retroviruses causing tumors that take a long time to develop are more common. These viruses lack oncogenes of their own and act instead by altering the expression of a cellular gene. For example, the avian leukosis virus (ALV) causes lymphomas in chickens. Hybridization of probes derived from the virus DNA isolated from tumor cells showed that the tumor cells' genomes contained integrated viral genomes. Closer examination of many different tumors revealed that these integrated proviruses were located within the same cellular gene. Furthermore, this gene was shown to be the cellular counterpart of the known retroviral oncogene, *myc*. An important model emerged to explain carcinogenesis by these retroviruses lacking oncogenes. As the viruses spread through the animal, they infect cells and integrate into cellular DNA essentially at random. In a rare cell, the virus drops into the c-*myc* gene ("c" for cellular) and perturbs the expression of this gene in a way that confers a growth advantage to the infected cell, which eventually multiplies to form a tumor. As we saw in Chapter 14, retroviruses used for gene therapy have generated cancer in this way.

Chromosome Translocations Can Cause Cancer

Physical rearrangements of human chromosomes can cause cancers by an analogous mechanism without any involvement of viruses (Fig. 15-3). Human B-cell lymphomas often contain chromosomal translocations in which the highly expressed immunoglobulin genes on chromosomes 2, 14, or 22 are joined to chromosome 8. Cloning the chromosome 8 DNA neighboring the translocated immunoglobulin locus revealed that the target was again the c-*myc* proto-oncogene. This rearrangement leads to the high-level expression of the c-*myc* gene from the powerful promoter in the immunoglobulin locus. A B-cell tumor results because the immunoglobulin promoter is active only in antibody-producing B cells. Similarly, human chronic myelogenous leukemia (CML) is caused by a chromosome rearrangement in which the end of chromosome 9 is attached to chromosome 22. The CML breakpoint on chromosome 9 is within the first intron of the large c-*abl* gene, and it is always joined to the then-previously unknown gene, *bcr*, on chromosome 22 (Fig. 15-3). The result is a hybrid mRNA that produces a novel protein, BCR-ABL, jointly encoded by both genes. Later in this chapter, we shall see how the successful drug Gleevec was developed specifically to inhibit this oncogene.

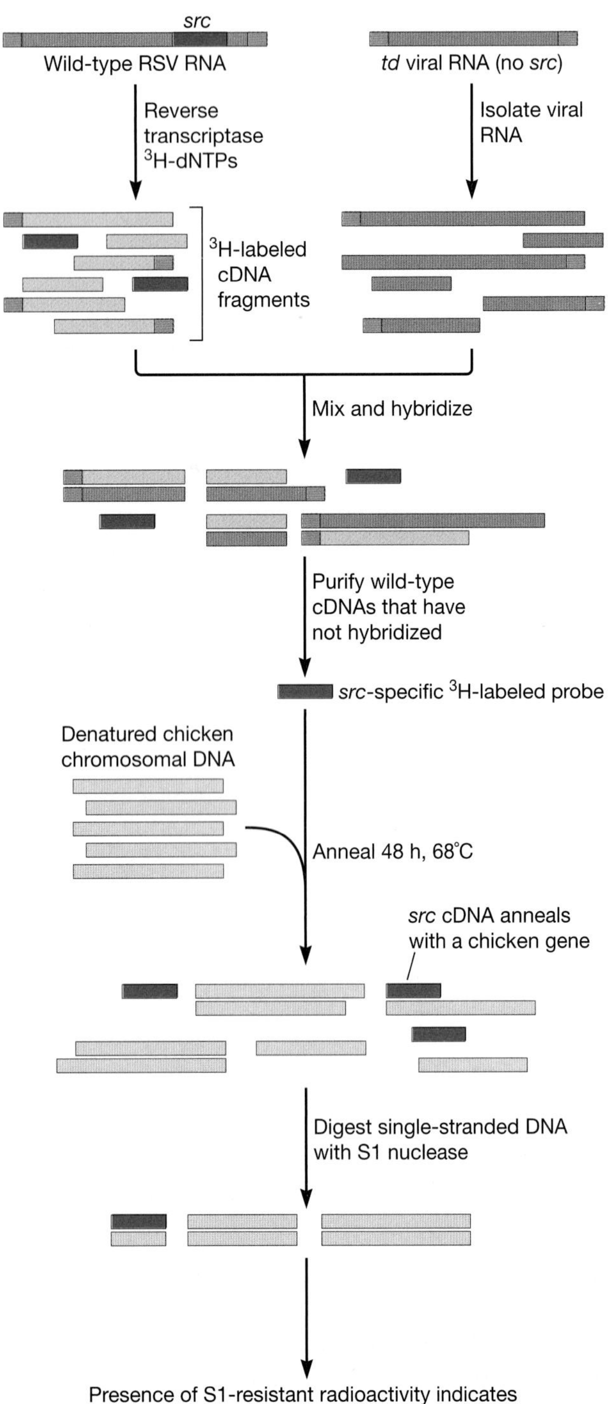

Recombinant DNA Techniques Are Used to Isolate the First Human Cancer Gene

At the same time that retrovirologists were cloning viral oncogenes, others were hunting directly for cellular oncogenes by using what were, at that time, newly developed methods for gene transfer into mammalian cells (Chapter 6). Over the next three years, four laboratories worked independently to isolate and identify human proto-oncogenes. All four laboratories followed the same four-stage strategy, although the experimental details differed. There were four steps. First, it was shown that transformed human cells contain oncogenes; second, human

FIGURE 15-2

Retroviral oncogenes have cellular counterparts. To determine the origin of the *src* transforming gene of Rous sarcoma virus (RSV), investigators needed to prepare a radiolabeled probe specific for *src* sequences. Cloning of tumor virus genomes was not possible at the time, because of both technical limitations and a voluntary moratorium on such experiments. Therefore, the investigators isolated RSV particles and, using the endogenous viral reverse transcriptase, transcribed the viral RNA genome into fragments of radiolabeled cDNA. They hybridized this cDNA to excess amounts of RNA from transformation-defective (*td*) RSV derivatives that lacked the *src* oncogene and isolated the labeled cDNA that failed to hybridize, thereby generating a probe highly enriched for *src* sequences. This *src* probe was in turn annealed to denatured chicken DNA (chicken is the host species for RSV). After annealing, the mixture was treated with S1 nuclease, an enzyme that degrades only single-stranded DNA. Some of the *src* DNA was resistant to S1 degradation, a finding indicating that it had formed double-stranded molecules with chicken DNA fragments. Thus, chicken DNA must possess a gene very similar to *src*.

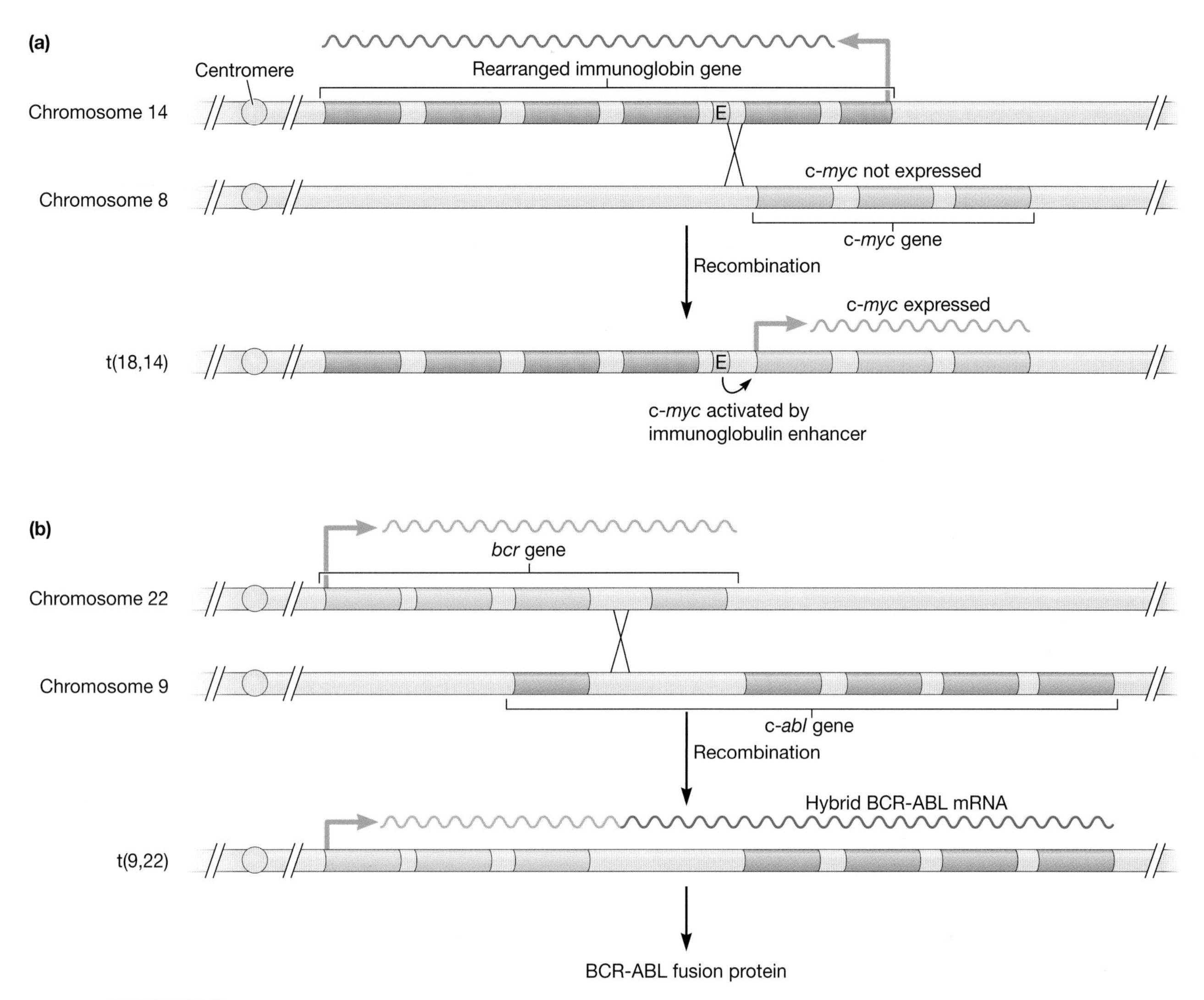

FIGURE 15-3

Some tumors have chromosomal rearrangements. (*a*) Many B-cell tumors carry chromosomal translocations that join the c-*myc* proto-oncogene (exons in *green*) to one of the immunoglobulin genes (exons in *blue*). The c-*myc* gene, normally off in terminally differentiated B cells, is kept on by the immunoglobulin gene enhancer (E). These translocated genes usually produce a normal c-*myc* protein. (*b*) Chronic myelogenous leukemia cells carry a hallmark chromosome translocation, which joins the c-*abl* proto-oncogene (exons in *red*) on chromosome 9 to chromosome 22 in the middle of a gene termed *bcr* (exons in *orange*). The translocation produces a novel BCR-ABL protein kinase no longer subject to its normal control. These translocations are *reciprocal,* which means that a second aberrant chromosome carrying the other pieces of each translocated chromosome is also formed. *Colored sections* represent gene exons.

DNA containing the oncogene was cloned; third, the oncogene was identified; and, fourth, the oncogene was compared with the proto-oncogene to determine what distinguished an active oncogene from its normal equivalent.

DNA was isolated from human cells transformed by chemical carcinogens on the assumption that these cells contained oncogenes activated by mutation. The DNA was transfected into 3T3 mouse fibroblasts, which have relatively normal growth properties. The cultures were nurtured for several weeks, and, on some of the dishes, foci of cells developed that overgrew their neighbors, presumably as a result of gene transfer. Very quickly these experiments were repeated with donor DNA from tumors of various tissues and species. About one-third of these tumor DNAs were capable of transforming recipient fibroblast cultures. It was evident that this gene transfer assay was detecting powerful cellular oncogenes present in a wide variety of tumors.

Cloning DNA from the region of the genome containing the oncogene was accomplished by tracking the human DNA associated with transformation through two or more rounds of transfection (Fig. 15-4). Cell cultures were transfected with DNA from a tumor cell line, and clones of transformed cells were isolated and grown in bulk culture. DNA was extracted from these cultures and used to transform a second set of cells. It was assumed that two rounds of transfection had selected the DNA containing the oncogene and that irrelevant DNA had been lost. The DNA was tracked either by using *Alu* repeats, the human repetitive sequence not found in the mouse genome, or by attaching a selective marker to fragments of DNA used for the first transfection. In this way, the oncogenes were localized to sections of cloned DNA a few kilobases long.

The most straightforward way of identifying the specific oncogene in the cloned DNA was to make use of the many DNA probes that reacted with the two dozen or so known retroviral oncogenes. Researchers used these probes to examine the transformed cultures for new DNA fragments related to the oncogenes. The patterns of hybridization were similar for DNA from normal cells and DNA from the transformed fibroblasts, except when a probe for the *ras* oncogene of Harvey sarcoma virus (a mouse retrovirus) was used. New *ras*-related DNA fragments were detected in several fibroblast clones independently transformed by the human tumor DNA, and the transformed phenotype correlated perfectly with the presence of new *ras* DNA in the cells. Furthermore, the cloned bladder oncogene also hybridized to the *ras* probe. The conclusion was inescapable: The oncogene in the human bladder cancer that had been passed from cell to cell by transfection was c-*ras*, the cellular counterpart of retroviral *ras*.

The next step was to determine the molecular change that had converted the normal c-*ras* into a virulent oncogene. Because of the complex phenotypes of cancers, it had been expected that they arose through changes in gene regulation, but comparison by restriction enzyme mapping of the activated *ras* gene from the tumor with its normal counterpart showed no differences. Investigators narrowed down the region containing the mutation by recombining pieces of the normal and transforming *ras* genes and testing the chimeric molecules for their ability to transform cells. The transforming mutation was rapidly localized to a 350-bp restriction fragment (Fig. 15-5). When this DNA was sequenced, it was found that a guanine nucleotide within the normal gene had been converted to a thymine in the oncogene, leading to a change from a codon encoding glycine to one encoding valine. This was a startling finding: The dramatic difference in growth and behavior of a cell having a normal or a mutant *ras* gene was due to a single-base-pair change within the thousands of base pairs in this gene among thousands of genes in the cell. In retrospect, it is not so surprising. After all, a single-amino-acid change in the β-globin molecule causes the many symptoms of sickle-cell anemia.

The Ras protein turned out to be a guanine nucleotide binding protein, a member of the large family of G proteins, molecular switches that regulate a variety of signal transduction pathways. In its inactive state, Ras is bound to GDP, but on activation by a signal from the prior step in the pathway, GDP is released, GTP binds, and Ras is activated, thereby transmitting the signal to the next step in the pathway (Fig. 15-6). Ras proteins possess an intrinsic enzymatic activity that hydrolyzes bound GTP to GDP, thereby returning them to their inactive state. The mutations in activated *ras* oncogenes destroy this GTP hydrolysis activity, thus locking the protein in its active conformation so that it transmits a signal even when it has not received one. However, as we shall see in the next section, subverting signal transduction pathways is not the only way in which oncoproteins wreak their havoc.

Cellular Proto-Oncogenes Affect Signal Transduction Pathways and Transcription

The identification of cellular proto-oncogenes told us that our genomes carry genes with the potential to kill us. What are these genes? What are their normal functions? How do viruses direct DNA damage and unleash their oncogenic potential? The fact that these genes have the potential to dramatically perturb cell growth, and the observation that cellular proto-oncogenes have been extraordinarily conserved in evolution, suggested that, in their normal context, these genes must have important functions in growth control. Indeed, most proto-oncogenes encode proteins that participate in signal transduction pathways through which signals to divide (or not divide) are relayed from outside the cell to the regulatory machinery within. Thus, the proto-onco-

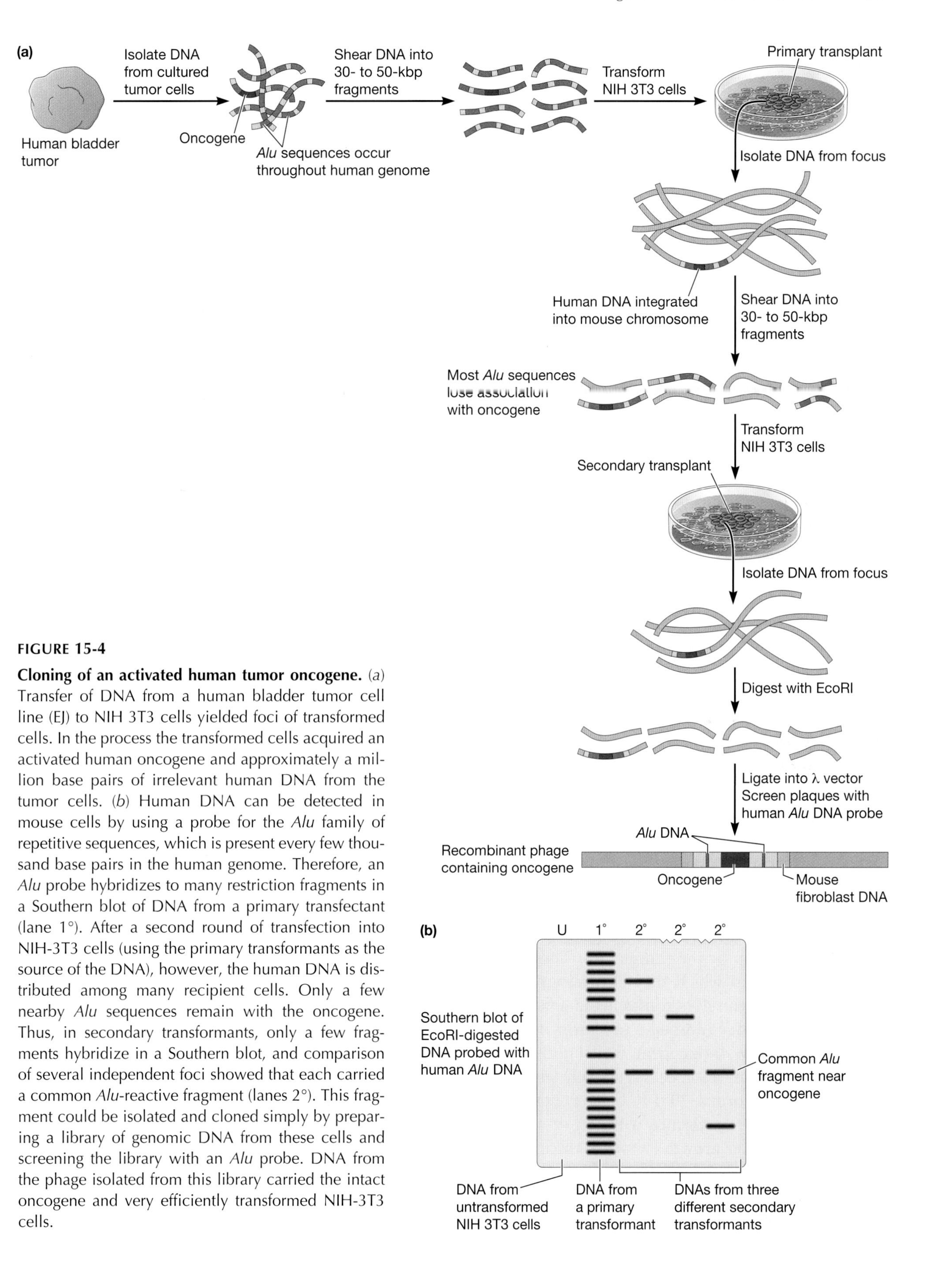

FIGURE 15-4

Cloning of an activated human tumor oncogene. (*a*) Transfer of DNA from a human bladder tumor cell line (EJ) to NIH 3T3 cells yielded foci of transformed cells. In the process the transformed cells acquired an activated human oncogene and approximately a million base pairs of irrelevant human DNA from the tumor cells. (*b*) Human DNA can be detected in mouse cells by using a probe for the *Alu* family of repetitive sequences, which is present every few thousand base pairs in the human genome. Therefore, an *Alu* probe hybridizes to many restriction fragments in a Southern blot of DNA from a primary transfectant (lane 1°). After a second round of transfection into NIH-3T3 cells (using the primary transformants as the source of the DNA), however, the human DNA is distributed among many recipient cells. Only a few nearby *Alu* sequences remain with the oncogene. Thus, in secondary transformants, only a few fragments hybridize in a Southern blot, and comparison of several independent foci showed that each carried a common *Alu*-reactive fragment (lanes 2°). This fragment could be isolated and cloned simply by preparing a library of genomic DNA from these cells and screening the library with an *Alu* probe. DNA from the phage isolated from this library carried the intact oncogene and very efficiently transformed NIH-3T3 cells.

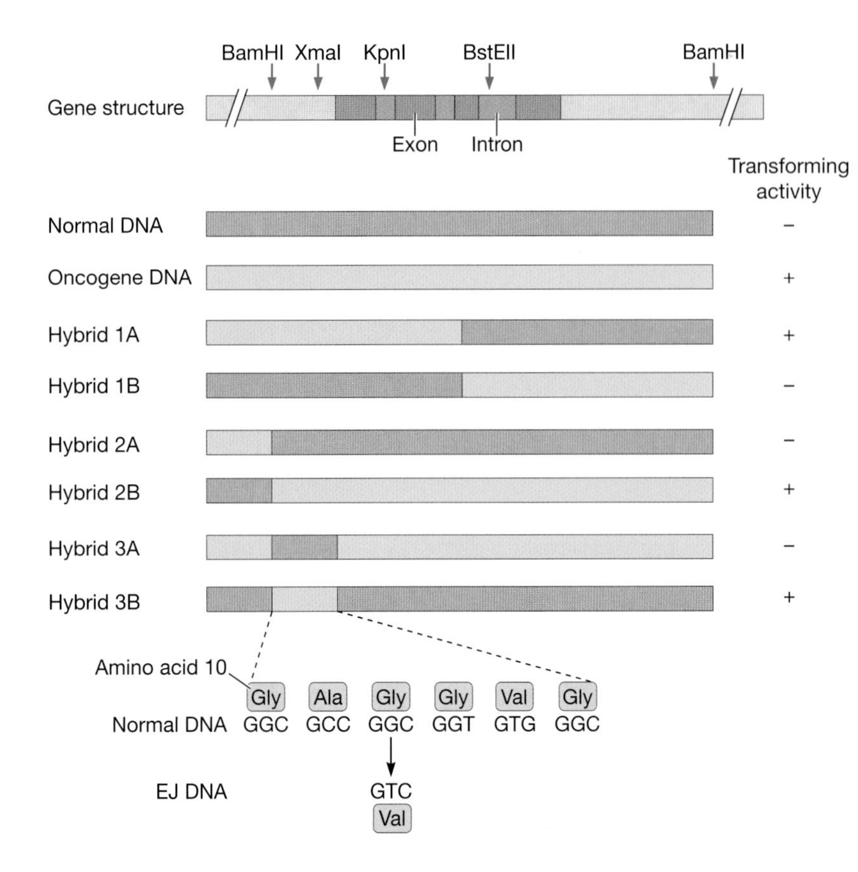

FIGURE 15-5

Locating the mutation that activated the *ras* gene—the mix-and-match experiment. Although a normal *ras* gene (*blue*) does not transform NIH-3T3 cells, the bladder carcinoma *ras* oncogene (*green*) does. To locate the sequence difference that accounted for the dramatic difference in the biological activity of these genes, investigators created chimeric genes from the two parent genes by using restriction sites common to both genes. Some chimeras transformed NIH 3T3 cells (indicated by +), whereas others did not (indicated by –). By making progressively finer chimeras, researchers quickly homed in on a short restriction fragment (XmaI to KpnI) that, upon sequencing, yielded a single-nucleotide difference, shown at the *bottom,* for the DNA strand corresponding to the sequence in the mRNA.

gene proteins are links in a molecular bucket brigade that pass growth signals into the cell. The oncogenic versions of these proteins are somehow able to pass buckets down the line even if they are not receiving them from their neighbors.

Some proto-oncogenes encode growth factors, the molecules that are themselves the signals to grow. One example is the *sis* oncogene, first identified in a monkey retrovirus. The *sis* oncogene encodes a form of platelet-derived growth factor (PDGF), a potent mitogen (an agent that induces mitosis) for mesenchymal cells such as fibroblasts. Cells infected with the *sis*-carrying virus become transformed via autocrine stimulation: They secrete a growth factor to which they also respond. Thus the cells bathe themselves constantly in a factor that makes them divide. Although the form of PDGF encoded by the v-*sis* oncogene is a slightly altered form of the natural factor, cells artificially engineered to overexpress normal PDGF behave exactly the same way as cells transformed with *sis.*

Next in line after growth factors in the molecular bucket brigade of signaling are growth factor receptors. Many growth factors act on cells through specific high-affinity cell surface receptor proteins endowed with protein-tyrosine kinase activity. Binding of growth factors to their receptors triggers a series of growth-promoting signals inside the cell. Some oncogenes encode altered growth factor receptors that get locked in the "on" position; that is, the receptor transmits its signal to the interior of the cell even when it is not bound to its growth factor. The viral *erbB* gene isolated from a chicken retrovirus, for example, encodes a form of the chicken epidermal growth factor (EGF) receptor that is shortened at both ends. In particular, its entire extracellular (EGF-binding) domain has been lopped off. This truncated receptor acts as if it is constantly bound to its ligand and therefore constitutively sends growth-promoting signals through continual tyrosine kinase activity.

Stable changes in the pattern of cellular gene expression must occur to change the growth state of cells. Therefore, transcription must be considered as an ultimate target of oncogene action. Indeed, many oncogenes act directly in the nucleus as transcription factors where they control the expression of cellular genes required for proliferation. A striking feature of these proto-oncogenes is that their expression is highly regulated. Genes like c-*fos*, c-*myc*, and c-*rel* are expressed at very low levels in quiescent cells but are quickly turned on in response to mitogenic signals that promote cell division. Because the regulation of expression of nuclear proto-oncogenes is

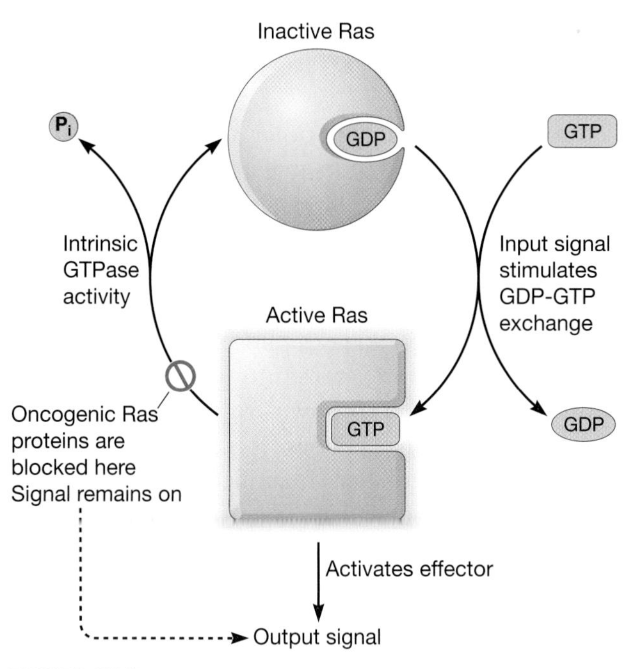

FIGURE 15-6

The Ras cycle. When bound to a molecule of GDP (*top*), Ras proteins are inactive for signaling. In response to an upstream signal, GDP is exchanged for GTP, thereby activating Ras. Normally Ras proteins remain in this state only briefly; they hydrolyze GTP to GDP and then return to the inactive state. Mutations that block this ability to hydrolyze GTP cause Ras to remain locked in the active conformation.

intrinsic to their function, activation of nuclear oncogenes often occurs by deregulated expression. We have already seen, for example, that the c-*myc* gene is a frequent target for integration of proviruses and for chromosomal rearrangements, both of which increase expression of c-*myc* or lead to its expression at an inappropriate time. Indeed, placing a normal c-*myc* gene in an expression vector that cannot be shut off by the cell can be sufficient to transform cells (Fig. 15-7).

In virtually every case where their functions are known, oncogenes lie along the signaling pathways by which cells receive and execute growth instructions. The mutations that activate these genes are either structural mutations that lead to the constitutive activity of a protein without an incoming signal (e.g., the protein kinases and Ras) or regulatory mutations that lead to the expression of the gene at the wrong place or time (e.g., the nuclear oncogene proteins and the growth factors themselves). Damage to oncogenes gives a cell a persistent internal growth signal in the absence of any external stimuli.

Herceptin Targets a Growth Factor Receptor Expressed on Many Breast Cancer Cells

Biologists used the viral *erbB* as a probe to identify the cellular equivalent and found it to be the gene encoding the epidermal growth factor receptor (*EGFR*). They also discovered a second related gene, *erbB2*, also named *HER2* for human EGF receptor 2. Furthermore, *HER2* was found to be the same gene as another known oncogene, called *neu*, that had been cloned from rat neuroblastomas. Even though it is closely related to the EGF receptor, the *HER2* receptor does not bind a growth factor but instead stimulates EGF signaling by interacting with the EGF receptor itself. That *HER2* was likely to be an important player in cancer became apparent when it was found to be overexpressed in 20–25% of breast

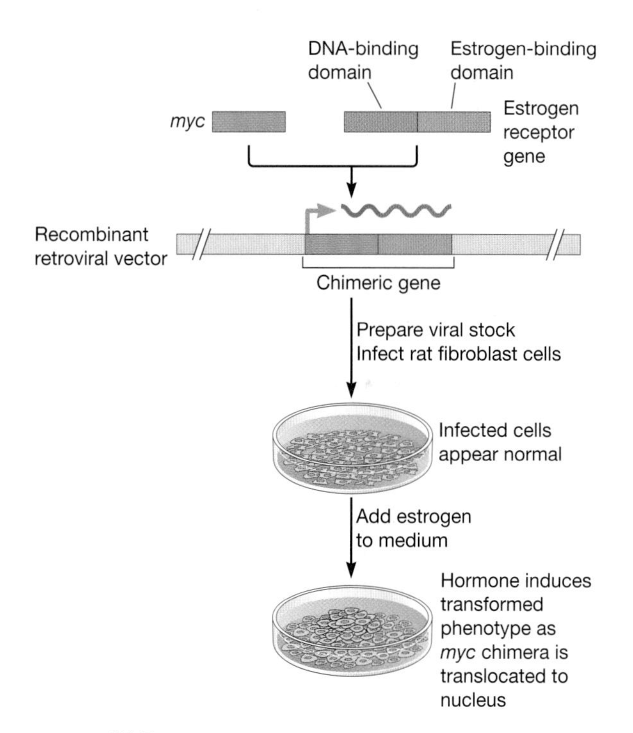

FIGURE 15-7

Deregulated production of a normal Myc protein is sufficient to transform cells. To study transformation by the Myc protein, investigators constructed a hybrid gene in which *myc* docking sequences were fused to the sequence encoding the hormone binding protein of the estrogen receptor. The hormone-binding protein domain acts as an intracellular switch, which keeps the linked protein domain inactive in the absence of ligand and active in its presence. The chimeric gene was placed in a retroviral vector, and the resulting virus stock was used to infect a culture of normal cells. In the absence of estrogen, the Myc protein was inactive, and the cells continued to grow normally. Estrogen treatment allowed active Myc protein to accumulate in the cell nuclei, and the cells became fully transformed.

cell cancers. Strikingly, tumors with *HER2* overexpression are much more aggressive and patients with *HER2* overexpressing tumors do not survive as well as patients whose tumors express *HER2* at normal levels. *HER2* is also overexpressed in many ovarian cancers, and, here too, *HER2* overexpression predicts a poor prognosis.

The overexpression of *HER2* suggested that the receptor might be a good target through which to destroy tumor cells. Mouse monoclonal antibodies were prepared that recognized the human HER2 extracellular domain, and some of these could block the proliferation of cultured cell lines that overexpressed *HER2*. However, these mouse antibodies could not be injected into humans, because the mouse IgG molecules would be recognized as foreign proteins and destroyed. The solution was to prepare a *humanized* antibody by inserting those portions of the mouse antibody molecule that recognized HER2 into a human IgG molecule (Fig. 15-8). These hybrid antibodies consist primarily of human antibody sequence; thus, they are not attacked by the patients' immune response system and remain in circulation for as long as a month.

The company Genentech developed a humanized anti-HER2 antibody, named Herceptin, which blocked the proliferation of *HER2* overexpressing cells in culture and was soon proved to inhibit tumors in vivo as well. Herceptin was approved by the U.S. FDA for the treatment of metastatic breast cancer. In these patients, Herceptin extended survival from 20 months to 25 months and reduced the death rate at one year from 33% to 22%. Herceptin appears to be particularly effective when used in combination with chemotherapy to treat women who have surgical removal of HER2-positive tumors. Although 33% of women receiving chemotherapy suffer relapse, the combined treatment reduced this number to 15%; presumably Herceptin is effective against small groups of cells that may have metastasized to other tissues.

How Herceptin causes tumor regression is not entirely clear. Part of its success is undoubtedly that the antibody molecules recruit cells of the immune system—natural killer cells and macrophages—to attack the tumor cells, and there is some evidence that it inhibits angiogenesis (growth of blood vessels) necessary for tumor growth. However, we have already seen that the antibody can block cell growth in culture and it seems to do this by direct inhibition of signaling through the tyrosine kinase domain of HER2.

Gleevec Inhibits a Kinase Mutated in Leukemia Cells

Protein-tyrosine kinases other than those that are transmembrane receptors make up the single largest family of oncogenes. These *nonreceptor kinases* occupy the shadowy area just inside the plasma membrane, where they are activated by binding to phosphotyrosine residues on activated membrane receptors. In turn, these kinases phosphorylate their protein targets, thereby transmitting signals from the upstream receptors to downstream molecules. This family of tyrosine kinases is encoded by genes including *src* (the first identified oncogene), *abl* (affected on the Philadelphia chromosome), and *kit* (a transmembrane tyrosine kinase receptor for stem cell factor required for hematopoiesis, melanogenesis, and gametogenesis). The very large number of oncogenic protein-tyrosine kinases suggests that tyrosine phosphorylation is a critical step in the signal transduction pathways controlling cell growth. Indeed, phosphorylated tyrosine is vanishingly rare in normal cells and is found only when cells are stimulated with growth factors or transformed with oncogenic growth factor receptors or when oncogenic intracellular tyrosine kinases are active.

As we have seen, tumor cells from patients with CML carry a reciprocal chromosome translocation that brings the tyrosine kinase gene *abl* from chromosome 9 next to the so-called breakpoint cluster region serine/threonine kinase gene (*bcr*) on chromosome 22. As a consequence of the translocation, a BCR-ABL fusion protein is made. In fact, three different fusion proteins are made depending on exactly where in the *BCR* gene the break has occurred. Each fusion protein is associated with a different form of CML: acute lymphoblastic leukemia (ALL), chronic lymphocytic leukemia (CLL), and chronic neutrophilic leukemia (CNL). The kinase activity of the protein encoded by c-*abl* is normally tightly controlled by cellular signals, but the fusion proteins have constitutively active kinase activity. Experimental evidence that the BCR-ABL protein is responsible for leukemia came when the BCR-ABL fusion protein was expressed in transgenic mice. These mice developed leukemias very similar to CML. Other transgenic mice were made carrying various mutated forms of BCR-ABL and one, with a single missense mutation that destroyed ABL kinase activity, never led to leukemia. This pointed to the ABL kinase as a possible target for leukemia therapy by blocking its activi-

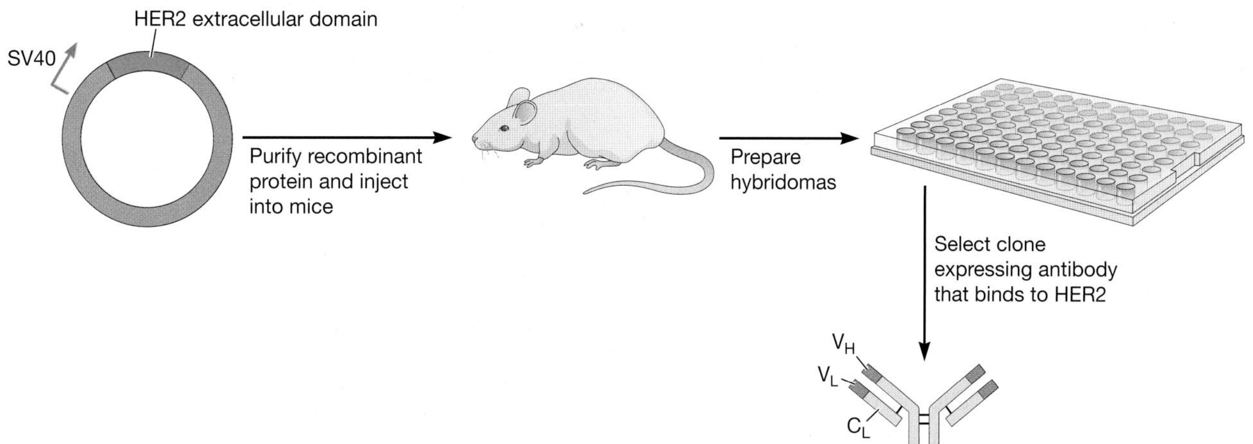

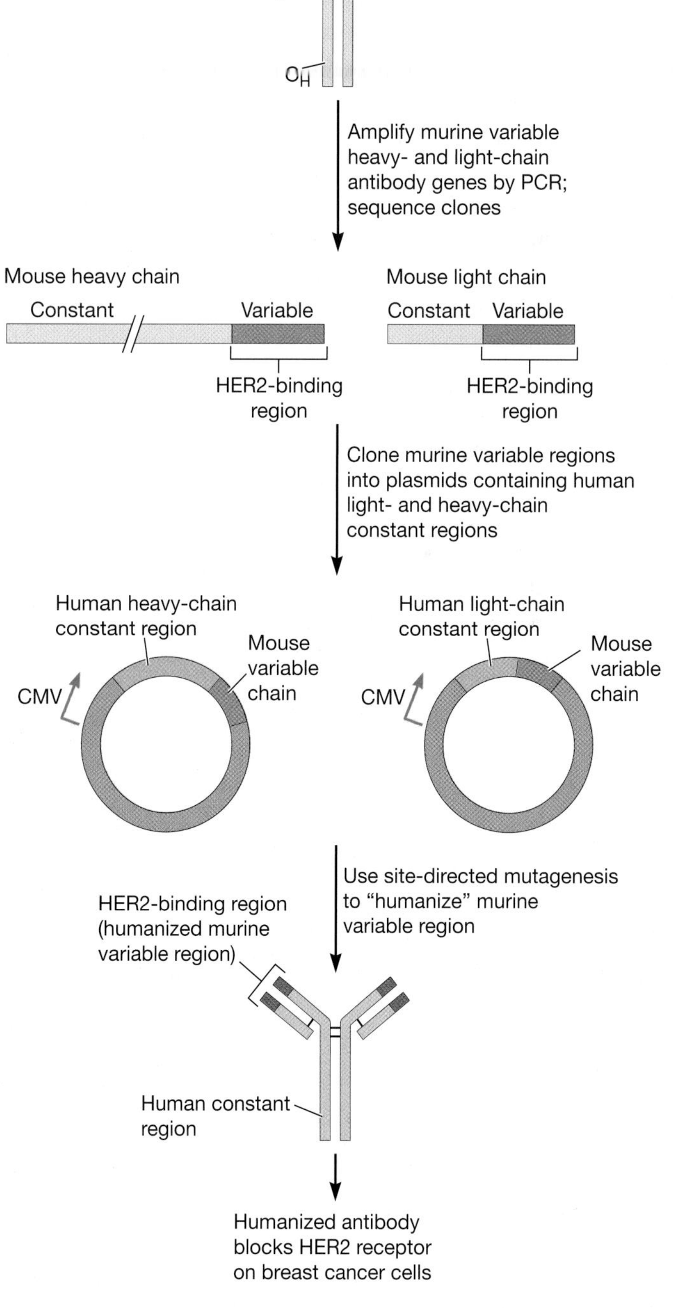

FIGURE 15-8

Herceptin is an example of a therapeutic humanized antibody. A bacterial expression construct was prepared that encoded the extracellular domain of HER2. This recombinant protein was purified, injected into mice, and used to prepare hybridomas. A clone expressing an antibody was found that not only bound to HER2 but could also inhibit the growth of HER2-overexpressing cancer cells. The variable regions of both the heavy chain and the light chain antibody genes were amplified by PCR and sequenced. These murine variable regions were cloned into a plasmid containing cDNA encoding the human light chain and heavy chain constant regions. Then, site-directed mutagenesis was used to convert much of the murine variable region to human sequences. Humanizing the light and heavy chains required 24 and 32 amino acid changes, respectively. The resulting antibody clones encoded a molecule that was almost entirely human, with the exception of the small portions of the variable region directly involved in antigen binding.

ty. However, because BCR-ABL is an intracellular protein, a therapeutic antibody (as for *HER2*) would have no effect. Another approach was needed to inhibit BCR-ABL within cells.

In the late 1980s, scientists at Ciba-Geigy (later merged with Sandoz to form Novartis) began screening through libraries of chemical compounds in a search for small molecules that could inhibit protein kinases. A molecule was identified that could weakly inhibit several kinases. In an attempt to enhance its activity, these scientists tried adding different chemical groups to the molecule. They developed one molecule, later named Gleevec, that inhibited BCR-ABL at very low concentrations and prevented the growth of CML cells in vitro. Remarkably, Gleevec has little effect on other kinases (Fig. 15-9). The cocrystal structure of Gleevec with BCR-ABL revealed why this is so. Although most kinases have very similar structures in their active state, the structures of the inactive state are much more varied. Gleevec achieves such striking specificity by binding to the inactive form of BCR-ABL and locking the "activation loop" that controls ABL kinase activity in this inactive conformation. Gleevec is an astonishingly successful treatment for CML; it achieves remission in more than 80% of newly diagnosed, early-stage patients. Previously, at most 50% of patients survived 5 years after diagnosis, but 89% of patients receiving Gleevec survive 5 years or more. Unfortunately, some tumors in some patients develop Gleevec resistance. BCR-ABL was sequenced from these resistant tumors, thereby revealing mutations that weakened or blocked Gleevec binding to BCR-ABL or led to overexpression of the fusion protein. Currently, chemists are trying to design new molecules capable of blocking the mutated kinases. Gleevec does inhibit other kinases, including two known to be involved in cancer: the Kit receptor in gastrointestinal tumors and the PDGF receptor in myeloid cancers. Gleevec is also effective in treating both types of cancer.

Evidence for Tumor Suppressor Genes Comes from Cells in Culture and the Familial Cancer Retinoblastoma

The first hints that there was more to the genetics of cancer than the activation of oncogenes came from cell culture experiments done in the 1970s. It had been assumed that oncogenes were dominant because, when introduced into cells, they induced changes even though the cells contained normal versions of the genes. However, researchers using somatic cell genetics to study cancer reached the opposite conclusion—tumorigenicity is recessive to normal growth. This was discovered through cell fusion experiments in which a culture containing cancer cells such as HeLa cells together with normal human diploid cells was treated with inactivated Sendai virus or polyethylene glycol. These agents cause the cell membranes to fuse so that hybrid cells are formed having nuclei from both cell types in the same cytoplasm (Fig. 15-10). The expectation, based on the studies with oncogenes, was that these hybrid cells would have a cancer phenotype, but they did not. Hybrid cells had normal characteristics, which suggests that normal cells must possess genes that can overpower the oncogenes and keep them in check. Sometimes hybrid cells changed and became cancer cells. Cytogenetic analysis revealed that these cells had lost chromosomes, commonly chromosomes 11 or 13. Demonstration that chromosome 11 indeed carried an anti-oncogene came from sophisticated chromosome transfer experiments in which isolated chromosomes from normal human cells were inserted into tumor cells (Fig. 15-10). Insertion of chromosome 11 was sufficient to reprogram the tumor cell into normal growth. This phenomenon is paralleled in vivo by the frequent observation of chromosome deletions in tumors, which are detectable by microscopy or by hybridization to molecular probes. These observations support the idea that cells have growth-suppressing genes that must be inactivated before tumors can develop.

Further evidence for the existence of tumor suppressor genes came from a very different source—studies of retinoblastoma, a rare inherited tumor of the eye that strikes young children. In its most common form, retinoblastoma occurs sporadically, as an isolated event in a family with no history of the disease, and affecting only one eye. But about one-third of retinoblastoma patients develop multiple tumors, usually in both eyes, and the children and siblings of such patients often develop the same disease, an observation suggesting that the susceptibility to retinoblastoma is inherited. The data are consistent with a mechanism involving a single gene.

In 1971, Alfred Knudson suggested a model to account for the sporadic and inherited forms of retinoblastoma. He supposed that the development of retinoblastoma required mutations in both copies

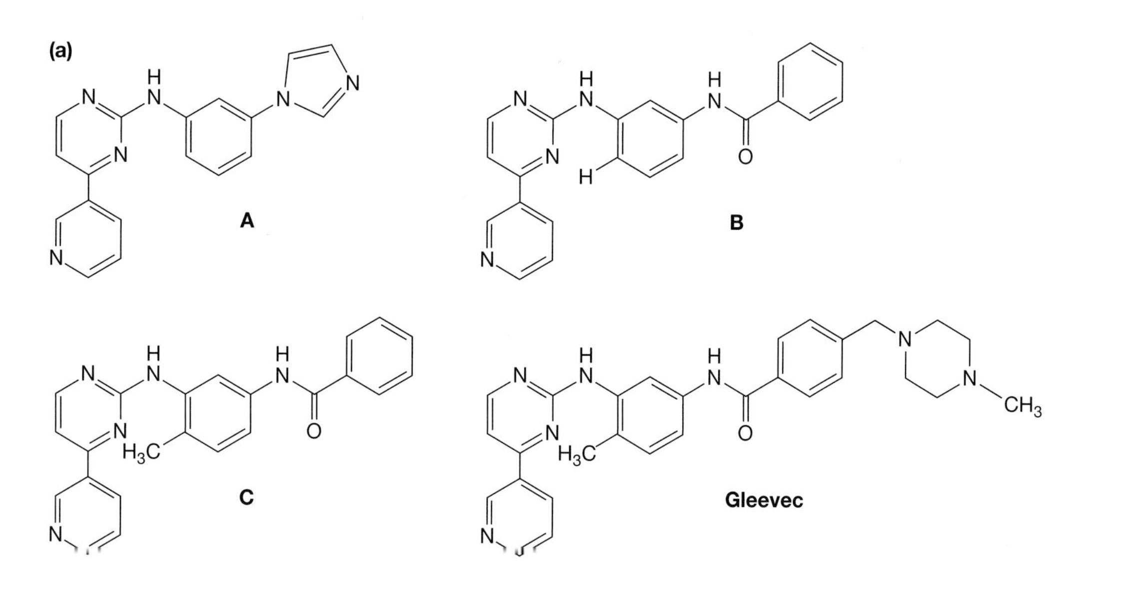

(b)

	Concentration of drug needed for 50% inhibition of kinase activity (IC_{50} nM)			
	A molecule	B molecule	C molecule	Gleevec
BCR-ABL	n.d.	n.d.	n.d.	25
c-Abl	3,300	2,800	361	188
c-Kit	1,100	1,100	785	413
PDGFR-β	390	870	400	386
VEGFR-2	1,400	1,300	>10,000	>10,000
EGFR (HER-1; ErbB)	>10,000	>10,000	>10,000	>10,000
FGFR-1	2,500	>10,000	>10,000	>10,000
c-Met	n.d.	>10,000	>10,000	>10,000
IGF-R	>10,000	>10,000	>10,000	>10,000
CDK1/cyclinB	92	200	>10,000	>10,000
c-Src	1,700	>10,000	>10,000	>10,000
PKC-α	1,000	1,200	2,000	>10,000

(c)

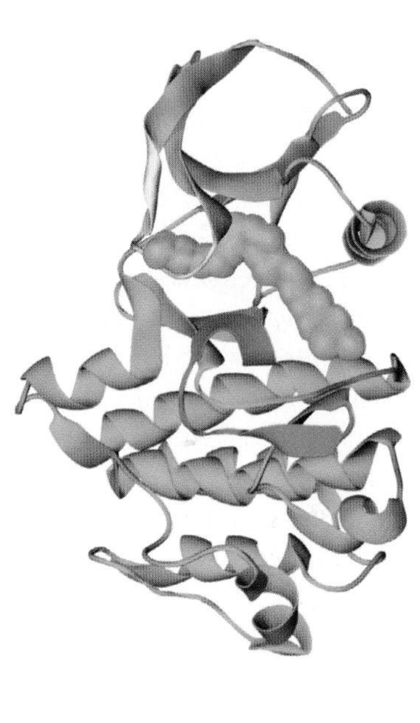

FIGURE 15-9

Gleevec was developed to specifically inhibit the BCR-ABL fusion kinase. (*a*) A library of chemical compounds was screened to identify small molecules that could inhibit the kinase activity of a receptor tyrosine kinase, the platelet-derived growth factor (PDGF) receptor. One molecule (A) had reasonable inhibitory activity. A succession of changes to this molecule were made—shown as B and C, culminating in STI-571, eventually named Gleevec. (*b*) Each new molecule could inhibit BCR-ABL at an even lower concentration. The indicated kinases were expressed and purified and mixed with ATP and a substrate to test their phosphorylation ability. Different amounts of each drug were tested to identify the concentration needed to inhibit 50% of the activity of each kinase—a concentration called the inhibitory $concentration_{50}$ (IC_{50}). The PDGF receptor and c-Kit are also inhibited, although a much higher concentration of the drug is required. n.d., not determined. (*c*) An X-ray crystal structure of BCR-ABL in complex with Gleevec was determined. The drug (*orange*) fits precisely into the active site of the kinase and locks the protruding activation loop in an inactive conformation.

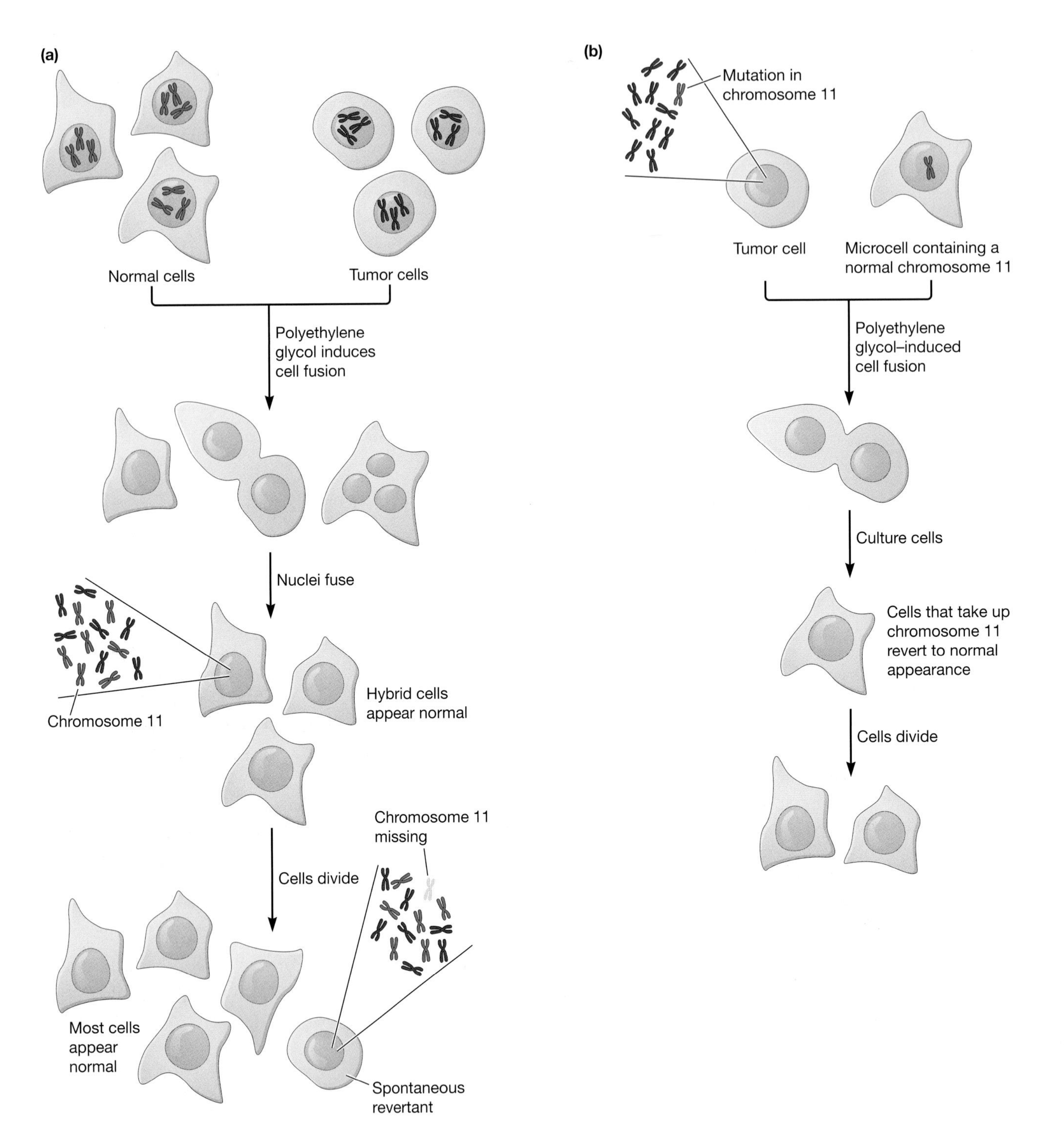

FIGURE 15-10

Normal growth phenotype is dominant to the transformed phenotype. (*a*) Investigators mixed normal cells and tumor cells and induced them to fuse by using the chemical polyethylene glycol (PEG). The resulting hybrids carry chromosomes from both parents, and they were uniformly normal in phenotype. Thus, genes from the normal partner are able to suppress the growth defect encoded in the genes of the transformed cell. But these hybrids are unstable, and they randomly shed chromosomes. By examining hybrids that spontaneously regained the transformed phenotype, investigators could identify the normal chromosome that was always lost from such cells. In certain crosses, it was chromosome 11. Indeed, transfer of chromosome 11 alone to tumor cells containing a mutated chromosome 11 was sufficient to revert their growth to normal (*b*). This chromosomal transfer was done by microcell transfer, which involves treating cells with drugs that disrupt the cytoskeleton and generate small cell fragments (microcells) containing single (or a few) chromosome(s) in their nuclei.

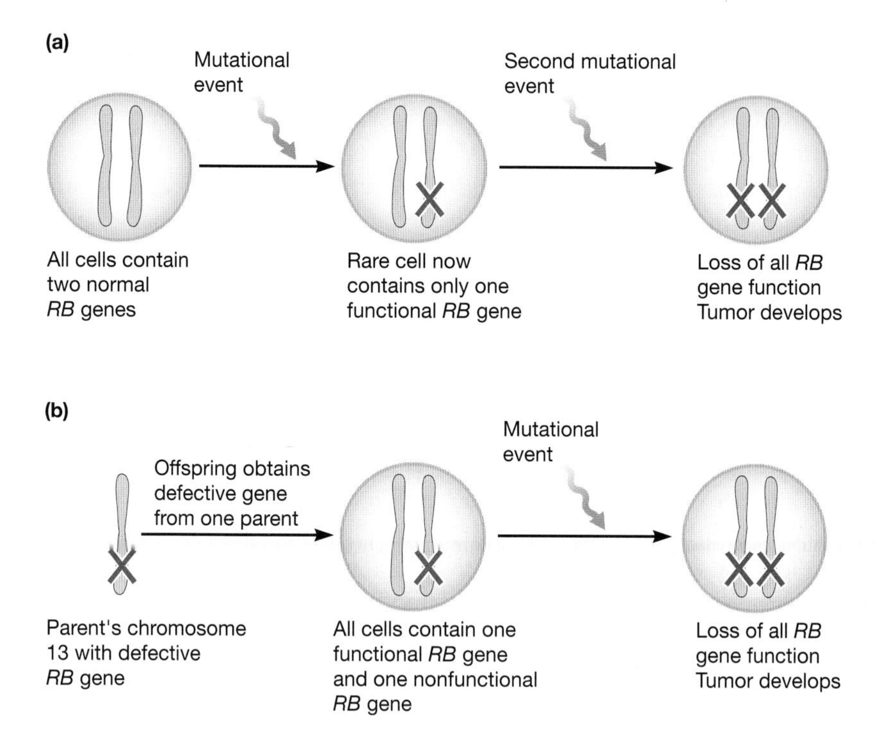

FIGURE 15-11

The development of sporadic and inherited retinoblastoma. (*a*) In the sporadic form of the disease, all cells in the body carry two functioning *RB* genes, one on each copy of chromosome 13. In a rare cell, a mutational event destroys one of the active genes. If one of the descendants of this cell suffers a second blow to its remaining *RB* gene, it develops into a tumor. The likelihood that two rare events will strike in the same cell is very low, hence sporadic retinoblastoma is a very uncommon tumor. (*b*) Some children, however, are born with one *RB* gene already damaged. This mutation can be inherited from a parent or can occur during development of the sperm or egg cell that gives rise to the child. Now, only a single event—damage to the remaining *RB* gene—is required for a tumor to develop. These children often develop multiple tumors in both eyes.

of a gene. In the sporadic form of the disease, both mutations would have to occur within a single retinoblast cell, an exceedingly infrequent circumstance that explains why these children never develop more than one tumor. In the inherited form of the disease, however, he suggested that one of the mutations is already present (inherited) in all retinal cells. Therefore, only a single additional mutation is required for a full-blown tumor, and tumors would be much more frequent in these individuals. A decade later, analysis of the chromosomes in tumor cells and normal tissues of retinoblastoma patients resoundingly confirmed Knudson's hypothesis (Fig. 15-11). Many retinoblastoma patients carried deletions in chromosome 13 and the gene inherited by afflicted children, termed *RB*, was mapped to this same chromosome. Most important, whereas unaffected tissues in these children could be shown to carry one mutant *RB* allele and one normal one, tumor DNA carried only mutant *RB* alleles. The *RB* gene must therefore be a tumor suppressor gene or anti-oncogene that normally functions to arrest the growth of retinal cells. The *RB* gene was cloned and its state determined in a large panel of retinoblastoma tumors. In many of these tumors, the *RB* gene was completely or partially deleted. In others, the gene was intact, but molecular analysis revealed simple point mutations, often in splice junctions, which caused exons to be skipped, thus leading to the production of an aberrant RB protein. Finally, it was shown that introducing the wild-type *RB* gene into tumor cells could revert their aberrant growth properties to normal growth patterns.

What does the RB protein do? How does it put the brakes on cell growth? Antibodies to the RB protein showed that it was located in the nucleus and that in newly dividing cells the protein was only lightly phosphorylated. As cells began to duplicate their DNA, the protein became very heavily phosphorylated. This observation suggested some connection to the clock mechanism by which cells regulate their cell division. Now we know that the RB protein inhibits DNA replication until the appropriate signals trigger RB phosphorylation, which releases it to allow transcription of genes needed for mitosis.

For a while it seemed that there was no link between these two classes of cancer genes, dominant oncogenes like *myc* and tumor suppressor genes like *RB*, but in one of the key discoveries in cancer biology, investigators made a connection between the two. Over the years, investigators had accumulated much information about the oncogenes of DNA tumor viruses and were particularly interested in the cellular proteins with which the viral onocogenic proteins interacted. These experiments were carried out by preparing an antibody to a particular viral oncoprotein, and using this to immunoprecipitate the oncoprotein and the cellular proteins bound to it, from lysates of infected cells. Shortly after the *RB* gene had been cloned, Ed Harlow recognized that the adenovirus E1A oncoprotein was associated with a cellular protein of a similar size to the RB protein. Further analysis proved that the E1A oncoprotein did, indeed, bind to and inactivate the RB tumor suppressor protein. Thus, the oncoproteins of the DNA tumor viruses, whose functions had proved so difficult to work out compared to their cousins in the RNA tumor viruses, were in fact working in a completely different fashion. They drove cell growth, not by pressing on the accelerator like conventional oncogenes, but by eliminating the brakes. Moreover, the RB protein led oncogene researchers beyond the primary signal transduction pathways they had been studying to the internal clock that regulates cell growth after the initial growth signal is received.

p53 Plays a Central Role in the Cell's Response to Cancer-inducing Damage

The protein, p53, named for its size of 53 kD, was first discovered as a cellular nuclear protein associated with and stabilized by the T-antigen protein of SV40. There seemed to be a strict correlation between the abundance of p53 and the oncogenic activity of the tumor virus. Uninfected cells and cells infected with nontumorigenic mutant SV40 viruses did not accumulate much p53. Shifting cells transformed with a mutant SV40 virus carrying a temperature-sensitive T antigen to the nonpermissive temperature resulted in simultaneous reversion of the cells to normal and loss of p53 protein. It seemed that p53 was a cellular oncogene used by the tumor viruses to transform cells. However, it was found that some p53 cDNAs did not transform cells and some tumors were found that lacked p53 entirely. Investigators then realized that the original p53 cDNAs, which had been isolated from tumor cell lines, were mutants that carried point mutations in their coding sequence. Wild-type p53 genes, isolated from normal cells, had the opposite effect of the original clones—they suppressed cell growth. In fact, inactivation of the p53 gene is now known to be the most common genetic defect in tumors; more than half of all tumors have loss-of-function mutations in p53. Inheriting a damaged p53 gene leads to dramatically elevated frequencies of cancer, a situation resembling the inheritance of retinoblastoma.

The Li–Fraumeni syndrome is an inherited susceptibility to a variety of cancers, in which 50% of affected individuals develop cancer by age 30 and 90% develop it by age 70. Conventional genetic mapping of the locus responsible for inheritance of the syndrome was not successful. But on the basis of the retinoblastoma precedent, in which patients are born with one inactivated allele of *RB*, investigators checked the DNA of Li–Fraumeni patients for changes in p53, the only other well-characterized tumor suppressor gene at that time. They amplified portions of the p53 gene from patient DNA by using PCR and then sequenced these samples. Indeed, in all five families tested, inheritance of the syndrome correlated with the presence of point mutations in the p53 gene. The rapid linking of this rare and mysterious syndrome to single-nucleotide changes in the genomes of affected individuals is yet another striking illustration of the power of modern technology to find the precise molecular defect underlying a complex disease. And, for p53, its contribution to the development of cancer and its role as a bona fide tumor suppressor were dramatically confirmed.

Nearly all inherited cancer syndromes are associated with one or a few types of tumors, but Li–Fraumeni patients develop nearly every known type of tumor. This wide-ranging effect is a testament to the role of p53: It is the central hub of a cellular network that senses a variety of stresses to the cell and coordinates a variety of responses.

When a cell is faced with DNA damage or excessive growth signals, p53 accumulates. Although p53 is continually transcribed and translated, under normal conditions the protein is degraded almost immediately at the direction of a partner protein, MDM2, which directs the ubiquitinylation of p53. Stress signals can stabilize p53 by way of several different mechanisms. For instance, DNA damage triggers phosphorylation of both p53 and MDM2 so that they can no longer bind. Without this interaction, p53 accumulates; it reaches levels sufficient to form tetramers that bind the promoters of its approximately 120 target genes, thereby activating transcription. These target genes direct a variety of cellular responses to stress, from the benign—inducing proteins needed for DNA repair—to the irrevocable—the programmed death of the cell. Biologists do not yet entirely understand why a stressful insult in one cell will cause p53 to merely halt growth, whereas the response to the same stress in a different cell could be programmed death.

The Several Steps to Colorectal Cancer: A Real-Life Tale of Oncogenes and Tumor Suppressors

The epidemiology of cancer has long suggested that cancer is a multistep disease. Statistical calculations, based on the increased frequency of cancer with age, estimate the number of steps as four to six. Molecular analyses have revealed the genetic changes underlying these steps in colorectal cancer, one of the most common human cancers.

Colorectal cancer occurs both sporadically and in an inherited form. One inherited form, a disease called familial adenomatous polyposis (FAP), is manifested as hundreds of precancerous polyps forming on the colon, some of which progress to fully malignant carcinomas. FAP is inherited as an autosomal-dominant trait, which suggests that the *FAP* locus encodes a tumor suppressor gene. The *APC* (adenomatous polyposis coli) gene was cloned from the region and shown to down-regulate the cytoplasmic protein β-catenin. When *APC* is lost, β-catenin accumulates and triggers cell proliferation. Thus loss of this tumor suppressor gene is one of the steps toward colorectal cancer. Conventional oncogenes are involved as well. Fully one-half of colon carcinomas and polyps larger than 1 cm carry activating mutations in one of the *ras* genes. The presence of these mutations in precancerous polyps suggests that activation of a *ras* gene is an early step in tumor development.

There are chromosomal deletions in colorectal cancer, thus implying the losses of tumor suppressor genes. Most common among these are losses in chromosomes 17 and 18. Chromosome 17 mutations delete the p53 gene. Deletion of p53 is a relatively late step in tumor development, found in 70% of carcinomas but rarely in polyps. The chromosome 18 region consistently lost in colorectal tumors (detectable in >70% of cases) is 18q21. This region contains two genes that may contribute to tumorigenesis. *DCC* (deleted in colorectal cancer) encodes a protein related to the cell adhesion molecules that mediate contact between cells and with the extracellular material within which cells grow. Perhaps loss of this gene allows tumor cells to escape their normal confines and invade surrounding tissue, a cardinal property of tumors. *DPC4/MADH4* is also in this region. It encodes Smad4, a member of the signal-transduction pathway leading from the transforming growth factor β (TGF-β) receptor.

The proposed order of events in the development of colorectal cancer is shown in Figure 15-12. Colorectal tumors have an average of one or two additional chromosomal losses, which may correspond to as-yet-unidentified tumor suppressor genes. Although there is a preferred order for these events—*ras* mutations occur early, for example, and p53 deletions late—it is the accumulation of these events rather than their order that matters in tumor development.

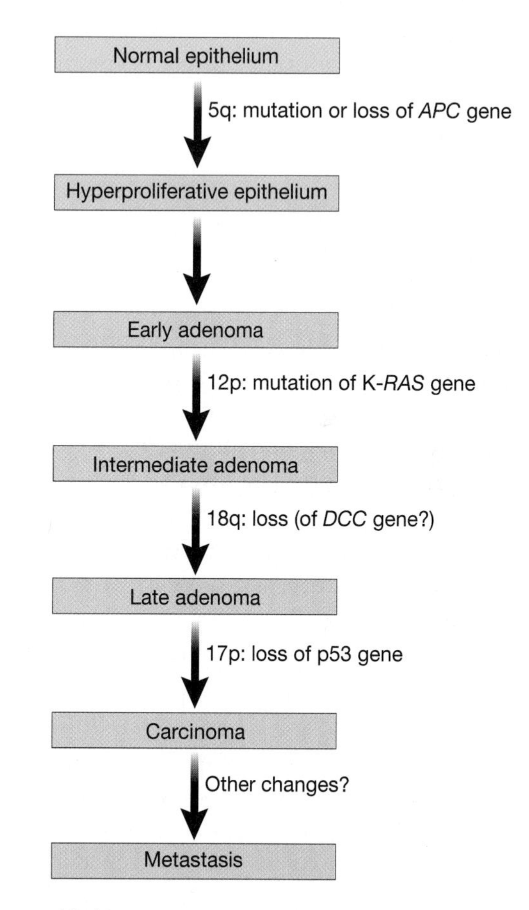

FIGURE 15-12

The road to colorectal cancer. The *arrows* indicate the stages in the development of a colorectal tumor, from normal colon epithelium through benign adenoma to malignant and metastatic carcinoma. Most steps are associated with a defined genetic change, which is indicated alongside each *arrow*. This order of genetic events does not appear to be obligatory. Rather, it is the cumulative effect of them all that leads to development of colorectal cancer.

Loss of p53 Contributes to the Warburg Effect

In 1931, the great German biochemist Otto Warburg won the Nobel Prize for Physiology or Medicine for his work on oxidative metabolism. In that same year, he observed that the metabolism of cancer cells is strikingly different from that of normal cells. Cancer cells produce more lactic acid and use less oxygen than normal cells. This difference is the consequence of an increased rate of glycolysis and decreased rate of aerobic respiration in cancer cells. This metabolic change, known as the Warburg effect, occurs in nearly every cancer. Very recently, scientists observed that simply deleting the p53 gene recapitulated the Warburg effect in cell lines. The decreased rate of respiration was evident even at the whole-organism level; in a swimming test, p53-null mice have half the endurance of wild-type mice.

To identify which target of p53 might be responsible for this change, biologists examined potential p53 target genes predicted from serial analysis of gene expression (SAGE) studies (see Chapter 13). One of the p53 target genes was *Synthesis of Cytochrome c Oxidase 2* (*SCO2*). As expected for a gene regulated by p53, *SCO2* promoter contains a p53 binding site, and expression of SCO2 is reduced in p53-null cells. Previous studies in yeast cells had shown that the SCO2 protein is required for the assembly of cytochrome oxidase, a mitochondrial protein complex essential for respiration. The level of such a protein might well be predicted to affect the respiratory capacity of cells, and indeed overexpression of *SCO2* increased the respiratory rate of cells. *SCO2* expression rescued respiration levels in p53-null cells, thereby demonstrating that the Warburg effect of respiratory inhibition can result from the loss of the tumor suppressor p53.

But why is decreased respiration an advantage in tumor development? At first glance, it would seem that switching from respiration to glycolysis would be a disadvantage, because glycolysis produces much less energy per sugar molecule than does respiration. However, respiration requires a steady supply of oxygen. As we will discuss later, growing tumors are limited by the availability of a blood supply. One advantage of the Warburg effect may be that cancer cells can survive better by producing energy through glycolysis. Also, some evidence suggests that respiratory inhibition can block cells from entering senescence, so the Warburg effect may also promote the longevity of cancer cells.

Failure of Apoptosis Contributes to Tumor Survival and Growth

The increase of glycolysis is just one of many effects of p53 loss. One of the most important roles of p53 in cancer is its effect on tumor growth. Tumor cells in which p53 has been deleted form tumors that grow much more rapidly than wild-type tumors. The aggressive growth of these p53-negative tumors is not due to more rapid cell division; the proliferative rate of p53-negative cells is identical to that of wild-type cells. However, when these tumors were examined for evidence of cell death, they had a much lower number of cells dying through *apoptosis*, a regulated process that sentences cells to death (Fig. 15-13).

Apoptosis was recognized as a distinct form of cell death through histological studies of tissues containing dying cells. Cells killed by violent trauma like temperature shock or abrupt pH change die by *necrosis*, in which cells swell and eventually burst. In contrast, cells dying by apoptosis proceed through an orderly process: They break down systematically into pieces by partitioning the contents of the cell into *blebs*, small packets enclosed by plasma membrane. Apoptosis is a normal, controlled process that occurs throughout the life of multicellular organisms. Morphogenesis of the hand, for example, requires cell death. The hand develops from a limb bud and the fingers are carved from the limb bud by the death of cells between the tissue that will become fingers. Later, in adult life, many tissues are continually renewed, with cells multiplying and dying in equal proportion. In the course of a year in the human body, approximately 3×10^{13} cells are born, and just as many die—a number equal to the total number of cells present at any one moment.

One of the key regulators of apoptosis was discovered by mapping the site of a frequent translocation in B-cell lymphomas. This translocation between chromosomes 14 and 18 resulted in the overexpression of a gene called *bcl-2* (B-cell lymphoma 2). *bcl-2* was unlike other oncogenes. Although its co-overexpression with the *myc* oncogene greatly accelerated the growth of lymphomas, *bcl-2* alone was not oncogenic. When transgenic mice overexpressing *bcl-2* in their blood cells were carefully examined, an unusually large number of cells were present. The rate of blood cell proliferation in wild-type and transgenic mice was the same, so the only explana-

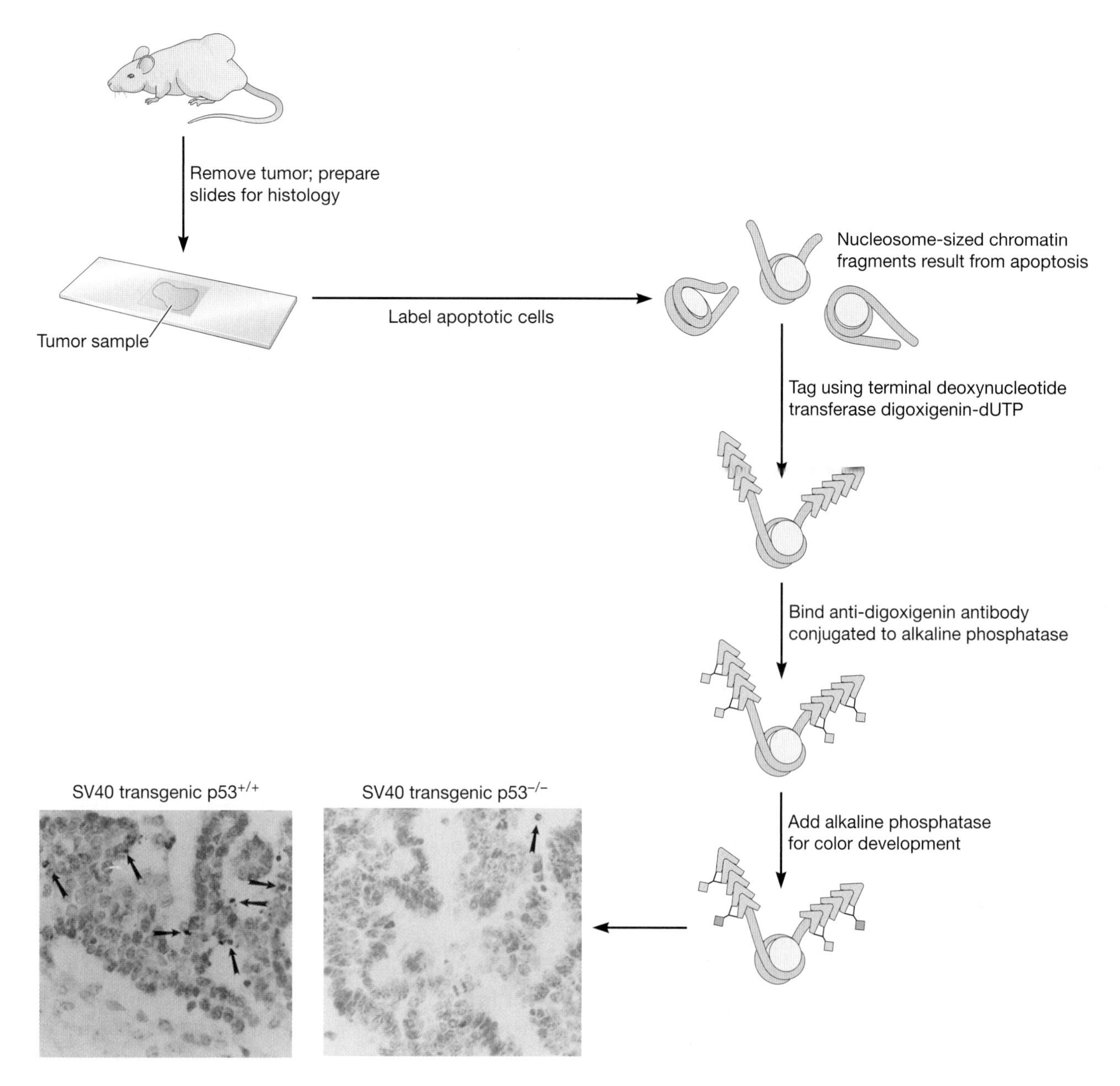

FIGURE 15-13

Loss of p53 makes cells resistant to apoptosis. A mouse cancer model was developed in which the tumor virus SV40 T antigen oncogene was overexpressed in the brain, thereby causing brain tumors. The SV40 transgenic mice were mated to p53 knockout mice to create mice lacking p53 and carrying the SV40 transgene. Tumors developed much more quickly in the p53-null SV40 transgenic mice than in those SV40 trangenic mice with wild-type p53. Therefore, biologists investigated the tumors for the rates of cell proliferation and death. The rates of cell division were identical in p53-null and p53 wild-type tumors. They then examined the number of cells undergoing programmed cell death, called *apoptosis*. Tumor slices were examined for apoptotic cells by using a method to label the double-stranded DNA breaks that occur late in apoptosis. Thin slices of the tumor tissue were prepared and attached to microscope slides. The slides were treated with terminal deoxynucleotidyl transferase (TdT) and dUTP conjugated to the small molecule digoxigenin. Apoptotic cells have many DNA ends that are the substrate for digoxigenin-dUTP addition by TdT. The labeled dUTP was detected by adding an antidigoxigenin antibody conjugated to the enzyme alkaline phosphatase. To detect the apoptotic cells, a chemical was added that was converted by alkaline phosphatase to a dark brown pigment (indicated by *arrows*). Wild-type tumors had high rates of apoptosis, but very few cells in p53-null tumors died.

tion for the increased number of cells was that *bcl-2* overexpression enabled cells to live longer.

When antibodies recognizing Bcl-2 were used to localize the protein in cells, Bcl-2 was found at an unexpected location—the outer membrane of the mitochondria. Previously, biologists believed the sole function of the mitochondria was energy production, but a series of elegant biochemical experiments proved that the abundant electron transport protein cytochrome *c* can trigger apoptosis. Cytochrome *c* is confined to the mitochondrion by the outer mitochondrial membrane, but this barrier is riddled with pores, the sites of a continuous tug-of-war between Bcl-2 and other related proteins. Bcl-2 tries to plug the pores, whereas other Bcl-2 family members, including Bax and Bak, try to pry open the pores. Cell signals influence the abundance and activity of all of these proteins, so that when a cell is subjected to stress, Bcl-2 is overcome by Bax and Bak, and the pores open and release cytochrome *c*. Once cytochrome *c* is released, it initiates a cascade of biochemical events that commit the cell to death.

The initiator of these events is the protein caspase-9, one of a family of cysteine aspartyl proteases. Transcribed and translated as inactive proenzymes, these proteases are activated by proteolytic activity. When cytochrome *c* is released from the mitochondria, it assembles with accessory proteins to form a complex that stimulates self-cleavage of the caspase-9 proenzyme. Liberated into its active state, caspase-9 frees the downstream executioner caspases-3, -6, and -7 that cleave a host of target proteins in the cytoskeleton, nucleus, and mitochondria. This cascade of proteases thus amplifies the small initial signal from the cytochrome *c* released from mitochondria into a biochemical response that destroys the cell.

Many of the genes that interact with p53 tilt the balance from cell survival to cell death. When p53 is activated, proapoptotic members of the *bcl-2* family of genes including *bax* are transcribed. Additionally, p53 activation leads to the repression of Bcl-2, lowering its levels and so making it more likely that the mitochondrial pores will open. p53 also affects a second apoptotic pathway, the extrinsic pathway, in which death receptor proteins such as tumor necrosis factor α are activated by their ligands. These proteins trigger a different set of initiator caspases, which activate the same executioner caspases as the intrinsic cytochrome *c* pathway. In addition, p53 activation dampens survival signals by activating the gene encoding insulin growth factor binding protein 3, which is secreted from cells to sequester the insulin growth factor 2 (IGF2). This blocks the survival signals normally transmitted from IGF2 through the Akt kinase that inhibit caspase-9 and other proapoptic proteins. The decrease in IGF2 signaling makes the cell more sensitive to apoptotic signals.

As tumors grow, it becomes increasingly difficult to acquire the oxygen and nutrients needed for survival, and so most tumors devise methods to resist apoptosis. About one-half of tumors mutate p53 itself and the rest undergo other changes, such as the overexpression of antiapoptotic genes like *bcl-2*, the hyperactivation of the Akt survival pathway, or the inactivation of proapoptotic genes like caspase-8. Overexpressing antiapoptotic proteins and repressing proapoptotic proteins helps tumors to resist death and stress signals. So, the loss of p53 function causes cells to be much more resistant to apoptosis. The loss of p53 in a tumor is an indicator of poor prognosis because the cells can continue growing even in the face of serious DNA damage caused by radiation treatment or chemotherapy.

Tumor Growth Requires New Blood Vessels

In the 1960s, the American surgeon Judah Folkman was working for the U.S. Navy and studying alternatives to blood transfusion. In the course of this research, Folkman observed that tumors transplanted into organs in isolated culture grew to a small size and halted, but if the growth-arrested tumors were then transplanted into animals, they grew rapidly once they had become supplied with blood vessels. It was not sufficient that the cancer cells were capable of growing uncontrollably; the tumor required a blood supply if it were to grow. Folkman went on to observe tumors transplanted into the eyes of rabbits. If a tumor was placed far away from any blood vessels, its growth was arrested just as if it had been put in organ culture. If, however, tumor transplants were placed near the blood vessel–rich iris, blood vessels grew toward the tumors that, now supplied with blood, began to grow. Folkman hypothesized that the tumors were releasing a factor that stimulated the growth of blood vessels and attracted them to the tumors. He homogenized tumors and isolated factors that could stimulate *angiogenesis*, the growth of blood vessels. Many of these angiogenic factors were proteins, and by purifying and sequencing these proteins, Folkman identified a number of genes encoding factors that promote the growth of blood vessels.

One of the most important angiogenic factors is vascular endothelial growth factor (VEGF). Harold Dvorak identified VEGF by biochemical purification of a factor that increased the permeability of blood vessels. Soon, biologists realized that VEGF was needed for the first steps of blood vessel formation and that mice with defects in VEGF function fail to sprout blood vessels. Because VEGF is so important for the early steps in blood vessel formation, it seemed that inhibiting its function might prevent the growth of blood vessels to tumors. To inhibit VEGF, researchers isolated a monoclonal antibody that specifically bound to VEGF, thereby signaling immune cells to take up and destroy the antibody–VEGF complex. A humanized version of one anti-VEGF antibody, called Avastin (or bevacizumab) has been approved by the U.S. FDA for use in cancer patients in combination with chemotherapy. This factor was approved after colon cancer patients receiving the combined treatment lived 5 months longer on average than patients receiving standard care.

After VEGF and other angiogenesis factors were discovered, biologists realized that these genes are expressed continually, but angiogenesis occurs only in very specific situations—in wound healing, in the development of the placenta, and, abnormally, in promoting the growth of tumors. Why do these factors not promote angiogenesis at all times? Clues came from the observations made by surgeons that secondary metastases often appear soon after a large, primary tumor had been removed. Folkman developed a mouse tumor model in which lung cancer cells were injected under the skin of mice, thus quickly forming a large, primary tumor. When the mice were examined, small secondary metastases were also apparent, but these had not acquired their own blood supply. Just as for cancer patients, when the primary tumors were surgically removed from the mice, the secondary metastases acquired a blood supply and grew quickly.

It seemed that the primary tumor was suppressing the development of the smaller tumors. Researchers collected the urine of tumor-bearing mice and found factors that inhibited angiogenesis. One of these antiangiogenic factors, called *angiostatin*, was found to be a fragment of a familiar protein, the blood clotting factor plasminogen. Full-length plasminogen does not affect the growth of blood vessels, but it is strongly inhibitory when proteolytically cleaved into angiostatin. This was but the first example of several common proteins that are proteolytically processed into angiogenesis inhibitors (Table 15-1). A second antiangiogenic factor, endostatin, was found to be a fragment of collagen type XVIII. This solved an old mystery—why do people with Down syndrome have much lower rates of cancer than the general population? The collagen XVIII gene, the source of endostatin, is on chromosome 21, and indeed Down patients have higher levels of endostatin in their bloodstream than do normal individuals, presumably because they have three copies of the collagen XVIII gene. This idea was explored by creating transgenic mice that expressed 50% more endostatin than normal mice. Tumors injected into these mice grew much more slowly than tumors in normal mice (Fig. 15-14). It was a remarkable finding that such a small increase in the level of an angiogenesis inhibitor could have so profound an effect on tumor growth. It was clear that antiangiogenesis factors might be potent suppressors of human cancers.

TABLE 15.1. Endogeneous inhibitors of angiogenesis

Antiangiogenic factor	Protein source	Source protein function
Endostatin	Collagen XVIII	Extracellular matrix
Arresten	Collagen IVα1	Extracellular matrix
Canstatin	Collagen IVα2	Extracellular matrix
Tumstatin	Collagen IVα3	Extracellular matrix
Anastellin	Fibronectin	Extracellular matrix
Angiostatin	Plasminogen	Blood clotting

There were high hopes that endostatin and angiostatin would prove to be effective therapies. They are natural molecules and so have few side effects, and their pharmacodynamics (e.g., their persistence in the body) were good. In these experimental situations, treatment with both proteins was effective, but FDA regulations require that the components of a combination therapy must individually be effective before they can be used in clinical trials. Furthermore, producing large quantities of proteins is difficult and expensive. Small, synthetic molecules can be made more easily and one has been approved by the U.S. FDA for use in patients. SU11248, directed against tyrosine kinases including the PDGF and VEGF receptors, is being used for treatment of gastrointestinal stromal tumor and renal cell carcinoma.

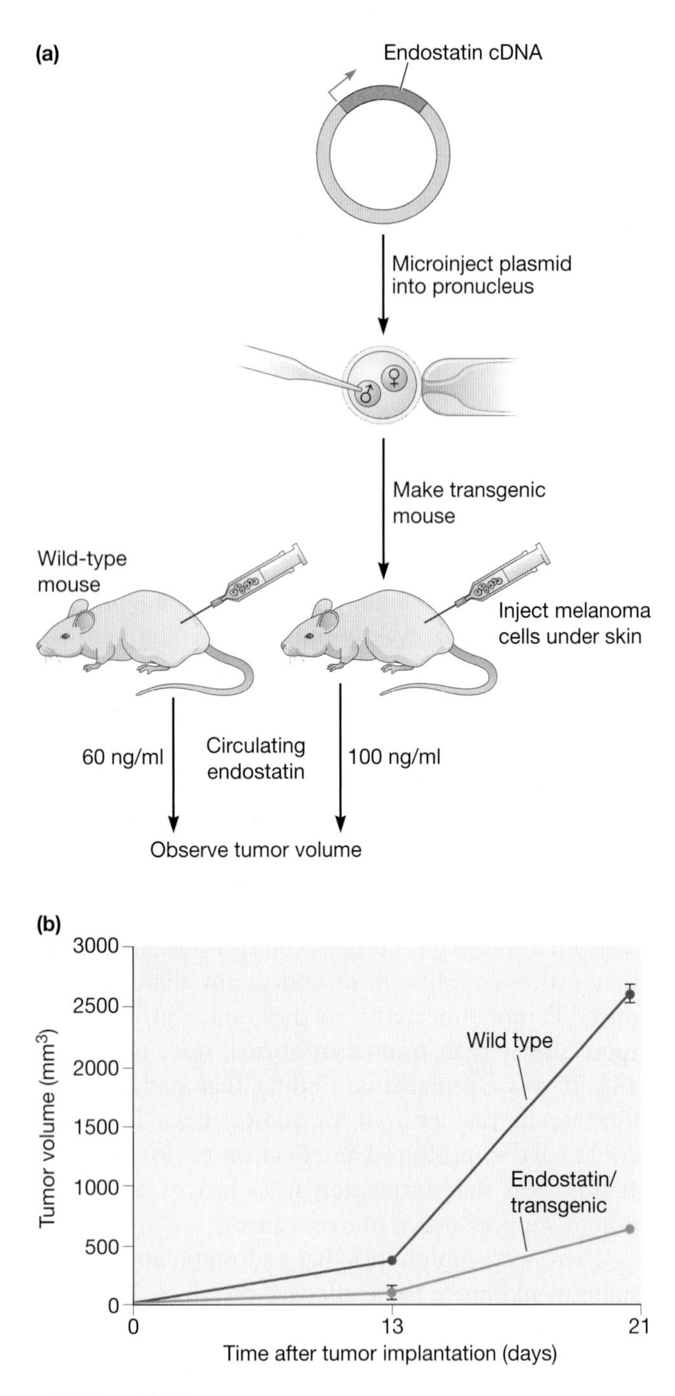

FIGURE 15-14

Small increases in the antiangiogenesis factor endostatin levels slow tumor growth. (*a*) A transgenic mouse was prepared that expressed endostatin (a fragment of collagen XVIII) under the control of a tetracycline-activated promoter. This promoter was activated by expressing the Tet promoter transactivator from a separate transgenic construct (not shown). The transgenic mice had increased levels of endostatin in their bloodstream as compared to wild-type mice. A melanoma tumor cell line was injected under the skin on the backs of both endostatin transgenic mice and control mice. The size of the resulting tumors was measured. (*b*) Tumors grew more slowly in the mice overexpressing endostatin.

A G Protein–coupled Receptor Interacts with the Extracellular Matrix to Regulate Metastasis

Metastasis, the spread of a cancer from its primary site to other parts of the body, is the main cause of death in cancer patients. Tumor cells must undergo a number of changes in order to colonize new sites. They must detach from the primary tumor, invade the circulatory system, migrate to another region of the body, invade the new site, survive, and proliferate in the new location. Clinical observations and studies of injected cancer cells show that very few tumor cells can accomplish all of these steps. To do so requires mutations in multiple genes that control these processes, and identifying these genes and understanding the underlying cell biology remain great challenges in cancer research. One strategy, exploiting the power of genomics-based techniques, used microarray analysis to look for all possible differences between metastatic and nonmetastatic cells.

Investigators derived several highly metastatic cell lines from a poorly metastatic melanoma cell line and injected 200,000 cells intravenously into mice. The gene expression profiles of the resulting tumors were compared, and one gene was identified (*GPR56*) whose expression was much lower in all tumors derived from the highly metastatic cell lines. The GPR56 protein is a member of a subfamily of G protein–coupled receptors (GPCRs). These proteins are also termed seven-transmembrane receptors, because they loop back and forth through the cell membrane seven times. When the extracellular domains of GPCRs are bound by molecules, the receptors transmit signals to the cell through a G protein.

To confirm the role of GPR56 in metastasis, investigators modified the melanoma cell line by overexpressing GPR56 from a transgene or by knocking it down using RNA interference (RNAi). Although modified GPR56 expression did not affect the in vitro growth rate of the cells, tumor development was profoundly changed when GPR56 expression levels were altered. The tumor cells with excess GPR56 created many fewer metastases, whereas tumor cells in which GPR56 was silenced by short hairpin RNAs (shRNAs) developed many more metastases. Clearly, GPR56 was important for the development of metastases, although it did not directly affect the growth of the tumor cells. So, what is the role of GPR56 in metastasis? What is its ligand?

A fusion protein was made that contained a part of the GPR56 extracellular domain fused to the Fc constant region of an antibody. This Fc-GPR56 fusion protein can be detected by histochemical analysis of tumor sections using a fluorescently tagged anti-Fc antibody. The GPR56 fusion protein was found to bind to a diffuse region outside of cells in a pattern consistent with the *extracellular matrix*, the network of carbohydrates and proteins, secreted by cells, that forms a scaffold within which cells grow. This finding suggested that the GPR56 ligand might be an ECM protein. To identify the protein, ECM fractions were prepared from mouse lungs by using detergents of increasing strength and fractionating the proteins so obtained by size. The proteins in the purified fraction that bound to GPR56 were characterized by mass spectrometry, and the major protein identified was tissue transglutaminase 2 (TG2). TG2 is widely expressed in the ECM and is implicated in wound healing. Furthermore, TG2 expression is often lost in aggressive tumors, and, strikingly, injection of TG2 into mouse tumors in TG2 knockout mice can slow tumor growth.

TG2, then, acts like a tumor suppressor and normally signals through GPR56 to inhibit cell growth. Down-regulation of GPR56 or the loss of other signaling components in the pathway enhances metastasis. TG2 also interacts with TGF-β, another tumor suppressor, and with other components of the ECM, including fibronectin and the integrins. It is clear that there is a complex network of interactions and signaling between tumor cells and their environment that opens up new possibilities for a better understanding of how metastasis can be controlled.

Microarrays Reveal Changes in the Genomes of Cancer Cells

Theodor Boveri recognized as long ago as 1914 that many tumors have chromosomal abnormalities, and he proposed that that these abnormalities were the cause of cancer. An alternative explanation is that chromosomal instability is a consequence of other genetic changes—for example, mutations in oncogenes. There has not yet been a resolution of this controversy and it is probable that neither chromosome instability nor oncogene mutation is the sole initiator of cancer. Nevertheless, chromosomal abnormalities are a key feature of cancer, as we saw in the translocations that lead to human B-cell lymphomas and CML. In these cases, an oncogene is moved from its usual position to a new position in the genome where it comes under the influence of a strong immunoglobulin promoter or produces a fusion protein with oncogenic properties.

In addition to these changes, cancer cells are frequently no longer diploid but are *aneuploid*; that is, they have abnormal numbers of chromosomes. Aneuploidy arises through errors during mitosis, and mutations have been found in the genes for spindle proteins and other proteins that regulate the movement of chromosomes. These may cause nondisjunction when chromatids fail to separate as they move to the opposite ends of the mitotic spindle. As a consequence, one daughter cell receives three copies of a chromosome (triploid) and the other daughter cell has only one copy (haploid). Such chromosome imbalance leads to further difficulties in subsequent mitoses.

Abnormalities like translocations, deletions, and aneuploidy can be seen in classical cytological chromosome preparations or by fluorescence in situ hybridization (FISH) using DNA probes. With the completion of the human genome sequence and the availability of microarray techniques, it has become possible to survey the entire genome for regions that have been deleted or amplified or regions that are far too small to be detected by light microscopy. These variations have become known as *copy number variants* or polymorphisms. Several methods have been developed to identify these variations, but all are based on preparing DNA from a tumor and normal tissue, labeling each with a different dye, and hybridizing both DNAs to the same microarray carrying either oligonucleotides or clones covering the whole human genome. Variations between the DNAs are revealed by differences in the ratios of the two labels. The strategy involved in one such approach, representational oligonucleotide microarray analysis (ROMA), is shown in Figure 15-15. It is possible to pinpoint regions as small as a few kilobases, where genes have been either amplified or deleted from tumor cells. As we shall see in the next section, these methods are revealing the panoply of genetic changes that occur in cancer cells and uncovering new oncogenes and tumor suppressors.

Mouse Models and Genomic Analysis Lead to the Discovery of Two New Oncogenes that Cooperate in Liver Cancer

Liver cancer (hepatocellular carcinoma or HCC) is the fifth most common cancer worldwide, and it is the third leading cause of cancer death because few treat-

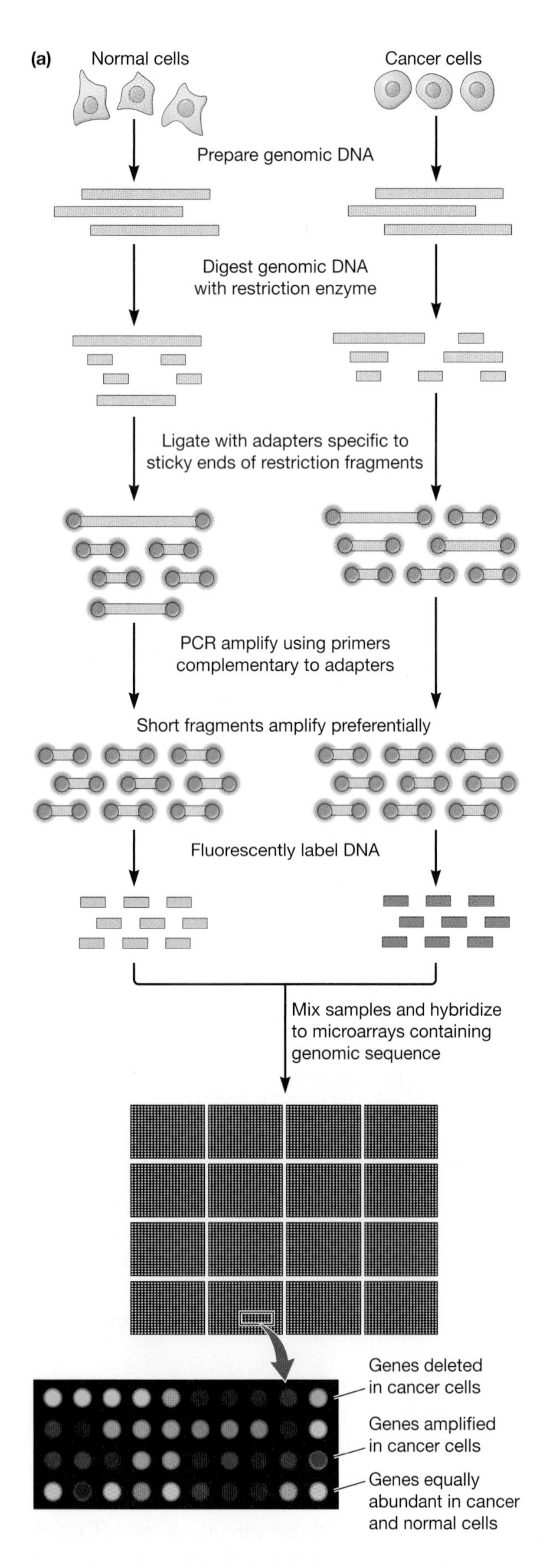

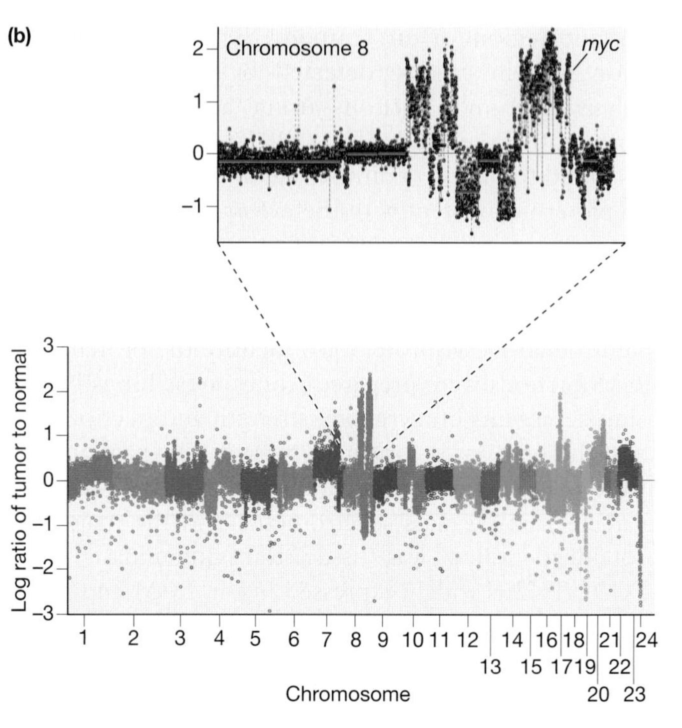

FIGURE 15-15

Representational oligonucleotide microarray analysis can detect copy number variation in tumor genomes. (*a*) Genomic DNA is prepared from both a tumor sample and a normal sample. Each sample of genomic DNA is digested with a restriction enzyme, and complementary oligonucleotide adapters are ligated to the ends of the DNA. These ligated adapters are then used as primers for PCR amplification. Shorter fragments are amplified preferentially over longer fragments, thus producing a *representation* of the genomic DNA. This representation contains the small fragments of the genome and excludes longer restriction fragments. Because only a portion of the genome is present, the complexity of the sample is greatly reduced and hybridization can proceed more efficiently. The PCR-amplified representation is fluorescently labeled—one sample with red fluorophores and the other with green. The labeled DNAs are mixed and hybridized to an oligonucleotide array that consists of 70-mer probes complementary to the genome. (*b*) Amplifications and deletions within the tumor genome can be detected by calculating the ratio of the two probes. The data shown compares a normal cell line to a breast tumor cell line. The amplification of the *myc* oncogene on human chromosome 8 is clearly visible, and a number of other changes are apparent.

ments other than surgery are available. It is known that HCC is initiated by liver damage, which can be caused by infection with hepatitis, by alcohol, or by toxins such as aflatoxin B1. These induce mutations in known oncogenes such as c-*met* and c-*myc*, and in the p53 tumor suppressor gene, but this is not the whole story. To understand this deadly cancer better, researchers undertook a carefully designed, comprehensive project that began with the development of a new mouse model for HCC, used microarrays to look for copy number variation, and used comparative genomics to discover two new oncogenes.

Hepatoblasts (liver stem cells) were isolated from the livers of embryonic mice by using antibodies to select cells expressing high levels of the transmembrane protein E-cadherin, a marker of the liver progenitor cells that have the potential to develop tumors. These cells were treated with retroviral vectors carrying the gene for fluorescent green protein to track the cells. Some vectors also carried shRNAs to knock down gene expression. Hepatoblasts were injected into mice that had been treated with a toxin to inhibit division of the host hepatocytes. Although the embryonic cells proliferated to contribute more than 1% of the cells in the recipient mouse livers, transplanted cells from wild-type embryos did not produce tumors. However, hepatoblasts that had been prepared from p53$^{-/-}$ knockout mice and had also been treated with retroviruses expressing c-*myc* or other activated oncogenes did form tumors. Importantly, the histology of these mouse tumors resembled that of human HCC.

Now that a reliable mouse model of human HCC was available, ROMA (Fig. 15-15) was used to search for DNA copy number changes that had occurred in the rare cells that gave rise to the tumors. The ROMA results revealed that a 1-Mb region at mouse chromosome 9qA1 was amplified in four of seven tumors initiated from p53$^{-/-}$ cells overexpressing c-*myc*. ROMA was also used to analyze 48 human HCCs, and two of these were shown to carry amplifications of a region at chromosome 11q22. That this region is syntenic with mouse chromosome 9qA1 strongly suggested that the region might carry an oncogene. Further comparative genomics of these syntenic regions identified 14 genes common to both, and the mRNA and protein expression levels of these genes were tested in mouse and human cancers. Only two of these genes were consistently overexpressed in all of the tumors—cellular inhibitor of apoptosis 1 (*cIAP1*) and Yes kinase–associated protein (*Yap*). Elevated levels of the proteins could be detected in tumor tissue sections.

To confirm that these proteins were important for carcinogenesis, *cIAP1* and *Yap* expression was altered in the p53$^{-/-}$*myc* hepatoblasts that had given rise to the tumors. Overexpression of cIAP or Yap proteins accelerated the appearance and increased the number and size of tumors, whereas silencing of either gene using shRNAs slowed tumor growth. Finally, simultaneous elevated expression of both genes resulted in even more potent promotion of tumor growth. Indeed, the effect was very much greater than simply the additive effects of each gene individually. It is not known whether the occurrence of two such oncogenes in the same amplification region has any special significance. A second example is known in breast cancer and more detailed studies of other amplified regions containing an oncogene may reveal an unsuspected companion. The discovery of these oncogenes is a tribute to the versatility of genetically manipulated mouse models and to the power of genomics-based analysis. Only a year and a half elapsed between the development of the tumor model to the full validation of these new oncogenes.

miRNAs Constitute a New Class of Oncogenes

Soon after microRNAs (miRNAs) were discovered (Chapter 9), biologists made miRNA microarrays to study miRNA expression in different tissues. It has become clear that some cancers have unusual miRNA expression patterns. In many B-cell lymphomas, five miRNAs called mir-17-92, expressed from a single poly-miRNA transcript on chromosome 13, are upregulated. To investigate their function, this cluster of miRNAs was expressed in a mouse model of lymphoma, where they accelerated both the development of lymphoma and the rate of death. The pattern of metastasis was altered so that the cancer was much more disseminated (Fig. 15-16). These experiments are the first example of oncogenic miRNAs, presumably acting by regulating other, as yet unknown, genes that alter cell growth or survival.

Telomerase Inhibition Can Kill Cancer Cells

As we learned earlier, there is an upper limit to the number of times a cell can divide before it stops growing and enters a phase called senescence. For

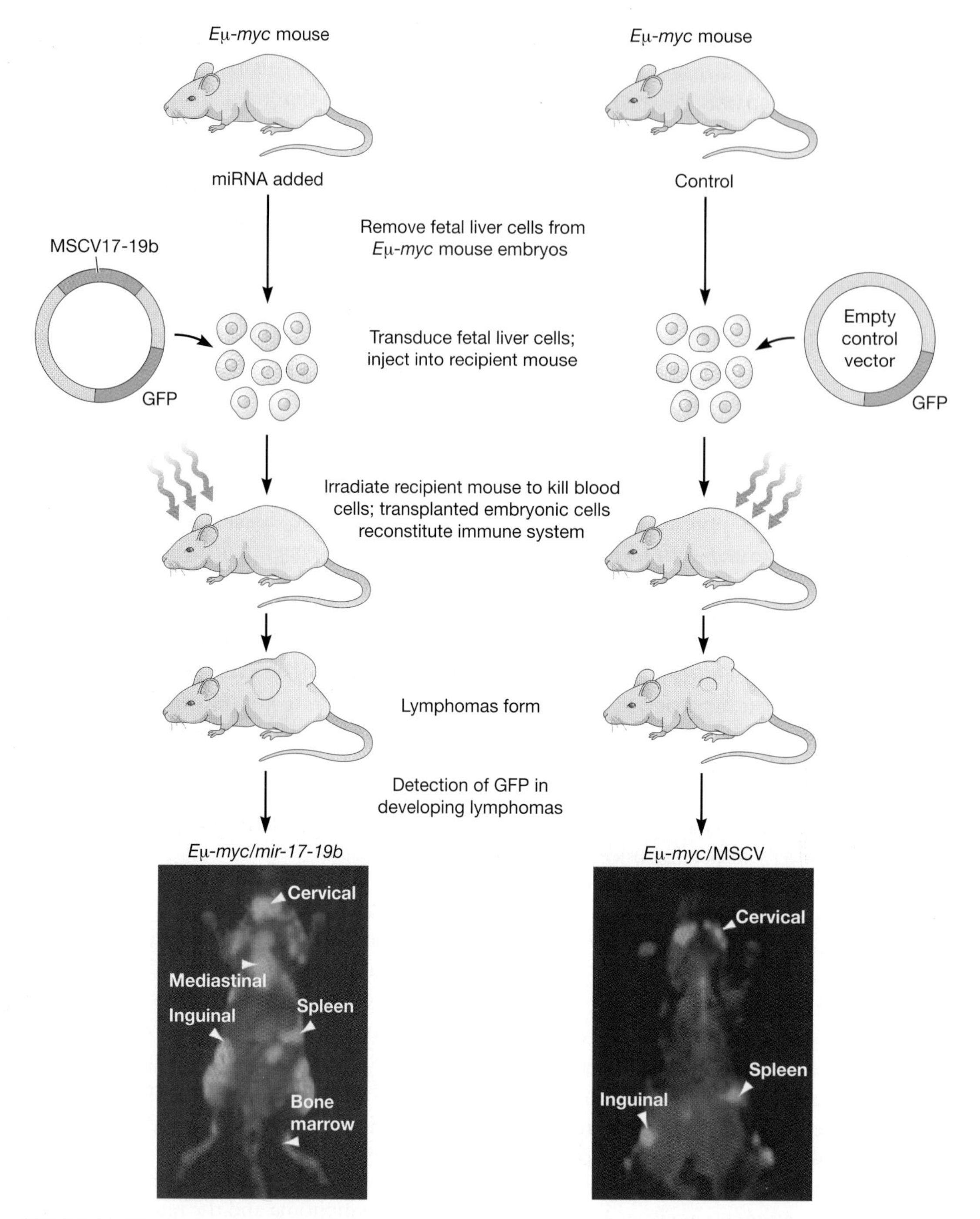

FIGURE 15-16

miRNA cluster 17-92 is an oncogene. The miRNA-17-92 polycistron is a cluster of miRNAs transcribed as a single RNA from a locus on human chromosome 13. The miRNAs are processed by Microprocessor and Dicer to release individual RNAs. An *E*μ-*myc* mouse model of B-cell lymphoma was used to test the effects on tumor growth of overexpression of the first two miRNAs in the cluster, miRNA-17-19b. As we saw in Fig. 15-2, the translocation of the *myc* oncogene to a position under control of the immunoglobulin promoter causes B-cell lymphomas. Liver cells are removed from *E*μ-*myc* mouse embryos. These *E*μ-*myc* transgenic cells, when transplanted into another mouse, grow to form B-cell lymphomas. (The recipient mouse is irradiated to kill its blood cells; this treatment stimulates the transplanted embryonic cells to proliferate and reconstitute the recipient's immune system.) The embryonic cells are transduced with a viral vector that contains GFP, which allows tracking of tumor growth, and miRNA-17-92. A control vector encoded GFP alone. Sensitive cameras are capable of detecting the GFP in developing lymphomas by imaging through the skin of the live mouse. The tumors expressing miRNA-17-92 were more invasive and killed the recipient mice more quickly than did the control Eμ-*myc* cells.

human embryonic diploid fibroblasts, the Hayflick limit is about 60 doublings. This limit is related to the end-replication problem faced by DNA polymerase as it moves along a template strand synthesizing a new strand (Chapter 7). DNA polymerase must initiate synthesis from an RNA primer and it synthesizes a new strand beginning with the primer's 3′ end. The need for the RNA primer means that the DNA polymerase cannot replicate the very 5′ end of the lagging strand because the primer occupies the last few bases of the template strand. As a consequence, the DNA of a chromosome shortens by 50–100 nucleotides each time the chromosome replicates. Eukaryotic chromosomes have telomeres at their ends, which are long sections made up of a repeat sequence, TTAGGG in human cells. These sequences protect the ends of the chromosomes, thus preventing the chromosomes from interacting with each other. Telomeres are eroded with each division until they no longer function properly, with catastrophic results for the cell.

What is it, then, that distinguishes a cell capable of indefinite growth from one that senesces? The answer is the ribonucleoprotein enzyme telomerase that catalyzes the addition of telomere repeats at the ends of eukaryotic chromosomes (Chapter 7). The RNA component of telomerase (hTer in human cells) anneals to the telomere repeat at the end of the chromosome and the protein component (hTERT), a reverse transcriptase, reverse transcribes the RNA component to add DNA repeats to the ends of chromosomes. Telomerase is silent in nearly all normal cells, except for germ cells, which must express telomerase so that eggs and sperm receive chromosomes ending in telomeres of the right length. If hTERT is overexpressed in cells, these will continue to proliferate past the point when they would normally reach the Hayflick limit and senesce. Strikingly, more than 90% of human tumors overexpress telomerase; no other characteristic is common among so many different tumors. The vast majority of tumors depend on telomerase for continued cell division, which makes this enzyme an attractive therapeutic target. Perhaps inhibiting telomerase activity would stop cell growth. Indeed, it does. Overexpressing dominant-negative forms of hTERT in tumor cell lines eventually leads the cells into crisis. (These mutant hTERTs probably work by "mopping-up" associated proteins so that the endogenous hTERT cannot work efficiently.)

In an alternative strategy, investigators first tried using a short interfering RNA (siRNA) to knock down the endogenous telomerase RNA component, hTer. As expected, hTer depletion led to telomere loss and cell senescence, in vitro and in vivo. Second, they replaced hTer with a mutant-template hTer (MT-hTer), which directed the addition of a nontelomeric sequence at the telomeres. This MT-hTer also inhibited cell growth, and it did it more rapidly than did the siRNA-induced depletion of hTer. Third, and unexpectedly, combining the siRNA with MT-hTer had a synergistic effect, which caused more rapid and greater inhibition of cell growth, accompanied by DNA damage foci. Interestingly, wild-type hTERT was needed, presumably so that MT-hTer/siRNA could interact with it and direct the addition of the nontelomeric sequences. The quick-acting toxic effect of MT-hTer/siRNA is quite distinct from any effect of telomere shortening, which would require many cell division cycles for serious shortening to occur. The mutant telomerase RNA affected even cells that had lost p53 function, and this, coupled with the near-universal telomerase activation in cancer cells, gives hope that this previously unsuspected pathway may provide a new method to target cancer cells.

Microarray Gene Expression Analysis Indicates Drug Response in Breast Cancer

Histopathologists currently recognize some 20 types of breast cancer based on the cells affected, the microscopic appearance of the cancers, and biochemical markers. We know, however, that there is considerable molecular heterogeneity in all cancers and that it is likely that these variations from cancer to cancer and person to person have a strong influence on the response of a patient to drug therapy. This variation should be reflected in gene expression—that is, which mRNAs are being made and in what quantities. Microarrays can be used to examine the expression of thousands of genes (Chapter 13), which is exactly what is needed to examine why different cancers respond differently to different treatments,

mRNA from the tumors from 78 women with breast cancer was labeled with the fluorescent dye, Cy5, and control, nontumor mRNA was labeled with Cy3. The labeled mRNA from each woman was hybridized to spotted microarrays containing more than 8000 human genes, and the data were analyzed to identify which genes were expressed at higher or lower levels in the tumors compared to the nontumor tissue (Fig. 15-17). Those genes showing the largest differences in expression levels are shown in horizontal rows in the display, and each column corresponds to the results from one woman's tumor comparison. To help interpretation, the genes are ordered according to similarity of expression levels, so that genes that behave similarly tend to form clusters.

Mathematical analysis, and even simple inspection of the data, identified five distinct clusters of women; the women in each cluster had similar gene expression patterns in their tumors (Fig. 15-17), although their cancers were not easily distinguishable on the basis of histology or other clinical parameters. The molecular variation between the tumors was immediately apparent. For example, genes known to be involved in regulating cellular proliferation were higher in one tumor subtype, whereas other genes involved in cell signaling pathways were expressed at higher levels in another subtype. But

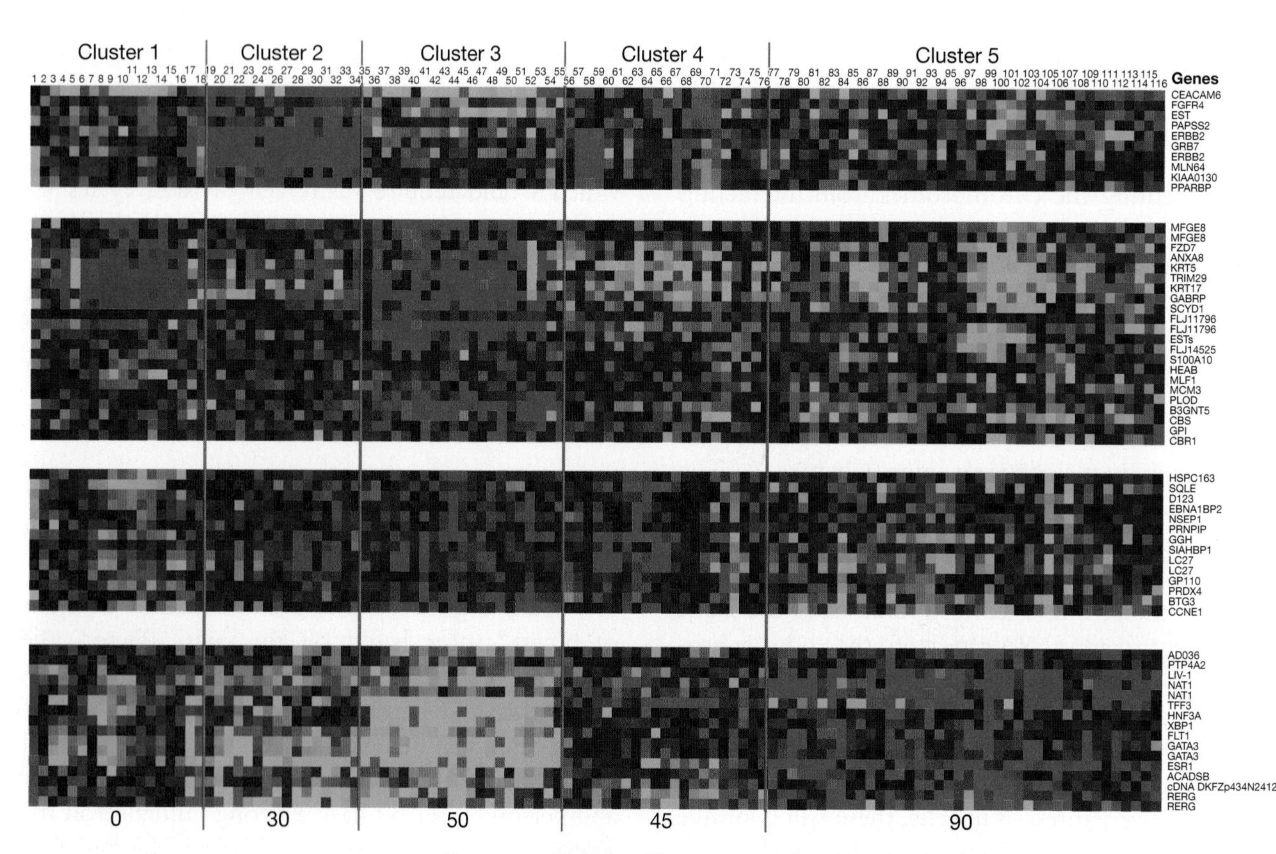

FIGURE 15-17

Heat maps are a useful tool for visualizing rankings of gene expression data from microarray experiments. In this example, mRNA was isolated from 78 women with breast cancer. For each woman in the study, the mRNA sample derived from the tumor was labeled with Cy5, and the control, nontumor mRNA was labeled with Cy3. The labeled mRNAs from each woman were hybridized to spotted microarrays containing more than 8000 human genes, and the data were analyzed to identify which genes were expressed at higher or lower levels in tumors compared to nontumor tissue. Each column contains the results of the 61 genes of the entire set that showed the most interesting patterns among the women. The sets of genes are arranged by using a method called hierarchical clustering that groups genes that have similar expression profiles across the different samples. The patterns of colored blocks generated by this clustering method are called heat maps. *Red* and *green* shading indicate higher or lower expression, respectively, in the tumor sample relative to the nontumor sample. As seen in the heat map, those genes with the most similar expression patterns are grouped together, revealing blocks (or clusters) of genes that behave similarly across many different tumors (clusters 1–5). Also, tumors showing the most similar expression patterns across the gene set are grouped together. In this case, such clustering reveals five distinct tumor subtypes, and the tumor subtype is a striking predictor of survival rate.

even more useful information was obtained from this experiment. The various tumor subtypes, defined solely on the basis of gene expression patterns, showed a strong correlation with clinical outcome in these women, all of whom had been treated with a drug commonly used in breast cancer. Those in cluster 5 in Figure 15-17 had the most favorable outcome—90% were still alive after 48 months following treatment—and those in cluster 1 had the poorest outcome—most not surviving even 1 month. Microarray experiments are too cumbersome to perform in the clinic, so it is not likely that they will be used to diagnose patients. However, once a small subset of the genes most relevant for predicting disease or treatment outcome is discovered, it may be possible to develop simpler tests that could be applied in clinical settings.

The Human Cancer Genome Project Will Identify All Mutations in Different Cancers

Cancers are complex genetic diseases that often take decades to develop and that require multiple independent mutations to turn a normal cell into a metastatic tumor. Some mutations remove the checks on cell growth; others alter the response of the cell to extracellular signals by altering growth factor receptors or by causing the cell to make its own growth factors. Still others subvert the fail-safe systems that would otherwise kill a cancer cell, and, finally, there are mutations that enable the cancer cell to metastasize. Cells may develop mutations in the DNA repair machinery—a *mutator* phenotype—which lead to a greatly increased accumulation of mutations. We now know many, but by no means all, of the genes involved in a particular cancer, and there are hundreds of different types of cancer, many of which have been little studied. The argument has been made that, just as the Human Genome Project (HGP) has provided us with a catalog of all the human genes and forms the basis for studies of human inherited disorders, so a catalog of all the changes found in cancers would be the foundation for a new, genomics-based attack on cancer. This is the goal of the proposed Human Cancer Genome Project (HCGP).

The largest genome studies to date have been carried out on protein kinases, which are frequently mutated in cancer. For example, 518 protein kinase genes were sequenced (~1.3 Mb of DNA per tumor) in samples of 16 breast, 26 lung, and 13 testicular primary tumors. The sequencing reveals that 37 of the tumors had no mutations in any kinase gene, 15 had fewer than ten mutations, and three cancers had more than ten, one of these having 52 mutations. The latter had probably acquired a mutator phenotype. The data from the lung cancers and the testicular cancers were strikingly different; one-half of lung cancers had at least one mutation, whereas only one of the 13 testicular cancers had a mutation. These analyses show that the commonly mutated oncogenes may be the exception rather than the rule.

What these results will mean for the HCGP is not clear. The proposal recognizes that many tumors will have to be sampled to detect genomic changes that are relevant to the cancer. It is estimated that about 250 samples of each tumor type will be sufficient, many more than in the studies just described. How many different cancer types will be studied? The HCGP proposes to analyze about 50, for a total of 12,500 samples. Each of these will be characterized for amplifications and deletions, chromosome rearrangements, mutations in the coding regions of all genes, regions of abnormal methylation, and expression profiles. The project will be spread over 10 years with a total cost of $1.5 billion. Just like the Human Genome Project (Chapter 11), the HCGP is controversial and for similar reasons: scientific merit and cost. A 3-year pilot project is proposed in which five cancer types will be targeted; it will provide data on a scale more suitable for assessing the value of the HCGP.

The increase in our understanding of cancer has been breathtaking in the 25 years since the first human cancer gene was cloned, and we are beginning to see applications of this knowledge reaching patients. The benefits of detecting cancer at the earliest stage can mean the difference between recovery and death. There are now four or five different molecular probes that are used to diagnose early-stage colorectal cancer, where death can be prevented if the disease is caught any time prior to the last step in progression of the cancer. Molecular diagnostics, determining the unique genetic profile of a patient's tumor, will lead to prediction of the course of the disease and suggest the appropriate treatment. But, the successes of Herceptin and Gleevec notwithstanding, we cannot be sanguine about the progress made thus far in developing new treatments. Our hope and expectation is that as our understanding of the underlying biology of cancer continues to grow, and as we learn more of the crucial differences between normal and cancer cells, we shall find new targets and new therapeutic strategies unlike any available at the present time.

Reading List

General

Hanahan D. and Weinberg R.A. 2000. The hallmarks of cancer. *Cell* **100:** 57–70.

Stillman B. and Stewart D., eds. 2006. *Molecular Approaches to Controlling Cancer.* Cold Spring Harbor Symposia on Quantitative Biology, Vol. 70. Cold Spring Harbor Laboratory Press, Cold Spring Harbor, New York.

Weinberg R.A. 2007. *The Biology of Cancer.* Garland Press, New York.

Environmental Factors Are the Primary Initiators of Cancer

Colditz G.A., Sellers T.A., and Trappido E. 2006. Epidemiology—Identifying the causes and preventabiltiy of cancer? *Nat. Rev. Cancer* **6:** 75–83.

Doll R. and Bradford Hill A. 1950. Smoking and carcinoma of the lung: Preliminary report. *Br. Med. J.* **2:** 739–748.

Wynder E.L. and Graham E.A. 1950. Tobacco smoking as a possible etiologic factor in bronchogenic carcinoma. *J. Am. Med. Assoc.* **143:** 329–336.

American Cancer Society. 2006 *Cancer Facts and Figures 2006.* American Cancer Society, Atlanta. http://www.cancer.org/downloads/STT/CAFF2006PWSecured.pdf

Rous P. 1911. A sarcoma of the fowl transmissible by an agent separable from the tumor cells. *J. Exp. Med.* **13:** 397–411.

Lowy D.R. and Schiller J.T. 2006. Prophylactic human papillomavirus vaccines. *J. Clin. Invest.* **116:** 1167–1173.

Tumor Cells Grow Abnormally in Cell Culture

Hayflick L. 1965. The limited in vitro lifetime of human diploid cell strains. *Exp. Cell Res.* **37:** 614–636.

Todaro G.J. and Green H. 1963. Quantitative studies of the growth of mouse embryo cells in culture and their development into established lines. *J. Cell Biol.* **17:** 299–313.

Jones H.W., Jr., McKusick V.A., Harper P.S., and Wuu K.D. 1971. George Otto Gey (1899–1970). The HeLa cell and a reappraisal of its origin. *Obstet. Gynecol.* **38:** 945–949.

Drayton S. and Peters G. 2002. Immortalisation and transformation revisited. *Curr. Opinion. Genet. Dev.* **12:** 98–104.

The First Cancer Genes Were Found Using Tumor Viruses

Varmus H. 1989. An historical overview of oncogenes. In *Oncogenes and the Molecular Origins of Cancer* (R.A. Weinberg, ed.). Cold Spring Harbor Laboratory Press, Cold Spring Harbor, New York.

Vogt M. and Dulbecco R. 1960. Virus–cell interaction with a tumor-producing virus. *Proc. Natl. Acad. Sci.* **46:** 365–370.

Vogt P.K. 1971. Spontaneous segregation of nontransforming viruses from cloned sarcoma viruses. *Virology* **46:** 939–946.

Martin G.S. 1970. Rous sarcoma virus: A function required for the maintenance of the transformed state. *Nature* **227:** 1021–1023.

Smith, A.E., Smith R., and Paucha E. 1979. Characterization of different tumor antigens present in cells transformed by simian virus 40. *Cell* **18:** 335–346.

Stehelin D., Varmus H.E., Bishop J.M, and Vogt P.K. 1976. DNA related to the transforming gene(s) of avian sarcoma viruses is present in normal avian DNA. *Nature* **260:** 170–173.

Spector D.H., Varmus H.E., and Bishop J.M. 1978. Nucleotide sequences related to the transforming gene of avian sarcoma virus are present in the DNA of uninfected vertebrates. *Proc. Natl. Acad. Sci.* **75:** 4102–4106.

Hayward W.S., Neel B.G., and Astrin S.M. 1981. Activation of a cellular *onc* gene by promoter insertion in ALV-induced lymphoid leukosis. *Nature* **290:** 475–480.

Dalla-Favera R.M., Bregni J., Erickson D., Patterson R., Gallo C., and Croce C.M. 1982. Human c-*myc onc* gene is located on the region of chromosome 8 that is translocated in Burkitt lymphoma cells. *Proc. Natl. Acad. Sci.* **79:** 7824–7827.

de Klein A., van Kessel A.G., Grosveld G., Bartram C.R., Hagemeijer A., et al. 1982. A cellular oncogene is translocated to the Philadelphia chromosome in chronic myelocytic leukaemia. *Nature* **300:** 765–767.

Recombinant DNA Techniques Isolate the First Human Cancer Gene

Rigby P.W.J. 1982. The oncogenic circle closes. *Nature* **297:** 451–453.

Shih C., Shilo B.Z., Goldfard M.P., Dannenberg A., and Weinberg R.A. 1979. Passage of phenotypes of chemically transformed cells via tansfection of DNA and chromatin. *Proc. Natl. Acad. Sci.* **76:** 5714–5718.

Goldfarb M., Shimizu K., Perucho M., and Wigler M. 1982. Isolation and preliminary characterization of a human transforming gene from T24 bladder carcinoma cells. *Nature* **296:** 404–409.

Shih C. and Weinberg R.A. 1982. Isolation of a transforming sequence from a human bladder carcinoma cell line. *Cell* **29:** 161–169.

Parada L.F., Tabin C.J., Shih C., and Weinberg, R.A. 1982. Human EJ bladder carcinoma oncogene is homologue of Harvey sarcoma virus *ras* gene. *Nature* **297:** 474–478.

Tabin C.J., Bradley S.M., Bargmann C.I., Weinberg R.A., Papageorge A.G., et al. 1982. Mechanism of activation of a human oncogene. *Nature* **300:** 143–149.

Taparowsky I., Suard Y., Fasano O., Shimizu K., Goldfarb M., and Wigler M. 1982. Activation of the T24 bladder carcinoma transforming gene is linked to a single amino acid change. *Nature* **300:** 762–765.

Cellular Proto-Oncogenes Affect Signal Transduction Pathways and Transcription

Collet M.S. and R.L. Erikson. 1978 Protein kinase activity associated with the avian sarcoma virus *src* gene product. *Proc. Natl. Acad. Sci.* **75:** 2021–2024.

Hunter T. and B.M. Sefton. 1980. Transforming gene product of Rous sarcoma virus phosphorylates tyrosine. *Proc. Natl. Acad. Sci.* **77:** 1311–1315.

Doolittle R.F., Hunkapiller M.W., Hood L.E., Deuare S.G., Robbins K.C., Aaronson S.A., and Antomades H.N. 1983. Simian sarcoma virus *onc* gene v-*sis* is derived from the gene (or genes) encoding a platelet derived growth factor. *Science* **221:** 275–276.

Downward J., Yarden Y., Mayes E., Scrace G., Totty N., et al. 1984. Close similarity of epidermal growth factor receptor and v-*erbB* oncogene protein sequences. *Nature* **307:** 521–527.

Bohmann D., Bos T.J., Admon A., Nishimura T., Vogt P.K., and Tjian R. 1987. Human proto-oncogene c-*jun* encodes a DNA-binding protein with structural and functional properties of transcription factor AP-I. *Science* **238:** 1386–1392.

Franza B.R., Jr., Rauscher F.J., III, Josephs S.F., and Curran T. 1988. The Fos complex and Fos-related antigens recognize sequence elements that contain AP-I binding sites. *Science* **239:** 1150–1153.

Herceptin Targets a Growth Factor Receptor Expressed on Many Breast Cancer Cells

Baselga J. 2006. Targeting tyrosine kinases in cancer: The second wave. *Science.* **312:** 1175–1178.

Schechter A.L., Hung M.C., Vaidyanathan L., Weinberg R.A., Yang-Feng T.L., et al. 1985. The *neu* gene: An *erbB*-homologous gene distinct from and unlinked to the gene encoding the EGF receptor. *Science.* **229:** 976–978.

King C.R., Kraus M.H., and Aaronson S.A. 1985. Amplification of a novel v-*erbB*-related gene in a human mammary carcinoma. *Science* **229:** 974–976.

Carter P., Presta L., Gorman C.M., Ridgway J.B., Henner D., et al. 1992. Humanization of an anti-p185HER2 antibody for human cancer therapy. *Proc. Natl. Acad. Sci.* **89:** 4285–4289.

Gleevec Inhibits a Kinase Mutated in Leukemia Cells

Druker B.J. and Lydon N.B. 2000. Lessons learned from the development of an Abl tyrosine kinase inhibitor for chronic myelogenous leukemia. *J. Clin. Invest.* **105:** 3–7.

Zhang X. and Ren R. 1998. BCR-ABL efficiently induces a myeloproliferative disease and production of excess interleukin-3 and granulocyte-macrophage colony-stimulating factor in mice: A novel model for chronic myelogenous leukemia. *Blood* **92:** 3829–3840.

Manley P.W., Cowan-Jacob S.W., Buchdunger E., Fabbro D., Fendrich G., et al. 2002. Imatinib: A selective tyrosine kinase inhibitor. *Eur. J. Cancer.* (suppl.) **5:** S19–S27.

Druker B.J., Tamura S., Buchdunger E., Ohno S., Segal G.M., et al. 1996. Effects of a selective inhibitor of the Abl tyrosine kinase on the growth of BCR-ABL positive cells. *Nat. Med.* **2:** 561–566.

Evidence for Tumor Suppressor Genes Comes from Cells in Culture and the Familial Cancer Retinoblastoma

Harris H. 1968. *Nucleus and Cytoplasm.* Clarendon Press, Oxford.

Harris H., Miller O.J., Klein G., Worst P., and Tachibana T. 1969. Suppression of malignancy by cell fusion. *Nature* **223:** 363–368.

Stanbridge E.J. 1976. Suppression of malignancy in human cells. *Nature* **260:** 17–20.

Weissman B.E., Saxon P.J., Pasquale S.R., Jones G.R., Geiser A.G., and Stanbridge E.J. 1987. Introduction of a normal human chromosome 11 into a Wilms' tumor cell line controls tumorigenic expression. *Science* **236:** 175–180.

Knudson A.G. Mutation and cancer: Statistical study of retinoblastoma. 1971. *Proc. Natl. Acad. Sci.* **68:** 820–823.

Friend S.H., Bernards R., Rogeli S., Weinberg R.A., Rapaport J.M., Albert D.M., and Dryja T.P. 1986. A human DNA segment with properties of the gene that predisposes to retinoblastoma and osteosarcoma. *Nature* **323:** 643–646.

Huang H.J., Yee J.K., Shew J.Y., Chen P.L., Bookstein R., et al. 1988. Suppression of the neoplastic phenotype by replacement of the *RB* gene in human cancer cells. *Science* **242:** 1563–1566.

Whyte P.K., Buchkovich J.M., Horowitz J.M., Friend S.H., Raybuck M., Weinberg R.A., and Harlow E. 1988. Association between an oncogene and an anti-oncogene: The adenovirus E1A proteins bind to the retinoblastoma gene product. *Nature* **334:** 124–129.

Whyte P., Williamson N.M., and E. Harlow. 1989. Cellular targets for transformation by the adenovirus E1A proteins. *Cell* **56:** 67–75.

p53 Plays a Central Role in the Cell's Response to Cancer-inducing Damage

Vogelstein B., Lane D., and Levine A.J. 2000. Surfing the p53 network. *Nature* **408:** 307–310.

Lane D.P. and L.V. Crawford. 1979. T-antigen is bound to host protein in SV40-transformed cells. *Nature* **278:** 261–263.

Mowat M., Cheng A., Kimura N., Bernstein A., and Benchimol S. 1985. Rearrangements of the cellular p53 gene in erythroleukemic cells transformed by Friend virus. *Nature* **314:** 633–636.

Finlay C.A., Hinds P.W., and Levine A.J. 1989. The p53 proto-oncogene can act as a suppressor of transformation. *Cell* **57:** 1083–1093.

Nigro J.M., Baker S.J., Presinger A.C., Jessup J.M., Hostetter R., et al. 1989. Mutations in the p53 gene occur in diverse human tumour types. *Nature* **342:** 705–708.

Symonds H., Krall L., Remington L., Saenz-Robles M., Lowe S., Jacks T., and Van Dyke T. 1994. p53-dependent apoptosis suppresses tumor growth and progression in vivo. *Cell* **78:** 703–711.

The Several Steps to Colorectal Cancer: A Real-Life Tale of Oncogenes and Tumor Suppressors

Fearon E.R. and Vogelstein B. 1990. A genetic model for colorectal tumorigenesis. *Cell* **61:** 759–767.

Baker S.J., Fearon E.R., Nigro J.M., Hamilton S.R., Preisinger A.C., et al. 1989. Chromosome 17 deletion and p53 gene mutation in colorectal carcinomas. *Science* **244:** 217–221.

Fearon E.R., Cho K.R., Nigro J.M., Kern S.E., Simons J.W.,

et al. 1990. Identification of a chromosome 18q gene that is altered in colorectal cancers. *Science* **247:** 49–56.

Woodford-Richens K.L., Rowan A.J, Gorman P., Halford S., Bicknell D.C., et al. 2001. SMAD4 mutations in colorectal cancer probably occur before chromosomal instability, but after divergence of the microsatellite instability pathway. *Proc. Natl. Acad. Sci.* **98:** 9719–9723.

Loss of p53 Contributes to the Warburg Effect

Matoba S., Kang J.-G, Patino W.D., Wragg A., Boehm M., et al. 2006. p53 regulates mitochondrial respiration. *Science* **312:** 1650–1653.

Failure of Apoptosis Contributes to Tumor Survival and Growth

Symonds H., Krall L., Remington L., Saenz-Robles M., Lowe S., Jacks T., and Van Dyke T. 1994. p53-dependent apoptosis suppresses tumor growth and progression in vivo. *Cell* **78:** 703–711.

Vaux D.L., Cory S., and Adams J.M. 1988. *Bcl-2* gene promotes haemopoietic cell survival and cooperates with c-*myc* to immortalize pre-B cells. *Nature* **335:** 440–442.

Oltvai Z.N., Milliman C.L., and Korsmeyer S.J. 1993 Bcl-2 heterodimerizes in vivo with a conserved homolog, Bax, that accelerates programmed cell death. *Cell* **74:** 609–619.

Li P., Nijhawan D., Budihardjo I., Srinivasula S.M., Ahmad M., Alnemri E.S., and Wang X. 1997. Cytochrome c and dATP-dependent formation of Apaf-1/caspase-9 complex initiates an apoptotic protease cascade. *Cell* **91:** 479–489.

Kluck R.M., Bossy-Wetzel E., Green D.R., and Newmeyer D.D. 1997. The release of cytochrome c from mitochondria: A primary site for Bcl-2 regulation of apoptosis. *Science* **275:** 1132–1136.

Yang J., Liu X.S., Bhalla K., Kim K.N., Ibrado A.M., et al. 1997. Prevention of apoptosis by Bcl-2: Release of cytochrome *c* from mitochondria blocked. *Science* **275:** 1129–1132.

Miyashita T. and Reed J.C. 1995. Tumor suppressor p53 is a direct transcriptional activator of the human bax gene. *Cell* **80:** 293–299.

Liu X.S., Kim C.N., Yang J., Jemmerson R., and Wang X. 1996. Induction of apoptotic program in cell-free extracts: Requirement for dATP and cytochrome c. *Cell* **86:** 147–157.

Butt A.J., Firth S.M., King M.A., and Baxter M.C. 2000. Insulin-like growth factor-binding protein-3 modulates expression of Bax and Bcl-2 and potentiates p53-independent radiation-induced apoptosis in human breast cancer cells. *J. Biol. Chem.* **275:** 39174–39181.

Tumors Need New Blood Vessels to Grow

Folkman J. 1971. Tumor angiogenesis: Therapeutic implications. *N. Engl. J. Med.* **285:** 1182–1186.

Folkman J. 2006. Angiogenesis. *Ann. Rev. Med.* **57:** 1–18.

Gimbrone M.A., Cotran R.S., Leapman S.B., and Folkman J. 1974. Tumor growth and neovascularization: An experimental model using the rabbit cornea. *J. Natl. Cancer Inst.* **52:** 413–427.

Senger D.R., Galli S.J., Dvorak A.M., Perruzzi C.A., Harvey V.S., and Dvorak H.F. 1983. Tumor cells secrete a vascular permeability factor that promotes accumulation of ascites fluid. *Science* **219:** 983–985.

O'Reilly M.S., Holmgren L., Shing Y., Chen C., Rosenthal R.A., et al. 1994. Angiostatin: A novel angiogenesis inhibitor that mediates the suppression of metastases by a Lewis lung carcinoma. *Cell* **79:** 315–328.

O'Reilly M.S., Boehm T., Shing Y., Fukai N., and Vasios G. 1997. Endostatin: An endogenous inhibitor of angiogenesis and tumor growth. *Cell* **88:** 277–285.

Hurwitz H., Fehrenbacher L., Novotny W., Cartwright T., Hainsworth J., et al. 2004. Bevacizumab plus irinotecan, fluorouracil, and leucovorin for metastatic colorectal cancer. *N. Engl. J. Med.* **350:** 2335–2342.

Sund M., Hamano Y., Sugimoto H., Sudhakar A., Soubasakos M., et al. 2005. Function of endogenous inhibitors of angiogenesis as endothelium-specific tumor suppressors. *Proc. Natl. Acad. Sci.* **102:** 2934–2939.

A Novel G Protein–coupled Receptor Interacts with the Extracellular Matrix to Regulate Metastasis

Mehlen P. and Puisieux A. 2006. Metastasis: A question of life or death. *Nat. Rev. Cancer* **6:** 449–458.

Xu L., Begum S., Hearn J.D., and Hynes R.O. 2006. GPR56, an atypical G protein–coupled receptor, binds tissue transglutaminase, TG2, and inhibits melanoma tumor growth and metastasis. *Proc. Natl. Acad. Sci.* **103:** 9023–9028.

Genome-wide Scans Detect Differences in Gene Copy Numbers in Cancer

Lucito R., Healy J., Alexander J., Reiner A., Esposito D., et al. 2003. Representational oligonucleotide microarray analysis: A high-resolution method to detect genome copy number variation. *Genome Res.* **13:** 2291–2305.

Sebat J., Lakshmi B., Troge J., Alexander J., Young J., et al. Large-scale copy number polymorphism in the human genome. *Science* **305:** 525–528.

Mouse Models and Genomic Analysis Lead to the Discovery of Two New Oncogenes that Cooperate in Liver Cancer

Zender L., Spector MS., Xue W., Flemming P., Cordon-Cardo C., et al. 2006. Identification and validation of oncogenes in liver cancer using an integrative oncogenomic approach. *Cell* **125:** 1253–1267.

miRNAs Constitute a New Class of Oncogenes

He L., Thomson J.M., Hemann M.T., Hernando-Monge E., Mu D., et al. 2005. A microRNA polycistron as a potential human oncogene. *Nature* **435:** 828–833.

Telomerase Is a Target for Therapy

Goldkorn A. and Blackburn E.H. 2006. Assembly of mutant-template telomerase RNA into catalytically active telomerase ribonucleoprotein that can act on telomeres is required for apoptosis and cell cycle arrest in human cancer cells. *Cancer Res.* **66:** 5763–5771.

Microarray Gene Expression Analysis Distinguishes Low- and High-Risk Breast Cancers

Sørlie T., Perou C.M., Tibshirani R., Aas T., Geisler S., et al. 2001. Gene expression patterns of breast carcinomas distinguish tumor subclasses with clinical implications. *Proc. Natl. Acad. Sci.* **98:** 10869–10874.

Bild A.H., Yao G., Chang J.T., Wang Q., Potti A., et al. 2006. Oncogenic pathway signatures in human cancers as a guide to targeted therapies. *Nature* **439:** 353–357.

Villuendas R., Steegmann J.L., Pollán M., Tracey L., Granda A., et al. 2006. Identification of genes involved in imatinib resistance in CML: A gene-expression profiling approach. *Leukemia* **20:** 1047–1054.

The Human Cancer Genome Project Will Identify All Mutations in Different Cancers

Hartwell L.H. and Lander E.S. (co-chairs). 2005. Report to the National Cancer Advisory Board. NCAB Working Group On Biomedical Technology. http://cancergenome.nih.gov/about/NCABReport_Feb05.pdf

Futreal P.A., Wooster R., and Stratton M.R. 2005. Somatic mutations in human cancer: Insights from resequencing the protein kinase gene family. *Cold Spring Harbor Symp. Quant. Biol.* **70:** 43–49.

Bignell G., Smith R., Hunter C., Stephens P., Davies H., et al. 2006. Sequence analysis of the protein kinase gene family in human testicular germ-cell tumors of adolescents and adults. *Genes Chromosomes Cancer* **45:** 42–46.

Davies H., Hunter C., Smith R., Stephens P., Greenman C., et al. 2005. Somatic mutations of the protein kinase gene family in human lung cancer. *Cancer Res.* **65:** 7591–7595.

Stephens P., Edkins S., Davies H., Greenman C., Cox C., et al. 2005. A screen of the complete protein kinase gene family identifies diverse patterns of somatic mutations in human breast cancer *Nat. Gen.* **37:** 590–592.

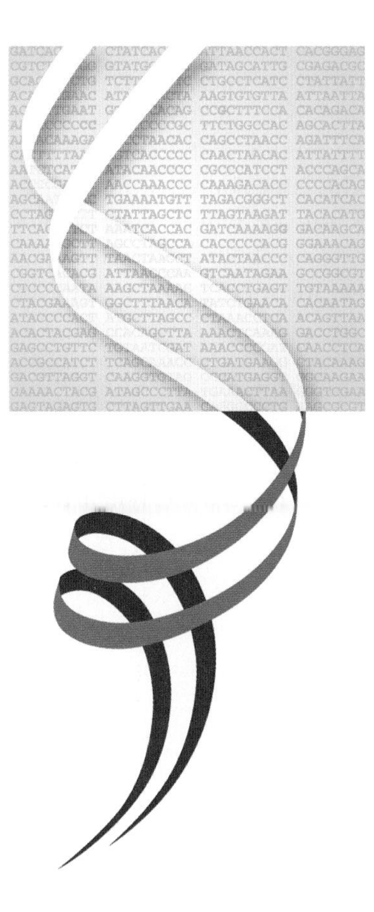

CHAPTER 16

DNA Fingerprinting and Forensics

Michael Anthony Williams is 41 years old. He has never learned to drive; he has never owned a cell phone, or a computer, or an iPod. Williams missed the experiences of most teenagers and young men because he was incarcerated in the Louisiana State Penitentiary, Angola, when he was 16 years old, locked away for 24 years for a crime he did not commit. That he did not serve life without parole is due to one of the most remarkable developments to come from our knowledge of DNA, human genetics, and the modern techniques of genomic analysis. Williams was freed because a comparison of his DNA with DNA isolated from crime scene evidence showed conclusively that he could not have been the perpetrator of the crime.

DNA profiling, popularly known as DNA fingerprinting, was developed from basic research on repetitive elements (Chapter 5) and has transformed personal identification, whether in forensic cases, missing persons, mass disasters, or paternity suits. The speed with which this has happened is extraordinary. DNA profiling was first used in 1985 and only 20 years later has become ubiquitous in law enforcement and is featured in television series and movies. During criminal investigations, it is used to exclude individuals suspected of crimes; in trials, it may convince a jury of an individual's guilt; and, in cases like that of Michael Anthony Williams, it can correct great injustices. DNA profiling is used to establish paternity and much more distant relationships by those tracing their ancestry. And by identifying victims, it can bring peace to those who lose relatives and friends in mass tragedies.

In this chapter, we will review the molecular basis and the techniques of DNA profiling and go on to look at some of the ways DNA profiling has been used. We will also discuss some of the potential problems that may arise as a consequence of the very rapid applications of DNA technology in human society.

Hypervariable or Variable Tandem Repeat Loci Can Be Used to Identify Individuals

Until about 15–20 years ago, techniques for identifying individuals made use of protein polymorphisms—for example, the familiar ABO and other blood groups. Such tests may show definitively that two samples are different—a smear of O positive blood cannot have come from an AB suspect—but these tests can rarely establish with certainty that two samples are from the same source or individual. DNA profiling makes use of highly polymorphic loci, chosen so that the probability is very low that two DNA samples with identical profiles could by chance have come from different individuals. The most useful polymorphisms for forensic purposes are found at so-called hypervariable loci.

The most informative hypervariable loci are those made up of a variable number of identical sequences joined together in tandem (Fig. 16-1). When DNA

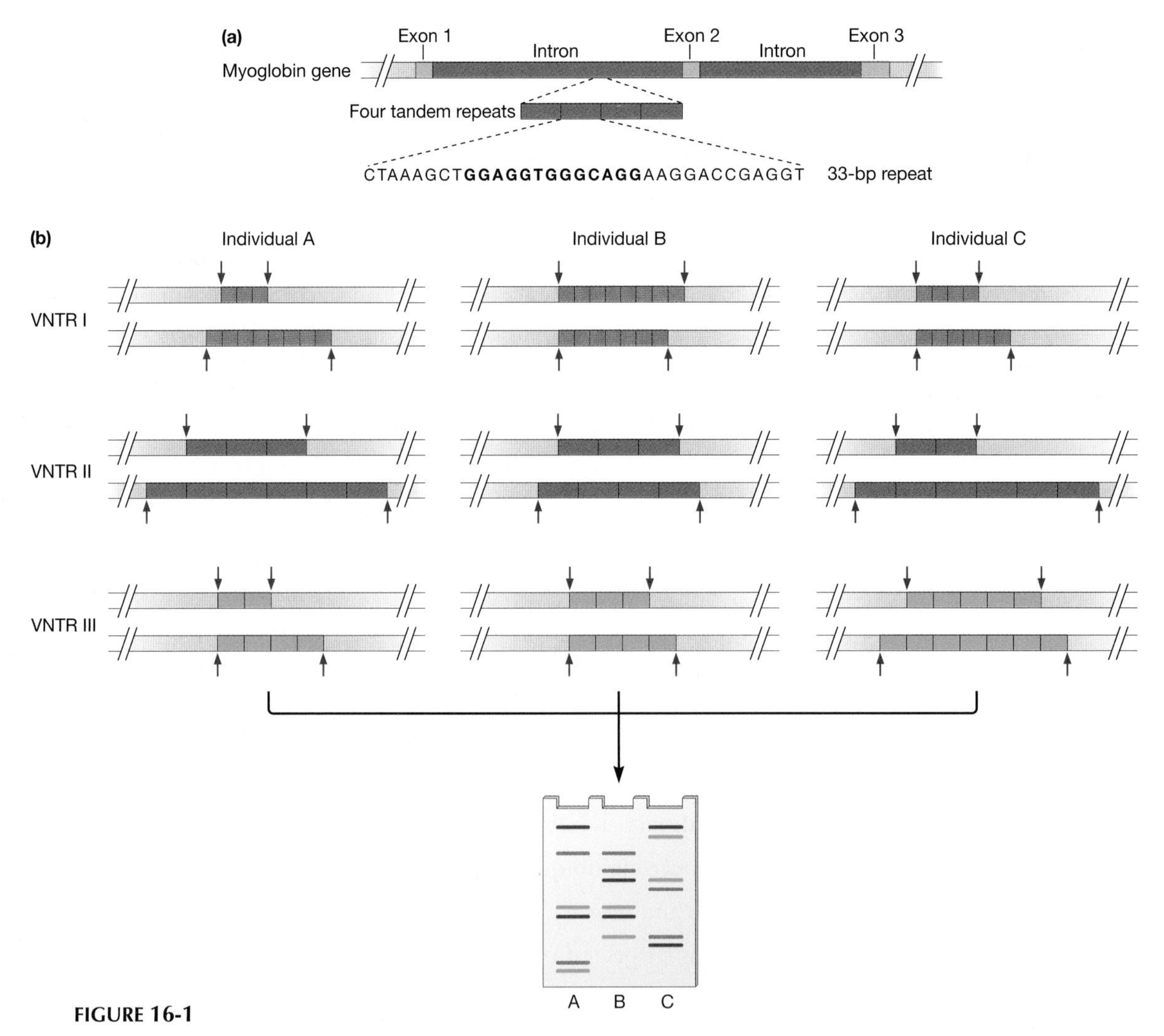

FIGURE 16-1

Variable number tandem repeat (VNTR) loci. (*a*) The VNTR locus in an intron of the myoglobin gene is composed of four repeats of a 33-bp sequence. The core sequence is shown in *bold type*. (*b*) The variability at a number of VNTR loci can generate diversity sufficient to identify individuals. Between two and eight tandem repeats of three different VNTR loci (I–III) have been detected by a hypervariable probe. Although different individuals may have some fragments in common, the chance that two individuals have all fragments in common is low. Restriction endonuclease sites are indicated by *red arrows*.

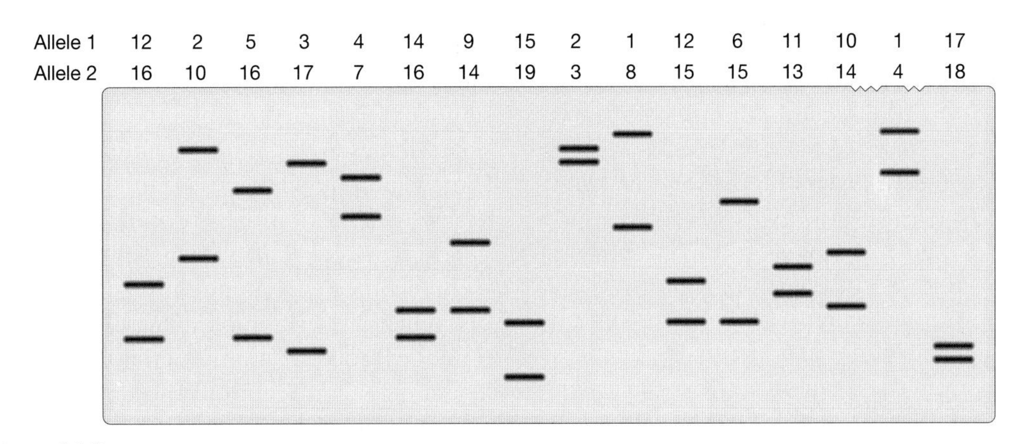

FIGURE 16-2

Fingerprinting, using variable number tandem repeat (VNTR) probes. The probe pYNH24 detects 19 alleles in 16 unrelated individuals. (The alleles are labeled from the largest, 1, to the smallest, 18.) No two individuals have the same pattern of fragments.

is digested with a restriction endonuclease that cuts the sequences flanking a hypervariable or variable number tandem repeat (VNTR) locus, the lengths of the DNA fragments produced in different individuals depend on the number of repeats at the locus; this number typically ranges from 15 to 70. There are many different VNTR loci in the genome, so the pattern of fragments from the VNTR loci in one individual is essentially unique for that individual.

The first highly variable probes to be used for individual identification were derived by Alec Jeffreys in the course of his studies of the human myoglobin gene. These first probes produced complicated patterns because they hybridized to fragments from many different VNTR loci in an individual's DNA. In contrast, a *single-locus probe* is specific for just one VNTR locus and detects only one (homozygous) or two (heterozygous) fragments per locus (or marker) in an individual's DNA. However, there is a very wide range of fragment sizes for each locus; for example, one of these probes detected a minimum of 77 alleles at one locus, with repeats ranging from 14 to more than 500 in number. The fragments detected by one VNTR probe and their pattern of inheritance are shown in Figure 16-2.

DNA probes were soon put to use, in 1985, in an immigration case where it was necessary to determine whether a boy was the son or the nephew of the woman who claimed him as her son. Conventional testing, using 17 protein polymorphisms, demonstrated that the boy and woman were related but could not determine their relationship more precisely. To do this, DNA samples were analyzed from the woman and two of her sisters and from the boy. Despite the lack of a sample from the father (there was some doubt about paternity), it was possible to show that the boy and the woman were related and that the relationship was that of son and mother (Table 16-1). The use of these probes in this way has led to the reuniting of many families who would otherwise have been kept apart.

The potential of these probes for establishing identities in forensic science cases was obvious, and Jeffreys chose the term "DNA fingerprinting" quite deliberately to draw an analogy between identification using conventional fingerprints and using DNA. The power of DNA profiling was demonstrated in the very first murder cases in which it was used, the two Narborough murders in England, committed in 1983 and 1986. A man had confessed to the second murder but not the first. Jeffreys was called in by the police and showed, just as the police had surmised, that one man was involved in both attacks. However, the police were astounded to learn that their suspect was not the perpetrator. This result was not believed, and the DNA fingerprinting process had to be repeated several times before the suspect

TABLE 16-1. DNA analysis is used to determine relationships

Question	Number of shared fragments	Probability[a]
Is the boy related to this family?	61	7×10^{-22}
Could an unrelated woman be the mother?	25	2×10^{-15}
Could the sister of the woman be the mother?	25	6×10^{-6}

[a]The probabilities are estimates of the likelihood that the relationships shown by the DNA profiles could have arisen by chance. For example, the probability that the boy could *by chance* have these 61 fragments in common with the family is 7×10^{-22}.

was released. Eventually the real murderer was caught, through careful police work, and DNA fingerprinting confirmed the identification.

Short Tandem Repeats Become the Standard for Forensic Applications

There are many other repetitive sequences in the human genome, and tandem repetitive sequences with repeats between two and seven base pairs in length are called *short tandem repeats* (STRs). These repeats are highly variable and, because of the short overall size of each allele, PCR can be used to amplify the alleles by using primers that hybridize to sequences on each side of the allele. Using PCR has the added advantage for forensic work that the very small samples, often of degraded DNA, typical of what might be found at a crime scene, can be analyzed successfully. However, not all STRs are suitable. Although they occur rarely, "microvariants," alleles that differ by more or less than the repeat length, can complicate the analysis. It is especially important that PCR amplification of the STRs does not produce artifacts that can confuse or complicate interpretation, given that the liberty (and sometimes life) of the accused depends on the outcome of the analysis.

The U.S. Federal Bureau of Investigation (FBI) implemented DNA profiling in 1988 and by 1997 had selected a set of 13 "core" STRs to be used as a standardized panel for investigative purposes (Table 16-2). These STRs are distributed across 12 autosomes and vary in repeat number from 5 to more than 30. An additional probe for the amelogenin gene is included in the panel. This gene codes for tooth enamel and is found in a region of the Y chromosome that has homology with the X chromosome but does not recombine with the X chromosome. The amelogenin gene on the X chromosome has a 6-bp deletion in intron 1. PCR amplification of the intron in male DNA produces two fragments 112 and 106 bp in length from the Y and X chromosomes, respectively; female DNA produces only a 106-bp fragment. Other law enforcement agencies around the world have adopted similar sets of markers (Table 16-2).

Multiplex PCR Amplification and Fluorescent Tags Are Used to Analyze Forensic STRs

STRs are PCR-amplified using pairs of primers that flank the STR sequences, one of each pair carrying one of four fluorescent tags. Following PCR, machines similar to those used for DNA sequencing determine the lengths of the STRs. The fragments

TABLE 16-2. Short tandem repeats (STRs) used for forensic DNA profiling by the Federal Bureau of Investigation (FBI; United States), the Forensic Science Service (FSS; United Kingdom), and Interpol (Europe)

	Repeat	No. of alleles	FBI	FSS	Interpol
D2S1338	[TGCC][TTCC]	20		✓	
D3S1358	[TCTG][TCTA]	10	✓	✓	✓
D5S818	AGAT	10	✓		
D7S820	GATA	11	✓		
D8S1179	[TCTA][TCTG]	10	✓	✓	✓
D13S317	TATC	8	✓		
D16S539	GATA	8	✓	✓	
D18S51	AGAA	15	✓	✓	✓
D19S433	AAGG	19		✓	
D21S11	[TCTA][TCTG]	69	✓	✓	✓
CSF1PO	TAGA	15	✓		
FGA	CTTT	19	✓	✓	✓
THO1	TCAT	7	✓	✓	✓
TPOX	GAAT	7	✓		
vWA	[TCTG][TCTA]	10	✓	✓	✓
Amelogenin	106 bp/112 bp	2	✓	✓	✓

The amelogenin gene is used for sex determination. The gene is on a homologous region of the X and Y chromosomes and so is present in two copies in both males and females. The length of the PCR product from the Y chromosome is 112 bp and the length from the X chromosome is 106 bp. Square brackets indicate STRs with complex repeat patterns.

are detected as they pass a laser, and software determines the type of fluorescent tags associated with each STR and produces tables and graphs of the data. By careful selection of the sizes of the amplified regions, multiple STRs can be analyzed in a single reaction and kits are commercially available that detect all 13 FBI STRs and the amelogenin gene, as well as the two additional STRs used in the United Kingdom, for a total of 16 regions (Fig. 16-3). In addition, "allelic ladders" are available that show the most common alleles at a locus. These are used to calibrate analyses and for quality control (Fig. 16-4). An allelic ladder is included in a separate lane on every slab gel and, because capillary gels change over time, allelic ladders are often injected every 10–15 samples for recalibration.

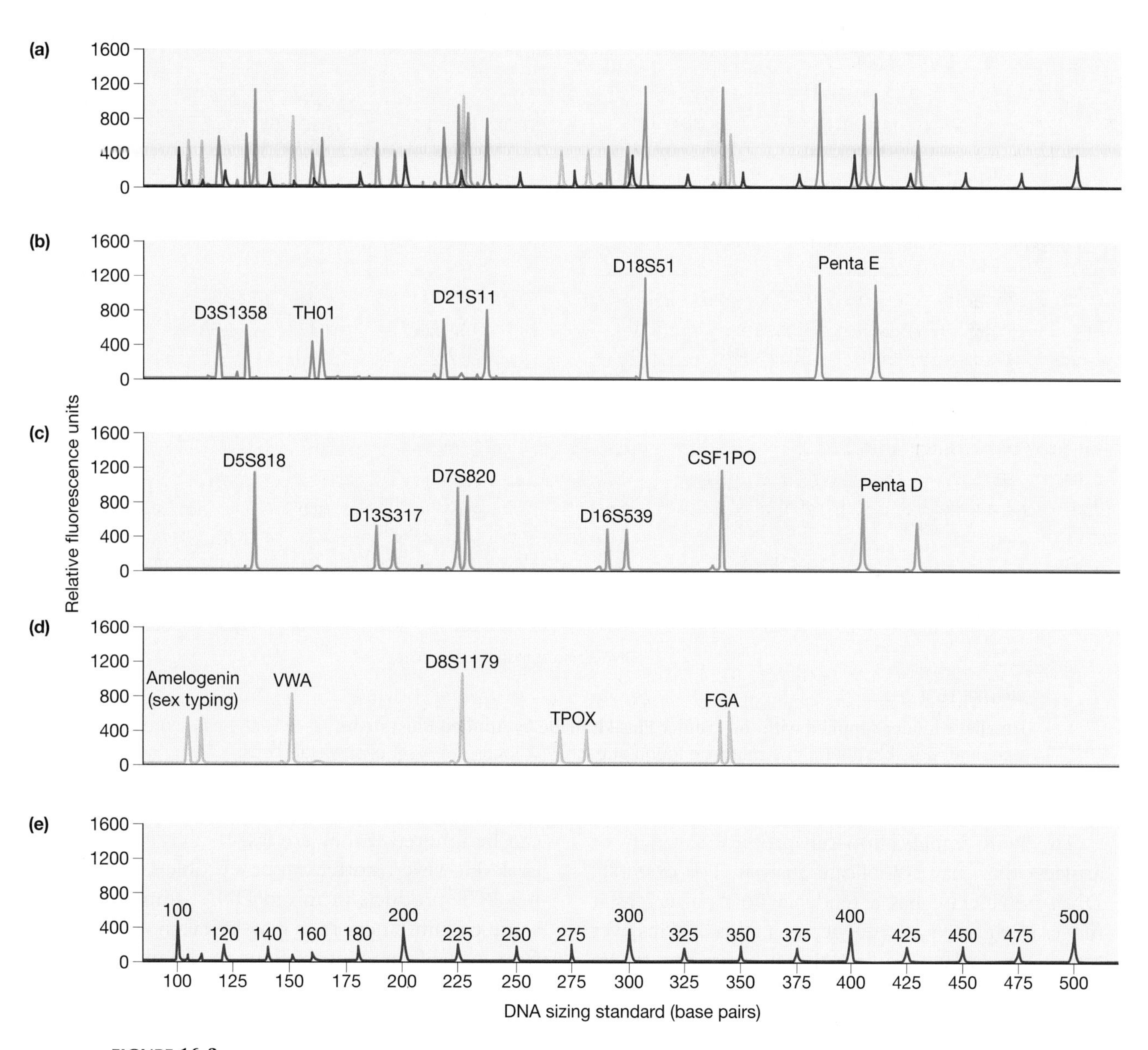

FIGURE 16-3

PowerPlex 16 (Promega Inc.) STR profile using the standard FBI panel. The traces are the output from a fluorescent analyzer. The *upper panel* (*a*) is an overlay of the four *lower panels* (*b–e*). Panel *e* is a set of size standards. Each peak is an allele of the locus. This individual is homozygous for those loci where there is only one peak (e.g., D18S51 and CSF1PO) and heterozygous for those loci with two peaks (e.g., D21S11 and FGA). The DNA sample came from a male: Both the 106- and 112-bp amelogenin alleles are present. Penta E and Penta D are not part of the FBI core STR panel.

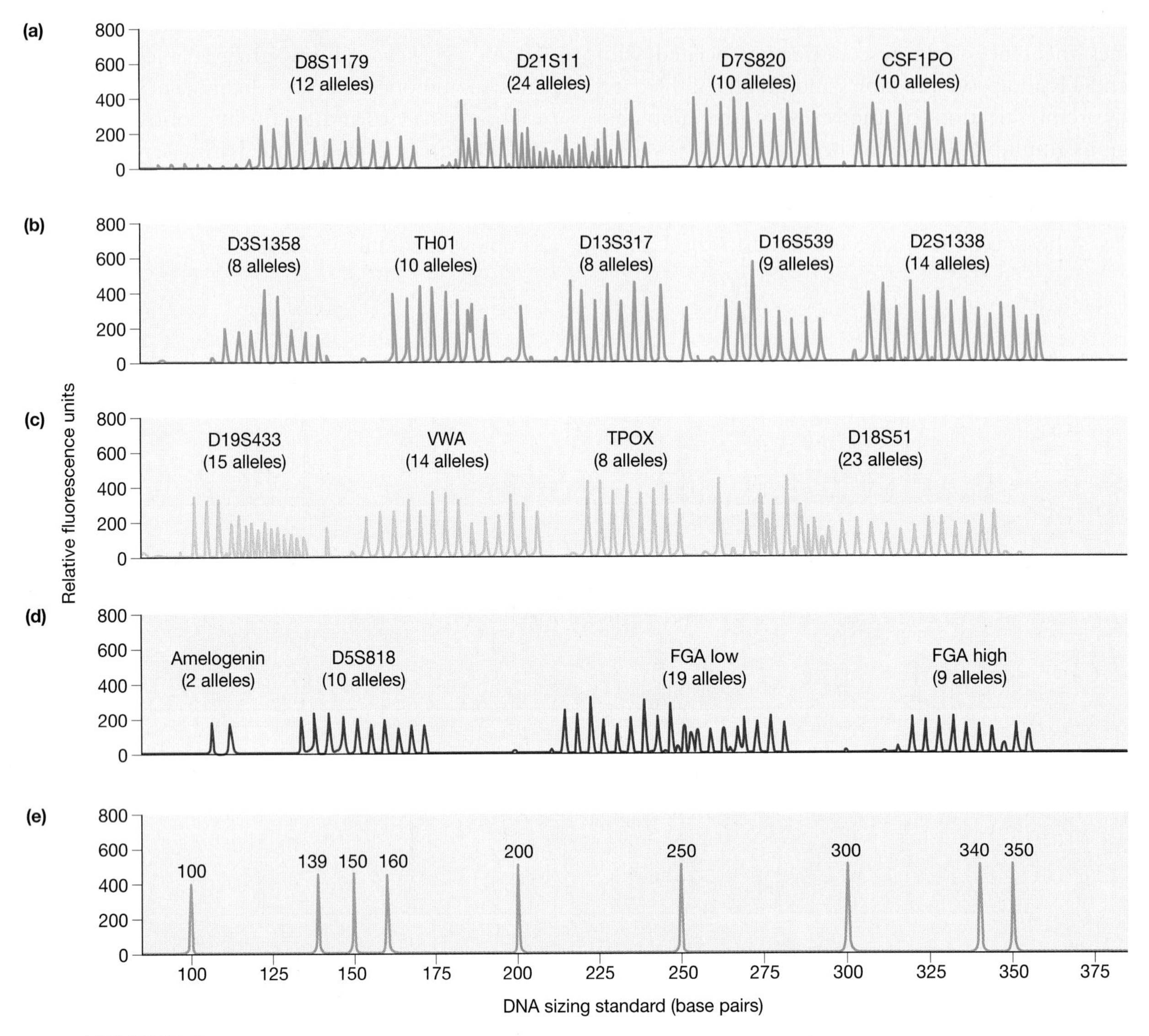

FIGURE 16-4

An allelic ladder supplied with the Profiler Plus kit made by Applied Biosystems. (*a–d*) Each peak represents one of all possible alleles that might be found at each STR locus. (*e*) A set of size standards.

The PCR amplification can produce a variety of artifacts that may complicate analysis. For example, DNA polymerase has a tendency to "stutter" as it moves along a repeat sequence so it adds or skips over a repeat to make a product that is either one repeat longer or shorter than the true length of the allele. These appear as small peaks flanking the large peak representing the true product (Fig. 16-5). For the STRs used in forensic analysis, stutter is typically present as one repeat shorter than the true allele length. For the standard set of STRs, stutter peaks range from less than 2% of the true peak to as much as 15%, depending on the STR, the length of the allele, and the amount of DNA analyzed. Generally these peaks can be ignored if they are less than 15% of the true peak. However, stutter can be a problem in interpreting PCR products in mixed DNA samples where in some circumstances they may be taken as true peaks from one of the contributions to the mixed sample. Other problems can be seen in the graphical output (trace) from a machine. For example, if a sample is overloaded, small peaks may occur in channels other than the one detecting the STRs (Fig. 16-5). This is because of overlap in the spectra of the fluorescent dyes. These "pull-up" peaks are usually small and, if large, can be manually excluded. (Any editing of a trace must be recorded.) Mixed samples are easy to detect by inspection of a trace. Instead of a maximum

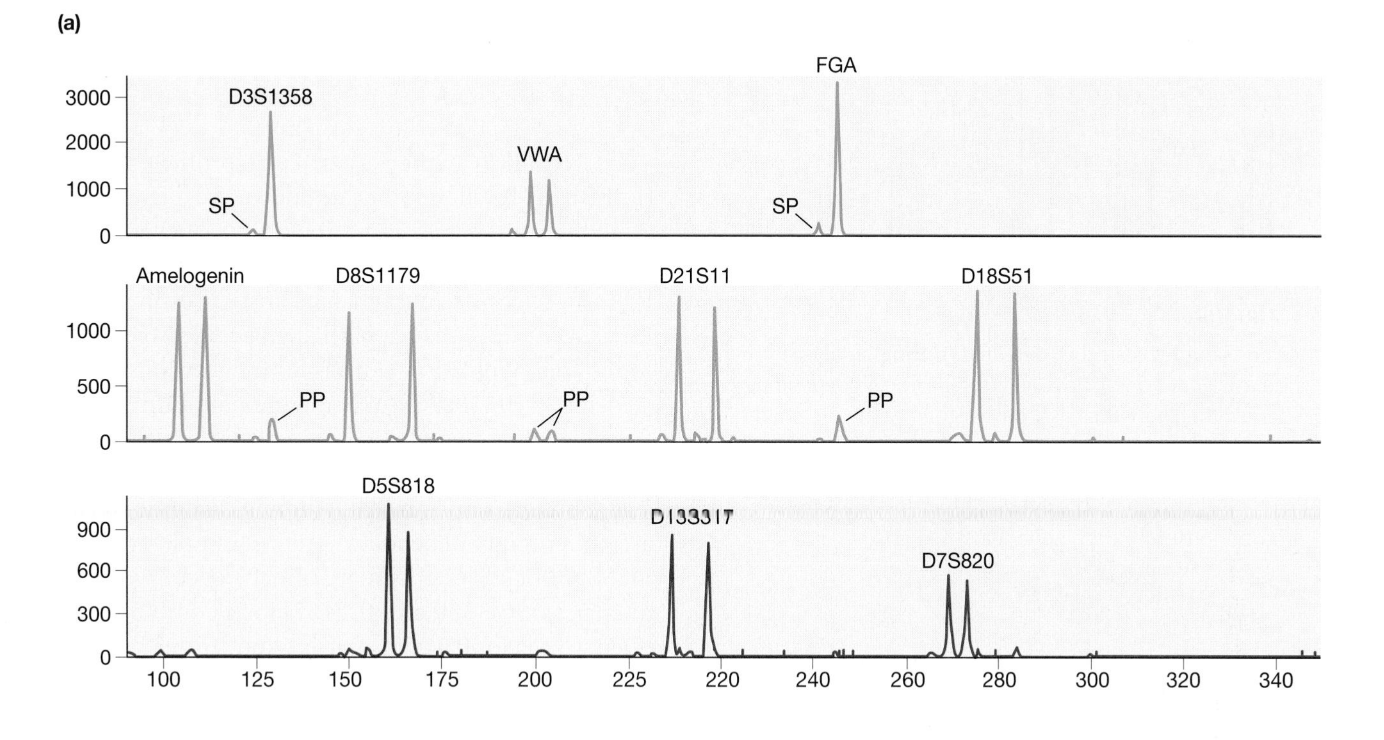

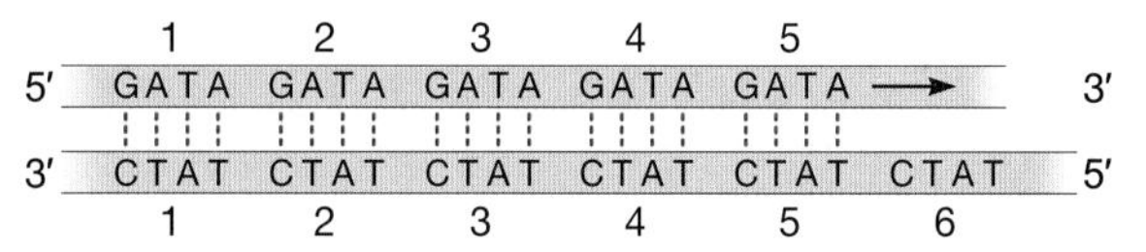

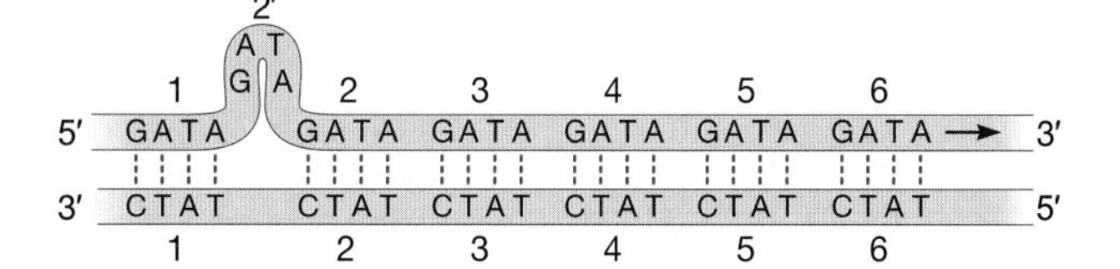

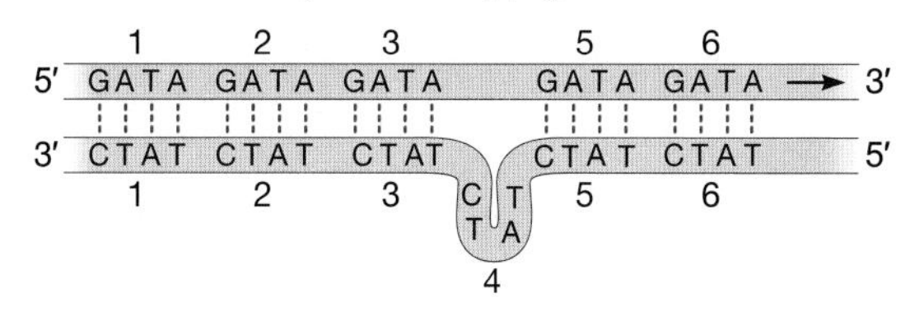

FIGURE 16-5

Stutter peaks and "pull-up" peaks. (*a*) The three traces are the outputs from the blue, green, and red channels of an analyzer (Applied Biosystems). The names of the loci are given above each set of peaks. The three small peaks labeled PP in the green channel are "pull-up" peaks. Each corresponds to a strong signal in blue channel. The peaks marked SP are examples of "stutter" peaks caused by the DNA polymerase slipping and synthesizing a PCR product one repeat shorter than the allele. (*b*) DNA polymerase is synthesizing the upper strand, using the lower strand with a CTAT repeat as the template strand. Occasionally, the polymerase slips backward, adding an extra repeat (2′). The polymerase may also slip forward, missing a repeat (4) and producing a PCR product one repeat shorter than the true allele length.

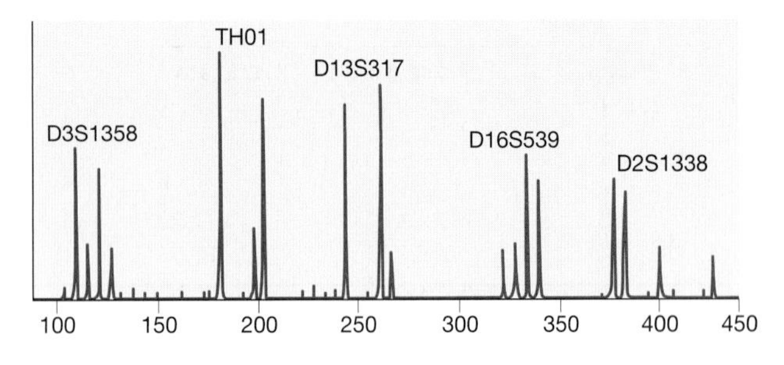

FIGURE 16-6

Mixed samples can be detected during analysis. There should be only one or two alleles at a locus, depending on whether the individual is homozygous or heterozygous at the locus, but in this case there are multiple alleles at some loci. For example, there are clearly four alleles at the D16S539 locus, thereby showing that this is a sample with contributions from two individuals.

of two peaks representing the locus on the two chromosomes, there may be three or four if two DNA genomes are present in the biological sample collected at the crime scene (Fig. 16-6).

Special care has to be taken with "low copy number" (LCN) DNA profiling in which more rounds of PCR are performed because the starting amount of DNA is typically less than 100 pg, which corresponds to the DNA extracted from about 16 cells (1 pg = 10^{-12} g). Where does this minute amount of DNA come from? When we touch a surface, we leave not only a fingerprint outlined by sweat and oils on our skin, we shed skin cells as well. It has proved possible to extract DNA samples from touched surfaces, even from fingerprints that have been treated with fingerprint powder or cyanoacrylate to make the prints more obvious. (Cyanoacrylate reacts with the biological materials of the fingerprint to produce a white substance.) However, amplification of such minute quantities of DNA requires using as many as 34 cycles of PCR (six more than recommended for routine analysis). This may lead to allelic imbalance at a heterozygote locus so that a heterozygous locus may appear to be homozygous. Just by chance, in the early cycles, one allele may be amplified more than the other and so come to dominate in the reaction mixture in the subsequent cycles. Contamination is a particular concern with any PCR amplification and particularly so for LCN analyses. It is possible for secondary transfers to occur; that is, skin cells may be passed from individual A to individual B, and then via B to a gun or other object. The presence of A's DNA on the gun would not be related to the crime.

The Uniqueness of a DNA Profile Is Calculated Using the Frequencies of the STR Alleles

To determine whether a particular profile—the set of FBI STR alleles for an individual—is unique would require determining the profile of every individual in the world. Because this is impossible, forensic scientists must make statistical estimates of how likely it is that a particular profile would be found in a population. This is done according to the principle of the Hardy–Weinberg equilibrium. If the frequencies of two alleles, A and a, of a marker are p and q, the frequencies of AA, Aa, and aa are p^2, $2pq$, and q^2, respectively, assuming that the alleles are inherited independently. For example, for the D5S818 locus, lengths of 12 and 13 occur with frequencies of 0.3539 and 0.1462, for a combined frequency, $2pq$, of 0.1035, or 1 in 9.66. If the 12 allele was homozygous, then the combined frequency would be p^2, 0.3539 x 0.3539, or 0.1252. The frequency for the complete profile of 13 STRs is found by multiplying together the frequencies of each of the STRs. In the example shown in Table 16-3, the overall probability of finding a matching profile from an unrelated individual by chance is 1 in 1.56 x 10^{15}, a vanishingly small possibility. With a global population of some 6.5 x 10^9, and assuming that the analyses have been performed correctly, the FBI panel of STRs provides very strong evidence of identification (excluding monozygotic twins).

Allele frequencies differ in different populations (Table 16-4). The frequency of allele 9 of the TH01 locus is similar in Caucasians, African–Americans, and Hispanics; the frequency of allele 9.3 is twofold higher in Hispanics and threefold higher in Caucasians than in African–Americans. These differences are being used to make predictions about the physical characteristics of an individual from a DNA profile. These predictions are highly controversial on ethical as well as scientific grounds as we shall discuss later in the chapter.

Databases Contain Millions of DNA Profiles

Computers and databases are essential for storing and retrieving DNA sequence data, and this is also true for DNA profile data. In the same way that a geneticist will compare an unknown sequence against GenBank, a forensic scientist will search one of the forensic DNA databanks for a match with the unknown profile. The FBI Combined DNA Index

TABLE 16-3. Calculation of match probability for a CODIS 13-STR profile

Locus	Allele	Frequency (p)	Allele	Frequency (q)	Combined ($2pq$; p^2)	Cumulative frequency	Cumulative (1 in)
D3S1358	16	0.2533	17	0.2152	0.1091	1.091×10^{-1}	9
vWA	17	0.2815	18	0.2003	0.1127	1.230×10^{-2}	81
FGA	21	0.1854	22	0.2185	0.081	9.963×10^{-4}	1,005
D8S1179	12	0.1854	14	0.1656	0.0614	6.117×10^{-5}	16,364
D21S11	28	0.1589	30	0.2782	0.0884	5.407×10^{-6}	184,922
D18S51	14	0.1374	16	0.1391	0.0382	2.086×10^{-7}	4,838,866
D5S818	12	0.3841	13	0.1407	0.1081	2.234×10^{-8}	44,762,757
D13S317	11	0.3394	14	0.0480	0.0326	7.282×10^{-10}	1,373,098,259
D7S820	9	0.1772			0.0314	2.286×10^{-11}	43,744,531,933
D16S539	9	0.1126	11	0.3212	0.0723	1.653×10^{-12}	6.05×10^{11}
TH01	6	0.2318			0.0537	8.876×10^{-14}	11.3×10^{13}
TPOX	8	0.5348			0.2860	2.537×10^{-14}	3.94×10^{13}
CSF1PO	10	0.2169			0.0470	6.192×10^{-15}	8.39×10^{14}

Based on data from STRBase http://www.cstl.nist.gov/biotech/strbase/

The alleles for an individual, typed with the 13 STRs used by the FBI, are listed. Where the second allele field has been left blank, the individual was homozygous for the allele listed. The frequency of finding each of the alleles is shown. The combined frequency of the two alleles is $2pq$ or p^2 for homozygous loci, and the cumulative frequency is calculated by multiplying together successive combined frequencies. The last column expresses the frequencies in terms of how often the observed set of alleles is likely to be observed. The complete set of these alleles might be found in 1 in 839 trillion individuals.

CODIS, Combined DNA Index System (United States); STR, short tandem repeat.

System (CODIS) has 3,275,000 profiles and the National DNA Database (NDNAD) of the Forensic Science Service (FSS) in the United Kingdom has 3,500,000 profiles. These sets of profiles represent approximately 1.0% and 5.2% of the U.S. and British populations, respectively, and these databases are growing rapidly (Fig. 16-7). CODIS contains five profile databases: those of convicted offenders, DNA samples from crime scenes, a missing person and human remains database, a relatives of missing persons database, and a population database for estimating the rarity of a DNA profile. CODIS is organized in a hierarchical fashion—the National DNA Index System at the highest level, followed by state and local level databases. The Armed Forces Repository of Specimen Samples for the Identification of Remains (AFRSSIR) does not keep a database of profiles, but stores the DNA samples of members of the services, in case at some time an identification has to be made. It makes sense to type these only when needed, as the AFRSSIR contains more than 4,800,000 samples (2006) but is called on to analyze less than 2500 military casualties each year.

TABLE 16-4. Frequencies of alleles of TH01 in Caucasian, African–American, and Hispanic groups in the United States

Allele	Caucasian (n = 302)	African–American (n = 258)	Hispanic (n = 140)
5	0.00166[a]	0.00388[a]	—
6	0.23179	0.12403	0.21429
7	0.19040	0.42054	0.27857
8	0.08444	0.19380	0.09643
9	0.11424	0.15116	0.15000
9.3	0.36755	0.10465	0.24643
10	0.00828	0.00194[a]	0.01429[a]
11	0.00166[a]	—	—

From Butler J.M. 2005. *Forensic DNA Typing: Biology, Technology, and Genetics of STR Markers,* Appendix 2. Elsevier, Amsterdam.

Not all alleles are found in all populations.

[a]Alleles whose frequency is considered too low for forensic use.

Mitochondrial DNA Variants Are Useful

The human mitochondrial genome was sequenced in 1981 by Fred Sanger and his colleagues in Cambridge. They determined that it is 16,569 nucleotides long and this sequence, deposited in GenBank, became the standard reference sequence for the mitochondrial genome. Interestingly, when the original 1981 material (human placenta) was resequenced in 1999, only 11 differences were found between the two sequences, a testimony to the high quality of the original sequencing carried out with what we would now regard as rather primitive means. This revised Cambridge Reference Sequence (from 1999) is now the standard against which other human mitochondrial sequences are compared.

Over evolutionary time, many of the original mitochondrial genes moved to the nuclear genome, but the human mitochondrion still retains 37 genes,

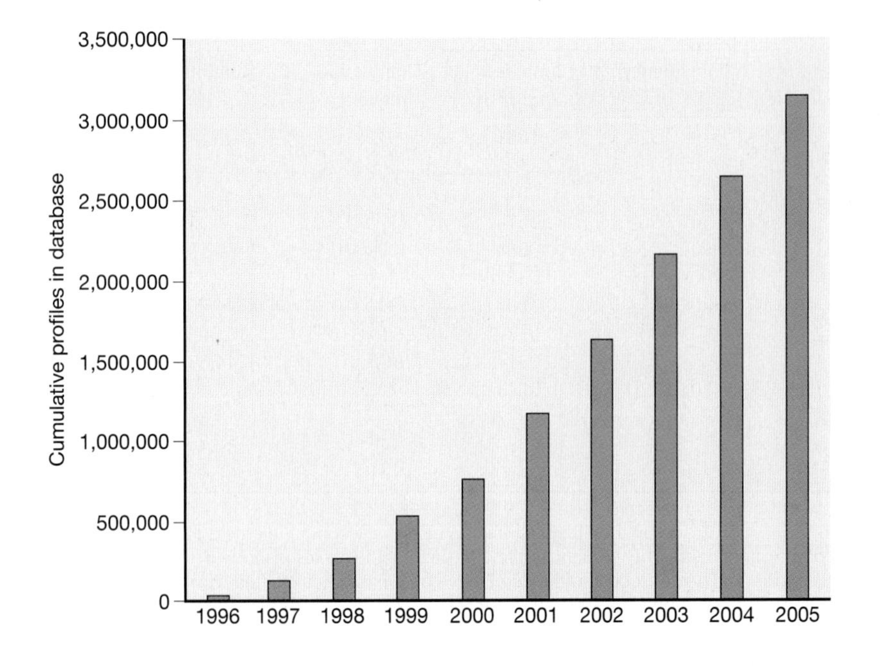

FIGURE 16-7

DNA profile databases have increased rapidly in size. The National DNA Database in the United Kingdom has grown from 36,000 profiles in 1996 to more than 3,000,000 profiles in 2005. The increase beginning in 2000 came about because of the implementation of the Home Office DNA Expansion Program and related legislation that broadened collection of DNA samples.

tightly packed in its circular chromosome. The mitochondrial genome has a 1200-bp region that does not code for protein. This "control region" contains the origin of replication for the chromosome, a promoter for each strand, and the D-loop region named for the characteristic structure that forms as replication begins. Within the D loop, there are two short hypervariable sequences, HV1 and HV2, 342- and 268-bp long, respectively. These sequences differ on average between unrelated individuals in about 7–14 bases out of a total of 610 bases, in contrast to an average of 1 per 1000 nucleotides for the human genome as a whole. These sequence variations are single-nucleotide polymorphisms (SNPs).

At present, the HV1 and HV2 nucleotide variations are analyzed by sequencing. Because the hypervariable sequences are short, templates for sequencing are efficiently generated by PCR. The amplified fragments are sequenced on both strands and compared with the Cambridge Reference Sequence. Sequencing is laborious, time consuming, and expensive, given that a majority of the sequence is constant between individuals so that several hundred base pairs are sequenced to determine changes in only a few. Although not yet adopted for routine analysis, a technique called allele-specific primer extension or "mini-sequencing" may make detection of mitochondrial SNPs faster and cheaper. Primers are synthesized that hybridize to the 19–21 nucleotides preceding the nucleotides that are polymorphic. Such a primer is added to the denatured DNA sample and the four dideoxynucleotides (ddNTPs), each labeled with a different fluorescent marker, and DNA polymerase are added. The ddNTP complementary to whichever nucleotide is present at the polymorphic site is added to the primer. For example, if the SNP is the G to A polymorphism at nucleotide 16,130, the polymerase will add a T to the primer and this can be distinguished by its fluorescence. Several SNPs can be assayed in the same reaction by adding a "tail" of nucleotides of differing length to the primer for each SNP. So, each SNP can be distinguished by length following electrophoresis, and the nucleotide incorporated at each SNP by its fluorescence.

The mitochondrial chromosome behaves as a single linkage group: All the loci are inherited as a block. Studies of large numbers of mitochondrial genomes have identified several mitochondrial haplotypes. These "mitotypes" are sets of SNPs that occur together on a single mitochondrial genome. If the SNPs are not independent of each other, and probabilities cannot be calculated by multiplying their individual frequencies, how are the data used? The simplest procedure is to count the differences between the two samples being compared and to follow the FBI's three-point classification (Fig. 16-8).

Exclusion. If there are two or more nucleotide differences between the questioned and known samples, the samples can be excluded as originating from the same person or maternal lineage.

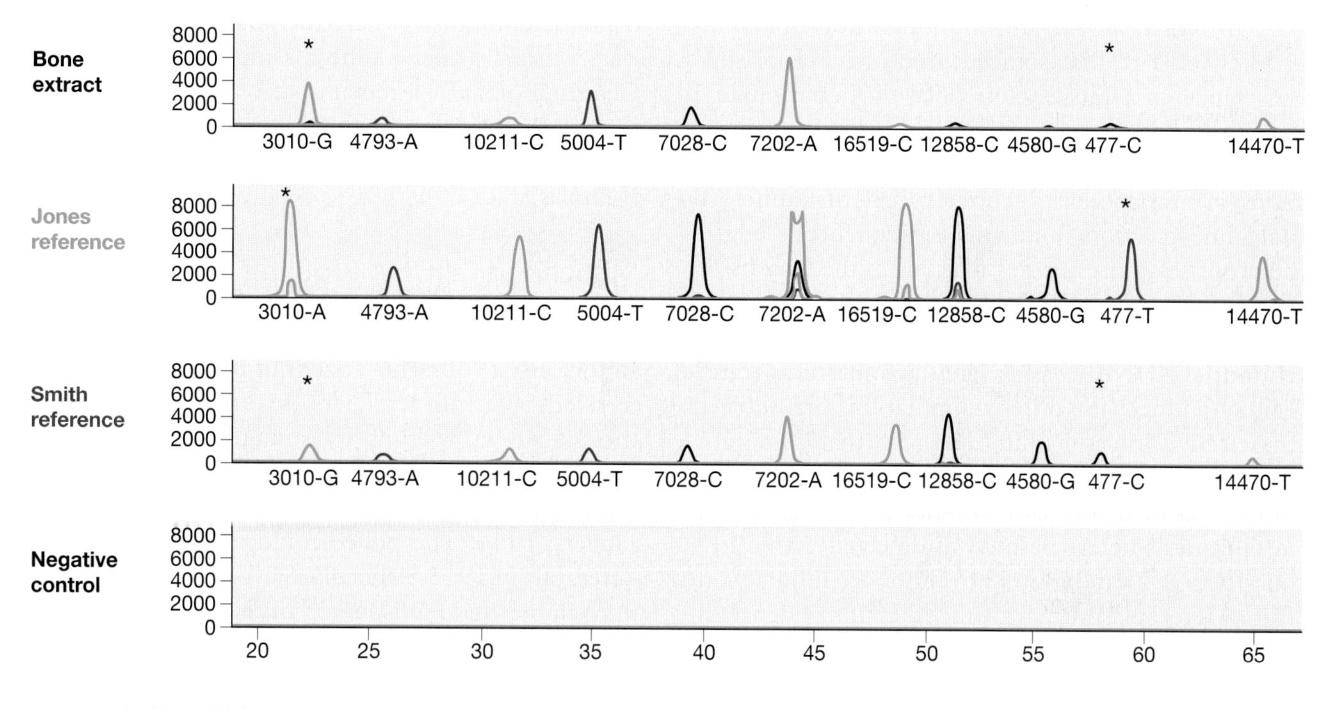

FIGURE 16-8

Mitochondrial SNPs are useful for making comparisons using degraded DNA. Skeletal remains have been found and might be of Jones or Smith. DNA was extracted from the bones and a DNA profile obtained using mitochondrial SNPs. (The degraded state of the bone DNA is evident from the small peaks at many loci.) Tissue samples known to have come from the two men were obtained, DNA was extracted, and the same SNP profile was determined. The label of each SNP is given, and base present is shown. Jones' sample differs from the bone sample at two loci, 477 and 3010 (*asterisks*). At these two sites, the skeletal and Smith samples are identical, having the same changes from the Cambridge reference sequence: T to C at 477 and A to G at 3010. For mitochondrial SNP analysis, samples differing at two loci are considered to be different, so the skeletal remains are of Smith and not Jones.

Inconclusive. If there is one nucleotide difference between the questioned and known samples, the result is inconclusive.

Cannot Exclude. If the sequences from questioned and known samples under comparison have a common base at each position, the samples cannot be excluded as originating from the same person or maternal lineage.

If the conclusion is "cannot exclude," a statistical estimate of the rarity of that mitochondrial profile can be made by counting the number of times that particular combination of SNPs (or mitoytpe) appear(s) in a population database. Using this number and the size of the database, the expected frequency of the observed profile can be calculated, together with upper and lower confidence limits. However, this calculation depends on the number of samples in the database, and the population frequencies of most mitochondrial DNA profiles cannot be calculated because they occur only once in the databases or are being detected for the first time. Although databases with very large amounts of mitochondrial sequence data exist, these are of variable quality and are not considered suitable for forensic studies. The FBI has developed a database of mitochondrial sequences, CODIS^{mt}, that contains sequences from only laboratories that have passed an accreditation process.

Mitochondrial Profiling May Be Complicated by Heteroplasmy

For many years it was thought that the genomes of all the mitochondria in a cell were identical. It was then discovered that a cell or tissue might have mitochondria with different genomes. This condition is known as *heteroplasmy*. It was thought to be rare but studies indicate that some degree of heteroplasmy exists in all individuals. Heteroplasmic mutations are of two kinds—sequence heteroplasmy and length heteroplasmy. The former occurs where two populations of mitochondria genomes exist that differ at only one position in the DNA sequence. Length heteroplasmy arises from variations in the lengths of tracts of cytosines in the HV1 and HV2 regions. It is believed that heteroplasmy arises because mitochondrial DNA has a high mutation rate. Mitochondria with different mutations are subject to selective forces within a cell, so that some are maintained, whereas others are lost. Different tissues differ in their degree of heteroplasmy. For example, in one study, heteroplasmy was detected with a frequency of 1.7% in blood, 8.6% in bone, and 9.7% in hairs. Heteroplasmy may be mistaken for evidence of mixed samples, but the mixed samples often have differences at numerous positions. In fact, rather than causing confusion, heteroplasmy can be of forensic value.

The most famous case involving heteroplasmic mitochondrial DNA was the identification of remains dug up in a forest in Siberia in 1991. These remains were suspected to be the bodies of Tsar Nicholas II and his family who had been murdered in the Bolshevik Revolution in 1918. Peter Gill undertook mitochondrial DNA profiling of the remains and showed that five of the recovered bodies were related and that three of these were females. One was identified as the Tsarina Empress Alexandra, as shown by comparing her mitochondrial DNA profile with that of a living relative. Prince Philip, Duke of Edinburgh, husband of Queen Elizabeth II, is Alexandra's grandnephew, and their profiles matched. The best source of reference DNA for comparison with the presumptive Tsar's body would have been the remains of his younger brother, Grand Duke Georgij Romanov, but the Russian authorities were reluctant to risk damaging the marble sarcophagus containing his body when opening the tomb. Instead, more distant relatives were found and their mitochondrial DNA profiles were shown to match one of the male bodies in the forest. However, the match was not exact. The Tsar's profile had a sequence variation—a C at position 16,169, instead of the T found in his living relatives. Moreover, further analysis found evidence of heteroplasmy; the Tsar's sample had mitochondrial DNA with both a C and a T. This might have been due to contamination but when the marble sarcophagus was eventually opened and a sample procured from Georgij Romanov's remains, it was found that he, too, was heteroplasmic for this sequence, thus providing strong evidence that the remains were of Tsar Nicholas (Fig. 16-9).

Mitochondrial DNA Profiling Reopens the Case of the Boston Strangler

In the early 1960s, the citizens of Boston were being terrified by a murderer who sexually assaulted and then killed his victims by strangulation. The first murder took place in June 1962, and the body of the 13th and last victim, Mary Sullivan, was found in January 1964. The police could not find the murderer, but in 1965, Albert DeSalvo, then imprisoned in Walpole State Prison for burglary and multiple rapes, confessed to the murder of Sullivan and, by association, the other murders ascribed to the Boston Strangler. However, he was never brought to trial for these murders and was himself murdered in 1973. Sullivan's sister and her son did not believe that DeSalvo killed Sullivan and persuaded James Starrs to reinvestigate the case. Starrs compared DeSalvo's confession with contemporary accounts and found several discrepancies. With the cooperation of Sullivan's sister and nephew and DeSalvo's brother, Starrs and David Foran, a forensic scientist, exhumed the body of Mary Sullivan. Examination of the body revealed traces of biological materials on head hair, pubic hair, and underwear that had been replaced on the body after the original autopsy.

The investigators were aware that only mitochondrial DNA was likely to have survived 36 years and that nested PCR amplification would have to be used to obtain a DNA profiile for Sullivan. Here amplification is first carried out for 20–30 cycles, and then a sample of this reaction is subjected to a further round of 20 cycles using primers internal to the first set. Contamination by extraneous sources of DNA is a serious matter when so many rounds of amplification are performed and thus extreme care was taken to avoid contamination of Sullivan's tissues. All members of the exhumation, autopsy, and laboratory teams were gowned, gloved, and masked. In the laboratory, all surfaces and solutions were treated with shortwave UV light to destroy DNA. Isolation of DNA and PCR amplification and analysis were carried out in different rooms to minimize any possibility of contamination.

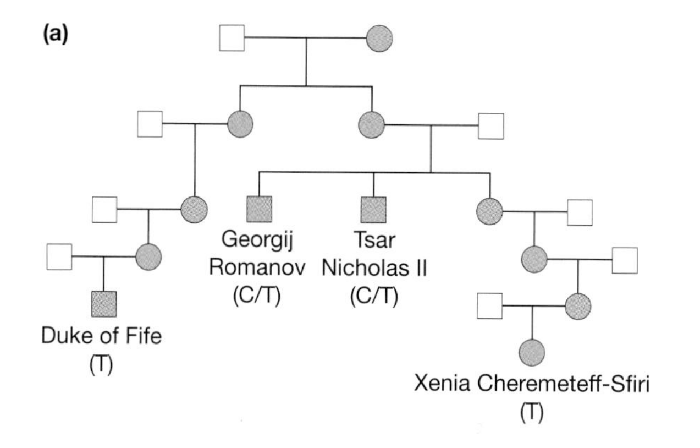

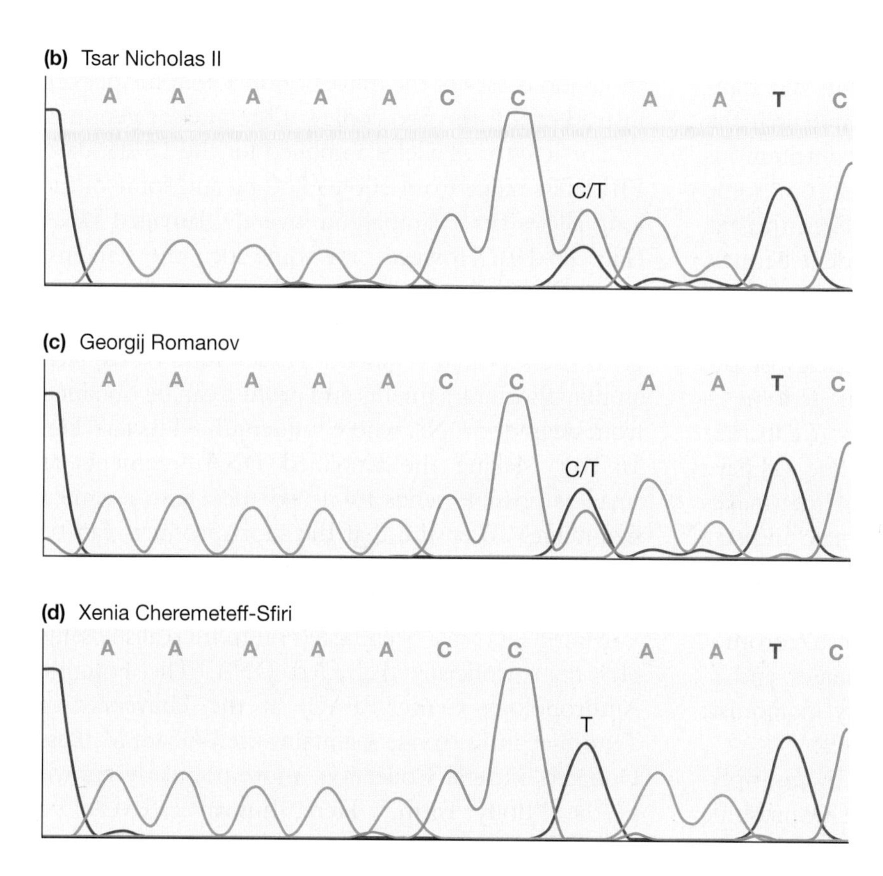

FIGURE 16-9

Heteroplasmy in the family of Tsar Nicholas II. (*a*) Family tree of Tsar Nicholas II. (*Orange* symbols indicate the maternal line.) (*b*) Sequencing revealed that the Tsar was heteroplasmic at nucleotide 16,169, having both a T and a C, thus showing that he had two populations of mitochondria. This was not seen in the distant relatives of the Tsar who were tested. Some used this finding to argue that the analysis should be disregarded. (*c*) A sample of mitochondrial DNA was later analyzed from Georgij Romanov, the Tsar's brother. He, too, had the same heteroplasmy, thereby confirming the identity of the Tsar. (*d*) Xenia Cheremeteff-Sfiri is the great-great-granddaughter of the Tsar's mother and, as she is in the direct maternal line, she should also show heteroplasmy. Similarly, the Duke of Fife, also in the maternal line, should show heteroplasmy. Neither do and it is assumed that over the course of four generations, the populations of mitochondria segregated to homoplasmy.

DNA was successfully amplified from the underwear and the pubic hair. These sequences differed: That from the underwear differed from the Cambridge Reference sequence at four places, whereas the pubic hair sequence differed at only one position. Furthermore, these DNAs did not come from Mary Sullivan and were not contaminants from any member of the investigative team. After comparison with sequences from DeSalvo's brother excluded DeSalvo as the source (the brother's DNA was kept in a separate building until all the testing had been completed on the evidential samples!), Albert DeSalvo's body was exhumed and the exclusion was confirmed by direct comparison with mitochondrial DNA isolated from DeSalvo's bone sample.

What conclusions can we draw from this information? All that can be said with certainty is that no DNA profiling evidence was found to associate Albert DeSalvo with the body of Mary Sullivan, thus making it extremely unlikely that he was her murderer. However, nothing useful can be said about the origins of the two mitochondrial DNAs. Many people could be the sources—the detectives who originally investigated the crime scene, the pathologists who conducted the autopsy, or the morticians who prepared the body for burial. The question still remains: Who was the Boston Strangler?

Y-Chromosome STRs Have the Advantage of Being Specific for Males

Because most crimes are committed by men, interpretation of DNA profiles in cases of rape and murder would be simpler if male-specific markers could be used. For example, in a case of rape where the assailant has a low sperm count, a minute amount of male sample may be mixed with a very large female sample, thereby possibly precluding analysis using the standard loci. In such a case, Y-chromosome markers could be analyzed independently of contributions from the female DNA. Unfortunately, Y-chromosome loci have some limitations for profiling. First, the loci on the Y chromosome are not independent because the chromosome does not undergo recombination, so that probabilities for Y chromosome markers cannot be calculated by multiplying the frequencies of the loci. Second, just as the mitochondrial genome is inherited through the maternal line, so the Y chromosome is patrilineal—all males in the same line will have the same Y chromosome. Thus the factor that makes the Y chromosome popular for tracing male ancestry in genealogical studies, also makes it impossible to distinguish between brothers or between sons and fathers. Nevertheless, the advantages of Y-chromosome-specific loci outweigh the disadvantages, and Y-chromosome markers are used routinely in forensic analysis in combination with other analyses.

If probabilities cannot be calculated by multiplying the frequencies of the different Y-chromosome STRs, then how are they calculated? The procedure is similar to that used for mitochondrial markers. That is, a comparison of an evidentiary profile with a profile determined at the same loci for a suspect will result in an exclusion, inclusion, or inconclusive result. However, inclusion, that the two samples are sufficiently similar that they might have come from the same individual, must be qualified with a statement to the effect that the result applies equally well to all patrilineal male relatives.

Degraded DNA Is Analyzed Using "Mini-STRs"

Blood stains scraped from a gatepost, semen from a decades-old rape kit, or tissue parts from the scene of a plane crash are not ideal sources for isolating DNA. Typically, such DNA is degraded to a greater or lesser degree so that only a very small amount of DNA may be isolated, and the DNA that is recovered is likely to be present in very short segments. Carrying out extra rounds of PCR amplification can compensate for the small amount of starting DNA, but this amplification increases both the proportion of DNA molecules with errors and the likelihood of amplifying contaminating DNA. Analyzing mitochondrial DNA can be helpful because a human cell has several hundred to several thousand mitochondria, each with several copies of its genome. Thus, there may be several thousand copies of the mitochondrial genome per cell instead of just the two copies of the nuclear genome.

The lengths of alleles amplified for the 13 standard FBI STRs range from 106 to 360 bp and some of the long alleles fail to amplify in severely damaged DNA (Fig. 16-10). However, for some loci, the standard primers lie some distance from the repeated segment, so it has proved possible to design primers much closer to the repeated segment. These "mini-STRs" can amplify shorter fragments and profiles can be obtained from degraded DNA where full-length STRs fail (Fig. 16-10). Making the amplified DNA fragments as small as possible tends to crowd them into a similar size range. To ensure that the short products can be distinguished from one another, fewer STRs are analyzed in a single reaction.

Mini-STRs have been tested on more realistic samples than artificially degraded DNA. The Forensic Anthropology Center (FAC) at the University of Tennessee in Knoxville maintains the William M. Bass Donated Skeletal Collection, more popularly known as the "Body Farm." Here, human cadavers are exposed to the environment where they decompose under a variety of environmental conditions, including being eaten by insects, birds, and other animals. The "Body Farm" provides data on the rates and nature of decomposition of human bodies—data that are used for interpreting real cases. Forensic scientists used samples from the "Body Farm" to determine whether mini-STR primers could amplify alleles that were missed by conventional primers using samples. They extracted DNA from a left femur that had been buried for 36 months and had been stored for a further month before processing. Conventional primers failed to detect alleles where the PCR product was longer than 275 bp. Those same alleles were detected with primers designed to amplify much shorter (<100 bp) PCR products (Fig. 16.11).

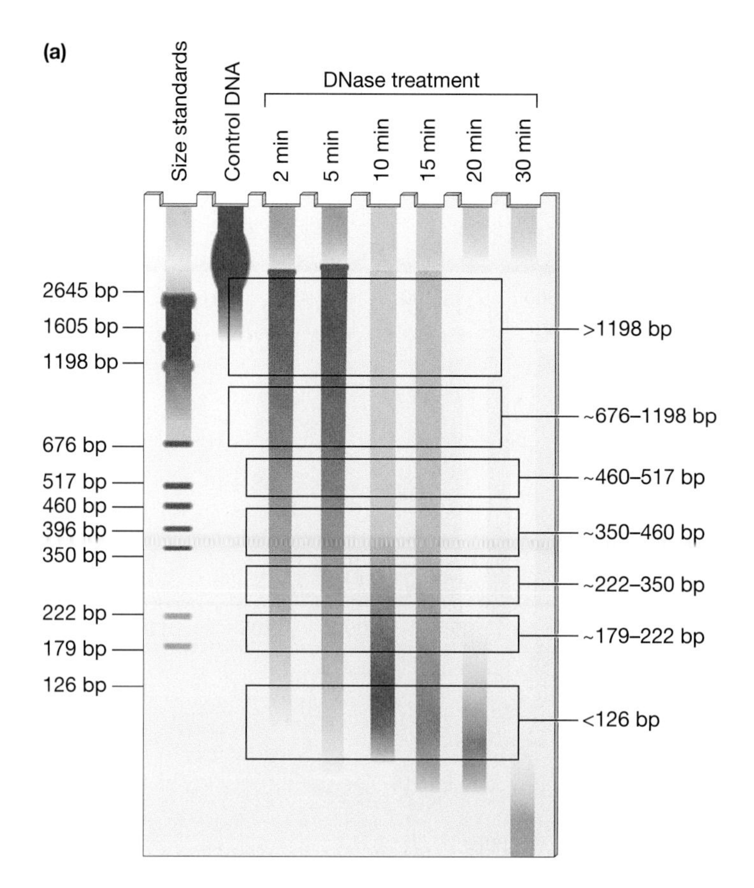

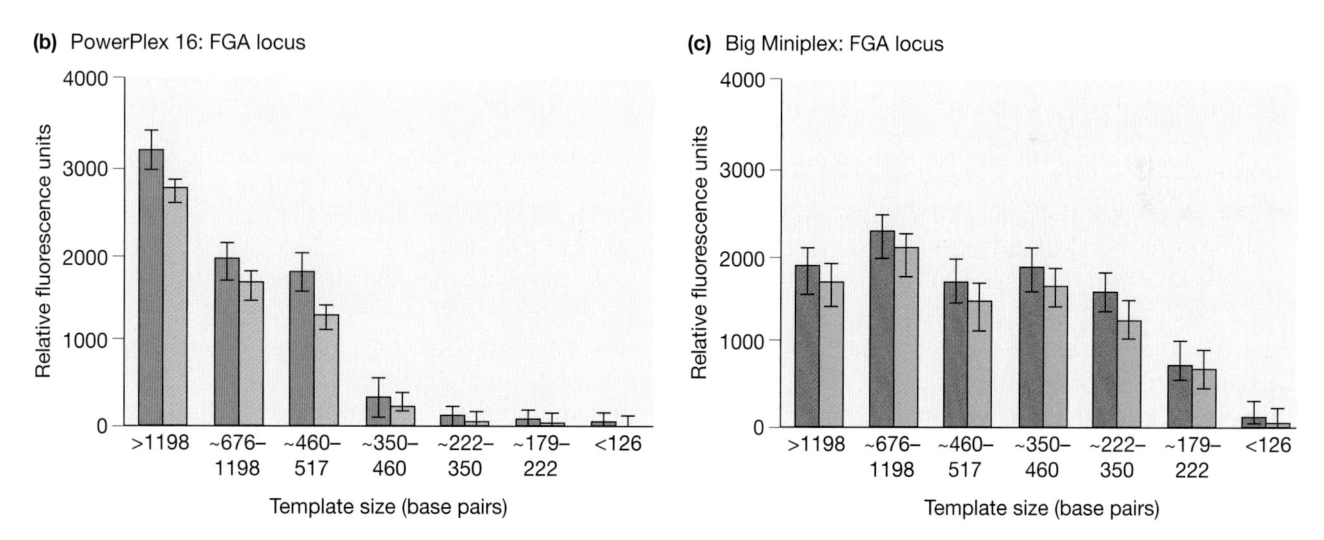

FIGURE 16-10

Mini-STRs are developed to amplify loci in degraded DNA. (*a*) Samples of DNA were incubated for varying lengths of time with DNase I to simulate degradation that occurs when tissues decay. The control lane contains a high-molecular-weight sample of untreated DNA that barely migrates through the gel. Samples treated with DNase I are cut into smaller and smaller fragments as shown by their increasing migration. Sections containing different sizes of DNA fragments were cut from the gel and the DNA extracted from each gel section. (*b*) These were tested with two sets of primers for amplification of the FGA locus. One, PowerPlex 16 (Promega Inc.), is a standard set used for forensic profiling, whereas the second set was for mini-STRs. The conventional primers make fragments between 308 and 464 bp, depending on the allele lengths. These primers fail to amplify the locus from DNA averaging 350–460 bp in length. (*c*) The mini-STRs, which amplify fragments in the range 125–281 bp, are successful except with the most degraded DNA. The two bars in *b* and *c* are the two alleles of the FGA locus in this sample.

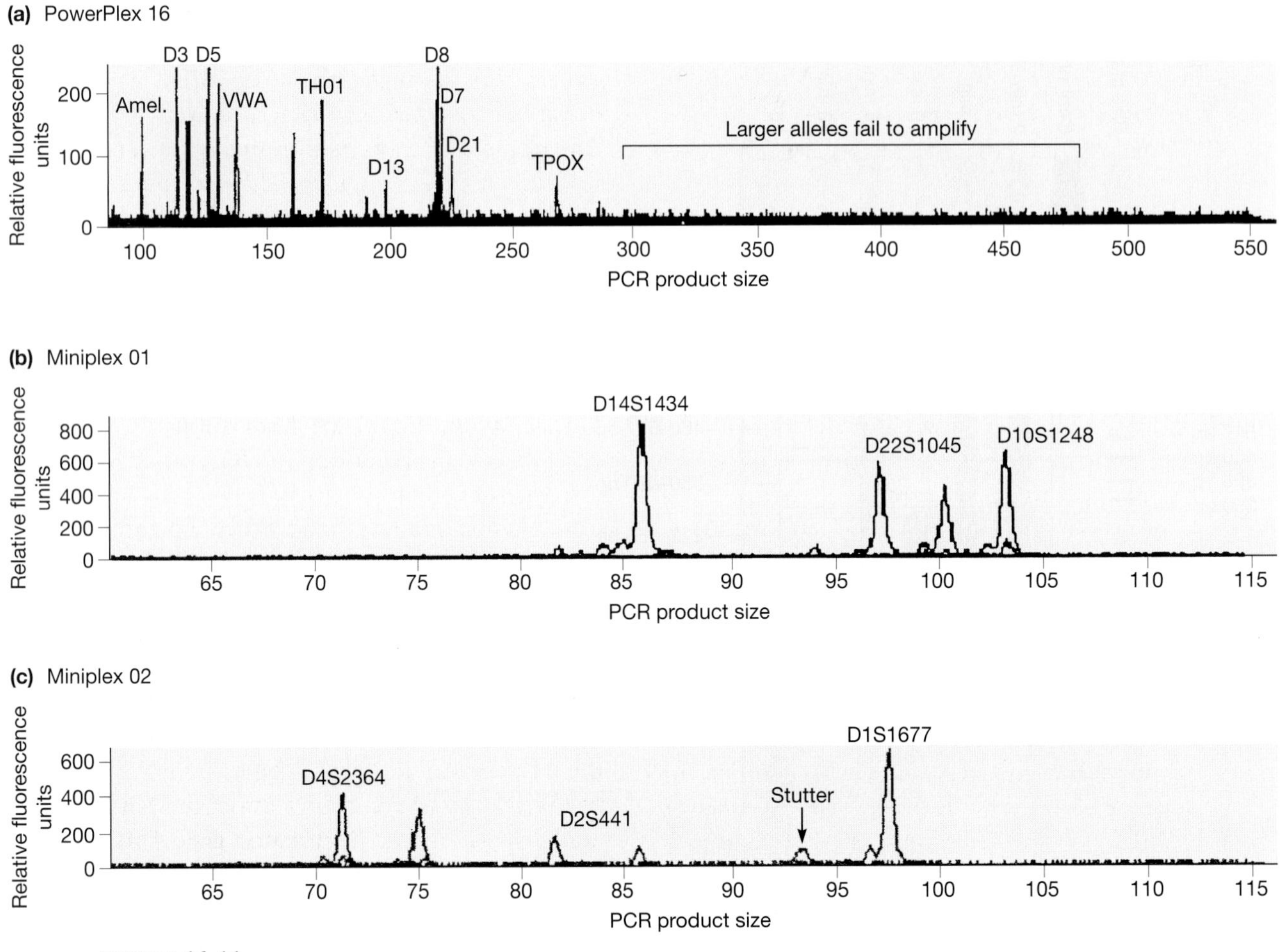

FIGURE 16-11

Testing mini-STRs on DNA from corpses buried for 36 months in University of Tennessee Forensic "Body Farm." (*a*) A sample of DNA prepared from a degraded bone sample was amplified using a standard PowerPlex 16 profiling kit (Promega Inc.). Alleles up to 260 bp in length are amplified but all those longer are lost. (The allele names have been abbreviated.) (*b,c*) Amplifications using the Miniplex 01 and Miniplex 02 primers, which produce very much shorter PCR products, amplify the "missing" alleles efficiently.

Single-Nucleotide Polymorphisms Show Promise for Forensic Applications

SNPs are the most common genetic variants in the human genome and provide a practically inexhaustible supply of markers. Just as for other types of genetic variability, SNPs can be used to distinguish between individuals, but they have some disadvantages for forensic applications. First, because SNPs have only two alleles in contrast to several alleles for STRs, at least 50 SNPs are needed to achieve the same statistical power as the set of 13 STRs in the FBI panel. Second, the presence of only two alleles for an SNP also makes it more difficult to detect mixtures of samples. Suppose that the two alleles at a given SNP are 1 and 2. If individual A is 1 and 2, but the sample has been mixed with DNA from an individual (B) who is 1 and 1, then the presence of the DNA from individual B cannot be detected except by quantitation, that is, by determining that that there are three copies of 1 and only one copy of 2. Third, fourfold more PCR amplifications are required to analyze 50 SNPs than are necessary for 13 STRs, thus increasing costs. Such an analysis also requires designing primers to amplify 50 SNPs in a multiplex PCR so that all 50 PCR products can be distinguished. This has not yet been achieved, but multiplex systems have been devised that carry out single base extension reactions for 52 SNPs in two batches, with 23 and 29 pairs of primers.

Why, then, with all these disadvantages, are SNPs attractive for forensic analysis? A major advantage is that SNP analysis can be carried out using degraded DNA where STR analysis would fail. The DNA to be amplified need contain only the SNP and a short section flanking it to which the primer anneals.

Deriving Physical Characteristics from DNA Profiles Is Controversial

DNA analysis has become popular in the genealogical world for suggesting ancestral origins; there are several companies offering Y-chromosome and mitochondrial DNA analyses to determine, for example, to which of the ancient tribes of Britain a man belongs or whether a man or woman has African, Native American, or Celtic haplotypes. It might be possible to use forensic DNA profiling in the same way to determine the ethnic or geographical origin of the individual from whom the DNA came. This would provide additional information that could be used to narrow the number of potential suspects.

However, there are scientific controversies about the extent to which genetic analysis using markers such as SNPs, *Alu* sequences, and microsatellites can distinguish groups of human beings. One study analyzing 100 *Alu* insertions in several hundred individuals found that the latter were assigned to their continent of origin with an accuracy of 99–100% when all 100 *Alu* polymorphisms were used. However, in another study of 1330 individuals that combined data from 21 *Alu* polymorphisms with data from Y-chromosome and β-globin polymorphisms, assignments could not be made with better than 70% accuracy. These conflicting results may be due in part to the ways in which samples are collected. One study, for example, found that worldwide sampling based on geographical location without regard to "race" demonstrated gradual changes in allele frequencies across the continents; that is, there is no significant grouping of alleles based on location.

There is considerable controversy about how to decide what constitutes a "population." In many cases, ethnic and racial identity is self-reported, and these populations contain individuals of diverse genetic background. Indeed, analyses show that the degree of admixture—the degree to which an individual's genetic background is derived from different groups—is considerable in most populations. And many individuals have been surprised by the extent to which analysis has revealed a much more mixed ancestry than they had suspected. The extent to which ethnicity can be used as a surrogate for underlying genetic variations is the subject of heated debate. For example, the drug BiDil is especially effective for treating heart failure in African–Americans and there is presumably a genetic basis for this. Critics decried FDA approval of BiDil specifically for treatment of African– Americans; they say that this implied government approval for a biological concept of race.

Analyses of SNPs within genes coding for physical characteristics may enable correlations to be made between SNPs and the appearance of an individual. One study examined the association of 754 SNPs with iris pigmentation in 851 individuals of European descent. Here, 335 of these SNPs were found in genes known to be associated with pigmentation, such as the *TYR* gene. This gene encodes tyrosinase, the enzyme that catalyzes the conversion of tyrosine to dopaquinone, the first step in melanin synthesis. Twenty of these SNPs were associated with iris color, but the study concluded overall that the iris color could not be predicted on the basis of the sequence of any single gene.

Inheritance of red hair is associated with variants in the melanocortin 1 receptor gene (*MC1R*) and resembles an autosomal recessive trait. However, there are multiple alleles of *MC1R* and different combinations lead to different shades. For example, one study found six variants at *MC1R* associated with red hair, and that individuals heterozygous for two of these, R151C and 537insC, were significantly more likely to have red hair. Another study examined 12 variants and showed that individuals heterozygous for any two were likely to have red hair, and, conversely, that the absence of any variant would indicate that the individual did not have red hair. At present, deriving physical characteristics from DNA samples is not standard practice and has been used only under carefully supervised conditions by the Forensic Science Service (FSS) in the United Kingdom.

DNA Evidence Must Be Collected Carefully at the Crime Scene

The importance of correct processing of a crime scene was demonstrated dramatically during the trial of O.J. Simpson. The defense team drew attention to how the police and other law enforcement agents had collected, handled, and stored evidence that yielded the DNA samples used to incriminate Simpson. Because the defense was able to cast doubt on the validity of the samples, thereby undermining the prosecution case, they were able to persuade the jury that there was a reasonable doubt that Simpson had committed the crimes despite his DNA profile matching evidence from the crime scene.

Special precautions must be taken when collecting biological evidence at a crime scene. Samples containing DNA may be widespread at a crime scene, because DNA is found in blood, saliva, skin cells, hair, dandruff, semen, and sweat. These may be found on objects as varied as the handle of a weapon, dirty laundry, the butt of a used cigarette, or the licked flap of an envelope. Small blood spots, even in cleaned areas, can be visualized by spraying with the chemiluminescent compound luminol dissolved in a solution containing hydrogen peroxide. The iron atom in hemoglobin acts as a catalyst for a reaction in which luminol is oxidized and, as it does so, luminesces, making blood spots visible. However, luminol can detect as little as 1 part in 10 million of blood, so that although luminol reveals the presence of blood, there may not be enough to extract DNA. DNA is susceptible to degradation by bacteria and molds, and therefore all samples are air-dried to minimize microbial growth. For the same reason, evidential samples are stored in paper bags rather than plastic bags, as the latter retain moisture.

Samples in addition to those from the perpetrator and the victim will be present at a crime scene. Just as for fingerprints, officers have to collect elimination samples so that DNAs from individuals not associated with the crime can be distinguished from DNA of a possible suspect. Officers investigating the scene must take care not to contaminate the scene with their own DNA by sneezing, handling samples without wearing gloves, or even talking. Such contamination can be detected during the laboratory investigation but adds unnecessary complexity to the analysis.

An essential component of all forensic work is the *chain of custody*. The integrity of evidence of any kind collected at a crime scene will be one of the major factors if an investigation goes to trial. Both prosecution and defense will use the evidence to bolster their own cases, and juries place great weight on physical evidence. Given the consequences of a guilty or not guilty verdict, evidence must be safeguarded at all stages to ensure that it is not lost or tampered with. Chain of custody is the process by which a sample of evidence is always under the care of a known individual. A form accompanies the sample, and when it is transferred from one individual to the next, the latter signs the form, noting the date and time of the transfer. Evidence is kept under seal and, when not being examined, is kept under lock and key.

DNA Profiling Enables "Cold Hit" Identification of Perpetrators

Databases of DNA profiles have now grown to include millions of profiles and this has led to increasing numbers of "cold hits." In these cases, an evidentiary sample is not compared with the profile of a particular suspect who has been identified through other investigations, for example, observed by eyewitnesses or by selling items stolen during a robbery. Instead, the evidentiary profile is compared with all the profiles in a computer database such as the U.S. National DNA Index System or the U.K. National DNA Database.

The success rate for identification has grown because the databases contain an increasing number of profiles of convicted felons, many of whom are in situations where they will commit a crime again. For example, the most recent survey of recidivism in the United States, based on all prisoners released in 1994 (272,111), showed that 67.5% were rearrested within 3 years. Given this regrettable pattern, many states have enacted laws that require all convicted felons to give a blood sample or a sample of buccal cells wiped from the inside cheek for DNA profiling. New York State takes samples from all individuals who commit any of more than 100 different offenses. The United Kingdom has adopted even more sweeping legislation; blood samples are required from all individuals arrested for a recordable offense, even before being charged with an offense. Individuals who are not charged may request that their DNA sample be destroyed, but the default state is that the sample is retained indefinitely. These measures have raised issues about civil liberties, but the increasing number of cases solved through "cold hits" demonstrates the value of these databases.

Because DNA samples are taken from individuals involved in so many different criminal activities, DNA profiling is proving useful in solving a wide range of crimes. Biological evidence capable of yielding DNA for profiling can be left, for example, on a broken window when a burglar forces entry into a house. Indeed, the greatest use of DNA profiling in the United Kingdom is in solving burglaries and vehicle thefts.

Familial Searches Identify Suspects

What can be done if an evidentiary profile is incomplete and cannot be matched to a profile in a database? Or if there is a complete profile, but there is not a match in the database? The FSS in the United Kingdom has exploited the fact that an individual who leaves behind a biological sample will share more alleles with a close relative than with someone drawn at random from the general population. Instead, therefore, of requiring a perfect match between evidential sample and a profile in the NDNAD, the stringency of the matching is reduced, one allele at a time. Profiles may be identified that have come from a parent, sibling, or child of the individual who left the biological evidence at the crime scene. As of 2005, the FSS has used this strategy to investigate some 20 cases and found useful information in about one-quarter of these. Take, for example, the cases of Pauline Floyd, Geraldine Hughes, and Sandra Newton.

These three young women were raped and murdered in 1973. At the time there was no apparent connection between the cases, and despite a massive police effort, no one was arrested. Twenty-seven years later, the FSS obtained a DNA profile from stains on Floyd's and Hughes' clothing. However, there was no match to the profile in the database in August 2000, and the police began collecting DNA samples from several hundred men. None of these matched the evidentiary profiles. In October 2001, a DNA profile from a biological sample from Newton showed that she had been killed by the same man as had Floyd and Hughes. The forensic scientists redoubled their efforts and decided to use the DNA profile from the evidentiary samples to see if they could identify individuals who might be *related* to the murderer. They compared the evidence profile to profiles already present in the NDNAD using less stringent matching and found 100 individuals who might be related to the murderer. One of these, Paul Kappen, immediately caught the attention of the detectives. His father, Joseph Kappen, had been on the list of suspects investigated the year before, but he had died in 1989 and so no DNA sample had been tested. Using DNA profiles from Joseph Kappen's wife and daughter, the forensic scientists were able to derive possible profiles for Kappen; those alleles present in the daughter's profile and absent from the mother's must have come from Joseph. The investigators determined that one of these derived profiles matched the profiles from the DNA recovered from the girls' bodies. On May 15, 2002, gravediggers began to exhume Joseph Kappen's body, and teeth and a femur were removed. Profiles of DNA extracted from these tissues showed conclusively that Joseph Kappen had been the murderer.

Arrest Warrants Are Issued Based on Only a DNA Profile

Detective Lori Gaglione and Assistant District Attorney Norman Gahn of Milwaukee County, Wisconsin, faced a dilemma. Detectives were convinced that three brutal rapes in 1993 had been carried out by the same man, and DNA profiling confirmed that this was indeed the case. However, the statute of limitations for rape in Wisconsin is 6 years and that deadline was rapidly approaching. (Statutes of limitation are intended to protect against cases being brought against individuals many years after a crime has been committed, when the memories of eyewitnesses have faded or become modified by subsequent experiences.) Gaglione and Gahn recognized that, in Wisconsin, a warrant had to carry either the name of the suspect or "...designate the person to be arrested by any description by which the person can be identified with reasonable certainty." Could a warrant based on a genetic profile be issued prior to the statute of limitations expiring? What, they reasoned, could identify a person with "reasonable certainty" if not a DNA profile? So, a warrant was issued in the name of "John Doe, unknown male" with matching DNA at "genetic locations D1S7, D2S44, D5S110, D10S28, and D17S79" (Fig. 16-12). This approach was soon challenged in a California case where the John Doe in that case was later identified, his name was entered in the warrant, and he was charged and convicted. A California Superior Court judge upheld the warrant, writing that DNA "...appears to be the best identifier of a person that we have."

Statues of limitations have subsequently undergone changes in response to such cases. Some states have even decided to eliminate the statute of limitations in cases where biological evidence is available. In 2004, the U.S. Congress passed the "Justice for All Act," which brought legislation up-to-date in relation to the implications of DNA profiling. The Act declares that the statute of limitations clock starts running once someone is identified by DNA profiling, instead of from the date when the crime

CASE NUMBER: WARRANT NUMBER:

01XF1363

The State of Wisconsin, Plaintiff
v.

Doe, John #13, Unknown Male with Matching Deoxyribonucleic Acid (DNA) Profile at Genetic Locations D3S1358 (15,16), vWA (17,19), FGA (22,26). D9S1179 (14), D21S11 (29,30), D18S51 (12, 16), D5S818 (12, 13), D13S317 (11, 12), D7S820 (8, 9), D16S539 (8, 13), THO1 (7, 9), TPOX (8), CSF1PO (7)
Defendant

AUTHORIZATION FOR EXTRADITION
(Check One)

☐ **Extradition is authorized from any location within the United States.**

☐ **Extradition is authorized from any adjoining state.**

☐ **Extradition is not authorized.**

STATE OF WISCONSIN CIRCUIT COURT MILWAUKEE COUNTY

FELONY WARRANT
(and AUTHORIZATION FOR EXTRADITION)

CRIME(S) AND STATUTE(S) VIOLATED:

Count 01: Sexual Assault, First Degree (Armed) In Violation Of Wisconsin Statutes Section 940.225(1)(b)

COMPLAINING WITNESS:

Detective Lori Gaglione

THE STATE OF WISCONSIN TO ANY LAW ENFORCEMENT OFFICER:

A Complaint, copy of which is attached, having been filed with me, accusing the Defendant(s) of committing the above stated crime(s) contrary to the above stated statutory sections of the Wisconsin Statutes, I find that probable cause exists that the crime(s) was/were committed by the Defendant(s).

You are, therefore, commanded to arrest and bring the named Defendant(s) before a presiding Judge of the Circuit Court, Criminal Division of Milwaukee County, or, if he is not available, before any acting Judge of the Circuit Court, Criminal Division of said County.

FIGURE 16-12
An example of a "John Doe" arrest warrant. This is the arrest warrant issued by the State of Wisconsin for "John Doe #13, Unknown Male" who is identified by his alleles for the 13 FBI STRs loci.

was committed. Thus the "Justice for All Act" acknowledges that the evidence provided by DNA profiling is quite different in nature from other forms of testimony. However, critics of the Act point out that other sources of evidence, including eyewitness accounts, will be used in court and that these are still subject to the concerns addressed by statute of limitations legislation.

DNA Profiling Frees the Wrongly Convicted

We think of DNA profiling being used to catch criminals, but, in fact, DNA analysis often excludes individuals from investigations and frees those wrongly convicted. Indeed, it has been estimated that about 30% of suspects in crimes are eliminated from an investigation based on DNA profiling, thus saving many hours of costly investigation.

Barry Scheck and Peter Neufeld are attorneys in New York City who have long been advocates for indigent defendants. In the late 1980s, when DNA fingerprinting was first being used in the United States, profiling was carried out in forensic laboratories associated with law enforcement agencies. Scheck and Neufeld were concerned that there was no independent analysis of DNA samples because many defendants could not afford to pay for it. Furthermore, at that time there were serious concerns about the quality of the analyses being performed in forensic laboratories unfamiliar with molecular genetic techniques. In 1992, as a consequence of their early experience, Scheck and Neufeld established the Innocence Project, which investigates convictions where DNA profiling might establish the innocence of the wrongly convicted. The example of Michael Anthony Williams that opens this chapter is typical. In more than one-half of the cases in which a conviction is overturned, the conviction was based on eyewitness testimony, often of a single individual, with no supporting evidence. A further one-third of convictions were based on false confessions, given when the judgment of the suspect is impaired by drugs or alcohol, or under duress, or because the suspect has a mental illness. There are now Innocence Projects or similar organizations in many states and, as of September 2006, 184 individuals have been freed, some after as long as 27 years in prison, by the actions of these organizations.

DNA profiling has also been used to resolve cases where doubt remained about the guilt of a murderer who had been executed. In England, James Hanratty had been found guilty of two murders and was hanged, the last execution in the United Kingdom. He died protesting his innocence and his family campaigned to have him pardoned. Celebrities were attracted to his cause and eventually DNA profiling was carried out on DNA extracted from semen on his victim's underwear and from cells on a handkerchief the attacker had used to disguise himself. These results were compared with profiles determined for Hanratty's brother and mother, and showed that Hanratty had been the assailant. The family insisted that the profiling was in error and had Hanratty's body exhumed so that a direct comparison could be done. But DNA profiling of Hanratty's tissues confirmed the previous findings. Lord Chief Justice Woolf declared that "...the DNA evidence establishes beyond doubt that James Hanratty was the murderer."

DNA Profiling Faced Challenges in the Courtroom

Scientific evidence has long held an ambiguous position in courtrooms in the United States; science and the law have different procedures and goals that may bring them into conflict. On one hand, scientific analysis appears to offer incontrovertible evidence as to the facts of a case, thereby resolving ambiguities or contradictions in the differing descriptions of a case as presented by prosecution and defense. At the same time, judges and lawyers have been wary of scientific evidence and expert testimony in general, in part because it can have disproportionate influence on juries. The law is concerned not only with whether a particular fact is true, but whether it is relevant to the case and admissible. It is the function of the judge to determine this and there may be occasions when evidence, even when relevant, is inadmissible. In the adversarial court system, it is the responsibility of both prosecution and defense to vigorously promote their own case and argue against the other side. This is not a process familiar to scientists and is one that makes them uncomfortable.

In the United States, the long-standing criteria for whether or not scientific evidence is admissible come from the 1923 *Frye v. United States* case. The court recognized that there were occasions when expert testimony was necessary and appropriate, and that there was a need to determine the quality of the science underlying the evidence. The trial judge had to determine whether a scientific principle was "...sufficiently established to have gained general acceptance in the particular field in which it belongs." In *Frye*, the court refused to admit evidence from a lie detector test by arguing that the "systolic blood pressure deception test" did not meet this standard. Seventy years after *Frye*, the Supreme Court reconsidered the admissibility of scientific evidence in *Daubert v. Merrell Dow Pharmaceuticals*. The Court suggested four questions that a judge should ask when considering whether to allow a jury hear scientific evidence: Has the scientific technique been tested to determine its validity? Has the theory or technique been subjected to peer review and publication? What is the known or potential rate of error of the scientific technique? What is the degree of acceptance within the scientific community? This increased the "gatekeeping" function of judges in deciding whether or not scientific evidence should be admitted in federal courts, and more than half of the states now follow the *Daubert* standard.

DNA profiling raised particular concerns and there were frequent challenges when it was first introduced. Some challenges were made on how well a particular sample had been analyzed, and it was clear that the performance of some laboratories left much to be desired. The FBI established working groups to examine laboratory practice and a system of accreditation and quality assessment was introduced. Other challenges dealt with variability within analyses. The first forensic profiling in the United States was done using single locus probes that detect restriction fragment polymorphisms. The variabilities in electrophoresis made it difficult to determine whether a band in an evidence sample and a band in a reference sample were identical. The change to the use of multiple STRs has alleviated this problem. Many challenges focused on the population genetics underlying the statistical calculations. Initially courts balked at the seemingly astronomical probabilities being quoted for matches between samples. It was also argued that classifications such as "Caucasian," "Black," and "Hispanic" as they appear on census forms, for example, are made up of diverse genetic populations. This genetic substructure, it was argued, invalidated estimates of allele frequencies. However, 20 years' experience and the continuing accumulation of data have validated DNA profiling.

Victims of Mass Disasters Are Identified by DNA Profiling

For many years, the identification of bodies relied on finding personal belongings associated with a body, on comparing fingerprints to the FBI or another database, and on dental records or other anthropological data. If the body was suspected to be that of a missing person, then, in the event an intact body was recovered, the skeletal remains, teeth, and bodily features such as scars could be compared with those of the missing person. However, there are circumstances in which there are many bodies and where death has come about through a catastrophic calamity such as a plane crash. Here the force of the impact causes severe fragmentation of bodies and mixing of the remains. Because DNA profiling can be performed on very small pieces of tissue, such profiles can be used to determine which tissue fragments have come from which individual.

Identification requires reference samples, that is, samples known to have come from the individuals who were involved in the disaster. These reference DNA samples might come from hairs on a hairbrush, skin cells on clothes or credit cards, or even the Guthrie cards carrying a spot of blood taken at birth to test for phenylketonuria and other inherited metabolic disorders. There may be occasions on which reference DNA samples are not available. In these circumstances, it may be possible to deduce the identity of the remains from the profiles of biological relatives. Requesting such samples must be done with the greatest sympathy and understanding for the relatives, and with the recognition that family analyses of this kind may discover unexpected relationships. Just as for linkage analyses in clinical genetics, adoptions and nonpaternity (and sometimes nonmaternity) may be revealed. It is important that privacy be maintained and that the answers provided be relevant to the question being asked.

DNA profiling was first used in this way in 1993 following the tragedy at Waco, Texas, where 82 members of the Branch Davidian religious group, including 25 children, died. Many of the bodies were damaged by the collapse of the buildings and the fires that followed. Forensic scientists from the U.S. Armed Forces Institute of Pathology and the FSS in the United Kingdom were called in to identify the remains. Using mitochondrial DNA sequencing, STRs, and other DNA markers, they made some 3000 comparisons between DNA profiles recovered from those killed and reference samples. The forensic scientists succeeded in identifying remains, many of which would have been impossible to identify using conventional means.

DNA profiling was used on an even larger scale in 1998 to identify people killed in the crash of Swissair flight 111 from New York to Geneva. Shortly after takeoff on September 2, the crew detected an abnormal smell but ascribed it to the air-conditioning system. When they realized that it was smoke, the crew turned to make a landing at Halifax, Nova Scotia, but the plane plunged into the sea by Peggy's Cove, Nova Scotia. Bodies were recovered from a depth of 190 feet. Forensic identification was necessary for all but one of the 229 people on board. The limitations of conventional fingerprinting were evident in this case. Although fingerprints were obtained from 1020 fingers, only 43 individuals had some form of fingerprint record that could be used to identify them. In the case of a plane disaster, there is a complete list of those on board so that a comprehensive set of reference samples, from the victims or from relatives, can be collected. Nevertheless, collecting samples was difficult because the passengers were from 21 different countries. More than 1200 human remains were subject to DNA profiling, and in this case 500,000 comparisons were made between remains and reference samples. Remarkably for such a complex operation, all 229 occupants were definitively identified within 3 months.

The greatest challenge faced by DNA profiling came with the terrorist attacks of September 11, 2001. It is hard to appreciate the monumental task that faced pathologists and forensic scientists in the aftermath of the attacks on the Pentagon and World Trade Center, and the crash of United Flight 93 in Somerset County, Pennsylvania. There were 189 murdered at the Pentagon and 45 in Somerset County, all of whom were identified (with the exception of five individuals from the Pentagon whose remains were never recovered). The responsibility for dealing with the collapse of the Twin Towers of the World Trade Center fell on the New York City Office of the Chief Medical Examiner (OCME). It was known that some 3000 bodies were intermingled with 1.5 million tons of rubble and these bodies had been subject to intense heat from the crashes and from the fires that burned beneath the rubble for a further 3 months. The OCME recognized that help was needed for the analysis of DNA samples and for curating the samples and managing the flow of data. The OCME therefore organized a consortium of public institutions and private companies that worked together throughout the project. As well as using the conventional set of FBI STRs, investigators used mitochondrial DNA sequencing, mini-STRs, and SNPs for profiling these degraded DNAs. In the attack on the World Trade Center, 2749 individuals are believed to have lost their lives; by September 2005, 20,120 tissue fragments had been recovered

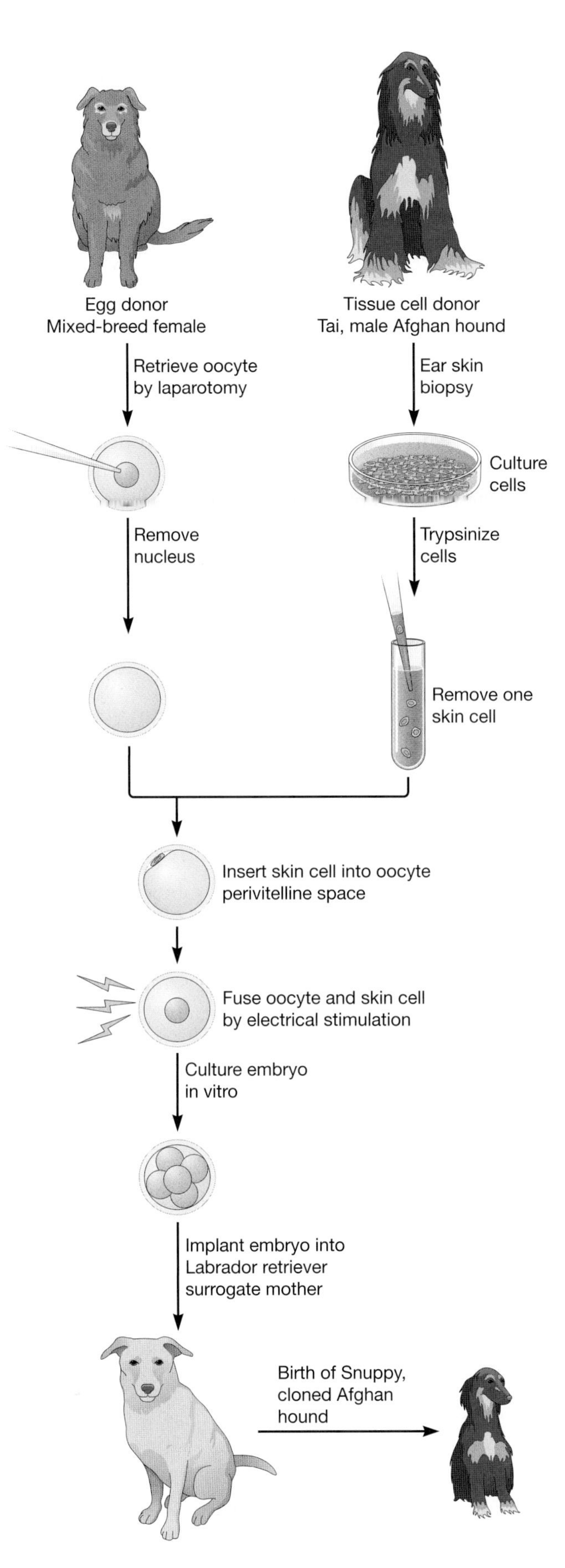

from the site and surrounding buildings. Of the 1594 individuals identified, 850 of the identifications were based on DNA alone, and altogether 52,000 STR, 44,000 mitochondrial DNA, and 17,000 SNP profiles were generated.

DNA Profiling Resolves a Case of Suspected Scientific Fraud

In 1997, the world was transfixed by the announcement that scientists in Scotland had succeeded in cloning a sheep, Dolly. It was a startling result for biologists because such cloning had been pronounced impossible, and it was a startling result for the world as it seemed to point the way to the Alphas, Betas, and Epsilons of Aldous Huxley's *Brave New World*. However, cloning by nuclear transfer has proved to be very difficult in most species and, to date, apparently impossible in some. So, Woo-Suk Huang and his colleagues in South Korea received much praise when, in 2001, they announced that they had produced a cloned dog, named Snuppy. In this case, a skin biopsy had been taken from Tai, an Afghan hound, and put in culture. A fibroblast from this culture was fused with an anucleated egg from a mixed-breed dog and after further culturing, the blastocyst was transferred to a surrogate mother (Fig. 16-13). Three years later, this South Korean group created a public sensation when they reported that they had created 11 human embryonic stem cell lines derived from blastocysts created by transferring cells from the egg donors to

FIGURE 16-13

How Snuppy was cloned. An oocyte was removed from a mixed breed female and its nucleus removed. An ear-skin biopsy was performed on Tai, an Afghan hound. The skin cells were grown in cell culture and one of these cells was injected below the perivitelline membrane of the enucleated egg. This was then subjected to a brief, high-voltage electrical pulse to force Tai's cell to fuse with the egg. This now has a diploid nucleus (as if it had been fertilized) and mitochondria from both the mother and from Tai's cell (in contrast to fertilization where the fertilized zygote would have only the mother's mitochondria). After culturing in vitro for a few days, the blastocyst was implanted into a female Labrador retriever that served as a surrogate mother. Nine weeks later, Snuppy was born.

TABLE 16-5. STRs profiles of Snuppy (clone), Tai (nucleus donor), and surrogate mother

	FH2010	FH2054	FH2079	Pez01	Pez05	Pez06	Pez08	Pez10	Pez11	Pez12
Tai	246	186	288	130	124	199	252	294	153	300
	246	194	288	138	132	199	252	298	153	326
Snuppy	246	186	288	130	124	199	252	294	153	300
	246	194	288	138	132	199	252	298	153	326
Surrogate	250	170	292	138	120	198	256	298	145	285
	254	170	296	138	124	203	256	298	153	297

From Parker et al. 2006. *Nature 440:* E1–E2.

This table shows the alleles for ten canine nuclear STRs tested in Snuppy (clone), Tai (donor of the nucleus), and the surrogate Labrador mother. It is clear that Tai and Snuppy are a perfect match for these STRs and that the Labrador mother is not related to Snuppy.

their own anucleated eggs. But only 1 year later, this work was shown to be fraudulent, which threw doubt on all research from that laboratory. As part of the investigation, two laboratories used DNA profiling to determine the origin of Snuppy.

There were three possibilities. Snuppy was simply an offspring of the supposedly surrogate mother and not a clone of the donor dog, Tai; Snuppy was genetically related to Tai because they were identical twins, a blastomere having been divided to produce Tai and Snuppy; or, as claimed, Snuppy was a clone of Tai, having nuclear DNA from Tai and mitochondrial DNA from the egg donor. First, the genetic relationship between Snuppy and Tai was determined using a combined total of 24 microsatellite repeat loci (Table 16-5). There was complete concordance between these markers for the two dogs. In addition, there was no relationship between Snuppy and his surrogate mother. Second, mitochondrial D-loop variation showed that Snuppy's mitochondria differed from those of Tai as would be expected if a nucleus alone had been removed from Tai's cells (Table 16-6). Third, the D-loop analysis also determined the origin of the egg that gave rise to Snuppy. Snuppy's mitochondrial DNA differed from the surrogate mother but was identical to that of the egg donor. The conclusion is that Snuppy is, in fact, a clone, created with a nucleus from Tai and an egg from the donor.

DNA profiling is the most significant development in forensics since the introduction of fingerprinting more than 100 years ago and is now in use throughout the world. However, DNA profiling is regarded with concern by many. The rapid increase in collecting DNA samples, the retention of samples, the sizes and interconnection of profile databases, and the likelihood that much personal information will be derived from DNA raise issues relating to civil liberties and genetic privacy. For example, ethnic composition of the DNA profile databases suggests to some that disadvantaged minorities will suffer further discrimination. Alec Jeffreys' solution is that the best way to avoid discrimination would be for *all* citizens to provide a DNA sample, provided that stringent rules were in place to prevent inappropriate use. DNA profiling is a most powerful forensic tool that can exonerate the innocent as well as convict the guilty. But our genetic makeup is such an intimate and private part of ourselves that we should be vigilant to guard against its misuse.

TABLE 16-6. Analysis of mitochondrial, cytochrome oxidase *b*, and 16S RNA STRs of Snuppy (clone), Tai (nucleus donor), egg donor, and surrogate mother

	Nucleotide position											
	Hypervariable region									Cyto. *b*	16S RNA	
	170	175	195	343	354	455	498	546	568	426	1117	1145
Tai	G	C	A	T	A	C	C	A	C	A	T	A
Snuppy	A	T	A	C	G	T	T	G	T	G	C	G
Egg donor	A	T	A	C	G	T	T	G	T	G	C	G
Surrogate	G	C	G	T	A	C	C	A	T	A	C	A

From Seoul National University Investigation Committee et al. 2006. *Nature 440:* E2–E3.

Snuppy and the egg donor have identical profiles as would be expected, and both are different from Tai and the surrogate mother. Taken together with the data from Table 16-5, these data show that Snuppy is a clone of Tai.

Reading List

Historical

Ballantyne J., Sensabaugh G., and Witkowski J.A., eds. 1989. *DNA Technology and Forensic Science*, Banbury Report 32. Cold Spring Harbor Laboratory, Cold Spring Harbor, New York. (Provides a broad review of the state of DNA profiling in its infancy.)

DNA Technology in Forensic Science. 1992. National Research Council, Washington, D.C. (A highly influential and controversial report.)

General

Jobling M.A. and Gill P. 2004. Encoded evidence: DNA on forensic analysis. *Nat. Rev. Genet.* **5:** 739–752.

Butler J.M. 2005. *Forensic DNA Typing. Biology, Technology, and Genetics of STR Markers.* Elsevier, Amsterdam. (A comprehensive and clear review of all aspects of STRs.)

Kobilinsky L., Liotti T.F., and Oeser-Sweat J. 2005. *DNA: Forensic and Legal Applications.* Wiley-Interscience, Hoboken, New Jersey. (A good source for legal issues relating to DNA profiling.)

Noble A.A. and Moulton B.W., eds. 2006. DNA fingerprinting and civil liberties: A symposium. *J. Law Med. Ethics* **34:** 141–414. (Many interesting articles covering societal issues arising from the use of DNA profiling.)

Hypervariable Loci

Wyman A.R. and White R. 1980. A highly polymorphic locus in human DNA. *Proc. Natl. Acad. Sci.* **77:** 6754–6758.

Jeffreys A.J., Wilson V., and Thein S.L. 1985. Hypervariable 'minisatellite' regions in human DNA. *Nature* **314:** 67–73.

Nakamura Y., Leppert M., O'Connell P., Wolff R., Holm T., et al. 1987. Variable number tandem repeat (VNTR) markers for human gene mapping. *Science* **235:** 1616–1622.

Weber J.L. and May P.E. 1989. Abundant class of human DNA polymorphisms which can be typed using the polymerase chain reaction. *Am. J. Hum. Genet.* **44:** 388–396.

Edwards A., Civitello A., Hammond H.A., and Caskey C.T. 1991. DNA typing and genetic mapping with trimeric and tetrameric tandem repeats. *Am. J. Hum. Genet.* **49:** 746–756.

Early Applications of DNA Profiling

Jeffreys A.J., Brookfield F.Y., and Semeonoff R. 1985. Positive identification of an immigration test-case using human DNA fingerprints. *Nature* **317:** 818–819.

Gill P., Jeffreys A.J., and Werrett D.J. 1985. Forensic application of DNA 'fingerprints'. *Nature* **318:** 577–579.

Wambaugh J. 1989. *The Blooding.* William Morrow and Company, New York.

Short Tandem Repeats (STRs)

Butler J.M. 2006. Genetics and genomics of core STR loci used in human identity testing. *J. Forensic Sci.* **51:** 253–265.

Short Tandem Repeat DNA Internet Database (STRBase) http://www.cstl.nist.gov/div831/strbase/ (An invaluable source for information on STRs and their forensic applications.)

Cotton E.A., Allsop R.F., Guest J.L., Frazier R.R.E., Koumi P., et al. 2000. Validation of the AMPFISTR SGM plus system for use in forensic case work. *Forensic Sci. Int.* **112:** 151–161.

Moretti T.R., Baumstark A.L., Defenbaugh D.A., Keys K.M., Smerick J.B., and Budowle B. 2001. Validation of short tandem repeats (STRs) for forensic usage: Performance testing of fluorescent multiplex STR systems and analysis of authentic and simulated forensic samples. *J. Forensic Sci.* **46:** 647–660.

Greenspoon S.A., Ban J.D., Pablo L., Crouse C.A., Kist F.G., et al. 2004. Validation and implementation of the PowerPlex 16 BIO System STR multiplex for forensic casework. *J. Forensic Sci.* **49:** 71–80.

Information on the multiplex STRs kits produced by Applied Biosystems (https://www2.appliedbiosystems.com/) and Promega (http://www.promega.com/).

Gill P., Whitaker J., Flaxmann C., Brown N., and Buckleton J. 2000. An investigation of the rigor of interpretation rules for STRs derived from less than 100 pg of DNA. *Forensic Sci. Int.* **112:** 17–40.

Gill P. 2001. Application of low copy number DNA profiling. *Croatian Med. J.* **42:** 229–232.

Databases Contain Millions of DNA Profiles

CODIS: Combined DNA Index System of the FBI. http://www.fbi.gov/hq/lab/codis/index1.htm

The National DNA Database Annual Report 2004–2005. http://www.acpo.police.uk/asp/policies/Data/NDNAD_AR_04_05.pdf

The National DNA Database. 2006. Parliamentary Office of Science and Technology, Postnote 258. http://www.parliament.uk/documents/upload/postpn258.pdf

Bieber F.R. 2006. Turning base hits into earned runs: Improving the effectiveness of forensic DNA data bank programs. *J. Law Med. Ethics* **34:** 2–13.

Williams R. and Johnson P. 2006. Inclusiveness, effectiveness and intrusiveness: Issues in the developing uses of DNA profiling in support of criminal investigations. *J. Law Med. Ethics* **34:** 234–247.

Mitochondrial DNA Profiling

Thompson W.E., Ramalho-Santos J., and Sutovsky P. 2003. Ubiquitination of prohibitin in mammalian sperm mitochondria: Possible roles in the regulation of mitochondrial inheritance and sperm quality control. *Biol. Reprod.* **69:** 254–260.

Anderson S., Bankier A.T., Barrell B.G., de Bruijn M.H., Coulson A.R., et al. 1981. Sequence and organization of the human mitochondrial genome. *Nature* **290:** 457–465.

Andrews R.M., Kubacka I., Chinnery P.F., Lightowlers R.N., Turnbull D.M., and Howell N. 1999. Reanalysis and revision of the Cambridge reference sequence for human mitochondrial DNA. *Nat. Genet.* **23:** 147.

Greenbergh B.D., Newbolda J.E., and Sugino A. 1983. Intraspecific nucleotide sequence variability surrounding the origin of replication in human mitochondrial DNA. *Gene* **21:** 33–49.

Morley J.M., Bark J.E., Evans C.E., Perry J.G., Hewitt C.A., and Tully G. 1999. Validation of mitochondrial DNA minisequencing for forensic casework. *Int. J. Legal Med.* **112:** 241–248.

Quintans B., Alvarez-Iglesias V., Salas A., Phillips C., Lareu M.V., and Carracedo A. 2004. Typing of mitochondrial DNA coding region SNPs of forensic and anthropological interest using SNaPshot minisequencing. *Forensic Sci. Int.* **140:** 251–257.

Budowle B., Allard M.W., Wilson M.R., and Chakraborty R. 2003. Forensics and mitochochondrial DNA: Applications, debates and foundations. *Annu. Rev. Genomics Hum. Genet.* **4:** 119–141.

Monson K.L., Miller K.W.P., Wilson M.R., DiZinno J.A., and Budowle B. 2002. The mtDNA population database: An integrated software and database resource for forensic comparison. *Forensic Sci. Commun.*, vol. **4**.

Mitochondrial Heteroplasmy and Tsar Nicholas II

Melton T. 2004. Mitochondrial DNA heteroplasmy. *Forensic Sci. Rev.* **16:** 1–20.

Gill P., Ivanov P.L., Kimpton C., Piercy R., Benson N., et al. 1994. Identification of the remains of the Romanov family by DNA analysis. *Nat. Genet.* **6:** 130–135.

Ivanov P.L., Wadham M.J., Roby R.K., Holland M.M., Weedn V.W., and Parsons T.J. 1996. Mitochondrial DNA sequence heteroplasmy in the Grand Duke of Russia Georgij Romanov establishes the authenticity of the remains of Tsar Nicholas II. *Nat. Genet.* **12:** 417–420.

Massie R.K. 1996. *The Romanovs: The Final Chapter.* Ballantine Books, New York.

The Case of the Boston Strangler

Kelly S. 2002. *The Boston Stranglers.* Pinnacle, New York.

Foran D.R. and Starrs J.E. 2004. In search of the Boston Strangler: Genetic evidence from the exhumation of Mary Sullivan. *Med. Sci. Law* **44:** 47–54.

Y Chromosome STRs

Budowle B., Wilson M.R., and DiZinno J.A. 1999. Mitochondrial DNA regions HVI and HVII population data. *Forensic Sci. Int.* **103:** 23–35.

Butler J.M. 2003. Recent developments in Y-short tandem repeat and Y-single nucleotide polymorphism analysis. *Forensic Sci. Rev.* **15:** 91–111.

Prinz M. 2003. Advantages and disadvantages of Y-short tandem repeat testing in forensic casework. *Forensic Sci. Rev.* **15:** 189–201.

Gusmao L. and Alves C. 2005. Y chromosome STR typing. *Methods Mol. Biol.* **297:** 67–82.

Roewer L., Krawczakb M., Willuweita S., Nagya M., Alvesc C., et al. 2001. Online reference database of European Y-chromosomal short tandem repeat (STR) haplotypes. *Forensic Sci. Int.* **118:** 106–113.

YHRD—Y Chromosome Haplotype Reference Database. http://www.ystr.org

"Mini- STRs" Are Used to Analyze Degraded DNA

Chung D.T., Drabek J., Opel K.L., Butler J.M., and McCord B.R. 2004. A study of the effects of degradation and template concentration on the amplification efficiency of the STR Miniplex primer sets. *J. Forensic Sci.* **49:** 733–740.

Coble M.D. and Butler J.M. 2005. Characterization of new mini-STR loci to aid analysis of degraded DNA. *J. Forensic Sci.* **50:** 43–53.

Single-Nucleotide Polymorphisms Show Promise for Forensic Applications

Sanchez J.J., Phillips C., Børsting C., Balogh K., Bogus M., et al. 2006. A multiplex assay with 52 single nucleotide polymorphisms for human identification. *Electrophoresis* **27:** 1713–1724.

Determining Population Origins Is Controversial

Bamshad M.J., Wooding S., Watkins W.S., Ostler C.T., Batzer M.A., and Jorde L.B. 2003. Human population genetic structure and inference of group membership. *Am. J. Hum. Genet.* **72:** 578–589.

Romualdi C., Balding D., Nasidze I.S., Risch G., Robichaux M., et al. 2002. Patterns of human diversity, within and among continents, inferred from biallelic DNA polymorphisms. *Genome Res.* **12:** 602–612.

Serre D. and Pääbo S. 2004. Evidence for gradients of human genetic diversity within and among continents. *Genome Res.* **14:** 1679–1685.

Wetton J.H., Tsang K.W., and Khan H. 2005. Inferring the population of origin of DNA evidence within the UK by allele-specific hybridization of Y-SNPs. *Forensic Sci. Int.* **152:** 45–53.

Vallone P.M. and Butler J.M. 2004. Y-SNP typing of U.S. African American and Caucasian samples using allele-specific hybridization and primer extension. *J. Forensic Sci.* **49:** 723–732.

Daly E. 2005. DNA tells students they aren't who they thought. *New York Times*, April 13, 2005.

Harmon A. 2006. Seeking ancestry in DNA ties uncovered by tests. *New York Times*, April 22, 2006.

Shriver M.D. and Kittles R.A. 2004. Genetic ancestry and the search for personalized genetic histories. *Nat. Rev. Genet.* **5:** 611–618.

Haga S.B. 2006. Policy implications of defining race and more by genome profiling. *Genomics Soc. Policy* **2:** 57–71.

Genome Profiling May Indicate Physical Characteristics

Ossorio P.N. 2006. About face: Forensic genetic testing for race and visible traits. *J. Law Med. Ethics* **34:** 277–292.

Frudakis T., Thomas M., Gaskin Z., Venkateswarlu K., Chandra K.S., et al. 2003. Sequences associated with human iris pigmentation. *Genetics* **165:** 2071–2083.

Rees J.L. 2004. The genetics of sun sensitivity in humans. *Am. J. Hum. Genet.* **75:** 739–751.

Flanagan N., Healy E., Ray A., Philips S., Todd C., et al. 2000. Pleiotropic effects of the melanocortin 1 receptor (*MC1R*) gene on human pigmentation. *Hum. Mol. Genet.* **9:** 2531–2537.

Grimes E.A., Noake P.J., Dixon L., and Urquhart A. 2001. Sequence polymorphism in the human melanocortin 1 receptor gene as an indicator of the red hair phenotype. *Forensic Sci. Int.* **122:** 124–129.

Applications of DNA-based Identification

National Institute of Justice. 1999. What every law enforcement officer should know about DNA evidence. National Institute of Justice, Washington, D.C. http://www.ncjrs.gov/pdffiles1/nij/bc000614.pdf

Langan P.A. and Levin D.J. 2002. Recidivism of prisoners released in 1994. Bureau of Justice Statistics Special Report, June 2002. U.S. Department of Justice. http://www.ojp.usdoj.gov/bjs/pub/pdf/rpr94.pdf

Greely H.T., Riordan D.P., Garrison N., and Mountain J.L. 2006. Family ties: The use of DNA offender databases to catch offenders' kin. *J. Law Med. Ethics* **34:** 248–262.

Haimes E. 2006. Social and ethical issues in the use of familial searching in forensic investigations: Insights from family and kinship studies. *J. Law Med. Ethics* **34:** 263–276.

Bieber F.R., Brenner C.H., and Lazer D. 2006. Finding criminals through DNA of their relatives. *Science* **312:** 1315–1316.

Gahn N. 2000. John Doe, D1S7, D2S44, D5S110, D10S28, D17S79, charged with rape. *Profiles in DNA* **3:** 8–9. http://www.promega.com/profiles/303/303_08.html

The Innocence Project. http://www.innocenceproject.org/

Scheck B., Neufeld P., and Dwyer J. 2000. *Actual Innocence: Five Days to Execution and Other Dispatches from the Wrongly Convicted.* Doubleday, New York.

Lord Woolf, Judgement in *Regina v. James Hanratty*, in the Supreme Court of Judicature, Court of Appeal (Criminal Division), 10 May 2002.

DNA Profiling Faced Challenges

Jasanoff S. 2006. Just evidence: The limits of science in the legal process. *J. Law Med. Ethics* **34:** 328–341.

Berger M.A. 2002. The Supreme Court's Trilogy on the Admissibility of Expert Testimony. In *Reference Manual on Scientific Evidence.* Federal Judicial Center, Washington, D.C., pp. 9–38.

Lander E.S. 1989. DNA fingerprinting on trial. *Nature* **339:** 501–505.

Lewontin R.C. and Hartl D.L. 1991. Population genetics in forensic DNA typing. *Science* **254:** 1745–1750.

Lander E.S. and Budowle B. 1994. DNA fingerprinting dispute laid to rest. *Nature* **371:** 735–738.

Victims of Mass Disasters Are Identified by DNA Profiling

Budowle B., Bieber F.R., and Eisenberg A.J. 2005. Forensic aspects of mass disasters: Strategic considerations for DNA-based human identification. *Legal Med.* 1–14.

Clayton T.M., Whitaker J.P., and Maguire C.N. 1995. Identification of bodies from the scene of a mass disaster using DNA amplification of short tandem repeat (STR) loci. *Forensic Sci. Int.* **76:** 7–15. (Waco.)

Clayton T.M., Whitaker J.P., Fisher D.L., Lee D.A., Holland M.M., et al. 1995. Further validation of a quadruplex STR DNA typing system: A collaborative effort to identify victims of a mass disaster. *Forensic Sci. Int.* **76:** 17–25. (Waco.)

Leclair B., Fregeau C.J., Bowen K.L., and Fourney R.M. 2004. Enhanced kinship analysis and STR-based DNA typing for human identification in mass fatality incidents: The Swissair Flight 111 disaster. *J. Forensic Sci.* **49:** 939–953.

Budimlija Z.M., Prinz M.K., Zelson-Mundorff A., Wiersema J., Bartelink E., et al. 2003. World Trade Center human identification project: Experiences with individual body identification cases. *Croat. Med. J.* **44:** 259–263.

Biesecker L.G., Bailey-Wilson J.E., Ballantyne J., Baum H., Bieber F.R., et al. 2005. DNA identifications after the 9/11 World Trade Center attack. *Science* **310:** 1122–1123.

DNA Profiling Resolves a Case of Suspected Scientific Fraud

Lee B.C., Kim M.K., Jang G., Oh H.J., Yuda F., et al. 2005. Dogs cloned from adult somatic cells. *Nature* **436:** 641.

Parker H.G., Kruglyak L., and Ostrander E.A. 2006. DNA analysis of a putative dog clone. *Nature* **440:** E1–E2.

Seoul National University Investigation Committee, Lee J.B., and Park C. 2006. Molecular genetics: Verification that Snuppy is a clone. *Nature* **440:** E2–E3.

Figure Credits

Every effort has been made to contact the copyright holders of figures in this test. Any copyright holders we have been unable to reach or for whom inaccurate information has been provided are invited to contact W. H. Freeman and Company.

Abbreviations: AAAS, American Association for the Advancement of Science; CSHLP, Cold Spring Harbor Laboratory Press; NAS, USA, National Academy of Sciences, U.S.A.

Section 1

Icon photograph, from http://spm.phy.bris.ac.uk.

Chapter 1

1.2, Reprinted from Stevens N.M., *Carnegie Inst. Wash. Publ. 36:* 1–33, © 1905; **1.3,** redrawn from Sturtevant A.H. and Beadle G.W., *An Introduction to Genetics,* © 1939 W.B. Saunders, Philadelphia; **1.5c,** redrawn from Bridges C.B., *J. Hered. 26:* 60–64, © 1935 Oxford University Press; **1.6,** based on data from Bridges C.B., 1919, *J. Exp. Zool. 28:* 265–305, and Beadle G.W., *An Introduction to Genetics,* © 1939 W.B. Sanders, Philadelphia; **1.7,** modified from Suzuki D.T. et al., *An Introduction to Genetic Analysis, 4e,* © 1989 W.H. Freeman; **1.11,** based on data from Avery O.T., *J. Exp. Med. 79:* 137–158, © 1944 Rockefeller University Press; **1.13b,** reprinted from Franklin R.E. and Gosling R.G., *Nature 171:* 740–741, ©1953 Macmillan; **1.17,** based on Meselson M. and Stahl F.W., *Proc. Natl. Acad. Sci. 44:* 671–682, © 1958 NAS, USA.

Chapter 2

2.12, Modified from Watson J.D. et al., *Molecular Biology of the Gene, 5e,* p. 423, © 2004 Pearson Education, Inc.; **2.13,** reprinted from Yusupov M.M. et al., *Science 292:* 883–896, © 2005 AAAS; **2.15,** modified from Batey R.T. et al., *Nature 432:* 411–415, © 2004 Macmillan.

Chapter 3

3.2b, Adapted from Bell C.E., *J. Mol. Biol. 312:* 921–926, © 2001 Elsevier; **3.5,** redrawn from Watson J.D. et al., *Molecular Biology of the Gene, 5e,* p. 495, © 2004 Pearson Education, Inc.; **3.8,** redrawn from Watson J.D. et al., *Molecular Biology of the Gene, 5e,* p. 533, © 2004 Pearson Education, Inc.; **3.9, photographs,** reprinted from Krizek B.A. and Fletcher J.C., *Nat. Rev. Genet. 6:* 688–698, © 2005 Macmillan; **3.9, diagram,** adapted from Gutierrez-Cortines M.E. and Davies B., *Trends Plant Sci. 5:* 471–476, © 2000 Elsevier; **3.11,** modified from Nusslein-Volhard C. and Wieschaus E., *Nature 287:* 795–801, © 1980 Macmillan; **3.13,** Fujioka M. et al., *Development 126:* 2527–2638, © 1999 Company of Biologists Ltd.; **3.14,** redrawn from Watson J.D. et al., *Molecular Biology of the Gene, 5e,* p. 605, © 2004 Pearson Education, Inc.; **3.17,** adapted from Watson J.D. et al., *Molecular Biology of the Gene, 5e,* p. 164, © 2004 Pearson Education, Inc.

Chapter 4

4.3, Courtesy of Stanley N. Cohen, Stanford University; **4.16a,b,c,** courtesy of N. Arnheim, University of Southern California; **4.17,** courtesy of D.M. Hunt, University of South Carolina School of Medicine, http://pathmicro.med.sc.edu/pcr/realtime-home.htm; **4.19,** redrawn from Vet J.A. et al., *Proc. Natl. Acad. Sci. 96:* 6394–6399, © 1999 NAS, USA.

Chapter 5

5.1, top, Courtesy of P. Chambon, modified from *Sci. Am. 244:* 60–71, © 1981 Scientific American; **5.1, bottom,** courtesy of Estate of Bunji Tagawa, modified from *Sci. Am. 244:* 60–71, © 1981 Scientific American; **5.3a,b,** adapted from Watson J.D. et al., *Molecular Biology of the Gene, 5e,* pp. 381–382, © 2004 Pearson Education, Inc.; **5.5,** courtesy of Lees-Miller J., adapted from *Mol. Cell Biol. 10:* 1729–1742, © 1990 American Society for Microbiology; **5.9,** redrawn from Sharpless N.E. and DePinho R.A., *Curr. Opin. Genet. Dev. 9:* 22–30, © 1999 Elsevier; **5.13,** redrawn from Watson J.D. et al., *Molecular Biology of the Gene, 5e,* p. 406, © 2004 Pearson Education, Inc.

Chapter 7

7.5, Redrawn from Watson J.D. et al., *Molecular Biology of the Gene, 5e,* p. 331, © 2004 Pearson Education, Inc.; **7.6,** adapted from Dewannieux M. et al., *Nat. Genet. 35:* 41–48, © 2003 Macmillan; **7.10a,** courtesy of the John Innes Centre; **7.10b,** adapted from Coen E.S. et al. *Cell 47:* 285–296, © 1986 Elsevier; **7.11,** modified from Kazazian H.H. et al., *Nature 332:* 164–166, © 1988 Macmillan; **7.14,** modified from Ivics Z. et al., *Cell 91:* 501–510, © 1997 Elsevier; **7.15,** based on data from Lehrman M.A. et al., *Science 227:* 140–146, © 1985 AAAS, and Lehrman M.A. et al., *Cell 48:* 827–835, © 1987 Elsevier; **7.16,** modified from Sijen T. and Plasterk R.H., *Nature 426:* 310–314, © 2003 Macmillan.

Chapter 8

8.1, Reprinted from Muller H.J., *Harvey Lect. 43:* 165–229, © 1950 Wiley-Liss; **8.2,** adapted from Henikoff S., Fig. 1a, p. 320, in *Epigenetic Mechanisms of Gene Regulation*, © 1996 CSHLP; **8.5, photograph,** reprinted from Brown C.J., *Nature 349:* 38–44, © 1991 Macmillan; **8.6,** modified from Carrel L. and Willard H.F., *Nature 434:* 400–404, © 2005 Macmillan; **8.7,** based on Kazazian H.H. et al., *Science 150:* 1601–1602, © 1965 AAAS; **8.8,** based on Chatterjee R.N. and Mukherjee A.S., *J. Cell Biol. 74:* 168–180, © 1977 Rockefeller University Press; **8.12,** modified from Yang A.S. et al., *Epigenetic Mechanisms of Gene Regulation*, © 1996 CSHLP, and Shibata A., Ph.D. thesis, University of California, 2005; **8.14,** modified from Engel N. et al., *Nat. Genet. 35:* 883–888, © 2004 Macmillan; **8.15,** adapted from Turner B.M., *Cell 111:* 285–291, © 2002 Elsevier, and Spotswood H.T. and Turner B.M., *J. Clin. Invest. 110:* 577–582, © 2002; **8.17, photographs,** reprinted from Fraga M.F. et al., *Proc. Natl. Acad. Sci. 102:* 10604–10609, © 2005 NAS, USA.

Chapter 9

9.1c, Redrawn from Napoli C. et al., *Plant Cell 2:* 279–289, © 1990 American Society of Plant Biologists; **9.3a,** adapted from Waterhouse P.M. et al., *Proc. Natl. Acad. Sci. 95:* 13959–13964, © 1998 NAS, USA; **9.6a,c,** adapted from Bernstein E. et al., *Nature 409:* 363–356, © 2001 Macmillan; **9.7, bottom,** redrawn from Elbashir S.M. et al., *EMBO J. 20:* 6877–6888, © 2001 Macmillan; **9.8b, graph,** redrawn from Palliser D. et al., *Nature 439:* 89–94, © 2006 Macmillan; **9.9,** modified from Ha I. et al., *Genes Dev. 10:* 3041–3150, © 1996 CSHLP; **9.10b, graph and gel,** modified from Zeng Y. et al., *Proc. Natl. Acad. Sci. 100:* 9779–9784, © 2003 NAS, USA; **9.13b,** redrawn from Allen E. et al., *Cell 121:* 207–221, ©2005 Elsevier; **9.14,** reprinted from Voinnet O. et al., *Nature 389:* 553, © 1997 Macmillan; **9.16a,b,** adapted from Liu J. et al., *Genes Dev. 18:* 2873–2878, © 2004 CSHLP.

Section 2

Icon photograph, courtesy of Stanley N. Cohen, Stanford University.

Chapter 10

10.1, Data from GenBank; **10.10,** modified from Rosenblum B.B. et al., *Nucleic Acids Res. 25:* 4500–4504, © 1997 Oxford University Press; **10.16,** courtesy of Robert Fleischmann and The Institute for Genomic Research, Rockville, MD, modified from *Science 269:* 496–512, © 1995 AAAS.

Chapter 11

11.7, from NCBI MapViewer http://www.ncbi.nlm.nih.gov/mapview/maps.cgi?taxid=9606&chr=4; **11.9b,** reprinted from McPherson J.D. et al., *Nature 409:* 934–941, © 2001 Macmillan.

Chapter 12

12.1a, Redrawn from Hu J. et al., *Mol. Cell 12:* 1379–1389, © 2003 Elsevier, **12.1b,** adapted from Pavletich N.P. and Pabo C.O., *Science 252:* 809–817, ©1991 AAAS; **12.1c,** adapted from Hung L.W. et al., *Nature 396:* 703–707, © 1998 Macmillan; **12.3,** courtesy of Greg Cooper, Stanford University; **12.6,** courtesy of Greg Cooper, Stanford University; **12.9,** from http://www.ncbi.nlm.nih.gov/BLAST; **12.10a,** adapted from Asif T. et al., *Nature 420:* 520–562 © 2002 Macmillan; **12.11,** adapted from Pennacchio L. et al., *Science 294:* 169–173 © 2001 AAAS; **12.12,** redrawn from Simon A.L. et al., *Proc. Natl. Acad. Sci. 99:* 2912–2917, © 2002 NAS, USA; **12.13,** adapted from Emison E.S. et al., *Nature 434:* 857–863, © 2005 Macmillan.

Chapter 13

13.2 and 13.4, Adapted from http://www.affymetrix.com/technology/manufacturing/index.affx; **13.6,** adapted from Iyer et al., 1999, *Science 283:* 83–87; **13.7,** adapted from Velculescu V.E. et al., *Trends Genet. 16:* 423–425, © 2000 Elsevier; **13.9b,** from Lieb J.D. et al., *Nat. Genet. 28:* 327–334, © 2001 Macmillan; **13.10b,** adapted from Rice J.C. and Allis C.D., *Curr. Opin. Cell Biol. 13:* 263–273, © 2001 Elsevier; **13.10c,** from http://genome.ucsc.edu/ENCODE; **13.11,** courtesy of Richard Myers, Alayne Brown, and Nathan Trinklein; **13.12,** based on principles in Trinklein N. et al., *Genome Res. 13:* 308–312, © 2003 CSHLP; **13.14a,b,** adapted from Rual J.F. et al., *Nature 437:* 1173–1178, © 2005 Macmillan; **13.15a,b,** reprinted from Zhu H. et al., *Science 293:* 2101–2105, © 2001 AAAS.; **13.15c,** modified from Zhu H. et al., *Science 293:* 2101, © 2001 AAAS; **13.16a,** based on Kumar A. and Snyder M., *Nature 415:* 123–124, © 2002 Macmillan; **13.16b,** redrawn from Gavin A. et al., *Nature 415:* 141–147, © 2002 Macmillan; **13.18a,** modified from Huh W. et al., *Nature 425:* 686–691, © 2003 Macmillan; **13.18b,** reprinted from Huh W. et al., *Nature 425:* 686–691, © 2003 Macmillan; **13.18c,** redrawn from Huh W. et al., *Nature 425:*686–691, © 2003 Macmillan.

Chapter 14

14.2, Modified from Marx J.L., *Science 255:* 923–925, ©1989 AAAS; **14.5,** modified from Zucher S. et al., *Am. J. Hum. Genet. 79:* 365–369, © 2006 University of Chicago Press; **14.12,** adapted from figure courtesy of Christa Lese Martin, Emory University; **14.13,** adapted from Syvanen A.C., *Nat. Genet. 37:* S5–S10, © 2005 Macmillan; **14.14,** modified from Fan J.B. et al., *Cold Spring Harbor Symp. Quant. Biol. 68:* 69–87, © 2003 CSHLP.

Chapter 15

15.1, Redrawn from *Death Rates: U.S. Mortality Use Tapes 1960–2002, U.S. Mortality Volumes, 1939–1950, National Center for Health Statistics, Centers for Disease Control and Prevention, 2005, Cigarette Consumption, U.S. Dept of Agriculture, 1900–2002*; **15.8,** based on data from Carter P. et al., *Proc. Natl. Acad. Sci. 89:* 4285–4289, © 1992 NAS, USA; **15.9a,b,** redrawn from Manley P.W. et al., *Eur. J. Cancer.* (suppl. 5) *38:* 519–527, © 2002 Elsevier; **15.9c,** redrawn from Deininger M. et al., *Blood 105:* 2640–2653, © 2005 American Society of Hematology; **15.13, right,** modified from *Chemicon International, ApopTag Peroxidase In Situ Apoptosis Detection Kit, S7100*, Fig. 1, p. 5; **15.13, photographs,** Symonds H. et al., *Cell 78:* 703–711, © 1994 Elsevier; **15.14 graph,** redrawn from Sund M. et al., *Proc. Natl. Acad. Sci. 102:* 2934–2939, © 2005 NAS, USA; **15.15, graphs** redrawn and **illustration** adapted from http://www.cshl.edu/publicreleases/revealing.html; **15.16, illustration** adapted from and **photographs** reprinted from He L. et al., *Nature 435:* 828–833, © 2005 Macmillan; **15.17,** modified from Sørlie T. et al., *Proc. Natl. Acad. Sci. 98:* 10869–10874, © 2001 NAS, USA.

Chapter 16

16.2, Adapted from Nakamura Y. et al., *Science 235:* 1616–1622, © 1987 AAAS; **16.3 and 16.4,** redrawn from Butler J.M., *Forensic DNA Typing, 2e*, © 2005 Elsevier; **16.5a,** courtesy of Mechthild Prinz, Office of the Chief Medical Examiner, New York City; **16.5b,** adapted from Butler J.M., *Forensic DNA Typing, 2e*, © 2005 Elsevier; **16.6,** redrawn from Butler J.M., 2004, *Curr. Protocols Human Genet.* (Unit 14.8, suppl. 41), 14.8.1–14.8.22; **16.7,** data from The National DNA Database Annual Report 2004–2005, http://www.acpo.police.uk/asp/policies/Data/NDNAD_AR_04_05.pdf; **16.8,** courtesy of Rebecca Just, Armed Forces DNA Identification Library; **16.9,** redrawn from Ivanov P.L. et al., *Nat. Genet. 12:* 417–420, © 1996 Macmillan; **16.10,** adapted from Chung D.T. et al., *J. Forensic Sci. 49:* 733–740, © 2004 ASTM International, West Conshohocken, PA; **16.11,** redrawn from Coble M.D. et al., *J. Forensic Sci. 50:* 43–53, © 2005 ASTM International, West Conshohocken, PA; **16.12,** reprinted from State of Wisconsin Circuit Court, Milwaukee County, Felony Warrant Case number 01XF1363; **16.13,** based on data in Lee B.C. et al., *Nature 436:* 641, © 2005 Macmillan.

Index

The letter "f" after a page number refers to a figure, "t" to a table.

E

H

I

J

K

L

Q

R